List of Integrals

22. $\int xe^{ax}\,dx = \frac{1}{a^2}(ax-1)e^{ax} + C, \quad a \neq 0$

23. $\int \ln x\,dx = x\ln x - x + C$

24. $\int x\ln x\,dx = \frac{x^2}{2}\ln x - \frac{x^2}{4} + C$

25. $\int \frac{dx}{x\ln x} = \ln|\ln x| + C$

BRIEF CALCULUS WITH APPLICATIONS

James W. Burgmeier
University of Vermont

Monte B. Boisen, Jr.
Virginia Tech

Max D. Larsen
The Gallup Organization

This book is a brief version of *Calculus with Applications.*

McGRAW-HILL PUBLISHING COMPANY

New York St. Louis San Francisco Auckland Bogotá Caracas
Hamburg Lisbon London Madrid Mexico Milan Montreal
New Delhi Oklahoma City Paris San Juan São Paulo
Singapore Sydney Tokyo Toronto

BRIEF CALCULUS WITH APPLICATIONS

1 2 3 4 5 6 7 8 9 0 VNH VNH 8 9 4 3 2 1 0 9

ISBN 0-07-557411-X

This book was set in Times Roman.
The editors were Robert A. Weinstein and John M. Morriss;
the designer was DeNee Skipper;
the production supervisor was Michael Weinstein.
Von Hoffmann Press, Inc., was printer and binder.

Library of Congress Cataloging-in-Publication Data

Burgmeier, James W.
Brief calculus with applications/James W. Burgmeier, Monte B. Boisen, Jr. Max D. Larsen.
p. cm.
Includes index.
ISBN 0-07-557411-X
1. Calculus. I. Boisen, Monte B., Jr. II. Larsen, Max D. III. Title.
QA303.B9259 1989b
515—dc20 89-7981

This text is dedicated to our families, for their patience, support, and love.

Pat, Glenn, and Robin

Helen, John, and Jennifer

Lillie, Mike, Paul, and Charlie

PREFACE

The purpose of this text is to develop the student's understanding of the basic concepts of calculus through realistic applications from the managerial, social, and life sciences. Numerous examples motivate the theoretical concepts and show how calculus can be applied in these areas. Realistic applications show that the mathematics presented in this course will have an impact on the student's future work. This approach ensures that the student learns the concepts, power, and usefulness of calculus.

The text presents the fundamentals of calculus intuitively and clearly without sacrificing accuracy. The balance of prose explanations and algebraic details found in the solutions to examples is the result of years of classroom experience. Instructors and students alike will find the text clear, concise, accurate, and readable.

Distinguishing Features

- ***Emphasis on examples:*** The many carefully explained examples illustrate procedures, show applications of the material, and elaborate on mathematical content. In many cases, the examples are somewhat idealized: we have adjusted the numerical constants to avoid tedious and distracting arithmetic. Students can thus focus attention on the calculus concept involved and grasp the mathematics more easily and quickly.
- ***Emphasis on applications:*** We have included a wealth of applications to the managerial, social, and life sciences adapted from the recent literature of these sciences. Each application has been chosen to represent faithfully the discussion from which it came as well as to provide a straightforward illustration of the calculus topic under study. Examples from each of the managerial, social, and life sciences are presented for most of the mathematical topics in this text, so instructors can emphasize one discipline or another according to the interests of the students. These examples require no specialized knowledge. The range of applications provides evidence that the mathematics learned in this course has relevancy and usefulness beyond the classroom.
- ***Exercises:*** Over 2400 exercises are included to give students ample opportunity to test their understanding and apply their

skills to real-world problems. Our experience shows that the student's understanding of calculus is linked directly to the number of problem-types solved successfully; we believe that the quality and variety of our exercises will foster this success. The answers to the odd-numbered exercises are given at the back of the book, and many of these include intermediate results so students can check their work. Answers to the even-numbered exercises are given in the Instructor's Manual.

- ***Figures and photographs:*** Wherever possible, we have used figures to motivate new concepts and as an aid in problem-solving. These figures are clearly drawn and carefully captioned and labeled to enhance student comprehension. In addition, the abundance of photographs throughout the text attests to the range of applied problems in this book and the diverse uses of mathematics. Every photo is carefully keyed to an applied example or topic in the text, illustrating the usefulness of calculus in the real world.
- ***Bibliography:*** The annotated bibliography lists sources for the applied problems and other works in which calculus is used at the level presented in this text. Readers who would like more information on the applications will find numerous source materials listed here.

Special Pedagogical Issues

- ***Limits and continuity:*** While our discussion of limits is intuitive, it does not present the material in a dishonest fashion. We have made a special effort to use applications to set the stage in Section 2.1 for the need for limits. Section 2.2 then discusses limits and how to calculate the limits of various functions. We have deliberately postponed continuity until Section 2.7 so that students can go immediately to the derivative in Section 2.3. Instructors who wish to omit Section 2.7 may do so without disruption. This organization should hold the student's attention and clarify why these concepts are important.
- ***Derivatives:*** Section 2.3 uses the informal treatment of limits given in Sections 2.1 and 2.2 to define the derivative and cover the simple algebraic rules for finding derivatives. The product, quotient, and chain rules are postponed until Section 3.1 in order to present some of the uses of the derivative immediately. Our experience has shown that this approach has more appeal for students.

  Sections 2.4 to 2.6 discuss several ways to interpret the derivative. Section 2.4 covers the slope of the tangent line and how it is related to the derivative. Section 2.5 discusses increasing and decreasing functions so students see a direct implementation of the tangent line concept. Section 2.6 uses applications to explore the derivative as a rate of change. These sections discuss various interpretations of the derivative and several of its applications.

- ***Curve sketching:*** In Section 4.2 we present a checklist for graphing a function. Sign charts and carefully labeled preliminary sketches accompany the examples; the final graphs are labeled with the important features of the functions. A detailed summary of the data is given under the caption of each figure. The examples and exercises in this section cover various types of functions to give students plenty of practice in graphing functions.
- ***Numerical methods:*** Computing technology has evolved enormously in the past several years and an increasing number of disciplines use numerical methods to approximate solutions to mathematical problems and to analyze data. Therefore, we introduce a variety of numerical methods throughout the text to acquaint students with the power and scope of these techniques. We discuss the total differential of several variables (Section 8.5), numerical integration (Sections 6.5 and 7.4), and the method of least squares (Section 8.6).

Features for the Student

- ***Algebra review:*** We have assumed that students taking this course will have had two years of high-school algebra or the equivalent. However, Appendix A includes a review of basic algebra. The numerous examples and exercises enable students to refresh their knowledge of algebra and to pinpoint those areas in which a deficiency may exist. They also give the instructor an option to cover as much review as needed.
- ***Algebraic steps:*** We know that students get discouraged if too many algebraic details are omitted. Therefore, in the exposition and in the examples, algebraic steps are clearly marked and explained. Many answers in the back of the book also include steps to the solution so students can check intermediate results. By providing intermediate steps we encourage students to persist in their problem-solving efforts.
- ***Table format:*** Except for some of the examples, the material in the text has been classroom-tested over a period of several years. Many pedagogical aspects of the text stem from this testing, including our special table format for many procedures. This format takes advantage of the fact that a compact array of information is more easily assimilated than a long list. These table formats have proved to be successful for learning and implementing many procedures, including integration by parts, the trapezoidal rule, Simpson's rule, the derivative tests for one and two independent variables, and the method of least squares.
- ***Plausibility arguments:*** The primary pedagogical tool we use is examples. Therefore, we have included only those proofs and theoretical discussions that contribute to an understanding of calculus and its applications. Even in these cases we often use

informal arguments since we believe students will find them more meaningful than formal mathematical proofs.

For Each Chapter

- ***Chapter introduction:*** Most of the chapter introductions discuss applications of the material; these applications and many others are then explored and developed in the examples and exercises of that chapter.
- ***Summary:*** Each chapter concludes with a concise summary highlighting the chapter's important concepts, definitions, formulas, and procedures. Students will find this summary an excellent study and review tool.
- ***Review exercises:*** In addition, at the end of each chapter, there is an extensive set of review exercises to test student comprehension and problem-solving skills. In keeping with our emphasis on applications, almost every review exercise set includes applied problems as well as numerous exercises to test skills and understanding.

Accuracy

To ensure accuracy, the exercises were solved independently by James W. Burgmeier and Wendy Fenwick. The examples have been solved and checked by several people, including Garret Etgen, Linda Holden, and Vincent P. Schielack. We are grateful for their keen eyes and attention to detail. If, however, you do find an error, please advise our publisher immediately so that it can be corrected in subsequent printings.

Flexibility

Since several chapters are independent of one another, there are various possible sequences of chapters, as the accompanying schema illustrates. For a one-semester course for business or managerial students, Sections 8.1 through 8.6 may be substituted for Chapter 6. Students who have a thorough precalculus background could skip Sections 1.1 through 1.3 and briefly study the material in the remainder of Chapter 1. The other optional sections in Chapters 1 through 7 are: 2.7; 4.3 through 4.6; 5.4; 6.5; 7.3; and 7.4 through 7.5.

Section Length

Generally, each section presents the amount of material that is appropriate for one fifty-minute lecture. Exceptions to this rule, such as Sections 4.2, 5.5, and 6.5, contain many examples and applications, rather than new mathematical material. Time devoted to these sections should depend on the disciplines represented in a particular class.

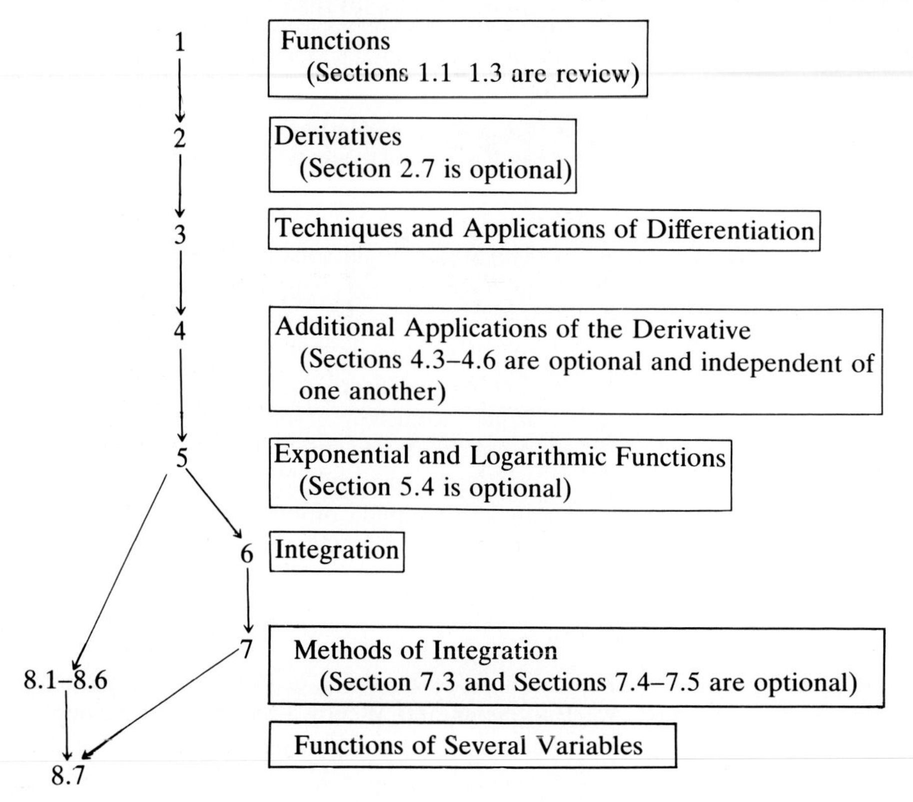

Supplements

This text is accompanied by a well-rounded supplements package, including Instructor's Manual, a Student Solutions Manual, Test Bank, Computerized Test Bank and software.

- *Student Solutions Manual:* The Student Solutions Manual was prepared by James W. Burgmeier and Wendy Fenwick and contains detailed solutions to selected odd-numbered exercises.
- *Instructor's Manual:* The Instructor's Manual was prepared by James W. Burgmeier and Wendy Fenwick, who prepared answers to even-numbered exercises, and Jan E. H. Johansson, who prepared sample tests. It includes three forms of a chapter test for each chapter, three forms for two mid-term exams, and three forms of two final exams to insure a plentiful supply of questions for instructors.
- *Test Bank:* The questions found in the sample tests, and additional test questions, are also available in a test bank form which corresponds to a computerized version for IBM-PC and compatibles.

The student supplements package also includes software. For more details, please contact your local McGraw-Hill Sales Representative.

Reviewers

We would like to thank the many reviewers who have read all or part of the manuscript. Their encouragement, comments, and suggestions have been most helpful.

Ann O'Connell	Providence College
Bruce H. Edwards	University of Florida
Garret Etgen	University of Houston
Brian D. Hassard	S.U.N.Y.-Buffalo
Kevin Hastings	Knox College
Robert Heal	Utah State University
Linda Holden	Indiana University
Lowell Leake	University of Cincinnati
Joyce Longman	Villanova University
Maurice Monahan	South Dakota State University
Hiram Paley	University of Illinois
Wayne Powell	University of Oklahoma
Vincent P. Schielack	Texas A&M
Daniel Shea	University of Wisconsin
Clifford W. Sloyer	University of Delaware
Robert E. White	North Carolina State University

Acknowledgments

We would like to express our appreciation to Wendy Fenwick at the University of Vermont for her work on the Student Solutions Manual and the Instructor's Manual; to Jan E. H. Johansson for preparing the Test Bank; to W. David Klemperer for his helpful suggestions on the forestry applications; and to Pat Burgmeier for typing the original manuscript.

Finally, we want to thank the editorial and production staff at McGraw-Hill and Random House: John Martindale, Senior Editor; Alexa Barnes, Development Editor; Margaret Pinette, Project Manager; her assistant, Julia Kerr; Michael Weinstein, Production Manager; his assistant, Susan Brown; and John Morriss, Editing Manager. Their help and support on this project have been invaluable.

James W. Burgmeier
Monte B. Boisen, Jr.
Max D. Larsen

INDEX OF APPLICATIONS

The following list is a selection from the approximately 600 applications in the text.

Business and Economics

Advertising in newspapers, 330, 364
Agricultural equipment sales, 179
Agricultural production, 332
Cable-television subscription campaign, 185
Cash management model, 177, 180
Class size related to fee, 80
Cobb–Douglas production function, 354–355
College fund, 21, 23, 200
Compound interest, 192–193, 194, 195, 214–216, 223, 224, 225
Constructing houses, 6
Consumer surplus, 264–266
Cost function, 25, 40, 167, 168, 173, 174, 177, 180, 181, 187, 188, 234, 238, 252, 306, 313, 330, 331, 340
Cost of building aquarium, 336–337
Crop yield, 80, 144, 306, 307, 320, 324, 330, 332
Demand function, 10, 27, 28, 29, 43, 112, 164, 174, 186, 188, 264, 265, 266, 323
Effect of tax on price and profit, 171–172, 173
Egg production rate on a chicken farm, 257
Electronics manufacturing, 102
Equilibrium price, 29
Farm profits related to choice of livestock, 331
Federal program to reduce unemployment, 120
Import fees, 173–174
Inspection for defective parts, 313
Interest-bearing account, 192–193, 194, 200–201, 214–216, 223, 224, 225, 226
Inventory model, 176
Investment rate of return, 304
Long-distance telephone costs, 30
Manufacturing batch size, 174–176, 288, 291
Manufacturing costs, 3, 10, 25–26, 80–81
Marginal, cost function, 168, 170, 172, 173, 187, 230, 234, 235, 238, 252, 257, 268, 277

demand, 323
price, 257
productivity of capital, 319, 323, 344, 346
productivity of labor, 319, 323, 344–345, 346
productivity of money, 336
profit, 268
revenue function, 169, 172, 173, 187, 234, 238, 239, 257, 277, 295, 299, 304
Maximizing, manufacturing profits, 84–85, 144, 331
production on a budget, 332, 333–334, 335–336, 340
profit, 74, 138, 142, 144, 169–170, 171, 173, 174, 187, 329, 331, 364
profits with limited resources, 142
rental income, 144
revenue, 140, 172, 173, 174, 364
Military spending versus welfare spending, 143
Minimizing, batching and storage costs, 174–177, 179, 180, 181, 188
inventory ordering costs, 180
production costs, 340
Newsletter design, 143
Optimal company size, 109, 112
Optimal order size, 176–177, 179, 180, 188
Optimal production level, 170, 171, 173
Penalty for price-fixing, 172–173
Plumbing costs, 30
Pollution from manufacturing plant, 323, 340
Postal rate function, 3, 4, 58, 86
Poverty level income, 31
Price elasticity, of demand, 181–183, 184, 185, 186, 188
of supply, 186, 188
Producer surplus, 267
Production as a function of shift, 331
Production cost, 25, 30, 40–41, 42, 73, 74, 112, 143, 145, 230, 235, 238, 252, 277
Production function, 39, 86, 170, 171–172, 252, 268, 295, 298–299, 304, 306–307, 313, 319, 323, 330, 331, 332, 333, 335, 340, 344, 346, 364
Production-inventory problem, 174, 179
Production-storage problem, 174, 179, 188
Productivity of manufacturing plant, 319, 323
Profit change, from advertising, 84, 85
from using fertilizer, 75
Profit from special edition of book, 144
Profit function, 42, 44, 74, 84, 85, 86, 138, 142, 144, 148, 169, 170, 171, 173, 187, 268, 328–329, 330, 331, 364
Publisher's profit from sale of novel, 330
Publishing, 143, 144, 313, 330
Rental car costs, 43
Restaurant inventory, 180
Restaurant tax, 184–185
Revenue function, 10, 31, 80, 167, 169, 172, 174, 187, 234, 238, 239, 264, 313, 330

Sales analysis, 225, 238, 239, 240, 247, 248, 257, 266, 286, 300, 323, 356
Sales from advertising, 225, 240, 306, 323, 356
Salesperson's income, 43
Setting advertising level, 166
Shipping costs, 364
Stock market investment strategy, 138, 144, 187
Store's profit, 328–329
Supply function, 29, 43, 265, 266
Tax effect on price, 171–172, 173
Television advertising, 323
Venture-capital investment profits, 74, 86
Worth of manufacturing company, 102

Life Sciences

Absorption rate of medicine, 220–221, 224
Animal diet cost, 340
Animal metabolism rate, 353
Athlete's workout, 284, 286
Bacteria population growth, 41, 85, 219, 223, 225, 268, 323, 351
Bacterial infection, 95, 225
Codling moth life cycle, 33, 37, 132
Cold medication to minimize symptoms, 330
Counting cricket chirps to measure temperature, 85
Decay of radioactive medicine, 223
Deer population program, 37
Drug concentration in bloodstream, 224
Earth's population, 223, 224
Field mice population growth, 257
Flu epidemic, 137–138, 143–144
Gerbil colony population, 42
Grasshopper population at various temperatures, 10
Growth rate of malignant tumor, 86
Insect infestation, 46–48, 51, 84
Landfill monitoring system, 41
Maze experiments, 2, 224
Mediterranean fruit fly infestation, 46–48, 51, 84
Mouse-to-elephant function (metabolism), 353
Park lands in city, 108–109
Pesticide/insecticide use, 46–48, 144, 330, 340
Photosynthesis rate of trees, 26–27, 30
Plant growth rate, 328, 331
Pollution index, 31, 320, 323, 343, 345
Population, density, 262–263, 266
  growth, 145, 216–217, 223, 224
  of United States, 224
Reaction rate to drugs, 306, 324

Reineke's equation, 222, 224
Rodent control, 41, 80
Runner's rate of respiration, 41
  training program, 43, 143
Sparrowhawk flight height, 277
Standard urea clearance function, 343–344
Tree, destruction by pests, 291
  diameter at maturity, 222–223, 224
Wildlife management approaches, 37
Wood production, 71, 74, 85, 300

Physical Sciences

Acceleration measurements, 83–84, 115, 239, 268, 277
Air-monitoring system, 221, 225–226
Animal holding pen, 364
Average speed of a car, 38–39, 81–82, 247, 248
Block of ice sliding down chute, 239
Capacity of livestock watering tank, 268
Car's speed before accident, 345
Chemical reaction rate, 80
Contamination of hay from Chernobyl explosion, 217–218
Deceleration measurements, 85, 236, 237, 239
Distance traveled, 85, 247–248, 253, 286–287, 291
Diving-pool depth, 85
Engine efficiency, 144
Fahrenheit-to-Celsius conversion function, 24–25
Fencing maximum pasture area, 139, 142, 146
Flight of arrow, 83–84, 239
Grain pile shape, 110
Highway safety, 208
Human cannonball, 85
Maximizing capacity, 141
Maximizing volume, 48–49, 51, 80, 129–130, 136–137
Measurement conversion, 24–25, 30
Minimizing surface area, 139, 142, 146
Missile's altitude recorder, 61
Oil slick, 112, 145
Projectile, 236, 239, 240, 248, 253, 268
Radioactive decay, 217–219, 223, 224
Rocket trajectory, 239
Stopping distances for various speeds, 237–238
Temperature measurements, 10, 24, 30
Trailer broken loose from car, 236–237
Velocity, measurements, 104, 253, 257, 277–278, 286, 291, 296
  of bomb dropped from airplane, 239
  of bullet, 85
  of hockey puck, 268

Water testing program, 253
Weather, 266, 312
Weather map showing isobars and isotherms, 311–312
Window design, 143

Social Sciences

Age of excavated bone, 218–219, 223
Burglaries during police strike, 300
Census, 93
Community cooperation index, 357
Crime rate, 330
Customer receptivity, 331
Education level and income level, 350–351, 358
Election campaign strategies, 113, 117, 120, 143
Final-exam grade versus study time, 223
Fundraising campaign, 43
Job-creation program, 34, 37
Job satisfaction, 340
Learning random sequences, 224
Mobility index, 357
Participation rank in foster-care center, 357
Popularity rank in foster-care center, 357
Psychology experiments, 2, 19–20, 23, 76–77, 80, 135, 143, 224, 277, 331, 339, 340
Quality-of-living index in rural area, 112
Reading proficiency test, 112
Researcher productivity, 257
Restaurant popularity, 268
Sociology, 30, 257, 300, 330, 349–350, 357, 358
  survey cost, 30
Spatial perception experiment, 19, 23
Standardized tests and amount of instruction, 6
Student achievement index, 339
Student receptivity to new concepts, 76–77, 80
Students' test scores, 356
Vacation planning, 340
Worker efficiency, 357

CONTENTS

1 ◆ FUNCTIONS 1
1.1 Introduction to Functions 2
1.2 Graphing Functions 11
1.3 Linear Functions 15
1.4 Applications of Linear Functions 24
1.5 Polynomials and the Absolute Value Function 31
1.6 Average Rate of Change 38
Summary 42
Review Exercises 43

2 ◆ THE DERIVATIVE 45
2.1 Motivation for The Derivative 46
2.2 Limits 52
2.3 Definition of the Derivative 61
2.4 Tangent Lines 64
2.5 Increasing and Decreasing Functions 75
2.6 Rate of Change 81
2.7 Continuity and Differentiability 86
Summary 94
Review Exercises 94

3 ◆ TECHNIQUES AND APPLICATIONS OF DIFFERENTIATION 97
3.1 Derivatives of Products and Quotients 98
3.2 The Chain Rule and Related Rates 105
3.3 The Second Derivative: Concavity 113
3.4 Derivative Tests for Relative Maxima and Relative Minima 120
3.5 Absolute Maxima and Absolute Minima 129
3.6 Applied Optimization Problems 135
Summary 144
Review Exercises 145

4 ◆ ADDITIONAL APPLICATIONS OF THE DERIVATIVE 147
4.1 Asymptotes 148
4.2 Graphing Functions 153
4.3 Implicit Differentiation 162
4.4 Applications to Economics 167
4.5 Inventory Problems 174
4.6 Relative Change and Elasticity 181
Summary 186
Review Exercises 187

5 • EXPONENTIAL AND LOGARITHMIC FUNCTIONS 189
5.1 Exponential Functions 190
5.2 The Function e^x 194
5.3 The Function ln x 200
5.4 Logarithmic Differentiation 210
5.5 Applications of the Exponential and Logarithmic Functions 214
Summary 224
Review Exercises 225

6 • INTEGRATION 227
6.1 Antidifferentiation 228
6.2 Applications of Antiderivatives 235
6.3 The Definite Integral 240
6.4 Area in the Plane 249
6.5 Riemann Sums and Applications 257
Summary 267
Review Exercises 268

7 • METHODS OF INTEGRATION 271
7.1 Integration by Substitution 272
7.2 Integration by Parts 278
7.3 Improper Integrals 286
7.4 The Trapezoidal Rule and Simpson's rule 291
7.5 Integration by the Use of Tables 300
Summary 303
Review Exercises 303

8 • FUNCTIONS OF MORE THAN ONE VARIABLE 305
8.1 Functions of Several Variables 306
8.2 Partial Derivatives 314
8.3 Optimization Problems 324
8.4 Constrained Optimization Problems and Lagrange Multipliers 331
8.5 The Total Differential 341
8.6 The Method of Least Squares 346
8.7 Double Integrals 358
Summary 363
Review Exercises 363

Appendix REVIEW OF ALGEBRA 365
Tables 377
Answers 393
Credits 429

Index I1

CREDITS

Chapter 1: page 1, R. B. Hoit/Photo Researchers; page 6, David Powers/Stock, Boston; page 24, Runk/Schoenberger/Grant Heilman
Chapter 2: page 45, Dean Abramson/Stock, Boston; page 46, Tom McHugh/Photo Researchers; page 71, George H. Harrison/Grant Heilman; page 77, Richard Wood/Taurus
Chapter 3: page 97, Jerry Howard, Positive Images; page 114, Arthur Grace/Stock, Boston; page 132, Grant Heilman; page 138, Mark Antman/The Image Works
Chapter 4: page 147, Barbara Rios/Photo Researchers; page 170, Courtesy Harley-Davidson Co.; page 176, The Photo Works/Photo Researchers
Chapter 5: page 189, Runk/Schoenberger/Grant Heilman; page 193, Hazel Hankin/Stock, Boston; page 214, Culver Pictures; page 219, Stan Levy/Photo Researchers
Chapter 6: page 227, Earth Observation Satellite Co.; page 241, Paul Conklin
Chapter 7: page 271, Barrera/TexaStock; page 278, Michael D. Sullivan, TexaStock; page 288, Topham/The Image Works
Chapter 8: page 305, Ellis Herwig/Stock, Boston; page 328, Grant Heilman; page 353, Culver Pictures
Chapter 9: page 365, Jack Spratt/The Image Works; page 395, John J. Krieger/Picture Cube
Chapter 10: page 405, George Gardner; page 410, Mark Antman/The Image Works; page 415, Wide World Photos; page 443, Leonard Lee Rue III/NAS/Photo Researchers
Chapter 11: page 449, Christopher Morrow/Photo Researchers; page 452, Paul Fortin/Stock, Boston; page 452, Bruce Iverson; page 457, Fred Lyon, Photo Researchers; page 474, Jerry Howard/Positive Images
Chapter 12: page 491, Alan Corey/The Image Works; page 500, Peter Menzel Photo Credits

***Facing page:* The rate of photosynthesis for a tree is a function of the amount of available light. Trees that thrive in the shade have a different type of function from those that do not. See Example 1.33.**

FUNCTIONS

· O · N · E ·

The first step in solving many applied problems is to express the problem in mathematical terms so that mathematical techniques can be used to solve it. Functions play an essential role in bridging the gap between a problem in its natural setting and its mathematical formulation because a function can describe the relationship between the quantities involved in a problem. Consider these situations:

- A manufacturer of automobiles wants to set the most profitable level of production for a particular model of car. A function can express the relationship between manufacturing costs and numbers of cars built.
- A psychologist studying the learning behavior of rats uses a function that expresses the relationship between the number of times a rat has run a particular maze and the number of seconds it takes the rat to go through the maze.
- A forester deciding whether a particular species of tree is suitable for reforesting a burned area uses a function that expresses the relationship between the amount of available light and the efficiency of photosynthesis for that species.
- A farmer trying to protect his apple orchard from a particular pest is interested in a function that relates the length of each stage of a pest's life to the average daily temperature.

1.1 Introduction to Functions

The list of possible applications of functions to problems faced by managerial, social, and life scientists is extensive. Throughout this book we will see many important applications of functions. In this section we discuss what a function is and develop mathematical notation for functions.

Each situation described in the introduction involves a relationship between two sets of numbers. For example, the automobile manufacturer's problem involves the set of all numbers of dollars that could be spent on manufacturing cars and all numbers of cars that could be manufactured. The manufacturer's decision about production levels would be easier if a rule were available that specifies the cost of manufacturing any given number of cars. This sort of rule is an essential part of a function.

Given two sets of numbers, A and B, a **function** from A to B expresses a relationship between these sets by assigning to *each* number in A *exactly one* number in B. The set A is called the **domain** of the function.

Consequently, in order to have a function, there must be a rule that specifies which element from B is assigned to each element chosen from A. If x is a variable that can assume values from A (that is, x can represent any number in A), and if y is a variable that can assume values from B, then a function from A to B associates exactly one value of y with each value of x.

Example 1.1 Consider the function that expresses the manufacturing cost corresponding to the number of automobiles manufactured. Here, A is the set of numbers of automobiles that could be manufactured and B is the set of all possible numbers of dollars. The function is given by the rule: "It costs y dollars to manufacture x automobiles."

Table 1.1 gives some hypothetical values for x and y. From these values, we see that it costs $100,000 to build just one automobile, $104,000 to build three, and $150,000 to build twenty automobiles. The high cost of building just one automobile reflects expenses incurred in setting up the production. Subsequent costs are due to materials and labor. ■

Table 1.1

Number of automobiles (x)	*Manufacturing cost* (y)
1	100,000
2	102,000
3	104,000
5	108,000
10	118,000
20	150,000

Because the value of y given by a function depends on the value of x "fed" into the function, we call y the **dependent variable**. Naturally, x is then called the **independent variable**. We will often emphasize the roles played by x and y by saying that *y is a function of x*. Occasionally, letters other than x and y are used as function variables; however, a function always has an independent variable and a dependent variable, and it will be clear from the context which letter stands for each type of variable.

Usually a formula or equation is used to define which value of the dependent variable is to be assigned to a given value of the independent variable.

Example 1.2 The formula

$$y = 3 + 2x$$

describes a function. For example, to the value $x = 1$ this function assigns the value $y = 3 + 2(1) = 5$.

The formula

$$y = (x^2 + 2)^3$$

describes another function; to the value $x = -2$ this function assigns the y-value 216 because $y = [(-2)^2 + 2]^3 = (4 + 2)^3 = 6^3 = 216$. ■

The next example illustrates that the rule specifying a function need not be a single formula.

Example 1.3 Consider a "postal rate" function: Let x represent the weight (in ounces) of a letter and let y be the amount (in cents) of postage needed to mail a letter by first class. In 1988 the rule was that if the weight x of a letter is 1 (ounce) or less, then $y = 25$; if x is more than one and less than or equal to 2, then $y = 45$; if x is more than two and less than or equal to 3, then $y = 65$, and so on. See Figure 1.1. ■

An essential part of the definition of a function from A to B is that to each number in A *exactly one* number from B is assigned. Therefore, not every equation defines a function, as shown in the next example.

Example 1.4 Let A and B denote the set of all real numbers. The rule that assigns to each value of x in A a number y in B such that $y^2 = x^2$ *does not* define a function because it does not determine a *unique* value

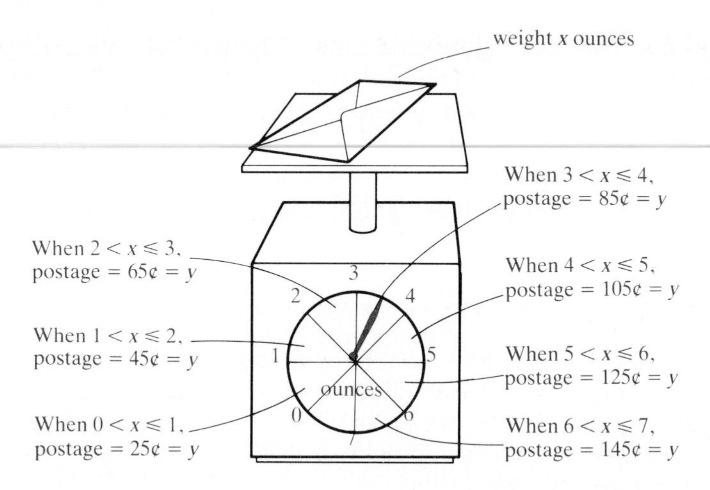

Figure 1.1
The postal rate function

of y corresponding to a given value of x. For example, for the value $x = 3$, two y-values, $y = 3$ and $y = -3$, satisfy the rule $y^2 = x^2$ (that is, both $(3)^2 = (3)^2$ and $(-3)^2 = (3)^2$). Having two values of y associated with one value of x violates the definition of a function. ■

Example 1.5 The rule relating y and x given by $y = x^2$ *does* define a function. Note, however, that if $x = 2$ (for example) then $y = 4$, and if $x = -2$ then again $y = 4$. This does not violate the definition of a function because only one value of y corresponds to each value of x (two values of x may yield the same value of y; this is not a violation of the definition of a function). ■

A formula such as $y = x^2 + 3x$ is interpreted as an abbreviation, or shorthand, for the rule relating the *given number* (independent variable) and the *assigned number* (dependent variable). In the formula $y = x^2 + 3x$, the assigned number is the sum of the square of the given number and three times the given number. The symbols x and y are only abbreviations for the given number and assigned number, respectively.

Instead of restating the rule that defines a function each time we wish to refer to the function, we usually denote the function by a letter. If f is used to denote a function, then the number y assigned to x will be denoted by $f(x)$ (read f *of* x). Note that $f(x)$ does *not* denote the product of f and x.

Example 1.6 If the function defined by the rule $y = x^2 + 3x$ is f, then we write

$$y = f(x) = x^2 + 3x$$

to denote the value of y corresponding to a value of x. Remember that x is a "placeholder" in this definition of $f(x)$, that is, the formula indicates that

$$f(\_) = (\_)^2 + 3(\_),$$

and any real number, or expression denoting a real number, may replace "_." Specifically, $f(2)$ denotes the value of y corresponding to

$x = 2$ and is obtained by substituting 2 for x in the rule for the function. Hence,

$$f(2) = (2)^2 + 3(2) = 4 + 6 = 10.$$

Similarly, for $x = -4$

$$f(-4) = (-4)^2 + 3(-4) = 16 - 12 = 4,$$

and for $x = a + 1$

$$f(a + 1) = (a + 1)^2 + 3(a + 1) = a^2 + 2a + 1 + 3a + 3 = a^2 + 5a + 4. \quad \blacksquare$$

If x is the independent variable of f, then we often refer to the function f as $f(x)$. The context will always clarify whether $f(x)$ denotes the function or the number assigned to x by f. However, there is a difference between the *function* f and its *value* $f(x)$ at a particular x. For the function f whose value at x is

$$f(x) = x^2 + 3x,$$

the rule of the function is "the assigned number is the sum of the square of the given number and three times the given number," and the formula specifies the assigned number associated with *each* given number. Virtually all the functions we will study may be defined by a formula, so we will not make any further distinction between the symbol, f, of the function and the symbol, $f(x)$, of function value.

The notation $f(x)$ is **functional notation**; it is a powerful tool in mathematics because it allows us to make precise statements easily. To remember what a function stands for in a given application, or to distinguish among several functions that are used at the same time, we will use a variety of letters for functional notation: $g(x)$, $w(x)$, $s(x)$, $c(x)$, and so on.

Example 1.7 Let $g(x) = 4 - x + x^2$. Find the following: $g(0)$, $g(2)$, $g(3)$, $g(a)$, $g(x + h)$, and $g(x) + g(h)$.

Solution

$$\begin{aligned}
g(0) &= 4 - 0 + 0^2 = 4 \\
g(2) &= 4 - 2 + (2)^2 = 6 \\
g(3) &= 4 - 3 + (3)^2 = 10 \\
g(a) &= 4 - a + a^2 \\
g(x + h) &= 4 - (x + h) + (x + h)^2 \\
&= 4 - x - h + x^2 + 2xh + h^2 \\
g(x) + g(h) &= (4 - x + x^2) + (4 - h + h^2) \\
&= 8 - x + x^2 - h + h^2.
\end{aligned}$$

It is important to note that the last two expressions are *not* equal, that is,

$$g(x + h) \neq g(x) + g(h). \quad \blacksquare$$

Example 1.8 When all other factors are constant, the average score attained by students in a particular school district on a standardized mathematics test is a function of the number of hours of mathematics instruction required for each student (to a maximum of 20 hours). We denote by x the number of hours of instruction required and by $s(x)$ the average score attained when x hours are required. Suppose we find (after careful analysis of all the available data) that, for $0 \leq x \leq 20$,

$$s(x) = 0.2x^2 + 0.5x + 5.$$

What is the average score if no instruction is required? What is the score if 10 hours of instruction are required? What is the score if 20 hours of instruction are required?

Solution The average score when no instruction is required is the value of $s(x)$ when $x = 0$. This is $s(0)$, and we find that

$$s(0) = 0.2(0)^2 + 0.5(0) + 5 = 5.$$

If 10 hours of instruction are required, the average score will be $s(10)$:

$$s(10) = 0.2(10)^2 + 0.5(10) + 5$$
$$= 20 + 5 + 5 = 30.$$

If 20 hours of instruction are required, the average score will be $s(20)$:

$$s(20) = 0.2(20)^2 + 0.5(20) + 5$$
$$= 80 + 10 + 5 = 95.$$

■

Example 1.9 A builder plans to build x houses in a subdivision that may contain no more than 100 houses. The total cost of building will be $C(x)$ dollars, where $C(x)$ is given by the rule

$$C(x) = -10x^2 + 20{,}000x + 50{,}000 \quad \text{for} \quad 0 \leq x \leq 100.$$

How much will it cost to build four houses? ten houses? Determine $C(0)$, and interpret its meaning.

Solution The cost of building four houses is

$$C(4) = -10(4)^2 + 20{,}000(4) + 50{,}000$$
$$= -160 + 80{,}000 + 50{,}000 = \$129{,}840$$

The cost of building ten houses is

$$C(10) = -10(10)^2 + 20{,}000(10) + 50{,}000$$
$$= -1{,}000 + 200{,}000 + 50{,}000 = \$249{,}000$$

Finally, $C(0) = -10(0)^2 + 20{,}000(0) + 50{,}000 = \$50{,}000$. This is the cost of building no houses. This amount is the *fixed cost* (cost occurred even if no houses are built), which may include such things as wages that have been promised to the employees, options on the land, and overhead for office facilities. ■

Sometimes the rule that defines a function $f(x)$ cannot be used for certain values of x, or some values of x may be unreasonable in the context of a particular situation. When this happens we must restrict the values that x is allowed to assume. The set of permissible values of x is the domain of $f(x)$.

Three types of restrictions are commonly placed on domains. First, the domain may be stated explicitly, as in the cost function for the subdivision houses in Example 1.9. Second, we must exclude from the domain any values of x that yield a fraction with a zero denominator, as shown in the next example.

Example 1.10 Consider the function

$$f(x) = \frac{x}{x-3}.$$

The rule can be executed for all values of x except 3 because we cannot divide by zero. Hence the domain of $f(x)$ is the set of all real numbers except 3. ■

A third restriction may be placed on the domain if the function involves a square root. (Note that by *square root* we mean the *principal* square root. For example $\sqrt{4} = 2$ not ± 2 nor -2.) Since the square root of a negative number is not a real number, we exclude from the domain of $f(x)$ any values of x that result in the square root of a negative number. Similarly, we exclude any even root (fourth root, sixth root, and so on) of a negative number.

Example 1.11 If $f(x)$ is defined by

$$f(x) = \sqrt{4-x},$$

then we must exclude from the domain of $f(x)$ any value of x that makes $4 - x < 0$. Therefore, we must exclude all $x > 4$, so the domain of $f(x)$ consists of all x such that $x \leq 4$. ■

A common-sense approach will usually be sufficient to determine the domain of the functions we will use in this book. Consequently, when given a rule for a function $f(x)$, we will assume without further comment that any value of x for which the rule cannot be executed is automatically excluded from the domain.

The domains we deal with are often intervals of real numbers, and it is convenient to use *interval notation* to describe them. Table 1.2 shows interval notations, their meanings, and graphical representations. (A more thorough discussion of intervals is given in Appendix A, Section A.3.)

For example, the domain of $s(x)$ in Example 1.8 is $[0, 20]$; the domain of $C(x)$ of Example 1.9 is $[0, 100]$; and the domain of $f(x)$ in Example 1.11 is $(-\infty, 4]$. The interval $[a, b]$ that includes the endpoints a and b is a **closed interval**. The interval (a, b) that excludes both a and b is an **open interval**.

Example 1.12 Determine the domain of

$$f(x) = \frac{1}{\sqrt{x+2}}.$$

Table 1.2
Interval notation

Interval	*Includes all x such that*	*Graphical representation*
$[a, b]$	$a \le x \le b$	a b
$(a, b]$	$a < x \le b$	a b
$[a, b)$	$a \le x < b$	a b
(a, b)	$a < x < b$	a b
$(-\infty, b]$	$x \le b$	b
$(-\infty, b)$	$x < b$	b
$[a, \infty)$	$a \le x$	a
(a, ∞)	$a < x$	a

Solution If $x < -2$, then $f(x)$ is not defined, since we cannot take the square root of a negative number. If $x = -2$, then $f(x)$ is not defined, since we cannot divide by zero. The rule can be used for all other values of x, so the domain is $(-2, \infty)$. ■

Example 1.13 Consider the function

$$s(t) = \frac{a\sqrt{t+b}}{t+c}$$

with certain constants a, b, and c. For what values of t is this function not defined, that is, what real numbers are not included in the domain?

Solution First, $s(t)$ is not defined when $t = -c$, since the denominator is then zero. Second, $s(t)$ is not defined when $t < -b$, since for $t < -b$ we would take a square root of a negative number. Consequently, $t = -c$ and all $t < -b$ are not in the domain of $s(t)$. ■

We are interested in expressions of the form

$$\frac{f(x) - f(a)}{x - a},$$

where a represents some constant value, because such expressions will be important for our later work. Note that $x = a$ yields a zero denominator, so this expression is not defined when $x = a$ even though the numerator is also zero at $x = a$.

Example 1.14 Let $f(x) = x^2 - 2x$. Form a new function $g(x)$ defined by

$$g(x) = \frac{f(x) - f(a)}{x - a}$$

for $a = 3$, and simplify.

Solution Substituting 3 for a in the expression yields

$$g(x) = \frac{f(x) - f(3)}{x - 3}$$

$$= \frac{(x^2 - 2x) - [(3)^2 - 2(3)]}{x - 3}$$

$$= \frac{x^2 - 2x - (9 - 6)}{x - 3} = \frac{x^2 - 2x - 3}{x - 3}.$$

The domain of $g(x)$ consists of all real numbers except 3. If we factor the numerator, we get $x^2 - 2x - 3 = (x - 3)(x + 1)$. Thus, if $x \neq 3$ then $x - 3 \neq 0$, and the nonzero factor can be cancelled to give

$$g(x) = \frac{(x - 3)(x + 1)}{x - 3} = x + 1.$$

Hence, $g(x) = x + 1$ for $x \neq 3$ and $g(x)$ is *not defined* for $x = 3$. Note that 3 is excluded from the final domain because it is not in the original domain of $g(x)$, even though the final expression $x + 1$ can be evaluated at 3. ∎

As the previous example demonstrates, the notion of *equal functions* depends on the domains as well as on the rule for assigning values. For example,

$$f(x) = \frac{(x + 3)(x - 2)}{x + 3} \quad \text{and} \quad g(x) = x - 2$$

are not equal functions, since -3 is in the domain of $g(x)$ but not in the domain of $f(x)$. However, if -3 is excluded from the domain of $g(x)$, then $f(x) = g(x)$.

In general, two functions $f(x)$ and $g(x)$ are **equal** if and only if they have the same domain and, for every value a in this domain, $f(a) = g(a)$.

Exercises

Exer. 1–10: Find the values of the function, and simplify where possible.

1. $f(x) = 3x + 11$; $f(2)$, $f(0)$, $f(-1)$, $f(-4)$, $f(h)$
2. $g(x) = 4x - 8$; $g(0)$, $g(-2)$, $g(2)$, $g(x + h)$
3. $f(x) = 2x^2 - 3x$; $f(0)$, $f(2)$, $f(-3)$, $f(h)$
4. $g(x) = 4 - x^2 + 3x^3$; $g(0)$, $g(2)$, $g(6)$, $g(x + h)$
5. $h(x) = \sqrt{x^2 - 3}$; $h(2)$, $h(3)$, $h(6)$, $h(1)$, $h(x + a)$, $h(x) + h(a)$
6. $f(x) = \dfrac{1}{x + 1}$; $f(0)$, $f(1)$, $f(3)$, $f(x + h)$, $f(x) + f(h)$
7. $f(x) = -17$; $f(0)$, $f(-1)$, $f(5)$, $f(x + h)$
8. $f(x) = 3x - \sqrt{x^2 - 3}$; $f(2)$, $f(3)$, $f(x + h)$
9. $f(x) = \dfrac{2x}{x^2 + 1}$; $f(0)$, $f(2)$, $f(-2)$, $f(x + h)$
10. $g(x) = (x^2 + 2)(3 - x)^2$; $g(0)$, $g(2)$, $g(-2)$, $g(3)$

Exer. 11–16: For the given functions, compute $f(2)$, $f(5)$, and $f(7)$. Compare the value of $f(7)$ with $f(2)+f(5)$.

11. $f(x)=x^2$
12. $f(x)=x^3$
13. $f(t)=\sqrt{t+3}$
14. $f(u)=\frac{1}{u}$
15. $f(x)=x^3-x$
16. $f(s)=s^4-4s^3$

Exer. 17–28: Determine if the equation defines y as a function of x. (For instance, the rule given in Exercise 20 associates with each value of x a value of y that satisfies $y^2=16-x^2$.)

17. $y=4x-3$
18. $\frac{y}{5}=2x$
19. $y=\sqrt{x-2}$
20. $x^2+y^2=16$
21. $x-2=\sqrt{y}$
22. $x=\sqrt{y-2}$
23. $x^2=y^3$
24. $y^2=x^3$
25. $y=\sqrt{x^2+1}$
26. $x=\sqrt{y^2+1}$
27. $x^2=y^4+1$
28. $x^3=y^3$

Exer. 29–31: A company finds that the cost (in hundreds of dollars) of manufacturing x trailers is given by $C(x)=80+45x$.

29. Find $C(10)$, $C(100)$, and $C(200)$
30. Find $C(0)$, and interpret its meaning.
31. Find $C(-10)$, and decide whether it is meaningful.

Exer. 32–35: Refer to Exercises 29–31. The number of trailers that would be sold if the trailers were priced at x hundred dollars each is the *demand*. Suppose the demand $D(x)$ is given by

$$D(x)=200-0.5x.$$

The money paid (in hundreds of dollars) to the company for the sale of trailers at x hundred dollars per trailer is x times the number of trailers sold. This money is the *revenue*, denoted by $R(x)$. Hence, if $D(x)$ trailers are sold at x hundred of dollars per trailer, the revenue (in hundreds of dollars) is

$$R(x)=x\,D(x)=x(200-0.5x).$$

32. Find $D(10)$, $D(100)$, $D(200)$, $R(10)$, $R(100)$, $R(200)$, and $R(300)$.
33. Find $D(0)$, and interpret its meaning.
34. Interpret the fact that $R(300)$ is less than $R(200)$.
35. How many trailers should be produced (that is, what would the demand be) if the selling price is \$5,000 per trailer?

Exer. 36–38: The number of grasshoppers in a certain field is given by

$$f(x)=300-0.1x^2,$$

where x is the number of days after the ground temperature reaches 15 °C in the fall.

36. How many grasshoppers will be in the field 10 days after the ground temperature reaches 15 °C?
37. How many grasshoppers will be in the field 40 days after the ground temperature reaches 15 °C?
38. How many days after the temperature reaches 15 °C will there be 260 grasshoppers in the field?

Exer. 39–46: Find the domain for the function and explain why each value excluded from the domain had to be excluded.

39. $f(x)=\frac{x+1}{x}$
40. $f(x)=3+\frac{4}{x-5}$
41. $g(x)=\frac{3}{(x-1)(x-2)}$
42. $Q(x)=\sqrt{x^2+5}$
43. $w(x)=\sqrt{3-x}$
44. $g(r)=\frac{1}{\sqrt{r-4}}$
45. $K(t)=\frac{3t}{t^2+1}$
46. $h(x)=\sqrt{x-5}$

Exer. 47–56: For the function $f(x)$ and value a, form

$$\frac{f(x)-f(a)}{x-a}$$

and simplify.

47. $f(x)=2x-3;\quad a=2$
48. $f(x)=x+6;\quad a=-3$
49. $f(x)=0.5+x;\quad a=-1$
50. $f(x)=x^2-1;\quad a=2$
51. $f(x)=x^2+x;\quad a=2$
52. $f(x)=3x^2-x+6;\quad a=-2$
53. $f(x)=6;\quad a=3$
54. $f(x)=1/x;\quad a=1$
55. $f(x)=\sqrt{x};\quad a=2$
56. $f(x)=\sqrt{3-x};\quad a=1$

Exer. 57–66: Find the largest domain for which $f(x)=g(x)$.

57. $f(x)=x+3;\quad g(x)=\frac{(x+3)(x-2)}{x-2}$
58. $f(x)=\frac{(x+3)(x+2)}{(x-1)(x+3)};\quad g(x)=\frac{x+2}{x-1}$
59. $f(x)=\frac{x^2+2x-8}{x-2};\quad g(x)=x+4$
60. $f(x)=\frac{x^2+3x+2}{x^2+4x+3};\quad g(x)=\frac{x^2+x-2}{x^2+x-6}$
61. $f(x)=\sqrt{(x-2)x};\quad g(x)=\sqrt{x-2}\sqrt{x}$
62. $f(x)=\frac{(x-2)(x+1)}{x+1};\quad g(x)=x-2$
63. $f(x)=\frac{(x+2)(x+1)}{x+1};\quad g(x)=\frac{(x+2)(x-3)}{x-3}$
64. $f(x)=x-3;\quad g(x)=\frac{x^2-x-6}{x+2}$
65. $f(x)=\frac{x^2+4x-5}{x^2+x-2};\quad g(x)=\frac{x^2+8x+15}{x^2+5x+6}$
66. $f(x)=\sqrt{(x-4)(x+2)};\quad g(x)=\sqrt{x-4}\sqrt{x+2}$

Graphing Functions

1.2

An illuminating way to analyze a function is to represent it graphically. Indeed, graphical representation of data is an indispensable device in almost all disciplines. To provide a framework for the graphical representation of a function, we construct two perpendicular lines and choose a unit of measurement to be used along each line (see Figure 1.2). We usually call the horizontal line the *x-axis* and the vertical line the *y-axis* (although we may sometimes use other letters to denote these axes). The point of intersection of these axes is the *origin*. The positive directions along the x-axis and the y-axis are to the right and upward, respectively; the negative directions along the x-axis and the y-axis are to the left and downward, respectively. This framework is called a **Cartesian coordinate system** of the plane.

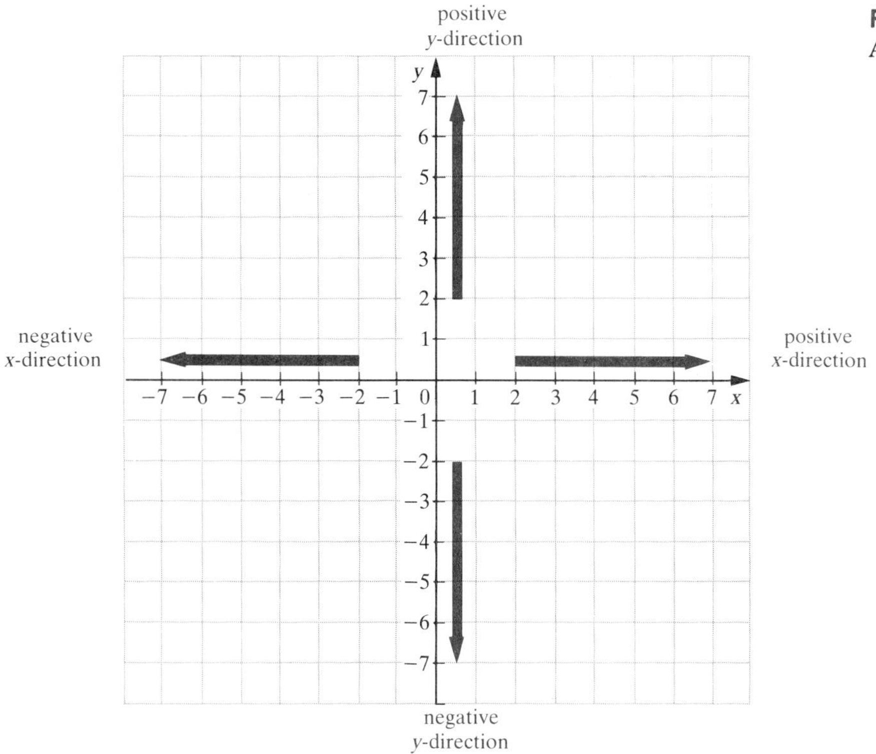

Figure 1.2
A Cartesian coordinate system

Points in the plane are labeled by *ordered pairs* of the form (a, b) such that a is the signed distance from the origin measured along the x-axis and b is the signed distance from the origin measured along the y-axis. See Figure 1.3. For the point (a, b), a is the *x-coordinate* and b is the *y-coordinate*.

The **graph of a function** is the set of points of the form (a, b) such that $b = f(a)$. Thus the graph of a function $f(x)$ is the set of points of the form $(a, f(a))$ for all a in the domain of $f(x)$. If $b = f(a)$, then we say that the graph of $f(x)$ *passes through* (a, b) or that the point (a, b) is

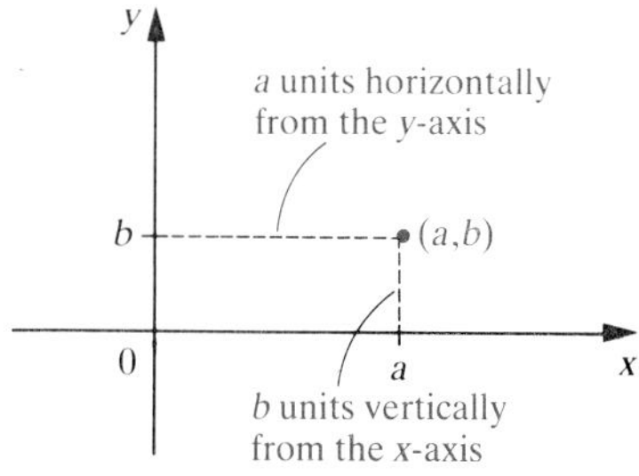

Figure 1.3
Location of the point (a, b) on a coordinate system

Table 1.3

Some values of $f(x) = 3x - 2$

x	-2	-1	0	1	2	3	4
$f(x)$	-8	-5	-2	1	4	7	10

on the graph of $f(x)$. Thus $(3, 6)$ is on the graph of $f(x) = x^2 - x$ because $f(3) = 3^2 - 3 = 6$. Similarly, the graph of $f(x) = x^3$ passes through $(2, 8)$ because $f(2) = 2^3 = 8$.

Example 1.15 Plot some points on the graph of the function $f(x) = 3x - 2$.

Solution We arbitrarily choose some values of a in the domain of f and plot $(a, f(a))$ for each of them. If $a = 2$, then $f(a) = f(2) = 3(2) - 2 = 4$, so $(2, 4)$ is a point on the graph. If $a = 3$, then $f(3) = 3(3) - 2 = 7$, so $(3, 7)$ is a point on the graph. We have shown the results of several such calculations in Table 1.3 and have plotted the corresponding points in Figure 1.4(a).

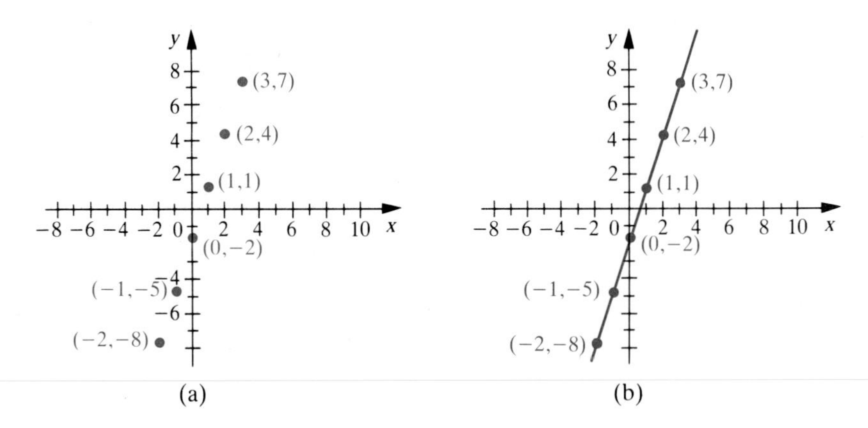

Figure 1.4
(a) Some points on the graph of $f(x) = 3x - 2$; (b) the graph of $f(x) = 3x - 2$

Of course, we cannot plot all of the points on the graph of $f(x) = 3x - 2$ because the graph contains an infinite number of points. However, the points we have plotted suggest that the graph is the line sketched in Figure 1.4(b). ■

In the preceding example, the graph of $f(x)$ crosses the y-axis at the point $(0, -2)$; therefore, -2 is the **y-intercept** of $f(x)$. A point is on the y-axis if and only if its x-coordinate is zero. So the **y-intercept of a function** $f(x)$ is $f(0)$, the y-coordinate of the point $(0, f(0))$ on the graph, provided that $f(x)$ is defined at $x = 0$.

A function can have at most one y-intercept. If it had two or more y-intercepts, then two or more y-values would correspond to $x = 0$, thus violating the definition of function.

An **x-intercept** of $f(x)$ is a number x in the domain of $f(x)$ for which $f(x) = 0$. The graph of $f(x)$ crosses the x-axis at each x-intercept. Hence, an x-intercept of a function is sometimes called a **zero** of the function. Since $3x - 2 = 0$ if and only if $x = \frac{2}{3}$, the function $f(x) = 3x - 2$ has one x-intercept, which is $\frac{2}{3}$. A function may have none, one, or many x-intercepts. See Figure 1.5.

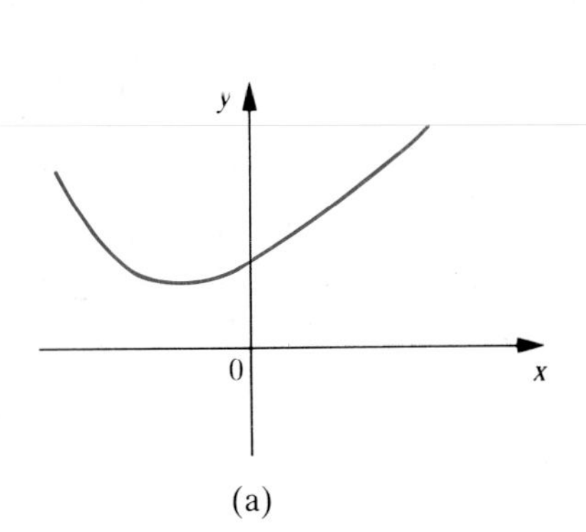

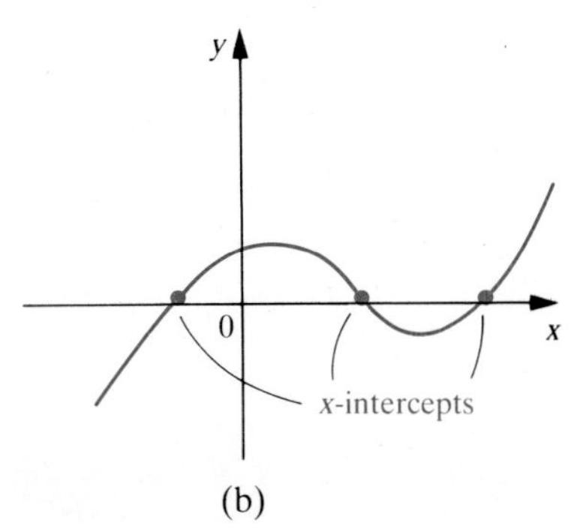

Figure 1.5
(a) A function with no x-intercept; (b) a function with three x-intercepts

Example 1.16 Determine a few points on the graph of $f(x) = -x^2 + 6x - 5$ and locate its y-intercept. Sketch the graph by drawing a smooth curve through the points, and find its x-intercepts.

Solution The y-intercept of $f(x)$ is found by evaluating $f(x)$ at $x = 0$. Since $f(0) = -0^2 + 6(0) - 5 = -5$, the y-intercept of $f(x)$ is -5. We then evaluate the function at a few selected values of x and construct Table 1.4 from the results. The value corresponding to $x = 3$ is found by calculating $f(3) = -3^2 + 6(3) - 5 = 4$. Other entries are found by similar calculations. The points corresponding to the table entries are plotted in Figure 1.6(a). The pattern of these points suggests the curve drawn in Figure 1.6(b). Since $f(1) = 0$ and $f(5) = 0$, we know that 1 and 5 are two of the x-intercepts. We will see later that these are the only x-intercepts of this function. ■

Table 1.4
Some values of $f(x) = -x^2 + 6x - 5$

x	−1	0	1	2	3	4	5	6
$f(x)$	−12	−5	0	3	4	3	0	−5

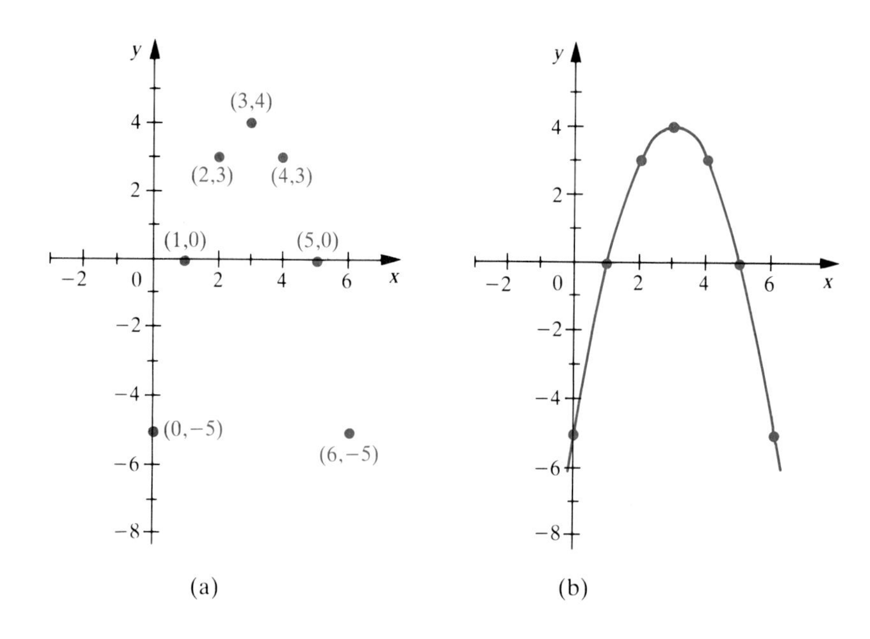

Figure 1.6
(a) Some points on the graph of $f(x) = -x^2 + 6x - 5$; (b) the graph of $f(x) = -x^2 + 6x - 5$

It is important to realize that the y-coordinate of a point (a, b) on the graph of a function $f(x)$ is computed by evaluating the function at $x = a$. Since $b = f(a)$, the value $f(a)$ is the signed vertical distance, measured at $x = a$, from the x-axis to the graph of the function at the point (a, b).

Suppose we are given the graph of some function $y = f(x)$. To find $f(a)$ for some number a in the domain of $f(x)$, we locate a on the x-axis, draw a line perpendicular to the x-axis through $x = a$, and determine the y-value (or y-coordinate) of the point where this line intersects the graph. This y-value is $f(a)$. See Figure 1.7(a). If this vertical line intersects the graph of $f(x)$ at more than one point, as in Figure 1.7(b), then a has more than one y-value assigned to it, thus violating the definition of function. We can easily determine at a glance whether a given curve in the plane is the graph of a function or not. *If any vertical line crosses a curve more than once, then the curve is not the graph of a function of x.* Otherwise, the curve *is* the graph of a function, and its domain is the set of all x-values a for which the vertical line passing through $x = a$ intersects the graph.

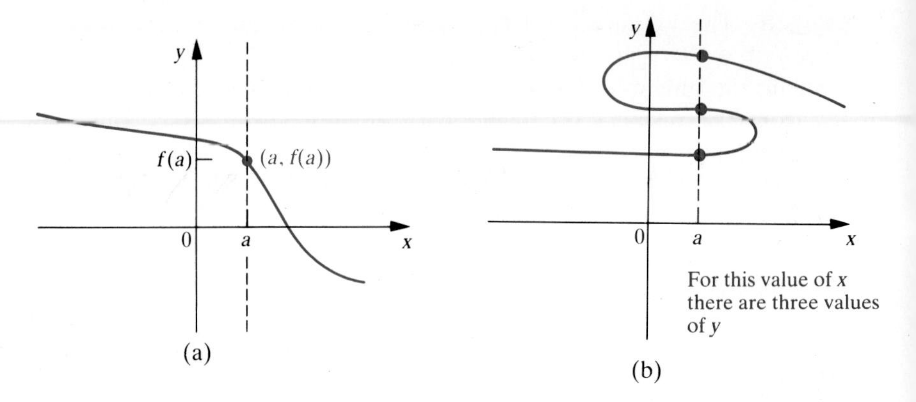

Figure 1.7
(a) A vertical line may intersect the graph of a function at only one point; (b) a curve that is not the graph of a function

Figure 1.8 shows the graphs of the important functions $f(x) = \sqrt{x}$ and $f(x) = 1/x$. Note that there are no points on the graph of $y = \sqrt{x}$ for $x < 0$ because these values are not in the domain of this function. Similarly, the graph of $y = 1/x$ does not cross the y-axis, since $x = 0$ (the value of x on the y-axis) is not in the domain of this function.

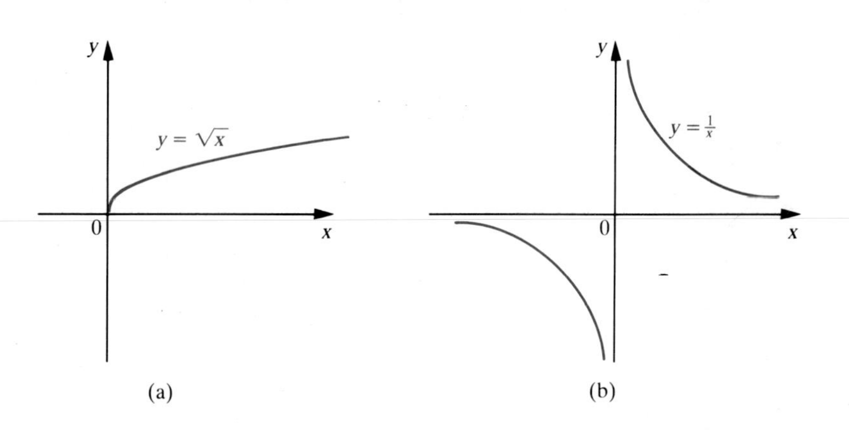

Figure 1.8
(a) The graph of $f(x) = \sqrt{x}$; (b) the graph of $f(x) = 1/x$

Exercises

Exer. 1–18: Plot these points on a common coordinate system.

1. $(2, 3)$
2. $(-2, -3)$
3. $(4, -7)$
4. $(6, -2)$
5. $(3, -4)$
6. $(-2, 0)$
7. $(2, 4)$
8. $(3, 4)$
9. $(-3, 4)$
10. $(0, 4)$
11. $(2, -3)$
12. $(2, 0)$
13. $(2, 5)$
14. $(2, -2)$
15. $(a, 2a)$; $a = 3$
16. $(a, 4a + 1)$; $a = -1$
17. $(b, b^2 - 2b)$; $b = 3$
18. $(b, b^2 + 2b - 6)$; $b = -4$

Exer. 19–22: Plot the following points. Use a new set of axes for each problem.

19. $(a, 3a)$; $a = -1, 0, 1, 2$
20. $(a, 2a + 1)$; $a = -3, -1, 0, 2$
21. $(a, a^2 - 2a + 1)$; $a = -2, -1, 0, 1, 2$
22. $(a, 2a^2 - 3a + 2)$; $a = -1, 0, 1, 2, 3$
23. Determine the coordinates of the third vertex of a right triangle that has points (a, b) and (c, d) as the ends of the hypotenuse and has legs parallel to the axes, as shown in the figure. (Note: there are two answers one of which is shown in the figure.)

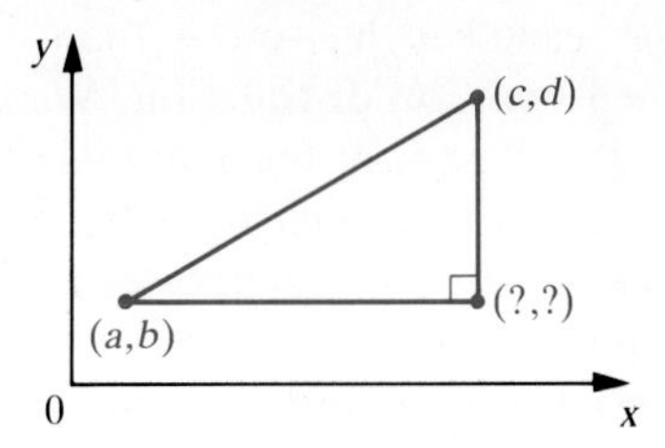

24. Determine the coordinates of the other two corners of the rectangle that has points (a, a^2) and $(b, 2b)$ as opposite corners and has sides parallel to the axes, as shown in the figure.

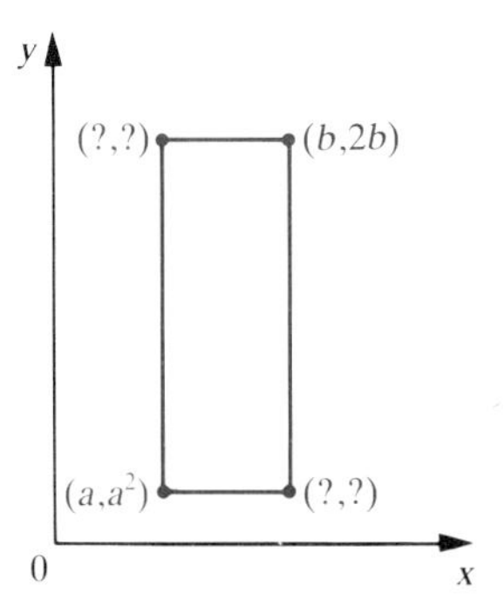

Exer. 25–40: Sketch the graph of the function by plotting several points and connecting them with a smooth curve. Find the y-intercept and the x-intercepts.

25. $f(x) = 4x$
26. $f(x) = -6x$
27. $f(x) = 2 - 4x$
28. $f(x) = 2x - 3$
29. $f(x) = 6$
30. $f(x) = -4$
31. $f(x) = x^2$
32. $f(x) = 2x^2$
33. $f(x) = x^2 + 3$
34. $f(x) = 2x^2 + 4$
35. $f(x) = 9 - x^2$
36. $f(x) = 4 - 2x^2$
37. $f(x) = \sqrt{x}, \quad x \geq 0$
38. $f(x) = \sqrt{3x}, \quad x \geq 0$
39. $f(x) = 1/x$
40. $f(x) = (1/x) - 2$

Exer. 41–44: Sketch the graph of the function by plotting several points and connecting them with a smooth curve. Find the y-intercept.

41. $f(x) = x^2 + x + 2$
42. $f(x) = x^2 - 4x + 6$
43. $f(x) = -x^2 + x + 2$
44. $f(x) = -x^2 - 4x + 6$

Exer. 45–50: Determine if the graph is the graph of y as a function of x. If not, explain why.

45.

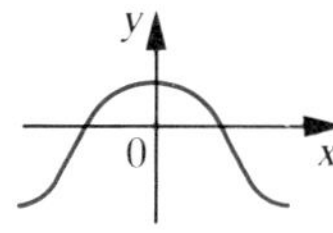

46.

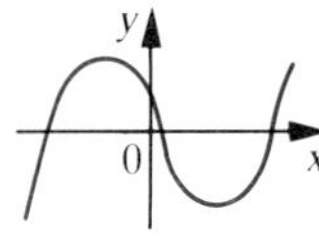

47.

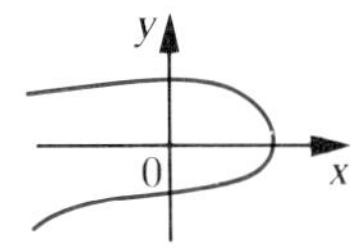

48.

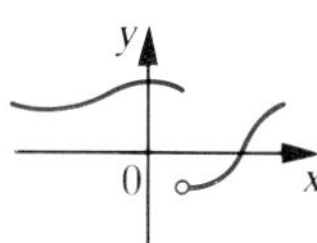

49.

50.

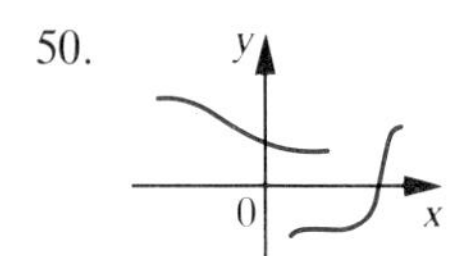

51. Consider the functions $f(x) = 2x - 2$ and $g(x) = -x + 7$. Their graphs are intersecting lines. Graph these functions on a common coordinate system by plotting a few points, and estimate where the lines intersect.
52. Work Exercise 51 with $f(x) = x + 3$ and $g(x) = 9 - 4x$.
53. Graph the functions $f(x) = x - 2$ and $g(x) = x^2 + x - 6$ and estimate where their graphs intersect.

Linear Functions

1.3

Now that we have introduced the general notion of functions and their graphs, we turn our attention to a detailed study of one type of function. In this section we will study **linear** functions—those whose graphs are straight lines. A function $f(x)$ is linear function if and only if it can be written in the form

$$f(x) = mx + b$$

for some constants m and b. Every function that can be written in this form has a straight line as its graph, and every function with a straight line as its graph can be written in this form. Functions such as $f(x) = 2x + 4$, $f(x) = -3x + 6$, and $f(x) = 5(2x + 6)$ are linear functions (the latter can be written as $f(x) = 10x + 30$). Functions such as $f(x) = x^2 + 3$ and $f(x) = 2\sqrt{x} + x + 6$ are not linear. Any nonvertical straight line is the graph of a function. A vertical line is not the graph of a function, since the x-value of the point where it crosses the x-axis has more than one y-value assigned to it (in fact, this x has infinitely many values assigned to it).

The constants m and b in the function $f(x) = mx + b$ have special

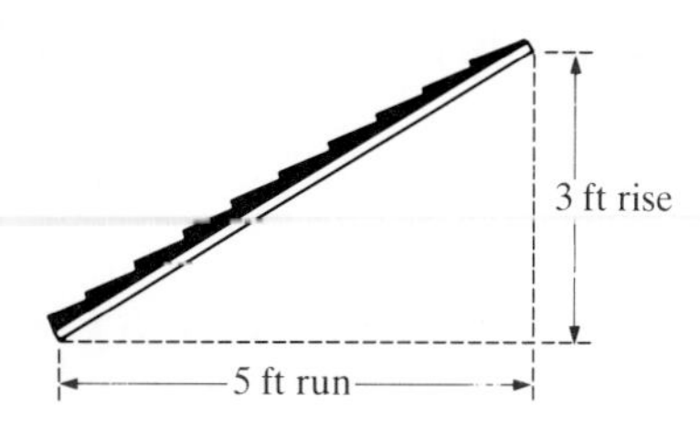

Figure 1.9
The "rise" and "run" of a pitched roof

meanings. Since $f(0) = m(0) + b = b$, we see that b is the y-intercept of $f(x)$. The constant m also gives important information about the graph of $f(x)$, namely, *m is the slope of its graph.* The definition of the slope of a line is related to an idea used in the construction industry. The slope, or pitch, of a roof is given by the ratio

$$\text{slope} = \frac{\text{rise}}{\text{run}}.$$

For example, saying that the roof shown in Figure 1.9 has a slope of $\frac{3}{5}$ indicates that the height of the roof increases 3 feet as we move 5 feet closer to the center of the building.

We can make a similar analysis of the linear function $f(x) = mx + b$. Suppose we choose two distinct arbitrary points on the graph, say $(x_1, f(x_1))$ and $(x_2, f(x_2))$, as shown in Figure 1.10. Because both points lie on the graph of $f(x) = mx + b$, the following two statements must be true:

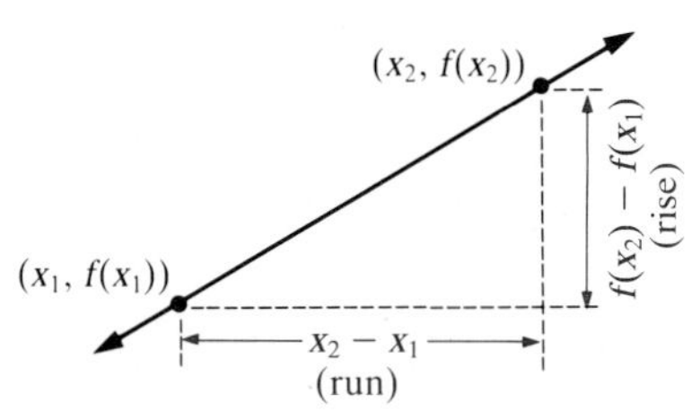

Figure 1.10
The "rise" and "run" of a line

$$f(x_1) = mx_1 + b \quad \text{and} \quad f(x_2) = mx_2 + b.$$

Thus, $\quad f(x_1) - mx_1 = b \quad \text{and} \quad f(x_2) - mx_2 = b.$

Since the expression on the left in both equations is equal to b, these expressions are equal to each other, so we have

$$f(x_2) - mx_2 = f(x_1) - mx_1.$$

We then solve for m:

$$f(x_2) - f(x_1) = mx_2 - mx_1 = m(x_2 - x_1).$$

Thus

$$m = \frac{f(x_2) - f(x_1)}{x_2 - x_1}.$$

Since $x_2 - x_1$ is the change in x, or the run, and since $f(x_2) - f(x_1)$ is the corresponding change in $f(x)$, or the rise, we conclude that m is the ratio rise/run. Hence m is the slope of the line $f(x) = mx + b$.

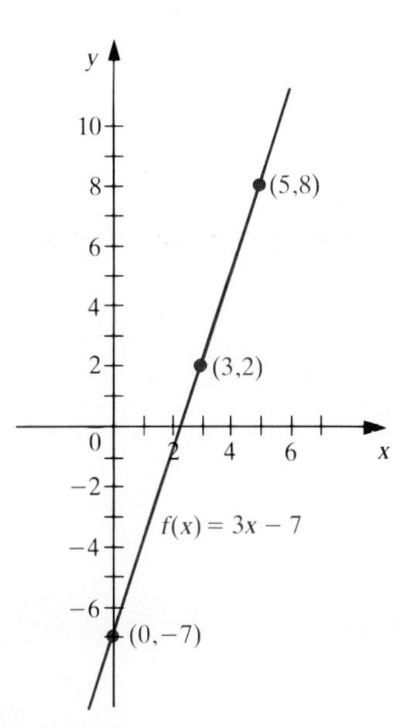

Figure 1.11
The graph of the linear function $f(x) = 3x - 7$

Example 1.17 Suppose that the graph of a linear function $f(x)$ passes through the points $(3, 2)$ and $(5, 8)$. Find m, b, and the equation describing $f(x)$.

Solution Let $x_1 = 3$ and $x_2 = 5$. Thus $f(x_1) = 2$ and $f(x_2) = 8$, so

$$f(x_2) - f(x_1) = 8 - 2 = 6 \quad \text{(rise)}$$

and

$$x_2 - x_1 = 5 - 3 = 2 \quad \text{(run)}$$

and so $m = \frac{6}{2} = 3$. Thus $f(x) = 3x + b$. To determine b, we use the fact that $f(5) = 8$, that is,

$$8 = 3(5) + b = 15 + b.$$

Hence, $b = 8 - 15 = -7$, and we can write $f(x)$ as

$$f(x) = 3x - 7.$$

The slope is 3 and the y-intercept is -7. See Figure 1.11. As a check, verify that points $(5, 8)$ and $(3, 2)$ both satisfy the equation. Also verify that we obtain the same equation if we choose $x_1 = 5$, $f(x_1) = 8$ and $x_2 = 3$, $f(x_2) = 2$. By considering similar triangles and using the fact that corresponding sides of similar triangles are proportional, we see that *any two points on the line will determine the same values for m and b.* ■

If the slope of a line is positive, then the line slants upward from left to right. The greater the slope, the more steeply the line rises. See Figure 1.12. If its slope is negative, then a line slants downward from left to right. The more negative the slope, the more steeply the line falls. When the slope is zero, the line slants neither up nor down, but is horizontal. (A line is horizontal if its rise is zero.) The only type of line that does not fit into one of these categories is a vertical line. The slope of a vertical line is undefined because the run for a vertical line is zero, and we cannot divide by zero.

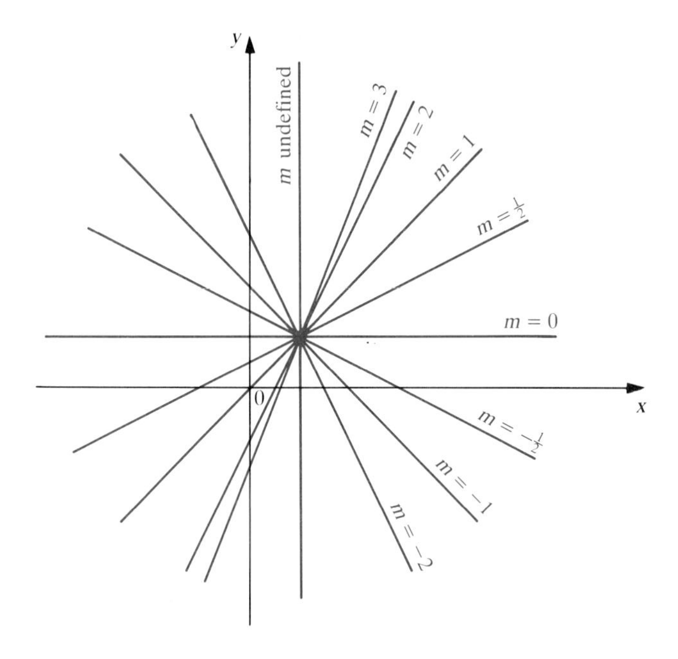

Figure 1.12
Several lines and their slopes

Example 1.18 Sketch the graph of $f(x) = 3x - 7$.

Solution One of two approaches can be taken to graph the line $f(x) = 3x - 7$. From Example 1.17 we know that $(5, 8)$ and $(3, 2)$ are points on the line. Hence, we could simply plot these two points and draw the line that passes through them. See Figure 1.11. Alternatively, we could use the facts that $m = 3 = 3/1$ and $b = -7$. Locate -7 on the y-axis. From that point we go three units up and one unit to the right (because the rise/run is $3/1$), and plot another point, then draw the line through the two points. See Figure 1.13. ■

The form of the equation of a linear function that we have used so far as the standard form, $y = mx + b$, is called the **slope-intercept form**, since m is the slope and b is the y-intercept of the line. Linear functions can

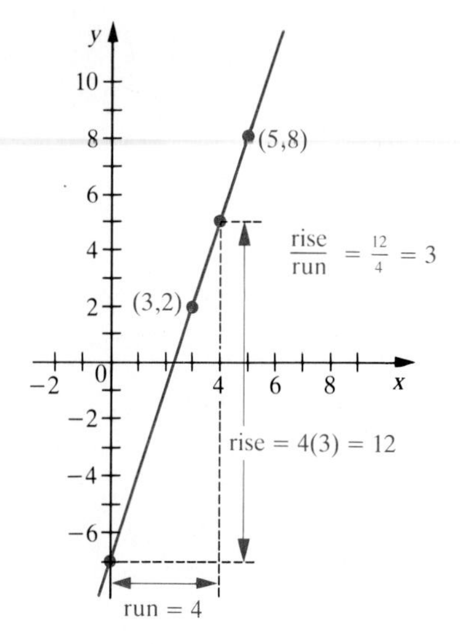

Figure 1.13
Two approaches to graphing $f(x) = 3x - 7$

be written in other forms; for example, if (x_1, y_1) is a point on the line $y = mx + b$, and m is the slope of the line, then the rise and run to another point (x, y) on the line are

$$\text{rise} = y - y_1 \qquad \text{run} = x - x_1;$$

then we can say

$$m = \frac{\text{rise}}{\text{run}} = \frac{y - y_1}{x - x_1}.$$

Hence, we can write the equation for the line as

$$y - y_1 = m(x - x_1).$$

This form is called the **point-slope form** because it uses one known point and the slope.

Example 1.19 Find the point-slope form of the equation of the line that passes through the point $(-2, 4)$ and has slope 3.

Solution In the point-slope form of the equation of the line, we have $m = 3$, $x_1 = -2$, and $y_1 = 4$. Thus an equation of the line is

$$y - 4 = 3[x - (-2)] = 3(x + 2).$$

Hence the point-slope form of the equation is

$$y - 4 = 3(x + 2).$$

■

Example 1.20 Find the slope-intercept form of the equation of the line that passes through the points $(-2, 4)$ and $(1, 2)$.

Solution We begin by finding the slope of the line. Taking (x_1, y_1) to be $(-2, 4)$ and (x_2, y_2) to be $(1, 2)$, we find

$$m = \frac{y_2 - y_1}{x_2 - x_1} = \frac{2 - 4}{1 - (-2)} = -\frac{2}{3}.$$

Using $(-2, 4)$ as the known point on the line, we obtain the point-slope form of the equation:

$$y - 4 = -\tfrac{2}{3}[x - (-2)] = -\tfrac{2}{3}(x + 2).$$

Simplifying the equation gives the slope-intercept form:

$$y = -\tfrac{2}{3}x + \tfrac{8}{3}.$$

Figure 1.14 shows the graph of this line.

Any linear equation in point-slope form can be rewritten in slope-intercept form in this manner. Verify that using $m = -\frac{2}{3}$ for the slope and $(1, 2)$ as the known point on the line produces the same equation.

■

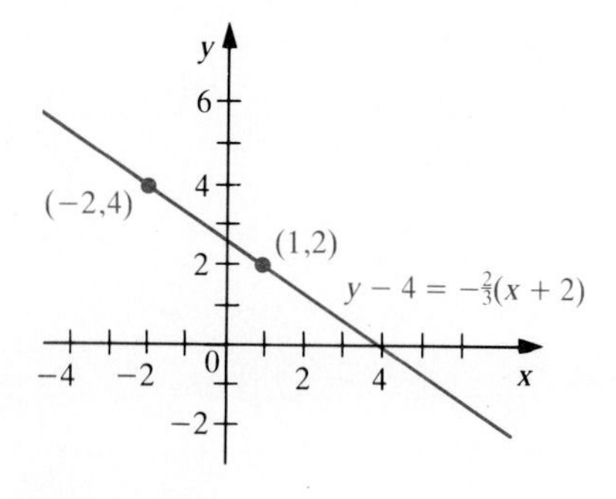

Figure 1.14
The graph of the line through $(-2, 4)$ and $(1, 2)$

The equation of a vertical line cannot be written in slope-intercept form or in point-slope form, since a verical line does not have a slope. For example, the equation of a vertical line that crosses the x-axis at 2 is $x = 2$; the equation of the vertical line that crosses the x-axis at -5 is $x = -5$. In general, the equation of the vertical line that crosses the x-axis at a is $x = a$.

Finding that two quantities are related by a linear function can sometimes lead to an important discovery. As reported in a recent study (L. A. Cooper and R. N. Shepard, "Turning something over in the mind," *Scientific American* 251, no. 6 [December 1984]: 106–115) on spatial perception, subjects were shown a series of pairs of pictures similar to those in Figure 1.15. Each pair showed the same object, or very similar objects, in different orientations. Subjects were asked to decide whether the pictures within each pair showed the same object. The researchers analyzed the time required to make a correct determination for each type of pair to determine how subjects think when making spatial perceptual comparisons. Example 1.21 is adapted from the results and analysis of that experiment.

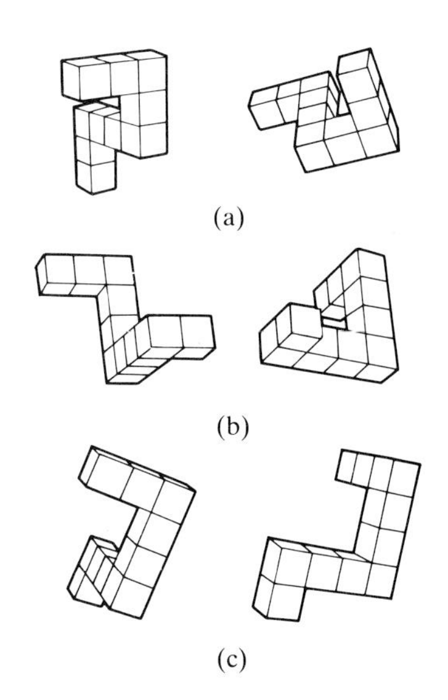

Figure 1.15
Are the two objects shown in each pair the same? (a) Yes—positions differ by a rotation within the plane; (b) yes—positions differ by a rotation out of the plane; (c) no—they are mirror images of each other

Example 1.21 Refer to Figure 1.15. Let $f(x)$ denote the average time (in seconds) that a subject requires to give a correct response when the angle between the orientations of the two pictures is x (in degrees). An analysis of the data shows that

$$f(x) = 0.02x + 1.2.$$

What is the average time required to give a correct response when the objects are in the same orientation? What is the average time when the angle between the orientations is 60°? When the angle is 180°? Suppose that each subject mentally rotates one object until both objects are in the same orientation before comparing them, and that the difference in response time from one orientation to another is due entirely to the time required for this rotating process. How many degrees per second does the (average) subject mentally rotate the object?

Solution When the pictures are in the same orientation, $x = 0$. Since $f(0) = 0.02(0) + 1.2 = 1.2$, it takes 1.2 sec, on the average, for a subject to give the correct response in this case. When the angle is 60°, the average response time is $f(60) = 0.02(60) + 1.2 = 2.4$ sec. When the angle is 180°, it takes $f(180) = 0.02(180) + 1.2 = 4.8$ sec. Plotting these values gives us a graph of the function (see Figure 1.16).

To answer the final question, note that from one data point to another on the graph, the rise of the line $y = f(x) = 0.02x + 1.2$ is the change in time, and the run is the change in orientation. Therefore, since the slope of the line is 0.02, we have

$$0.02 = \frac{\text{change in time}}{\text{change in orientation}}.$$

When the change in time is 1 sec,

$$0.02 = \frac{1}{\text{change in orientation}}$$

and so

$$\text{change in orientation} = \frac{1}{0.02} = 50.$$

Hence a subject mentally rotates the picture 50° per second. Thus, the average time it takes a subject to decide whether two objects are the

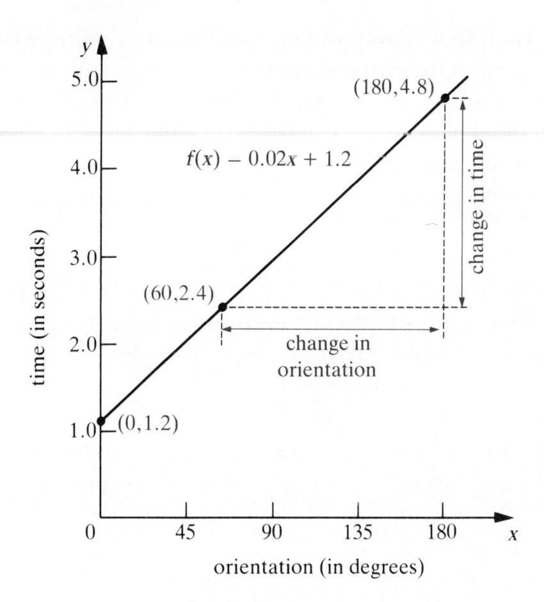

Figure 1.16
The graph of $f(x) = 0.02x + 1.2$

same is equal to the time it takes the subject mentally to rotate one object 50° per second until the objects are in the same orientation, followed by a 1.2-sec period to make the comparison. Because this analysis fits the "rotating model" so well and seems to contradict other proposed models, the researchers are persuaded that this is how the subject accomplishes this type of spatial perceptual comparisons. If the function $f(x)$ had not been linear, then the number of degrees that the objects were rotated per second would not have been a constant and the rotating model would not have been supported by the data. ■

We now consider the relationship between the equations of lines and the geometric properties of lines. Given any two lines in the plane, the lines either intersect at a single point or are parallel. Since the slope of a line determines the degree to which the line is slanted relative to the x-axis (its rise over run), two lines are parallel if and only if they have the same slope. Hence, the lines whose equations are $y = m_1x + b_1$ and $y = m_2x + b_2$ are **parallel** if and only if $m_1 = m_2$.

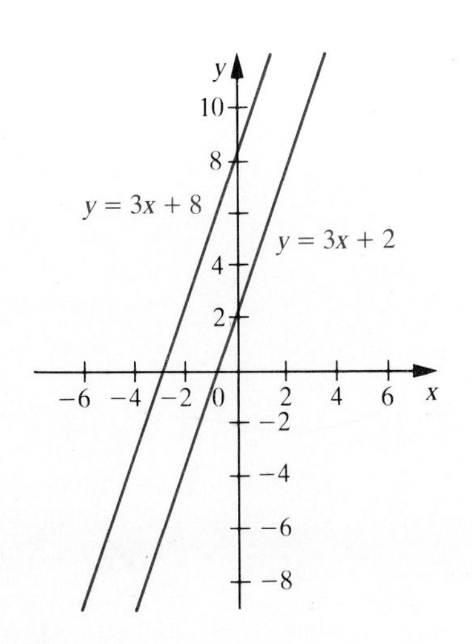

Figure 1.17
Graphs of parallel lines $y = 3x + 8$ and $y = 3x + 2$

Example 1.22 Find the equation of the line that is parallel to $y = 3x + 2$ and passes through the point (2, 14).

Solution The line we seek is parallel to the line given by $y = 3x + 2$, which has slope 3; therefore, the slope of the line we seek must also be 3. Its equation must be of the form

$$y = 3x + b$$

for some b. Since the line passes through (2, 14), we have

$$14 = 3(2) + b.$$

Solving for b, we get $b = 8$. Therefore the equation of the line is $y = 3x + 8$. See Figure 1.17. ■

When two lines are not parallel, we find their intersection point by solving their equations simultaneously.

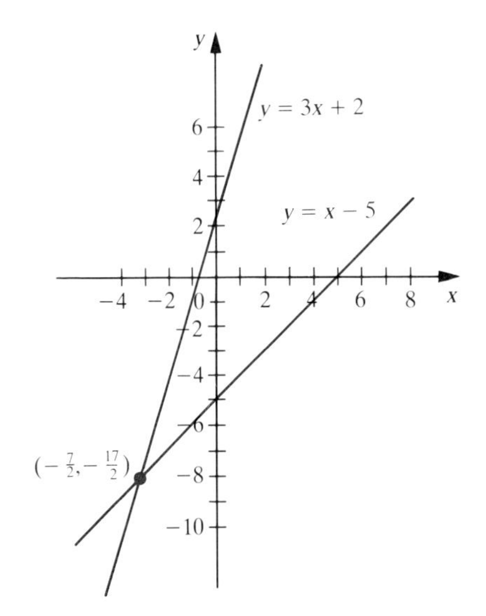

Figure 1.18
Graphs of the lines $y = 3x + 2$ and $y = x - 5$

Example 1.23 Determine the point (x, y) where the graphs of $y = 3x + 2$ and $y = x - 5$ intersect.

Solution We seek an x-value and a y-value such that the point (x, y) is on both lines. That is, we seek x and y so that $y = 3x + 2$ and $y = x - 5$. Since both expressions are equal to y, we can write

$$3x + 2 = x - 5$$
$$3x - x = -5 - 2$$
$$2x = -7$$
$$x = -\tfrac{7}{2}.$$

To determine the value of y, we use either of the original equations and substitute the value we have found for x:

$$y = x - 5 = -\tfrac{7}{2} - 5 = -\tfrac{17}{2}.$$

Thus, the lines intersect at $(-\frac{7}{2}, -\frac{17}{2})$. See Figure 1.18. ■

Example 1.24 Suppose that a college fund is set up for a three-year-old child so that the value of the fund in t years will be $V(t)$ dollars, and suppose that the cost of the average college education in t years will be $C(t)$ dollars, where these amounts are specified by the formulas

$$V(t) = 8000 + 5000t \qquad \text{and} \qquad C(t) = 52{,}000 + 2000t.$$

Determine the point in time when the value of the college fund will equal the cost of an average college education.

Solution To determine when the value of the fund will equal the cost of education, we need to find the value of t such that $V(t) = C(t)$, that is, the value of t where the graphs of $V(t)$ and $C(t)$ intersect. Setting $V(t) = C(t)$ and solving for t, we have

$$8000 + 5000t = 52{,}000 + 2000t$$
$$3000t = 44{,}000$$
$$t = 14\tfrac{2}{3}.$$

Hence in 14 years eight months, the fund will have the money necessary to pay for the child's college education. The three-year-old child will be nearly 18 years old when the college fund and education costs become equal. Consequently, the college fund was properly established to meet the anticipated education expenses. ■

The slopes of two lines can be used to determine if the lines are perpendicular. The lines whose equations are

$$y = m_1x + b_1 \qquad \text{and} \qquad y = m_2x + b_2$$

with $m_1 \neq 0$ are **perpendicular** if and only if $m_2 = -1/m_1$. (See Exercise 80 for an outline of a proof of this fact.)

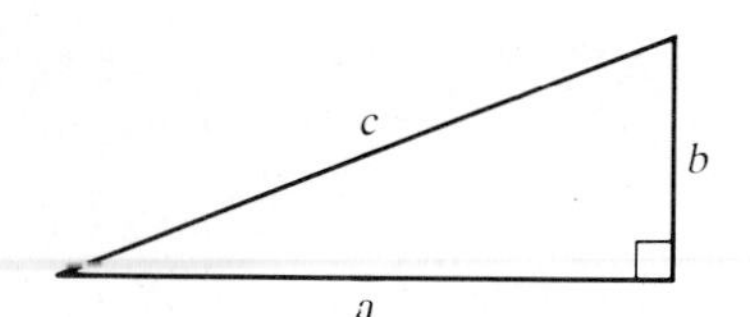

Figure 1.19
The Pythagorean theorem: $c^2 = a^2 + b^2$

Example 1.25 Find the equation of a line that passes through (8, 1) and is perpendicular to the line $y = 4x + 5$.

Solution Since the slope of the line $y = 4x + 5$ is 4, the slope of a line perpendicular to it is $-\frac{1}{4} = -0.25$. Hence its equation is of the form

$$y = -0.25x + b$$

for some b. Since the line passes through (8, 1), we have

$$1 = -0.25(8) + b,$$

or $b = 3$. Therefore the equation is $y = -0.25x + 3$. ■

The distance between two points in a Cartesian coordinate system can be calculated using a formula adapted from the Pythagorean theorem, which states that in a right triangle with legs of length a and b and hypotenuse c, we have $c^2 = a^2 + b^2$. See Figure 1.19.

In Figure 1.20 a right triangle is shown with (x_1, y_1) and (x_2, y_2) as two of its vertices. The length of the hypotenuse of the right triangle is d, the distance between the given points; and the legs of the triangle have lengths $x_2 - x_1$ and $y_2 - y_1$. Hence, by the Pythagorean theorem,

$$d = \sqrt{(x_2 - x_1)^2 + (y_2 - y_1)^2}.$$

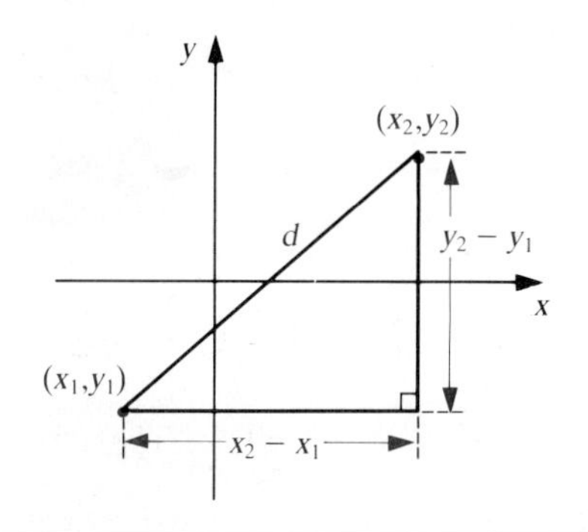

Figure 1.20
Geometry for determining the distance between two points

Note that in Figure 1.20 $x_2 - x_1$ and $y_2 - y_1$ are both positive. However, since $(x_2 - x_1)^2$ and $(y_2 - y_1)^2$ appear in the formula, and since the square of a number is never negative, the formula works when either $(x_2 - x_1)$ or $(y_2 - y_1)$ or both are negative, as the next example illustrates.

Example 1.26 Find the distance between the points:

(a) (2, 3) and (5, 7) (b) (−3, 5) and (2, 8) (c) (4, 3) and (−1, 5).

Solution (a) According to the distance formula, the distance between (2, 3) and (5, 7) is

$$\sqrt{(5-2)^2 + (7-3)^2} = \sqrt{3^2 + 4^2} = \sqrt{25} = 5.$$

(b) The distance between (−3, 5) and (2, 8) is

$$\sqrt{[2-(-3)]^2 + (8-5)^2} = \sqrt{5^2 + 3^2} = \sqrt{34}.$$

(c) The distance between (4, 3) and (−1, 5) is

$$\sqrt{(-1-4)^2 + (5-3)^2} = \sqrt{(-5)^2 + (2)^2} = \sqrt{29}.$$ ■

Exercises

Exer. 1–8: Sketch the graph of the function $f(x) = mx + b$ for the given conditions. Use a common coordinate system for each problem, and label each line carefully.

1. $m = 0$; $b = -4, -2, -1, 0, 1, 2, 4$
   (Seven separate lines are required.)
2. $b = 2$; $m = 1, 2, 3, 4, 10$
3. $b = 2$; $m = \frac{1}{2}, \frac{1}{5}, \frac{1}{10}, \frac{4}{5}$
4. $b = 0$; $m = -1, -\frac{1}{2}, -\frac{1}{5}, -2, -5$
5. $m = 2$; $b = 1, 3, 6, 8$
6. $b = 1$; $m = 2, -\frac{1}{2}$
7. $b = -1, 1$; $m = -1, 1$
   (Four separate lines are required.)
8. $b = -2, 2$; $m = 2, -\frac{1}{2}$
   (Four separate lines are required.)

Exer. 9–20: Find the equation of the form $f(x) = mx + b$ for the linear function satisfying the given conditions and sketch its graph.

9. Passes through $(1, 2)$ and has slope 3
10. Passes through $(3, 3)$ and has slope -3
11. Passes through $(4, 1)$ and $(7, 4)$
12. Passes through $(4, 1)$ and $(6, 9)$.
13. Slope 2 and y-intercept 4
14. Slope -4 and y-intercept 1
15. Passes through $(4, -2)$ and has slope 7
16. Passes through $(1, 0)$ and has slope $\frac{1}{2}$
17. Slope $\frac{2}{3}$ and y-intercept 10
18. Slope $-\frac{3}{2}$ and y-intercept 5
19. Slope 2 and x-intercept -1
20. Slope -3 and x-intercept 1

Exer. 21–36: Write the equation in point-slope form for the linear function satisfying the given conditions.

21. Passes through $(2, 1)$ and has slope $\frac{1}{2}$
22. Passes through $(3, -4)$ and has slope -3
23. Passes through $(4, 4)$ and $(1, 1)$
24. Passes through $(3, -2)$ and $(2, -3)$
25. Slope -2 and y-intercept 2
26. Slope 3 and y-intercept 3
27. Passes through $(1, 2)$ and has slope -2
28. Passes through $(-1, 2)$ and has slope 4.
29. Slope 3 and x-intercept 1
30. Slope -2 and x-intercept 6
31. Slope 4 and x-intercept 3
32. y-intercept 2 and x-intercept 6
33. y-intercept 6 and x-intercept 2
34. y-intercept -2 and x-intercept 1
35. y-intercept -3 and x-intercept -2
36. y-intercept 0 and x-intercept 2

Exer. 37–46: Determine if the equation is a linear function. If not, explain why.

37. $7 - 3x = 2y$
38. $y = 1/x$
39. $2x + 4y - 2 = 0$
40. $x = 3y + 7$
41. $f(x) = (x + 1)(x + 3)$
42. $y = (x + 1)(7 + x)$
43. $y = 0$
44. $f(x) = 3x^2 + 2$
45. $y = -3x$
46. $3xy = 2$

Exer. 47–49: Suppose that the average time (in seconds) required by a subject to give a correct response in Example 1.21 is

$$f(x) = 0.03x + 1.5.$$

47. How long does it take a subject to give a correct response when the pictures have the same orientation?
48. How long does it take a subject to give a correct response when the angle between the orientations is 140°?
49. Assuming that the subjects are mentally rotating one of the pictures, at how many degrees per second are they performing this rotation?

Exer. 50–61: Find the linear function that satisfies the given conditions.

50. Parallel to the graph of $y = 4x + 8$ and passes through $(5, 1)$
51. Parallel to the graph of $y = -2x + 3$ and passes through $(2, 8)$
52. Parallel to the graph of $y = 0.5x - 2$ and passes through $(-1, 6)$
53. Parallel to the graph of $y = 1.2x - 15$ and passes through $(-4, -3)$
54. Parallel to the graph of $3y + 9x = 7$ and passes through $(7, -4)$
55. Parallel to the graph of $-2y + 8x = 3$ and passes through $(-5, 3)$
56. Perpendicular to the graph of $y = x + 3$ and passes through $(1, 9)$
57. Perpendicular to the graph of $y = 0.5x - 6$ and passes through $(-2, 4)$
58. Perpendicular to the graph of $y = \frac{2}{3}x + 3$ and passes through $(-3, 5)$
59. Perpendicular to the graph of $y = -5x + 2$ and passes through $(-3, 6)$
60. Perpendicular to the graph of $y = 10x - 8$ and passes through the origin
61. Perpendicular to the graph of $6y - 3x = 5$ and passes through $(-3, 6)$.

Exer. 62–69: Find the intersection point of the lines.

62. $y = 3x$; $y = \frac{3}{2}x + 1$
63. $y = x - 3$; $y = 3 - x$
64. $y = 4$; $y = 2x + 3$
65. $y = -2x + 4$; $y = 2x - 8$
66. $y - 3x = -5$; $y - 4x = 23$
67. $3y - 4x = -9$; $4y + 2x = 10$
68. $-5y + 3x = -11$; $3y + 8x = -13$
69. $2y + 25x = 81$; $14y - 12x = 6$

Exer. 70–77: Find the distance between the points

70. $(1, 3)$ and $(4, 7)$
71. $(-2, 4)$ and $(1, -3)$
72. $(2, 8)$ and $(5, 14)$
73. $(4, -3)$ and $(2, 5)$
74. $(-1, 6)$ and $(4, 8)$
75. $(3, 8)$ and $(-1, 4)$
76. $(-2, -5)$ and $(-1, 2)$
77. $(-2, 5)$ and $(-5, 7)$

78. Refer to Example 1.24. Suppose the value of the college fund discussed is expected to be $V(t) = 6000 + 4000t$ dollars in t years, and the cost of an average college education is expected to be $C(t) = 60{,}000 + 3000t$ dollars. Determine the point in time when the value of the college fund will equal the cost

of an average college education.

79. Work Exercise 78 with $V(t) = 12{,}000 + 3000t$ and $C(t) = 50{,}000 + 3500t$. What conclusion can you draw about this financial plan?
80. Use the figure to show that $m_1 m_2 = -1$. (*Hint:* Start by writing the Pythagorean theorem for the three right triangles in the figure.)

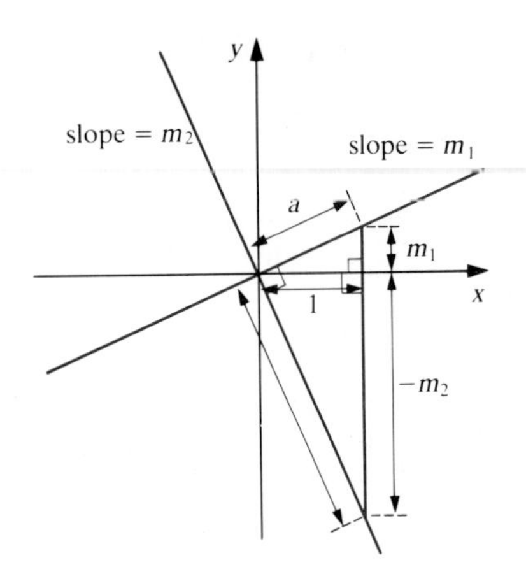

1.4 Applications of Linear Functions

We have seen that in many applications the relationship between the quantities involved can be described or approximated by a linear function. In this section we will discuss more applications. The first example deals with the conversion from one system of measurement to another, a process that is almost always linear.

Example 1.27 Let F denote Fahrenheit temperature and C denote Celsius temperature. The graph of the linear function converting F to C must pass through the point $(32, 0)$, since 32 °F and 0° C both correspond to the freezing point of water. Similarly, using the boiling point of water, the graph must pass through $(212, 100)$.

(a) Find a linear equation that gives Celsius temperature as a function of Fahrenheit temperature.

(b) What Celsius temperature corresponds to 100 °F?

(c) Verify that -40 °F $= -40$ °C.

(d) Is there any other temperature that is the same on both scales?

Solution **(a)** We write $C = mF + b$ and determine m and b. Since the graph passes through $(32, 0)$ and $(212, 100)$, we have

$$m = \frac{100 - 0}{212 - 32} = \frac{100}{180} = \frac{5}{9}.$$

Thus $C = \frac{5}{9}F + b$. Since the graph passes through $(32, 0)$, we see that $0 = (\frac{5}{9})32 + b$; thus $b = -(\frac{5}{9})32 = -\frac{160}{9}$. Hence, the slope-intercept form of the equation is

$$C = \tfrac{5}{9}F - \tfrac{160}{9}.$$

It is, however, more convenient to remember the formula in the form

$$C = \tfrac{5}{9}F - \tfrac{5}{9}(32) = \tfrac{5}{9}(F - 32).$$

This linear function is graphed in Figure 1.21.

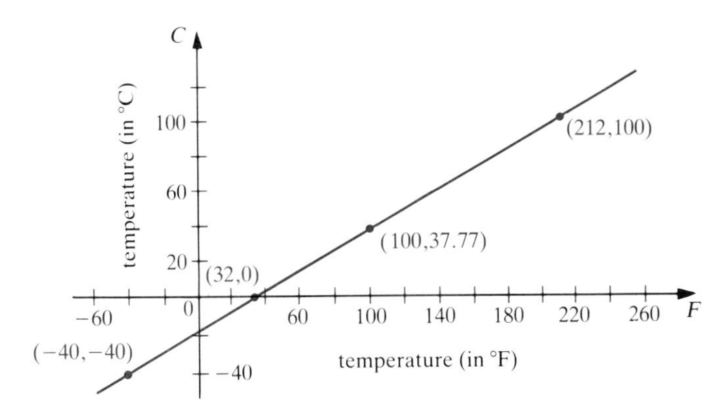

Figure 1.21
Graph of the Fahrenheit-to-Celsius conversion function

(b) To find the value of C corresponding to $F = 100$, we compute

$$C = \tfrac{5}{9}(100 - 32) = \tfrac{5}{9}(68) = 37.77.$$

(c) When $F = -40$ we compute C to be $C = \tfrac{5}{9}(-40 - 32) = -40$. Thus $-40\,°F = -40\,°C$.

(d) To decide if there is any other temperature that is the same on both scales, we set $F = C$ and solve for F. Thus

$$F = C = \tfrac{5}{9}(F - 32)$$

or

$$9F = 5(F - 32) = 5F - 160.$$

Hence $4F = -160$, and we get $F = -40$, which indicates that only -40 is the same on both scales. ∎

A manufacturing company sets the production level of a certain item after determining the cost of producing a given number of the item. To analyze this situation, cost should be expressed as a function of the number of items produced. One of the most powerful applications of calculus for the managerial sciences deals with cost and related functions. A cost function is linear if the extra cost required to increase production by a certain amount, say 10 items, is a constant regardless of the current production level; for example, when a production increase from 20 to 30 items costs the same as a production increase from 1000 to 1010 items. In practice, cost functions are seldom linear; however, for this section we will consider only linear cost functions.

Note that the cost of producing a certain number of items is the sum of the cost of manufacturing the items and the fixed cost (the cost incurred even if no items are produced).

Example 1.28 Suppose the daily production cost is \$23,000 for 150 video cassette recorders (VCRs) and \$29,000 for 210 VCRs. Find (a) the linear cost function, (b) the fixed cost, and (c) the production cost for 400 VCRs.

Solution (a) Let x denote the number of VCRs produced per day and let C denote the cost in dollars. We are given two points on the linear cost function: $(150, 23{,}000)$ and $(210, 29{,}000)$. The slope of the cost function is

$$\text{slope} = \frac{29{,}000 - 23{,}000}{210 - 150} = \frac{6000}{60} = 100.$$

The equation of the line that has slope 100 and passes through (150, 23,000) is

$$C - 23{,}000 = 100(x - 150)$$

or

$$C = 100x + 8000.$$

This is the cost function. (b) The fixed cost is the cost incurred even if no VCR is produced. Setting $x = 0$ yields $C = \$8000$ for the fixed cost. (c) The production cost for $x = 400$ VCRs is $C = 100(400) + 8000 = \$48{,}000$. ■

Example 1.29 A consumer plans to buy one of two cars. The first car costs $11,000 and gets 40 miles per gallon of gasoline. The second car costs $9,500 and gets 30 miles per gallon. Considering only the price of the car and gasoline consumed, determine the cost of buying and driving each car x miles, assuming that gasoline costs $1.20 per gallon. How many miles must the consumer drive the first car until the greater initial cost is completely offset by lower fuel cost?

Solution Let $C_1(x)$ represent the cost of driving the first car x miles, that is, the sum of the cost of the car and the cost of the gasoline consumed. The number of gallons consumed by the first car in traveling x miles is $x/40$; thus the cost of the gasoline consumed in driving the car x miles is $(1.20)(x/40)$. Adding the costs of the car and the gasoline, we have

$$C_1(x) = 11{,}000 + (1.20)\left(\frac{x}{40}\right) = 11{,}000 + 0.03x.$$

Similarly, the cost $C_2(x)$ of driving the second car x miles is

$$C_2(x) = 9500 + (1.20)\left(\frac{x}{30}\right) = 9500 + 0.04x.$$

We wish to find the number of miles x such that $C_1(x) = C_2(x)$. Thus

$$11{,}000 + 0.03x = 9500 + 0.04x$$
$$1500 = 0.01x$$
$$x = 150{,}000.$$

Therefore, the consumer would need to drive the first car 150,000 miles before the initial price differential is offset by lower fuel costs. ■

Linear functions can often be applied to the life sciences. The following example is based on actual data from B. Zeide, "Tolerance and self-tolerance of trees," *Forest Ecology and Management* 13 (1985): 149–166.

Example 1.30 The rate of photosynthesis of a tree generally increases as the amount of available light increases. Hence assuming that all other environmental factors are constant, the rate of photosynthesis is a function of the amount of available light. This function varies from species to species, but most species fall into one of two categories. Species that become saturated at a moderate level of light so that any additional light results in only very small increases in the rate of photosynthesis are *shade-tolerant* since they perform at near peak efficiency in the shade. Species in the second category are said to be *shade-intolerant.* Let $T(x)$ denote the rate of photosynthesis for shade-tolerant trees when the available light is x thousand footcandles, a measure of light intensity. Let $I(x)$ denote the corresponding function for shade-intolerant trees. In both cases, the rate of photosynthesis is measured as a percentage of the greatest rate observed for all species. (Thus both $T(x)$ and $I(x)$ are between 0 and 100.) When x is between 2 and 10 (thousand footcandles), $T(x)$ and $I(x)$ are accurately given by the linear functions

$$T(x) = 0.5x + 84 \quad \text{and} \quad I(x) = 5x + 49.$$

(a) What do the slopes of these functions tell us about the nature of the trees they describe?
(b) Determine the rate of photosynthesis for each type of tree at $x = 2$ and at $x = 10$.
(c) At what level of light intensity do shade-tolerant and shade-intolerant trees have the same rate of photosynthesis?

Solution **(a)** Since the slope of $T(x)$ $[m = 0.5]$ is very small compared to that of $I(x)$ $[m = 5]$, we conclude, as expected, that the rate of photosynthesis for shade-tolerant trees will increase much less as x increases than will the rate for shade-intolerant trees.
(b) Since $T(2) = 85$ and $I(2) = 59$, we see that at 2000 footcandles the shade-tolerant trees have a much higher rate of photosynthesis than the shade-intolerant trees. Since $T(10) = 89$ and $I(10) = 99$, the opposite is true at 10,000 footcandles.
(c) The two categories of trees have the same rate of photosynthesis when $T(x) = I(x)$—that is, when

$$\begin{aligned} 0.5x + 84 &= 5x + 49 \\ 35 &= 4.5x \\ 7.8 &= x. \end{aligned}$$

Hence the rate of photosynthesis is the same for trees in both categories at 7800 footcandles. ■

An important concept in economics and business management is that of supply and demand for a product or service. A **demand function** expresses the relationship between the price of a product and the quantity of the product that will sell. In a free marketplace, a change in price will affect the quantity that can be sold (presumably, the higher the price, the fewer the sales); a change in the quantity available will affect the price (generally, the scarcer an object is, the more expensive it becomes).

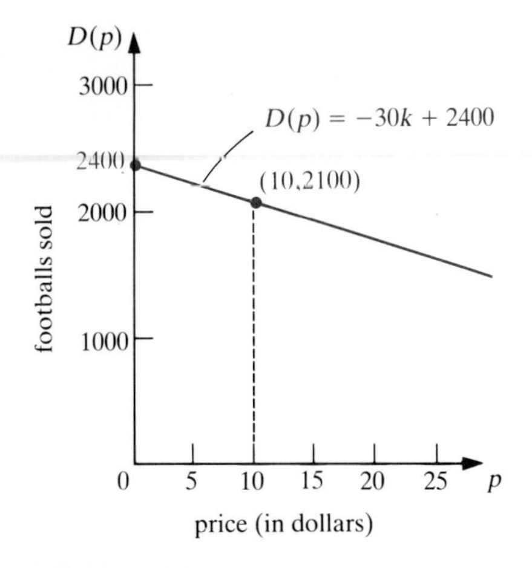

Figure 1.22
Graph of the demand function $D(p) = -30p + 2400$

Example 1.31 Suppose that the demand function for a brand of football is given by

$$D(p) = -30p + 2400$$

for the price p in dollars and the number of footballs $D(p)$ that can be sold at that price. See Figure 1.22.
(a) What is the demand for this brand of football when the price is \$10.00?
(b) What is the effect on demand when the price is increased by \$2.00?

Solution **(a)** Since

$$\begin{aligned} D(10) &= (-30)10 + 2400 \\ &= -300 + 2400 = 2100, \end{aligned}$$

we see that 2100 footballs will be sold if the price of each football is set at \$10.00.
(b) Assume the present price is k dollars. The demand at k dollars is

$$D(k) = -30k + 2400.$$

If the price is increased by \$2.00 the new price is k + \$2.00, and the demand is

$$\begin{aligned} D(k+2) &= -30(k+2) + 2400 \\ &= -30k + (-60) + 2400 \\ &= -30k + 2340. \end{aligned}$$

Hence the change in demand when the price is increased by \$2.00 is

$$D(k+2) - D(k) = -30k + 2340 - (-30k + 2400) = -60.$$

Consequently, each increase of \$2.00 results in a demand of 60 fewer footballs. We could have reached the same conclusion by noting that since the slope m of the graph is

$$m = \frac{\text{rise}}{\text{run}} = \frac{\text{change in } D(p)}{\text{change in } p}$$

and since the change in p is 2 and $m = -30$, we have

$$-30 = \frac{\text{change in } D(p)}{2}.$$

Again we find that corresponding to an increase of 2 for p

$$\text{change in } D(p) = (-30)(2) = -60.$$

Before leaving this example, we note that $D(0) = 2400$ footballs, that is, if the price is zero dollars, the demand will be only 2400 footballs! Phrased another way, this demand function predicts that if footballs are given away, only 2400 customers will want one! This is unrealistic, of course, and explains why few demand functions are linear. ■

Just as the quantity of a product that will be sold is related to the price, the quantity that manufacturers will produce is related to the price. A **supply function** expresses the relationship between the price of a product and the quantity manufacturers are willing to produce at that selling price. In a free marketplace, a change in price will affect the quantity produced. Usually, low prices correspond to low production levels, and vice versa.

Example 1.32 A certain candy bar has a demand function $D(p)$ and a supply function $S(p)$ given by

$$D(p) = -2p + 275 \quad \text{and} \quad S(p) = 3p + 50$$

for the price p in cents and both $D(p)$ and $S(p)$ in thousands of candy bars. Compare the supply and the demand when $p = 20$ and when $p = 50$. At what price does supply equal demand?

Solution Since $D(20) = 235$ and $S(20) = 110$, the demand is greater than the supply when the price is 20 cents. Since $D(50) = 175$ and $S(50) = 200$, the supply is greater than the demand when the price is 50 cents. To find the price at which the supply equals the demand, we solve for the intersection point of the graphs $y = D(p)$ and $y = S(p)$, that is, we must find the value of p where $D(p) = S(p)$. See Figure 1.23. Thus

$$\begin{aligned} D(p) &= S(p) \\ -2p + 275 &= 3p + 50 \\ -5p &= -225 \\ p &= 45. \end{aligned}$$

Hence supply will equal demand when the price is 45 cents. ■

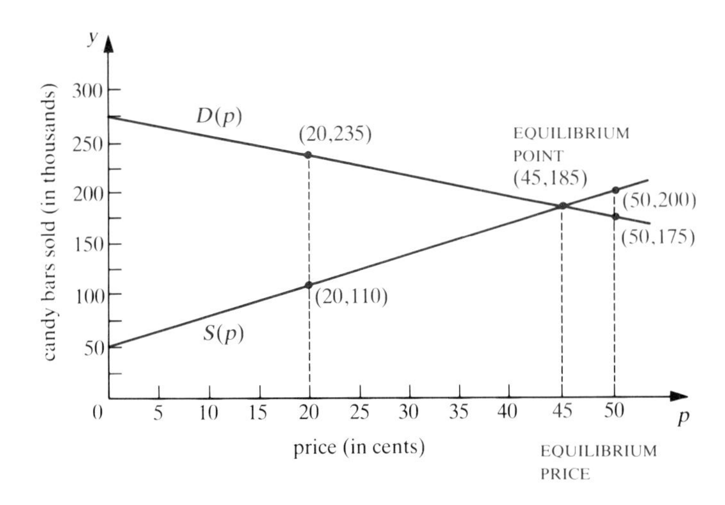

Figure 1.23
Demand and supply functions for the candy bar in Example 1.32

When graphs of demand and supply functions intersect, the point of intersection is called the **equilibrium point** and the price corresponding to the point of intersection is the **equilibrium price**, the price at which consumers will purchase the same quantity of a product that manufacturers will produce at that selling price. When the price of a product is less than equilibrium price, more of the product is demanded than produced, resulting in a shortage in the marketplace. When the price is greater than the equilibrium price, more of the product is produced than bought, creating a surplus. Consequently, determining the equilibrium price accurately is essential if a company is to devise a reasonable production strategy. In Example 1.32 the equilibrium price for the candy bar is 45 cents, and since $D(45) = -2(45) + 275 = 185$, the equilibrium point is (45, 185), which indicates that 185,000 candy bars can be produced and sold at 45 cents each.

Exercises

1. Refer to Example 1.27. (a) Find the linear function that expresses the conversion of C (Celsius) to F (Fahrenheit) by finding two points on the line that is the graph of this function. What Fahrenheit temperature corresponds to 50 °C?
(b) The conversion of F to C is given by the linear function $C = \frac{5}{9}(F - 32)$. Solve this equation for F as a function of C and compare this with your function in part (a).
2. One yard is the same length as 0.9144 meter. (a) Find a function that changes yards to meters. (b) How many meters equal 100 yards? (c) Find a function that changes meters to yards. (d) How many yards equal 1500 meters? (e) Which is longer, the 1500-meter race or the 1-mile (1760-yard) race?
3. Given that 1 mile is 1.609 kilometers, find a function that converts kilometers to miles and a function that converts miles to kilometers. If the speed limit is 65 mi/hr, and you are traveling at 100 km/hr, are you speeding?
4. Suppose that the cost function of interviews for a sociology survey is linear. If the fixed cost is \$300.00 and \$540.00 is the cost of conducting 20 interviews, find the cost function $C(x)$. Find the cost of interviewing a sample population of 120.
5. The production cost is \$10,000 for 90 cameras and \$20,000 for 200 cameras. Find the linear cost function. Find the production cost for 300 cameras. What is the fixed cost?
6. Example 1.30 discusses the relationship between the rate of photosynthesis for shade-tolerant and shade-intolerant trees when the available light x, in thousands of footcandles, is between 2 and 10. When x is between 0.5 and 2, $T(x)$ and $I(x)$ are

$$T(x) = 33x + 19, \qquad I(x) = 27x + 5.$$

What do the slopes of these functions tell us about the nature of the trees in this low-light situation? Determine the rate of photosynthesis for each type of tree at $x = 0.5$ and at $x = 2$. Find the point of intersection of the two linear functions and explain why, in this low-light situation, no amount of light yields the same rate of photosynthesis for trees in both categories.
7. Refer to Example 1.30. Suppose that a particular species of tree has a rate of photosynthesis of

$$S(x) = 0.7x + 48 \quad \text{when} \quad 2 \le x \le 10.$$

Comparing $S(x)$ to the functions $T(x)$ and $I(x)$ of the example, decide whether this species is shade tolerant or shade intolerant. Explain how you made your decision.
8. Refer to Example 1.29. Suppose that the first car costs \$12,000 and gets 50 miles per gallon of gasoline and the second car costs \$9,000 and gets 30 miles per gallon. If the price of fuel is \$1.30 per gallon, find the cost of buying and driving each car x miles. How far must the first car be driven before the additional initial cost is completely offset by the lower fuel cost?
9. A long-distance telephone company charges a connection fee of \$3.00 for an international call plus \$1.50 per minute. Determine the function $C(t)$ that gives the cost (in dollars) of an international call lasting t minutes. How much would a 12-minute call cost?
10. A long-distance telephone company, which is in competition with the company described in Exercise 9, charges a connection fee of \$5.00 for an international call plus \$1.25 per minute. Determine the function $F(t)$ that gives the cost of an international call lasting t minutes. For what length of call would the two companies charge the same amount?
11. A plumber charges a \$30 fee plus \$16 per hour for making a house call. Find the function $C(t)$ that gives the plumber's charge for a house call of t hours. If the plumber's charge for a certain house call is \$158 (excluding materials), how long was the house call?
12. In the inner city of a large metropolitan area, the average annual income for the head of household of a family of six is \$8,800 and is increasing at a rate of

\$500 each year. Find a function that gives the average annual income in x years. If \$12,000 is the poverty level for a family of six use this function to determine how many years it will take until the average annual income of an inner-city head of household of a family of six is greater than the poverty level (assuming a constant poverty level).

13. Suppose \$55,000 is the cost of producing 2000 hand-held calculators and \$100,000 is the cost for 4000. If the cost function is linear, what is the cost of producing 6500 calculators?
14. A large farm cooperative sells pinto beans at 30 cents per pound with a delivery charge of 5 cents per pound and a handling charge of 75 cents per order. Find a function $C(x)$ for the cost of an order of x pounds.
15. The pollution index $p(t)$ in a certain city is a linear function of the time t. If the pollution index is 4.0 at 2:00 P.M. and is up to 7.5 at 6:00 P.M., predict what it will be at 9:00 P.M.
16. Suppose that the pollution index, a linear function of time, is 4.2 at both 3:00 P.M. and 8:00 P.M. What is the pollution index at 5:30 P.M.?

Exer. 17–24: Find the equilibrium point for the given demand and supply functions:

17. $D(p) = -4p + 665; \quad S(p) = 5p + 80$
18. $D(p) = -2p + 8; \quad S(p) = p + 2$
19. $D(p) = -2p + 600; \quad S(p) = 2p + 40$
20. $D(p) = -(5/12)p + 18; \quad S(p) = (1/4)p + 2$
21. $D(p) = -(2.5)p + 40.5; \quad S(p) = 3p + 19$
22. $D(p) = -(0.1)p + 20; \quad S(p) = (0.5)p + 16$
23. $D(p) = -1.8p + 57; \quad S(p) = 0.4p + 44$
24. $D(p) = -(3.4)p + 78; \quad S(p) = (0.05)p + 9$

Exer. 25–27: The demand for home-baked pies is given by $D(x) = 350 - 2x$ where x is the price of each pie (in cents).

25. What is the effect on sales of pies if the price is increased by 10 cents?
26. What is the effect of a 5-cent price decrease?
27. Suppose the current price of 150 cents is increased to 175 cents. What will be the effect on revenue? (Revenue is the product of demand and price.)

Exer. 28–30: The demand for a certain item is given by $D(x) = 1000 - 29x$ where x is the price per unit (in dollars) and $D(x)$ is the number (in hundreds) at price x.

28. How many units will be sold when the price is \$25?
29. What is the effect of a \$5 increase in price?
30. What is the effect on revenue of lowering the price from \$30 to \$20? (Revenue is the product of demand and price.)

1.5 Polynomials and the Absolute Value Function

Many applications involve functions that are more complicated than linear functions. In this section we will study *polynomial* functions and the *absolute value* function. A **polynomial** is a function of the form

$$y = f(x) = a_n x^n + a_{n-1}x^{n-1} + \cdots + a_1 x + a_0, \qquad a_n \neq 0$$

for a nonnegative integer n, called the **degree** of the polynomial. The quantities $a_n, a_{n-1}, \ldots, a_0$ represent constants and are the **coefficients** of the polynomial; a_n is the **leading coefficient**. Table 1.5 gives several examples of polynomials.

Every linear function is a polynomial of degree 1, except for those of the form $y = constant$. If the constant does not equal zero, a polynomial of the form $y =$ constant has degree zero. The polynomial $y = 0$ has no degree assigned to it.

A **quadratic** polynomial is a polynomial of degree two and may be written

$$ax^2 + bx + c$$

where a, b, and c are constants with $a \neq 0$. For convenience we are using a, b, c rather than a_2, a_1, a_0.

Table 1.5

Polynomial	*Degree*	*Leading coefficient*
$4x^2 - 2x + 1$	2	4
$6x - 3$	1	6
$-x^3$	3	−1
$-12x^5 + x - 1$	5	−12
$1 + 2x + 3x^7$	7	3

Example 1.33 Sketch a graph of the quadratic function $f(x) = -2x^2 - 4x + 3$ by plotting several points.

Solution Several points on the graph are shown in Table 1.6. These points were ploted to give us the graph shown in Figure 1.24. ■

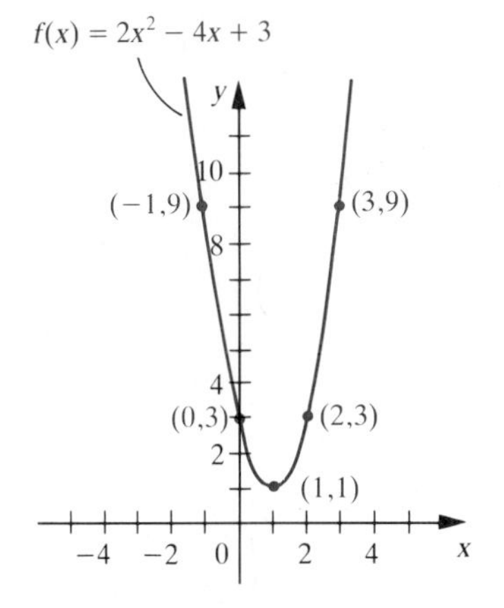

Figure 1.24
Graph of $f(x) = 2x^2 - 4x + 3$

Table 1.6
Some values of $f(x) = 2x^2 - 4x + 3$

x	$f(x)$
−1	9
0	3
1	1
2	3
3	9

The graph of a quadratic function is called a **parabola**, which may be shaped like one of the two typical graphs in Figure 1.25. In Figure 1.25(a), the graph opens upward and looks like a valley (see also Figure 1.24 in Example 1.33). The bottommost point is called the **vertex**. In Figure 1.25(b), the graph opens downward and looks like a hill; in this case the topmost point is the vertex. If we can determine, by inspecting the function, the coordinates of the vertex and whether the graph opens upward or downward, then we can draw a reasonable sketch by plotting only a few points. The y-intercept, for example, is always a good point to plot.

We can determine the direction the parabola opens and the location of the vertex by the following rules.

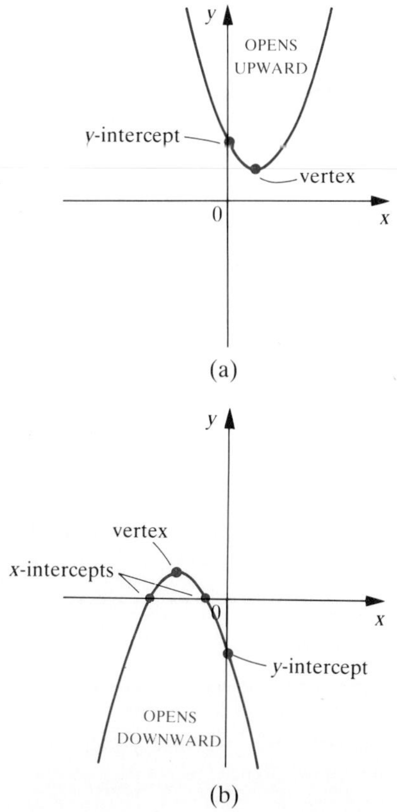

Figure 1.25
(a) Typical graph of $ax^2 + bx + c$, $a > 0$; (b) typical graph of $ax^2 + bx + c$, $a < 0$

The graph of $y = f(x) = ax^2 + bx + c$, $a \neq 0$ is a parabola.
Rule 1: If $a > 0$, then the parabola opens upward (that is, shaped like a valley).
Rule 2: If $a < 0$, then the parabola opens downward (that is, shaped like a hill).
Rule 3: The vertex is the point $(d, f(d))$ where $d = -b/(2a)$.

These rules can be verified without the use of calculus (see any precalculus book). Later we will show how they may be verified easily using techniques of calculus.

Example 1.34 Sketch the graph of $y = f(x)$ for

$$f(x) = 2x^2 + 12x + 11.$$

Solution Since $a = 2 > 0$, from Rule 1 we know that this parabola opens upward. To use Rule 3 we need to calculate d. Since $d = -b/(2a)$ with $a = 2$ and $b = 12$, we have

$$d = -\frac{12}{2(2)} = -3$$

so the vertex of the parabola is $(-3, f(-3))$. To find $f(-3)$ we substitute -3 for x and get

$$f(-3) = 2(-3)^2 + 12(-3) + 11 = -7.$$

Therefore, the vertex is $(d, f(d)) = (-3, -7)$. The y-intercept is $f(0) = 11$. We should plot a few more points to sketch a good graph. Since $f(1) = 25$ and $f(-6) = 11$, the points $(1, 25)$ and $(-6, 11)$ lie on the graph. Plotting these points, we obtain the sketch of the graph in Figure 1.26. ■

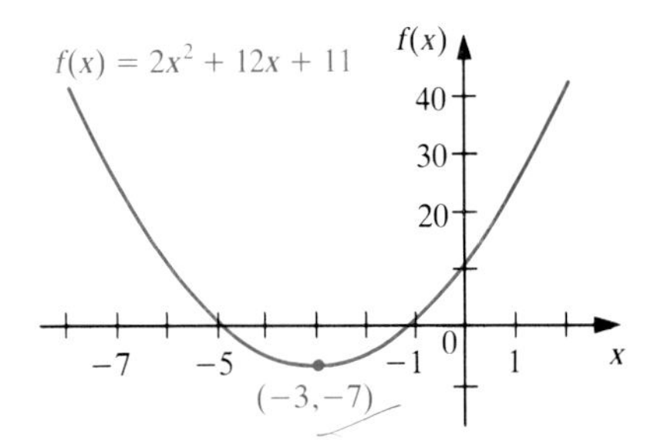

Figure 1.26
Graph of $f(x) = 2x^2 + 12x + 11$

Suppose that the graph of a quadratic function $f(x) = ax^2 + bx + c$ opens downward (that is, $a < 0$). Then the vertex is the highest point on the graph, and the value of the function at the vertex is the maximum value that the function can attain. Finding such a maximum value is very important in many applications. The following example is adapted from P. L. Shaffer and H. J. Gold, "A simulation model of population dynamics of the codling moth *cydia pomonella*," (*Ecological Modelling* 30 [1985]: 247–274.)

Example 1.35 In the life cycle of the codling moth, the searching stage is the period after the larva has hatched from its egg until it penetrates an apple. The larva is particularly vulnerable during this period. If all other environmental factors are constant and no pesticide is present, the survival rate of the larva during this stage is a function of the air temperature. Let $N(T)$ denote the percentage of larva that survives this stage for air temperature T (in °C). This study found that when $20 \le T \le 30$,

$$N(T) = -0.85T^2 + 45.4T - 546.$$

(a) What percentage of larva survives the searching stage when the temperature is 20 °C?
(b) What percentage survives at 30 °C?
(c) At what temperature is the survival rate the greatest? What percentage survives at that temperature?

Solution **(a)** Since $N(20) = -0.85(20)^2 + 45.4(20) - 546 = 22$, only 22% of the larva survives when the temperature is 20 °C (68 °F).
(b) Since $N(30) = -0.85(30)^2 + 45.4(30) - 546 = 51$, 51% survives at 30 °C (86 °F).
(c) Note that since $a = -0.85 < 0$, the graph of $N(T)$ is shaped like a hill; thus the vertex is a maximum point. By the rules stated earlier, the vertex occurs when $T = \frac{-b}{2a} = \frac{-45.4}{2(-0.85)} = 26.7$ °C. Since $N(26.7) = 60$, we know that 60% of the larva will survive the searching stage at this optimal temperature. ■

$\frac{-b}{2a}$ vertex

In many applications, such as the codling moth example, we find the maximum or minimum value of a function. In the next few chapters we develop efficient techniques for finding maxima and minima of a variety of functions. These techniques will eliminate the need for specialized rules such as those used for the parabola.

Recall that the x-intercepts of a function $f(x)$ occur when $f(x)=0$. If $f(x)$ is a quadratic function, then it may be possible to factor $f(x)$ to easily find the x-intercepts. For example, if $f(x)=x^2-7x+12$, then we can factor $f(x)$ as $f(x)=(x-3)(x-4)$. Thus $f(x)=0$ when $x=3$ or $x=4$. Similarly, we can factor $f(x)=2x^2+8x+6$ as $f(x)=2(x^2+4x+3)=2(x+1)(x+3)$. From this expression we see that $f(x)$ is zero when $x=-1$ or $x=-3$. However, the quadratic function

$$f(x)=ax^2+bx+c, \qquad a\neq 0$$

may not be easily factored. In such cases the quadratic formula (see Appendix A, Section A.2) gives the solutions to $f(x)=0$:

$$x=\frac{-b\pm\sqrt{b^2-4ac}}{2a}.$$

Since we cannot take the square root of a negative number, if $b^2-4ac<0$ the equation has no real solutions and $f(x)$ has no x-intercepts. If $b^2-4ac=0$, we have only one solution and $f(x)$ has only one x-intercept. If $b^2-4ac>0$, the equation has two solutions and $f(x)$ has two x-intercepts. Consequently, we conclude that a quadratic function may have either none, one, or two x-intercepts.

Example 1.36 Find the x-intercepts of $f(x)=2x^2+12x+11$.

Solution Using the quadratic formula, we find that $f(x)=0$ when

$$x=\frac{-12\pm\sqrt{(12)^2-4(2)(11)}}{2(2)}$$
$$=\frac{-12\pm\sqrt{144-88}}{4}=\frac{-12\pm\sqrt{56}}{4}.$$

Hence, the x-intercepts are the two values

$$x=\frac{-12+\sqrt{56}}{4}\approx-1.129 \qquad \text{and} \qquad x=\frac{-12-\sqrt{56}}{4}\approx-4.871.$$

Refer to Figure 1.26 to see that these are indeed reasonable x-values for the points where the graph crosses the x-axis. ■

Example 1.37 Two programs have been designed to create jobs in a midwestern city. One sociologist's study concludes that Program 1 will have produced $f(x)$ thousands of jobs x months after the program begins, and Program 2 will have produced $g(x)$ thousands of jobs x months after the program begins, where

$$f(x)=-0.1x^2+2x \qquad \text{and} \qquad g(x)=0.5x.$$

(a) Compare the effectiveness of the programs (that is, how many jobs

each creates) at the end of 2 months, 4 months, 10 months, and 20 months.
(b) If the program is scheduled to last 12 months, which program should the city implement?
(c) Is there a point in time at which the two programs will have created the same number of jobs?
(d) Graph the two functions on the same coordinate system.
(e) If the program is scheduled to last indefinitely, which of the two programs should the city implement?

Solution **(a)** To compare the effectiveness of the programs at the end of 2 months, we compare $f(2)$ with $g(2)$. Since

$$f(2) = -0.1(2)^2 + 2(2) = 3.6 \quad \text{and} \quad g(2) = 0.5(2) = 1,$$

we see that the Program 1 will have created a total of 3600 jobs by the end of 2 months while the Program 2 will have produced only 1000 jobs. At the end of 4 months we have

$$f(4) = -0.1(4)^2 + 2(4) = 6.4 \quad \text{and} \quad g(4) = 2.$$

Hence, at the end of four months Program 1 will have created 6400 jobs while Program 2 will have produced only 2000 jobs. Since $f(10) = 10$ and $g(10) = 5$, at the end of 10 months Program 1 will have produced 10,000 jobs while the total for Program 2 is 5000 jobs. At the end of 20 months, $f(20) = 0$ and $g(20) = 10$: Program 1 will have created no jobs at all (possibly because it overtaxes the business community, eventually causing a loss of jobs, which offsets the program's gains), while Program 2 will have produced 10,000 jobs.
(b) If the program is scheduled to last only 12 months then, since $f(12) = 9.6$ and $g(12) = 6$, Program 1 produces 3600 more jobs than Program 2, and the first program should be adopted.
(c) To determine at what point in time the two programs will have created the same number of jobs, we need to find the value of x for which $f(x) = g(x)$. Setting these equal and solving for x, we have

$$-0.1x^2 + 2x = 0.5x$$

$$0 = 0.1x^2 - 1.5x.$$

Factoring yields

$$0 = 0.1x(x - 15).$$

Hence, the solutions are $x = 15$ and $x = 0$. Therefore, $f(0) = g(0)$ and $f(15) = g(15)$. Now $f(0) = g(0)$ merely tells us that the programs start at the same point, but $f(15) = g(15)$ tells us that the programs will have created the same number of jobs at the end of 15 months.
(d) We graph $y = f(x)$ (Figure 1.27) by observing that, according to Rule 2, the graph is a parabola with the shape of a hill; by Rule 3 its vertex is the point (10, 10) (because $-b/(2a) = -2/[2(-0.1)] = 10$). We have already determined that the points (2, 3.6), (4, 6.4), and (20, 0) are on the graph. The graph of $y = g(x)$ is a line that passes through the two points (0, 0) and (20, 10).

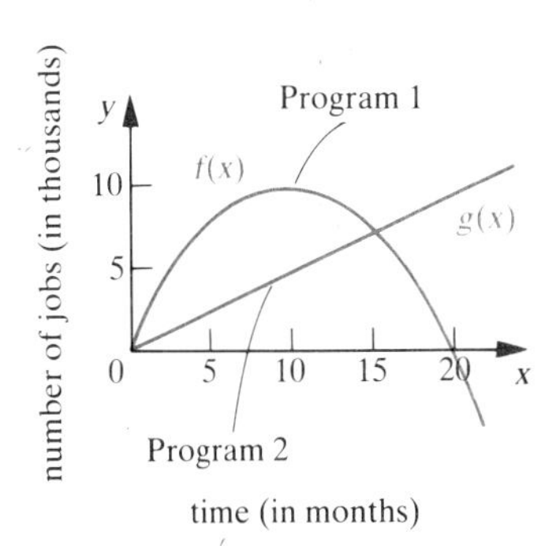

Figure 1.27
Graphs of the job programs in Example 1.37

(e) Note how obvious the relationship between the two functions appears when they are graphed! We can easily see how Program 1 betters Program 2 for the first 15 months, but in the long run, Program

2 is more sound. Consequently, if the program is to continue indefinitely, Program 2 is the better choice. ■

The x-intercepts of a quadratic function can be found easily by the quadratic formula; however, finding the x-intercepts of polynomials of higher degree is usually much more difficult and requires techniques beyond the scope of this text. The graph of a polynomial of degree higher than two can also be more complicated than that of a quadratic function. In Chapters 3 and 4 we will discuss methods for determining the graph of polynomials without extensive evaluation of the polynomial.

We now briefly consider a function that is quite different from polynomials. The absolute value of a real number is defined and discussed in Appendix A, Section A.3. We denote the absolute value of a number a by $|a|$; for example, the absolute value of 6 is denoted by $|6|$. The absolute value leaves positive numbers and zero unchanged but changes the sign of negative numbers. Since each real number x has exactly one corresponding absolute value, the absolute value defines a function:

$$f(x) = |x| = \begin{cases} x & \text{if } x \geq 0 \\ -x & \text{if } x < 0 \end{cases}$$

Example 1.38 Sketch the graph of the function $f(x) = |x|$.

Solution If $x \geq 0$, then $f(x) = x$, which has as its graph a line that passes through the origin and has slope 1. Some points on this line are $(1, 1)$, $(4, 4)$, and $(7, 7)$. If $x < 0$, then $f(x) = -x$, which has as its graph a line that passes through the points $(-1, 1)$, $(-4, 4)$, and $(-7, 7)$. These points are plotted in Figure 1.28, which shows the graph of $f(x) = |x|$. ■

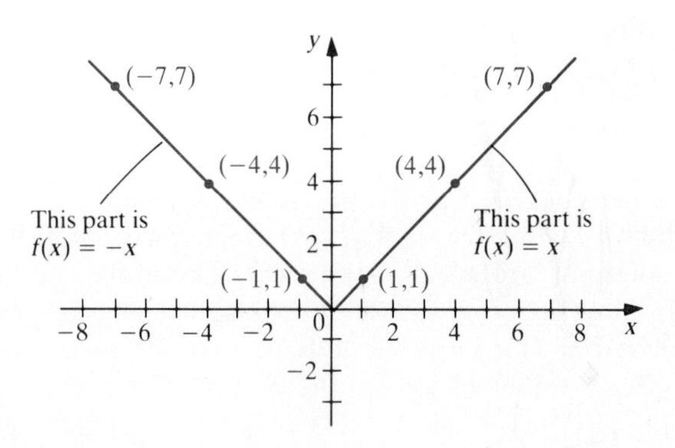

Figure 1.28
Graph of the absolute value function, $f(x) = |x|$

Example 1.39 Find a rule for the function $f(x) = |x - 4|$ that does not use absolute value notation and graph $f(x)$.

Solution We apply the definition of the absolute value to get

$$f(x) = |x - 4| = \begin{cases} x - 4 & \text{if } x - 4 \geq 0 \\ -(x - 4) & \text{if } x - 4 < 0 \end{cases}$$

or, equivalently, $f(x) = \begin{cases} x-4 & \text{if } x \geq 4 \\ 4-x & \text{if } x < 4 \end{cases}$

(We have used the fact that $-(x-4) = 4-x$.) The graph of $f(x)$ for $x < 4$ is the line $f(x) = 4 - x$, which passes through $(0, 4)$ and $(3, 1)$; the graph of $f(x)$ for $x \geq 4$ is the line $f(x) = x - 4$, which passes through $(5, 1)$ and $(6, 2)$. These points are plotted in the graph in Figure 1.29. ■

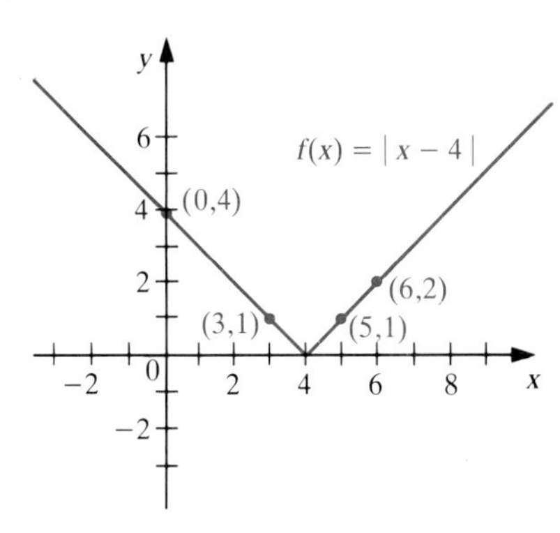

Figure 1.29
Graph of $f(x) = |x - 4|$

Exercises

Exer. 1–14: Graph the quadratic function and find its y-intercept and any x-intercepts.

1. $f(x) = 3x^2$
2. $f(x) = -4x^2$
3. $f(x) = x^2 + 6x + 5$
4. $f(x) = -x^2 + 4x + 3$
5. $g(x) = 2x^2 + 4x - 8$
6. $g(x) = -4x^2 + 8x + 6$
7. $f(x) = 3x^2 + 12x + 5$
8. $D(x) = 3x^2 + 9x + 9$
9. $w(x) = -2x^2 - 8x + 12$
10. $f(x) = -2x^2 - 6x - 2$
11. $M(x) = -2x^2 + 12x - 3$
12. $S(x) = -5x^2 + 7x - 4$
13. $p(x) = 0.2x^2 + 0.4x + 8$
14. $C(x) = -0.01x^2 + 0.06x - 12$

Exer. 15–18: Find the points of intersection of the graphs of $y = f(x)$ and $y = g(x)$ and graph both functions on a common coordinate system. (See Example 1.37.)

15. $f(x) = x^2 + 2x + 1$; $g(x) = 4x + 4$
16. $f(x) = -2x^2 + 6x - 5$; $g(x) = 4x - 7$
17. $f(x) = 2x^2 + 8x + 3$; $g(x) = 3x + 1$
18. $f(x) = -3x^2 + 12x - 6$; $g(x) = x + 9$

19. Refer to Example 1.37. Suppose the sociologist finds that Program 1 is expected to yield $f(x)$ thousands of jobs in x months where $f(x) = -0.1x^2 + 3x$, and Program 2 is expected to yield $g(x)$ thousands of jobs in x months where $g(x) = x$.
(a) Compare the two programs at the end of 4 months, 10 months, and 30 months.
(b) At what point in time will the two programs have created the same number of jobs?
(c) If the program is scheduled to last 12 months, which program should be adopted?
(d) Graph $y = f(x)$ and $y = g(x)$ on one coordinate system.
(e) If the program is to be implemented for an indefinite length of time, which of the two programs should be adopted? Explain.

20. A forest ranger has a choice between two wildlife management approaches. The first approach will produce a population of $f(x)$ deer at the end of x years, where

$$f(x) = 20x^2 - 80x + 500.$$

The second approach will produce a population of $g(x)$ deer at the end of x years, where

$$g(x) = -50x^2 + 400x + 500.$$

(a) Which approach yields the larger deer population by the end of 2 years? 4 years?
(b) At what point in time will the two approaches yield the same deer population?
(c) Graph $y = f(x)$ and $y = g(x)$ on one coordinate system.
(d) Which approach would you recommend to produce the greatest possible number of deer over an indefinite length of time.

Exer. 21–22: Refer to Example 1.37.

21. The percentage of codling moth eggs that survives (hatches) is $N(T)$ when the temperature is T °C and is given by

$$N(T) = -0.53T^2 + 25T - 209 \quad \text{for} \quad 15 \leq T \leq 30.$$

What percentage of eggs survives when the temperature is 15 °C? What percentage survives at 30 °C? At what temperature is the survival rate of eggs the greatest, and what percentage survives at the optimal temperature?

22. The percentage of codling moths that survives the pupa stage when the temperature is T (°C) is

$$N(T) = -1.42T^2 + 68T - 746 \quad \text{for} \quad 20 \leq T \leq 30.$$

What percentage survives the pupa stage when the temperature is 20 °C? What percentage survives at 30 °C? At what temperature is the survival rate the greatest during the pupa stage, and what percentage survives at the optimal temperature?

Exer. 23–32: Evaluate the expression.

23. $|-2|$
24. $|3|$
25. $|0|$
26. $|x-2|;\quad x=1$
27. $|x^2|;\quad x=-3$
28. $|-4|$
29. $|4-6|$
30. $|6-4|$
31. $|2+x|;\quad x=-4$
32. $|x+4|;\quad x=-3$

Exer. 33–38: Find a rule for the function that does not use absolute value notation, then sketch the graph of $y=f(x)$.

33. $f(x)=|x-1|$
34. $f(x)=|2-x|$
35. $f(x)=|-x|$
36. $f(x)=|2x-1|$
37. $f(x)=|2x+3|$
38. $f(x)=3+|x-1|$

1.6 Average Rate of Change

If an automobile travels 150 miles in three hours, its average speed for that time period is

$$\frac{150}{3}=50 \text{ miles per hour}$$

(because distance divided by time is rate). Suppose that the motion of this automobile is described by the function

$$s(t)=10t^2+20t \quad \text{with} \quad 0\le t\le 3$$

where t represents time measured in hours and $s(t)$ represents the number of miles traveled by time t. To find the car's average speed over the last two hours we use the time period that begins at $t=1$ and ends at $t=3$. Therefore, the distance traveled during this period is $s(3)$, the distance traveled over the full three hours, minus $s(1)$, the distance traveled during the first hour. Hence the distance traveled is $s(3)-s(1)$, and the time elapsed is $3-1$. Therefore, the average speed over the last two hours of the trip is

$$\frac{s(3)-s(1)}{3-1}=\frac{150-30}{2}=60.$$

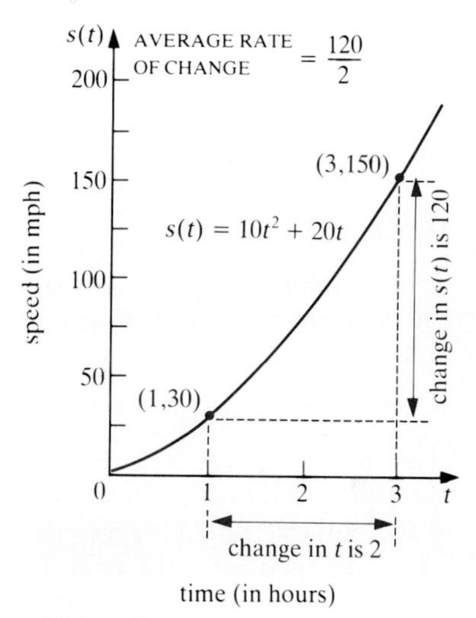

Figure 1.30
Average speed of a car over a 2-hour period

See Figure 1.30. The average speed over the 2-hour period is 60 mph. The expression $s(3)-s(1)$ represents the change in $s(t)$ from $t=1$ to $t=3$, and $3-1$ represents the change in t. Hence,

$$\text{average rate of speed}=\frac{\text{change in } s(t)}{\text{change in } t}$$

In general, given a function $f(x)$ and two values of x, say a and b, the change in $f(x)$ as x changes from a to b is $f(b)-f(a)$. Thus, the average rate of change of $f(x)$ with respect to x as x changes from a to b is

$$\frac{f(b)-f(a)}{b-a}.$$

In other words,

$$\text{average rate of change of } f(x) = \frac{\text{change in } f(x)}{\text{change in } x}.$$

Note that the average rate of change of $f(x)$ is the slope of the line that passes through the points $(a, f(a))$ and $(b, f(b))$. Consequently, if $f(x)$ is a linear function, then the average rate of change of $f(x)$ over any interval is the slope of the graph of $f(x)$.

Example 1.40 Show that the average rate of change of the linear function $f(x) = 3x - 4$ (a) from $x = 2$ to $x = 4$ and (b) from $x = 3$ to $x = 7$ is the slope of its graph.

Solution (a) The slope of the graph of $f(x)$ is 3. The average rate of change of $f(x)$ from $x = 2$ to $x = 4$ is

$$\frac{f(4) - f(2)}{4 - 2} = \frac{8 - 2}{2} = 3,$$

which equals the slope.
(b) The average rate of change of $f(x)$ as x goes from 3 to 7 is

$$\frac{f(7) - f(3)}{7 - 3} = \frac{17 - 5}{4} = 3,$$

which also equals the slope. ■

Example 1.41 A small company has determined that the number $p(x)$ of finished music stands it can produce each day is given by the function

$$p(x) = 10x - 0.1x^2 \quad \text{for} \quad 0 \le x \le 50,$$

where x is the number of employees. What is the average rate of change of production (number of stands produced) per additional employee as the workforce increases from 10 to 15 employees? What is the average rate of change of production as the workforce changes from 20 to 25?

Solution The change in production as the workforce changes from $x = 10$ to $x = 15$ is $p(15) - p(10)$, so the average rate of change in production as the workforce changes from 10 to 15 is

$$\frac{p(15) - p(10)}{15 - 10} = \frac{127.5 - 90}{5} = \frac{37.5}{5} = 7.5$$

stands per additional worker per day. The average rate of change of production as the workforce increases from 20 to 25 is

$$\frac{p(25) - p(20)}{25 - 20} = \frac{187.5 - 160}{5} = \frac{27.5}{5} = 5.5$$

stands per additional worker per day. The manager interprets these numbers in the following way. When there are 10 employees and another 5 are added, the number of stands produced per day will increase an average of 7.5 stands for each new employee; when 5 employees are added to a staff of 20, the average number of stands produced per day will increase 5.5 stands for each new employee. This information will help the manager decide whether adding more employees will increase or decrease profits. ■

Example 1.42 Suppose that the cost C in dollars of producing x wool hats is given by the quadratic function

$$C(x) = 20 + 1.6x - 0.002x^2.$$

(a) What is the change in cost between producing 50 hats and 100 hats?
(b) What is the average rate of change in cost per additional hat between producing 50 hats and producing 100 hats?
(c) Is the average rate of change in cost between producing 150 hats and producing 200 hats higher or lower than that between 50 and 100 hats?

Solution **(a)** The costs of producing 50 hats is

$$C(50) = 20 + 1.6(50) - 0.002(50)^2 = 95$$

and the cost of producing 100 hats is

$$C(100) = 20 + 1.6(100) - 0.002(100)^2 = 160,$$

for a change in cost of \$65.
(b) The average rate of change in cost between producing 50 hats and producing 100 hats is

$$\frac{\text{change in cost}}{\text{change in number produced}} = \frac{C(100) - C(50)}{100 - 50} = \frac{160 - 95}{50} = 1.3.$$

Over the range 50-100, on the average, the cost for producing each additional hat is \$1.30.
(c) The average rate of change in cost between producing 150 hats and

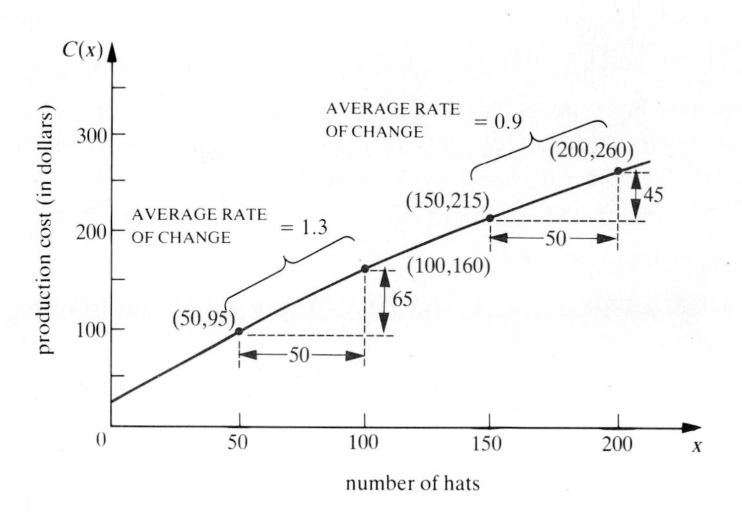

Figure 1.31
Average rate of change of the cost function in Example 1.42

producing 200 hats is

$$\frac{C(200) - C(150)}{200 - 150} = \frac{260 - 215}{50} = 0.9.$$

Thus, over the range 150-200, on the average, the cost for producing each additional hat is \$0.90.

The average rate of change over the interval $x = 150$ to $x = 200$ is lower than the average rate of change over the interval $x = 50$ to $x = 100$. Figure 1.31 illustrates this result. The curve appears to rise more quickly on the interval [50, 100] than on [150, 200]. ■

Exercises

Exer. 1–6: For the function $f(x) = 2x - 7$, find the following:

1. Change in $f(x)$ from $x = 2$ to $x = 4$
2. Change in $f(x)$ from $x = 5$ to $x = 8$
3. Change in $f(x)$ from $x = -1$ to $x = 4$
4. Average rate of change of $f(x)$ from $x = 2$ to $x = 4$
5. Average rate of change of $f(x)$ from $x = 5$ to $x = 8$
6. Average rate of change of $f(x)$ from $x = -1$ to $x = 4$

Exer. 7–8: A runner's rate of respiration is a function of the amount of carbon dioxide in the lungs. The function $f(x) = 0.59x - 10.39$ gives the number of breaths per minute as a function of the amount x of carbon dioxide in the lungs. Find the average rate of change of respiration when the amount of carbon dioxide changes as indicated.

7. From $x = 10$ to $x = 20$
8. From $x = 40$ to $x = 50$

Exer. 9–14: For the function $f(x) = 2x^2 - x + 1$, find the following.

9. Change in $f(x)$ from $x = 1$ to $x = 4$
10. Change in $f(x)$ from $x = -2$ to $x = 0$
11. Change in $f(x)$ from $x = 3$ to $x = 6$
12. Average rate of change of $f(x)$ from $x = 1$ to $x = 4$
13. Average rate of change of $f(x)$ from $x = -2$ to $x = 0$
14. Average rate of change of $f(x)$ from $x = 3$ to $x = 6$

Exer. 15–17: The number of mice in a landfill t hours after a monitoring system is activated is given by $m(t) = 1000(1 + 0.35t + t^2)$. Find the following.

15. The change in the number of mice between $t = 2$ and $t = 5$
16. The average rate of change in the number of mice between $t = 2$ and $t = 5$
17. The average rate of change in the number of mice between $t = 1$ and $t = 5$

Exer. 18–22: For the function $f(x) = 6 - 3x$, find the following.

18. Change in value of $f(x)$ from $x = 1$ to $x = 4$
19. Change in value of $f(x)$ from $x = 2$ to $x = 6$
20. Average change in value of $f(x)$ from $x = 1$ to $x = 4$
21. Average change in value from $f(x)$ from $x = 2$ to $x = 6$
22. What conclusion can be made about the average change of this funtion over any interval?

Exer. 23–25: A population of 1000 bacteria is introduced into an environment in which it grows according to the equation

$$P(t) = 1000 + \frac{1000t}{100 + t^2}$$

for time t measured in hours. Find the average rate of change of the population during the specified time period.

23. The first two hours ($t = 0$ to $t = 2$)
24. The fifth hour (from $t = 4$ until $t = 5$)
25. As t changes from 3 to 5

Exer. 26–28: The cost of producing x canvas backpacks is $C(x) = 10 + 25x + \sqrt{x}$.

26. Find the cost of producing the twenty-fifth backpack.
27. Find the average rate of change in cost from producing 25 backpacks to producing 49.
28. Find the average rate of change in cost from producing 100 backpacks to producing 225.

Exer. 29–31: A student group decides to produce and sell "survival kits" (consisting of pizza, soda, candy, and so on) before final exams. The profit from producing and selling x kits is $p(x) = 2x + (3/x)$ dollars.

29. Find the profit from selling the tenth kit.
30. Find the average rate of change in profit from selling 15 kits to selling 30.
31. Find the average rage of change in profit from selling 100 kits to selling 150.

Exer. 32–33: Suppose that the number of gallons $G(t)$ of water in a tank that is being drained for cleaning is given by the formula

$$G(t) = 400(60 - t^2)$$

for the number of minutes t after the draining starts. Find the average rate of change of the number of gallons of liquid in the tank for the specified time period.

32. During the first four minutes.
33. For the first four minutes after $t = 2$.

34. Refer to Example 1.42. Suppose that the cost function for the hat manufacturer is

$$C = 300 + 12x - 0.03x^2.$$

What is the average change in cost per additional hat between producing 20 hats and 60 hats?

35. A corporation finds that its profit P (in millions of dollars) as a function of the number of years t that the corporation has been in business is given by

$$P(t) = -2t^2 + 32t - 66.$$

$P(t)$ is the total profit for the first t years. What is the average yearly profit between the fourth and seventh year? Between the eighth and twelfth year?

36. The population P (in hundreds) of a gerbil colony living under restricted conditions is given by

$$P(t) = (t + 2)^2 + 0.1$$

at time t years. Find the average rate of change in the size of the population (a) between 1 year and 2 years; (b) between 2 years and 3 years; and (c) between 1 year and 3 years.

Summary

A **function** $f(x)$ from a set A to a set B expresses a relationship between the two sets by assigning to each number in A exactly one number in B. The number in B to which a number x in A is assigned is denoted by $f(x)$. The **domain** of a function $f(x)$ is the set A.

The **graph** of a function $f(x)$ is the set of points $(a, f(a))$ for all a in the domain of $f(x)$. The **y-intercept** of a function $f(x)$ is the number $f(0)$ and corresponds to the point $(0, f(0))$ where the graph intersects the y-axis. An **x-intercept** is a value of x for which $f(x) = 0$ and corresponds to a point where the graph crosses the x-axis. If a vertical line crosses a curve more than once, then the curve is not the graph of a function.

A function $f(x)$ is a **linear function** if and only if it can be written in the form $f(x) = mx + b$ for constants m and b. The graph of the linear function $f(x) = mx + b$ is a line that has **slope** m and y-intercept b. Therefore,

$$y = mx + b$$

is called the **slope-intercept** form of the equation of the line. An equation of the line that has slope m and passes through the point (x_1, y_1) is

$$y - y_1 = m(x - x_1).$$

This form of the equation of the line is called the **point-slope form**.

If two lines are given by equations $y = m_1x + b_1$ and $y = m_2x + b_2$, then they are **parallel** if $m_1 = m_2$ or **perpendicular** if $m_1 = -1/m_2$. If the lines are not parallel, then they intersect at exactly one point, which is found by solving the two equations simultaneously for x and y. The **distance** d between two points (x_1, y_1) and (x_2, y_2) is

$$d = \sqrt{(x_2 - x_1)^2 + (y_2 - y_1)^2}.$$

A **demand** function expresses the relationship between the price of a product and the quantity of a product that will sell at that price. A **supply** function relates the quantity produced to the price of the product. If the graphs of functions representing demand and supply intersect, then the price corresponding to the point of intersection is the **equilibrium price**, the price at which customers will purchase the same quantity of a product that manufacturers wish to produce at that selling price.

A **polynomial** is a function of the form

$$y = f(x) = a_nx^n + a_{n-1}x^{n-1} + \cdots + a_1x + a_0, \qquad a_n \neq 0,$$

for a nonnegative integer n and constants a_i. When $n = 2$, the polynomial is a **quadratic function**. The graph of the quadratic $f(x) = ax^2 + bx + c$ is a **parabola** that opens upward if $a > 0$, or downward if $a < 0$. The **vertex** of the parabola is the point $(d, f(d))$ for $d = -b/(2a)$.

The **absolute value** function $f(x) = |x|$ is defined by

$$f(x) = |x| = \begin{cases} x & \text{if } x \geq 0 \\ -x & \text{if } x < 0 \end{cases}$$

The **average rate of change** of a function $f(x)$ with respect to x as x changes from $x = a$ to $x = b$ is

$$\frac{f(b) - f(a)}{b - a}.$$

Review Exercises

1. If $f(x) = 9 - x^2$, find $f(2)$ and $f(a - 3)$.
2. If $f(x) = \dfrac{1}{x+3}$, determine $\dfrac{f(x) - f(2)}{x - 2}$, and simplify.
3. Find the domain of the function defined by $f(x) = \sqrt{x + 3}/(x + 2)$.
4. Find the domain of the function $f(x) = \sqrt{x^2 + 4x + 3}$.

Exer. 5–10: Find the equation of the line that satisfies the given conditions. Write the equation in the form $y = mx + b$.

5. Slope -5 and y-intercept 3
6. Slope 3 and x-intercept 4
7. Slope 2 and passes through $(-1, 3)$
8. Passes through $(2, 4)$ and $(3, 9)$
9. Passes through $(2, 4)$ and is horizontal
10. With y-intercept 4 and x-intercept 2

11. Find the point-slope form of the equation of the line that passes through the points $(1, -2)$ and $(-3, 6)$.
12. Find the equation of the line that is parallel to the line $y = 3x + 8$ and passes through the point $(4, 5)$.
13. Find the equation of the line that is perpendicular to the line $y = 0.5x + 6$ and passes through the origin.
14. Find the equation of the line that is perpendicular to the line $3x + 4y = 5$ and passes through the point of intersection of the lines $y = 2x - 3$ and $y - x = 1$.
15. A salesman earns \$150 per week plus a 15% commission on his weekly gross sales. Write a function that gives his weekly income in terms of his gross sales. What will his income be for a week when he has gross sales of \$800?
16. A rental car costs \$35 per day plus 22¢ per mile. Find a function $C(x)$ that gives the cost of renting such a car for a week if x miles are driven.

Exer. 17–20: Find the y-intercept and any x-intercepts of the graph of $y = f(x)$.

17. $f(x) = 3x - 2$
18 $f(x) = -4x + 1$
19. $f(x) = 2x^2 + x - 1$
20. $f(x) = -3x^2 + 2x - 4$

Exer. 21–24: Find the points of intersection of the graphs of $y = f(x)$ and $y = g(x)$.

21. $f(x) = 4x + 1$; $g(x) = -2x + 13$
22. $f(x) = 3x + 5$; $g(x) = x^2 + 4x + 3$
23. $f(x) = -x + 7$; $g(x) = x^2 + 3x - 5$
24. $f(x) = 3x^2 + 2x - 5$; $g(x) = 2x^2 + x + 7$

Exer. 25–30: Sketch the graph of $y = f(t)$.

25. $f(t) = 2t - 3$
26. $f(t) = 2t^2 - t + 1$
27. $f(t) = 2t^2 + t - 1$
28. $f(t) = 4$
29. $f(t) = -3t + 7$
30. $f(t) = -4t^2 + 8t + 1$

31. Of two proposed fundraising drives, the first is expected to raise $f(x)$ thousands of dollars in an x-week campaign, where

$$f(x) = 5x.$$

The second is expected to raise $g(x)$ thousands of dollars in an x week campaign, where

$$g(x) = 0.5x^2.$$

Which would raise the most money in a 5-week campaign? Which would raise the most money in a 12-week campaign? For what length campaign would the two types raise the same amount of money?

32. A long-distance runner who wants to increase endurance through an exercise and diet program is considering two programs. One program will allow the runner to run at top performance for $f(x)$ miles x weeks after the program begins, where

$$f(x) = -x^2 + 7x + 6.$$

The second program will allow the runner to run at top performance for $g(x)$ miles after x weeks of training, with

$$g(x) = x + 6.$$

If a big race is coming up in four weeks, which program should the runner follow? How many miles will the runner be able to run at top performance in this race? At what point in time will the two programs produce the same effect? If continued for several months, which program would be better?

33. The daily demand for radios is given by $D(x) = 500 - 10x$ for the price x (in dollars) of one radio. What will be the effect on the number of radios sold daily if the selling price is decreased by \$5?
34. Refer to Exercise 33. Suppose that the supply function for radios is $S(x) = 4x + 220$ radios per day. What is the equilibrium price? How many radios will be sold daily at the equilibrium price?

Exer. 35–40: Find the average rate of change of $f(x)$ over the given interval.

35. $f(x) = 3x + 7$; from $x = 2$ to $x = 9$

36. $f(x) = 2x^2 - 4x + 6;$ from $x = 5$ to $x = 8$
37. $f(x) = -3x^2 + 2x + 3;$ from $x = -3$ to $x = 2$
38. $f(x) = \dfrac{2x + 1}{x^2 + 6x - 1};$ from $x = 1$ to $x = 4$
39. $f(x) = |4x - 3|;$ from $x = 2$ to $x = 6$
40. $f(x) = |6x + 9|;$ from $x = -4$ to $x = 1$

Exer. 41–42: A high-school soccer club is selling caps to raise money for uniforms, and their profit is $p(x) = 2x - 5/x$ dollars for selling x caps.

41. What is the profit for selling the twentieth cap?
42. Find the average rate of change in profit from selling 20 caps to selling 40.

Facing page: **The skis of the skier are tangent to the hill. At the top of the hill, they are horizontal.**

· T · W · O ·

The *derivative* is an important concept in calculus. Two major uses of the derivative are (1) the measurement of the rate at which some quantity changes with time and (2) the measurement of the steepness of a graph. The nature and usefulness of the derivative will be illustrated in Section 2.1 by two examples. In this chapter we define the derivative and discuss some of its applications.

2.1 Motivation for the Derivative

In this section we will explore two situations to illustrate important uses of the derivative. The first use, measurement of the rate of change of a quantity over time, is demonstrated by a fruit fly infestation.

The Fruit Fly Population: Rate of Growth

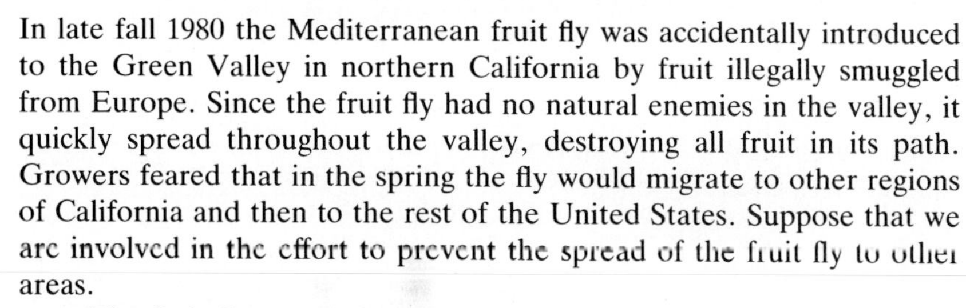

In late fall 1980 the Mediterranean fruit fly was accidentally introduced to the Green Valley in northern California by fruit illegally smuggled from Europe. Since the fruit fly had no natural enemies in the valley, it quickly spread throughout the valley, destroying all fruit in its path. Growers feared that in the spring the fly would migrate to other regions of California and then to the rest of the United States. Suppose that we are involved in the effort to prevent the spread of the fruit fly to other areas.

The fruit fly population has been analyzed over a 4-week period, and the number of Mediterranean fruit flies t weeks after their discovery is found to be $P(t)$ million fruit flies where

$$P(t) = \tfrac{1}{4}t^2 + 3t + 1.$$

For instance, the population of fruit flies in the Green Valley four weeks after discovery ($t = 4$) is

$$P(4) = \tfrac{1}{4}(4)^2 + 3(4) + 1 = 17,$$

that is, 17 million fruit flies are present at the end of four weeks. This formula is expected to be accurate for many more weeks.

At this time (four weeks after the discovery of the flies) two methods are proposed to control the fruit fly population: (1) The entire area could be sprayed from the air with a strong insecticide, and (2) sterile male fruit flies could be introduced into the valley and as much fruit as possible could be harvested early, eliminating some food supply for the flies. The majority of fruit growers favor the first method, but it poses a health risk to humans and animals; therefore, environmentalists and most of the residents of Green Valley favor the second method. However, the second method has two drawbacks: complete implementation will require about four more weeks, and, if the fruit fly population is growing too rapidly when implementation begins, the method will be unsuccessful. The growers and residents compromise by deciding that method 2 will be used only if it can be determined now that four weeks from now ($t = 8$) the fruit fly population will be growing at the relatively slow rate of 7.5 million flies per week. Otherwise, the first method should be applied immediately.

Our choice of method thus depends on determining the rate of growth of the fruit fly population. In fact we need the rate of growth at the end of the eighth week after discovery. Furthermore, this rate must be determined from current data because if method 2 is implemented but proves to be unsuccessful, then in four more weeks the fruit fly population will be well established and even spraying may be ineffective.

In Section 1.6. we introduced a method for calculating the *average* rate of growth over a given period of time. For example, the average rate of growth of the fruit fly population between the eighth and the twelfth week is

$$\frac{P(12) - P(8)}{12 - 8} = \frac{73 - 41}{4} = \frac{32}{4} = 8 \text{ million fruit flies per week.}$$

Figure 2.1 shows the graph of $P(t)$ along with the average rate of change during this time. This average growth is larger than the established limit of 7.5 million flies per week. However, the population may be growing at a faster rate in the twelfth week than in the eighth week, so this average rate for the interval [8, 12] may be greater than the rate of growth at $t = 8$.

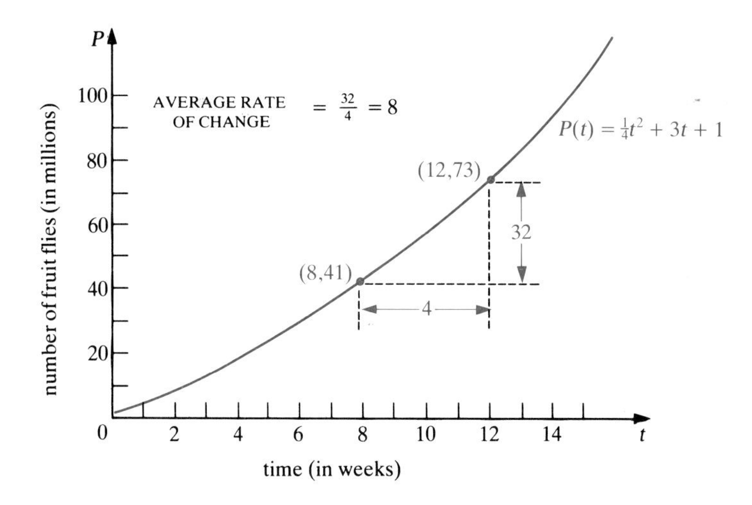

Figure 2.1
Average rate of growth of the fruit fly population from week 8 to week 12

The number we seek is the growth rate at the instant $t = 8$ rather than an average rate of growth over an interval of time. How can we define such a notion? To answer this question, let s be a point in time greater than $t = 8$; then the average rate of growth between time $t = 8$ and time $t = s$ is

$$\frac{P(s) - P(8)}{s - 8}.$$

In Figure 2.2 we have shown the average rate of change over this time interval. If s is far away from $t = 8$, then the average is not an accurate

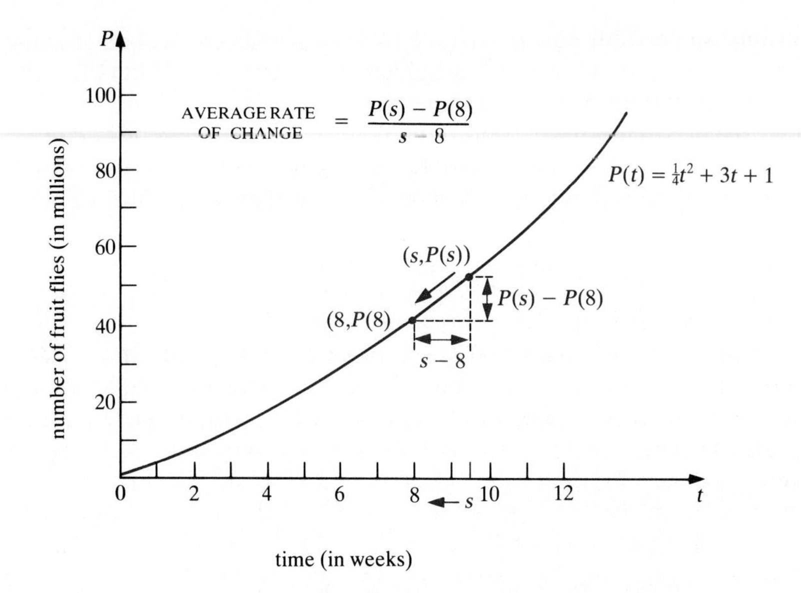

Figure 2.2
Average growth rate from week 8 to week *s*

estimate of the rate of growth near $t = 8$. The closer s is to $t = 8$, the better this average growth rate approximates the actual rate of growth at the instant $t = 8$. If the average growth rate approaches a particular value as s gets closer and closer to $t = 8$, then that particular value is the rate of growth at the instant $t = 8$. We call the process involving "closer and closer values" a *limiting process*, and we denote the result of the process by the expression

$$\lim_{s \to 8} \frac{P(s) - P(8)}{s - 8}.$$

This limiting process is successful for our function $P(t)$, but it is not successful for all functions. When it is successful, this particular *limit* will be called the *derivative* of $P(t)$ at $t = 8$. When we have learned how to calculate such limits, we will find that the rate of growth of the fruit fly population at $t = 8$ weeks is 7 million flies per week. Consequently, the Mediterranean fruit fly population can be controlled effectively by method 2, and the residents of Green Valley will be spared the hazards of spraying.

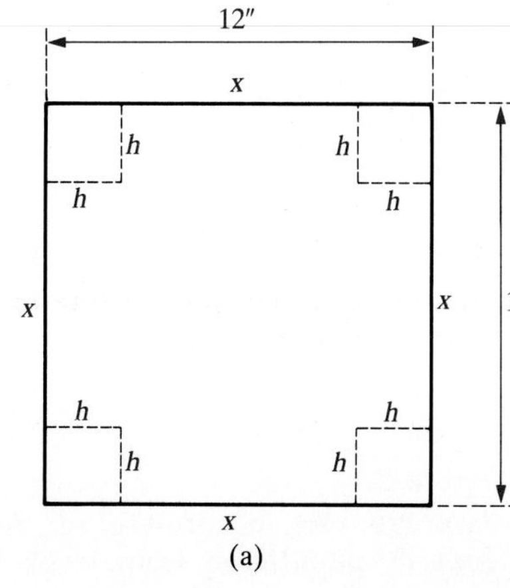

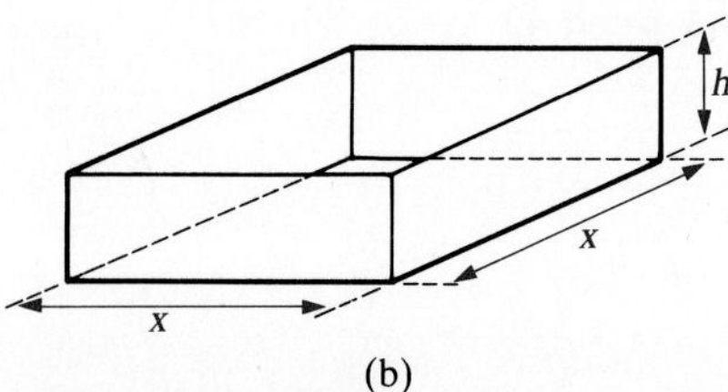

Figure 2.3
(a) A square piece of plastic with corners cut out; (b) a cake pan formed by bending up sides of the piece of plastic in (a)

The Cake Pan: Optimization

Our second example appears to be quite different from the fruit fly example; however, as we will see, both examples embody a similar idea. A manufacturing company plans to make microwave-safe cake pans by cutting squares out of the corners of a 12-inch by 12-inch piece of plastic, and then bending the sides up as shown in Figure 2.3(a). The seam along each corner will be fused to finish the pan. The manufacturer wishes to determine what size square should be cut from the corners to make a pan of the greatest possible volume.

The resulting pan has a square base, with sides of length x inches and is h inches deep. See Figure 2.3(b). Consequently, its volume (in cubic inches) is given by

$$V = x^2h.$$

Each side of the original piece of plastic is 12 inches long and consists of two sections of length h and one section of length x, so that $2h + x = 12$, that is, $h = 6 - \frac{1}{2}x$. Thus, the volume V can be expressed as a function $f(x)$ of x as

$$V = f(x) = x^2(6 - \tfrac{1}{2}x) = 6x^2 - \tfrac{1}{2}x^3 \qquad \text{when } x \geq 0.$$

The graph of this function is shown in Figure 2.4.

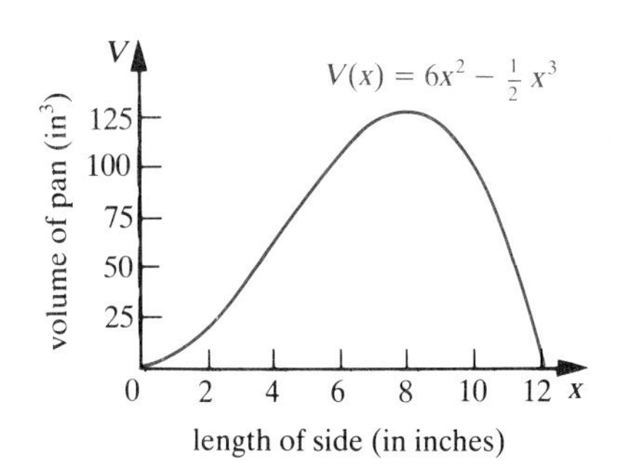

Figure 2.4
Graph of $f(x) = 6x^2 - \frac{1}{2}x^3$

The maximum volume, that is, the maximum height of the curve in the graph will occur at the "top of the hill." How can we locate the value of x that corresponds to the top of the hill without guessing from the graph? The concept of a **line tangent to the curve** will help us determine this value. We want to define the line tangent to a curve so that it indicates the "slope of the curve" at the point of tangency. Consider the skier illustrated in Figure 2.5. The slope of the line formed by the skis represents the steepness of the hill and whether it is uphill or downhill. We wish to define mathematically the line corresponding to each position of the skis.

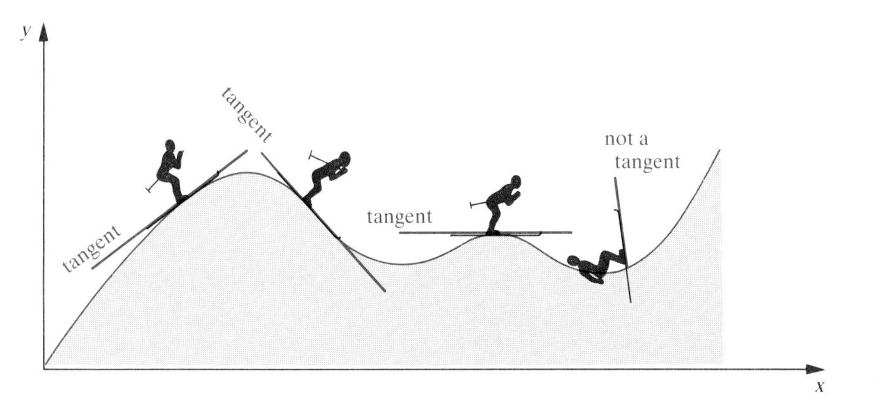

Figure 2.5
Skis indicate the tangent line to a curve at several points

To define the line tangent to the graph of a function $f(x)$ at a point P, we construct the *secant line* through P and a nearby point Q (that is, a line crossing the graph at points P and Q). Let a denote the x-coordinate of P and let x denote the x-coordinate of Q. Then the y-coordinates of these points are $f(a)$ and $f(x)$, respectively. See Figure 2.6. As Q moves closer to P, the slope of the secant line through P and Q approximates more accurately the slope of the curve at P. Figure 2.6 shows several secant lines with x successively closer to a. We define the *line tangent to the curve at P* to be the **limit** *of these secant lines as Q gets closer and closer to P.*

The slope of the line through P and Q is

$$\text{slope of secant line} = \frac{f(x) - f(a)}{x - a}.$$

Thus, the slope of the line tangent to the curve at P is the limit of this

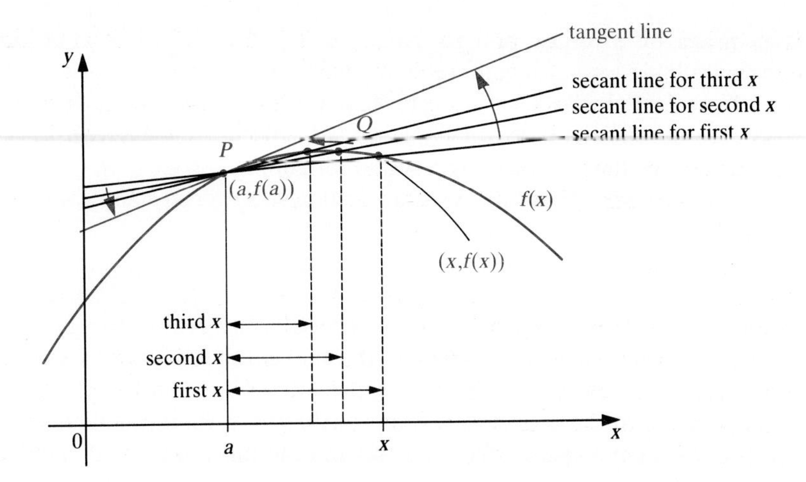

Figure 2.6
Geometric definition of the tangent line as the limit of secant lines

expression as x gets closer and closer to a, denoted by

$$\lim_{x \to a} \frac{f(x) - f(a)}{x - a},$$

where we are using *limit* as an informal, intuitive concept. (As with the fruit fly example, this limiting process is successful for some functions and fails for others. We will study limits in more detail in the next section.)

Just as the skier's skis are horizontal at the top of a hill, the tangent line is horizontal at the maximum point of a graph. Since the slope of a horizontal line is zero, the slope of the tangent line to a function is zero at the maximum points of the function.

To finish the cake pan problem, we need to find the value of x where the graph of the function in Figure 2.4 attains its maximum value; that is, where the slope of the tangent line to the function $f(x) = 6x^2 - \frac{1}{2}x^3$ is zero. In Section 2.3 we will find that the maximum of this function occurs at $x = 8$ (see Figure 2.7). Consequently, to construct a cake pan with the greatest possible volume from a 12-inch square piece of plastic, we let x equal 8 inches (see Figure 2.3). The base of the pan

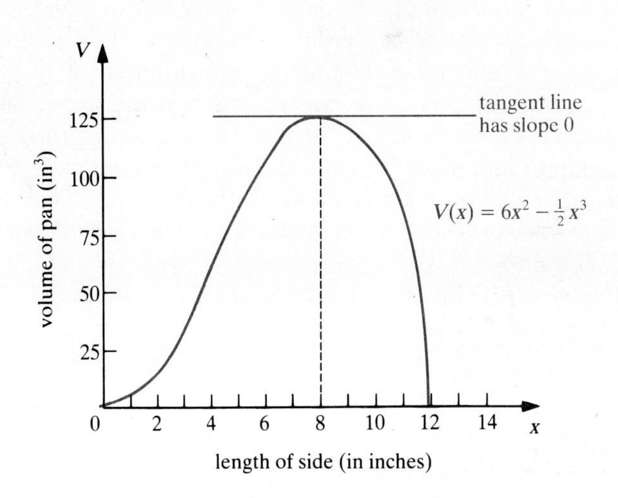

Figure 2.7
Maximum value of $f(x) = 6x^2 - \frac{1}{2}x^3$ for $0 < x < 12$ occurs at $x = 8$

is 8 inches by 8 inches and its height is 2 inches. These dimensions correspond to those of a commonly available microwave cakepan.

Note that the limit we used to find the slope of the tangent line is similar to the limit used to solve the fruit fly problem. In Section 2.3 we will define this limit to be the *derivative* of $f(x)$ at $x = a$.

To summarize, both of our examples were solved by the process of taking limits. A particular limit yields the derivative of a function at a particular value. The derivative may represent the instantaneous rate of change of a function (the rate of growth of a population of fruit flies, for example). It also represents the slope of the line tangent to the graph of a function, from which we determine maximum (or minimum) values of the function (such as the maximum volume of the cake pan). An understanding of the concepts of limits and derivatives is essential to the solution of many applications. We will discuss these topics in detail in Section 2.2.

Exercises

Exer. 1–2: If the population of the Mediterranean fruit fly in the Green Valley t weeks after discovery is

$$P(t) = \tfrac{1}{4}t^2 + 3t + 1,$$

calculate the average rate of growth for $P(t)$ between $t = 8$ and $t = s$ for the given values of s—that is, evaluate the expression

$$\frac{P(s) - P(8)}{s - 8}$$

for each value of s. Examine your answers, and suggest a reasonable value for

$$\lim_{s \to 8} \frac{P(s) - P(8)}{s - 8}.$$

1. $s = 16, 10, 9, 8.5, 8.1, 8.01$
2. $s = 4, 6, 7.9, 7.99$

Exer. 3–8: The volume of the cake pan to be constructed from a square piece of plastic is given by

$$V = f(x) = 6x^2 - \tfrac{1}{2}x^3.$$

Use the function values for $f(x)$ given in the following table to find the slope of the secant line through $(8, f(8))$ and the given point.

x	$f(x)$
9	121.5
8.5	126.4375
8.2	127.756
8.1	127.9395
8.01	127.9993995
8.001	127.999994
8	128

3. $(9, f(9))$
4. $(8.5, f(8.5))$
5. $(8.2, f(8.2))$
6. $(8.1, f(8.1))$
7. $(8.01, f(8.01))$
8. $(8.001, f(8.001))$
9. The slope of the tangent line at $x = 8$ is the limit of the slope of these secant lines near $x = 8$. Use this fact and your answers to Exer. 3–8 to suggest a value for the slope of the tangent line at $x = 8$.

Exer. 10–14: Use the data in the following table for $f(x) = 6x^2 - \frac{1}{2}x^3$ to find the slope of the secant line through $(8, f(8))$ and the given point.

x	$f(x)$
7	122.5
7.5	126.5625
7.7	127.4735
7.9	127.9405
7.99	127.9994005
7.999	127.999994
8	128

10. $(7.5, f(7.5))$
11. $(7.7, f(7.7))$
12. $(7.9, f(7.9))$
13. $(7.99, f(7.99))$
14. $(7.999, f(7.999))$
15. Use the answers to Exer. 10–14 to rework Exercise 9.
16. Explain how your answers to Exercises 9 and 15 support the assertion made in the text (see page 50) that the tangent to $f(x) = 6x^2 - \frac{1}{2}x^3$ at $x = 8$ is horizontal.

2.2 Limits

The two motivating examples discussed in the previous section illustrate the usefulness of calculating limits of functions. In this section we will discuss what a limit is and how we find a limit.

Example 2.1 Consider the values of the function $f(x) = 3x + 5$ as x gets very close to $x = 2$. To see what is happening to $f(x)$, we have evaluated $f(x)$ at $x = 1$, 1.5, 1.9, 1.99, and 1.999 and at $x = 3$, 2.5, 2.1, 2.01, and 2.001. The results are shown in Table 2.1.

Table 2.1
Values of $f(x) = 3x + 5$ near $x = 2$

x	$f(x)$
1	8
1.5	9.5
1.9	10.7
1.99	10.97
1.999	10.997
2.001	11.003
2.01	11.03
2.1	11.3
2.5	12.5
3	14

We see that as x gets closer and closer to 2, $f(x)$ gets closer and closer to 11. Therefore we write

$$\lim_{x\to 2} f(x) = 11. \qquad \blacksquare$$

Note that in Example 2.1 we might have guessed that the limit of $f(x) = 3x + 5$ as x approaches 2 is 11 by observing that $f(2) = 11$. In the following example, however, it is not possible to evaluate the function at the value being approached.

Example 2.2 Consider the values of the function

$$f(x) = \frac{x^2 + 3x + 2}{x + 1}$$

as x approaches $x = -1$. Note that -1 is not in the domain of $f(x)$ because the denominator is zero when $x = -1$. However, we are not interested in $x = -1$, but only in x-values *close* to -1. Table 2.2 shows evaluations of $f(x)$ for several x-values as x approaches -1. From the values in the table we guess that $f(x)$ gets increasingly close to 1 as x approaches -1. We write this as

$$\lim_{x\to -1} f(x) = 1. \qquad \blacksquare$$

Table 2.2
Values of $f(x) = \dfrac{x^2 + 3x + 2}{x + 1}$ near $x = -1$

x	$f(x)$
−2	0
−1.5	0.5
−1.1	0.9
−1.01	0.99
−1.001	0.999
−0.999	1.001
−0.99	1.01
−0.9	1.1
−0.5	1.5
0	2

Let a denote a real number and let $f(x)$ denote a function. We say that L is the **limit of $f(x)$ as x approaches a** if $f(x)$ gets closer and closer to L as x approaches a. In this case, we write

$$\lim_{x\to a} f(x) = L.$$

Note that we do *not* consider what happens when x is *equal* to a. This is very important because many interesting applications of the limit will involve functions or expressions that are not defined at a. The solution to each of the two examples discussed in the previous section requires finding the limit of an expression at a value where the expression is undefined. If there is no single real number L that $f(x)$ gets close to as x approaches a, then no limit exists. This concept is illustrated in the following example.

Example 2.3 Find $\lim_{x\to 6} f(x)$ where $f(x) = \dfrac{1}{x - 6}$, if it exists.

Solution Note that $f(x)$ is not defined at $x=6$. From Table 2.3, which shows the behavior of $f(x)$ as x approaches 6, we can see that $f(x)$ does not approach any specific number L. As x approaches 6 with $x<6$, $f(x)$ gets very large in the negative direction; as x approaches 6 with $x>6$, $f(x)$ gets very large in the positive direction. Therefore $\lim_{x\to 6} f(x)$ does not exist.

Figure 2.8 illustrates the behavior of $f(x)=1/(x-6)$ near $x=6$. ■

Table 2.3
Values of $f(x)=\dfrac{1}{x-6}$ near $x=6$

x	$\dfrac{1}{x-6}$
4	−0.5
5	−1
5.5	−2
5.9	−10
5.99	−100
6.01	100
6.1	10
6.5	2
7	1
8	0.5

In Examples 2.2 and 2.3 we examined $\lim_{x\to a} f(x)$ for functions that are not defined at a. In one case the limit existed, and in the other case the limit did not exist. Both of these functions are rational functions (quotients of polynomials). To see why these functions behave so differently, we observe what happens to the numerator and denominator of each function as x approaches a. (The following observations may be confirmed by forming a table of values as in the previous examples.)

We consider $f(x)=1/(x-6)$ first. As x approaches 6, the numerator is always 1 and the denominator approaches 0. Hence, as x gets closer and closer to 6, we are dividing 1 by a smaller and smaller number, and so the absolute value of $1/(x-6)$ gets larger and larger. Thus the function does not get close to any single number L.

On the other hand, for the fraction

$$\frac{x^2+3x+2}{x+1}$$

both numerator and denominator approach zero as $x\to -1$. In such a case we cannot immediately determine if the limit exists. Sometimes the limit of a quotient can be simplified using the following observation. Suppose that $f(x)$ and $g(x)$ are functions that are equal to each other for all x except $x=a$. Then, since the limits of $f(x)$ and of $g(x)$ as x approaches a depend only on the function values near $x=a$ but not at a, these limits are equal. Consequently, if we have a complicated function $f(x)$, and wish to find its limit as x approaches a, and if we can find a simpler function $g(x)$ that equals $f(x)$ for all x except $x=a$, we can then find the limit of $f(x)$ by finding the limit of the simpler function $g(x)$.

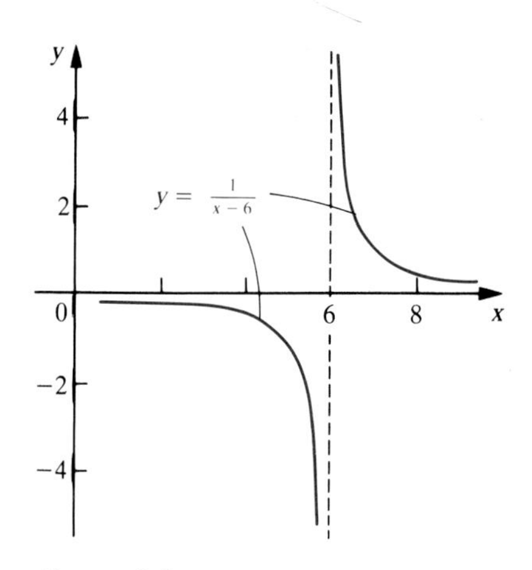

Figure 2.8
Graph of $f(x)=\dfrac{1}{x-6}$

Example 2.4 Find $\lim_{x\to -1} f(x)$ with $f(x)=(x^2+3x+2)/(x+1)$ by finding the limit of a simpler function $g(x)$ such that $g(x)=f(x)$ for all values of x except -1.

Solution The numerator of $f(x)$ can be factored as

$$x^2+3x+2=(x+2)(x+1).$$

Hence
$$f(x)=\frac{(x+2)(x+1)}{x+1}.$$

Since any nonzero factor that occurs in both the numerator and denominator can be cancelled, this quotient is equal to the function

$$g(x) = x + 2$$

for all values of x except $x = -1$. By the remarks preceding this example,

$$\lim_{x \to -1} f(x) = \lim_{x \to -1} g(x).$$

For the simple function $g(x) = x + 2$ we can easily see that its limit is 1 as x approaches -1, that is, $\lim_{x \to -1} g(x) = 1$. Thus $\lim_{x \to -1} f(x) = 1$, as we determined in Example 2.2. Figure 2.9 shows the graph of both functions. Note that the *only* difference between these functions is that $f(x)$ is not defined at $x = -1$ while $g(x)$ is defined for $x = -1$. At all other values of x, $f(x) = g(x)$. ■

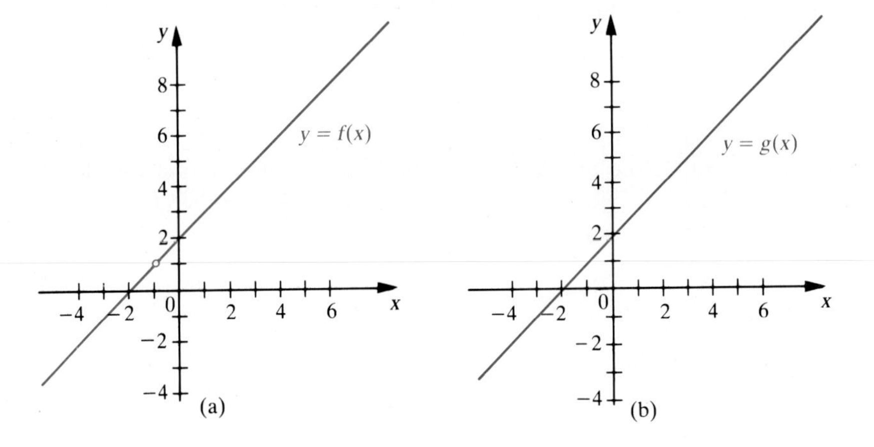

Figure 2.9
(a) Graph of $f(x) = \dfrac{x^2 + 3x + 2}{x + 1}$; (b) graph of $g(x) = x + 2$

To solve the examples in Section 2.1 we need to find

$$\lim_{x \to a} \frac{f(x) - f(a)}{x - a}$$

where $f(x)$ is a given function. We will use the technique illustrated in the previous example to simplify our work in finding this limit.

Example 2.5 Let $f(x) = x^2$ and let $a = 2$. If the limit exists, find the value of

$$\lim_{x \to a} \frac{f(x) - f(a)}{x - a}.$$

Solution First note that

$$\frac{f(x) - f(a)}{x - a} = \frac{x^2 - 4}{x - 2} = \frac{(x + 2)(x - 2)}{x - 2} = x + 2$$

for every value of x except $x = 2$. Since the limit of $x + 2$ as x

approaches 2 is 4, we have

$$\lim_{x\to 2}\frac{x^2-4}{x-2}=\lim_{x\to 2}\frac{(x+2)(x-2)}{x-2}=\lim_{x\to 2}(x+2)=4.$$ ■

To simplify the task of finding limits, we now present some helpful rules for determining limits. We may use these rules to find the limits of most functions of interest to us without making tables of values.

Limit Theorems

Suppose that both $\lim_{x\to a} f(x)$ and $\lim_{x\to a} g(x)$ exist. Then the following rules are valid:

A. $\lim_{x\to a} c = c$ for all constants c.

B. $\lim_{x\to a} x^n = a^n$ for all positive numbers n.

C. $\lim_{x\to a} [cf(x)] = c \lim_{x\to a} f(x)$ for all constants c.

D. $\lim_{x\to a} [f(x) \pm g(x)] = \lim_{x\to a} f(x) \pm \lim_{x\to a} g(x)$

E. $\lim_{x\to a} [f(x)g(x)] = \left[\lim_{x\to a} f(x)\right]\left[\lim_{x\to a} g(x)\right]$

F. $\lim_{x\to a} \frac{f(x)}{g(x)} = \frac{\lim_{x\to a} f(x)}{\lim_{x\to a} g(x)}$ provided $\lim_{x\to a} g(x) \neq 0$.

G. $\lim_{x\to a} [f(x)]^r = \left[\lim_{x\to a} f(x)\right]^r$ for all positive numbers r.

The next example illustrates these rules.

Example 2.6 Evaluate each limit:

(a) $\lim_{x\to 2} x^3$ **(b)** $\lim_{x\to 3} 5x^2$ **(c)** $\lim_{x\to 2} (x^3 + 4x^5)$

(d) $\lim_{x\to 2} x^3(x^2+4)$ **(e)** $\lim_{x\to 4} \frac{x-5}{x^2}$ **(f)** $\lim_{x\to -1} \sqrt{x^2+8}$

Solution **(a)** Applying rule B with $a = 2$ and $n = 3$ gives

$$\lim_{x\to 2} x^3 = 2^3 = 8.$$

(b) Rule C states that a constant may be "pulled out" of a limit. Thus 5 can be pulled out of this expression, and then rule B may be used with

$a = 3$ and $n = 2$:

$$\lim_{x\to 3} 5x^2 = 5 \lim_{x\to 3} x^2 = 5(3^2) = 45.$$

Note that only a constant, not an expression involving x, may be pulled out of a limit.

(c) Rule D states that the limit of a sum (or difference) is the sum (or difference) of the limits. Thus,

$$\lim_{x\to 2} (x^3 + 4x^5) = \lim_{x\to 2} x^3 + \lim_{x\to 2} 4x^5 = (2)^3 + 4(2)^5 = 136.$$

(d) Rule E states that the limit of a product is the product of the limits, so we obtain

$$\lim_{x\to 2} x^3(x^2+4) = \left[\lim_{x\to 2} x^3\right]\left[\lim_{x\to 2} (x^2+4)\right]$$

In the last expression we can use the limit of sums rule (D) and the fact that, by rule A, $\lim_{x\to 2} 4 = 4$ to conclude that

$$\lim_{x\to 2} x^3(x^2+4) = (2^3)(2^2+4) = 8(8) = 64.$$

(e) The stipulation that $\lim_{x\to a} g(x) \neq 0$ is important in rule F for the limit of a quotient. Since $g(x) = x^2$ in this example and $\lim_{x\to 4} g(x) = \lim_{x\to 4} x^2 = 16 \neq 0$, we can apply rule F:

$$\lim_{x\to 4} \frac{x-5}{x^2} = \frac{\lim_{x\to 4} (x-5)}{\lim_{x\to 4} x^2} = \frac{4-5}{16} = \frac{-1}{16}.$$

(f) First recall that $\sqrt{x^2+8} = (x^2+8)^{1/2}$. Then,

$$\lim_{x\to -1} (x^2+8)^{1/2} = \left[\lim_{x\to -1} (x^2+8)\right]^{1/2} \qquad \text{(rule G)}$$

$$= \left[\lim_{x\to -1} x^2 + \lim_{x\to -1} 8\right]^{1/2} \qquad \text{(rule D)}$$

$$= [(-1)^2 + 8]^{1/2} = 9^{1/2} = 3.$$

■

These limit theorems provide easy methods for finding limits of polynomials and rational functions. Given a polynomial

$$f(x) = b_0 + b_1x + b_2x^2 + \cdots + b_nx^n,$$

the limit of $f(x)$ as x approaches a can be calculated by first applying rule D on the limit of sums:

$$\lim_{x\to a} f(x) = \lim_{x\to a} (b_0) + \lim_{x\to a} (b_1x) + \lim_{x\to a} (b_2x^2) + \cdots + \lim_{x\to a} (b_nx^n)$$

then rules A and C:

$$\lim_{x\to a} f(x) = b_0 + b_1 \lim_{x\to a} (x) + b_2 \lim_{x\to a} (x^2) + \cdots + b_n \lim_{x\to a} (x^n)$$

and finally rule B:

$$\lim_{x \to a} f(x) = b_0 + b_1 a + b_2 a^2 + \cdots + b_n a^n = f(a).$$

Therefore, we find the limit of a polynomial as x approaches a by substituting a for x in the polynomial, that is, evaluating the polynomial at $x = a$.

Example 2.7 Find $\lim_{x \to 3} (4x^3 - 8x^2 + 2x - 5)$.

Solution According to the preceding discussion, the value of this limit can be obtained by evaluating the polynomial at $x = 3$. Thus

$$\lim_{x \to 3} (4x^3 - 8x^2 + 2x - 5) = 4(3)^3 - 8(3)^2 + 2(3) - 5 = 37. \quad \blacksquare$$

Now consider the limit as x approaches a of the rational function $f(x)/g(x)$ where $f(x)$ and $g(x)$ are polynomials. Rule F on quotients can be applied to this quotient, provided the limit of $g(x)$ is not zero. Since $f(x)$ and $g(x)$ are polynomials their limits as $x \to a$ are $f(a)$ and $g(a)$, respectively. Hence, by rule F we have the following rational function rule.

Rational Function Rule

$$\lim_{x \to a} \frac{f(x)}{g(x)} = \frac{f(a)}{g(a)}$$

where $f(x)$ and $g(x)$ are polynomials and $g(a) \neq 0$.

Example 2.8 For $f(x) = 1/x^2$ and $a = 3$, find

$$\lim_{x \to a} \frac{f(x) - f(a)}{x - a}.$$

Solution Substituting $f(x) = 1/x^2$ and $a = 3$ into this expression and simplifying, we obtain

$$\begin{aligned}
\frac{f(x) - f(a)}{x - a} &= \frac{f(x) - f(3)}{x - 3} \\
&= \frac{\frac{1}{x^2} - \frac{1}{3^2}}{x - 3} \qquad \text{(now we find a common denominator)} \\
&= \frac{\frac{9 - x^2}{9x^2}}{x - 3} \\
&= \frac{-(x^2 - 9)}{9x^2(x - 3)}.
\end{aligned}$$

We cannot yet use the rational function rule since the denominator is zero at $x = 3$. We factor the numerator of the last expression to get

$$\frac{f(x) - f(3)}{x - 3} = \frac{-(x-3)(x+3)}{9x^2(x-3)} = \frac{-(x+3)}{9x^2}$$

except at $x = 3$. Therefore,

$$\lim_{x \to 3} \frac{f(x) - f(3)}{x - 3} = \lim_{x \to 3} \frac{-(x+3)}{9x^2}$$
$$= \frac{-(3+3)}{9(3)^2} = \frac{-6}{81} = -\frac{2}{27},$$

since $\lim_{x \to 3} 9x^2 \neq 0$. Hence, the desired limit is $-\frac{2}{27}$. ■

To find the limit of a function $f(x)$ as x approaches a, we must consider x-values both on the right side and on the left side of a. If we consider values on just one side of a, we get a *one-sided limit.* We define L to be the **left-hand limit** of $f(x)$ as x approaches a if $f(x)$ gets closer and closer to L as x gets closer to a with x less than a (on the left side of a). A left-hand limit involves only x-values *less than a.* Similarly, we define L to be the **right-hand limit** of $f(x)$ as x approaches a if $f(x)$ gets closer and closer to L as x gets closer and closer to a with x greater than a (on the right side of a). A right-hand limit involves only x-values *greater than a.* We denote the left-hand limit of $f(x)$ as x approaches a by

$$\lim_{x \to a^-} f(x).$$

The minus sign on the a indicates that we are considering only values of x less than a, and close to a. Similarly, the notation for the right-hand limit is

$$\lim_{x \to a^+} f(x).$$

The plus sign on the a indicates that we are considering only values of x greater than a, and close to a.

Of course, if $f(x)$ has a limit L as x approaches a, then it has a left-hand limit and a right-hand limit, and both are equal to L. Conversely, if both one-sided limits exist and are equal, then the limit exists, and is this common value. A function may have one-sided limits at some a but fail to have a limit at a; that is, if the one-sided limits exist but are not equal, or if one of the one-sided limits does not exist, then $\lim_{x \to a} f(x)$ does not exist.

Example 2.9 Consider the postal rate function $f(x)$ where $f(x)$ is the number of cents required to mail a first-class letter weighing x ounces. (This function was discussed in Example 1.3.) Find

$$\lim_{x \to 3^-} f(x) \quad \text{and} \quad \lim_{x \to 3^+} f(x).$$

Solution From Example 1.3 we see that for all x in the interval $(2, 3]$, we have $f(x) = 65$. Therefore, as x gets closer and closer to 3 from the left ($x<3$), $f(x)$ will be a constant 65, and so

$$\lim_{x\to 3^-} f(x) = 65.$$

Similarly, since $f(x)=85$ for all x in the interval $(3, 4]$, as x gets closer and closer to 3 from the right ($x>3$), $f(x)$ will be a constant 85, and so

$$\lim_{x\to 3^+} f(x) = 85.$$

Figure 2.10 shows the behavior of $f(x)$ near $x=3$. Note that since $\lim_{x\to 3^-} f(x) \neq \lim_{x\to 3^+} f(x)$, we conclude that $\lim_{x\to 3} f(x)$ does not exist. ■

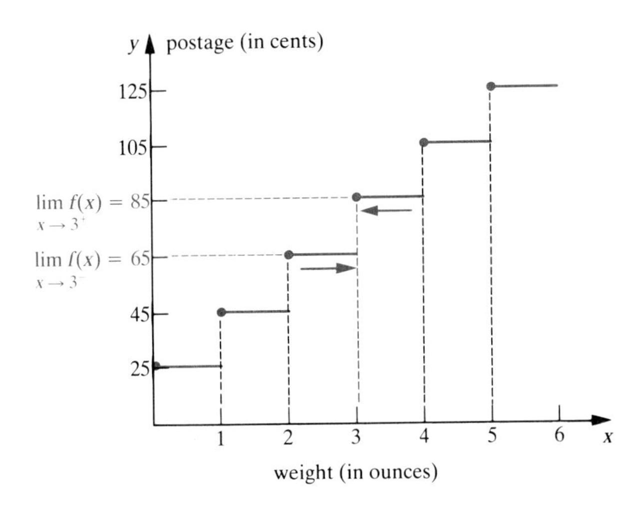

Figure 2.10
The postal scale function near $x=3$: the right-hand limit is 85; the left-hand limit is 65

Example 2.10 Let

$$f(x) = \begin{cases} 2x & \text{for } x \le 4 \\ 3x & \text{for } x > 4 \end{cases} \quad \text{and} \quad g(x) = \begin{cases} x^2 - 1 & \text{for } x < 2 \\ 4x - 5 & \text{for } x > 2 \end{cases}$$

Find the one-sided limits:

$$\lim_{x\to 4^-} f(x), \qquad \lim_{x\to 4^+} f(x), \qquad \lim_{x\to 2^-} g(x), \qquad \lim_{x\to 2^+} g(x).$$

Find $\lim_{x\to 4} f(x)$ and $\lim_{x\to 2} g(x)$.

Solution For $\lim_{x\to 4^-} f(x)$ the values of x we are considering are always less than 4, so $f(x)=2x$. Thus

$$\lim_{x\to 4^-} f(x) = \lim_{x\to 4^-} 2x = 8.$$

Similarly since $f(x)=3x$ for $x>4$,

$$\lim_{x\to 4^+} f(x) = \lim_{x\to 4^+} 3x = 3(4) = 12.$$

The two one-sided limits are different, so $\lim_{x\to 4} f(x)$ does not exist. Figure 2.11 shows the graph of $f(x)$.

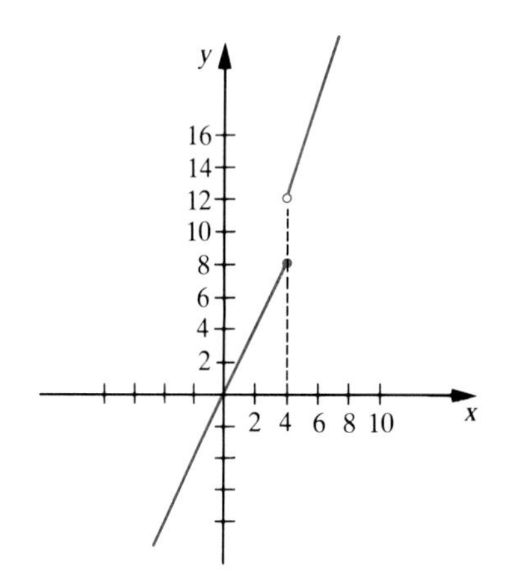

Figure 2.11
Graph of $f(x) = \begin{cases} 2x & \text{for } x \le 4 \\ 3x & \text{for } x > 4 \end{cases}$

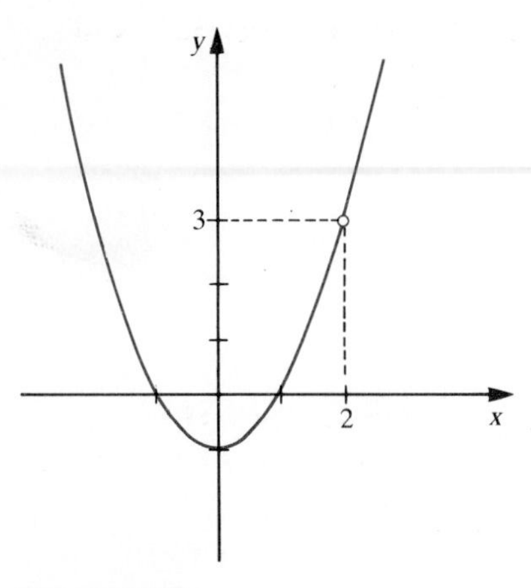

Figure 2.12

Graph of $g(x) = \begin{cases} x^2 - 1 & \text{for } x < 2 \\ 4x - 5 & \text{for } x > 2 \end{cases}$

For the function $g(x)$ we have

$$\lim_{x \to 2^-} g(x) = \lim_{x \to 2^-} (x^2 - 1) = 2^2 - 1 = 3,$$

and

$$\lim_{x \to 2^+} g(x) = \lim_{x \to 2^+} (4x - 5) = 4(2) - 5 = 3.$$

Since the two one-sided limits are equal, $\lim_{x \to 2} g(x)$ exists and equals 3. Note, however, that $g(x)$ is not defined at $x = 2$. Figure 2.12 shows the graph of $g(x)$. ■

One-sided limits are helpful in describing the behavior of a function near a point at which the function is not defined or has no limit. For example, the function $1/(x - 6)$, studied in Example 2.3, becomes arbitrarily large in the negative direction when x approaches 6 from the left, and arbitrarily large in the positive direction when x approaches 6 from the right. We express these facts as follows: $f(x) \to -\infty$ as $x \to 6^-$ and $f(x) \to \infty$ as $x \to 6^+$, respectively. (Refer to Figure 2.8.)

Exercises

Exer. 1–8: Find the limit by making a table of function values near the x-value at which the limit is taken. If the limit does not exist, explain why.

1. $\lim_{x \to 2} (x + 2)$
2. $\lim_{x \to -3} (x - 4)$
3. $\lim_{t \to 3} (2t + 5)$
4. $\lim_{t \to 4} \dfrac{1}{t + 3}$
5. $\lim_{u \to 4} \dfrac{1}{u - 4}$
6. $\lim_{x \to 2} \dfrac{x - 2}{x^2}$
7. $\lim_{x \to 3} \dfrac{x}{(x - 3)^2}$
8. $\lim_{p \to 0} \dfrac{p^2 + 3p}{p}$

Exer. 9–26: Evaluate the limit using the limit theorems on page 55. If the limit does not exist, explain why.

9. $\lim_{x \to 2} (x^2 - 4)$
10. $\lim_{x \to -1} (2x + 4)$
11. $\lim_{x \to 4} (x^3 - 2x)$
12. $\lim_{x \to 0} (x^3 + 2x - 6)$
13. $\lim_{t \to 2} \dfrac{t^2 - 4}{t - 2}$
14. $\lim_{t \to 5} \dfrac{5t^2 + 25}{t - 5}$
15. $\lim_{x \to 2} \dfrac{2x}{(x - 2)^2}$
16. $\lim_{x \to 0} \dfrac{x^3 - 2x}{x}$
17. $\lim_{p \to 0} \dfrac{p^3}{p^3 - 2p}$
18. $\lim_{p \to 0} \dfrac{p^3 - 2p}{p^2}$
19. $\lim_{x \to 1} \dfrac{(x - 3)^2}{3x}$
20. $\lim_{x \to -4} \dfrac{x^2 + 2x - 8}{x + 4}$
21. $\lim_{u \to 2} [(u + 3) + (u^2 + 4)]$
22. $\lim_{r \to -1} [(3 - r) + (2r^2 + 1)]$
23. $\lim_{x \to 2} [(x^2 + 2) - (x^3)]$
24. $\lim_{x \to 3} \left(x^2 + 3 + \dfrac{x^2 - 9}{x - 3}\right)$
25. $\lim_{x \to 0} \left(\dfrac{x^2 + x}{2x} + \dfrac{2x - 3}{x}\right)$
26. $\lim_{x \to 2} x^2 \left(\dfrac{x - 4}{x + 3}\right)^3$

Exer. 27–30: Given $\lim_{x \to a} f(x) = -4$ and $\lim_{x \to a} g(x) = 7$, evaluate the limit.

27. $\lim_{x \to a} f(x)g(x)$
28. $\lim_{x \to a} [f(x) + f(x)g(x)]$
29. $\lim_{x \to a} [(f(x))^2 + 3f(x)g(x) + 5g(x)]$
30. $\lim_{x \to a} [(f(x))^2 - 2(g(x))^2 + 3f(x)g(x)]$

Exer. 31–34: Given $\lim_{x \to a} f(x) = 5$ and $\lim_{x \to a} g(x) = -2$, evaluate the limit.

31. $\lim_{x \to a} 4f(x)$
32. $\lim_{x \to a} [2f(x) - 12g(x)]$
33. $\lim_{x \to a} (4f(x) + 3g(x))$
34. $\lim_{x \to a} (6f(x) + 3g(x))$

Exer. 35–44: For $f(x)$ and a as specified, evaluate

$$\lim_{x \to a} \frac{f(x) - f(a)}{x - a}.$$

35. $f(x) = 2x;\quad a = 3$
36. $f(x) = 1/x^2;\quad a = 1$
37. $f(x) = x^2;\quad a = 4$
38. $f(x) = x^2 + 1;\quad a = 2$

39. $f(x)=\dfrac{1}{x-2}$; $a=-1$

40. $f(x)=3x-4$; $a=-1$

41. $f(x)=1-2x^2$; $a=3$

42. $f(x)=x^3-4$; $a=2$

43. $f(x)=x^3$; $a=2$

44. $f(x)=1/x^3$; $a=3$

45. Suppose that when a missile is recovered after a test flight, the tape that records the missile's altitude during the flight has been damaged so that the altitude at $t=24$ seconds after launch cannot be read. By analyzing the data, the altitude $a(t)$ (in feet) at t seconds after launch is determined to be

$$a(t)=\frac{18t^3-408t^2-576t}{3t-72}.$$

What was the altitude at $t=24$?

46. Refer to Exercise 45. Suppose the altitude recorder is damaged at $t=16$, and $a(t)$ is determined to be

$$a(t)=\frac{7t^3-110t^2-32t}{2t-32}.$$

What was the altitude at $t=16$?

Exer. 47–55: Find the one-sided limit, provided it exists.

47. $\lim_{x\to 4^-} f(x)$ where $f(x)=\begin{cases} x^2-3 & \text{for } x<4 \\ 5x+1 & \text{for } x>4 \end{cases}$

48. $\lim_{r\to 3^+} \sqrt{r-3}$

49. $\lim_{t\to 3^-} \sqrt{t-3}$

50. $\lim_{x\to 1^-} \dfrac{x^2-1}{x-1}$

51. $\lim_{x\to 0^+} f(x)$ where $f(x)=\begin{cases} 3x-2 & \text{for } x<0 \\ 7x-2 & \text{for } x>0 \end{cases}$

52. $\lim_{x\to 1^+} f(x)$ where $f(x)=\begin{cases} 5x^2-2 & \text{for } x<1 \\ 7x-2 & \text{for } x\geq 1 \end{cases}$

53. $\lim_{t\to 3^-} f(t)$ where $f(t)=\begin{cases} t^3-2t & \text{for } t<3 \\ t^2+12 & \text{for } t>3 \end{cases}$

54. $\lim_{x\to 2^+} \dfrac{x^3-8}{x^2-4}$

55. $\lim_{x\to 1^-} f(x)$ where $f(x)=\begin{cases} 2x^2-2x & \text{for } x<1 \\ -x+4 & \text{for } x>1 \end{cases}$

2.3 Definition of the Derivative

In the first two sections of this chapter we have seen the quotient

$$\frac{f(x)-f(a)}{x-a}$$

several times, and we have examined the limit of this expression for several functions $f(x)$ and numbers a. In this section we study this limit in detail.

The **derivative** of a function $f(x)$ at a value a is defined to be

$$\lim_{x\to a}\frac{f(x)-f(a)}{x-a},$$

provided the limit exists. If the limit does not exist, then we say that $f(x)$ has no derivative at a. Note that the expression $[f(x)-f(a)]/(x-a)$ whose limit we seek is not defined at $x=a$.

If a function $f(x)$ has a derivative at $x=a$, we say that $f(x)$ is **differentiable at *a***. If $f(x)$ is differentiable for all a in its domain, then $f(x)$ is said to be a **differentiable function**. The process of finding the derivative of a function is **differentiation**.

The derivative is an essential tool in applications because we can use it to describe important properties of a function. In the cake-pan example in Section 2.1 we used the derivative to determine the *maximum-size* pan that could be constructed from a square piece of plastic. We found that the derivative yields the slope of the line tangent to the graph of the function. This geometric interpretation of the derivative will be discussed in Section 2.4. In the fruit fly example in Section 2.1 we used the derivative to measure the *rate of change*, or rate of growth, of the fruit fly population. This physical interpretation of the derivative will be explored in Section 2.6. In Section 4.4 we will discuss a *financial* interpretation of the derivative. In this section, we will examine the algebraic aspects of differentiation.

Example 2.11 Find the derivative of $f(x) = x^2$ (a) at $a = 1$ and (b) at an arbitrary value of a.

Solution **(a)** The derivative of $f(x)$ at $x = 1$ is

$$\begin{aligned}\lim_{x\to 1}\frac{f(x)-f(1)}{x-1} &= \lim_{x\to 1}\frac{x^2-1}{x-1}\\ &= \lim_{x\to 1}\frac{(x+1)(x-1)}{x-1}\\ &= \lim_{x\to 1}(x+1) = 2.\end{aligned}$$

Hence, the derivative of $f(x) = x^2$ at 1 is 2.

(b) We proceed in a similar fashion to find the derivative of $f(x)$ at $x = a$:

$$\begin{aligned}\lim_{x\to a}\frac{f(x)-f(a)}{x-a} &= \lim_{x\to a}\frac{x^2-a^2}{x-a}\\ &= \lim_{x\to a}\frac{(x+a)(x-a)}{x-a}\\ &= \lim_{x\to a}(x+a) = 2a.\end{aligned}$$

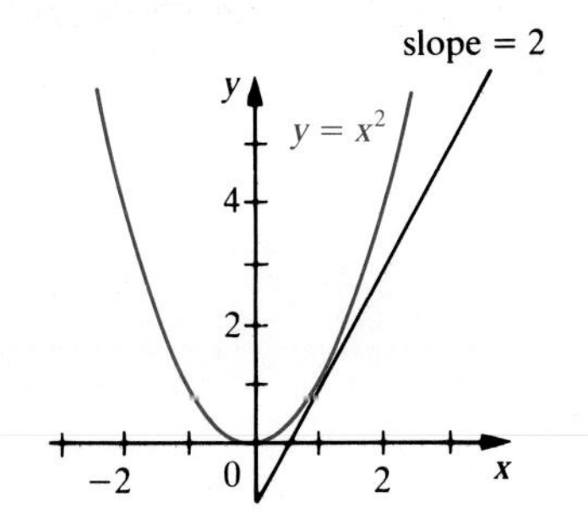

Figure 2.13
Graph of $f(x) = x^2$ with tangent line at $x = 1$

Thus, the derivative of $f(x) = x^2$ at any point a is $2a$. In particular, the derivative of $f(x) = x^2$ at $a = 1$ is $2(1) = 2$, as we found in part (a). Figure 2.13 shows the graph of $f(x) = x^2$ and the tangent line at $x = 1$. ∎

In the previous example we derived a rule for finding the derivative of $f(x) = x^2$ at any value $x = a$. In fact, the derivative at a is $2a$. If we write x in place of a, then the derivative at x is $2x$. Hence, the process of differentiating $f(x) = x^2$ defines a new function, which we denote by $f'(x)$, so that $f'(x) = 2x$ for the function $f(x) = x^2$.

In general, if $f(x)$ is any differentiable function, then the **derivative of $f(x)$** is the function $f'(x)$ that gives the value of the derivative of $f(x)$ at x. At $x = a$ we define $f'(a)$ by

$$f'(a) = \lim_{x\to a}\frac{f(x)-f(a)}{x-a}.$$

An alternative notation can be used to define the derivative of a function. If we let $h = x - a$ in the above definition, then $x = a + h$, and as x approaches a, h approaches 0 (and conversely, as h approaches 0, x approaches a). Hence, for any value a,

$$f'(a) = \lim_{h\to 0}\frac{f(a+h)-f(a)}{h} \qquad (1)$$

is an equivalent expression for the derivative. The numerator of equation (1) is the change in $f(x)$ from $x=a$ to $x=a+h$, or the rise, and the denominator is the run of a secant line. Figure 2.14 shows three secant lines obtained with progressively smaller values of h.

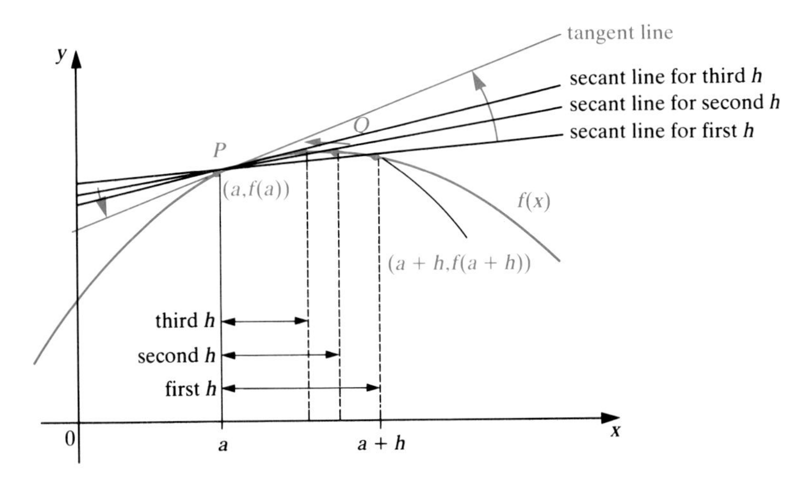

Figure 2.14
Geometry for the definition of the derivative

For the function $f(x)=x^2$ in Example 2.11, $f(a+h)=(a+h)^2$, so

$$\begin{aligned} f'(a) &= \lim_{h\to 0}\frac{f(a+h)-f(a)}{h} \\ &= \lim_{h\to 0}\frac{(a+h)^2-a^2}{h} \\ &= \lim_{h\to 0}\frac{a^2+2ah+h^2-a^2}{h} && \text{(expand the squared term)} \\ &= \lim_{h\to 0}\frac{2ah+h^2}{h} && \text{(cancel first and last terms)} \\ &= \lim_{h\to 0}(2a+h)=2a. && \text{(cancel } h\text{)} \end{aligned}$$

Of course, this result is the same as that obtained in the example, but the algebra in this method is perhaps easier. Note that the equations $f'(a)=2a$ for any a and $f'(x)=2x$ for any x both indicate that the derivative of $f(x)$ is two times the value of the independent variable. Since x is a more familiar variable, when writing formulas such as these we usually write the derivative in terms of x, such as $f'(x)=2x$. If we write x instead of a in the alternative definition of $f'(a)$ given in equation (1), we have the following definition.

Definition of the Derivative

$$f'(x)=\lim_{h\to 0}\frac{f(x+h)-f(x)}{h}$$

Example 2.12 Find $f'(x)$ for each function:

(a) $f(x)=1$ **(b)** $f(x)=x$ **(c)** $f(x)=x^3$.

Solution **(a)** For $f(x)=1$ we have

$$\begin{aligned} f'(x) &= \lim_{h\to 0}\frac{f(x+h)-f(x)}{h} \\ &= \lim_{h\to 0}\frac{1-1}{h} = \lim_{h\to 0} 0 = 0. \end{aligned}$$

Hence the derivative of $f(x)=1$ is $f'(x)=0$.
(b) Now consider the function $f(x)=x$. We have

$$\begin{aligned} f'(x) &= \lim_{h\to 0}\frac{f(x+h)-f(x)}{h} \\ &= \lim_{h\to 0}\frac{(x+h)-x}{h} \\ &= \lim_{h\to 0}\frac{h}{h} = \lim_{h\to 0} 1 = 1. \end{aligned}$$

Thus the derivative of $f(x)=x$ is $f'(x)=1$.
(c) For $f(x)=x^3$, we have

$$\begin{aligned} f'(x) &= \lim_{h\to 0}\frac{f(x+h)-f(x)}{h} \\ &= \lim_{h\to 0}\frac{(x+h)^3-x^3}{h} \\ &= \lim_{h\to 0}\frac{x^3+3x^2h+3xh^2+h^3-x^3}{h} && \text{(expand the cubed term)} \\ &= \lim_{h\to 0}\frac{3x^2h+3xh^2+h^3}{h} && \text{(cancel cubed terms)} \\ &= \lim_{h\to 0}\,[3x^2+3xh+h^2] = 3x^2 && \text{(cancel } h\text{)} \end{aligned}$$

because both $3xh$ and h^2 approach zero as h approaches zero. Therefore, $f'(x)=3x^2$. ■

To summarize the derivative formulas we have developed, note the following.

If $f(x)=1$ then $f'(x)=0$.
If $f(x)=x$ then $f'(x)=1$.
If $f(x)=x^2$ then $f'(x)=2x$.
If $f(x)=x^3$ then $f'(x)=3x^2$.

It appears that when $f(x)=x^r$ then $f'(x)=rx^{r-1}$, that is, the exponent in $f(x)$ becomes the coefficient of x raised to a new exponent that is one less than the original exponent in $f(x)$. This observation, which is true for any real number r, provides a powerful tool for finding derivatives.

Power Rule for Derivatives

If $f(x) = x^r$, and r is any real number, then

$$f'(x) = rx^{r-1}.$$

Example 2.14 Find $f'(x)$ for each function:

$$\textbf{(a)}\ f(x) = x^4 \qquad \textbf{(b)}\ f(x) = x^{2/3} \qquad \textbf{(c)}\ f(x) = \frac{1}{x^2}$$

Solution **(a)** For $f(x) = x^4$ the exponent is $r = 4$, and thus $r - 1 = 3$. Hence $f'(x) = 4x^3$.
(b) In $f(x) = x^{2/3}$ the exponent is $r = \frac{2}{3}$ and $r - 1 = -\frac{1}{3}$, so $f'(x) = \frac{2}{3}x^{-1/3}$.
(c) Since $f(x) = 1/x^2 = x^{-2}$, the exponent is $r = -2$; thus $f'(x) = -2x^{-2-1} = -2x^{-3}$. ■

Example 2.15 Determine $f'(8)$ when $f(x) = \dfrac{1}{\sqrt[3]{x^2}}$.

Solution Recall that $1/\sqrt[3]{x^2} = x^{-2/3}$, so

$$f'(x) = -\frac{2}{3}x^{(-(2/3)-1)} = -\frac{2}{3}x^{-5/3} = -\frac{2}{3}\frac{1}{x^{5/3}}.$$

Therefore, $$f'(8) = -\frac{2}{3}\frac{1}{8^{5/3}} = -\frac{2}{3}\left(\frac{1}{32}\right) = -\frac{1}{48}.$$

Notice that we found the derivative $f'(x)$ *before* evaluating it at the specific value $x = 8$. ■

The derivative of an expression such as $x^2 + 6x + 2$ can also be denoted by $(x^2 + 6x + 2)'$. Other notations for the derivative of $f(x)$ are

$$\frac{d}{dx}f(x) \quad \text{or} \quad \frac{df(x)}{dx}.$$

If $f(x)$ is given by an explicit expression such as $f(x) = x^2 + 6x + 2$, we may write the derivative as

$$\frac{d}{dx}(x^2 + 6x + 2).$$

If a function is defined by an equation such as $y = x^2 + 6x + 2$, then the derivative can be denoted by dy/dx. In summary, if $y = f(x)$ is a differentiable function, any of the following notations can be used interchangeably for the derivative of $f(x)$:

$$f'(x) \qquad \frac{df(x)}{dx} \qquad \frac{dy}{dx} \qquad y'.$$

The value of the derivative at a is given by any of these notations:

$$f'(a) \qquad \left.\frac{df(x)}{dx}\right|_{x=a} \qquad \left.\frac{dy}{dx}\right|_{x=a} \qquad y'\,|_{x=a} \qquad y'(a).$$

Two useful results concerning derivatives are similar to rules C and D of the limit theorems given in Section 2.2. This similarity arises because the derivative is defined by a limit. We state the corresponding rules here.

Derivative Rules

Let $f(x)$ and $g(x)$ be differentiable functions and let c denote a constant.

A. Derivative of a Constant Multiple

$$\frac{d}{dx}(cf(x)) = c\frac{df(x)}{dx}$$

B. Derivative of a Sum (or Difference)

$$\frac{d}{dx}[f(x) \pm g(x)] = \frac{d}{dx}f(x) \pm \frac{d}{dx}g(x)$$

We present a proof of the sum rule here and leave the remaining proofs as exercises.

Proof of the Sum Rule Let $S(x)$ denote the sum of the functions $f(x)$ and $g(x)$, that is, $S(x) = f(x) + g(x)$. Then, by definition of the derivative,

$$\begin{aligned}
S'(x) &= \lim_{h\to 0}\frac{S(x+h) - S(x)}{h} \\
&= \lim_{h\to 0}\frac{[f(x+h) + g(x+h)] - [f(x) + g(x)]}{h} \\
&= \lim_{h\to 0}\frac{[f(x+h) - f(x)] + [g(x+h) - g(x)]}{h} && \text{(rearrange terms)} \\
&= \lim_{h\to 0}\left[\frac{f(x+h) - f(x)}{h} + \frac{g(x+h) - g(x)}{h}\right] && \text{(rewrite as two fractions)} \\
&= \lim_{h\to 0}\frac{f(x+h) - f(x)}{h} + \lim_{h\to 0}\frac{g(x+h) - g(x)}{h} && \text{(by limit theorem D)} \\
&= f'(x) + g'(x).
\end{aligned}$$

Hence,

$$S'(x) = f'(x) + g'(x) = \frac{d}{dx}f(x) + \frac{d}{dx}g(x).$$

Therefore $$\frac{d}{dx}[f(x)+g(x)]=\frac{d}{dx}f(x)+\frac{d}{dx}g(x).$$ ■

The following examples illustrate use of the derivative rules.

Example 2.16 Find each derivative:

(a) $\frac{d}{dx}(4x^3)$ **(b)** $\frac{d}{dx}(x^2+x^5)$ **(c)** $\frac{d}{dx}(3x^2+4x^{1/2})$ **(d)** $\frac{d}{dx}(4)$

Solution **(a)** Rule A states that a constant may be "pulled out" of a derivative. Thus,

$$\frac{d}{dx}(4x^3)=4\frac{d}{dx}(x^3)=4(3x^2)=12x^2.$$

(b) Rule B may be used to differentiate each term in a sum separately:

$$\frac{d}{dx}(x^2+x^5)=\frac{d}{dx}(x^2)+\frac{d}{dx}(x^5)=2x+5x^4.$$

(c) This derivative combines rules A and B:

$$\begin{aligned}\frac{d}{dx}(3x^2+4x^{1/2})&=\frac{d}{dx}(3x^2)+\frac{d}{dx}(4x^{1/2}) && \text{(rule B)}\\ &=3\frac{d}{dx}(x^2)+4\frac{d}{dx}(x^{1/2}) && \text{(rule A)}\\ &=3(2x)+4(\tfrac{1}{2})(x^{-1/2})\\ &=6x+2x^{-1/2}=6x+\frac{2}{\sqrt{x}}.\end{aligned}$$

(d) We saw in Example 2.12 that $d(1)/dx=0$. Hence,

$$\frac{d}{dx}4=4\frac{d}{dx}1=0.$$ ■

Part (d) of the previous example illustrates that *the derivative of any constant is zero.*

Example 2.17 **(a)** If $y=2x^3-x^2$, find y'.
(b) If $f(x)=4x-x^3$, find $f'(2)$.

(c) If $y=\sqrt{x}+4x$, find $\left.\frac{dy}{dx}\right|_{x=1}$.

Solution **(a)** Applying the difference and constant multiple derivative rules, we have

$$\begin{aligned}y'&=(2x^3-x^2)'\\ &=(2x^3)'-(x^2)' && \text{(rule B)}\\ &=2(x^3)'-(x^2)' && \text{(rule A)}\\ &=2(3x^2)-2x=6x^2-2x.\end{aligned}$$

(b) Similarly, the derivative of $f(x) = 4x - x^3$ is

$$\begin{aligned} f'(x) &= (4x - x^3)' \\ &= 4(x)' - (x^3)' \qquad \text{(rules A and B)} \\ &= 4(1) - (3x^2) = 4 - 3x^2. \end{aligned}$$

Hence $$f'(2) = 4 - 3(2)^2 = -8.$$

(c) The derivative of $y = \sqrt{x} + 4x$ is

$$\begin{aligned} \frac{dy}{dx} &= y' = (x^{1/2})' + (4x)' \\ &= \tfrac{1}{2}x^{-1/2} + 4 = \frac{1}{2\sqrt{x}} + 4. \end{aligned}$$

Thus, $$y'\,|_{x=1} = \tfrac{1}{2} + 4 = \tfrac{9}{2}.$$ ■

Using the power rule and derivative rules A and B, we can find the derivative of any polynomial function. Thus, *all polynomial functions are differentiable functions.*

The limit theorems for constant multiples and for sums can be translated into similar theorems for the derivative, but the limit theorems for products and quotients have no corresponding derivative formulas. We will introduce the more complicated rules for the derivatives of products and quotients in Section 3.1.

Exercises

Exer. 1–10: Find $f'(a)$ using $f'(a) = \lim_{x \to a} \dfrac{f(x) - f(a)}{x - a}$.

1. $f(x) = 2x;\quad a = 1$
2. $f(x) = 1 - 3x;\quad a = 2$
3. $f(x) = x^2;\quad a = -1$
4. $f(x) = x - x^2;\quad a = 2$
5. $f(x) = x^{1/2};\quad a = 2$
6. $f(x) = 1/x^2;\quad a = 1$
7. $f(x) = 6;\quad a = 1$
8. $f(x) = \dfrac{1}{x + 1};\quad a = 2$
9. $f(x) = \dfrac{x}{x - 1};\quad a = 3$
10. $f(x) = x^2 - 7;\quad a = 7$

Exer. 11–16: Identify $f(x)$ and a so that the limit of the expression is the derivative of $f(x)$ at $x = a$.

11. $\lim_{x \to 4} \dfrac{(x^2 - 3) - (4^2 - 3)}{x - 4}$
12. $\lim_{x \to 3} \dfrac{2x + 5 - 11}{x - 3}$
13. $\lim_{x \to 2} \dfrac{(x^2 - 1) - 3}{x - 2}$
14. $\lim_{x \to -2} \dfrac{x + 4 - 2}{x + 2}$
15. $\lim_{x \to 2} \dfrac{x^2 + 2 - 6}{x - 2}$
16. $\lim_{x \to 2} \dfrac{x^3 - 8}{x - 2}$

Exer. 17–32: Use the definition $f'(x) = \lim_{h \to 0} \dfrac{f(x + h) - f(x)}{h}$ to find $f'(x)$.

17. $f(x) = 7x$
18. $f(x) = 5x^3$
19. $f(x) = x - 2$
20. $f(x) = -3x$
21. $f(x) = 1/x$
22. $f(x) = -3/x$
23. $f(x) = 2x + 7$
24. $f(x) = -x^2 + 6x$
25. $f(x) = 5x^2 - 3x$
26. $f(x) = (5x + 8)^2$
27. $f(x) = 1/x^5$
28. $f(x) = 1/x^2$
29. $f(x) = (2x + 1)^2$
30. $f(x) = \dfrac{5x}{2x + 1}$
31. $f(x) = \dfrac{x}{x + 1}$
32. $f(x) = \dfrac{x + 2}{x - 6}$

Exer. 33–46: Find $f'(x)$ using the power rule, sum rule, and constant multiple rule for derivatives.

33. $f(x) = x^3$
34. $f(x) = x^4$
35. $f(x) = 4x^5$
36. $f(x) = -6x^3$
37. $f(x) = x^{-3/4}$
38. $f(x) = \frac{1}{2}$
39. $f(x) = 6x^{2/3} + 4x$
40. $f(x) = 7/\sqrt{x}$
41. $f(x) = 3x^2 + 2x - 7$
42. $f(x) = 16x^4 + 3x^2 + 2x$
43. $f(x) = (x^4 - 4) + 3(x^3 - x)$
44. $f(x) = 8x^{1/3} + 12(x^2 + 6)$
45. $f(x) = 3x^2 - 16 + 7x^{-2/3}$
46. $f(x) = 7x^6 - 3x^4 + 6$

Exer. 47–54: Find dy/dx for the function using the power rule, sum rule, and constant multiple rule for derivatives.

47. $y = 16x - 2$
48. $y = -16$
49. $y = x^2 + 14x^6$
50. $y = x^{17} + 3x - 17$
51. $y = 117$
52. $y = 4/\sqrt[3]{x}$
53. $y = 2/x^2$
54. $y = x^2 + (2/x^2)$

Exer. 55–72: Given the functions

$$y = f(x) = 4x^{1/3} + 6x,$$

$$u = g(x) = -3x^3 + 4,$$

and
$$v = h(x) = \frac{3}{x^2} - \frac{x^2}{3},$$

find the indicated function or number.

55. $\dfrac{dy}{dx}$
56. $g'(4)$
57. $\left.\dfrac{dy}{dx}\right|_{x=8}$
58. $h'(1)$
59. y'
60. $\dfrac{du}{dx}$
61. $g'(x)$
62. $\left.\dfrac{dv}{dx}\right|_{x=4}$
63. $f'(0)$
64. $\left.\dfrac{d}{dx}g(x)\right|_{x=1}$
65. $\dfrac{d}{dx}h(x)$
66. v'
67. $h'(x)$
68. $g'(2)$
69. $\left.\dfrac{du}{dx}\right|_{x=-1}$
70. $v'|_{x=2}$
71. $\left.\dfrac{dv}{dx}\right|_{x=-2}$
72. $f'(2)g'(x)$

Exer. 73–82: Determine the values of x for which $f'(x) = 0$.

73. $f(x) = x^2 - 2x$
74. $f(x) = x^3 - 3x^2 + 3x + 3$
75. $f(x) = 3x$
76. $f(x) = x^3 - 12x + 16$
77. $f(x) = 4 - x^2$
78. $f(x) = 16$
79. $f(x) = x^3 - 16x^2$
80. $f(x) = x^4 + 8x^2 + 12$
81. $f(x) = 4x^3 - x^4$
82. $f(x) = x^3 - 9x^2 + 27x - 82$

83. We observed that if $f(x) = c$ for a constant c, then $f'(x) = 0$, since $f(x)$ can be written as $f(x) = cx^0$. Show that $f'(x) = 0$ using the definition of the derivative.

Exer. 84–85: Use the definition of the derivative and the limit theorems from Section 2.2 to prove the derivative rule.

84. $\dfrac{d}{dx}(f(x) - g(x)) = \dfrac{d}{dx}f(x) - \dfrac{d}{dx}g(x).$

85. $\dfrac{d}{dx}cf(x) = c\dfrac{d}{dx}f(x).$

Tangent Lines 2.4

In Section 2.1 we found the size of the largest cake pan that could be constructed from a square piece of plastic by determining where the slope of the tangent line to the graph of a certain function was zero. We observed (as shown in Figure 2.6 and repeated here in Figure 2.15) that the line tangent to the graph of a function $f(x)$ at $x = a$ is the limit of secant lines through the points $(a, f(a))$ and $(x, f(x))$ as x approaches a. Therefore, the slopes of these secant lines approach the slope of the tangent line as x approaches a. Since the slope of any secant line has the form

$$\text{slope of secant line} = \frac{f(x) - f(a)}{x - a},$$

$$\text{the slope of the tangent line} = \lim_{x \to a} \frac{f(x) - f(a)}{x - a},$$

provided the limit exists.

Suppose that the limit does exist at $x = a$. Then the limit is the derivative of $f(x)$ at a, that is, the slope of the tangent line to the graph of $f(x)$ at $x = a$ is $f'(a)$. Given that $f'(a)$ is the slope and that the line passes through the point $(a, f(a))$, the equation of this tangent line is

$$y - f(a) = f'(a)(x - a).$$

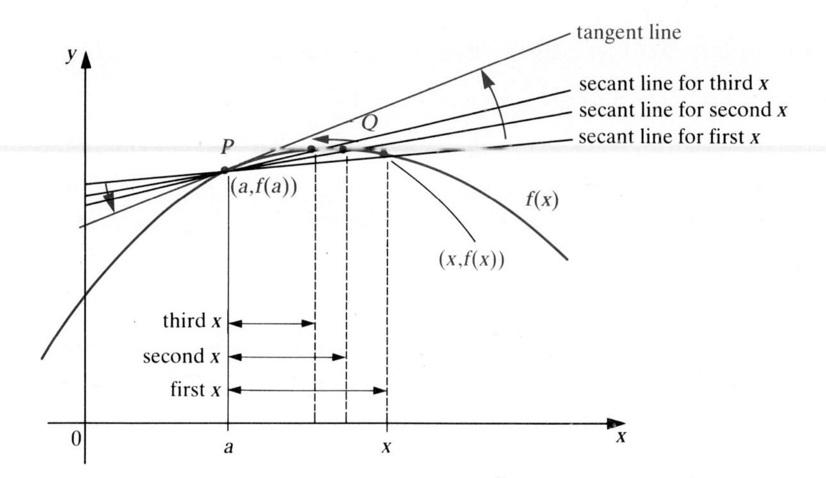

Figure 2.15
Geometric definition of the tangent line as the limit of secant lines

When $f(x)$ is a function such that $f'(a)$ does not exist, then either the tangent line has an undefined slope (is a vertical line) or there is no tangent line at all.

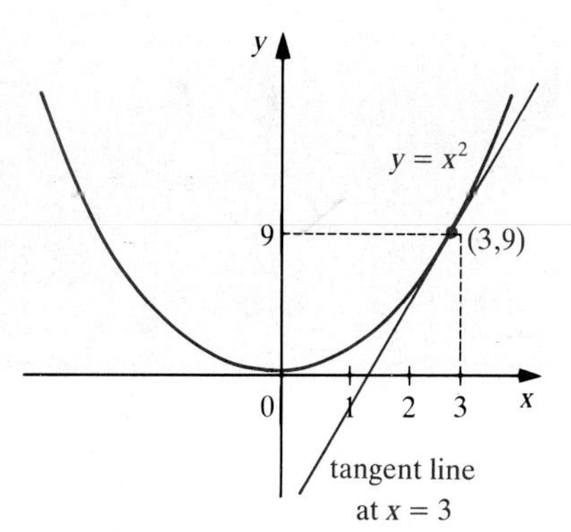

Figure 2.16
Tangent line to the graph of $f(x) = x^2$ at $x = 3$

Example 2.18 Consider the graph of the function $f(x) = x^2$, shown in Figure 2.16. Write the equation of the tangent line at $(3, 9)$.

Solution We know that $f'(x) = 2x$. Since $f'(3) = 6$, the line tangent to the graph at $(3, 9)$ has slope 6. Using the point-slope form, we find the equation of the tangent line is

$$y - 9 = 6(x - 3) \quad \text{or} \quad y = 6x - 9. \quad ■$$

A word of caution is needed here. The slope of a line is a number, not a function of x; therefore, the slope of the tangent line at $(a, f(a))$ cannot be $f'(x)$; rather it is the number $f'(a)$. In the above example the slope at $x = 3$ is not $2x$; the slope is $2x$ *evaluated* at the specific value of x: the x-coordinate of the point of tangency.

Example 2.19 Given $y = f(x) = x^4 + 2x + 2$, determine where the line tangent to the graph at the point $(-1, 1)$ crosses the x-axis.

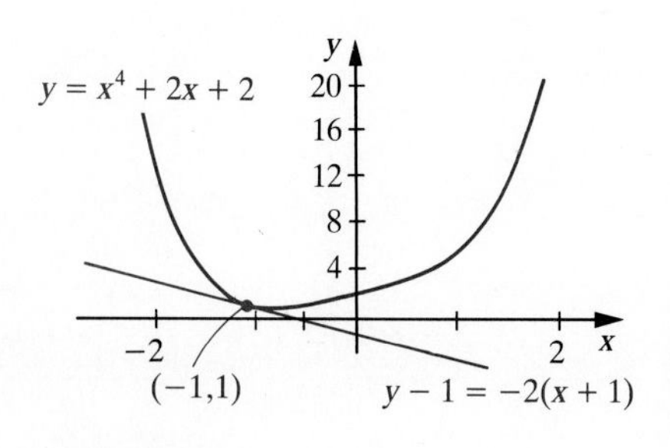

Figure 2.17
Tangent line to the graph of $f(x) = x^4 + 2x + 2$ at $x = -1$

Solution First we find the slope of the tangent line at $(-1, 1)$. Since $f'(x) = 4x^3 + 2$, we have $f'(-1) = 4(-1)^3 + 2 = -2$. Thus, the slope of the tangent line is -2. The line passes through $(-1, 1)$, so its equation is

$$y - 1 = -2[x - (-1)] \quad \text{or} \quad y - 1 = -2(x + 1).$$

This line crosses the x-axis when $y = 0$, that is, when $0 - 1 = -2(x + 1)$. Solving for x yields $x = -\frac{1}{2}$. Thus, the tangent line crosses the x-axis at the point $(-\frac{1}{2}, 0)$. See Figure 2.17. ■

Example 2.20 For what value of x does the graph of $f(x) = x^3 + 3x^2 - 9x$ have a horizontal tangent line?

Solution Since the slope of a horizontal line is zero, we can rephrase this question to ask where the tangent line has slope zero, that is, where is $f'(x)=0$. We first differentiate $f(x)$ to obtain

$$f'(x)=3x^2+6x-9=3(x^2+2x-3)$$
$$=3(x+3)(x-1).$$

We then set $f'(x)=0$ and solve for x. From $3(x+3)(x-1)=0$, we find that $x=-3$ and $x=1$ are values of x where $f'(x)=0$. Thus, the slope of the tangent line to the graph of $f(x)$ is zero when $x=-3$ and when $x=1$. See Figure 2.18. ■

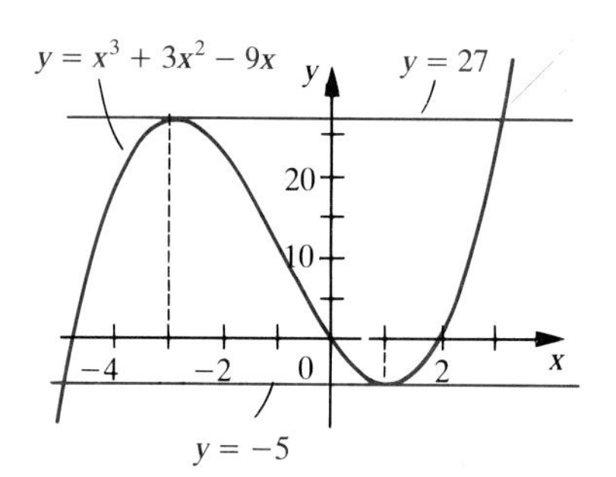

Figure 2.18
The horizontal tangent lines of $f(x)=x^3+3x^2-9x$ are at $x=-3$ and $x=1$

The following example, which is based on actual data, shows how a forester may decide when to harvest a forest to maximize the average annual wood production.

Example 2.21 A forester wishes to maximize the average annual yield from acreage devoted to the growth of shortleaf pines. The procedure is to plant 750 seedlings per acre, harvest the wood by clearcutting after x years, and begin the cycle again. Research has shown that for this acreage, wood production $W(x)$ (in ft³/acre) is given by

$$W(x)=-2.5x^2+240x-2200$$

after x years of growth for $15 \leq x \leq 40$.
(a) Find the average annual wood production for $x=20$.
(b) Find the average annual wood production for $x=40$.
(c) Find the value of x that maximizes the average annual wood production.

Solution **(a)** For $x=20$, we find

$$W(20)=-2.5(20)^2+240(20)-2200=1600.$$

Since this production is over a 20-year period, the *average* annual wood production is $W(20)/20=1600/20=80$ ft³/acre.
(b) At $x=40$, $W(40)=3400$, which gives an average annual production of $W(40)/40=3400/40=85$ ft³/acre.
(c) The average annual production at time x is the average rate of change in $W(x)$ over the time interval from 0 to x, that is,

$$\text{average of } W(x) \text{ from 0 to } x=\frac{W(x)-0}{x-0}.$$

This expression is the slope of a line through the origin, $(0,0)$, and $(x, W(x))$, a point on $W(x)$. To find the largest possible average, we want to find the steepest line that passes through the origin and a point on the graph of $W(x)$. Figure 2.19 shows that the required line is tangent to $W(x)$ because any steeper line will not intersect $W(x)$.

To find the equation of the desired line, we begin by finding the equation of the tangent line to the graph of $W(x)$ at a specific, but as yet unknown, point $(a, W(a))$. Since such a line passes through $(a, W(a))$ with slope $W'(a)$, its equation is

$$y-W(a)=W'(a)(x-a).$$

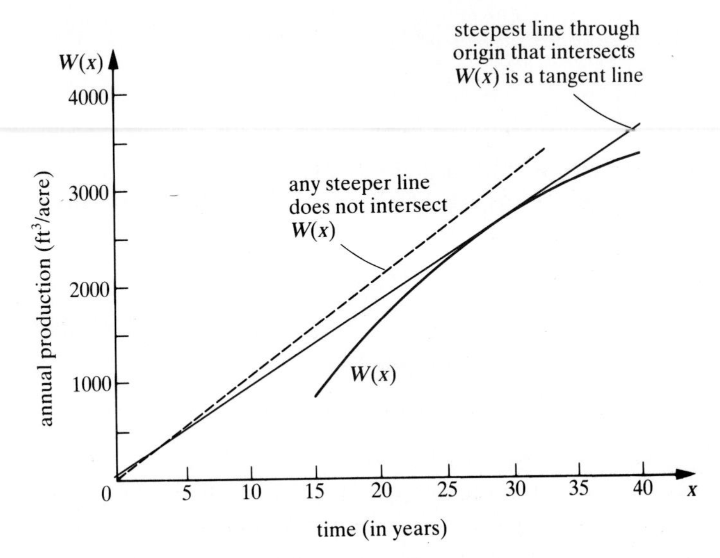

Figure 2.19
The maximum annual wood production is the slope of the tangent line to $W(x)$ that passes through the origin

The line must pass through $(0, 0)$, so we set x and y to zero, obtaining

$$0 - W(a) = W'(a)(0 - a).$$

Hence we need to find a value of a such that

$$W(a) = aW'(a).$$

From our original function $W(x)$, we find that $W'(x) = -5x + 240$, so $W(a) = aW'(a)$ can be written as

$$-2.5a^2 + 240a - 2200 = a(-5a + 240)$$

or

$$2.5a^2 - 2200 = 0$$

so that

$$a = 29.67.$$

Hence the maximum average annual production will occur when the pines grow for about 30 years and will be $W(30)/30 = 2750/30 \approx 91.67$ ft³/acre. (Note: the function for this example was approximated from actual values; the actual function is $W(x) = -2.489x^2 + 238.608x - 2178.85$, which yields $a = 29.59$ and a maximum average annual production of about 91.3 ft³/acre.) ■

We may use a tangent line to predict the change in the value of a function over a small interval. Figure 2.20 shows the relation between the change in the functional value from x to $x + h$ and the change in the tangent line over the same interval. Since the slope of the tangent line at x is $f'(x)$ and the run from x to $x + h$ is h

$$f'(x) = \frac{\text{rise along the tangent line}}{\text{run}}$$

$$= \frac{\text{rise along the tangent line}}{h}$$

Therefore, rise along the tangent line $= f'(x)h$

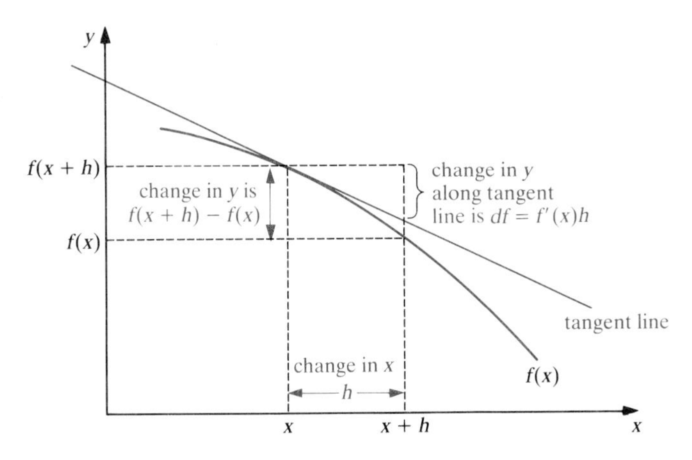

Figure 2.20
Geometric definition of df, the differential of $f(x)$

and since

$$f(x+h)-f(x) = \text{change in the function value}$$
$$\approx \text{rise along the tangent line,}$$

we have

$$f(x+h)-f(x) \approx f'(x)h.$$

Since the slope is also denoted by df/dx, we frequently denote the run parallel to the x-axis by dx and the rise along the tangent line by

$$df = \frac{df}{dx}\,dx = f'(x)\,dx$$

The quantity $df = f'(x)\,dx$ is called the **differential** of $f(x)$.

Example 2.22 For a small manufacturing company, the cost $C(x)$ (in dollars) of producing x items of a particular product is found to be

$$C(x) = 0.1x^3 + 5x^2 + 125x + 300.$$

Compute the increase in cost incurred by increasing production from 30 items to 31 items, and compare this amount with the value of the differential dC at $x = 30$.

Solution The cost of producing 30 items is $C(30) = \$11{,}250$. The cost of producing 31 items is $C(31) = \$11{,}959.10$. Thus the increase in cost is \$709.10. On the other hand, the value of the differential is $dC = C'(x)\cdot dx$, or $dC = C'(30)\cdot 1$ because the change in production from $x = 30$ to $x = 31$ is $dx = 1$. Since

$$C'(x) = 0.3x^2 + 10x + 125,$$

we have $C'(30) = 695$ so that $dC = 695 \cdot 1 = \$695$. The quantity dC differs from the actual increase in cost by \$14.10, or about 2% of the actual cost. ■

Exercises

Exer. 1–8: Write the equation of the line tangent to the graph at the given point.

1. $y = f(x) = x^2 - 2x$; $(1, -1)$
2. $y = f(x) = x^3 - 3x$; $(1, -2)$

3. $y = f(x) = 4 - x^3$; $(2, -4)$
4. $y = f(x) = x^4 - 3x^2 + 2$; $(0, 2)$
5. $y = f(x) = 7x - 6$; $(1, 1)$
6. $y = f(x) = x - 2\sqrt{x}$; $(4, 0)$
7. $y = f(x) = 4\sqrt[3]{x}$; $(8, 8)$
8. $y = f(x) = 1/x^2$; $(\frac{1}{2}, 4)$

Exer. 9–16: Write the equation of the tangent line to the graph of $f(x)$ at the given x-value

9. $y = f(x) = x^4 - 2x$; $x = 1$
10. $y = f(x) = x^2 - (1/x^2)$; $x = 2$
11. $y = f(x) = 3x^{-3/2}$; $x = 4$
12. $y = f(x) = 3x^2 - x$; $x = 4$
13. $y = f(x) = x^4 - x^3$; $x = 1$
14. $y = f(x) = 3x^4 - 4x^3$; $x = 2$
15. $y = f(x) = 4\sqrt[4]{x}$; $x = 16$
16. $y = f(x) = x^{2/3} - 2$; $x = 8$

Exer. 17–26: For the graph of the given function, determine the value(s) of x at which the tangent line has the indicated slope.

17. $y = f(x) = x^2 - 2x + 0.5$; slope 4
18. $y = f(x) = x^2 - 3x$; slope -1
19. $y = f(x) = x^3 - 8$; slope 4
20. $y = f(x) = 6 - 0.3x$; slope $\frac{1}{2}$
21. $y = f(x) = x^4 - 4x^3 - 4$; slope 0
22. $y = f(x) = x^3 - 4$; slope 0
23. $y = f(x) = \frac{1}{3}x^3 - x^2 + 1$; slope -1
24. $y = f(x) = 2x^3 - 9x^2 + 12x$; slope 0
25. $y = f(x) = x^3 + 12x$; slope -4
26. $y = f(x) = x^3 + 3x^2 + 3x + 9$; slope 0

Exer. 27–28: Consider the curve $y = x - \frac{1}{4}x^2$.

27. Find the slope of the line tangent to the curve when $x = -1, 0, 2, 4, 5$. Plot the corresponding points on the curve and draw the tangent lines at each point. Sketch the curve.
28. Find two points on the curve, each with the property that the tangent line at the point also passes through the point $(-\frac{1}{2}, 0)$.

Exer. 29–36: Find the value(s) of x for which the tangent line to the graph of the given function satisfies the stated condition.

29. $y = 4 - x - x^2$; parallel to $y = 3x + \frac{1}{2}$
30. $y = 3x^2 - 6$; parallel to $y = -x$
31. $y = 16x - 13$; perpendicular to $y = 3x$
32. $y = x^3 - 3x + 4$; parallel to $y = 9x - 7$
33. $y = \frac{1}{3}x^3 - 2x^2 - 1$; perpendicular to $y = \frac{1}{4}x + \frac{1}{4}$
34. $y = x^2 - 4x + 4$; perpendicular to $y = 2x$
35. $y = x^4 - 4x^3 + 12$; parallel to x-axis
36. $y = 2x^4 + 8x - 22$; perpendicular to the y-axis

Exer. 37–38: For the given function, find an equation of the line that is tangent to the graph and passes through the origin.

37. $y = x^2 - 2x + 9$
38. $y = -200x^2 + 4000x - 9800$

39. Refer to Example 2.21. Suppose that for a particular site the wood production $W(x)$ (in ft^3/acre) of the shortleaf pine is

$$W(x) = -3.546x^2 + 270.12x - 2406.56$$

after x years of growth for $15 \le x \le 40$. Find the average annual wood production if $x = 20$ and if $x = 35$. Find the value of x that maximizes the average annual wood production by finding the line that is tangent to the graph of $W(x)$ and passes through the origin.

40. A specialist in venture-capital investments makes a career of starting new companies, keeping them for a few years, and then selling them. As a company grows, its rate of growth slows, which causes a decrease in the average annual profit that the investor will make from the sale of the company. Suppose that the profit the investor will make from the sale of a particular company after holding it for x years is $P(x)$ thousands of dollars, where

$$P(x) = -23x^2 + 976x - 828$$

for $0 \le x \le 20$. Find the average annual profit that the investor will make if the company is held for (a) 4 years; (b) 10 years. Explain why the greatest average annual profit will be equal to the slope of the line that is tangent to the graph of $P(x)$ and passes through the origin. Find this tangent line. Determine how long the investor should hold the company, and find the greatest average annual profit.

Exer. 41–44: Use the differential df to estimate the change in $f(x)$.

41. $f(x) = x^2$ at $x = 3$; $dx = 0.5$
42. $f(x) = 1/x$ at $x = 3$; $dx = 0.5$
43. $f(x) = 2x^2 - x$ at $x = 2$; $dx = 0.2$
44. $f(x) = 4x^2 + x$ at $x = 2$; $dx = 0.5$

45. For a small manufacturing company, the cost $C(x)$ (in dollars) of producing x items of a particular product is

$$C(x) = 0.1x^2 + 25x + 250.$$

Compute the cost of increasing production from 10 items to 11 items, and compare this cost with the value of the differential dC at $x = 10$.

Increasing and Decreasing Functions

2.5

To determine whether to hire more employees a manager needs to know if profit would be increased or decreased by an increase in the labor force. A city planner who evaluates requests for new traffic lights needs to determine whether congestion will increase or decrease as the number of lights is increased. To order a sufficient amount of fertilizer, a farmer needs to know whether applying more fertilizer will increase or decrease the profit earned from the crop. The discussion in this section will develop a method for solving these types of problems and will provide a basis for finding maxima and minima discussed in Chapter 3.

We say that a function $f(x)$ is **increasing** on an interval if $f(x_2) > f(x_1)$ for every pair of values x_1 and x_2 in the interval with $x_2 > x_1$. In other words, if x_2 is to the right of x_1, then the function value at x_2 is greater (higher on the graph) than the function value at x_1. Similarly, we say that a function $f(x)$ is **decreasing** on an interval if $f(x_2) < f(x_1)$ for every pair of values in the interval with $x_2 > x_1$; that is, if x_2 is to the right of x_1, then the function value at x_2 is less (lower on the graph) than the function value at x_1. Thus, the graph of an increasing function appears to rise "uphill" as x moves to the right. See Figure 2.21(a). The graph of a decreasing function falls "downhill" as x moves to the right, as shown in Figure 2.21(b).

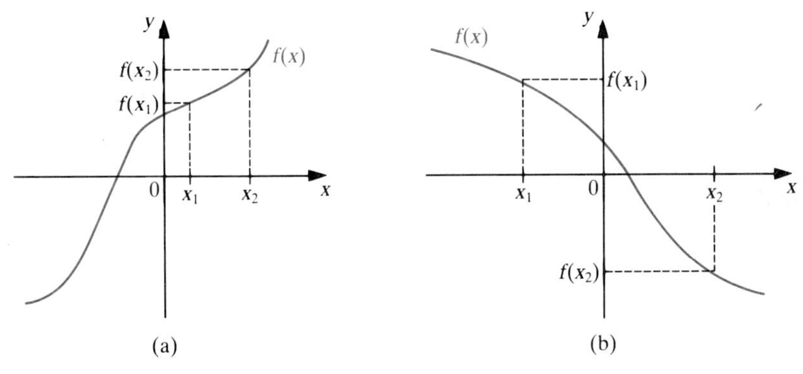

Figure 2.21
(a) An increasing function; (b) a decreasing function

A function may be increasing for some values of x and decreasing for other values of x. Figure 2.22 shows the graph of a function that is increasing for $x < 1$, decreasing for $1 < x < 3$, and increasing for $x > 3$.

The concept of increasing and decreasing functions is closely related to the derivative. First suppose that $f(x)$ is increasing and differentiable on some interval. Let x_1 and x_2 be points in that interval wth $x_2 > x_1$. The fact that $f(x)$ is an increasing function indicates that $f(x_2) > f(x_1)$. Consequently, $f(x_2) - f(x_1) > 0$ and $x_2 - x_1 > 0$, so

$$\frac{f(x_2) - f(x_1)}{x_2 - x_1} > 0.$$

If we let x_2 approach x_1, this ratio becomes the one-sided limit

$$\lim_{x_2 \to x_1^+} \frac{f(x_2) - f(x_1)}{x_2 - x_1}.$$

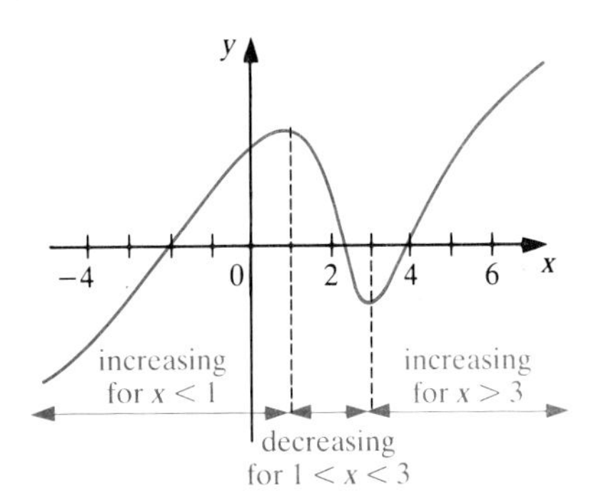

Figure 2.22
The graph of a function that decreases for $1 < x < 3$ and increases for all other x

Since $f(x)$ is differentiable, $f'(x_1)$ exists and this one-sided limit must equal $f'(x_1)$. The ratio in the limit is positive, but may decrease to zero as we let x_2 approach x_1; it could not, however, become negative.

> If $f(x)$ is an increasing function on an interval *and* $f'(x)$ exists there, then $f'(x) \geq 0$ on that interval.
>
> If $f(x)$ is a decreasing function on an interval *and* $f'(x)$ exists there, then $f'(x) \leq 0$ on that interval.

The condition that $f'(x) \geq 0$ for all values of x in an interval does not guarantee that $f(x)$ is increasing on that interval. For example, $f(x) = 4$ is a constant function and hence is neither increasing nor decreasing on any interval, even though $f'(x) \geq 0$ for all x (in fact, $f'(x) = 0$). On the other hand $f(x) = x^3$ is increasing on every interval, and yet $f'(x) = 0$ when $x = 0$. Consequently, the case where $f'(x) = 0$ is inconclusive. By avoiding the case $f'(x) = 0$, we can make the following statements.

> Whenever $f'(x) > 0$ on an interval, $f(x)$ is an increasing function on that interval.
>
> Whenever $f'(x) < 0$ on an interval, $f(x)$ is a decreasing function on that interval.

Example 2.23 Determine the values of x for which $f(x) = 2x^2 + 8x - 1$ is (a) increasing and (b) decreasing.

Solution **(a)** To find the values of x for which $f(x)$ is increasing, we first find the values of x for which $f'(x) > 0$. Since $f'(x) = 4x + 8$, we see that $f'(x) > 0$ is equivalent to $4x + 8 > 0$. We then solve for x:

$$4(x + 2) > 0$$

$$x + 2 > 0 \qquad \text{(since } 4 > 0\text{)}$$

$$x > -2.$$

Hence, $f(x)$ is increasing when $x > -2$.

(b) Similarly, $f'(x) = 4x + 8 < 0$ when $x < -2$, so $f(x)$ is decreasing when $x < -2$. Note that $f(x)$ is decreasing to the left of -2 and increasing to the right of -2. A graph of $f(x)$ is shown in Figure 2.23. ■

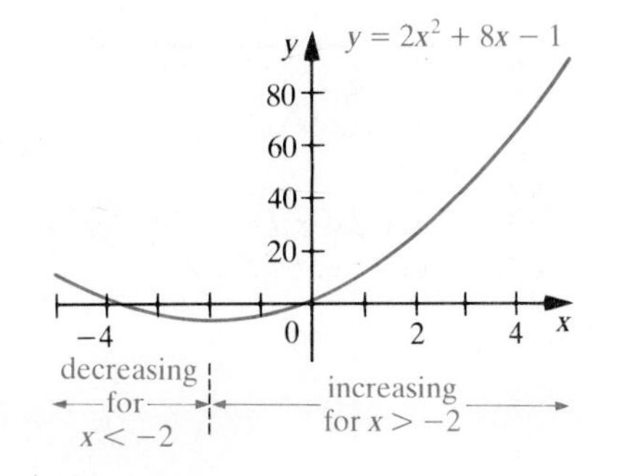

Figure 2.23
$f(x) = 2x^2 + 8x - 1$ is decreasing for $x < -2$ and increasing for $x > -2$

Example 2.24 By studying the learning behavior of a group of students a psychologist determines that receptivity, the ability of the students to grasp a difficult concept, is dependent on the number of minutes of the teacher's presentation that have elapsed before the concept is introduced. At the beginning of a lecture a student's interest is stimulated, but as time passes, attention becomes diffused. Analysis of this group's results indicate that the ability of a student to grasp a difficult concept is given by the function

$$G(x) = -0.1x^2 + 2.6x + 43$$

where the value $G(x)$ is a measure of receptivity after x minutes of presentation.

(a) Determine the values of x for which student receptivity is increasing and decreasing.

(b) Is student interest being stimulated after 10 minutes or is attentiveness falling off?

(c) Where in the presentation should the most difficult concept be placed?

(d) For the intelligence level of the students in this group, a certain concept requires a receptivity of 55. Is it possible to teach the students this concept?

Solution **(a)** Student interest is being stimulated when $G(x)$ is increasing, and attentiveness is falling off when $G(x)$ is decreasing. To determine where $G(x)$ is increasing and where it is decreasing, we use the derivative of $G(x)$ that is, $G'(x) = -0.2x + 2.6$. We know that $G(x)$ is increasing where $G'(x) > 0$, so we solve the inequality

$$-0.2x + 2.6 > 0,$$

which is equivalent to

$$-0.2x > -2.6$$

$$0.2x < 2.6 \qquad \text{(inequality sign is reversed)}$$

$$x < 13.$$

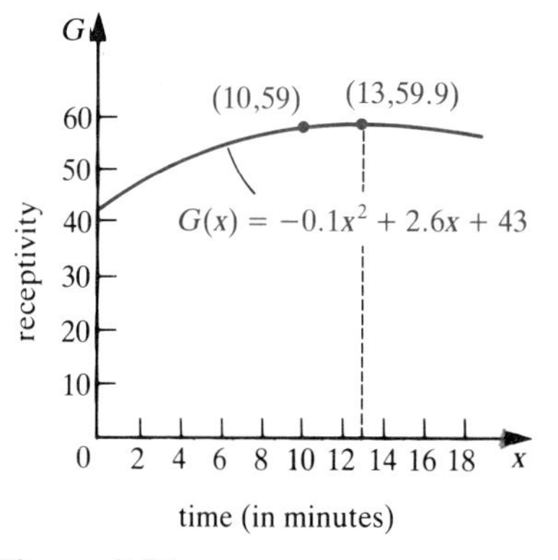

Figure 2.24
Graph of student receptivity function $G(x)$ in Example 2.24

Thus, student receptivity $G(x)$ is increasing for $x < 13$ and decreasing for $x > 13$. (See Figure 2.24.)

(b) Since $G(x)$ is increasing at $x = 10$, student interest is still being stimulated 10 minutes into the presentation.

(c) Receptivity is increasing during the first 13 minutes and decreasing after the first 13 minutes, maximum receptivity occurs 13 minutes into the presentation. Thus, the most difficult concept should be discussed 13 minutes after the presentation begins.

(d) Since $G(13) = 59.9$ and the concept we want to present requires a receptivity of 55, it is possible to teach this concept to these students. ■

To systematically locate the intervals in which a function $f(x)$ is increasing or decreasing, we find the values where $f'(x)$ equals zero or does not exist. Any point on the graph of $f(x)$ where $f'(x)$ equals zero or does not exist is called a **critical point**. At such a point a function may switch from increasing to decreasing or from decreasing to increasing. The x-coordinate of a critical point is referred to as a **critical value**.

A **critical value of $f(x)$** is a value $x = a$ such that

(i) $f(x)$ is defined at $x = a$,

and (ii) either $f'(a) = 0$ or $f'(a)$ does not exist.

Example 2.26 Let $f(x) = x^3 - 3x^2 - 9x + 10$. Find the intervals for which $f(x)$ is increasing and decreasing. Sketch a graph of the function.

Solution Note that

$$f'(x) = 3x^2 - 6x - 9 = 3(x - 3)(x + 1),$$

so $f'(x) = 0$ for $x = -1$ and for $x = 3$. Hence the critical values are $x = -1$ and $x = 3$. We consider the three regions on the x-axis determined by these two critical values, namely, $x < -1$, $-1 < x < 3$, and $x > 3$. To determine whether $f'(x)$ is positive or negative on each interval we consider a *test value t* in the interval. Since the sign of $f'(x)$ cannot change *within* these intervals, the sign of $f'(x)$ at *any* value of x in each interval gives the sign throughout the interval.

The use of test values is demonstrated in Table 2.4. We have (arbitrarily) selected $t = -2$ for the interval $(-\infty, -1)$, $t = 0$ for the interval $(-1, 3)$, and $t = 4$ for the interval $(3, \infty)$. For example, since $f'(x)$ is positive at the test value $t = -2$ $[f'(-2) = 15]$, the sign of $f'(x)$ is positive for all x in the interval $(-\infty, -1)$; thus $f(x)$ is increasing on this interval.

Table 2.4

The sign of $f'(t)$ at test values determines the sign of $f'(x) = 3x^2 - 6x - 9$ on each interval

Interval	$(-\infty, -1)$	$(-1, 3)$	$(3, \infty)$
test value t	-2	0	4
$f'(t)$	15	-9	15
sign of $f'(t)$	+	$-$	+
on the interval $f(x)$ is	increasing	decreasing	increasing

The function values corresponding to the critical values are

$$f(-1) = (-1)^3 - 3(-1)^2 - 9(-1) + 10 = 15$$

and

$$f(3) = (3)^3 - 3(3)^2 - 9(3) + 10 = -17.$$

Plotting the two critical points $(-1, 15)$ and $(3, -17)$, and using the information from Table 2.4 about the increasing and decreasing nature of $f(x)$, we obtain a rough idea of the graph of the function. A sketch of the graph of $f(x)$ is shown in Figure 2.25. ■

Example 2.27 Let $f(x) = \frac{1}{3}x - x^{2/3}$. Find the values of x for which $f(x)$ is increasing and decreasing. Sketch a graph of the function.

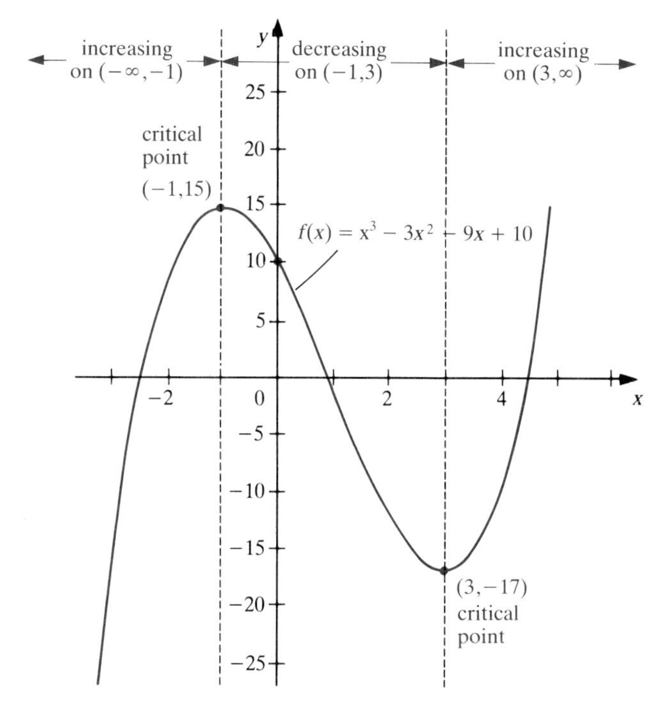

Figure 2.25
Graph of $f(x) = x^3 - 3x^2 - 9x + 10$

Solution The derivative of $f(x)$ is

$$f'(x) = \frac{1}{3} - \frac{2}{3}x^{-1/3} = \frac{1}{3} - \frac{2}{3\sqrt[3]{x}}. \tag{1}$$

This derivative is zero when

$$\frac{1}{3} = \frac{2}{3\sqrt[3]{x}}$$

$$\sqrt[3]{x} = 2. \qquad \text{(multiply by } 3\sqrt[3]{x}\text{)}$$

Thus $x = 8$ is a critical value. Note that at $x = 0$, $f'(x)$ is not defined but $f(x)$ is defined. Consequently, $x = 0$ is also a critical value, so we consider the sign of $f'(x)$ on the intervals $(-\infty, 0)$, $(0, 8)$, and $(8, \infty)$. Table 2.5 shows the test values chosen and the conclusions obtained.

Table 2.5
The sign of $f'(t)$ at test values determines the sign of $f'(x) = \frac{1}{3} - 2/(3\sqrt[3]{x})$ on each interval

Interval	$(-\infty, 0)$	$(0, 8)$	$(8, \infty)$
test value t	-1	1	27
$f'(t)$	1	$-\frac{1}{3}$	$\frac{1}{9}$
sign of $f'(t)$	$+$	$-$	$+$
on the interval $f(x)$ is	increasing	decreasing	increasing

The function values corresponding to the critical values are $f(0) = 0$ and $f(8) = -\frac{4}{3}$. Note that the term $-2/(3\sqrt[3]{x})$ in equation (1) indicates that $f'(x) \to -\infty$ as $x \to 0^+$ and $f'(x) \to +\infty$ as $x \to 0^-$. This information has been used to sketch the graph of $f(x)$, shown in Figure 2.26. ■

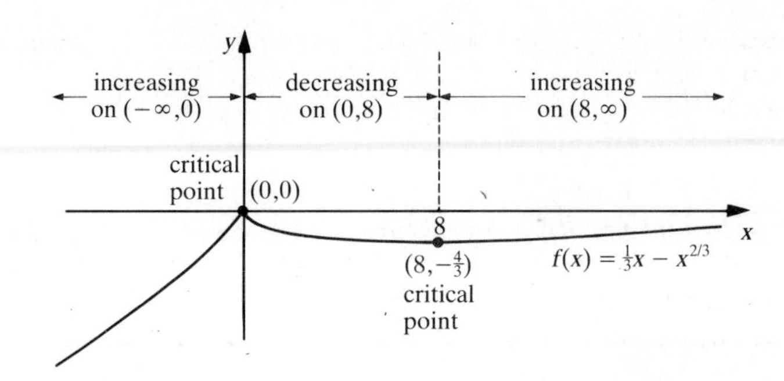

Figure 2.26
Graph of $f(x) = \frac{1}{3}x - x^{2/3}$

Exercises

Exer. 1–12: Find the critical values of $f(x)$, and then determine the values of x for which the function is increasing and decreasing.

1. $f(x) = x^2 + 6x - 10$
2. $f(x) = 8 - 14x + x^2$
3. $f(x) = 3x^2 + 4$
4. $f(x) = 6 - 2x$
5. $f(x) = 6x - 2$
6. $f(x) = x^3 - 4x$
7. $f(x) = 8x + 4x^3$
8. $f(x) = 2x^3 - 6x^2 - 36x + 7$
9. $f(x) = x^3 - \frac{3}{2}x^2 - 6x + 12$
10. $f(x) = x^3 - 9x^2 + 24x - 113$
11. $f(x) = x^3 - 12$
12. $f(x) = 12 + 9x - 6x^2 + x^3$

Exer. 13–22: Find the critical values of $f(x)$ and determine the values of x for which the function is increasing and decreasing. Use this information to sketch the graph of the function.

13. $f(x) = x^2 - 3x + 4$
14. $f(x) = 4x^2 - x + 4$
15. $f(x) = x^3 - 3x + 4$
16. $f(x) = (1/x) + x$
17. $f(x) = -4x^3 - 6x^2$
18. $f(x) = x^4 - 8x^2 + 4$
19. $f(x) = \frac{1}{6}x^2 - \sqrt[3]{x}$
20. $f(x) = \frac{1}{4}x - \sqrt{x}$
21. $f(x) = x - \sqrt{x}$
22. $f(x) = x^2 - 3\sqrt{x}$

23. If the rate $R(x)$ of a chemical reaction is given by

$$R(x) = 0.0004(6x - x^2),$$

where x is the amount of catalyst present $(0 < x < 6)$, find the values of x for which R is increasing.

24. If t days after the application of a pesticide, the number of rodents $r(t)$ in a particular area is

$$r(t) = 100 - 15t + 2.5t^2,$$

is $r(t)$ increasing (a) when $t = 2$? (b) when $t = 4$? (c) when $t = 6$?

25. A diving instructor finds that 20 pupils will enroll when the fee is \$60. To increase the class size, the fee is decreased by \$3 per additional pupil.

(a) Determine the amount $R(x)$ of revenue when x pupils are enrolled for $x \geq 20$.

(b) Is the revenue increasing when 25 pupils are enrolled? when 30 pupils are enrolled?

26. Consider the cake-pan example in Section 2.1. For a pan made from a 12-inch square of plastic, the volume (in cu. in.) is $V(x) = 6x^2 - \frac{1}{2}x^3$. Is the volume increasing when the length of the square being removed is 5 inches?

27. Refer to Example 2.24. In a follow-up study the psychologist finds that when visual aids are used, student receptivity is given by

$$G(x) = -0.1x^2 + 4.4x + 46.$$

Graph $G(x)$ and answer the following questions.

(a) During what time interval is student interest stimulated?

(b) During what time interval is attentiveness declining?

(c) Where in the presentation should the most difficult concept be placed?

(d) If a difficult concept requires a receptivity level of 80, can it be taught to students in this group?

28. A farmer finds that if a corn crop is sprayed with x hundred-gallon barrels of liquid fertilizer per acre, then the yield (in bushels per acre) will be

$$y = -3x^2 + 42x + 150.$$

Graph the yield function and determine where the graph is increasing and decreasing. The farmer currently sprays 5 barrels per acre. Should more or less spray be applied? Suppose the farmer currently sprays 9 barrels per acre. Should more or less spray be applied?

29. A tennis racket manufacturing company finds that the cost C (in dollars) of manufacturing each racket is a function of the number x of rackets manufactured per hour, according to the rule

$$C = 0.2x^2 - 32x + 1295.$$

For what values of x is the cost increasing? Will the cost per racket increase or decrease if production is increased slightly from a current level of 70 rackets per hour? from 95 rackets per hour?

30. Refer to Exercise 29. Suppose that the company opens a new factory and finds that the cost C of manufacturing each racket is now given by

$$C = 0.25x^2 - 62x + 3860.$$

If the company currently produces 100 rackets, determine whether the cost of each racket increases or decreases if production is increased. For what production level is the cost of each racket decreasing, and for what level is the cost increasing?

Rate of Change

2.6

The solution of the fruit fly example of Section 2.1 involves calculating the rate at which the fruit fly population is changing at a particular point in time. This type of problem is encountered frequently. For example, a manager of a company wants to know at what rate the company's profits are increasing (or decreasing); a sociologist wants to know at what rate crime is changing in a particular city. In this section we will explore how we can use the derivative to calculate the rate of change of a function.

Consider the calculation of the average speed of a car. Suppose that the function $f(t)$ gives the distance we have driven since the start of a trip. If t is measured in hours, then $f(1)$ will be the distance we traveled in the first hour, $f(2)$ will be the distance we traveled in the first two hours, and so on. To find the distance traveled between the fourth hour and fifth hour of the trip, we subtract: $f(5) - f(4)$. Suppose the values of $f(t)$ are as shown in Table 2.6. Then the distance driven between the fourth hour and fifth hour is $220 - 170 = 50$ miles. Since this distance was covered in one hour, the average speed of the car for this hour is obtained from

Table 2.6
Function values for $f(t)$

t	$f(t)$ miles
1 hr	30
2 hr	80
3 hr	120
4 hr	170
4 hr 1 sec	170.0125
4 hr 1 min	170.75
4 hr 6 min	175
5 hr	220

$$\text{average speed for 1 hour} = \frac{f(5) - f(4)}{5 - 4} = \frac{50 \text{ miles}}{1 \text{ hour}} = 50 \text{ mph}.$$

Now it can easily happen that this average speed is *not* the speedometer reading for most of the hour. The 50 mph average could have been attained by traveling at 60 mph for a while and then at 40 mph for a while. The speedometer reading may actually have been 50 for only a fraction of a second.

Next, suppose we calculate the average speed for the time interval from 4 hours to 4 hours 6 minutes. From the table, we see that the distance traveled is 5 miles; the time elapsed is $\frac{1}{10}$ hour, so

$$\text{average speed for 6 minutes} = \frac{f(4.1) - f(4)}{4.1 - 4} = \frac{5}{1/10} = 50 \text{ mph}.$$

Again, the speedometer may not have been on 50 for most of the six minutes.

If we repeat this calculation for the time interval of 1 minute ($=\frac{1}{60}$ hr) that begins at the four-hour mark, we get

$$\text{average speed for 1 minute} = \frac{f(4.01667) - f(4)}{4.01667 - 4} \approx \frac{0.75}{1/60} = 45 \text{ mph}.$$

The speedometer reading probably stayed near 45 for the one-minute time interval, but it could still have varied.

Finally, suppose we compute the average speed over the time interval of one second ($\frac{1}{3600}$ hr) that begins at the four-hour mark:

$$\text{average speed for 1 second} = \frac{f(4.0002778) - f(4)}{4.0002778 - 4} \approx \frac{0.0125}{1/3600} = 45 \text{ mph.}$$

For this case the speedometer will probably be on 45 for the entire second; that is, the speedometer reading and the average speed for this (short) time interval are virtually the same. The instantaneous speed, as noted on the speedometer, and the average speed over an extremely short time interval are essentially equal. Notice that if we let t denote any one of the times (5 hours), (4 hours, 6 minutes), (4 hours, 1 minute) or (4 hours, 1 second), then each average is computed by

$$\text{average speed} = \frac{f(t) - f(4)}{t - 4}.$$

Also note that these values of t are approaching 4. We thus have

$$\text{instantaneous speed} = \lim_{t \to 4} \frac{f(t) - f(4)}{t - 4}.$$

This process is exactly the same process that we used to find the derivative of $f(t)$ at $t = 4$.

At any point a and for any function $f(t)$, we define the **instantaneous rate of change at *a*** as

$$\lim_{t \to a} \frac{f(t) - f(a)}{t - a},$$

provided the limit exists; that is,

$$f'(a) = \text{instantaneous rate of change of } f(t) \text{ with respect to } t \text{ at } t = a.$$

Consequently, for a function $f(t)$, the rate at which the functional value $f(t)$ changes with respect to t at a given value $t = a$ is given by $f'(a)$. We have just seen that the speedometer on an automobile measures the rate at which the distance traveled is changing with respect to time. If $s(t)$ denotes the distance traveled in t hours, then $s'(t)$ is the rate of change of $s(t)$ with respect to t; that is, $s'(t)$ *is* the *velocity* of the car at time t. We usually denote velocity by $v(t)$.

Example 2.28 If $s(t) = 12t^2 + 3t$ denotes the distance traveled (in feet) by a dragster after t seconds have elapsed, how fast is the dragster traveling at $t = 5$ seconds?

Solution The velocity $v(t)$ of the dragster (in ft/sec) is given by

$$v(t) = s'(t) = 24t + 3.$$

Hence, the velocity at $t = 5$ seconds is $v(5) = 123 \text{ ft/sec} = 83.9 \text{ mi/hr}$. ■

Acceleration is the rate of change of velocity with respect to time.

For example, if we drop a rock from the top of a building, the acceleration due to gravity is 32 ft/sec/sec, which means that the velocity of the rock increases by 32 ft/sec each second that the rock is falling. In general, given the velocity $v(t)$ as a function of time t, $v'(t)$ represents the rate of change of $v(t)$ with respect to t; that is, $v'(t)$ is the *acceleration* of the object at time t. We usually denote acceleration by $a(t)$.

Since $v(t) = s'(t)$, where $s(t)$ is the distance traveled, we find $a(t)$ by taking the derivative of $s(t)$, and then taking the derivative of the result, $s'(t)$. Thus, $a(t)$ is called the *second derivative of* $s(t)$, denoted by $s''(t)$. We will study second derivatives in Section 3.3.

Example 2.29 If $v(t) = 2t^2 - 6t + 8$ ft/sec denotes the velocity of a train after t seconds, what is the acceleration of the train at $t = 4$ seconds?

Solution The acceleration $a(t)$ is given by

$$a(t) = v'(t) = 4t - 6,$$

so the acceleration after 4 seconds is $a(4) = 10\text{ ft/sec}^2$. ■

Example 2.30 Suppose that an arrow is shot vertically into the air and its distance from the ground after t seconds in flight is given by

$$s(t) = -16t^2 + 80t.$$

(a) What is the arrow's velocity after $t = 2$ seconds?
(b) How high is the arrow when it begins to fall toward the earth? (That is, what is the highest point in the trajectory of the arrow?)
(c) How many seconds after the arrow is shot does the arrow hit the earth?
(d) Describe the acceleration of the arrow.

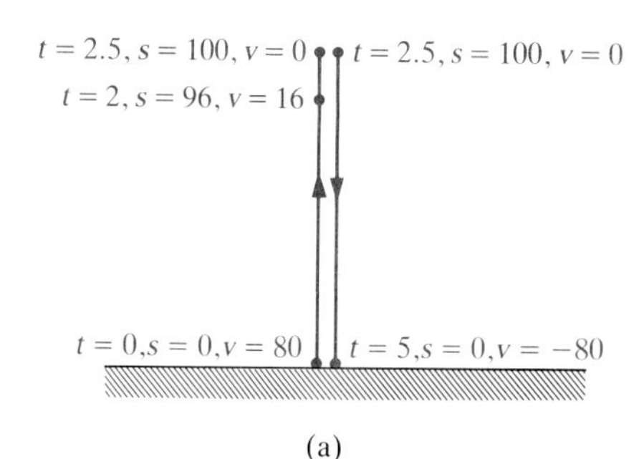

Solution In this problem, $s(t)$ denotes the distance above the ground instead of distance traveled (see Figure 2.27(b)). Hence, when the arrow is rising, $s(t)$ will be increasing, so the velocity $v(t) = s'(t)$ will be positive. When the arrow is falling, $s(t)$ will be decreasing and the velocity $v(t) = s'(t)$ will be negative. Therefore, due to the way that $s(t)$ is defined in this problem, upward is the positive direction and downward is the negative direction.

(a) To find the arrow's velocity we use $v(t) = s'(t) = -32t + 80$, so $v(2) = 16$ ft/sec.

(b) We first ask "What is special about the arrow when it is at its highest point?" At the highest point, the arrow is no longer rising nor has it begun to fall, so its velocity is zero. We set $v(t) = 0$ and solve for t:

$$v(t) = -32t + 80 = 0$$

$$t = \frac{-80}{-32} = 2.5\text{ sec.}$$

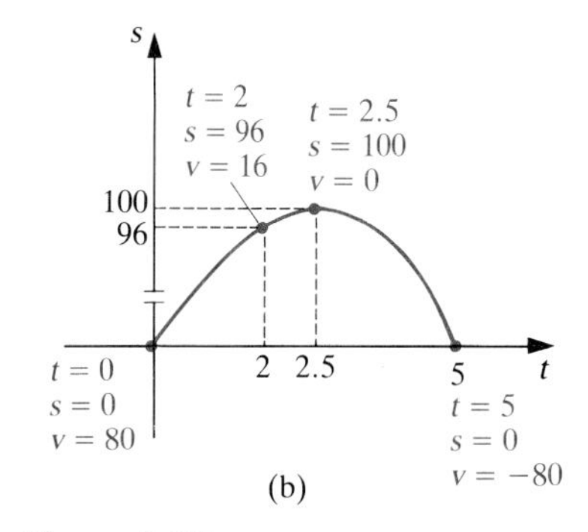

Figure 2.27
(a) The path of an arrow shot into the air; (b) the graph of $s(t)$, the height of the arrow at time t

A quick check of values before and after 2.5 (for example, $v(2) = 16$ and $v(3) = -16$) confirms that $v(t)$ is positive before $t = 2.5$ and negative after $t = 2.5$. That means that the arrow is rising before $t = 2.5$ and falling after $t = 2.5$. To find the maximum height, we evaluate $s(t)$ at $t = 2.5$. Since $s(2.5) = 100$, the highest point in the trajectory of the arrow is 100 feet.

(c) We note that when the arrow hits the ground, $s(t) = 0$. Solving for t gives

$$\begin{aligned} s(t) = -16t^2 + 80t &= 0 \\ 16t(-t+5) &= 0 \qquad \text{(factor)} \\ t = 0 \quad \text{or} \quad t &= 5. \end{aligned}$$

Since $t = 0$ is the time when the arrow is shot, $t = 5$ is the time when the arrow hits the ground. The arrow's flight lasts 5 seconds.

(d) The acceleration is $a(t) = v'(t) = -32$, and so the arrow has a constant acceleration of -32 ft/sec^2 (in the downward direction) due to gravity. ■

Example 2.31 Refer to Section 2.1 (page 46). At what rate will the fruit fly population be growing 8 weeks after the discovery of the infestation? What is the rate 10 weeks after discovery?

Solution Recall that $P(t) = \frac{1}{4}t^2 + 3t + 1$ is the predicted number of fruit flies (in millions) t weeks after their discovery. Since

$$P'(t) = \tfrac{1}{2}t + 3,$$

we have

$$P'(8) = \tfrac{1}{2}(8) + 3 = 7,$$

so the rate of growth of the population 8 weeks after discovery is 7 million fruit flies per week. Ten weeks after discovery, the rate of growth is $P'(10) = 8$ million fruit flies per week. ■

Example 2.32 A manufacturer of nonprescription drugs finds that the weekly profit P depends on the number of dollars x spent each week on advertising. The relationship is given by the formula $P(x) = 1000 + 2000x - x^2$ for $0 \le x \le 2000$. If the manufacturer is spending \$700 per week for advertising at the present time and plans to increase this amount, at what rate will the profit increase or decrease?

Solution The rate of change of P with respect to x is given by

$$\frac{dP}{dx} = 2000 - 2x.$$

When $x = 700$, the rate of change is $dP/dx = 600$, which means that profit will increase at a rate of \$600 for each additional dollar spent on advertising. In fact, $dP/dx > 0$ whenever $x < 1000$, so when the amount spent for advertising is less than \$1000 and is increasing, profit increases. However, if the amount spent for advertising is greater than \$1000, increasing the amount spent for advertising will decrease profit. ■

Exercises

Exer. 1–4: Find the rate of change of $f(t)$ for the given value of t.

1. $f(t) = t^3 + 2t^2$; $t = 2$
2. $f(t) = t + \frac{1}{t}$; $t = 1$
3. $f(t) = t^4 - 3t + 4$; $t = 3$
4. $f(t) = t^2 + 4t^{1/2}$; $t = 4$

5. If the distance an object travels after t seconds is given by $s(t) = 20t^2 - 3t + 5$, what is the velocity of the object after 3 seconds? What is the acceleration of the object after 8 seconds?
6. If the distance an object travels after t seconds is given by $s(t) = 12t^3 - 3t^2 + 5$, what is the velocity of the object after 3 seconds? What is the acceleration of the object after 5 seconds?
7. If the distance an object travels after t seconds is given by $s(t) = 2t^4 - 2t^3 + 5t$, what is the velocity of the object after 2 seconds? What is the acceleration of the object after 4 seconds?
8. If the distance an object travels after t seconds is given by $s(t) = \sqrt{t} - 3t$, what is the velocity of the object after 4 seconds? What is the acceleration of the object after 9 seconds?

Exer. 9–12: A ball is thrown into the air and its distance from the ground after t seconds is given by the distance function $s(t)$.

(a) What is the ball's velocity when it reaches its highest point and begins to fall toward the earth?
(b) How high is the ball when it begins to fall toward the earth?
(c) How many seconds after the ball is thrown does it hit the ground?
(d) Describe the acceleration of the ball.

9. $s(t) = -16t^2 + 51.2t$
10. $s(t) = -16t^2 + 96t$
11. $s(t) = -16t^2 + 96t + 108$
12. $s(t) = -t^2 + 16t$

13. Pat joins the circus as a human cannonball. It is calculated that t seconds after blastoff Pat will be $s(t) = -16t^2 + 64t$ feet from the floor of the civic center. How high must the ceiling of the civic center be in order for Pat to fall gently into the net, rather than crash into the ceiling? What is Pat's muzzle velocity, that is, at $t = 0$?
14. Work Exercise 13 if Pat's distance function is $s(t) = -16t^2 + 32t$.

Exer. 15–18: The size of a bacteria population at time t (in hours) is given by the function $p(t)$. Determine the growth rate when (a) $t = 1$ hour; (b) $t = 5$ hours; (c) $t = 10$ hours.

15. $p(t) = 10^6 + 10t - 1000t^2$
16. $p(t) = 10t^3 - 5t^2 + 1000$
17. $p(t) = 10t^{3/2} - 5t + 5000$
18. $p(t) = -10t^3 + 50t^2 + 1000$

19. Refer to Example 2.32. Suppose that the drug manufacturer has a profit function $P(x) = 3000 + 6000x - 2x^2$ for $0 \le x \le 3500$, where x is the number of dollars spent per week on advertising. If \$1400 per week is currently being spent on advertising, what is the rate at which profit is changing with respect to advertising expenditures? Would you recommend spending more on advertising?
20. Work Exercise 19 using \$2700 as the amount currently spent on advertising.
21. On a warm summer evening, the temperature can be estimated by counting the number of chirps a cricket makes in 15 seconds and adding 40 to this number. This thermometer is reasonably accurate between 55 °F and 100 °F. Calculate the rate of change of chirps (per 15 seconds) per Fahrenheit degree. (*Hint:* First determine the functional relationship between the number of chirps and the temperature.)
22. A bullet travels $s(t) = 700t - 4t^2$ feet t seconds after it is fired. What is its initial velocity? How fast is the bullet traveling after 6 seconds? Calculate the deceleration (negative acceleration) of the bullet after 6 seconds.
23. Work Exercise 22 for $s(t) = 1000t - 5t^{3/2}$.
24. A diver dives into a swimming pool and plunges $s(t) = 20t - 5t^2$ feet under the surface t seconds after entering the water. How deep does the pool have to be so that the diver doesn't hit bottom?
25. In Exercise 24, if the diver plunges $s(t) = 30t - 6t^2$ feet under the surface t seconds after entering the water, how deep must the pool be?
26. A forester finds that if 750 shortleaf pine seedlings are planted per acre at a particular site, then x years later the wood production $W(x)$ (in ft^3/acre) will be

$$W(x) = -2.489x^2 + 238.608x - 2178.85$$

for $15 \le x \le 40$. At what rate is the wood production growing 20 years after planting? At what rate is the wood production growing after 30 years?
27. Suppose that at a particular site the wood production $W(x)$ (in ft^3/acre) of shortleaf pine planted x years earlier is

$$W(x) = -3.546x^2 + 270.12x - 2406.56$$

for $15 \le x \le 40$. At what rate is the wood production growing 25 years after planting? At what rate is the wood production growing after 35 years?

28. Suppose that the profit $P(x)$ (in thousands of dollars) an investor makes from the sale of a particular company after holding it for x years is

$$P(x) = -23x^2 + 976x - 828 \quad \text{for} \quad 0 \le x \le 30.$$

Find the rate at which the profit is increasing or decreasing when the investor has held the company for 2 years. At what rate is the profit increasing or decreasing when the company has been held for 25 years? How long should the investor hold the company in order to realize the maximum profit?

29. The daily output of a factory assembly line is $f(t) = 60t + t^2 - \frac{1}{12}t^3$ units after t hours of operation for $0 \le t \le 8$. What is the rate of production when $t = 2$?

30. Suppose the weight (in grams) of a malignant tumor is given by $w(t) = 0.01t^3$ for time t (in months). What is the rate of growth of the tumor when $t = 5$?

31. Suppose that t hours after being placed in a freezer, the temperature $T(t)$ of a package of vegetables is

$$T(t) = 65 - 11t + \frac{4}{t} \quad \text{for} \quad 1 \le t \le 5.$$

How fast is the temperature changing when $t = 1$?

2.7 Continuity and Differentiability

We have analyzed functions whose graphs exhibit all sorts of behavior from gaps (the postal rate function) to corners (the absolute value function). In this section we will study two types of functions whose graphs are particularly well behaved: *continuous* functions and *differentiable* functions.

Consider the graphs shown in Figure 2.28. Intuitively, we may sense that the functions whose graphs are shown in parts (a), (b), and (c) are not continuous at $x = a$, but the function in part (d) is continuous at $x = a$. A few comments on why the first three graphs do not seem to be continuous will be helpful in understanding the definition of continuity given later. The function in Figure 2.28(a) is not even defined at $x = a$. The graph of the function in (b) has a "jump" at $x = a$; also note that $\lim_{x \to a} f(x)$ does not exist because $f(x)$ approaches one value from the left and another value from the right. The function in (c) is defined at $x = a$ and $\lim_{x \to a} f(x)$ exists (denoted by L in the figure). However, $f(a)$ is not equal to this limit; $f(a)$ is "too large." The graph in (d) exhibits none of these problems; the function is defined at $x = a$. Also, $\lim_{x \to a} f(x)$ exists and equals $f(a)$.

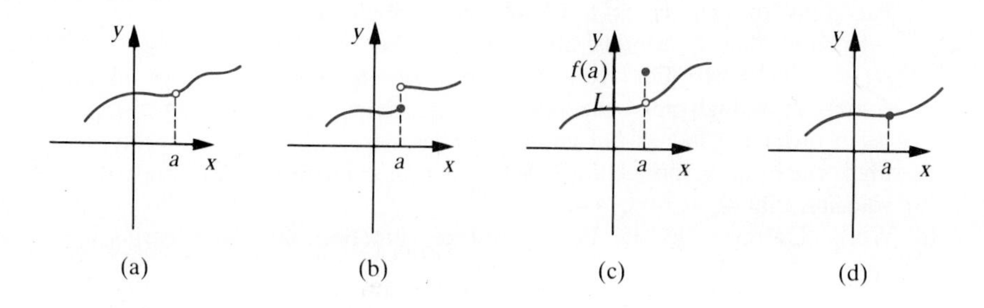

Figure 2.28
(a) $f(a)$ is not defined; (b) a "jump" at $x = a$ ($\lim_{x \to a} f(x)$ does not exist); (c) $\lim_{x \to a} f(x)$ exists but $\lim_{x \to a} f(x) \ne f(a)$; (d) $f(x)$ is continuous at $x = a$

Motivated by the preceding observations, we state the following definition.

> A function $f(x)$ is **continuous at $x = a$** if and only if
>
> (1) $f(x)$ is defined at $x = a$;
>
> (2) $\lim_{x \to a} f(x)$ exists; and
>
> (3) $\lim_{x \to a} f(x) = f(a)$

In Section 2.2 we showed that if $f(x)$ is a polynomial, then

$$\lim_{x \to a} f(x) = f(a)$$

for any real number a. Hence every polynomial is continuous at every real number. A function that is continuous at every real number is called a **continuous function**. We also showed that for a rational function $f(x)/g(x)$ (where $f(x)$ and $g(x)$ are polynomials)

$$\lim_{x \to a} \frac{f(x)}{g(x)} = \frac{f(a)}{g(a)}$$

except when $g(a) = 0$. Consequently, any rational function is continuous except for values where its denominator is zero.

An intuitive way to visualize a continuous function is to think of a function whose graph can be drawn without lifting the pencil from the paper. Hence, the graph of a function that is continuous for all values of x has no "gaps". (Figure 2.29 shows graphs of several continuous functions.) This visualization is useful, but fails sometimes. In fact, there exists a function that is continuous at $x = 0$, and yet its graph cannot be drawn near $x = 0$. Thus this characterization of continuous functions is not quite accurate, but it does provide an intuitive sense of continuity.

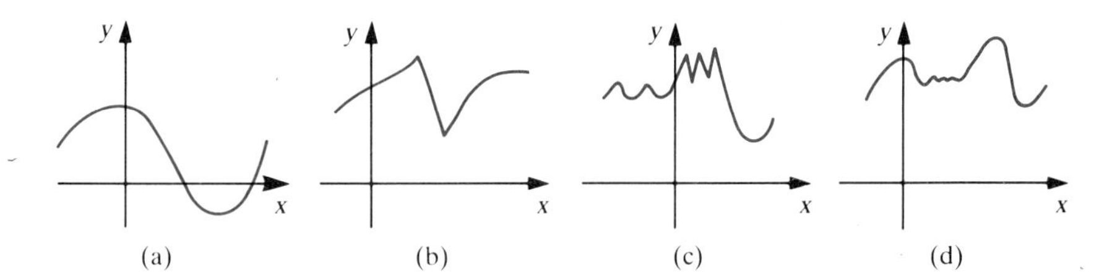

Figure 2.29
Several continuous functions

Example 2.33 Explain why each function $f(x)$ graphed in Figure 2.30 is not continuous at a.

Solution We summarize our observations in Table 2.7, in which each column corresponds to a condition in our definition of continuity. Note that the last column, headed by $\lim_{x \to a} f(x) = f(a)$, applies only when both $\lim_{x \to a} f(x)$ and $f(a)$ exist, and so we have left it blank in the other cases.

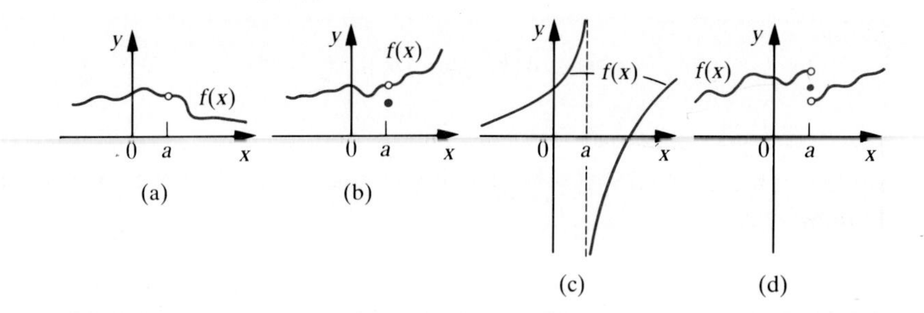

Figure 2.30
Graphs of several functions $f(x)$ that are not continuous at $x = a$

A function is continuous only if the conditions described in all three columns are fulfilled (indicated by a "yes").

Table 2.7

Graph	*$f(a)$ exists*	*$\lim_{x\to a} f(x)$ exists*	*$\lim_{x\to a} f(x) = f(a)$*
(a)	no	yes	—
(b)	yes	yes	no
(c)	no	no	—
(d)	yes	no	—

(a) Although $\lim_{x\to a} f(x)$ exists, $f(a)$ does not exist.

(b) Both $\lim_{x\to a} f(x)$ and $f(a)$ exist, but they are not equal.

(c) Neither $\lim_{x\to a} f(x)$ nor $f(a)$ exists.

(d) Although $f(a)$ exists, the two one-sided limits are different; therefore, $\lim_{x\to a} f(x)$ does not exist. ■

Example 2.34 Let $f(x) = [x] =$ the largest integer $\le x$. Find all points where $f(x)$ is not continuous.

Solution This function is called the **greatest integer function**. To get some experience with this function we first evaluate it for several values of x. If $x = 3.6$, then $f(x) = f(3.6) = 3$, because 3 is the largest integer that does not exceed 3.6. Similarly, $f(3.1) = 3$, $f(3.98) = 3$, and in fact, for all x such that $3 < x < 4$, we have $f(x) = 3$. Also note that $f(3) = 3$, because the largest integer not exceeding 3 is 3 itself. In a similar fashion, we find that for all x between 1 and 2, $f(x) = 1$, and for $2 \le x < 3$, $f(x) = 2$. Note, however, that $f(-2.6) = -3$, because -3 is the largest integer not larger than -2. Thus $f(x)$ is constant for all values of x between integers and jumps from one integer value to the next as x crosses an integer value.

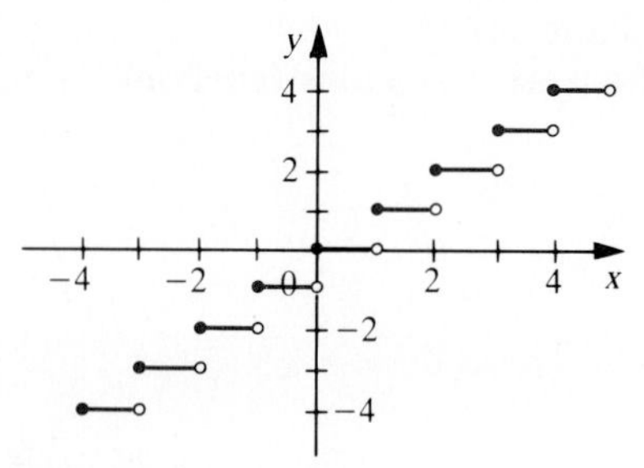

Figure 2.31
Graph of the greatest integer function, $f(x) = [x]$

Figure 2.31 shows the graph of $f(x)$, which certainly gives the impression that the function is not continuous at the integers. Is this true, or is the graph drawn incorrectly? To determine continuity or discontinuity at a point $x = a$, we must ascertain whether or not $\lim_{x\to a} f(x) = f(a)$. For example, consider $a = 3$. We will find the one-sided limits

$$\lim_{x\to 3^+} [x] \quad \text{and} \quad \lim_{x\to 3^-} [x].$$

First consider values of x that are close to 3 but slightly larger than 3, say $3 < x < 3.1$. Then $[x]$ will be 3 for all such x, and so the right-hand limit is

$$\lim_{x\to 3^+} [x] = 3.$$

Next, consider values of x near 3 but slightly less than 3, say $2.9 < x < 3$. For all such values of x, $[x] = 2$, so the left-hand limit is

$$\lim_{x\to 3^-} [x] = 2.$$

Consequently, the one-sided limits are different, so $\lim_{x\to 3} [x]$ does not exist, and $[x]$ is discontinuous at 3. The same computations could have been done at any integer, with the same conclusion: the two one-sided limits are different. Consequently, the function is not continuous at any integer. ■

Note the similarity between the greatest integer function and the postal rate function; the graph of each function looks like steps. Such a function is called a **step function**.

Table 2.8 shows an analysis of several functions for continuity at $a = 3$. Each function satisfies some combination of the three properties in the definition of a continuous function.

Function	***$f(3)$ exists***	***$\lim_{x\to 3} f(x)$ exists***	***$\lim_{x\to 3} f(x) = f(3)$***
$f(x) = \dfrac{1}{x-3}$	no	no	—
$f(x) = \dfrac{x^2-9}{x-3}$	no	yes	—
$f(x) = [x]$	yes	no	—
$f(x) = \begin{cases} 5 & \text{for } x = 3 \\ \dfrac{x^2-9}{x-3} & \text{for } x \neq 3 \end{cases}$	yes	yes	no
$f(x) = \dfrac{x^2+5}{x-2}$	yes	yes	yes

Table 2.8

A *no* in any column indicates that the function is discontinuous at $a = 3$. A continuous function has a *yes* in all three columns.

Recall that a function $f(x)$ is *differentiable at* $x = a$ if it has a derivative at a, and a function is *differentiable* if it has a derivative for all values of x in its domain. We have already observed that if $f(x)$ is continuous, then its graph has no gap or break. If $f(x)$ is differentiable, then its graph not only has no gap or break but must also be smooth with no corners or cusps. Since the conditions of differentiability of $f(x)$ are stricter than those implied by continuity, it is natural to ask whether every differentiable function is continuous.

■ **Theorem** If $f(x)$ is differentiable at $x = a$, then it is continuous at $x = a$.

Proof Suppose that $f(x)$ is differentiable at $x = a$. To show that $f(x)$ is continuous at $x = a$, we will show that $\lim_{x \to a} f(x) = f(a)$. Note that

$$\begin{aligned}\lim_{x \to a} [f(x) - f(a)] &= \lim_{x \to a} \left[\frac{f(x) - f(a)}{x - a}(x - a)\right] \\ &= \lim_{x \to a} \frac{f(x) - f(a)}{x - a} \lim_{x \to a} (x - a).\end{aligned}$$

Since $f(x)$ is differentiable at $x = a$, the first limit exists and equals $f'(a)$. Thus

$$\lim_{x \to a} [f(x) - f(a)] = f'(a) \cdot 0 = 0,$$

that is, $\lim_{x \to a} [f(x) - f(a)] = 0$. Since $f(a)$ does not depend on x, $\lim_{x \to a} f(a) = f(a)$. Therefore,

$$\begin{aligned}\lim_{x \to a} f(x) &= \lim_{x \to a} [f(x) - f(a) + f(a)] \\ &= \lim_{x \to a} [f(x) - f(a)] + \lim_{x \to a} f(a) \\ &= 0 + f(a) = f(a),\end{aligned}$$

as we wanted to show. ■

In Example 2.37, we will show that the absolute value function has no derivative at 0 and thus is not a differentiable function. Since the absolute value function is a continuous function, we see that a function can be continuous at a point without being differentiable there; that is, the converse of the preceding theorem is not true: *continuity does* ***not*** *imply differentiability.*

If a function has a "corner," where its graph seems to have two tangent lines, then the function is not differentiable at that point. None of the functions shown in Figure 2.32 are differentiable at $x = 3$.

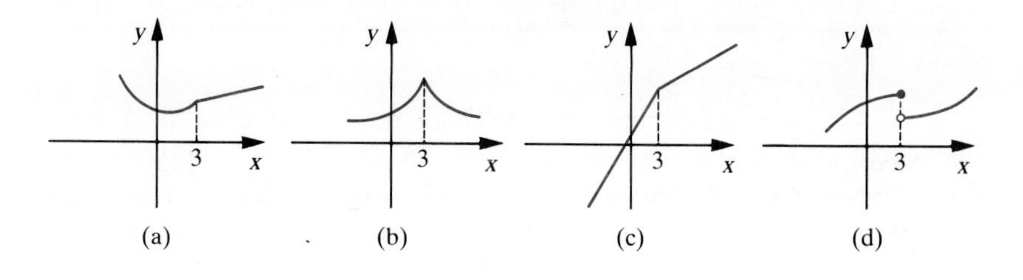

Figure 2.32 2.32
Several functions that are not differentiable at $x = 3$

Another way of stating the continuity theorem is that if a function is not continuous at a point, then it cannot be differentiable at that point. Since continuity is often easier to show than differentiability, we can use this result, discontinuity at a point, to determine that a function is not differentiable at that point.

Example 2.35 Determine if $f(x)$ is differentiable at $x = 2$ where

$$f(x) = \begin{cases} 2x^2 + 4x + 3 & \text{for } x \le 2 \\ x^2 - 2x + 5 & \text{for } x > 2 \end{cases}$$

Solution We first analyze the function for continuity at $x = 2$ by taking the one-sided limits there. Since the limit theorems apply to one-sided limits as well as to two-sided limits, we have

$$\lim_{x\to 2^-} f(x) = \lim_{x\to 2^-} (2x^2 + 4x + 3) = 2(2)^2 + 4(2) + 3 = 19$$

and $$\lim_{x\to 2^+} f(x) = \lim_{x\to 2^+} (x^2 - 2x + 5) = 2^2 - 2(2) + 5 = 5.$$

Since the one-sided limits do not agree, $\lim_{x\to 2} f(x)$ does not exist, thus $f(x)$ is not continuous at $x = 2$. Therefore, by the theorem, $f(x)$ is not differentiable at $x = 2$. See Figure 2.33. ■

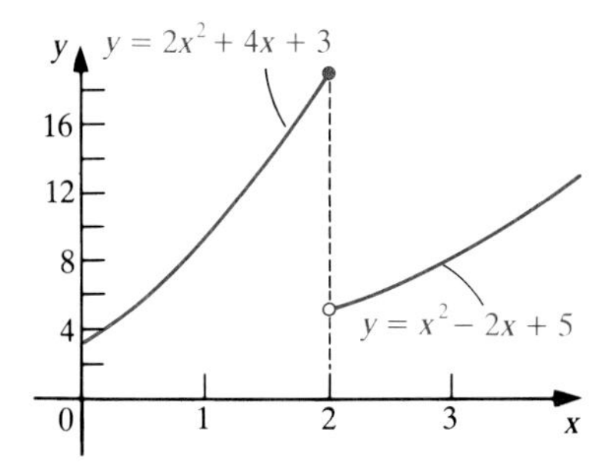

Figure 2.33
Graph of $f(x)$ in Example 2.35

Example 2.36 Determine if $f(x)$ is differentiable at $x = 2$ where

$$f(x) = \begin{cases} 2x^2 + 3x + 1 & \text{for } x \le 2 \\ 3x^2 - 5x + 13 & \text{for } x > 2 \end{cases}$$

Solution We first analyze the function for continuity at $x = 2$ by taking the one-sided limits there. Since

$$\lim_{x\to 2^-} f(x) = 2(2)^2 + 3 \cdot 2 + 1 = 15,$$

and $$\lim_{x\to 2^+} f(x) = 3(2)^2 - 5 \cdot 2 + 13 = 15,$$

and since $f(2) = 15$, we see that $f(x)$ is continuous at $x = 2$. Therefore, $f(x)$ *may* be differentiable there. Consider the two one-sided limits of the "difference quotient"

$$\frac{f(x) - f(2)}{x - 2}.$$

To be differentiable at $x = 2$, both one-sided limits of this expression must exist and be equal. Since $f(x) = 2x^2 + 3x + 1$ when $x < 2$, the left-hand limit of this expression will equal the derivative of $2x^2 + 3x + 1$ at $x = 2$; that is,

$$\lim_{x\to 2^-} \frac{f(x) - f(2)}{x - 2} = (4x + 3)\Big|_{x=2} = 4 \cdot 2 + 3 = 11.$$

Hence the left-hand limit is 11. Since $f(x) = 3x^2 - 5x + 13$ when $x > 2$ the right-hand limit of the difference quotient will equal the derivative

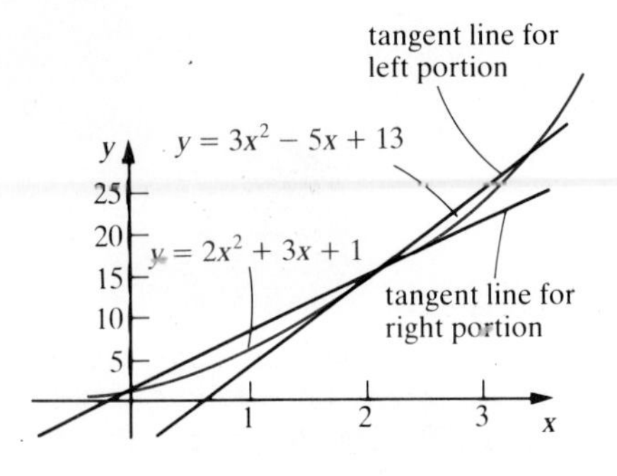

Figure 2.34
Graph of $f(x)$ in Example 2.36. The lines have slopes equal to the left- and right-hand limits of the difference quotient of $f(x)$ at $x = z$.

of $3x^2 - 5x + 13$ evaluated at $x = 2$. This evaluation yields $6(2) - 5 = 7$. Since the one-sided limits do not agree, the limit

$$\lim_{x \to 2} \frac{f(x) - f(2)}{x - 2}$$

does not exist, and hence $f(x)$ is not differentiable at $x = 2$. See Figure 2.34. ■

Example 2.37 Let $f(x) = |x|$. Find $f'(x)$ for $x \neq 0$ and show that $f(x)$ is not differentiable at $x = 0$.

Solution Recall from the definition of absolute value given in Section 1.5 that if $x > 0$, then we have $f(x) = x$ so $f'(x) = 1$ for $x > 0$. Similarly, if $x < 0$, then $f(x) = -x$ and $f'(x) = -1$. Since $f'(x) = 1$ for all $x > 0$,

$$\lim_{x \to 0^+} f'(x) = 1,$$

while $f'(x) = -1$ for all $x < 0$ implies

$$\lim_{x \to 0^-} f'(x) = -1.$$

These two one-sided limits do not agree, so we conclude that $f(x) = |x|$ is *not* differentiable at $x = 0$. Figure 2.35 shows the graphs of $f(x)$ and $f'(x)$. The "corner" at the origin in the graph of $f(x)$ accounts for the lack of differentiability there. In Section 2.4 we showed that the derivative of $f(x)$ equals the slope of the line tangent to the graph of $f(x)$ at x. This function apparently has *two* tangent lines at the origin; if it were differentiable at 0, then the function would have only one tangent line (with slope $f'(0)$). ■

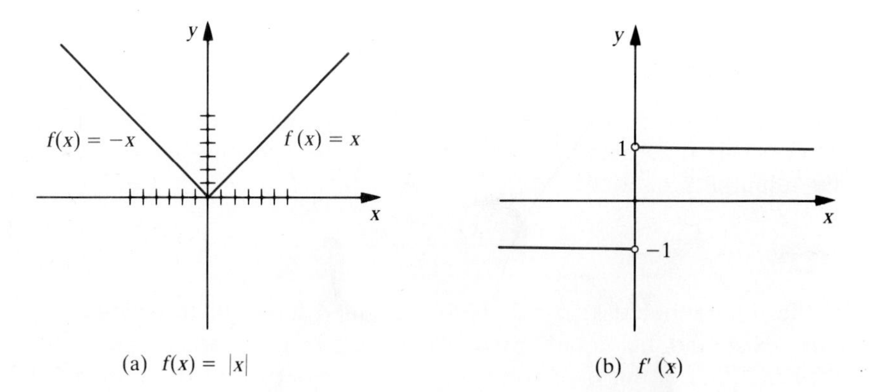

Figure 2.35
Graphs of $f(x)$ and $f'(x)$ for $f(x) = |x|$

We have mentioned that a function which has a corner at some point is not differentiable there. However, a function could have a tangent line at a point and still fail to be differentiable there. The graphs in Figure 2.36 show two examples of a function that fails to have a derivative at $x = a$ because the tangent line is vertical at $x = a$, and so has an undefined slope.

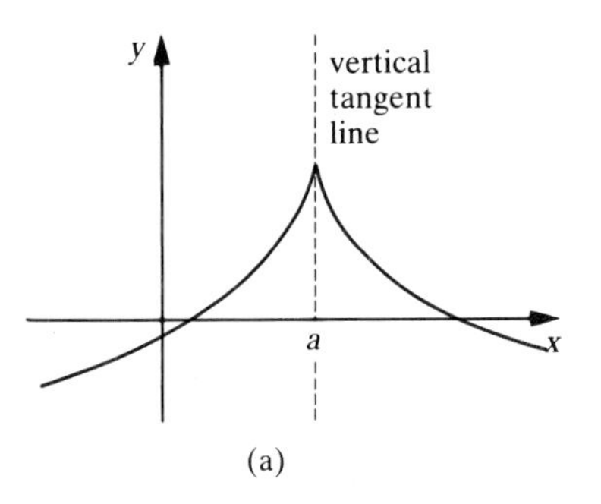

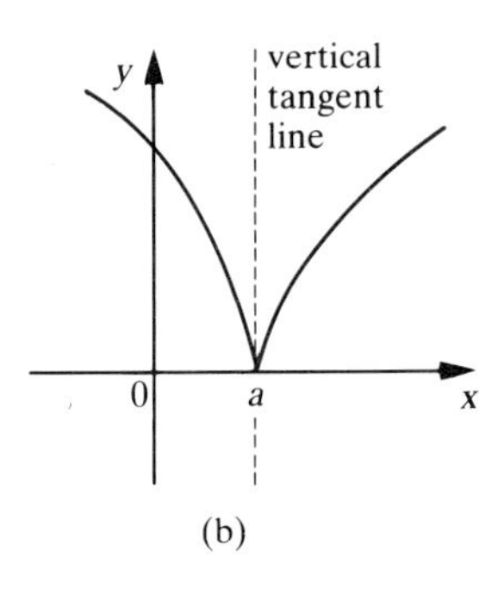

Figure 2.36
Graphs of functions that have vertical tangent lines at $x = a$ and are not differentiable at a

Exercises

1. Suppose that the 18 B.C. census records were lost for the ancient city of Nesral, and the population $P(x)$ in thousands during the year x B.C. is given by

$$P(x) = \frac{x^2 - 14x - 72}{2x - 36}.$$

Assuming that nothing extraordinary occurred in 18 B.C., what was the population of Nesral in that year?

2. A missile is recovered after a test flight, and the tape that records the missile's altitude during flight has been damaged so that the altitude at $t = 24$ seconds after launch cannot be read. By analyzing the data, the altitude $a(t)$ (in feet) at t seconds after launch is determined to be

$$a(t) = \frac{18t^3 - 408t^2 - 576t}{3t - 72}.$$

Find the altitude at $t = 24$.

Exer. 3–15: Determine if the polynomial or rational function is continuous at the given value of a. If the function is continuous, find the limit as x approaches a.

3. $f(x) = x^4 - 3x^2 + 2x + 5;\quad a = 2$
4. $f(x) = 5x^3 - 7x^2 + 2x + 1;\quad a = 3$
5. $f(x) = \dfrac{2x}{x+5};\quad a = 7$
6. $f(x) = \dfrac{3x^2 + 5x - 1}{x + 3};\quad a = 4$
7. $f(x) = \dfrac{x^2 - 4x + 3}{x - 3};\quad a = 3$
8. $f(x) = \dfrac{x^2 + 2x - 4}{x^2 + 3};\quad a = 2$
9. $f(x) = \dfrac{x^2 - 4x - 5}{x^2 - 3x - 10};\quad a = 5$
10. $f(x) = \dfrac{x^2 + 3x - 10}{x^2 + x + 12};\quad a = 3$
11. $f(x) = \dfrac{x^2 + 2x - 8}{x^2 + 5x + 4};\quad a = -4$
12. $f(x) = \dfrac{x^3 - x + 1}{x^2 + x - 6};\quad a = 2$
13. $f(x) = \dfrac{x^2 + 5x + 6}{x^2 + 4x + 3};\quad a = -3$
14. $f(x) = \dfrac{x^3 - 8}{x - 2};\quad a = -2$
15. $f(x) = \dfrac{x^4 - 16}{x^2 + 4};\quad a = -2$

Exer. 16–30: Determine if the function is differentiable at the indicated value. If the function is not differentiable, determine if it is continuous there.

16. $f(x) = \begin{cases} 6x^2 - 3x + 4 & \text{for } x \le 2 \\ x^2 + 3x + 6 & \text{for } x > 2 \end{cases}$ at $x = 2$
17. $f(x) = \begin{cases} 4x^2 - 2x - 4 & \text{for } x < 1 \\ 2x^2 - 6x + 2 & \text{for } x \ge 1 \end{cases}$ at $x = 1$
18. $f(x) = \begin{cases} 2x^2 + 5x - 6 & \text{for } x \le -1 \\ -x^2 - x - 9 & \text{for } x > -1 \end{cases}$ at $x = -1$ and $x = 1$
19. $f(x) = \begin{cases} -2x^2 + 12x - 8 & \text{for } x \le 3 \\ 3x^2 + x - 20 & \text{for } x > 3 \end{cases}$ at $x = 3$ and $x = 4$
20. $f(x) = \begin{cases} 3x^2 - 6x - 9 & \text{for } x < -2 \\ x^2 - 14x - 17 & \text{for } x \ge -2 \end{cases}$ at $x = 2$ and $x = -2$
21. $f(x) = \begin{cases} 5x^2 + 4x - 8 & \text{for } x \le 2 \\ 2x^2 + 16x - 20 & \text{for } x > 2 \end{cases}$ at $x = 0$ and $x = 2$
22. $f(x) = \begin{cases} x^2 + x & \text{for } x < 2 \\ x + 4 & \text{for } x \ge 2 \end{cases}$ at $x = 2$
23. $f(x) = \begin{cases} 2x^2 - x & \text{for } x \le 1 \\ -2x + 3 & \text{for } x > 1 \end{cases}$ at $x = 1$

24. $f(x) = \dfrac{x^2 + 4x + 4}{x^2 - 4}$ at $x = -2$

25. $f(x) = \dfrac{x^2 + 4x + 4}{x^2 - 4}$ at $x = 2$

26. $f(x) = \begin{cases} \dfrac{x^2 - 2x - 15}{x^2 - 3x - 10} & \text{for } x \neq 5 \\ 8/7 & \text{for } x = 5 \end{cases}$ at $x = 5$

27. $f(x) = \dfrac{x^2 - 2x - 5}{x^2 - 3x - 10}$ at $x = -2$

28. $f(x) = |x - 2|$ at $x = 2$

29. $f(x) = \dfrac{|x|}{x + 5}$ at $x = 0$

30. $f(x) = |x^2 + 4|$ at $x = -2$

Summary

Given a function $f(x)$ and a real number a, we say that L is the **limit** of $f(x)$ as x approaches a, and we write

$$\lim_{x \to a} f(x) = L,$$

provided $f(x)$ gets closer and closer to L as x approaches a. If $f(x)$ gets closer and closer to L as x approaches a from the left, then L is the **left-hand limit**

$$\lim_{x \to a^-} f(x) = L.$$

If $f(x)$ gets closer and closer to L as x approaches a from the right, then L is the **right-hand limit**

$$\lim_{x \to a^+} f(x) = L.$$

The limit does not always exist at $x = a$ even if the function is defined at a; however, the limit may exist even if $f(x)$ is not defined at a.

The **derivative** of $f(x)$ at a is defined by

$$f'(a) = \lim_{x \to a} \frac{f(x) - f(a)}{x - a}.$$

An equivalent definition of the derivative $f'(x)$ is given by the limit

$$f'(x) = \lim_{h \to 0} \frac{f(x + h) - f(x)}{h}.$$

Other notations for the derivative of $y = f(x)$ include

$$\frac{df(x)}{dx} \qquad \frac{dy}{dx} \qquad f'(x) \qquad y'$$

The power rule states that if $f(x) = x^r$, then $f'(x) = rx^{r-1}$. Other important rules of derivatives include:

$$\frac{d}{dx} cf(x) = c \frac{d}{dx} f(x), \quad c \text{ a constant;}$$

$$\frac{d}{dx}[f(x) \pm g(x)] = \frac{df(x)}{dx} \pm \frac{dg(x)}{dx}.$$

The equation of the **line tangent to the graph** of $f(x)$ at $x = a$, assuming that such a line exists, is

$$y - f(a) = f'(a)(x - a).$$

Whenever $f'(x) > 0$ on an interval, $f(x)$ is **increasing** on that interval. Whenever $f'(x) < 0$ on an interval, $f(x)$ is **decreasing** on that interval. Points where $f'(x)$ equals zero or does not exist are **critical points**.

Many applications of the derivative are based on the observation that the **instantaneous rate of change** of the function $f(t)$ with respect to t at $t = a$ is $f'(a)$. For example, if $s(t)$ is the distance-traveled function, then $s'(t)$ is the velocity function.

A function $f(x)$ is **continuous** at $x = a$ if and only if $f(a)$ exists, $\lim_{x \to a} f(x)$ exists, and $\lim_{x \to a} f(x) = f(a)$. If $f'(a)$ exists, then $f(x)$ is **differentiable** at $x = a$. If a function is differentiable at $x = a$, then it must be continuous there.

Review Exercises

Exer. 1–3: Find the limit or explain why it does not exist.

1. $\lim_{x \to -2} \dfrac{x + 2}{3x - 5}$

2. $\lim_{t \to 3} \dfrac{t + 3}{t - 3}$

3. $\lim_{u \to -2} \dfrac{u^2 + u - 2}{u^2 + 3u + 2}$

Exer. 4–6: Find the one-sided limit or explain why it does not exist.

4. $\lim_{x \to 2^-} (x - 2)$

5. $\lim_{u \to 4^+} \frac{u-4}{u-4}$

6. $\lim_{t \to 3^-} \frac{|t-3|+t}{t+2}$

Exer. 7–9: Given $\lim_{x \to a} f(x) = 3$ and $\lim_{x \to a} g(x) = -5$, find the indicated limit.

7. $\lim_{x \to a} [f(x)g(x) + 4f(x)]$
8. $\lim_{x \to a} ([f(x)]^2 + 3g(x) - 2)$
9. $\lim_{x \to a} ([f(x) + g(x)]^3 + 12)$

10. Given $f(x) = x^2 + x$ and $a = 2$, find $\frac{f(x) - f(a)}{x - a}$ and simplify.
11. Given $f(x) = 4 - x^2$, use the definition of the derivative to find $f'(-1)$.
12. Given $f(x) = \frac{3}{x}$, find $\frac{f(x+h) - f(x)}{h}$ and simplify.
13. Given $f(x) = 4/(x+1)$ and $a = 4$, find $f'(a)$ by evaluating $\lim_{h \to 0} \frac{f(a+h) - f(a)}{h}$.

Exer. 14–19: Find $f'(x)$.

14. $f(x) = 3x^4 - 7$
15. $f(x) = 3x^{5/3}$
16. $f(x) = 9x^{-2}$
17. $f(x) = 3x^{-2/3} + 6x^{5/3}$
18. $f(x) = 6/\sqrt{3x} + x\sqrt{x}$
19. $f(x) = (x+3)/x$
20. Given $f(x) = 2x^{-3} + 3x^2$, find $f'(4)$.
21. Given $f(x) = 3x^2 - x + 7$, find $f'(-2)$.

22. Find the equation of the tangent line to the curve $y = 2 - x^2$ at the point $(2, -2)$.

Exer. 23–24: Find the value of x for which the tangent line to the graph of $f(x)$ satisfies the given condition.

23. $f(x) = 5x^2 + 20x - 6$; horizontal
24. $f(x) = 3x^2 + 15x - 2$; slope = 18

Exer. 25–26: Find the equation of the line tangent to the graph of $f(x)$ that passes through the origin.

25. $f(x) = -2x^2 + 5x - 6$
26. $f(x) = 3x^2 + 10x + 12$

27. For what values of x is the graph of $f(x) = -3x^2 + 12x + 6$ increasing, and for what values of x is it decreasing? Find the critical point(s) of $f(x)$.
28. For what values of x is the graph of $f(x) = x^3 - 3x - 1$ increasing, and for what values is it decreasing? Find the critical points of $f(x)$.

Exer. 29–34: Find the critical values of the function.

29. $f(x) = x^2 - 4x + 5$
30. $f(x) = x^2 - 5x + 1$
31. $f(t) = t^2 + 6t - 3$
32. $g(t) = t^4 - 2t^2$
33. $C(x) = x^3 + 3x^2 - 9x + 1$
34. $W(x) = x^3 + 3x$

35. The number of people (in thousands) infected by bacterial infection is $v = 2t^2 + 38t + 60$, where t is the number of days after a treatment campaign is initiated. Is the number of people infected increasing or decreasing 12 days after the campaign is started?

Exer. 36–42: Determine if the function is continuous at the indicated value of a. If so, determine if it is differentiable there.

36. $f(x) = 3x^2 - 2x + 5$ at $a = 2$
37. $f(x) = \frac{x^2 - 4}{x - 2}$ at $a = 2$
38. $f(t) = \begin{cases} \frac{t^2 - 4}{t - 2} & \text{for } t \neq 2 \\ 4 & \text{for } t = 2 \end{cases}$ at $a = 2$
39. $f(u) = |u - 4|$ at $a = 4$
40. $f(x) = \begin{cases} 3x^2 - 4x + 6 & \text{for } x \geq 3 \\ -2x^2 + 3x + 5 & \text{for } x < 3 \end{cases}$ at $a = 3$
41. $f(x) = \begin{cases} 2x^2 + 5x - 6 & \text{for } x > -2 \\ -x^2 - 7x - 18 & \text{for } x < -2 \end{cases}$ at $a = -2$
42. $f(t) = \begin{cases} -4t^2 + 3t - 2 & \text{for } t > -1 \\ 5t^2 + 10t - 4 & \text{for } t \leq -1 \end{cases}$ at $a = -1$

Facing page: **The rate of growth of a city's park system should be related to the rate of growth of a city's population. See Example 3.12.**

Techniques and Applications of Differentiation

· T · H · R · E · E ·

We have discussed the definition of the derivative, how to differentiate relatively simple functions, and applications of the derivative in a few situations. In this chapter, we will expand our ability to compute derivatives by introducing the product rule, the quotient rule, and the chain rule. This completes the rules we will need to differentiate the functions used in this text. In the remainder of the chapter we will extend the applications we have already discussed to more complicated functions and to more realistic situations.

3.1 Derivatives of Products and Quotients

Many functions are products or quotients of simpler functions that can be differentiated easily. In this section we present methods for finding the derivatives of these more complicated functions from the derivatives of simpler ones.

First consider the derivative of the product of two functions. Unfortunately, the derivative of the product of two functions is *not* the product of their derivatives. This fact may be observed by noting that $f(x)=x^2$ is the product of x and x; that is, $f(x)=x^2=(x)(x)$. Note that the derivative of each factor is 1. If we could find the derivative of a product by multiplying the derivatives of the factors, then the derivative of $f(x)=x^2$ would be $f'(x)=(1)(1)$, instead of the correct result of $f'(x)=2x$. The correct method is given by the product rule, as follows.

The Product Rule

If $f(x)$ and $g(x)$ are differentiable functions and

$$P(x)=f(x)g(x),$$

then $P(x)$ is differentiable and

$$P'(x)=f(x)g'(x)+g(x)f'(x).$$

We can also write this product rule using the d/dx notation for the derivative:

$$\frac{d}{dx}f(x)g(x)=f(x)\frac{d}{dx}g(x)+g(x)\frac{d}{dx}f(x)$$

Thus *the derivative of the product of two functions is the first function times the derivative of the second, plus the second function times the derivative of the first.* We will give a proof of this rule at the end of this section.

Example 3.1 Compute the derivative of each function:
(a) $P(x) = x^2(x^3 - 2)$
(b) $P(x) = (x + 2)(x^3 - x)$
(c) $P(x) = (3x^4 + 2x^2 + 1)(5x^3 + 6)$

Solution **(a)** Here $P(x)$ is the product of x^2 and $x^3 - 2$. If we let $f(x) = x^2$ and $g(x) = x^3 - 2$, then $P(x) = f(x)g(x)$. The product rule states that

$$\begin{aligned} P'(x) &= f(x)g'(x) + g(x)f'(x) \\ &= x^2(3x^2) + (x^3 - 2)2x \\ &= 3x^4 + 2x^4 - 4x \\ &= 5x^4 - 4x. \end{aligned}$$

To check that the product rule has given the right answer, we can find the product of x^2 and $x^3 - 2$ first, and then differentiate:

$$\begin{aligned} P(x) &= x^2(x^3 - 2) = x^5 - 2x^2 \\ P'(x) &= 5x^4 - 4x. \end{aligned}$$

(b) It is not necessary to introduce $f(x)$ and $g(x)$ as in the solution to part (a). We can write the derivative of $P(x) = (x + 2)(x^3 - x)$ as

$$\begin{aligned} P'(x) &= \frac{d}{dx}[(x + 2)(x^3 - x)] \\ &= (x + 2)\frac{d}{dx}(x^3 - x) + (x^3 - x)\frac{d}{dx}(x + 2) \\ &= (x + 2)(3x^2 - 1) + (x^3 - x)(1) \\ &= 3x^3 + 6x^2 - x - 2 + x^3 - x \qquad \text{(multiply factors)} \\ &= 4x^3 + 6x^2 - 2x - 2. \end{aligned}$$

(c) Proceeding as in part (b), we obtain

$$\begin{aligned} P'(x) &= \frac{d}{dx}[(3x^4 + 2x^2 + 1)(5x^3 + 6)] \\ &= (3x^4 + 2x^2 + 1)\frac{d}{dx}(5x^3 + 6) + (5x^3 + 6)\frac{d}{dx}(3x^4 + 2x^2 + 1) \\ &= (3x^4 + 2x^2 + 1)(15x^2) + (5x^3 + 6)(12x^3 + 4x) \\ &= 105x^6 + 50x^4 + 72x^3 + 15x^2 + 24x. \qquad \text{(multiply)} \end{aligned}$$

■

Example 3.2 Determine the equation of the tangent line to the graph of $f(x) = (x^2 + x)(x^2 - 2)$ at $x = 1$.

Solution We first determine $f'(x)$:

$$\begin{aligned} f'(x) &= \frac{d}{dx}[(x^2 + x)(x^2 - 2)] \\ &= (x^2 + x)\frac{d}{dx}(x^2 - 2) + (x^2 - 2)\frac{d}{dx}(x^2 + x) \\ &= (x^2 + x)2x + (x^2 - 2)(2x + 1). \end{aligned}$$

Thus the slope of the tangent line at $x = 1$ is

$$f'(1) = (1^2 + 1)2(1) + (1^2 - 2)[2(1) + 1] = 1.$$

Since $f(1) = -2$, the point of tangency is $(1, -2)$ and so the equation of the tangent line is

$$y + 2 = 1(x - 1).$$

In slope-intercept form, this equation is $y = x - 3$. ∎

Suppose we have the product of three differentiable functions $u(x)$, $v(x)$, and $w(x)$, and we want to find the derivative of this product. We write the product $u(x)v(x)w(x)$ as $u(x)[v(x)w(x)]$ and considering $f(x)$ to be $u(x)$ and $g(x)$ to be $v(x)w(x)$ in the product rule. Then

$$\begin{aligned}\frac{d}{dx}[u(x)v(x)w(x)] &= u(x)\frac{d}{dx}[v(x)w(x)] + [v(x)w(x)]\frac{d}{dx}u(x)\\ &= u(x)[v(x)w'(x) + w(x)v'(x)] + v(x)w(x)u'(x)\end{aligned}$$

or, expanding and rearranging terms,

$$\begin{aligned}\frac{d}{dx}[u(x)v(x)w(x)] = u'(x)v(x)w(x) &+ u(x)v'(x)w(x)\\ &+ u(x)v(x)w'(x). \qquad (1)\end{aligned}$$

Note that the order is "$u\,v\,w$" in each term and that the differentiated factor is the first factor in the first term, the second factor in the second term, and the third factor in the third term. Generalizations to more than three factors also follow this scheme (as does the product of two factors).

Example 3.3 Use the result in equation (1) to find the derivative of $f(x) = (4x^5 + 3x^2 + 4)(7x^4 - x^3)(x^2 + 1)$.

Solution Let $u(x)$ be the factor $(4x^5 + 3x^2 + 4)$, let $v(x)$ be the factor $(7x^4 - x^3)$, and let $w(x)$ denote the third factor $(x^2 + 1)$. Using equation (1) and using d/dx instead of prime notation, we have

$$\begin{aligned}f'(x) &= \frac{d}{dx}[(4x^5 + 3x^2 + 4)(7x^4 - x^3)(x^2 + 1)]\\ &= \frac{d}{dx}[(4x^5 + 3x^2 + 4)](7x^4 - x^3)(x^2 + 1)\\ &\quad + (4x^5 + 3x^2 + 4)\frac{d}{dx}[(7x^4 - x^3)](x^2 + 1)\\ &\quad + (4x^5 + 3x^2 + 4)(7x^4 - x^3)\frac{d}{dx}[(x^2 + 1)]\end{aligned}$$

or

$$\begin{aligned}f'(x) = (20x^4 + 6x)(7x^4 - x^3)(x^2 + 1) &+ (4x^5 + 3x^2 + 4)(28x^3 - 3x^2)(x^2 + 1)\\ &+ (4x^5 + 3x^2 + 4)(7x^4 - x^3)(2x)\end{aligned}$$

Since this derivative was obtained for illustrative purposes only, we will not simplify the result. ■

Now suppose that we need to differentiate a function $f(x)$ that is a quotient of two functions; for example,

$$f(x) = \frac{5x^3 + 2x - 4}{2x^2 + 8x + 5}.$$

As in the case of the product rule, the most natural (but incorrect) guess is that if $f(x) = g(x)/h(x)$, then $f'(x) = g'(x)/h'(x)$. However, if this were correct then the derivative of $f(x) = 1/x$ would be $0/1 = 0$ instead of $-1/x^2$.

The Quotient Rule

If $f(x)$ and $g(x)$ are differentiable functions and

$$Q(x) = \frac{f(x)}{g(x)},$$

the $Q(x)$ is differentiable and

$$Q'(x) = \frac{g(x)f'(x) - f(x)g'(x)}{[g(x)]^2}.$$

We may also write this rule using the d/dx notation:

$$\frac{d}{dx}Q(x) = \frac{g(x)\frac{d}{dx}f(x) - f(x)\frac{d}{dx}g(x)}{[(g(x)]^2}.$$

Thus the derivative of a quotient of two differentiable functions is *the denominator times the derivative of the numerator, minus the numerator times the derivative of the denominator, all divided by the square of the denominator.*

Example 3.4 Compute the derivative of each function:

(a) $Q(x) = \dfrac{x^2}{x-3}$ **(b)** $Q(x) = \dfrac{x^2 - 4}{6x}$ **(c)** $Q(x) = \dfrac{5x^3 + 2x - 4}{2x^2 + 8x + 5}$

Solution **(a)** Here $f(x)=x^2$ and $g(x)=x-3$. Hence

$$Q'(x)=\frac{g(x)\frac{d}{dx}f(x)-f(x)\frac{d}{dx}g(x)}{[g(x)]^2}$$

$$=\frac{(x-3)\frac{d}{dx}(x^2)-(x^2)\frac{d}{dx}(x-3)}{(x-3)^2}$$

$$=\frac{(x-3)2x-x^2(1)}{(x-3)^2}=\frac{2x^2-6x-x^2}{(x-3)^2}=\frac{x^2-6x}{(x-3)^2}.$$

(b)

$$Q'(x)=\frac{6x\frac{d}{dx}(x^2-4)-(x^2-4)\frac{d}{dx}(6x)}{(6x)^2}$$

$$=\frac{(6x)(2x)-(x^2-4)6}{(6x)^2}$$

$$=\frac{12x^2-6x^2+24}{36x^2}=\frac{6(x^2+4)}{36x^2}=\frac{x^2+4}{6x^2}.$$

(c) $Q'(x)$

$$=\frac{(2x^2+8x+5)\frac{d}{dx}(5x^3+2x-4)-(5x^3+2x-4)\frac{d}{dx}(2x^2+8x+5)}{(2x^2+8x+5)^2}$$

$$=\frac{(2x^2+8x+5)(15x^2+2)-(5x^3+2x-4)(4x+8)}{(2x^2+8x+5)^2}$$ ■

Example 3.5 An electronics manufacturing company has a total worth of $w(x)$ million dollars x years after incorporation, where

$$w(x)=\frac{20x^{3/2}}{x+16}.$$

What is the rate of change of its total worth after 4 years?

Solution The rate of change is the derivative of $w(x)$ with respect to x:

$$\frac{dw(x)}{dx}=\frac{(x+16)\frac{d}{dx}(20x^{3/2})-20x^{3/2}\frac{d}{dx}(x+16)}{(x+16)^2}$$

$$=\frac{(x+16)[20(\frac{3}{2})x^{1/2}]-20x^{3/2}(1)}{(x+16)^2}$$

$$=\frac{(x+16)(30x^{1/2})-20x\cdot x^{1/2}}{(x+16)^2} \qquad \text{(use } x^{3/2}=x\cdot x^{1/2}\text{)}$$

$$=\frac{x^{1/2}(30x+480-20x)}{(x+16)^2} \qquad \text{(factor out } x^{1/2}\text{, simplify)}$$

$$=\frac{x^{1/2}(10x+480)}{(x+16)^2}=\frac{10\sqrt{x}(x+48)}{(x+16)^2}.$$

Thus, the rate of change of $w(x)$ when $x = 4$ is

$$\left.\frac{dw}{dx}\right|_{x=4} = \frac{10(2)(4+48)}{20^2}$$

$$= 2.6 \text{ million dollars per year.}$$

■

Proof of the Product Rule Let $P(x) = f(x)g(x)$ and suppose that both $f(x)$ and $g(x)$ are differentiable functions. Recall that this means that

$$f'(x) = \lim_{h\to 0} \frac{f(x+h) - f(x)}{h} \quad \text{and} \quad g'(x) = \lim_{h\to 0} \frac{g(x+h) - g(x)}{h}.$$

From the definition of the derivative, we have

$$P'(x) = \lim_{h\to 0} \frac{P(x+h) - P(x)}{h}$$

$$= \lim_{h\to 0} \frac{f(x+h)g(x+h) - f(x)g(x)}{h}.$$

We now subtract and add the term $f(x)g(x+h)$ to the terms in the numerator. (We will see momentarily why this is useful.) The above expression for $P'(x)$ becomes

$$P'(x) = \lim_{h\to 0} \frac{f(x+h)g(x+h) - f(x)g(x+h) + f(x)g(x+h) - f(x)g(x)}{h}.$$

We factor $g(x+h)$ from the first two terms and $f(x)$ from the last two terms in the numerator, obtaining

$$P'(x) = \lim_{h\to 0} \left[\frac{f(x+h) - f(x)}{h} g(x+h) + f(x) \frac{g(x+h) - g(x)}{h}\right].$$

(The terms we subtracted and added made this factoring possible.) Splitting the limit into two terms, and using the fact that the limit of a product is equal to the product of the limits, we have

$$P'(x) = \left(\lim_{h\to 0} \frac{f(x+h) - f(x)}{h}\right)\left(\lim_{h\to 0} g(x+h)\right)$$

$$+ \left(\lim_{h\to 0} f(x)\right)\left(\lim_{h\to 0} \frac{g(x+h) - g(x)}{h}\right).$$

As noted at the start of this proof, the first limit is $f'(x)$. Since $g(x)$ is differentiable, it is also continuous, so $g(x+h) \to g(x)$ as $h \to 0$. Thus, the second limit is $g(x)$. For the third limit, $f(x) \to f(x)$ as $h \to 0$. The fourth limit is $g'(x)$, as noted at the beginning of the proof. Therefore,

$$P'(x) = f'(x)g(x) + f(x)g'(x).$$

This expression is a rearrangement of that given in the product rule. ■

Exercises

Exer. 1–14: Compute the derivative of the function using the product rule and power rule.

1. $f(x) = 2x^2(x^3 - 4)$
2. $f(x) = 4x(x - x^3 + 2)$
3. $f(x) = (2x^2 - 1)x^3$
4. $f(x) = (x - 1)(3x^2 - 2)$

5. $f(t) = (t + 1)(t^2 + 1)$
6. $f(x) = (x^2 + 2x - 1)(x^3 + 3)$
7. $f(x) = (3x + \frac{1}{3})\left(x + \frac{1}{x}\right)$
8. $f(t) = (2t^2 + t^{2/3})\left(3t + \frac{1}{t}\right)$
9. $f(u) = \sqrt[5]{u}\,(u^5 + 3u)$
10. $f(x) = (x^{3/5} + x^{2/3})(2x^4 - 3x)$
11. $f(r) = (2r^3 + 4r^5)(r^{-2} - 3r^{-1})$
12. $f(u) = (4u^2 - 5u^{1/3})(3u^{-3} + 5u^2)$
13. $f(x) = (2x^2 - 1)(2x^2 - 1)$
14. $f(x) = (3x^3 + 2x + 5)^2$

Exer. 15–28: Compute the derivative of the function.

15. $f(x) = \dfrac{x}{x + 1}$
16. $f(x) = \dfrac{x + 1}{x}$
17. $f(x) = \dfrac{3x^{-1}}{x + 3}$
18. $f(x) = \dfrac{\sqrt[3]{x}}{x^2 + 2}$
19. $f(t) = \dfrac{7t^2 + 6}{1 - 2t^2}$
20. $f(t) = \dfrac{t + \sqrt{t}}{t^2 - 1}$
21. $f(x) = 6x + \dfrac{3x^2}{2x - 1}$
22. $f(x) = \left(x + \dfrac{1}{x}\right)(3 - x^2)$
23. $f(r) = 7\left(\dfrac{r^2 - 1}{2r + 3}\right) + 3$
24. $f(p) = \dfrac{(3p + 4)(p^2 - 1)}{p^3 - 3p}$
25. $f(x) = \left(\dfrac{x^2}{x - 3}\right)\left(\dfrac{\sqrt{x}}{2 + x^2}\right)$
26. $f(x) = x + 3\left(\dfrac{\sqrt{x}}{x^2 - 1}\right)$
27. $f(u) = 3u^{2/3}\left(\dfrac{u + 14}{3 - u^2}\right)$
28. $f(t) = \left(t^2 + \dfrac{2}{t^2}\right)\left(\dfrac{t}{t^2 - 1}\right)$

Exer. 29–40: Use the function to determine the indicated derivative.

29. $y = x^2(x^{1/3} - 2);\quad \left.\dfrac{dy}{dx}\right|_{x=1}$
30. $y = (x^2 - 1)(2x + 3);\quad \left.\dfrac{dy}{dx}\right|_{x=2}$
31. $y = \sqrt[3]{t}\,(1 - t^3);\quad \left.\dfrac{dy}{dt}\right|_{t=8}$
32. $y = (2x^2 - 1)(x + 4);\quad \left.\dfrac{d^2y}{dx^2}\right|_{x=0}$
33. $f(x) = (3x^2 - 4)(x - x^4);\quad f'(2)$
34. $f(p) = (p - \sqrt{p})(3p^2 + 1);\quad f'(4)$
35. $y = \dfrac{u - 1}{u + 4};\quad \left.\dfrac{dy}{du}\right|_{u=4}$
36. $y = \dfrac{x^2 - 4}{3x};\quad \left.\dfrac{dy}{dx}\right|_{x=-1}$
37. $y = \dfrac{x\sqrt[3]{x}}{x - 4};\quad \left.\dfrac{dy}{dx}\right|_{x=1}$
38. $f(s) = \left(\dfrac{s^2}{3 - s^2}\right)\left(\dfrac{s - 3}{4}\right);\quad f'(-2)$
39. $f(t) = \dfrac{3t^2 - 6}{4t} + 2;\quad f'(1)$
40. $y = \dfrac{(x^3 - x)(3 - \sqrt[3]{x})}{x^2 + 4};\quad \left.\dfrac{dy}{dx}\right|_{x=1}$

Exer. 41–46: Find the equation of the tangent line to the function corresponding to the given value of x.

41. $f(x) = \dfrac{\sqrt[3]{x}}{x - 4};\quad x = 8$
42. $f(x) = \dfrac{\sqrt[3]{x}}{3 - x^2};\quad x = 1$
43. $f(x) = 2x(x^3 - 2x + 1);\quad x = 1$
44. $f(x) = \dfrac{x^2 - 1}{x + 3};\quad x = 5$
45. $f(x) = x^2(2x^{1/4} - 4);\quad x = 1$
46. $f(x) = \dfrac{x^2(\sqrt{x} - 3)}{x + 1};\quad x = 9$

Exer. 47–50: Use the product rule to find the derivative of the function.

47. $f(x) = (x - 3)(2x + 5)(4x + 1)$
48. $w(x) = x(x^3 + 3)(x^4 - x^2 + 3)$
49. $f(x) = (x^2 - 1)(3x - 5)(x^3 - x)$
50. $f(x) = (x - 3)(x - 4)(x^3 + x^2 - x + 4)$

51. Prove the quotient rule, using the proof of the product rule as a model.
52. The position of a particle moving along the x-axis is given by $x(t) = t^2/(t + 5)$ for $t \geq 0$. Find the velocity of the particle at $t = 3$.
53. The total assets of a company is $A(t)$ million dollars t years after new management took control, where

$$A(t) = \frac{3t^{4/3} + 4}{2t + 3}.$$

What is the rate of change of the assets at $t = 3$?

3.2 The Chain Rule and Related Rates

Many functions are "built" from simpler functions. For example, the function $f(x) = \sqrt{3x + 1}$ uses the square root function and the function $3x + 1$. If we let $g(x) = \sqrt{x}$ and $h(x) = 3x + 1$, then $f(x)$ is $g(h(x))$ because

$$g(h(x)) = \sqrt{h(x)} = \sqrt{3x + 1} = f(x).$$

Both $g(x)$ and $h(x)$ can be differentiated easily. In this section we will discuss how to find the derivative of a function "composed" of simpler functions that can be differentiated easily.

The function $g(h(x))$ is called the **composition** of functions g and h. Essentially, we are substituting $h(x)$ for every occurrence of the variable x in $g(x)$.

Example 3.6 If $g(x) = \sqrt[3]{x}$ and $h(x) = x^3 - 4x$, find the compositions

(a) $g(h(x))$ and **(b)** $h(g(x))$.

Solution **(a)** We obtain the composition $g(h(x))$ by replacing every occurrence of x in $g(x)$ with $h(x)$:

$$g(h(x)) = \sqrt[3]{h(x)}.$$

Thus

$$g(h(x)) = \sqrt[3]{x^3 - 4x}.$$

(b) For the composition $h(g(x))$, we have

$$\begin{aligned} h(g(x)) &= [g(x)]^3 - 4g(x) \\ &= [\sqrt[3]{x}]^3 - 4\sqrt[3]{x} \end{aligned}$$

or, equivalently,

$$h(g(x)) = x - 4\sqrt[3]{x}.$$ ∎

Note that in the previous example $g(h(x)) \neq h(g(x))$. In fact, it is rare that $g(h(x)) = h(g(x))$. See Exercises 57–60 for examples of compositions that are equal.

Example 3.7 Express $f(x) = (x^2 + 3)^4$ as the composition of simpler functions.

Solution This function can be regarded as some expression raised to the fourth power. Thus, we let $g(u) = u^4$ and let $u = h(x) = x^2 + 3$. Then

$$g(h(x)) = [h(x)]^4 = (x^2 + 3)^4 = f(x).$$ ∎

As we have noted, composition of functions enables us to view a complicated function as one composed of simpler functions. Our final differentiation rule, called the chain rule, relates the derivative of a composite function to the derivatives of its components.

The Chain Rule

Let $g(u)$ and $h(x)$ be differentiable functions. Setting $u = h(x)$ produces the function $f(x) = g(h(x))$. Perhaps the two most natural guesses for the derivative of $f(x) = g(h(x))$ are $g'(h(x))$ and $g'(h'(x))$. Neither of these is correct, as we can see from the example

$$\frac{d}{dx}(2x^2)^3 = \frac{d}{dx}8x^6 = 48x^5.$$

Here $g(x) = x^3$ and $h(x) = 2x^2$, so that $g(h(x)) = [h(x)]^3 = (2x^2)^3$. Since $g'(x) = 3x^2$ and $h'(x) = 4x$, we can compute our two "guesses" for the derivative:

$$g'(h(x)) = 3[h(x)]^2 = 3(2x^2)^2 = 12x^4$$

and

$$g'(h'(x)) = 3[h'(x)]^2 = 3(4x)^2 = 48x^2.$$

The correct derivative is $(d/dx)8x^6 = 48x^5$, which indicates that both guesses are wrong. The correct rule for finding the derivative of $g(h(x))$ is as follows.

The Chain Rule

Let $g(u)$ and $u = h(x)$ be differentiable functions and let $f(x) = g(h(x))$. Then $f(x)$ is a differentiable function of x, and

$$f'(x) = g'(u)h'(x) = g'(h(x))h'(x),$$

where $g'(u)$ is the derivative of $g(u)$ with respect to u.

The right side of the chain rule is frequently paraphrased as *the derivative of the outside times the derivative of the inside.* The terms *outside* and *inside* refer to the functions $g(u)$ and $h(x)$, respectively. In the composition $f(x) = g(h(x))$, $g(u)$ is the outside function and $h(x)$ is the inside function. A proof of the chain rule is given at the end of this section.

Example 3.8 Find $f'(x)$ where $f(x) = (2x^2)^3$.

Solution Since $f(x) = g(h(x))$, where $u = h(x) = 2x^2$ and $g(u) = u^3$, we have, by the chain rule,

$$\begin{aligned} f'(x) &= g'(u)h'(x) \\ &= (3u^2)(4x) \\ &= [3(2x^2)^2](4x) \qquad (u = 2x^2) \\ &= 12x(4x^4) = 48x^5. \end{aligned}$$

Since $f(x)$ may also be written as $f(x) = 8x^6$, we see that $f'(x) = 48x^5$ is the correct derivative. ■

Example 3.9 Find $f'(x)$ for each function:

(a) $f(x) = (x^2 + 3x)^5$ **(b)** $f(x) = \sqrt[3]{x^2 + 2x - 1}$ **(c)** $f(x) = x^2(4 - 3x)^{4/3}$

Solution **(a)** This function can be viewed as an expression to the fifth power. Letting $g(u) = u^5$ and $u = h(x) = x^2 + 3x$ in the notation of the chain rule, we have

$$\begin{aligned} f'(x) &= 5u^4 h'(x) \\ &= 5(x^2 + 3x)^4(2x + 3). \end{aligned}$$

(b) We may think of this function as $u = h(x) = x^2 + 2x - 1$ raised to the 1/3 power. Let $g(u) = u^{1/3}$. By the chain rule, the derivative is

$$\begin{aligned} f'(x) &= \tfrac{1}{3}u^{-2/3}h'(x) \\ &= \tfrac{1}{3}(x^2 + 2x - 1)^{-2/3}(2x + 2). \end{aligned}$$

(c) This function is a product of the functions x^2 and $(4 - 3x)^{4/3}$, so we use the product rule:

$$f'(x) = x^2 \frac{d}{dx}[(4 - 3x)^{4/3}] + (4 - 3x)^{4/3}(2x).$$

Now the chain rule is used to find the derivative of $(4 - 3x)^{4/3}$, with $u = h(x) = 4 - 3x$ and $g(u) = u^{4/3}$:

$$\frac{d}{dx}(4 - 3x)^{4/3} = \tfrac{4}{3}(4 - 3x)^{1/3}(-3) = -4(4 - 3x)^{1/3}.$$

Thus $$f'(x) = x^2(-4)(4 - 3x)^{1/3} + 2x(4 - 3x)^{4/3}.$$

For completeness, we simplify this expression by factoring out the common factor $(4 - 3x)^{1/3}$:

$$\begin{aligned} f'(x) &= (4 - 3x)^{1/3}[-4x^2 + 2x(4 - 3x)] \\ &= (4 - 3x)^{1/3}(8x - 10x^2). \end{aligned}$$ ■

In most uses of the chain rule, it is not necessary to write down formally the definitions of $u = h(x)$ and $g(u)$. The next example illustrates this.

Example 3.10 Find the derivative of each function:

(a) $f(x) = (x^3 + x)^5$ **(b)** $w(s) = 4\sqrt{5s^4 - 3s + 5}$

Solution **(a)** The derivative of an expression to the fifth power is 5 times (the expression to the fourth power) times (the derivative of the expression). Thus

$$f'(x) = \frac{d}{dx}(x^3 + x)^5 = 5(x^3 + x)^4(3x^2 + 1).$$

(b) Similarly,

$$\begin{aligned} w'(s) &= \frac{d}{ds} 4\sqrt{5s^4 - 3s + 5} \\ &= 4\frac{d}{ds}(5s^4 - 3s + 5)^{1/2} \\ &= 4 \cdot \tfrac{1}{2}(5s^4 - 3s + 5)^{-1/2}(20s^3 - 3) \\ &= \frac{2(20s^3 - 3)}{\sqrt{5s^4 - 3s + 5}}. \end{aligned}$$ ■

A special case of the chain rule is used so much that an explicit statement is warranted. Suppose that $f(x) = g(h(x))$, where $g(u) = u^r$ for some real number r. By the power rule, given in Chapter 2, $g'(u) = r u^{r-1}$. Thus

$$f'(x) = \frac{d}{dx}[h(x)]^r = [r u^{r-1}]h'(x) = r[h(x)]^{r-1}h'(x).$$

We can state this formula as a general power rule, as follows.

The General Power Rule

Let $h(x)$ be a differentiable function and let r be a real number (a constant). Then

$$\frac{d}{dx}[h(x)]^r = r[h(x)]^{r-1}\frac{dh(x)}{dx}.$$

We now return to the statement of the chain rule given earlier and restate it using different notation. We begin with $y = f(x)$ as the composition of $y = g(u)$ and $u = h(x)$. If both functions are differentiable, then the chain rule states that

$$\frac{dy}{dx} = \frac{dy}{du}\frac{du}{dx}.$$

Example 3.11 Compute dy/dx if $y = 5u^3$ and $u = x^2 - x^3$.

Solution First, $dy/du = 15u^2$ and $du/dx = 2x - 3x^2$. Thus

$$\frac{dy}{dx} = \frac{dy}{du}\frac{du}{dx} = (15u^2)(2x - 3x^2)$$
$$= 15(x^2 - x^3)^2(2x - 3x^2).$$

In the last step, $x^2 - x^3$ replaces u so that dy/dx is expressed entirely in terms of x. ■

Example 3.12 According to a national quality-of-living scale the minimum number n of acres of park land in a city rated "acceptable" is

$$n(P) = 4 + 5P + \tfrac{1}{2}P^{3/2}$$

where P is the population in thousands. Suppose that Eastridge, a city of 100,000, is currently rated as an acceptable city and its population is increasing at the rate of 2100 people per year. At what rate should the Eastridge City Council be purchasing park land to keep its acceptable rating?

Solution Since the population P is measured in thousands, the "current conditions" are $P = 100$ thousand people and $dP/dt = 2.1$ thousand people per year. We are asked to find dn/dt, the rate of change of the amount of park land. Here n is a function of P and P is a function of t, so using the chain rule we can determine dn/dt by

$$\frac{dn}{dt} = \frac{dn}{dP}\frac{dP}{dt}.$$

Since $n(P) = 4 + 5P + \frac{1}{2}P^{3/2}$,

$$\frac{dn}{dP} = 5 + \tfrac{3}{4}P^{1/2} = 5 + 0.75\sqrt{P}.$$

Thus,

$$\frac{dn}{dt} = (5 + 0.75\sqrt{P})\frac{dP}{dt}.$$

Consequently, when $P = 100$,

$$\left.\frac{dn}{dt}\right|_{P=100} = (5 + 7.5)(2.1) = 26.25.$$

Thus, the Eastridge City Council should purchase an additional 26.25 acres of park land this year to keep its rating. ■

Related Rates

The previous example illustrates how we can use the chain rule to study the rate at which a variable—call it y—is changing with respect to time. If we can express y directly as a function of time t, then, as we have seen, dy/dt gives the rate at which y is changing with respect to time. However, in some applications the only information available is the rate at which a third variable, say x, is changing with respect to t and a functional relationship between y and x. We refer to a problem involving this situation as a **related rates** problem. Since y is a function of x and we know the rate dx/dt at which x changes with respect to time, we can find dy/dt using the chain rule:

$$\frac{dy}{dt} = \frac{dy}{dx}\frac{dx}{dt}.$$

Example 3.13 A milling company is a subcontractor for a manufacturing firm. According to the milling company's accountants, the size of the company is optimal when its total assets are A hundred dollars, where

$$A = 0.000001x^2 + 0.1x + 5000,$$

and the total value of the contracts currently held by the manufacturing firm is x thousand dollars. At the present time, $x = 70{,}000$ (that is, 70 million dollars) and the company is at its optimal size. New contracts outnumber contracts being completed, in such a way that x is currently increasing at a rate of 80 (thousand dollars) per month. At what rate should the milling company increase its total assets to maintain its optimal size?

Solution We are asked to find dA/dt, but we are not given a direct functional relationship between A and t (time, in months). However, we do have A as a function of x and we know dx/dt. In particular, at the present time $A = 0.000001x^2 + 0.1x + 5000$ and the rate of change of x is $dx/dt = 80$. We find dA/dt by the chain rule:

$$\frac{dA}{dt} = \frac{dA}{dx}\frac{dx}{dt}.$$

Consequently,

$$\frac{dA}{dt} = (0.000002x + 0.1)\frac{dx}{dt}.$$

Since $x = 70{,}000$ and $dx/dt = 80$, we calculate $dA/dt = 19.2$. Recall that $A(t)$ denotes the total assets in hundred dollars, so the milling company should increase its assets at a rate of \$1920 per month. ■

Example 3.14 After an abundant harvest of corn, farmers store the grain in large conical piles. The corn piles always maintain a shape in which the height is one-half the radius of the base. If corn is being added to the pile at 120 ft³/min, how fast is the height of the pile increasing when the height is 5 feet?

Solution The formula for the volume of a cone is

$$V = \tfrac{1}{3}\pi r^2 h$$

where r is the radius of the base and h is the height. See Figure 3.1. The shape of the corn pile indicates that $h = r/2$, or $r = 2h$. Thus,

$$V = \tfrac{1}{3}\pi(2h)^2 h = \frac{4\pi}{3}h^3.$$

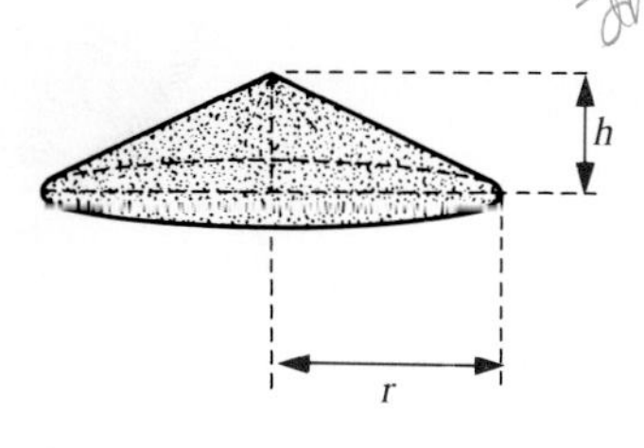

Figure 3.1
A conical pile of corn, discussed in Example 3.14

From the description of the situation we know that $dV/dt = 120$. We are asked to find dh/dt when $h = 5$. By the chain rule

$$\frac{dV}{dt} = \frac{dV}{dh}\frac{dh}{dt}$$

so

$$\frac{dh}{dt} = \frac{dV/dt}{dV/dh} = \frac{120}{4\pi h^2}. \qquad \text{(differentiate } \tfrac{4}{3}\pi h^3\text{)}$$

When $h = 5$,

$$\frac{dh}{dt} = \frac{120}{4\pi(25)} \approx 0.38.$$

When the height of the conical pile is 5 feet, the height is increasing at approximately 0.38 ft/min. ■

We now give a proof of the chain rule. As in the statement of the rule, we are assuming that $f(x)$ and $g(x)$ are differentiable.

Proof of the Chain Rule Let $F(x) = f(g(x))$. We wish to show that $F'(x) = f'(g(x))g'(x)$. By definition,

$$F'(x) = \lim_{h\to 0}\frac{F(x+h) - F(x)}{h}$$

$$= \lim_{h\to 0}\frac{f(g(x+h)) - f(g(x))}{h}.$$

To simplify the notation, let $u = g(x)$ and let $H = g(x+h) - g(x)$. Note that since $g(x)$ is continuous, $H \to 0$ as $h \to 0$. (To be precise, we must assume that H is never zero. This assumption could be avoided if we used a slightly different and more technical approach.) Since $g(x+h) = g(x) + H = u + H$,

$$F'(x) = \lim_{h\to 0} \frac{f(u+H) - f(u)}{h}.$$

Using u and H in the definition of $f'(u)$,

$$f'(u) = \lim_{H\to 0} \frac{f(u+H) - f(u)}{H}.$$

Consequently, if the above limit for $F'(x)$ had an H instead of h in the denominator, then that limit would be $f'(u)$. For this reason we "force" an H in the denominator as follows:

$$\begin{aligned} F'(x) &= \lim_{h\to 0} \left(\frac{f(u+H) - f(u)}{h} \frac{H}{H} \right) \\ &= \lim_{h\to 0} \left(\frac{f(u+H) - f(u)}{H} \frac{H}{h} \right) \\ &= \left(\lim_{H\to 0} \frac{f(u+H) - f(u)}{H} \right) \left(\lim_{h\to 0} \frac{H}{h} \right). \end{aligned}$$

(We have replaced $h \to 0$ in the left factor with $H \to 0$, since $H \to 0$ as $h \to 0$.) As noted above, the limit of the left factor in the above equation is $f'(u)$; the second limit is

$$\lim_{h\to 0} \frac{H}{h} = \lim_{h\to 0} \frac{g(x+h) - g(x)}{h} = g'(x).$$

Thus the limits in $F'(x)$ may be replaced with $f'(u)$ and $g'(x)$. Finally, since $u = g(x)$, we have

$$F'(x) = f'(u)g'(x) = f'(g(x))g'(x),$$

as required. ■

Exercises

Exer. 1–10: Express the function as a composition of two simpler functions.

1. $f(x) = (3x^2 - 4)^{-3}$
2. $f(x) = (4.00 - 2.15x)^{2/3}$
3. $f(x) = \sqrt{x^2 + 4}$
4. $f(x) = 4\sqrt[3]{3x^2 + 1}$
5. $f(t) = 2(t^2 - 4)^{1/3} + 5$
6. $f(t) = (t^2 - 2t + 1)^{3/4}$
7. $y = \sqrt{4s^2 + 1} + 1.03$
8. $y = (8 - 6x^{1/3})^2$
9. $y = 14/(3x^2 - 4)^2$
10. $y = (x^{16} - 2)^{16}$

Exer. 11–16: For functions $f(x)$ and $g(x)$, find the compositions $f(g(x))$ and $g(f(x))$. Note that $f(g(x)) \neq g(f(x))$.

11. $f(x) = x^2 + x$; $g(x) = 3x$
12. $f(x) = x^3 + 4x$; $g(x) = x + 4$
13. $f(x) = \dfrac{x}{x+3}$; $g(x) = x^2$
14. $f(x) = \dfrac{x^2 + x}{x+2}$; $g(x) = \dfrac{3}{x}$
15. $f(x) = \dfrac{2x+3}{x-2}$; $g(x) = \sqrt{x}$
16. $f(x) = (x+5)^3$; $g(x) = 3x^2 + 2$

Exer. 17–38: Find the derivative of the function.

17. $f(x) = (3x - 2)^5$
18. $f(x) = (x^{14} - 2)^6$
19. $f(x) = (x^2 - 4)^2$
20. $f(x) = (2x^2 + x)^8$
21. $f(x) = (x - 3)^{3/4}$
22. $f(p) = 2(p - p^4)^{2/3}$
23. $f(x) = (x^2 - 3)^{3/4}$
24. $f(x) = 3/(x^2 - 1)^2$
25. $f(t) = (4.00 - 2.15t)^{2/3}$
26. $f(t) = (2t^2 - 3t^3)^{4/3}$
27. $f(x) = 2(x^2 - 3x + 1)^{-1/2}$
28. $f(x) = (x^2 + 3)^2$
29. $f(r) = (\sqrt{r} - 2r)^3$
30. $f(x) = (3 - 2x)^{-1/3}$
31. $f(x) = (x^2 + 2)^2(x^2 + 6)$
32. $f(x) = [(x + 2)(3 - x)]^3$
33. $f(t) = \sqrt{(t^2 + 6)(6 - 3t)}$
34. $f(x) = \sqrt[3]{x^2 + 4}/(3 - x)$
35. $y = x^2(3 - x)^2$
36. $y = (3x^2 + 2)^4(3 - x)$
37. $y = \sqrt{(r - 3)(3r + 2)^3}$
38. $y = (t^2 + 4)/(3 - t)^2$

Exer. 39–44: Determine the equation of the tangent line to the graph of the function at the specified point.

39. $f(x) = (2 - 3x)^2$; $(1, 1)$
40. $f(x) = (2 - x)^2(x^2 + 1)$; $(2, 0)$
41. $f(u) = \sqrt{u^2 - 2u + 6}$; $(-1, 3)$
42. $f(x) = (x^2 - 2x)^{4/3}$; $(1, 1)$
43. $y = (2x - 1)^3$; $(2, 27)$
44. $y = (2x)^{2/3} - 1$; $(\frac{1}{2}, 0)$

45. The demand function for premium quality guitars made by a company is

$$D(p) = 6000 - 64\sqrt{0.01p^2 + 250}$$

where $D(p)$ is the number of guitars sold if the price is p dollars. The revenue function for these guitars is $R(p) = pD(p)$. Determine $R'(p)$.

46. The population of a rural area is projected to be $P(t) = 100t^2 + 2500t + 100{,}000$ people, t years in the future. A quality-of-living index Q is given by $Q(P) = 75 - \sqrt{P/10}$. Determine the rate of change of Q with respect to t when $t = 10$.

47. The radius of a spherical balloon is increasing at the rate of 2 in./sec, How fast is the volume increasing when the radius is 6 inches? (The volume of a sphere of radius r is given by $V = \frac{4}{3}\pi r^3$.)

48. A spherical balloon is being inflated so that its volume is increasing at a rate of 3 in.^{3}/sec. After 5 seconds (that is, when the volume of the balloon is 15 in.^{3}), at what rate is the radius of the balloon increasing?

49. The average score $s(x)$ on a reading proficiency test given to third graders is found to be a function of the number x (in hundreds) of books in the school library. If

$$s(x) = \sqrt{x^2 + 5x + 6},$$

and if the size of the school library is 2000 books and is increasing at a rate of 300 books per year, at what rate is the average score on the reading test increasing?

50. Refer to Example 3.13. Suppose that the milling company finds that its size is optimal when

$$A = 0.0001x^2 + 0.2x + 4000.$$

At what rate should it decrease its total assets when $x = 80{,}000$ and x is decreasing at a rate of \$40,000 per month?

51. Oil is leaking from an offshore tank at the rate of 10 m^3/hr (or 2778 cm^3/sec). Assume that the oil forms a uniform circular layer 0.5 cm thick on the water. How fast is the area of the oil slick increasing? How fast is the radius of the oil slick increasing?

52. The Gotham City Cab Company finds that the number of cabs C needed is dependent on the population P (in thousands) of Gotham City according to the rule

$$C = 0.01P^{3/2} + 0.3P + 5.$$

If the population of Gotham City is currently 1,600,000 (or $P = 1600$) and if the population is increasing at a rate of 4000 per month, at what rate should the company increase its cab fleet?

53. A ladder 15 feet long is leaning against a vertical wall. The bottom of the ladder is being pulled away from the wall at a rate of 3 ft/min. How fast is the top of the ladder moving when the bottom is 9 feet from the wall? (*Hint:* Draw a diagram, letting x be the distance of the bottom of the ladder from the wall and y be the vertical distance of the ladder top from the ground. Use the Pythagorean theorem ($c^2 = a^2 + b^2$ in a right triangle) to obtain a relationship between x and y.)

54. The cost $C(x)$ of producing a certain product is $C(x) = 2x^{3/2} + 60x^{1/2} + 7000$, where x is the number of items produced. At the production level $x = 100$, the cost of items produced is increasing at the rate of \$600 per month. Find the monthly rate of increase in production.

55. Oil drips onto a hot pavement and forms a circular slick. The radius is increasing at a rate of 2 cm/min. How fast is the area changing when the radius of the slick is 10 cm?

56. The population P in a rural area is increasing at an annual rate of 3%. The present population is 40 (thousand). The pollution index I is determined by $I(P) = 30 + \sqrt{P^2 - 40}$. Find the rate of change of I with respect to time t (in years), measured from the present.

Exer. 57–60: Two functions $f(x)$ and $g(x)$ are *inverses* of one another if $f(g(x)) = x$ and $g(f(x)) = x$ for all x in the domain of both $f(x)$ and $g(x)$. Verify that the given functions are inverses of each other.

57. $f(x) = \dfrac{1}{x + 2}$; $g(x) = \dfrac{1}{x} - 2$

58. $f(x) = \dfrac{x}{x+1}$; $g(x) = \dfrac{-x}{x-1}$

59. $f(x) = x^3$; $g(x) = \sqrt[3]{x}$

60. $f(x) = 3x + 2$; $g(x) = \dfrac{x-2}{3}$

61. Suppose the function $Q(x)$ is such that $Q'(x) = 1/x$. Find $(d/dx)Q(x^3 + x)$.

62. Suppose the function $Q(x)$ is such that $Q'(x) = 1/(x^3 + 1)$. Find $(d/dx)Q(4x^2 + 2x)$.

63. For some function $f(x)$, suppose that $f'(1) = 4$. If $g(x) = f(x^3 - 3x - 1)$, find $g'(2)$.

64. For some function $f(x)$, suppose that $f'(3) = 2$. If $g(x) = f(2x^2 - 3x + 1)$, find $g'(2)$.

65. For some function $f(x)$, suppose that $f(2) = -25$ and $f'(2) = 5$. If $g(x) = xf(x^2 + x - 10)$, find $g'(3)$.

The Second Derivative: Concavity

3.3

In Section 2.5 we used the derivative of a function $f(x)$ to study the increasing and decreasing behavior of $f(x)$. Although this information is useful in understanding and predicting events described by the function, it is not sufficiently refined for some applications. For example, suppose we analyze the performance of two political candidates' election campaigns. The graphs in Figure 3.2 show the percentages of potential voters $A(t)$ and $B(t)$ that support candidates A and B after t weeks of campaigning. Note that after five weeks ($t = 5$) the percentage of potential votes for candidate A is growing at the rate of 5% per week, while that for candidate B is growing at the rate of 3% per week. Since at $t = 5$ both candidates have 35% of the vote, we might conclude that candidate A is better off than candidate B. However, the rate of increase of voter support for A is diminishing while the rate of increase of voter support for B is growing. We say that B's campaign has a "snowballing" effect, while A's does not. In fact, note that on election day ($t = 10$), candidate B has 55% of the vote while A has only 45%.

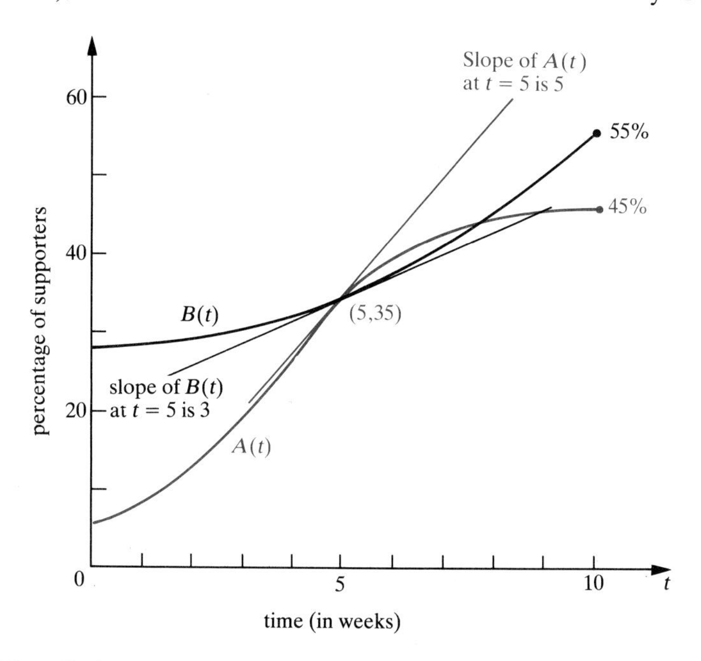

Figure 3.2
Percentage of potential voters that support candidates A and B

The distinction between the functions $A(t)$ and $B(t)$ lies in the rates at which their slopes are changing. Since the slope of a function is

the derivative, and since the rate of change is also a derivative, the rate of change of the slope is actually the "derivative of the derivative."

Let $f(x)$ be a function such that $f'(x)$ exists. If the function $f'(x)$ is also differentiable, then the derivative of $f'(x)$ is called the **second derivative** of $f(x)$ and is denoted by $f''(x)$ or, sometimes, by $f^{(2)}(x)$.

Example 3.15 Let $f(x) = x^3 + 2x^2 + 5x - 6$. Find $f''(x)$.

Solution Since $f'(x) = 3x^2 + 4x + 5$, we have $f''(x) = 6x + 4$. ■

In a similar fashion, $f'''(x)$ or, equivalently, $f^{(3)}(x)$ denotes the **third derivative** of $f(x)$ and is defined to be the derivative of $f''(x)$. Continuing in this way, $f^{(n)}(x)$ denotes the ***n*th derivative** of $f(x)$ and is defined to be the derivative of $f^{(n-1)}(x)$. In general, derivatives other than $f'(x)$ are called **higher-order derivatives**.

Example 3.16 Find $f'(x)$, $f''(x)$, $f^{(3)}(x)$, and $f^{(4)}(x)$ where

$$f(x) = x^9 + 2x^7 + 5x^4 - 2x^3 + 6x + 2.$$

Solution By taking derivatives successively, we have

$$f'(x) = 9x^8 + 14x^6 + 20x^3 - 6x^2 + 6$$

$$f''(x) = 72x^7 + 84x^5 + 60x^2 - 12x$$

$$f^{(3)}(x) = 504x^6 + 420x^4 + 120x - 12$$

$$f^{(4)}(x) = 3024x^5 + 1680x^3 + 120.$$ ■

When a function is given in the form $y = f(x)$, then, as we have seen, dy/dx denotes $f'(x)$. Sometimes dy/dx is written as $(d/dx)(y)$ and, in general, $(d/dx)(\ )$ represents the derivative of any expression within the parentheses. Using this notation, the second derivative of y with respect to x is

$$\frac{d}{dx}\frac{d}{dx}(y).$$

This suggests the notation $(d^2/dx^2)(y)$ or d^2y/dx^2 for the second derivative of y with respect to x. In general, d^ny/dx^n will be used to denote the nth derivative of y with respect to x. An alternative notation for d^2y/dx^2 is y''.

Example 3.17 If $y = x^3 - 12x^2$, find all x such that $\dfrac{d^2y}{dx^2} = 0$.

Solution Since

$$\frac{dy}{dx} = 3x^2 - 24x,$$

we have

$$\frac{d^2y}{dx^2} = 6x - 24 = 0$$

only when $x = 4$. ■

Example 3.18 An airplane travels $s(t)$ feet t seconds after takeoff where

$$s(t) = \tfrac{5}{3}t^3 + 270t.$$

What is the plane's acceleration after 8 seconds?

Solution In Section 2.6 we saw that acceleration is the derivative of velocity which, in turn, is the derivative of the distance-traveled function. Consequently, the acceleration after t seconds is $a(t) = s''(t)$. Therefore,

$$\begin{aligned} a(t) &= s''(t) \\ &= (\tfrac{5}{3}t^3 + 270t)'' \\ &= (5t^2 + 270)' = 10t. \end{aligned}$$

Since $a(8) = 80$, the acceleration of the plane after 8 seconds is 80 ft/sec/sec. ∎

At the beginning of this section we noted that if we are considering a function $f(x)$, then $f''(x)$ is the rate of change of $f'(x)$. Consider the graph of $f(x)$ in Figure 3.3. Several tangent lines have been sketched and their slopes noted. Note that as x moves to the right, the slope of the tangent line decreases. In other words, $f'(x)$ is a decreasing function, so its derivative is negative: $[f'(x)]' < 0$. Thus, any graph shaped like the curve in Figure 3.3 has a negative second derivative. When the curve is "bending downward" in this manner, we say that the graph of the function is **concave downward**. The graph of a function $f(x)$ is concave downward whenever $f''(x) < 0$.

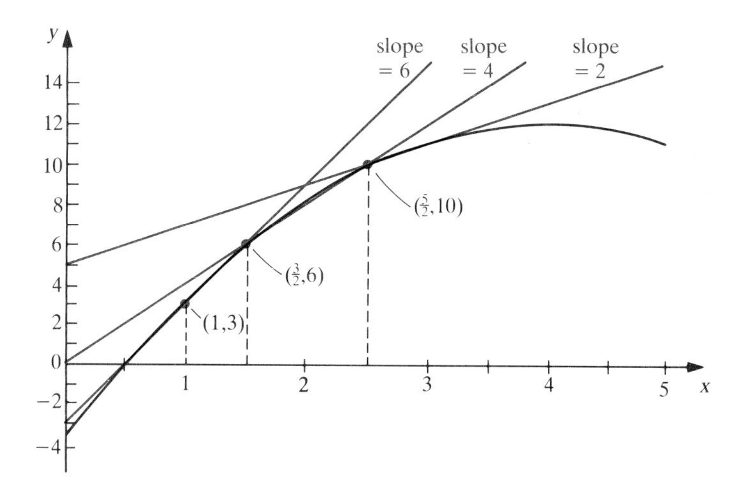

Figure 3.3
Since the slope decreases as x moves to the right, this curve is concave downward and $f''(x) < 0$

Similarly, if the graph of a function $f(x)$ is "bending upward," the graph is said to be **concave upward**; this occurs whenever $f''(x) > 0$. The graph in Figure 3.4(a) shows a function that is concave downward, so its second derivative is always negative; Figure 3.4(b) shows a function that is concave upward; its second derivative is always positive.

A point on the graph of a function at which the function changes

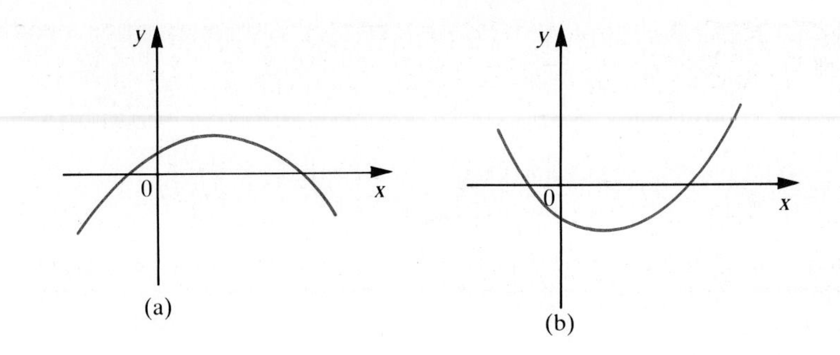

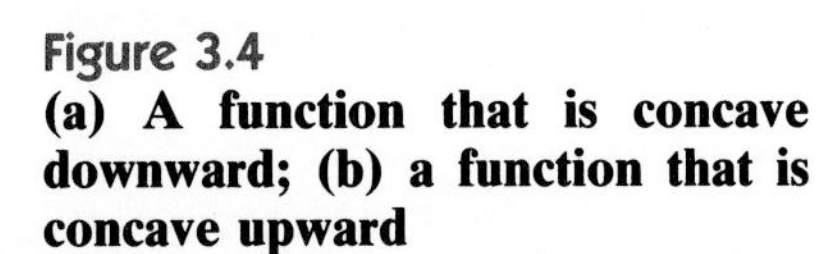

Figure 3.4
(a) A function that is concave downward; (b) a function that is concave upward

from concave upward to concave downward, or vice versa, is called an **inflection point**. The notions of concavity and inflection points are illustrated in Figure 3.5. Since $f''(x)$ has different signs on opposite sides of an inflection point, an inflection point can occur only where $f''(x) = 0$ or where $f''(x)$ is not defined.

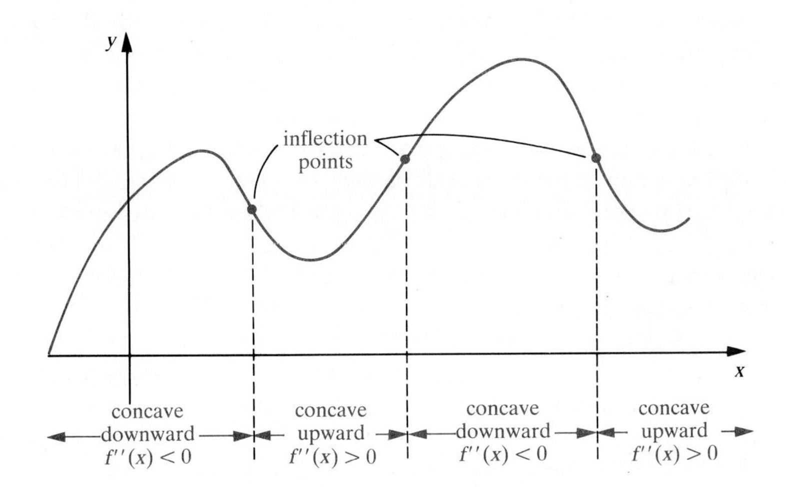

Figure 3.5
An inflection point is a point of transition between concave upward and concave downward

Example 3.19 Consider the function

$$f(x) = 2x^3 - 3x^2 - 12x + 2.$$

Determine the values of x where $f(x)$ is increasing, decreasing, concave upward, and concave downward. Locate all inflection points. Sketch the graphs of $f(x)$, $f'(x)$, and $f''(x)$ using the information obtained.

Solution To determine where $f(x)$ is increasing and decreasing, we compute $f'(x)$:

$$f'(x) = 6x^2 - 6x - 12 = 6(x + 1)(x - 2).$$

This function is defined everywhere and equals zero at $x = -1$ and $x = 2$ (these are critical values). Since $f'(x)$ is continuous, it can change sign only at $x = -1$ and at $x = 2$. We determine the sign of $f'(x)$ in each of the intervals $(-\infty, -1)$, $(-1, 2)$, and $(2, \infty)$ from an arbitrarily chosen test value in each interval. Table 3.1 shows the results and the conclusions about the behavior of $f(x)$.

Interval	$(-\infty, -1)$	$(-1, 2)$	$(2, \infty)$
test value t	-2	0	3
$f'(t)$	24	-12	24
sign of $f'(x)$ on interval	+	−	+
behavior of $f(x)$	increasing	decreasing	increasing

Table 3.1
Behavior of $f(x)$ near critical values

Thus $f(x)$ is increasing for $x < -1$ and for $x > 2$, and it is decreasing for $-1 < x < 2$.

To determine regions where $f(x)$ is concave upward and concave downward, we compute $f''(x)$:

$$f''(x) = 12x - 6 = 12(x - \tfrac{1}{2}).$$

Therefore $f''(x) = 0$ only when $x = \frac{1}{2}$. Since $f''(x)$ is positive when $x > \frac{1}{2}$ and negative when $x < \frac{1}{2}$, the graph is concave upward for $x > \frac{1}{2}$ and concave downward for $x < \frac{1}{2}$. The concavity changes when $x = \frac{1}{2}$, so the point $(\frac{1}{2}, f(\frac{1}{2})) = (\frac{1}{2}, -\frac{9}{2})$ is a point of inflection.

We have discovered enough information about $f(x)$ to sketch its graph. We know that it is increasing on $(-\infty, -1)$, decreasing on $(-1, 2)$, and increasing on $(2, \infty)$. It has horizontal tangent lines at $x = -1$ $(y = f(-1) = 9)$ and at $x = 2$ $(y = f(2) = -18)$. It is concave downward for $x < \frac{1}{2}$ and concave upward for $x > \frac{1}{2}$. We can also observe that its y-intercept is $y = 2$.

We also have enough information to sketch the graph of $y = f'(x) = 6x^2 - 6x - 12$. This is a parabola that opens upward (because of the positive coefficient of x^2) and has x-intercepts $x = -1$ and $x = 2$ (the critical values of $f(x)$). The y-intercept of $f'(x)$ is $y = -12$. Finally, the graph of $y = f''(x)$ is a line with slope 12 and y-intercept -6.

The three functions $f(x)$, $f'(x)$, and $f''(x)$ are graphed in Figure 3.6. Notice that the graph of $f'(x)$ is positive (above the x-axis) when $f(x)$ is increasing, and $f'(x)$ is negative (below the x-axis) when $f(x)$ is decreasing. Also note that the graph of $f''(x)$ is negative (below the x-axis) when the graph of $f(x)$ is concave downward, and the graph of $f''(x)$ is positive (above the x-axis) when the graph of $f(x)$ is concave upward. ■

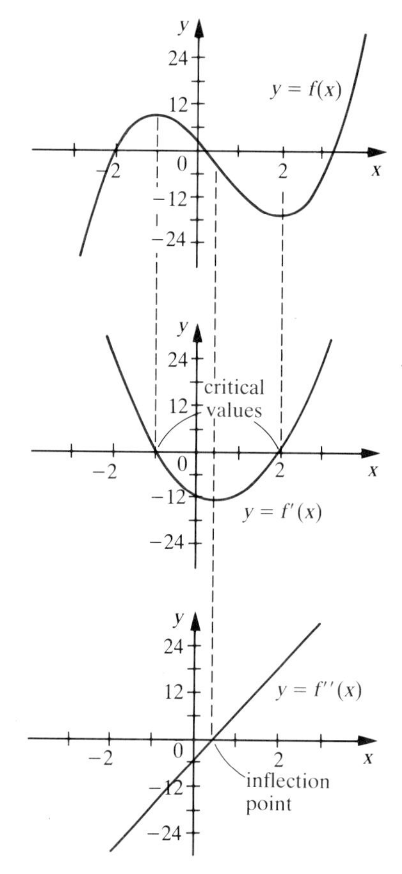

Figure 3.6
Graphs of $f(x)$, $f'(x)$, and $f''(x)$ for $f(x) = 2x^3 - 3x^2 - 12x + 2$

Example 3.20 A newcomer enters the primary election for state senator. Suppose the campaign manager believes that the campaign must have a snowballing effect (that is, the rate of growth of the number of potential votes must be increasing) for the candidate to win. Poll results show that the number of potential votes (in millions) in the xth week of the campaign is given by

$$s(x) = 3x + 2\sqrt{x} + 4.5.$$

What is the prognosis for this candidate's campaign?

Solution The rate of change of the number of potential votes is given by

$$s'(x) = 3 + \frac{1}{\sqrt{x}},$$

which is positive for all $x > 0$. Thus, the number of potential votes is increasing. However, the rate at which this number is growing is given by

$$s''(x) = \frac{-1}{2\sqrt{x^3}},$$

which is negative for all $x > 0$. Hence the graph of the function $s(x)$ is concave downward, indicating that the *rate of increase* of potential votes is tapering off. Consequently, the poll results predict that this candidate's campaign will not have the desired snowballing effect unless the manager adopts a new strategy. ■

Example 3.21 Find the inflection points of the function

$$f(x) = \frac{(x^2 - 4)^6}{12}.$$

Solution Interpret the function as $f(x) = \frac{1}{12}(x^2 - 4)^6$ and use the constant multiple rule and the chain rule to find $f'(x)$. The result is

$$f'(x) = \tfrac{1}{12}[6(x^2 - 4)^5(2x)] = x(x^2 - 4)^5.$$

To find the second derivative, we must use the product rule to differentiate $f'(x)$. We obtain

$$\begin{aligned} f''(x) &= x\frac{d}{dx}[(x^2 - 4)^5] + (x^2 - 4)^5\frac{d}{dx}(x) \\ &= x[5(x^2 - 4)^4(2x)] + (x^2 - 4)^5 \\ &= 10x^2(x^2 - 4)^4 + (x^2 - 4)^5 \\ &= (x^2 - 4)^4(10x^2 + x^2 - 4) \qquad \text{(factor out } (x^2 - 4)^4\text{)} \\ &= (x^2 - 4)^4(11x^2 - 4). \end{aligned}$$

This final expression for $f''(x)$ is zero when

$$x^2 - 4 = 0 \quad \text{or} \quad 11x^2 - 4 = 0.$$

These equations imply that

$$x = -2, \qquad x = 2, \qquad x = \frac{-2}{\sqrt{11}}, \quad \text{or} \quad x = \frac{2}{\sqrt{11}}.$$

The sign of $f''(x)$ does not change at either $x = 2$ or $x = -2$ because the first factor in $f''(x)$ has an exponent of 4 and, hence, is never negative; near $x = \pm 2$, (when the first factor is zero) the second factor $(11x^2 - 4)$ is positive (close to 40, in fact). Consequently, $f''(x)$ is positive on both sides of $x = 2$ and on both sides of $x = -2$, so neither $x = -2$ nor $x = 2$ corresponds to an inflection point.

The sign of $f''(x)$ does change at $x = \pm 2/\sqrt{11} \approx \pm 0.603$. In fact,

$f''(\pm 0.6) \approx -7.0$ and $f''(\pm 0.61) \approx 16$ so the values $x = \pm 20/\sqrt{11}$ correspond to inflection points. Figure 3.7 shows the graph of $f(x)$ with its inflection points. ■

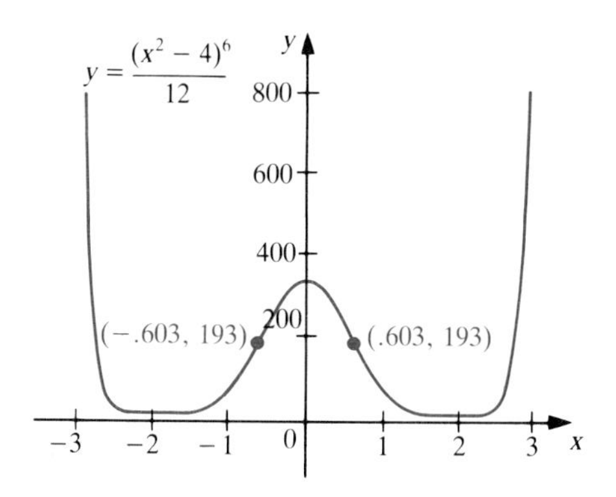

Figure 3.7
Graph of $f(x) = \frac{1}{12}(x^2 - 4)^6$ with inflection points

The previous example illustrates that *not all values of x* where $f''(x) = 0$ are inflection points. The next example shows that an inflection point may occur where $f''(x)$ is not zero!

Example 3.22 Find the inflection points of $f(x) = \sqrt[3]{x}$.

Solution We have $f(x) = x^{1/3}$, so that $f'(x) = \frac{1}{3}x^{-2/3}$, and therefore

$$f''(x) = -\tfrac{2}{9}x^{-5/3} = \frac{-2}{9\sqrt[3]{x^5}}.$$

Even though $f''(x)$ is never equal to zero, it is positive for all $x < 0$ and negative for all $x > 0$. Hence $x = 0$ is a value of x where the concavity of $f(x)$ changes; $(0, 0)$ *is* an inflection point of this function. The graph of $f(x)$ is shown in Figure 3.8. Note that the graph has a vertical tangent line at $x = 0$ because $f'(x) \to \infty$ as $|x| \to 0$. Also note that neither $f'(x)$ nor $f''(x)$ is defined at $x = 0$. ■

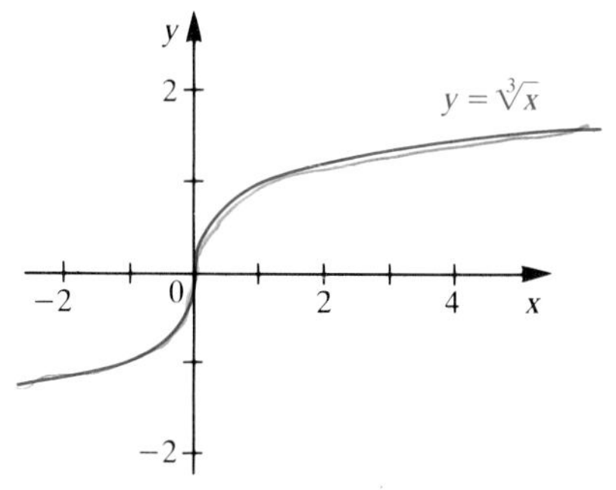

Figure 3.8
Graph of $f(x) = \sqrt[3]{x}$

The previous two examples provide guidelines for determining the inflection points of a function $f(x)$.

1. Determine the values of x in the domain of $f(x)$ where $f''(x)$ equals zero or does not exist.
2. Determine which of the values of x found in step 1 have the property that the sign of $f''(x)$ is different on opposite sides of that x-value. These values correspond to inflection points.

Exercises

Exer. 1–20: For the given function, compute the indicated derivatives.

1. $f(x) = x^4 - 3x^3$; $f''(x)$, $f''(2)$
2. $f(x) = 6x^5 + 13x^2$; $f'''(x)$, $f''(-1)$
3. $f(t) = t^3 - 2t^2 + 4$; $f''(t)$, $f''(3)$
4. $f(x) = \frac{3}{x} + 16x^2$; $f''(x)$, $f^{(3)}(2)$
5. $f(x) = x^{1/2} - 4x^3$; $f''(x)$, $f''(4)$
6. $f(u) = \frac{2}{u^3} + \frac{u^3}{2}$; $f'''(u)$, $f^{(3)}(1)$
7. $y = (x^3 - 62)^5$; $\frac{d^2y}{dx^2}$, $\left.\frac{d^2y}{dx^2}\right|_{x=4}$
8. $y = (3x^3 - 2x)^{1/2}$; $\frac{d^2y}{dx^2}$, $\left.\frac{d^2y}{dx^2}\right|_{x=1}$
9. $y = \frac{3t}{t+1}$; $\frac{d^2y}{dt^2}$, $\left.\frac{d^2y}{dt^2}\right|_{t=3}$
10. $y = 4x^3 + 5x^2 - 7x + 2$; $\frac{d^3y}{dx^3}$, $\left.\frac{d^3y}{dx^3}\right|_{x=2}$
11. $y = 3x^{1/2} + 5x^{2/3}$; $\frac{d^2y}{dx^2}$, $\left.\frac{d^2y}{dx^2}\right|_{x=1}$
12. $y = 4\sqrt{z^3} + 7z^{1/2}$; $\frac{d^2y}{dz^2}$, $\left.\frac{d^2y}{dz^2}\right|_{z=3}$.
13. $y = 6x^4 - 3x^2 + 5$; y'', $y''|_{x=3}$
14. $y = 3x^5 - 2x^3 + 5$; y''', $y'''|_{x=-2}$
15. $y = \frac{3}{t^{2/5}} - 5t^3$; y''', $y'''|_{t=-1}$
16. $y = \sqrt[3]{r^2} - 2r^4$; y''', $y'''|_{r=8}$
17. $f(x) = 2x^7 + 5x^4 + 200x$; $f^{(4)}(x)$, $f^{(7)}(x)$
18. $f(x) = 2x^{8/3}$; $f^{(3)}(x)$, $f^{(4)}(x)$, $f^{(7)}(x)$
19. $y = 3s^5 - 2s^{1/2}$; $y^{(4)}$, $y^{(5)}$, $y^{(5)}|_{s=4}$
20. $y = 3x^{2/5} - 2x^8$; $\frac{d^4y}{dx^4}$, $\frac{d^6y}{dx^6}$

Exer. 21–38: Determine regions where the graph of the function is concave upward and concave downward. Locate all points of inflection.

21. $f(x)=x^3-3x^2+3$
22. $f(x)=x^3+12x-12$
23. $f(t)=t^3-12t+6$
24. $f(x)=x^3-8$
25. $f(x)=2x^2+5x-6$
26. $f(p)=-5p^2+2p+18$
27. $f(s)=s^4-4s^3+4s^2-6$
28. $f(x)=x^4+4x^3+4$
29. $f(r)=r^4-8r^3+18r^2+36$
30. $f(x)=x^4-2x^2+1$
31. $f(x)=3x^4-4x^3$
32. $f(w)=w^5-5w^3$
33. $f(x)=x/(x^2+1)$
34. $f(x)=x^{2/3}-\frac{1}{5}x^{5/3}$
35. $f(z)=z^4-8z^2+12$
36. $f(x)=x^4-4x^3-3x^2+48x$
37. $f(t)=3t^5-20t^3+15$
38. $f(x)=6x^5-15x^4+10x^3-12$

39. Let $y=f(x)=ax^2+bx+c$ with $a\neq 0$. Show that $f(x)$ is concave upward for all x if $a>0$ and concave downward for all x if $a<0$. Show that $x=-b/2a$ is a critical value. (See the rules given in Section 1.5 (page 32) for sketching a parabola.)

Exer. 40–46: Sketch a graph of a continuous function that satisfies the given conditions.

40. Increasing and concave upward for $x<-1$;
increasing and concave downward for $-1<x<2$;
increasing and concave upward for $x>2$;
$f'(2)=0$.
41. Increasing for all x;
concave downward for $x<-2$;
concave upward for $x>-2$;
$f'(-2)=1$; $f(0)=0$.
42. Increasing and concave downward for $x<3$;
increasing and concave upward for $x>3$;
$f'(3)=0$; $f(3)=0$.
43. Decreasing for $x<-4$;
concave upward for $x<-1$;
increasing for $-4<x<2$;
concave downward for $x>-1$;
decreasing for $x>2$;
$f(0)=0$.
44. $f'(-2)=f'(4)=0$;
increasing for $x<4$;
decreasing for $x>4$.
45. Concave upward for $x<-6$ and for $x>6$;
concave downward for $-6<x<6$;
$f'(x)>0$ for all x.
46. Concave downward for $x<4$ and for $x>12$;
concave upward for $4<x<12$;
decreasing for $x<8$;
increasing for $x>8$;
$f(0)=4$.
47. Refer to Example 3.18. Suppose that the plane has traveled $s(t)=\frac{5}{3}t^3+16t^2+27t$ feet t seconds after takeoff $(0\leq t\leq 4)$. What is its acceleration after 4 seconds?
48. Refer to Example 3.20. Suppose the candidate has adopted a new strategy, and the potential voter support x weeks into the campaign is now predicted to be $s(x)=2x^{3/2}+5x+4.5$. Will the candidate win?
49. Federal economic advisors note that not only is the number of unemployed people rising, but the rate at which unemployment is rising is also rising. They recommend increasing federal spending to stimulate the economy in order to generate jobs. To prevent federal overspending and the resulting rise in inflation, the government should cut back the spending program as soon as the rate of increase of unemployment begins to decrease. Let $u(t)$ denote the number of unemployed people (in thousands). Economists predict that t months after the spending program is implemented, $u(t)$ will be given by

$$u(t)=5000+1092t+42t^2-t^3.$$

(a) How many months of federal spending will be necessary to achieve the desired decrease in the rate of increase in unemployment? (*Hint:* du/dt represents the rate at which unemployment is changing, and d^2u/dt^2 represents the rate at which this rate is changing. Consequently, the government should continue its spending program as long as $d^2u/dt^2>0$.)
(b) How many months of spending will elapse before the number of unemployed begins to decrease?
(c) How many people are unemployed when the government cuts back the spending program?
(d) What is the maximum number of people unemployed?

3.4 Derivative Tests for Relative Maxima and Relative Minima

We discussed the problem of finding the optimal size for a cake pan in Section 2.1 and saw that its solution involved determining the location of the maximum value of a function. We have also discussed the role of the derivative in finding critical values of a function. However, a critical value may correspond to a "maximum" value of the function, or a "minimum" value, or neither. In this section we present two *tests* for determining the behavior of a function near a critical value.

A function $f(x)$ has a **relative maximum** $f(a)$ at the point $(a, f(a))$ if $f(a) \geq f(x)$ for all x in some open interval containing a. In other words, $f(a)$ is a maximum for the function if we consider only values of x *near* a. The graphs in Figure 3.9 illustrate several relative maxima.

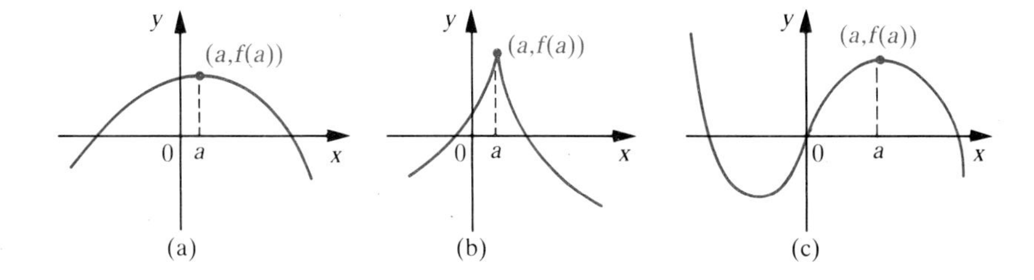

Figure 3.9
Some ways a relative maximum can occur

A function $f(x)$ has a **relative minimum** $f(a)$ at the point $(a, f(a))$ if $f(a) \leq f(x)$ for all x in some open interval containing a. Graphs in Figure 3.10 illustrate several relative minima.

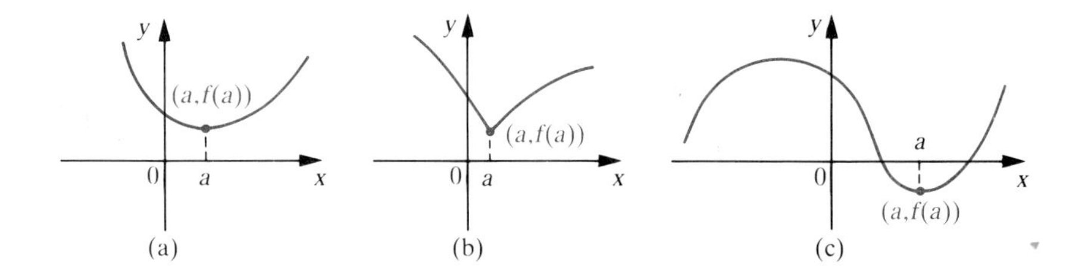

Figure 3.10
Some ways a relative minimum can occur

The derivative provides the key to determining the existence of relative maxima and minima. First suppose that $f(x)$ has a relative maximum at $x = a$ and that the derivative $f'(a)$ exists. Then $f(a) \geq f(a+h)$ for all (small) values of h. Hence $f(a+h) - f(a) \leq 0$, so

$$\frac{f(a+h) - f(a)}{h} \leq 0 \quad \text{for } h > 0$$

(a negative numerator, positive denominator), and

$$\frac{f(a+h) - f(a)}{h} \geq 0 \quad \text{for } h < 0$$

(a negative numerator, negative denominator). As $h \to 0$, both of the above ratios approach $f'(a)$, and therefore $f'(a) \leq 0$ and $f'(a) \geq 0$. Consequently, we see that $f'(a) = 0$ at a relative maximum. A similar argument can be made at a relative minimum.

At every relative maximum or relative minimum of a function $f(x)$, $f(x)$ has either a zero derivative or no derivative. Recall from Section 2.5 that values of a in the domain of $f(x)$ where $f'(a)$ equals zero or does not exist are critical values of $f(x)$. Thus, all relative maxima and relative minima occur at critical points.

Figure 3.11 shows critical points for several functions. Although all relative maxima and relative minima occur at critical values, not all critical values are at relative maxima or relative minima. The derivative $f'(a)$ may be zero but the critical value $x = a$ may not be a relative maximum or relative minimum, as shown in Figure 3.11(c), (d), and (f). If $f'(a) = 0$, then the tangent line is horizontal at $x = a$, as we anticipated from the skier illustration in Section 2.1.

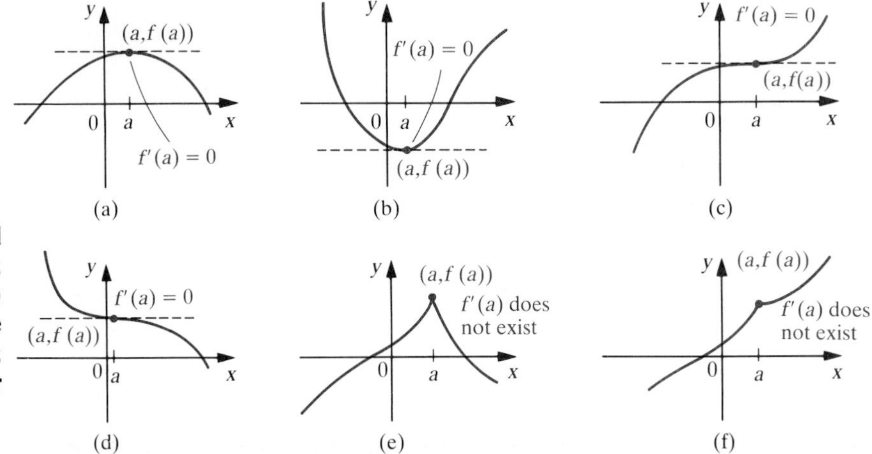

Figure 3.11
Critical values $x = a$ of several functions: (a) relative maximum; (b) relative minimum; (c) and (d) not relative maxima or relative minima; (e) relative maximum; (f) not a relative maximum or relative minimum

The following test can be used to determine when a critical value corresponds to a relative maximum or relative minimum.

The First Derivative Test

Let a be a critical value of $f(x)$ and suppose $f(x)$ is differentiable for all values of x near a (but not necessarily at a). For values of x near a,

A. if $f'(x) \geq 0$ for $x < a$ and $f'(x) \leq 0$ for $x > a$, then $f(a)$ is a relative maximum;
B. if $f'(x) \leq 0$ for $x < a$ and $f'(x) \geq 0$ for $x > a$, then $f(a)$ is a relative minimum;
C. if $f'(x)$ has the same sign on both sides of $x = a$, then $f(a)$ is neither a relative minimum nor a relative maximum.

The condition A of the test says that if $f(x)$ is increasing immediately to the left of $x = a$ and decreasing immediately to the right of $x = a$, then $f(a)$ is a relative maximum. Condition B says that if $f(x)$ is decreasing to the left of $x = a$ and increasing to the right of $x = a$, then $f(a)$ is a relative minimum. Reexamine Figure 3.11 to see how the First Derivative Test applies in each case.

Example 3.23 Find all relative maxima and relative minima of $f(x) = x^2 - 6x + 5$.

Solution Since $f(x)$ is differentiable everywhere, the only critical values occur when $f'(x) = 2x - 6 = 2(x - 3) = 0$, that is, when $x = 3$. If $x < 3$, then $f'(x) < 0$ and if $x > 3$, then $f'(x) > 0$. Thus, by the First Derivative Test, $f(x)$ has a relative minimum at $x = 3$, and that relative minimum value is $f(3) = 9 - 18 + 5 = -4$. Note that $f(x)$ has no relative maximum in this example. The graph of $f(x)$ and the relative minimum at $(3, -4)$ is shown in Figure 3.12. ■

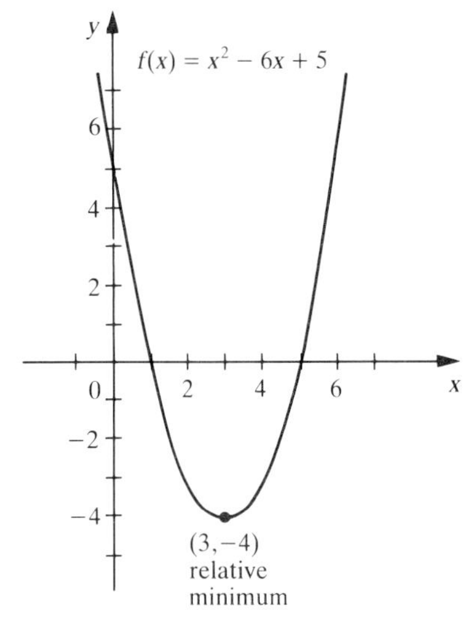

Figure 3.12
Graph of $f(x) = x^2 - 6x + 5$

Example 3.24 Find all relative maxima and relative minima of $f(x) = 2x - 5x^{4/5} + 3$. Sketch the graph of $f(x)$.

Solution We first locate the critical values to determine where the relative maxima and relative minima can occur. Since

$$\begin{aligned} f'(x) &= 2 - 4x^{-1/5} \\ &= \frac{2x^{1/5}}{x^{1/5}} - \frac{4}{x^{1/5}} \\ &= \frac{2x^{1/5} - 4}{x^{1/5}} = \frac{2\sqrt[5]{x} - 4}{\sqrt[5]{x}}, \end{aligned} \tag{1}$$

we note that $f'(x)$ is undefined at $x = 0$, although $x = 0$ is in the domain of $f(x)$. Hence $x = 0$ is a critical value. The remaining critical values of $f(x)$ are found by setting $f'(x) = 0$ and solving for x, which yields

$$\begin{aligned} 2x^{1/5} - 4 &= 0 \\ \sqrt[5]{x} &= 2 \\ x = 2^5 &= 32, \qquad \text{(raise to 5th power)} \end{aligned}$$

so $x = 32$ is also a critical value. To determine the nature of these critical values, we need to find the sign of $f'(x)$ on each side of each critical value. We consider three intervals: $(-\infty, 0)$, $(0, 32)$, and $(32, \infty)$. Since $f'(x)$ cannot change sign *within* any of these intervals, we can find whether $f'(x)$ is positive or negative throughout an interval by computing $f'(x)$ at any arbitrary test value t in the interval. We choose test values $t = -1$, $t = 1$, and $t = 3^5$ (so that $\sqrt[5]{t} = 3$), because $f'(x)$ can be evaluated easily at each of these numbers. Table 3.2 shows our results; these are summarized in Figure 3.13.

Our calculations indicate that $f(x)$ has a relative maximum at $x = 0$ and a relative minimum at $x = 32$. From equation (1), note that

Interval	$(-\infty, 0)$	$(0, 32)$	$(32, \infty)$
test value t	-1	1	3^5
$f'(t)$	6	-2	$\frac{2}{3}$
sign of $f'(x)$ on the interval	$+$	$-$	$+$
behavior of $f(x)$	increasing	decreasing	increasing

Table 3.2
Behavior of $f(x)$ near critical values

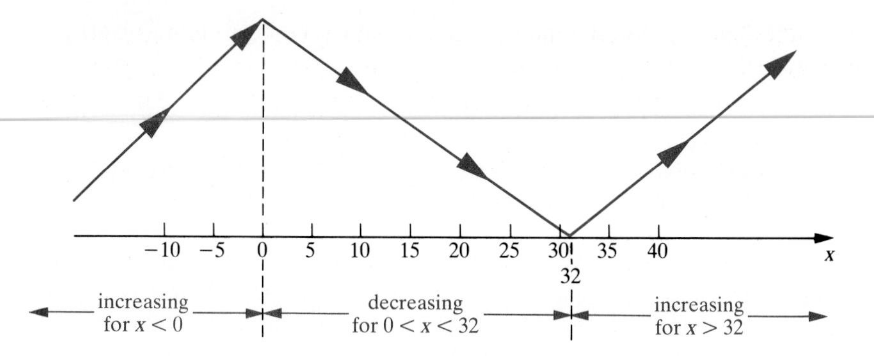

Figure 3.13
Increasing and decreasing regions for $f(x)=2x-5x^{4/5}+3$

$|f'(x)|\to\infty$ as $x\to 0$. In fact, from Table 3.2 we know that $f'(x)$ is positive when $x<0$, so $f'(x)\to+\infty$ as $x\to 0^-$. Similarly, since $f'(x)<0$ for $0<x<32$, we know that $f'(x)\to-\infty$ as $x\to 0^+$. The concavity of $f(x)$ may be determined by computing $f''(x)$. From equation (1),

$$f''(x)=\tfrac{4}{5}x^{-6/5}=\frac{4}{5\sqrt[5]{x^6}} \quad \text{for } x\neq 0.$$

Since $f''(x)$ is positive, the graph of $f(x)$ is concave upward for all $x\neq 0$. The graph of $f(x)$ is shown in Figure 3.14. ■

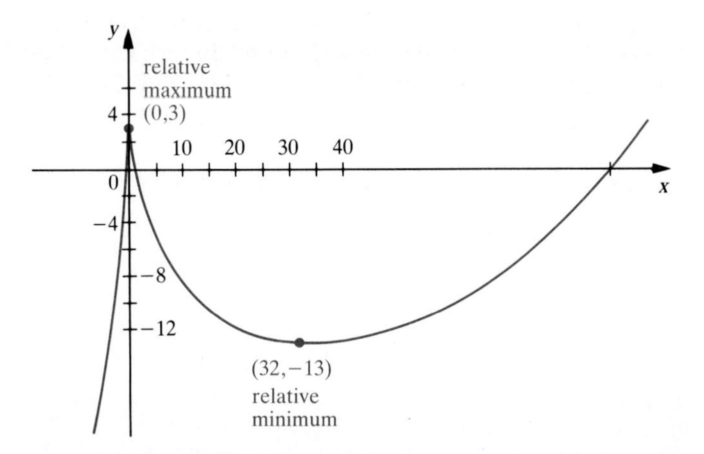

Figure 3.14
Graph of $f(x)=2x-5x^{4/5}+3$

In cases where $f'(a)=0$, another test can be used to determine if $f(a)$ is a relative maximum or a relative minimum. Although this test is often easier to use than the First Derivative Test, it may be inconclusive or may not be applicable in some cases.

The Second Derivative Test

Suppose that $f(x)$ is a twice-differentiable function in some interval about a, and suppose that $f'(a)=0$.

A. If $f''(a)<0$, then $f(a)$ is a relative maximum of $f(x)$.
B. If $f''(a)>0$, then $f(a)$ is a relative minimum of $f(x)$.
C. If $f''(a)=0$, then the test is inconclusive.

Condition A of the test says that the graph of $f(x)$ is concave downward, similar to the graph in Figure 3.15(a). Since $f''(x)$ is negative, the function $f'(x)$ is a decreasing function; that is, the slope of the tangent line is decreasing. Since the slope of $f(x)$ is zero at the critical value, the slope must be positive to the left and negative to the right of the critical value. Thus, according to the First Derivative Test, the critical value corresponds to a relative maximum.

Condition B says that the graph is concave upward, similar to the graph in Figure 3.15(b). In this case $f''(x) > 0$ implies that $f'(x)$ is an increasing function. Since $f'(x)$ is increasing and is zero at the critical value, it must be negative to the left and positive to the right of the critical value. According to the First Derivative Test, $f(x)$ has a relative minimum at the critical value.

If $f''(a)$ equals zero or does not exist, then we can make no conclusion about the concavity of $f(x)$. The First Derivative Test should be used in such cases. The failure of the Second Derivative Test when $f''(x)$ is zero at a critical value may be observed for the two functions $f_1(x) = x^4$ and $f_2(x) = -x^4$. The graph of $f_1(x)$, shown in Figure 3.16(a), has a critical value and a relative *minimum* at $x = 0$. Furthermore, $f_1''(0) = 0$. The graph of $f_2(x)$ is shown in Figure 3.16(b). This function has a critical value and a relative *maximum* at $x = 0$. Again $f_2''(0) = 0$. Although each of these functions has a second derivative equal to zero at a critical value, they behave oppositely at the critical value. These facts confirm that it is impossible to reach a conclusion about the nature of the critical value when $f''(x) = 0$ without further investigation.

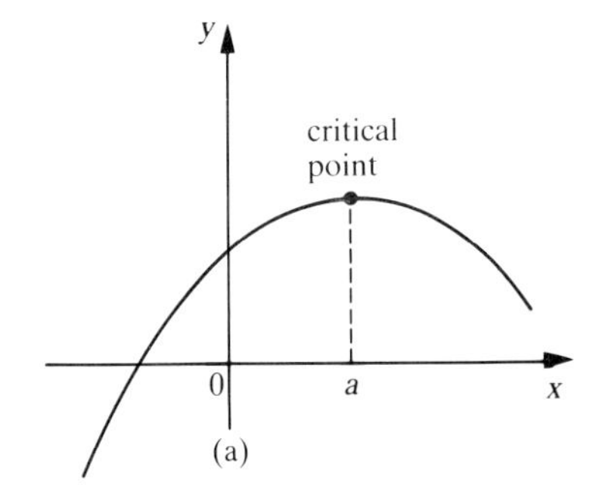

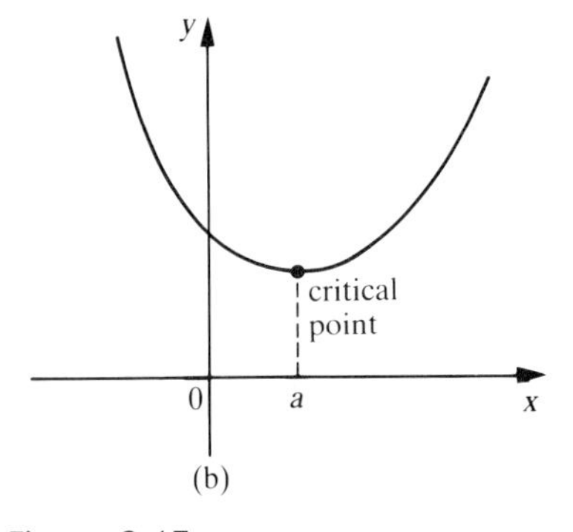

Figure 3.15
(a) A graph with $f''(x) < 0$ near the critical value a; (b) a graph with $f''(x) > 0$ near the critical value a

Example 3.25 Find all relative maxima and relative minima of $f(x) = 2x^3 + 3x^2 - 36x$.

Solution First find $f'(x)$ and $f''(x)$:

$$f'(x) = 6x^2 + 6x - 36 = 6(x^2 + x - 6)$$
$$f''(x) = 12x + 6 = 6(2x + 1).$$

To determine values of x for which a relative maximum or relative minimum may occur, we solve $f'(x) = 0$ for x:

$$x^2 + x - 6 = 0$$
$$(x + 3)(x - 2) = 0$$
$$x = -3 \quad \text{or} \quad x = 2$$

Hence $x = -3$ and $x = 2$ are critical values. We use the Second Derivative Test to determine if these values are relative maxima or relative minima:

$$f''(-3) = 6[2(-3) + 1] = -30 < 0$$
$$f''(2) = 6[2(2) + 1] = 30 > 0.$$

Since $f'(-3) = 0$ and $f''(-3) < 0$, we conclude that $f(x)$ has a relative maximum at $x = -3$. The relative maximum value is $f(-3) = 81$. Since $f'(2) = 0$ and $f''(2) > 0$, the function has a relative minimum value of $f(2) = -44$ when $x = 2$. A graph of $f(x)$ is sketched in Figure 3.17. ∎

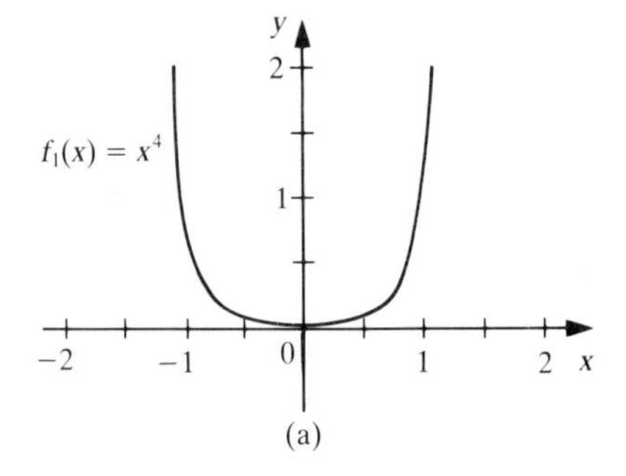

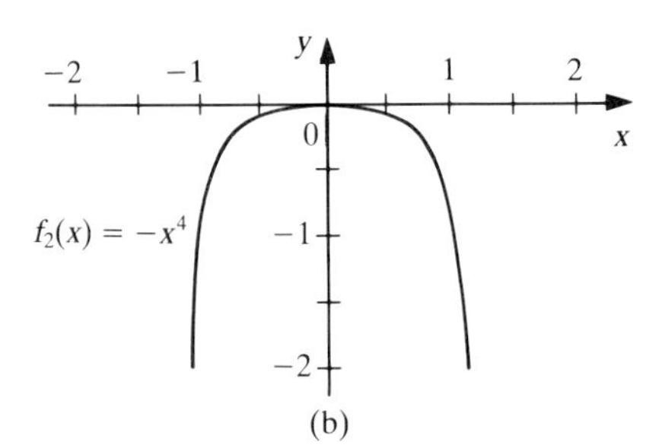

Figure 3.16
(a) Graph of $f_1(x) = x^4$; (b) graph of $f_2(x) = -x^4$

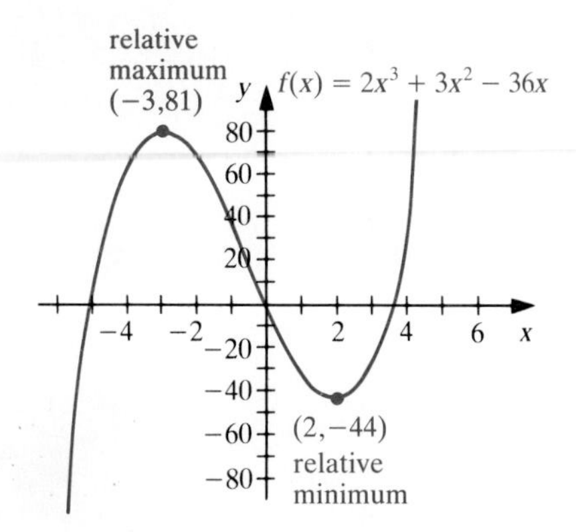

Figure 3.17
Graph of $f(x) = 2x^3 + 3x^2 - 36x$

In the preceding example, the functional value at $x = -3$ is 81 and the functional value at $x = 2$ is -44. The former is a relative maximum, and the latter, a relative minimum. Since $81 > -44$, we may be tempted to conclude from this information alone that a relative maximum occurs at $(-3, 81)$ and that $(2, -44)$ is a relative minimum. The next example shows that we cannot rely on the functional *value* to distinguish between relative maxima and relative minima.

Example 3.26 Find the critical values of

$$f(x) = x + \frac{1}{x}$$

and classify them as corresponding to relative maxima or relative minima.

Solution We first write $f(x)$ as

$$f(x) = x + x^{-1}$$

and then use the power rule to find $f'(x)$ and $f''(x)$. The results are

$$f'(x) = 1 - x^{-2} = 1 - \frac{1}{x^2},$$

$$f''(x) = 2x^{-3} = \frac{2}{x^3}.$$

We set $f'(x)$ to zero and solve for x, obtaining

$$1 - \frac{1}{x^2} = 0$$

or

$$\frac{x^2 - 1}{x^2} = 0$$

so that $\qquad x^2 = 1. \qquad$ (numerator must equal zero)

Therefore, critical values occur at

$$x = 1 \quad \text{and} \quad x = -1.$$

Even though $x = 0$ is a value where $f'(x)$ is not defined, $x = 0$ is not in the domain of $f(x)$, and is therefore not a critical value.

Since $f'(-1) = 0$ and $f'(1) = 0$, we may use the Second Derivative Test to classify these critical values. We evaluate $f''(1)$ and $f''(-1)$:

$$f''(1) = 2 > 0 \quad \text{and} \quad f''(-1) = -2 < 0.$$

Thus, the point $(1, 2)$ is a relative minimum and the point $(-1, -2)$ is a relative maximum. Note that the functional value 2 at the minimum is *greater* than the functional value -2 at the maximum. See Figure 3.18. ■

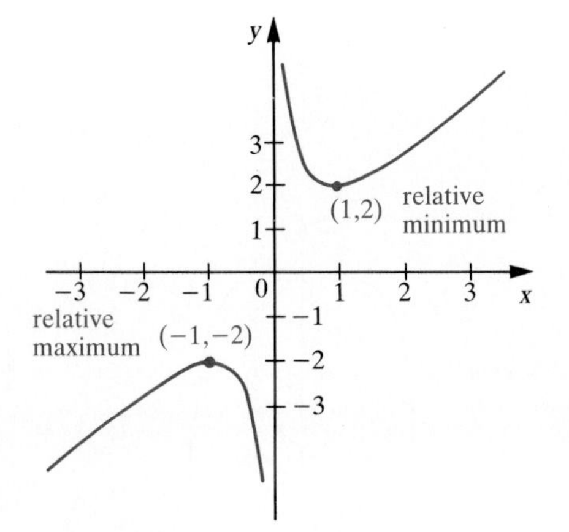

Figure 3.18
Graph of $f(x) = x + \dfrac{1}{x}$

Example 3.27 Find all critical values of $f(x) = x^2 - 3x^{4/3}$. Determine if each critical value corresponds to a relative maximum, a relative minimum, or neither.

Solution The derivative is $f'(x) = 2x - 4x^{1/3}$. There are no values of x where $f'(x)$ is not defined, and it is zero when

$$2x - 4x^{1/3} = 0,$$

that is, when $$x = 2\sqrt[3]{x}.$$

Cubing both sides gives $$x^3 = 8x.$$

Hence either $x = 0$ or $x^2 = 8$. The function $f(x)$ therefore has critical values at $x = 0$, $x = +\sqrt{8}$, and $x = -\sqrt{8}$. The corresponding y-coordinates are computed from the formula for $f(x)$ to be 0, -4, and -4, respectively.

To classify the critical values as relative maxima, relative minima, or neither, we compute $f''(x)$ and then apply the Second Derivative Test:

$$f''(x) = 2 - \tfrac{4}{3}x^{-2/3} = 2 - \frac{4}{3\sqrt[3]{x^2}}.$$

At $x = \pm\sqrt{8}$, we have $x^2 = 8$ so

$$f''(\pm\sqrt{8}) = 2 - \frac{4}{3\sqrt[3]{8}} = 2 - \frac{4}{3(2)} = \frac{4}{3} > 0.$$

Consequently, we have relative minima at $x = -\sqrt{8}$ and at $x = +\sqrt{8}$.

Since $f''(x)$ is not defined at the critical value $x = 0$, the Second Derivative Test is not applicable. Therefore, we must use the First Derivative Test to determine the nature of the critical value at the origin. Since $-\sqrt{8} < -1 < 0$ and $0 < 1 < \sqrt{8}$, we can determine the sign of $f'(x)$ to the left and to the right of $x = 0$ by evaluating $f'(-1)$ and $f'(1)$, respectively; that is, $t = 1$ and $t = -1$ are acceptable test values for determining the sign of $f'(x)$. From $f'(x) = 2x - 4\sqrt[3]{x}$, we compute $f'(-1) = -2 + 4 = 2$ and $f'(1) = 2 - 4 = -2$. Hence, $f(x)$ is increasing to the left of $x = 0$ and decreasing to the right; we conclude that a relative maximum occurs at $x = 0$. Figure 3.19 shows the graph of this function. ■

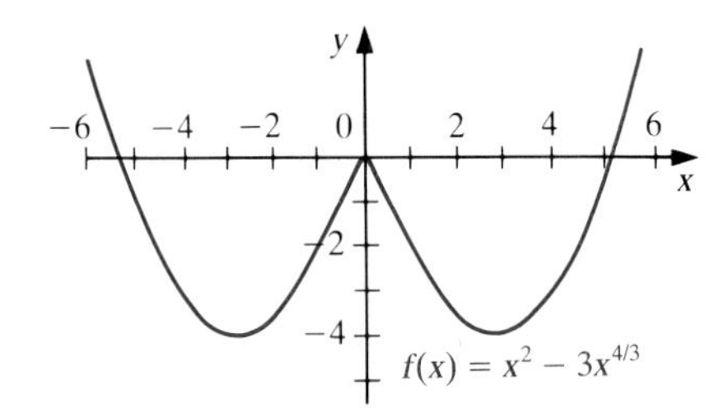

Figure 3.19
Graph of $f(x) = x^2 - 3x^{4/3}$

The following example illustrates that $f'(a) = 0$ does not guarantee that $f(x)$ has a relative maximum or relative minimum at $x = a$.

Example 3.28 Find all relative maxima and relative minima for $f(x) = 3x^4 - 16x^3 + 24x^2$.

Solution We first find the derivative

$$\begin{aligned} f'(x) &= 12x^3 - 48x^2 + 48x \\ &= 12x(x^2 - 4x + 4) = 12x(x-2)^2 \end{aligned}$$

and the second derivative:

$$\begin{aligned} f''(x) &= 36x^2 - 96x + 48 \\ &= 12(3x^2 - 8x + 4) = 12(3x - 2)(x - 2). \end{aligned}$$

Since $f'(x)$ is defined for all x and $f'(x) = 0$ only when $x = 0$ or $x = 2$, these are the only values of x for which $f(x)$ can have a relative maximum or relative minimum. From $f''(0)=48>0$, the Second Derivative Test implies that $f(x)$ has a relative minimum of $f(0)=0$ when $x=0$. However, $f''(2)=0$, so the Second Derivative Test is inconclusive, and we must apply the First Derivative Test. Since $f'(x)=12x(x-2)^2$,

$$f'(x)>0 \quad \text{for } x \text{ near 2 but } x<2$$

and

$$f'(x)>0 \quad \text{for } x \text{ near 2 but } x>2.$$

This information is recorded in Figure 3.20. Thus, the graph of $f(x)$ is increasing to the left *and* to the right of $x=2$, so there is no relative

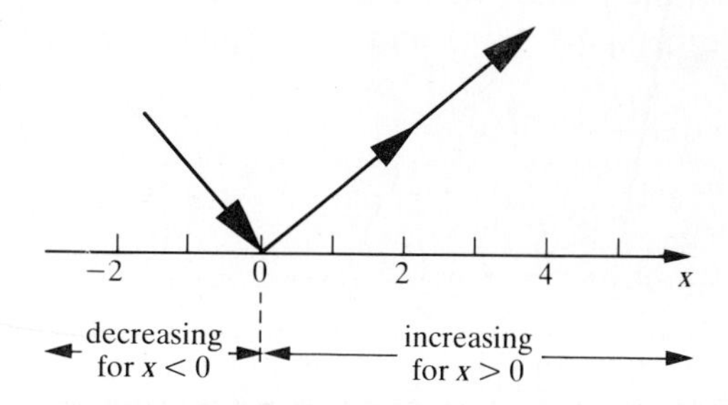

Figure 3.20
Increasing and decreasing regions for $f(x)=3x^4-16x^3+24x^2$

maximum or relative minimum at $x=2$. The graph of $f(x)$ is shown in Figure 3.21. ■

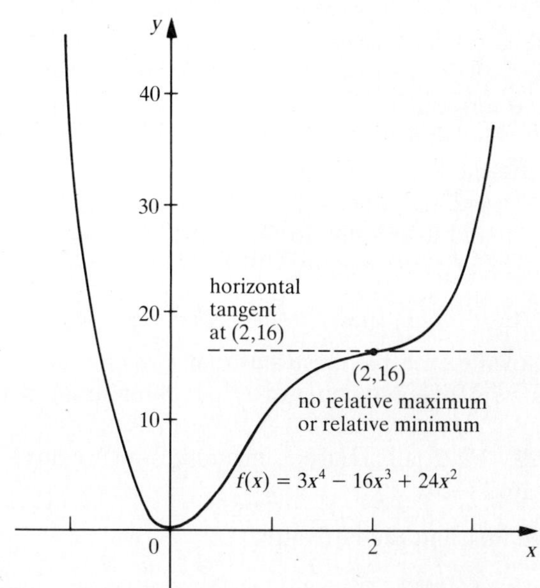

Figure 3.21
The graph of $f(x)=3x^4-16x^3+24x^2$

Exercises

Exer. 1–20: Find the critical value(s) for the function. Use the First Derivative Test to classify the results as relative maxima, relative minima, or neither. Sketch a graph of the function by plotting points corresponding to relative maxima and relative minima, observing where the function is increasing, decreasing, concave upward, and concave downward.

1. $f(x) = x^2 - 2x + 7$
2. $f(x) = 3x^2 + 4$
3. $f(x) = -x^2 + 6x + 9$
4. $f(x) = -4x^2 + 3x + 6$
5. $f(t) = -4t^2 + 2t + 1$
6. $f(x) = 2x^2 - 3x + 4$
7. $f(x) = x^3 - 6x^2 + 9x - 12$
8. $Q(r) = r^3 - 12r - 16$
9. $f(u) = u^3 - 9$
10. $f(x) = x^3 - 3x^2$
11. $f(x) = \dfrac{x-1}{x^2-1}$
12. $f(t) = \dfrac{t^2+1}{t-4}$
13. $w(x) = 1 - a\sqrt[3]{x^2}, \quad a > 0$
14. $f(x) = x\sqrt{x^2-1}$
15. $f(s) = \dfrac{1}{s^2+1}$
16. $g(s) = \dfrac{(s+2)^{4/3}}{s-3}$
17. $f(x) = \dfrac{x-2}{(x-1)^2}$
18. $f(x) = x - 3x^{1/3}$
19. $g(p) = \dfrac{2p+5}{p^2+2p-1}$
20. $f(x) = \dfrac{x-1}{x^2-x+1}$

Exer. 21–34: Find the critical value(s) for the function. Use the Second Derivative Test when possible to classify the results as corresponding to relative maxima or relative minima. When necessary, use the First Derivative Test.

21. $f(x) = x^2 - 2x + 1$
22. $f(x) = -x^3 + 4x^2 - 2$
23. $f(x) = -2x^2 - 5x + 7$
24. $f(x) = x^3 - 3x^2 + 4$
25. $f(t) = -3t^2 + 2t + 1$
26. $g(t) = t^3 - 3t + 6$
27. $f(x) = -x^3 + 12x - 16$
28. $f(x) = x^3 + 12x + 16$
29. $g(s) = -s^3 + s$
30. $f(p) = -p^3 - 6p^2 - 9p + 1$
31. $f(x) = \dfrac{x^2-x}{x-2}$
32. $f(x) = x\sqrt[3]{x^2-1}$
33. $Q(r) = r\sqrt{r^2-4}$
34. $f(x) = x(x^2-1)^3$

Exer. 35–48: Locate all relative maxima and relative minima for the function.

35. $f(x) = 2x - 5x^{4/5} + 3$
36. $f(x) = 4x^2 - \frac{3}{8}x^{8/3} + 1$
37. $f(x) = x^{-1/2}$
38. $f(x) = 16x - 3x^{2/3} + 2$
39. $g(t) = t^{1/3}$
40. $w(r) = r^{2/5} + r^{7/5}$
41. $f(x) = x^2 - 5x^{2/5}$
42. $f(x) = x^{3/2} - x^{1/2} + 4$
43. $f(x) = x^2 - 6x^{1/3} - 1$
44. $f(x) = 3x^{8/3} - 32x$
45. $f(u) = u - u^{5/3}$
46. $f(s) = s(s^3 - 1)^4$
47. $f(x) = x^2 - 4x^{4/3}$
48. $f(x) = x^{5/3} - x^{2/3}$

49. Show that $f(x) = x^3 + (3/x)$ is such that its relative minimum is greater than its relative maximum.

Absolute Maxima and Absolute Minima — 3.5

Many applications involve determining the maximum or minimum value of a function $f(x)$ on a closed interval. The variable x frequently denotes a quantity of some physical entity (dollars, speed, time, items produced) and hence only certain values of x make sense in the context of the application. We seek the maximum or minimum of $f(x)$ when x is restricted to "reasonable" values.

The **absolute maximum** of a function is a value $f(a)$ such that $f(a) \geq f(x)$ for *all* values of x in the domain of $f(x)$. Similarly, the **absolute minimum** of a function is a value $f(a)$ such that $f(a) \leq f(x)$ for **all** values of x in the domain of $f(x)$. The graphs in Figure 3.22 demonstrate that absolute maxima and absolute minima do not always occur.

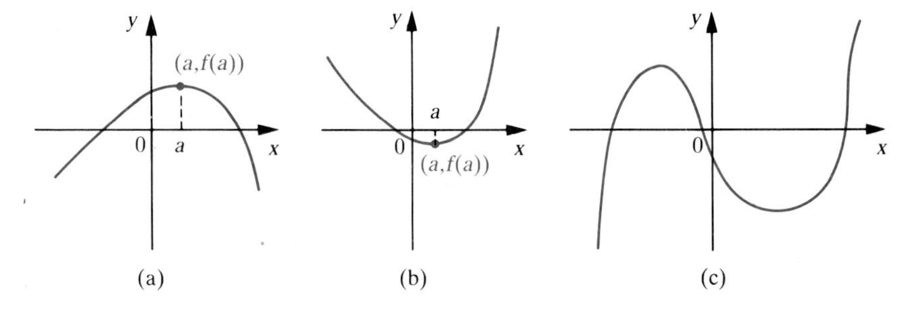

Figure 3.22
Some functions have absolute maxima and/or minima while others do not: (a) $f(a)$ is the absolute maximum, no absolute minimum; (b) $f(a)$ is the absolute minimum, no absolute maximum; (c) no absolute maximum, no absolute minimum

Each function in Figure 3.22 has the set of all real numbers as its domain. In most applications, however, we are interested in functions with more restricted domains. Consider the example (in Section 2.1) about maximizing the volume of a cake pan constructed from a sheet of plastic, 12 inches square (see Figure 3.23). This problem makes sense only for values of x such that $0 \le x \le 12$. Hence, to find the maximum volume, we must examine the values of the volume for x between 0 and 12, and determine the absolute maximum for those values of x.

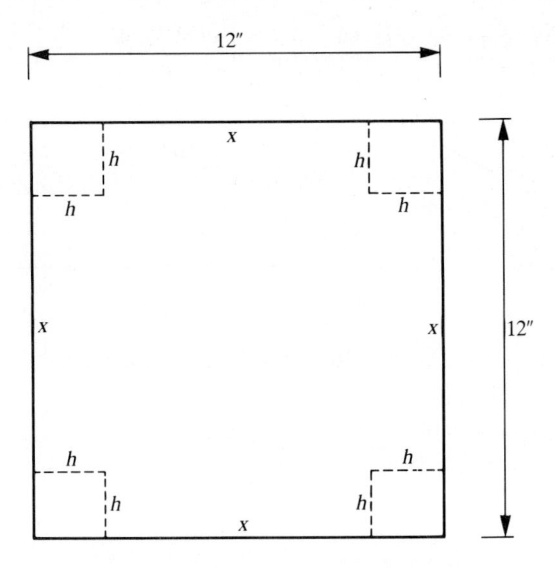

Figure 3.23
A cake pan formed by cutting squares from the corners of a piece of plastic

Since the function $f(x) = 6x^2 - \frac{1}{2}x^3$ represents the volume of the cake pan in Figure 3.23, and since x must lie in the closed interval $[0, 12]$, we seek the absolute maximum of $f(x)$ on the domain $[0, 12]$. The following result is useful.

Theorem For a *continuous* function $f(x)$, the absolute maximum and the absolute minimum on a *finite closed* interval $[b, c]$ always exist. The absolute maximum occurs either at an *endpoint* or at a *relative maximum* within the interval. The absolute minimum occurs either at an *endpoint* or at a *relative minimum* in the interval.

Figure 3.24 illustrates the absolute maximum and absolute minimum for several functions indicating the variety of possible locations for these values. By the theorem, we see that the absolute maximum value of a continuous function $f(x)$ on $[b, c]$ is the *greatest* number among $f(b)$, $f(c)$, and the relative maxima. The absolute minimum value of $f(x)$ on $[b, c]$ is the *least* number among $f(b)$, $f(c)$, and the relative minima. The determination of absolute maxima and absolute minima does not require that we know in advance if the number $f(a)$ (for which

Figure 3.24
Ways that an absolute maximum and absolute minimum can occur on a closed interval: (a) maximum and minimum at endpoints; (b) maximum at critical point and minimum at endpoint; (c) maximum at endpoint and minimum at critical point; (d) maximum and minimum at critical points

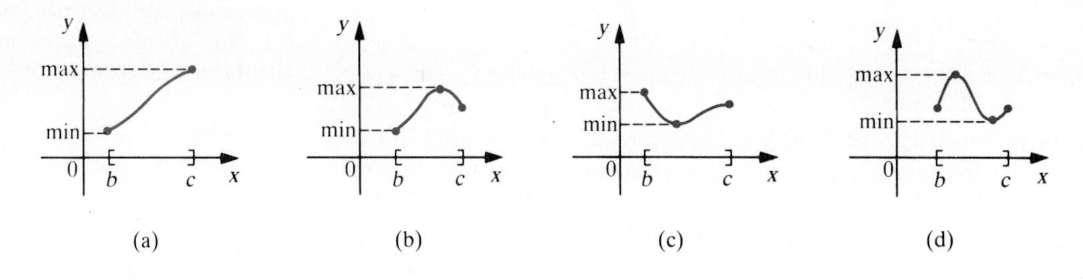

a is a critical value) is a relative maximum, a relative minimum, or neither. The procedure is as follows.

Finding Absolute Maxima and Absolute Minima

Let $f(x)$ be a continuous function on the finite closed interval $[b, c]$. Find all critical values a in (b, c). Of the values $f(b)$, $f(c)$, and these $f(a)$'s, the largest is the absolute maximum of $f(x)$ on $[b, c]$ and the smallest is the absolute minimum of $f(x)$ on $[b, c]$.

Example 3.29 Find the absolute maximum and absolute minimum of $f(x) = -3x^2 + 12x + 7$ on $[1, 6]$.

Solution Since $f'(x) = -6x + 12$ is defined for all x, we have critical values wherever $f'(x) = 0$. Since $f'(x) = 0$ when $x = 2$, we compare the numbers

$f(1) = 16$	$f(2) = 19$	$f(6) = -29$
endpoint	critical value	endpoint

Thus, the absolute maximum of $f(x)$ on $[1, 6]$ is 19, which occurs when $x = 2$, and the absolute minimum of $f(x)$ on $[1, 6]$ is -29, which occurs when $x = 6$. See Figure 3.25. ■

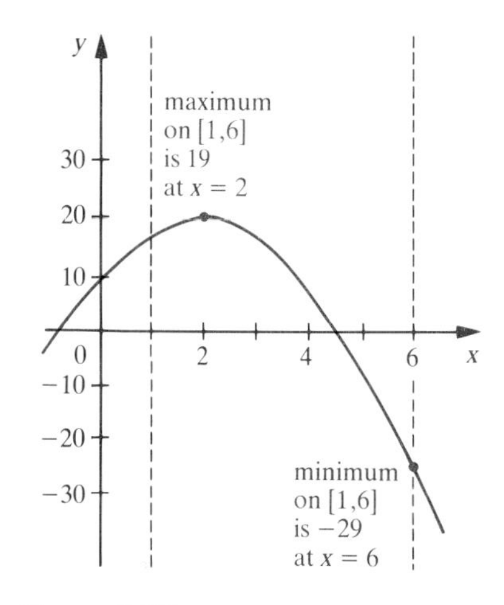

Figure 3.25
For $f(x) = -3x^2 + 12x + 7$ on $[1, 6]$ the absolute maximum is at $x = 2$ and the absolute minimum is at $x = 6$

Example 3.30 Find the absolute maximum and absolute minimum of $f(x) = \frac{1}{4}x^4 - 3x^3 + 10x^2$ on the interval $[1, 3]$.

Solution The derivative is $f'(x) = x^3 - 9x^2 + 20x = x(x^2 - 9x + 20)$, or $f'(x) = x(x-4)(x-5)$. Thus the critical values are $x = 0$, $x = 4$, and $x = 5$. None of these values lies in the interval $[1, 3]$ on which we seek the maximum and minimum. Thus we need to consider only the functional values at the endpoints. A calculation gives

$$f(1) = 7.25 \quad \text{and} \quad f(3) = 29.25.$$

Hence on the interval $[1, 3]$, the absolute maximum value of $f(x)$ is $f(3) = 29.25$ and the absolute minimum of $f(x)$ is $f(1) = 7.25$. ■

Example 3.31 Find the absolute maximum and absolute minimum of $f(x) = \frac{1}{4}x^4 - x^3 + x^2 + \frac{3}{4}$ on the interval $[-\frac{3}{2}, \frac{3}{2}]$.

Solution The derivative $f'(x) = x^3 - 3x^2 + 2x = x(x-1)(x-2)$ is zero at $x = 0$, $x = 1$, and $x = 2$. The first two critical values are in the given interval, but $x = 2$ is not. We therefore choose the largest and smallest of the values $f(-\frac{3}{2})$, $f(0)$, $f(1)$, and $f(\frac{3}{2})$. We compute

$$f(-\tfrac{3}{2}) = \tfrac{489}{64} = 7.64063 \qquad f(0) = \tfrac{3}{4} = 0.75$$
$$f(1) = 1 \qquad f(\tfrac{3}{2}) = \tfrac{57}{64} = 0.890625$$

Hence the absolute maximum value on $[-\frac{3}{2}, \frac{3}{2}]$ occurs at $x = -\frac{3}{2}$ and is 7.64063; the absolute minimum occurs at $x = 0$ and is 0.75. ■

Example 3.32 Refer to Example 1.36 (on page 34) about the life cycle of the codling moth. The brief period of time after the larva hatches and before it penetrates an apple is referred to as the searching stage. When all other factors are held constant, the average amount of time in days, $A(t)$, that it takes for the larva to pass through this stage is a function of the air temperature t (in degrees Celsius). It has been determined that for $20 \le t \le 30$, $A(t)$ is approximately

$$A(t) = \frac{1}{-0.03t^2 + 1.67t - 13.65} = (-0.03t^2 + 1.67t - 13.65)^{-1}.$$

(a) Find the average duration of the searching stage when $t = 22$ °C.
(b) Find the average duration of the stage when $t = 30$ °C.
(c) For temperatures in the range [20, 30], what is the shortest searching stage and what is the longest? At what temperatures do these occur?

Solution **(a)** Since

$$A(22) = \frac{1}{-0.03(22)^2 + 1.67(22) - 13.65} = \frac{1}{8.57} \approx 0.117,$$

it takes about 0.117 days, or 2.81 hours, on the average, for a larva to pass through the searching stage when the temperature is 22 °C.
(b) Since $A(30) = 1/9.45 \approx 0.106$, it takes, on the average, 0.106 days, or 2.54 hours for a larva to pass through the searching stage when $t = 30$ °C.
(c) The shortest and the longest searching stages correspond to the absolute minimum and absolute maximum of $A(t)$ on the interval [20, 30], respectively. Using the chain rule, we obtain

$$A'(t) = -(-0.03t^2 + 1.67t - 13.65)^{-2}(-0.06t + 1.67)$$
$$= \frac{-(-0.06t + 1.67)}{(-0.03t^2 + 1.67t - 13.65)^2}.$$

By the quadratic formula, it may be verified that the denominator of $A'(t)$ is zero when $t \approx 9.95$ and when $t \approx 45.71$. Since both of these values are outside the domain [20, 30], we conclude that $A'(t)$ is defined for all values of t in [20, 30]. Hence the only critical values in the interval [20, 30] occur where $A'(t) = 0 = -0.06t + 1.67$, that is, at $t \approx 27.8$ °C.

Evaluating $A(t)$ at $t = 27.8$ and at the endpoints, we find that $A(20) \approx 0.129$ days (3.10 hours), $A(27.8) = 0.104$ days (2.50 hours), and, as calculated in part (b), $A(30) = 0.106$ days (2.54 hours). Consequently, the shortest searching stage occurs at 27.8 °C and lasts, on the average, 2.5 hours, while the longest searching stage occurs at 20 °C and lasts, on the average, 3.1 hours. Hence, at the least advantageous temperature (27.8 °C), the larva can be killed only during the 2.5 hours between its hatching and penetrating an apple. However, at 20 °C, larva are vulnerable to pesticide for more than three hours. ■

The examples in this section have dealt with continuous functions on finite closed intervals. Three conditions—(i) a continuous function, (ii) a closed interval, and (iii) a finite interval—are necessary to *guarantee* the existence of an absolute maximum and absolute minimum on the interval. Violation of any of these conditions may result in a situation where an absolute maximum, an absolute minimum, or both fail to exist. For example, the function $f(x) = 1/x$ is *not* continuous on the finite closed interval $[-1, 1]$ and has neither an absolute maximum nor an absolute minimum on that interval. See Figure 3.26(a).

The function $f(x) = x$ is continuous on the finite *open* interval $(-1, 1)$. Even though $f(x)$ takes on every value between -1 and 1 on this interval, 1 is not an absolute maximum for $f(x)$ on this interval because there is no number a in the interval $(-1, 1)$ such that $f(a) = 1$. The only value of a for which $f(a) = 1$ is $a = 1$, and this lies *outside* the interval $(-1, 1)$. This function has neither an absolute maximum nor an absolute minimum on the open interval $(-1, 1)$. See Figure 3.26(b).

Finally, an *infinite* interval may result in no absolute maximum or absolue minimum. Again, consider the function $f(x) = x$. On the infinite interval $(-\infty, \infty)$, $f(x)$ has neither an absolute maximum nor an absolute minimum. On $[0, \infty)$ it has an absolute minimum of 0 at $x = 0$, but it has no absolute maximum.

Consequently, violation of one or more of the three conditions that guarantee existence of absolute maxima and absolute minima allows the construction of functions and intervals with no absolute maxima or absolute minima. However, a particular function on a particular interval may still have an absolute maximum, an absolute minimum, or both without satisfying the three conditions. We will consider the case of a continuous (in fact, differentiable) function on an infinite interval in later applications. The next two examples illustrate a method for finding an absolute maximum and absolute minimum, provided that they exist.

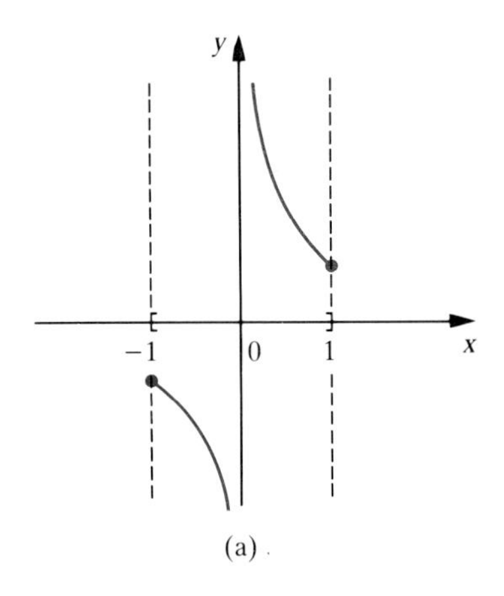

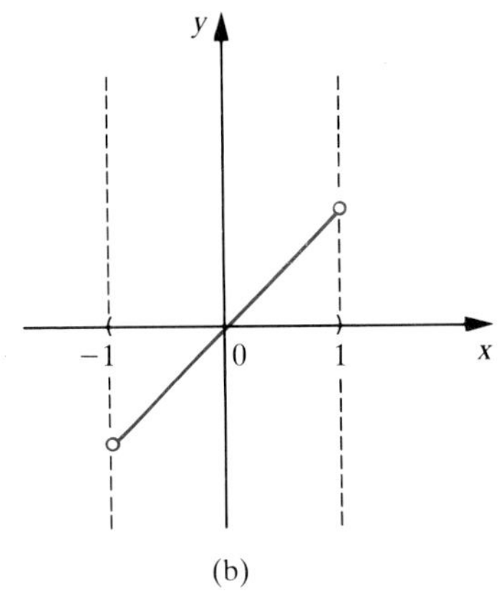

Figure 3.26
(a) $f(x) = 1/x$ on $[-1, 1]$;
(b) $f(x) = x$ on $(-1, 1)$; neither has an absolute maximum or minimum

Example 3.33 Determine the absolute maximum and absolute minimum of $f(x) = 1/x$ on $[2, \infty)$.

Solution We begin by finding all critical values. The derivative of $f(x)$ is

$$f'(x) = -\frac{1}{x^2}.$$

The derivative is never zero and is undefined at $x = 0$. Since 0 is not in the domain of the function, $x = 0$ is not a critical value. Therefore, the function has no critical values. However, note that on the given interval $[2, \infty)$, $f'(x) < 0$, indicating that $f(x)$ is a decreasing function on this interval. Hence the value of the function at the left endpoint is its absolute maximum value on the interval $[2, \infty)$. The absolute maximum of $f(x) = 1/x$ on $[2, \infty)$ occurs at $x = 2$ and is $\frac{1}{2}$. Even though $f(x) \to 0$ as $x \to \infty$, $f(x) > 0$ for all $x > 0$, so there is no number a such that $f(a) = 0$. Consequently, this function has no absolute minimum on $[2, \infty)$. ■

Example 3.34 Determine the absolute maximum and absolute minimum of $f(x) = x/(x^2+1)$ on $[0, \infty)$.

Solution We begin by determining the critical values of $f(x)$. We find the derivative by the quotient rule:

$$f'(x) = \frac{d}{dx}\frac{x}{x^2+1} = \frac{(x^2+1)(1) - x(2x)}{(x^2+1)^2}$$

$$= \frac{1-x^2}{(x^2+1)^2}. \qquad (1)$$

This expression is zero when the numerator is zero, that is, when $x = \pm 1$. Since we are interested in the interval $[0, \infty)$, the critical value $x = -1$ is ignored. The y-coordinate corresponding to $x = 1$ is $y = \frac{1}{2}$.

If we were interested in a finite closed interval such as $[0, 2]$, then we would compare $f(0)$, $f(1)$, and $f(2)$. (The interval $[0, 2]$ was chosen because it contains the critical value; other intervals such as $[0, 3]$, $[0, 4]$, and so on would be equally acceptable.) Of the numbers $f(0)$, $f(1)$, and $f(2)$, the largest value is the absolute maximum and the smallest value is the absolute minimum on $[0, 2]$. In fact, $f(0) = 0$, $f(1) = \frac{1}{2}$, and $f(2) = \frac{2}{5}$, so that, *on the interval* $[0, 2]$, the absolute maximum is $\frac{1}{2}$ and the absolute minimum is 0.

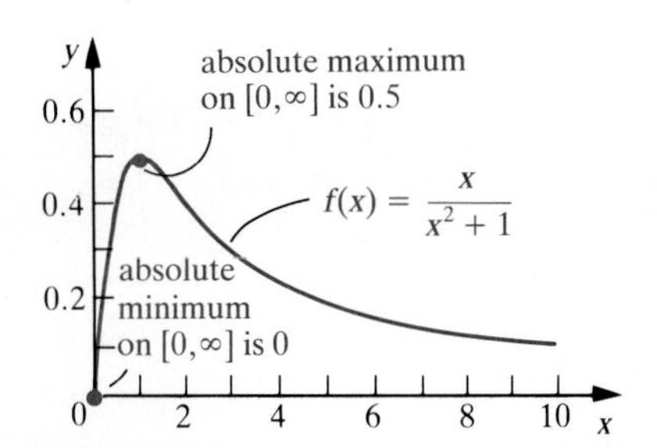

Figure 3.27

Graph of $f(x) = \dfrac{x}{x^2+1}$

To complete our solution, we must analyze what happens to $f(x)$ on $[2, \infty)$. For x in the interval $[2, \infty)$, equation (1) shows that $f'(x) < 0$. This indicates that $f(x)$ is decreasing on $[2, \infty)$, so $f(x) \leq f(2) = \frac{2}{5}$ for all $x \geq 2$. Furthermore, both the numerator and denominator of $f(x)$ are positive on this interval, so $f(x) > 0$. Thus $0 < f(x) \leq \frac{2}{5}$ on the interval $[2, \infty)$.

Combining our conclusions on the intervals $[0, 2]$ and $[2, \infty)$, we see that the absolute maximum of $f(x)$ on $[0, \infty)$ is $\frac{1}{2}$, which occurs at $x = 1$, and the absolute minimum is 0, which occurs at $x = 0$. The graph of $f(x)$ is shown in Figure 3.27. ■

As illustrated in Examples 3.33 and 3.34, the problem of determining the absolute maximum and absolute minimum of a continuous function $f(x)$ on $[a, \infty)$ can often be solved as follows. Determine an interval $[a, b]$ containing all critical values of $f(x)$ which are smaller than b. Find the absolute maximum and absolute minimum on $[a, b]$ by the method described at the beginning of this section. Analyze the behavior of $f(x)$ on $[b, \infty)$; (note that $f(x)$ is always increasing or always decreasing on this interval). Combine the information on $[a, b]$ and $[b, \infty)$.

Exercises

Exer. 1–26: Find the absolute maxima and absolute minima for the function on the closed interval.

1. $f(x) = x^2 + 2x - 1$; $[-3, 2]$
2. $f(x) = x^2 + 2x - 1$; $[3, 5]$
3. $f(x) = -2x^2 + 8x + 5$; $[1, 5]$
4. $f(x) = 2x^3 - 3x^2 - 36x + 12$; $[-4, 3]$
5. $f(s) = 2s^3 - 6s + 23$; $[-3, \frac{3}{2}]$
6. $f(t) = -2t^3 - 3t^2 + 12t - 5$; $[-3, 5]$

7. $f(x) = \frac{x^2 - 2}{x - 4}$; $[0, 3]$
8. $f(x) = x(x^3 - 1)^4$; $[-2, 2]$
9. $g(r) = \frac{r - 5}{r + 2}$; $[0, 6]$
10. $M(s) = \frac{s^2 - 3}{s - 4}$; $[1, 3]$
11. $f(x) = \frac{(x + 2)^{4/3}}{x - 3}$; $[0, 2]$
12. $f(x) = \frac{1}{x^2 + 1}$; $[-3, 3]$
13. $Q(s) = s - 3s^{1/3}$; $[-1, 1]$
14. $f(x) = \frac{x - 2}{(x - 1)^2}$; $[2, 4]$
15. $f(x) = \frac{x - 1}{x^2 - x + 1}$; $[-2, 2]$
16. $N(t) = \frac{2t + 5}{t^2 + 2t - 1}$; $[1, 6]$
17. $f(x) = \frac{x^2 + 1}{x - 4}$; $[0, 3]$
18. $f(x) = x^3 - 12x - 16$; $[-3, 4]$
19. $f(u) = u\sqrt{u^2 - 1}$; $[1, 4]$
20. $f(x) = x^3 - 3x^2$; $[-1, 4]$
21. $f(x) = -x^3 + x$; $[0, 1]$
22. $P(t) = -t^3 - 6t^2 - 9t + 1$; $[-4, 0]$
23. $f(x) = \frac{x^2 - x}{x - 2}$; $[-1, 1]$
24. $f(x) = x\sqrt[3]{x^2 - 1}$; $[-1, 1]$
25. $w(t) = t\sqrt{t^2 - 4}$; $[3, 4]$
26. $f(x) = x(x^2 - 1)^3$; $[0, 1]$

Exer. 27–40: Find the absolute maximum and absolute minimum of the function on the interval.

27. $f(x) = 4/x$; $[1, \infty)$
28. $f(x) = 3/x$; $(-\infty, -1]$
29. $f(x) = 2x + 1/x$; $[1, \infty)$
30. $f(x) = 2x^2 + 4/x$; $(0, \infty)$
31. $g(t) = \frac{t - 2}{(t + 1)^2}$; $[0, \infty)$
32. $f(w) = \frac{w + 2}{w^2 + 4w + 5}$; $[0, \infty)$
33. $f(x) = \frac{x}{x^2 + 1}$; $[4, \infty)$
34. $f(x) = \frac{x}{x^2 + 1}$; $(-\infty, \infty)$
35. $f(s) = \frac{s}{s^4 + 1}$; $(-\infty, \infty)$
36. $f(x) = \frac{x}{x^4 + 1}$; $(2, \infty)$
37. $f(x) = x^3 - x$; $[0, \infty)$
38. $M(p) = p^3 - p^2$; $[0, \infty)$
39. $Q(r) = r + 1/r$; $[\frac{1}{2}, \infty)$
40. $f(x) = \frac{x + 2}{x^2 + 4}$; $[0, \infty)$

Applied Optimization Problems

3.6

We began our discussion of derivatives by considering how to design a cake pan of maximum volume from a square piece of plastic. Other maximum–minimum (max–min) problems include the following:

- A manager of a company wants to determine how to obtain the most profit or spend the least amount of money per item manufactured.
- A politician wants to decide how much television advertising to buy to get the most votes.
- A sociologist wants to determine when a certain government program is most effective or least costly.
- A psychologist wants to know under which experimental conditions a subject reacts to a stimulus with the most surprise.

Individuals frequently face questions of how to obtain the most or the least of some entity. Since we are trying to find optimal values, such questions are usually called **optimization problems**. To solve an optimization problem we need to study the behavior of a function in order to determine its maximum or minimum value. The principal component for success in working applied optimization problems is good organization. The following steps will usually lead to solutions of these problems.

1. **Understand the problem.** These problems are normally in narrative form. Read the problem carefully (perhaps several times) and restate it in your own words. Assign a letter variable to each of the quantities discussed. Wherever possible, sketch a diagram to show the relationships between quantities. Identify the quantity that is to be maximized or minimized.
2. **Form a mathematical statement of the problem.** Use the information obtained in Step 1 to find a function of one variable that represents the quantity to be maximized or minimized. This may require substitutions involving known relationships among the variables. Write this function explicitly and determine its domain.
3. **Determine the maximum or minimum.** Apply the techniques presented in Sections 3.4 and 3.5 to determine the desired maximum or minimum: locate critical values; use one of the derivative tests to identify maxima and minima; if the domain is a closed interval, check the endpoints.
4. **Answer the question.** Review the question asked in the narrative to be certain that you have answered it. The answers to some questions may require a computation after the maximum or minimum has been determined.

Example 3.35 A manufacturer plans to form a cake pan by removing squares from the corners of a square piece of plastic and bending up the sides as shown in Figure 3.28. What are the dimensions of the pan of maximum volume?

Solution This is the cake pan example in Section 2.1.

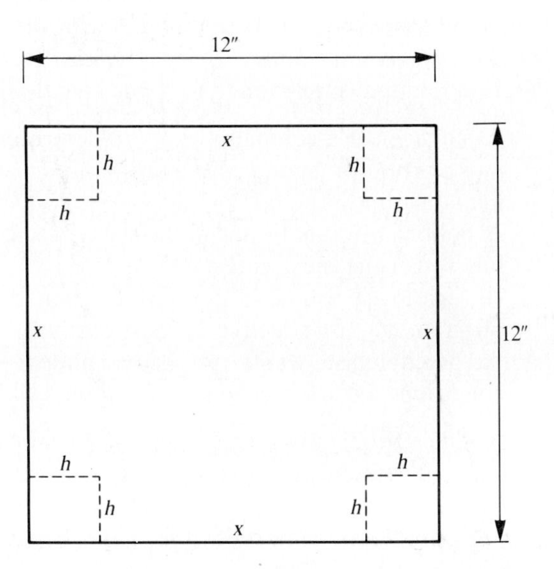

Figure 3.28
Relationship between x and h in Example 3.35

Step 1 We assign the following letters to the unknown quantities involved:

$$V = \text{volume of the pan}$$
$$x = \text{the length of a side of the square base}$$
$$h = \text{the length of a side of the corner to be cut out}$$
$$= \text{the height of the pan.}$$

We then sketch a diagram illustrating the relationships between the quantities (see Figure 3.28). The following relationships are observed from the diagram:

$$V = x^2h \quad \text{and} \quad 2h + x = 12.$$

Step 2 We now obtain V, the quantity to be maximized, as a function of one variable. Since

$$h = \frac{12 - x}{2},$$

we have

$$V = x^2h = x^2\left(\frac{12 - x}{2}\right)$$

or

$$V = 6x^2 - \tfrac{1}{2}x^3.$$

Note that only values of x on the closed interval $[0, 12]$ make sense in this problem, so we seek the maximum value of V for $0 \le x \le 12$. To emphasize the fact that V is a function of x, we will write $V(x)$ for the volume.

Step 3 We find the absolute maximum of $V(x)$ on $[0, 12]$. Applying the techniques of the previous section, we find the derivative

$$\frac{dV}{dx} = 12x - \tfrac{3}{2}x^2 = \tfrac{3}{2}x(8 - x).$$

Since $V'(x) = 0$ when $x = 0$ and $x = 8$, and since $x = 0$ and $x = 12$ are endpoints of the domain, we compare the values of $V(x)$ at $x = 0$, $x = 8$, and $x = 12$:

$$V(0) = 0, \qquad V(8) = 128, \qquad V(12) = 0.$$

We conclude that $V(x)$ has a maximum value of 128 in.^{3}, which occurs at $x = 8$ inches. See Figure 3.29.

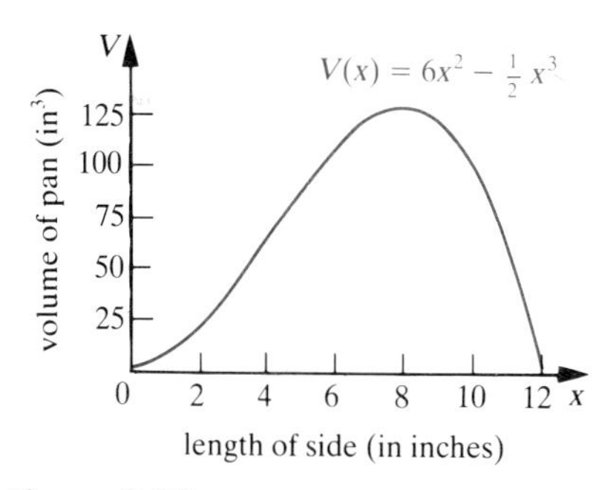

Figure 3.29
Graph of the volume function $V(x) = 6x^2 - \frac{1}{2}x^3$

Step 4 We review the problem to ensure that the question asked has been answered. In this case we are asked to find the dimensions of the pan. From our picture, we see that the dimensions of the box are x by x by h. Since $x = 8$ and

$$h = \frac{12 - x}{2} = \frac{12 - 8}{2} = 2,$$

the dimensions of the largest tray are 8 inches by 8 inches by 2 inches, a very common size for microwave trays and cake pans. ■

Example 3.36 For the first 15 days of a flu epidemic, public health workers predict that t days after the start of the epidemic the number of people I (in hundreds) contracting the infection each day is given by

$$I = -3t^2 + 54t + 73 \quad \text{for } 0 \leq t \leq 15.$$

What is the worst day of the epidemic? On the worst day, how many people will be infected?

Figure 3.30
Graph of $I(t) = -3t^2 + 54t + 73$

Solution Since we wish to find the maximum value of I, which is already expressed as a function of t, steps one and two are done for us in the statement of the problem. To emphasize the dependence of I on t, we will write $I(t)$. The domain of $I(t)$ is the interval $[0, 15]$. We want to find the absolute maximum of $I(t)$ for $0 \leq t \leq 15$. From $I'(t) = -6t + 54$, we see that $I(t)$ has a critical value at $t = 9$; therefore, we compare $I(0)$, $I(9)$, and $I(15)$. Since $I(9) = 316$, $I(0) = 73$, and $I(15) = 208$, the absolute maximum of $I(t)$ on $[0, 15]$ is 316, which occurs at $t = 9$. Thus the ninth day is the worst day of the epidemic, and 316 people are infected on this day. A review of the problem verifies that we have answered all of the questions. The graph of $I(t)$ is sketched in Figure 3.30. ◾

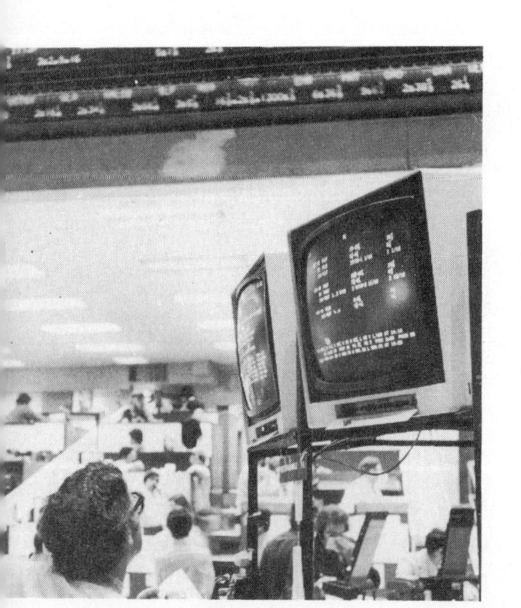

Example 3.37 Suppose we plan to invest in a particular stock on the New York Stock Exchange. After studying the behavior of the stock, we predict that, over the next 30 days, the price per share of the stock on the dth day of the period is given by

$$P(d) = 0.01d^3 - 0.39d^2 + 3.15d + 22.$$

To make the most money, when should we buy the stock and when should we sell? How much money would be made per share purchased (ignoring brokerage fees)?

Solution For maximum profit, we want to buy at the minimum price and sell at the maximum price over the 30-day interval. Note that on the New York Stock Exchange, we are allowed to sell a stock before buying it (this is called selling a stock short). Hence, we can buy and sell in any order. We begin by finding the critical values in the interval. Since

$$\begin{aligned} P'(d) &= 0.03d^2 - 0.78d + 3.15 \\ &= 0.03(d^2 - 26d + 105) \\ &= 0.03(d - 21)(d - 5), \end{aligned}$$

we see that $P'(d) = 0$ when $d = 5$ and $d = 21$. Thus $P(5)$ and $P(31)$ are candidates for maxima and minima. We compare the prices at $d = 0$, 5, 21, and 30.

$$P(0) = \$22.00 \qquad P(5) = \$29.25$$
$$P(21) = \$8.77 \qquad P(30) = \$35.50$$

We conclude that we should buy on day 21 and sell on day 30 to realize a profit of $\$35.50 - \$8.77 = \$26.73$ per share. (Since the Exchange deals only in eighths of dollars, the actual profit would be \$26.75.) See Figure 3.31. ◾

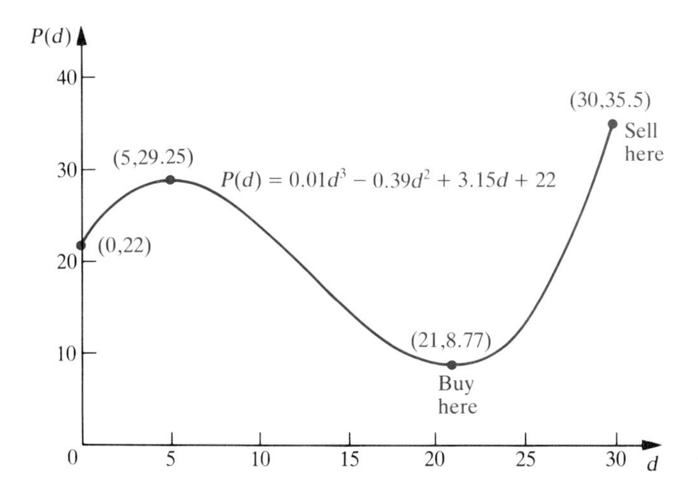

Figure 3.31
Graph of the price function $P(d) = 0.01d^3 - 0.39d^2 + 3.15d + 22$

Example 3.38 A farmer wishes to enclose a rectangular pasture, using a river as one boundary and adding a fence for the other boundaries. If 400 feet of fencing is available, find the dimensions of the largest pasture that can be enclosed?

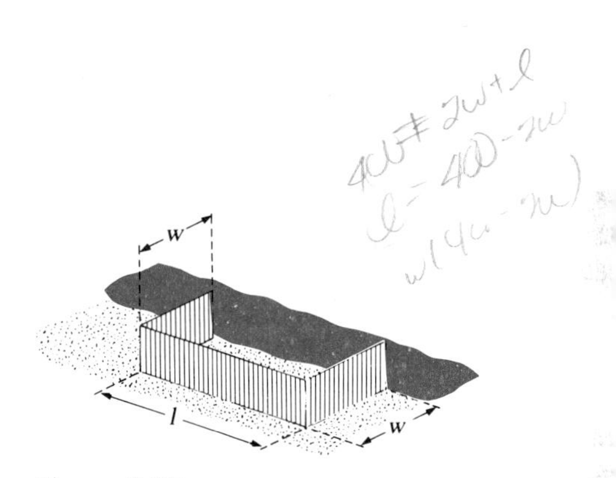

Figure 3.32
A fenced area using a river as one boundary

Solution Consider the diagram in Figure 3.32. Let w be the width of the pasture and let l be the length. Then the area A of the pasture is given by $A = wl$. Since the fence is $2w + l$ feet long and the farmer has 400 feet of fencing, we have $400 = 2w + l$, or $l = 400 - 2w$. Thus,

$$A = wl = w(400 - 2w) = 400w - 2w^2.$$

This gives A as a function of w, so we write A as $A(w)$. Note that since $2w + l = 400$ and $l \geq 0$, values for w must lie on the interval $[0, 200]$. We want to find the maximum of $A(w)$ on this interval. Since

$$\frac{dA}{dw} = 400 - 4w,$$

we see that $A'(w) = 0$ when $w = 100$; thus $A(100)$ is possibly a maximum. We compare

$$A(0) = 0, \qquad A(100) = 20{,}000, \qquad A(200) = 0$$

and conclude that the maximum area is 20,000 ft^2. To answer the question asked, we must find the *dimensions* of the fenced pasture. We know that $w = 100$; since $400 = 2w + l$, we get $l = 200$. The maximum area is obtained when the width is 100 feet and the length is 200 feet. ■

Example 3.39 Find the dimensions of a crate with square base and top such that its volume is 8000 in^3 and its surface area is a minimum (see Figure 3.33).

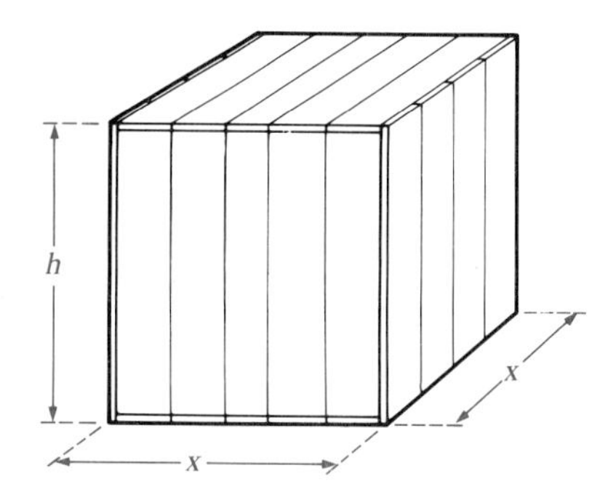

Figure 3.33
Crate with square top and bottom

Solution As noted in Figure 3.33, x denotes the length of each side of the bottom and h denotes the height of the crate. Then the volume is $V = x^2h = 8000$. The area of the top (and of the bottom) is x^2. The area of each of the four sides is xh. Hence the total area A of the sides, top, and bottom is $A = 2x^2 + 4xh$. Since $x^2h = 8000$, we have $h = 8000/x^2$. Using this in the expressions for area, we obtain

$$A(x) = 2x^2 + 4x\left(\frac{8000}{x^2}\right) = 2x^2 + \frac{32,000}{x}. \qquad (1)$$

We cannot have a length of $x = 0$ for the top or bottom, but any positive value of x is permissible. The crate with the least area may have small top and bottom and be quite tall (to have a volume of 8000 in.^{3}) or it may be very short with a large top and bottom; that is, we seek the minimum of $A(x)$ for $0 < x < \infty$.

We next find the critical values of $A(x)$:

$$A'(x) = 4x - \frac{32,000}{x^2}.$$

This expression is zero when

$$4x = \frac{32,000}{x^2}$$

or

$$4x^3 = 32,000.$$

Thus

$$x^3 = 8000.$$

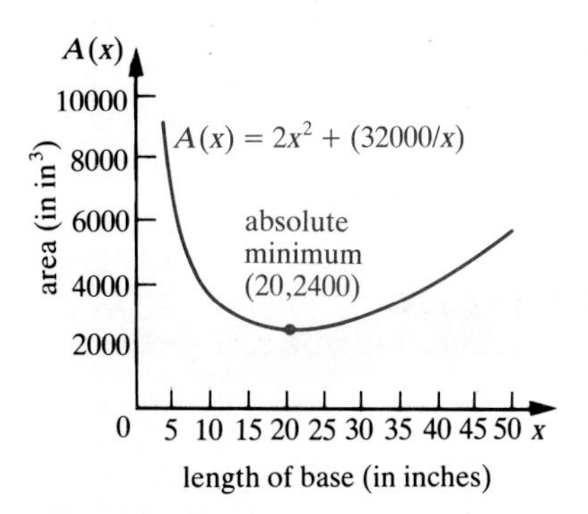

Figure 3.34
Graph of $A(x)$ in Example 3.39, showing the absolute minimum at (20, 2400)

Hence $A'(x)$ is zero when $x = 8000^{1/3} = 20$. A computation shows that $A''(x) = 4 + (64,000/x^3)$ is positive at $x = 20$, so $x = 20$ corresponds to a relative minimum. Note that the term $32,000/x$ in equation (1) causes $A(x)$ to approach infinity as $x \to 0^+$ and the term $2x^2$ causes $A(x)$ to approach infinity as $x \to \infty$. Consequently, the relative minimum at $x = 20$ is an absolute minimum. (See Figure 3.34). The corresponding value for h is $h = 8000/x^2 = 8000/400 = 20$. The crate with least surface area is a cube measuring 20 inches by 20 inches by 20 inches. ■

Example 3.40 The distributor of a new electronic game finds that to sell x thousand games per week the price must be set at $p(x) = \sqrt{900 - 2x^2}$ dollars. Determine how many thousand games per week must be sold to produce the maximum revenue $R(x)$, which is given by $R(x) = xp(x)$.

Solution We limit consideration of x to $[0, 15\sqrt{2}]$, since the price is undefined if $900 - 2x^2 < 0$. Since $R(x) = x(900 - 2x^2)^{1/2}$, we apply the product rule and the chain rule to compute the derivative:

$$R'(x) = x(\tfrac{1}{2})(900 - 2x^2)^{-1/2}(-4x) + (900 - 2x^2)^{1/2}(1).$$

We can factor out $(900 - 2x^2)^{-1/2}$ by noting that $(900 - 2x^2)^{1/2} = (900 - 2x^2)^{-1/2}(900 - 2x^2)$. This gives

$$R'(x) = (900 - 2x^2)^{-1/2}[-2x^2 + (900 - 2x^2)]$$
$$= \frac{900 - 4x^2}{(900 - 2x^2)^{1/2}}.$$

Thus, $R'(x) = 0$ when $900 - 4x^2 = 0$, that is, when $x = 15$. To find the maximum of $R(x)$, we compare $R(0)=0$, $R(15)=225\sqrt{2}$, and $R(15\sqrt{2})=0$. The maximum value of $R(x)$ is $225\sqrt{2}$, which occurs when $x=15$. The distributor receives maximum revenue by selling 15 thousand games at $p(15)=15\sqrt{2}=\$21.21$ each. ■

Example 3.41 A company manufacturers rain gutters shaped as shown in Figure 3.35. For aesthetic reasons, the bottom and the slanted side are each 4 inches wide. The company plans to advertise the gutters as "designed for maximum carrying capacity" and, of course, must adhere to truth-in-advertising laws. Find the height of the vertical side and the width of the top to obtain maximum carrying capcity?

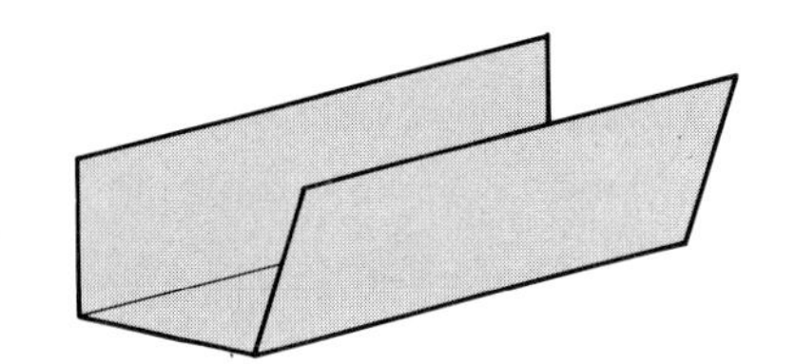

Figure 3.35
A section of rain gutter

Solution The carrying capacity of the rain gutter will be greatest if the dimensions are chosen to yield maximum cross-sectional area. From Figure 3.36, this area is the area of the rectangular portion plus the area of the triangular portion. Thus

$$\text{area} = 4h + \tfrac{1}{2}xh = (4 + \tfrac{1}{2}x)h.$$

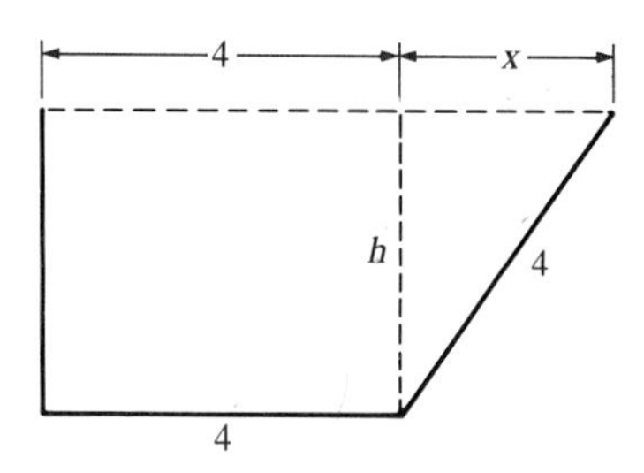

Figure 3.36
Cross section of a rain gutter

The quantities x and h are related by the Pythagorean Theorem for right triangles, namely

$$x^2 + h^2 = 16.$$

Solving this equation for h and substituting the result into the equation for area yields

$$\text{area} = f(x) = (4 + \tfrac{1}{2}x)\sqrt{16 - x^2}.$$

From the description of the problem, we know that $x \geq 0$, and from the factor $\sqrt{16-x^2}$, we must have $x \leq 4$. Hence the value of x that maximizes this function is between 0 and 4. We use the product rule to find $f'(x)$. We first write the square root factor in exponent form:

$$f(x) = (4 + \tfrac{1}{2}x)(16 - x^2)^{1/2}.$$

Then $$f'(x) = (4 + \tfrac{1}{2}x)\frac{d}{dx}(16 - x^2)^{1/2} + (16 - x^2)^{1/2}\frac{d}{dx}(4 + \tfrac{1}{2}x).$$

We use the chain rule to find $(d/dx)(16-x^2)^{1/2}$. The result is

$$\frac{d}{dx}(16 - x^2)^{1/2} = \tfrac{1}{2}(16 - x^2)^{-1/2}(-2x) = -x(16 - x^2)^{-1/2}.$$

Consequently,

$$\begin{aligned} f'(x) &= (4 + \tfrac{1}{2}x)(-x(16 - x^2)^{-1/2}) + (16 - x^2)^{1/2}(\tfrac{1}{2}) \\ &= \frac{-x(4 + \tfrac{1}{2}x)}{\sqrt{16 - x^2}} + \tfrac{1}{2}\sqrt{16 - x^2} \\ &= \frac{-2x(4 + \tfrac{1}{2}x) + (16 - x^2)}{2\sqrt{16 - x^2}}. \qquad \text{(common denominator)} \\ &= \frac{16 - 8x - 2x^2}{2\sqrt{16 - x^2}}. \end{aligned}$$

This last expression is zero when the numerator is zero, that is, when

$$16 - 8x - 2x^2 = 0.$$

This equation may be solved by the quadratic formula (see Section 1.5 or Appendix A.2). The results are

$$x = \frac{8 \pm \sqrt{64 + 128}}{2(-2)} = -2 \pm 2\sqrt{3}.$$

Since the domain of $f(x)$ is the interval $[0, 4]$, we discard the negative root. The positive root is $2(-1 + \sqrt{3}) \approx 1.46$. Another critical value may occur where the denominator is zero, namely at $x = \pm 4$. Since the domain is $[0, 4]$, $x = +4$ *is* in the domain and is a critical value. (Note that -4 is *not* in the domain of the function; it is therefore not a critical value.) Hence we must compare the values of $f(x)$ at $x = 0$, $x = 1.46$, and $x = 4$. We find $f(0) = 16$, $f(4) = 0$, and $f(1.46) = 17.6$. Thus, $x = 1.46$ produces the maximum carrying capacity. The corresponding value of h is $h = \sqrt{16 - x^2} \approx 3.72$. The company should manufacture rain gutters that are 3.72 inches deep and 5.46 inches across at the top. ■

Exercises

Exer. 1–30: Use the guidelines at the beginning of this section to solve each problem.

1. Find the dimensions of the crate shown in the figure, with square base and no top such that its volume is 500 ft^3 and its surface area is a minimum. (*Hint:* Show that you wish to minimize $S(x) = (2000/x) + x^2$ on $(0, \infty)$.)

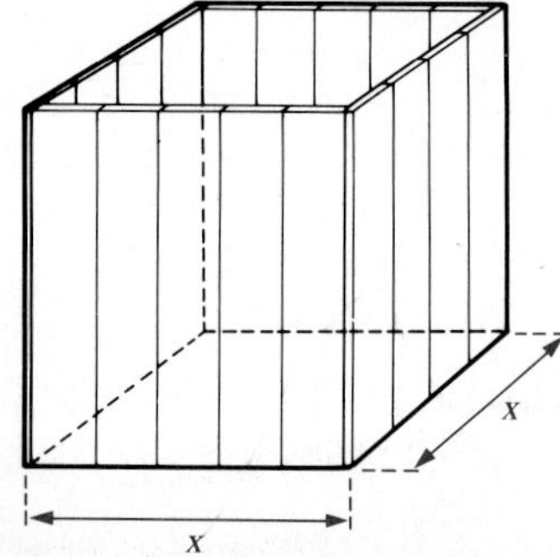

2. A farmer wants to enclose a rectangular pasture in a corner of a field as shown in the figure. The neighbors already have a fence on the boundary lines, so only two sides of the area must be fenced. If 500 feet of fencing are available, what are the dimensions of the pasture of largest area that can be enclosed?

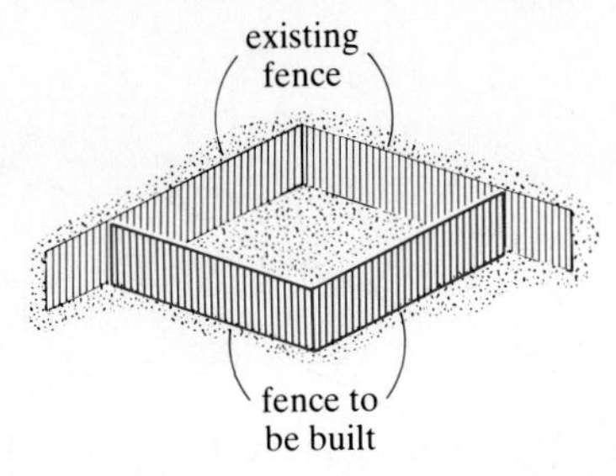

3. A manufacturer of electrical wire finds that the daily profit P (in dollars) from the manufacture of x hundred spools of wire is given by $P(x) = 4x^3 - 24x^2 + 36x - 5$. Because of a scarcity of copper, the maximum number of spools the company can make per day is 100. How many spools should the company make in order to maximize its profit?
4. Suppose that the daily profit $P(x)$ (in dollars) of a manufacturing company is given by $P(x) = -2x^2 + 168x - 1800$ when the company makes x motor bikes per day for $0 \le x \le 50$. Find the production level that yields the maximum profit. What is the maximum profit?
5. Suppose that the weekly profit made by a refrigerator manufacturing company follows the rule $P = -x^2 + 4x + 6$, where P is profit (in thousand dollars per week) and x is the number of refrigerators made (in hundreds per week) for $0 \le x \le 5$. How many hundreds of refrigerators should be made each week by the company in order to maximize the weekly profit? What is the maximum weekly profit?
6. Refer to Example 3.38. Suppose that the farmer has to double the fence (thereby using twice as much fencing per foot) along the portions perpendicular to the river (designated by w in Figure 3.32). What are the dimensions of the largest pasture that can be enclosed?
7. Find the two positive numbers with sum 20 having the maximum product. (*Hint:* Let x and y be the numbers; then $x + y = 20$. If $N = xy$, then $N = x(20 - x)$. Find x such that N is a maximum.)
8. Find two numbers whose difference is 40 and whose product is a minimum.

9. Find two positive numbers whose sum is 40 such that the sum of their squares is as small as possible.
10. Find two numbers such that their sum is 30 and the sum of twice the square of the smaller one and the square of the larger one is a minimum.
11. A building designer knows that a certain window requires two boards for the top and one board each on the sides and bottom. The window must have an area of 6 ft^2 to meet lighting specifications. Find the dimensions of the window that satisfies these requirements and uses the least amount of wood. Ignore the width of the boards.
12. When designing a newsletter, the publisher decides that the text page will have a printed area of 96 in.^2 with margins of 1 inch on each side and $1\frac{1}{2}$ inches at the top and bottom. What page dimensions will allow the maximum amount of printed area?
13. Find the most economical proportions (that is, minimize the amount of metal needed) for a closed cylindrical can that will hold 16 cubic inches. Recall that if a cylinder has radius r and height h, its volume is $\pi r^2 h$ and its lateral area is $2\pi rh$. The area of a circle of radius r is πr^2.
14. Work Exercise 13 for a cylindrical can with no top.
15. Suppose that the trajectory of a rocket is given by $y = -10x^2 + 40x$, where x denotes the distance (in miles) downrange and y denotes the altitude (in thousand feet). Find the maximum altitude attained by the rocket.
16. Suppose that you are running for President of the United States: if you promise to spend too much money on welfare, the conservatives won't vote for you, and if you promise to spend too little, the liberals won't vote for you. Your opponent has taken a firm stand on the issue. You feel that compromising your position on welfare is worth the gain in votes, and hence you are willing to adjust your stance on this issue to attract more votes. After extensive analysis of the mood of the voters, you determine that if you decide to spend x billion dollars on welfare, then the percentage of the electorate P that will vote for you is given by $P = -0.6x^2 + 6.36x + 35.146$. How much should you promise to spend on welfare to attract as many votes as possible in the election? If you adopt that stance, will you win?
17. Refer to Exercise 16. You find that your stance on military spending rather than welfare spending is crucial. If you propose to change the current budget for defense by x billion dollars, then the percentage of the electorate $P(x)$ that will vote for you is given by $P = -0.2x^2 - 5.6x + 8.6$. What is your optimum stance on this issue? If you adopt the optimum stance, will you win the election?
18. A psychologist sets up the following experiment for a large number of college freshman volunteers. Each volunteer performs a complicated manual task repeatedly. Each time the task is performed, an observer counts the number of errors committed. The psychologist finds that the volunteers improve their performance as they learn the task, but after a while their performance deteriorates as they become bored. After analyzing the data carefully, the psychologist finds that the number of errors E committed by the average volunteer the xth time the task is performed is given by the rule $E = 2x^2 - 32x + 131$. On which repetition do the volunteers as a group demonstrate the highest level of proficiency (that is, commit the fewest errors)? How proficient does the group become (that is, what is the smallest number of errors committed)?
19. Refer to Exercise 18. Suppose the psychologist repeates the experiment with a group of seniors. If the average number of errors committed on the xth attempt is now given by $E = 3x^2 - 30x + 82$, then on which repetition do the volunteers demonstrate the highest level of proficiency? How proficient does the group become?
20. A company finds that if it manufactures x hundred calculators each day, the production cost $C(x)$ is given by $C(x) = 2x^3 - 3x^2 - 12x + 500$. How many hundreds of calculators should the company produce per day to minimize the cost of production? Verify that you have indeed found a minimum. What is the minimum cost?
21. A track coach experiments with a star cross-country runner and finds that when the runner is on a special endurance and speed training program, the runner's endurance E and speed S are both functions of the runner's weight w as follows:

$$E = -w^2 + 280w - 19{,}500;$$
$$S = -0.4w^2 + 128w - 10{,}090,$$

where endurance and speed have been measured in comparable units. Note that the graphs of E and S are both parabolas.
(a) Sketch a graph of E and S.
(b) At what weight does the runner have maximum endurance?
(c) At what weight does the runner have maximum speed?
(d) It has been determined that a cross-country runner needs more endurance than speed and is at peak performance level when the sum $2E + S$ is maximum. Sketch a graph of $2E + S$.
(e) At what weight is the runner at peak performance level?
22. At the beginning of a flu epidemic it is predicted that t days after the start of the epidemic the number of people I (in hundreds) infected that day is given by

$I = -2t^2 + 48t + 62$. What is the worst day of the epidemic? On the worst day, how many people will be infected?

23. An automobile timing dictates at what point in the piston stroke the spark plug fires. The timing is set relative to a reference point called top dead center TDC). Most cars should be set at some specified value between $-10°$ TDC and $+10°$ TDC (meaning between 10 degrees before and 10 degrees after TDC). A manufacturer finds that the efficiency of a certain engine at d degrees TDC is E (in percentage), where $E = 80 + 2d^{4/7} - 0.5d$. What value of d will maximize the engine efficiency?
24. A farmer finds that if x barrels of a pesticide are applied per acre, the yield (in bushels of oats per acre) will be $y = \frac{1}{3}x^3 - 4x^2 + 15x + 1$. Due to environmental considerations, the government prohibits the use of more than 8 barrels of pesticide per acre. How many barrels of pesticide per acre should the farmer use to maximize the yield? What is the maximum yield?
25. Work Exercise 24, assuming the government has tightened restrictions to allow only 6 barrels of pesticide per acre. Now how many barrels of pesticide should be used per acre to maximize yield? What is the maximum yield?
26. Suppose that we wish to invest in three stocks during the next thirty days. The price functions giving our predicted price for each stock on the dth day are:

For the first stock:

$$P_1(d) = 0.0002d^3 + 0.096d^2 - 1.98d + 30$$

For the second stock:

$$P_2(d) = -0.0001d^3 + 0.029d^2 - 1.04d + 25$$

For the third stock:

$$P_3(d) = -0.0001d^3 - 0.046d^2 + 1.96d + 20$$

Determine when we should buy and sell each stock. Recall that on the New York Stock Exchange we are allowed to sell a stock before buying it.

27. Crowding decreases yield in an orange grove. If 36 trees are planted per acre, production per tree will be 8 bushels. For each additional tree per acre, production is decreased by 0.15 bushel. How many trees should be planted per acre to maximize production?
28. The Des Moines Housing Authority manages 1800 apartments. If the rent is set at \$400 per month, all apartments will be rented. However, for each \$5 increase in rent, the number of apartments rented decreases by 20. What rent should be charged to obtain the maximum rental income?
29. A book company found that if it produced x copies of a special edition, then its profit would be $P(x) = -0.001x^2 + 7x - 800$. If the maximum number of books the company can produce is 2000, how many should it produce to obtain the maximum profit? What if the maximum number of books the company can produce is 5000? In each case, what is the company's maximum profit? How much would the company make per book? (*Hint:* the profit per book is $P(x)/x$.)
30. A game manufacturer can produce x games per day at a cost (in dollars) of $C(x) = 10 + 5x + \frac{1}{4}x^{3/2}$. Each game sells for \$8. How many games should be produced per day to maximize profit?

Summary

Product Rule: If $g(x)$ and $h(x)$ are differentiable functions and $f(x) = g(x)h(x)$, then $f(x)$ is differentiable and

$$f'(x) = g(x)h'(x) + h(x)g'(x).$$

Quotient Rule: If $g(x)$ and $h(x)$ are differentiable functions and $f(x) = g(x)/h(x)$, then $f(x)$ is differentiable (except where $h(x) = 0$) and

$$f'(x) = \frac{h(x)g'(x) - g(x)h'(x)}{[h(x)]^2}.$$

Chain Rule: If $h(x)$ and $g(u)$ are differentiable functions and $f(x) = g(h(x))$ (composition of g with h), then $f(x)$ is differentiable and

$$f'(x) = g'(h(x))h'(x).$$

General Power Rule: If $f(x) = [h(x)]^r$ for any real number r, then

$$f'(x) = r[h(x)]^{r-1}h'(x).$$

If $y = f(x)$ is the composition of $y = g(u)$ with $u = h(x)$, then the chain rule may be written as

$$\frac{dy}{dx} = \frac{dy}{du}\frac{du}{dx}$$

and can be used to solve **related rates problems.**

A function $f(x)$ has a **relative maximum** $f(a)$ at the point $(a, f(a))$ if $f(a) \geq f(x)$ for all x in some open interval containing a. Similarly, $f(x)$ has a **relative minimum** $f(a)$ at $(a, f(a))$ if $f(a) \leq f(x)$ for all x in some open interval containing a.

Values of a in the domain of a function $f(x)$ such that $f'(a)$ equals zero or does not exist are **critical values** of $f(x)$. Since relative maxima and relative minima occur only at critical values, the first step in locating relative maxima and relative minima of a function is to determine its critical values.

There are two tests for determining the critical values that are actually relative maxima or relative minima:

First Derivative Test: Let $f(x)$ be a function with a critical value at $x = a$, and suppose that $f(x)$ is differentiable for all values of x near a, but not necessarily at a itself.

A. If $f'(x) > 0$ for $x < a$ and $f'(x) < 0$ for $x > a$, then $f(a)$ is a relative maximum.

B. If $f'(x) < 0$ for $x < a$ and $f'(x) > 0$ for $x > a$, then $f(a)$ is a relative minimum.

Second Derivative Test: Let $f(x)$ be a function that is differentiable in some interval about $x = a$. Suppose that $f'(a) = 0$ and that $f''(a)$ exists.

A. If $f''(a) < 0$, $f(a)$ is a relative maximum of $f(x)$.

B. If $f''(a) > 0$, $f(a)$ is a relative minimum of $f(x)$.

C. If $f''(a) = 0$, the test is inconclusive.

The **absolute maximum** of a function $f(x)$ is a value $f(a)$ such that $f(a) \geq f(x)$ for all values of x in the domain of $f(x)$. Similarly, $f(a)$ is an **absolute minimum** of $f(x)$ if $f(a) \leq f(x)$ for all values of x in the domain of $f(x)$. The absolute maximum of a continuous function $f(x)$ on a finite closed interval $[b, c]$ is the greatest of $f(b)$, $f(c)$, and the values of $f(x)$ at its critical values within the interval. The absolute minimum of $f(x)$ on $[b, c]$ is the least of $f(b)$, $f(c)$, and the values at the critical values within the interval.

The graph of a function $f(x)$ is **concave downward** when $f''(x) < 0$ and **concave upward** when $f''(x) > 0$. A point on the graph of a function at which the concavity of the function changes is an **inflection point**. Inflection points occur only where $f''(x)$ equals zero or does not exist.

Good organization is extremely important in solving optimization problems. The following steps will usually be helpful. Refer to page 136 for detailed descriptions.

1. Understand the problem.
2. Form a mathematical statement of the problem.
3. Determine the maximum or the minimum.
4. Answer the question.

Review Exercises

Exer. 1–16: Find the derivative of each function.

1. $f(x) = 2x^6(3x - 5)$
2. $y = \sqrt{1 - 2x^3}$
3. $f(x) = \dfrac{6x}{x^2 - 5}$
4. $y = x^2(4 - x^2)^3$
5. $f(x) = 2x - \dfrac{x - 4}{\sqrt{x}}$
6. $f(x) = \dfrac{(x^2 - 2)^{1/3}}{3x}$
7. $y = \sqrt{x + (1/x)}$
8. $f(x) = \dfrac{3x}{4x - 1}$
9. $y = \sqrt[3]{x^4 + 5x + 3}$
10. $f(x) = \dfrac{2x^3}{x - 4}$
11. $f(t) = \dfrac{\sqrt[3]{3t + 5}}{t^2 - 1}$
12. $f(u) = \dfrac{3u^2 - 5u}{u + 7}$
13. $f(s) = \dfrac{(s^2 + 3)(4s - 5)}{s^4 + 3s}$
14. $f(t) = \dfrac{t + 1}{(4t - 5)(2t + 3)}$
15. $f(x) = \dfrac{\sqrt{x} + \sqrt[3]{x}}{x^3 + 1}$
16. $f(x) = \dfrac{\sqrt[3]{x}}{x + 2}$

17. The cost of producing a certain product is $C(x) = 2x^{3/2} + 60x^{1/2} + 7000$ (in dollars), where x is the number of items produced. The current production level is 100 items. The company plans to increase production. Use the chain rule to estimate the increase in production (for next year) corresponding to a \$3000 cost increase.
18. Oil dripping onto a hot pavement forms a circular pattern. The radius of the slick is increasing at a rate of 2 cm/min. How fast is the area changing when the radius is 10 cm?
19. The population in a rural area is increasing at an annual rate of 3%. The present population is 40,000. The pollution index I is determined by $I(P) = 30 + \sqrt{P^2 - P}$, where P is measured in thousands. What is the rate of change of I with respect to t, with time t given in years from the present?
20. A spherical balloon is inflated at a rate of 2 in³/sec. How fast is the diameter changing when the diameter is 4 inches? Note: the volume of a sphere of radius r is $\frac{4}{3}\pi r^3$.

Exer. 21–32: Find all relative maxima and relative minima for the function.

21. $f(x) = x^2 - 6x + 5$
22. $f(x) = x^3 - 6x^2 + 11x - 6$
23. $f(x) = x^3 - 6x^2 + 9x$
24. $g(s) = \dfrac{s^2 - 3}{s - 2}$

25. $f(t) = 5 + 9t - \frac{1}{3}t^3$

26. $f(x) = \dfrac{\sqrt[3]{x}}{x^2 - 1}$

27. $f(x) = \dfrac{x^2 - 3}{x^2 - 5}$

28. $f(x) = \dfrac{x}{x^2 - 1}$

29. $Q(r) = r^2(r^3 - 1)^5$

30. $f(p) = (p - 1)^3(p - 3)^4$

31. $f(x) = x - 6\sqrt[3]{x^2}$

32. $f(x) = 2x - 6x^{5/4}$

Exer. 33–42: Find the absolute maximum and absolute minimum of the function on the given interval.

33. $f(x) = x^2 - 3x + 2;\ [0, 5]$
34. $f(x) = x^2 - 3;\ [-2, 3]$
35. $f(x) = x^3 - 3x + 2;\ [-1, 3]$
36. $f(s) = \frac{1}{4}s^4 - \frac{1}{2}s;\ [-1, 1]$
37. $f(t) = \dfrac{t - 3}{t - 5};\ [0, 2]$
38. $f(x) = 3x - 4;\ [0, 3]$
39. $f(x) = x(x^3 - 1)^4;\ [-1, 1]$
40. $f(x) = \dfrac{x^2 - 3}{x^2 + 1};\ [-2, 1]$
41. $g(z) = \dfrac{\sqrt[3]{z}}{z + 2};\ [0, \infty)$
42. $f(t) = t + \dfrac{4}{t};\ [1, \infty)$

43. Suppose $L(x)$ is a function such that $L'(x) = \dfrac{x}{x + 3}$. Find $\dfrac{d}{dx}L(x^2 + x)$.

44. A section of a river that makes a right angle turn is shown in the figure. An area is to be fenced at this turn, using the river as an unfenced boundary on two sides. The amount of fence available is 200 feet, and the fenced area is divided into two sections by a length of fencing. Find the dimensions of the rectangular region of largest area that can be fenced.

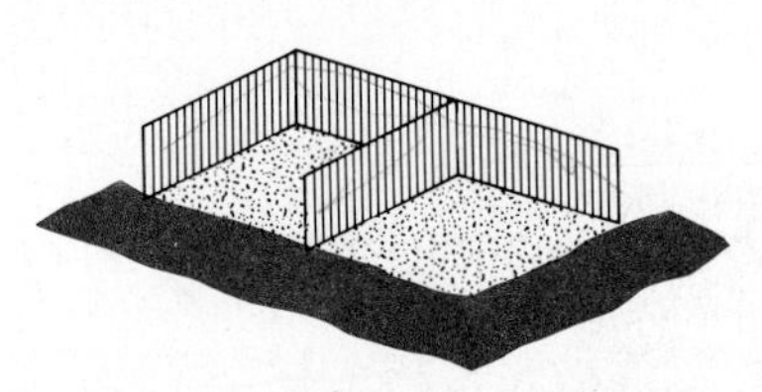

45. Consider an ice-cream cone of height h, and let r denote the radius of the top of the cone. The volume of such a cone is $V = \frac{1}{3}\pi r^2 h$ and the surface area is $S = \pi\sqrt{r^4 + r^2h^2}$. Suppose an ice-cream parlor owner wants to order cones that hold $V = \frac{1}{3}\pi\sqrt{2} \approx 1.48\ \text{in}^3$ (in the cone itself, excluding the ice-cream piled on top). Find r and h to obtain this volume and still yield a minimum surface area.

***Facing page:* The advisability of levying a tax on restaurant meals can be determined using the principle of elasticity of demand. See Example 4.30.**

Additional Applications of the Derivative

· F · O · U · R ·

In the previous chapters we have discussed applications of the derivative to velocity-acceleration problems, related rates problems, and optimization problems. The first and second derivatives of a function are useful for determining the graph of the function. In this chapter we give additional applications of the derivative in graphing functions and present applications of the derivative to economics. These applications include inventory control, batch manufacturing decisions, and price elasticity.

4.1 Asymptotes

We have seen how the graph of a function illustrates some of the important facts about the function and how understanding these facts can aid in the solution of many applied problems. We have learned how to determine where a graph is increasing, decreasing, concave upward, and concave downward, and where its relative maxima and relative minima occur. These facts have enabled us to graph many functions. In this section we will discuss two more features of graphs: horizontal and vertical asymptotes. In Section 4.2 we will suggest a check list of things to do when graphing functions.

Horizontal and vertical **asymptotes** provide information about the behavior of the graph of a function. Horizontal asymptotes give information about the behavior of a graph for values of x that are large in the positive or negative direction. They will help answer certain types of questions: Is there an amount of profit that a particular company cannot exceed, no matter how high its level of production? Can the number of new markets for a product increase indefinitely? What is the largest population of deer that a particular forest can support?

Example 4.1 The profit function for a particular company is given by $P(x) = (5000x - 200)/x$, where x is the number of 100-item lots produced. Are there any bounds to the profit this company can make?

Solution By writing the function as $P(x) = 5000 - (200/x)$, we observe that $P'(x) = 200/x^2$. Since $P'(x) > 0$ for all $x > 0$, we conclude that $P(x)$ is an increasing function for all $x > 0$. Hence, the company's profit always increases as production increases. We might therefore conclude that the company's profits can grow without bound; however, no matter how many lots are produced, the company's profits can never exceed \$5000. Observe that since $x > 0$, then $200/x > 0$; hence, to compute $P(x)$ from

$$P(x) = 5000 - \frac{200}{x},$$

we subtract a positive number from 5000 to get $P(x)$. Thus, $P(x) < 5000$ for all values of $x > 0$. Note that $200/x$ gets very close to zero as x gets large, so the company can realize a profit very close to \$5000 if it produces a large quantity. See the graph of $P(x)$ in Figure 4.1. The line $y = 5000$ is a horizontal asymptote because $P(x)$ gets close to 5000 as x gets larger. ■

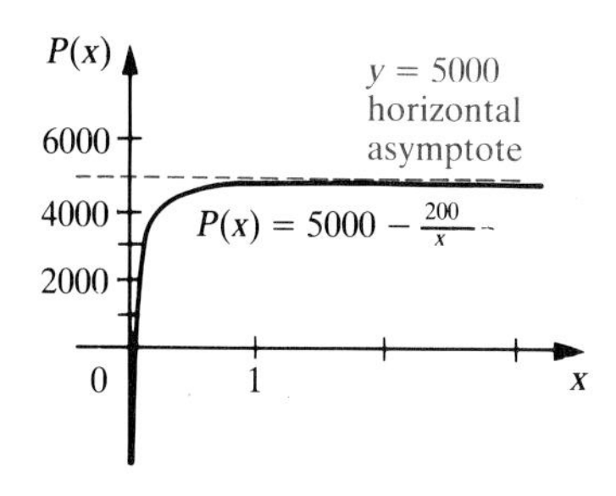

Figure 4.1
Graph of the profit function $P(x) = 5000 - \dfrac{200}{x}$ and its horizontal asymptote $y = 5000$

As the previous example illustrates, it is important to determine if a function has a horizontal asymptote and, if so, to locate it. The line $y = b$ is a **horizontal asymptote** for $f(x)$ if $f(x)$ gets close to b as x gets large in either the positive or negative direction.

To facilitate our discussion of asymptotes, we will use the notation $x \to \infty$, read *as x approaches infinity,* to mean x increases without bound in the positive direction. If a function $f(x)$ has the property that it approaches a particular number b as x approaches infinity, we will write this property as $f(x) \to b$ as $x \to \infty$, or as

$$\lim_{x\to\infty} f(x) = b.$$

For the function $P(x)$ in Example 4.1, we have $\lim\limits_{x\to\infty} P(x) = 5000.$ In a similar manner, we define $x \to -\infty$ to mean that x is negative and $|x|$ is increasing without bound. If a function $f(x)$ approaches a number c as $x \to -\infty$, then we write $\lim\limits_{x\to-\infty} f(x) = c$.

The definition of horizontal asymptote can be stated as *$y = b$ is a horizontal asymptote for $f(x)$ if $f(x) \to b$ as $x \to \infty$* or *as $x \to -\infty$*; that is, the horizontal line $y = b$ is a horizontal asymptote of $f(x)$ if either

$$\lim_{x\to\infty} f(x) = b \quad \text{or} \quad \lim_{x\to-\infty} f(x) = b.$$

Note that the definition implies that the limit exists. If neither limit exists, then $f(x)$ does not have a horizontal asymptote. The procedure for identifying horizontal asymptotes is demonstrated in the following example.

Example 4.2 Find the horizontal asymptotes, provided they exist, for each function:

$$\textbf{(a)}\ f(x) = \frac{4x+1}{x} \qquad \textbf{(b)}\ f(x) = \frac{4x+1}{x^2}$$

$$\textbf{(c)}\ f(x) = \frac{4x^2+1}{x} \qquad \textbf{(d)}\ f(x) = \frac{3x^2+6x}{2x^2-5}$$

Solution **(a)** We divide each term in the numerator and denominator by x, the highest power of x in the denominator:

$$f(x) = \frac{4 + \frac{1}{x}}{1} = 4 + \frac{1}{x}.$$

As $x \to \infty$, the term $1/x \to 0$, so $f(x) \to 4$. Therefore, $y = 4$ is a horizontal asymptote for $f(x)$. Also, as $x \to -\infty$, the term $1/x \to 0$, so $f(x) \to 4$. Thus $y = 4$ is the only horizontal asymptote for $f(x)$.

(b) Following the same procedure as in part (a), we divide numerator and denominator by x^2:

$$f(x) = \frac{4}{x} + \frac{1}{x^2}.$$

As $x \to \infty$, both terms approach 0. Thus, $\lim_{x\to\infty} f(x) = 0$ and hence $y = 0$ is a horizontal asymptote for $f(x)$. Also, as $x \to -\infty$, $f(x) \to 0$. The only horizontal asymptote is $y = 0$, which is the x-axis.

(c) We divide each term in the numerator and denominator by x to obtain

$$f(x) = 4x + \frac{1}{x}.$$

As $x \to \infty$, the term $4x \to \infty$ and the term $1/x \to 0$, so $f(x) = 4x + (1/x) \to \infty$. Similarly, as $x \to -\infty$, $f(x) \to -\infty$. Hence, $f(x)$ has no horizontal asymptote.

(d) We divide numerator and denominator by x^2 to obtain

$$f(x) = \frac{3 + \frac{6}{x}}{2 - \frac{5}{x^2}}.$$

As $x \to \pm\infty$, the terms $6/x$ and $5/x^2$ approach zero, so $f(x) \to \frac{3}{2}$ as $x \to \pm\infty$. Consequently, the line $y = \frac{3}{2}$ is a horizontal asymptote for $f(x)$. The graphs of the first three functions are shown in Figure 4.2. ■

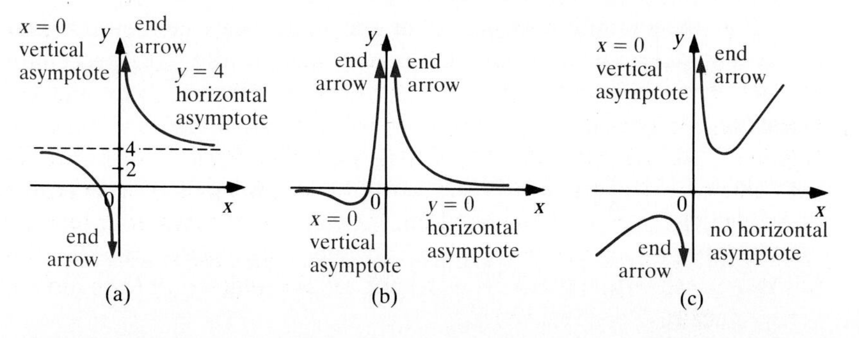

Figure 4.2
Behavior of $f(x)$ near the horizontal and vertical asymptotes: (a) $f(x) = (4x + 1)/x$; (b) $f(x) = (4x + 1)/x^2$; (c) $f(x) = (4x^2 + 1)/x$

As noted in the previous example, to find horizontal asymptotes of a rational function we divide the numerator and denominator by the highest power of x that appears in the denominator, and let $x \to \pm\infty$. We may then use the limits

$$\lim_{x\to\infty} \frac{c}{x^n} = 0 \quad \text{and} \quad \lim_{x\to-\infty} \frac{c}{x^n} = 0 \quad (\text{for positive } n),$$

to determine any horizontal asymptotes.

A **vertical asymptote** is a vertical line $x = a$ such that $f(x) \to +\infty$ or $f(x) \to -\infty$ as x approaches a from the left or from the right. The line $x = a$ is a vertical asymptote of $f(x)$ if either

$$|f(x)| \to \infty \quad \text{as } x \to a^- \quad \text{or} \quad |f(x)| \to \infty \quad \text{as } x \to a^+.$$

We will frequently use the notation $x \to a^-$ and $x \to a^+$ in discussions involving vertical asymptotes. Recall from Chapter 2 that $x \to a^-$ is an abbreviation for x approaching a with $x < a$ (x is on the left side of a); $x \to a^+$ denotes that x is approaching a with $x > a$ (x is on the right side of a).

Generally, $f(x)$ is undefined at any of its vertical asymptotes; consequently, these asymptotes are usually easy to find. For a rational function, a value of x that makes the denominator zero while the numerator is not zero corresponds to the location of a vertical asymptote. In other words, if

$$f(x) = \frac{p(x)}{q(x)},$$

where $p(x)$ and $q(x)$ are polynomials, and if $p(a) \neq 0$ while $q(a) = 0$, then $x = a$ is a vertical asymptote for $f(x)$.

Example 4.3 Each of the following functions has a vertical asymptote at $x = 0$:

$$\text{(a) } f(x) = \frac{4x+1}{x} \quad \text{(b) } f(x) = \frac{4x+1}{x^2} \quad \text{(c) } f(x) = \frac{4x^2+1}{x}$$

Determine the behavior of the graph of each function near $x = 0$.

Solution These functions were considered in Example 4.2.

(a) If x is near zero but greater than zero, then both the numerator and the denominator are positive. Since we are considering x quite close to zero, the value of the function is a large positive number. Table 4.1 supports this claim. Thus, $f(x) \to +\infty$ as $x \to 0^+$.

For x quite close to zero but negative, the numerator, $4x + 1$, is still positive (close to 1) but the denominator is negative. We have $f(x) \to -\infty$ as $x \to 0^-$, as shown in Table 4.2. This behavior and the results on the horizontal asymptote from Example 4.2(a) are shown in Figure 4.2(a). We will refer to the marks near the vertical asymptote as *end arrows*. The graph of the function continues in the indicated direction. The heavy segments along the horizontal asymptote indicate that the graph of $f(x)$ is close to this line. The entire graph of this function is shown; in Section 4.2 we will obtain sufficient information to verify that the graph is correct.

(b) The function $f(x) = (4x + 1)/x^2$ has the same numerator as the function in part (a). For x near zero, $4x + 1$ is close to 1 and is positive. The denominator x^2 is always positive. Thus, $f(x) \to +\infty$ as $x \to 0$ from either side. Figure 4.2(b) shows the end arrows that indicate this behavior.

(c) This function, $f(x) = (4x^2 + 1)/x$, is very similar to part (a) near the asymptote at $x = 0$. The only difference between these functions is the term $4x$ versus $4x^2$ in the numerator. For x close to zero, both $4x$ and $4x^2$ are quite small, so we have $(4x + 1) \approx 1$ and $(4x^2 + 1) \approx 1$. Consequently, *near* $x = 0$, $f(x) = (4x^2 + 1)/x$ behaves like $f(x)$ in part (a). Figure 4.2(c) shows this behavior. We will soon verify that the functions in parts (a) and (c) have the graphs shown and that their graphs are similar *only* near $x = 0$. ■

Table 4.1
Behavior of $f(x) = \dfrac{4x+1}{x}$ as $x \to 0^+$

x	$f(x)$
0.1	14
0.01	104
0.001	1004
0.0001	10004

Table 4.2
Behavior of $f(x) = \dfrac{4x+1}{x}$ as $x \to 0^-$

x	$f(x)$
−0.1	−6
−0.01	−96
−0.001	−996
−0.0001	−9996

Example 4.4 For each function, identify the vertical asymptotes, provided they exist, and indicate on a graph the behavior of the function near each asymptote:

$$\textbf{(a)}\ f(x)=\frac{1}{x-4} \quad \textbf{(b)}\ f(x)=\frac{1}{(x-2)^2} \quad \textbf{(c)}\ f(x)=\frac{x-x^2}{x}$$

Solution **(a)** Since we have zero in the denominator for $x=4$, and since the numerator is not zero at 4, the line $x=4$ is a vertical asymptote. Note that $f(x)<0$ when $x<4$, so $f(x)\to-\infty$ as $x\to 4^-$. Since $f(x)>0$ when $x>4$, $f(x)\to+\infty$ as $x\to 4^+$. Hence, the behavior of $f(x)$ near $x=4$ is as shown in Figure 4.3(a).

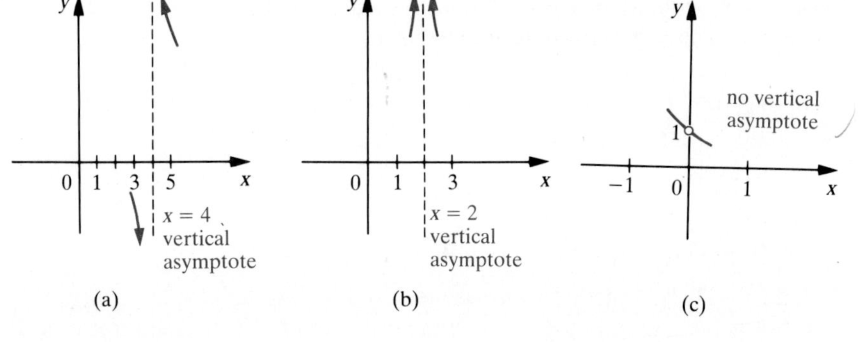

Figure 4.3
End arrows near the vertical asymptotes:
(a) $f(x)=(1)/x-4$;
(b) $f(x)=(1)/(x-2)^2$;
(c) $f(x)=(x-x^2)/x$

(b) In this case, at $x=2$, the denominator is zero and the numerator is not zero. Thus $x=2$ is a vertical asymptote. Because the denominator is $(x-2)^2>0$ for $x\neq 2$, we see that $f(x)\to+\infty$ when $x\to 2^-$ and $f(x)\to+\infty$ when $x\to 2^+$. The behavior of $f(x)$ near $x=2$ is shown in Figure 4.3(b).

(c) At $x=0$ we have the denominator zero but the numerator is also zero. To see if $x=0$ is an asymptote, we must determine if $f(x)\to\pm\infty$ as $x\to 0$. Note that

$$\lim_{x\to 0} f(x) = \lim_{x\to 0}\frac{x-x^2}{x}$$

$$= \lim_{x\to 0}(1-x) = 1.$$

Since $f(x)$ does not approach $\pm\infty$ as x approaches 0, the line $x=0$ is not a vertical asymptote, and $f(x)$ has no vertical asymptote. See Figure 4.3(c). ■

Note that polynomials of positive degree have no horizontal or vertical asymptotes. A polynomial is defined for all values of x (hence, no vertical asymptote) and approaches $\pm\infty$ as $x\to\infty$ or as $x\to-\infty$ (hence, no horizontal asymptote).

Exercises

Exer. 1–12: Find all horizontal asymptotes for the function.

1. $f(x)=\dfrac{2x-3}{x}$

2. $f(x)=\dfrac{2x^2+5}{x}$

3. $f(x) = \dfrac{-5x + 2}{x}$

4. $f(x) = \dfrac{3 - 2x}{4x}$

5. $f(x) = \dfrac{2x + 1}{3x - 2}$

6. $f(x) = \dfrac{x - 2}{4x + 1}$

7. $f(x) = \dfrac{3x + 8}{-2x + 5}$

8. $f(x) = \dfrac{8x^2 + 2x - 1}{4x^2 + 2}$

9. $f(x) = \dfrac{x - 2}{2 - x}$

10. $f(x) = \dfrac{x^3}{4 - x^3}$

11. $f(x) = \dfrac{x^3 + x}{1 - x^3}$

12. $f(x) = \dfrac{x^2 + x}{3x^2 - 1}$

Exer. 13–30: Find all vertical asymptotes for the function. Describe the behavior of the function near each vertical asymptote; that is, position the end arrows.

13. $f(x) = \dfrac{2x + 3}{x}$

14. $f(x) = \dfrac{3x - 5}{x}$

15. $f(x) = \dfrac{x}{x - 2}$

16. $f(x) = \dfrac{2x}{4x - 12}$

17. $f(x) = \dfrac{x}{x + 3}$

18. $f(x) = \dfrac{2x + 1}{x - 1}$

19. $f(x) = \dfrac{x^2 + x}{x}$

20. $f(x) = \dfrac{3x + 1}{x + 2}$

21. $f(x) = \dfrac{x^2}{x^2 - 4}$

22. $f(x) = \dfrac{x^3}{x^2 + x}$

23. $f(x) = \dfrac{2x}{x^2 - 5x + 4}$

24. $f(x) = \dfrac{2x - 1}{4x^2 - 1}$

25. $f(x) = \dfrac{2 - x}{x^2 - 4x}$

26. $f(x) = x^2 - \dfrac{1}{x^2}$

27. $f(x) = 2x + \dfrac{2}{x}$

28. $f(x) = \dfrac{x^2}{x^2 + 4x + 3}$

29. $f(x) = \dfrac{1}{x - 2} + \dfrac{5}{x - 6}$

30. $f(x) = \dfrac{2}{3 - x} - \dfrac{5}{3x - 5}$

Graphing Functions

4.2

We have now discussed all the graphing features we need to consider to accurately sketch many functions and now present a checklist for graphing a function.

Checklist for Graphing a Function

1. Find the y-intercept by evaluating the function at $x = 0$. Find x-intercepts when practical.
2. Find any vertical asymptotes and locate their end arrows.
3. Find any horizontal asymptotes.
4. Calculate $f'(x)$.
5. Determine and plot all critical values.
6. Calculate $f''(x)$.
7. Determine if the critical values correspond to relative maxima or relative minima.
8. Determine where $f(x)$ is increasing or decreasing, that is, where $f'(x) > 0$ or $f'(x) < 0$.
9. Determine where $f(x)$ is concave upward or downward, that is, where $f''(x) > 0$ or $f''(x) < 0$.
10. Determine and plot all inflection points.

We do not necessarily have to do the items in the order listed and, in some cases, it may not be necessary to do all of them. If the first five items yield adequate information about the graph to make an accurate sketch, then the other items need not be done.

Example 4.5 Sketch the graph of $f(x) = 3x^4 + 8x^3 + 6x^2 + 4$.

Solution Setting $x = 0$, we find that the y-intercept is $y = 4$. Since $f(x)$ is a polynomial, it has no asymptotes. We now compute and simplify the derivatives:

$$f'(x) = 12x^3 + 24x^2 + 12x$$
$$= 12x(x^2 + 2x + 1) = 12x(x + 1)^2$$

and

$$f''(x) = 36x^2 + 48x + 12$$
$$= 12(3x^2 + 4x + 1) = 12(3x + 1)(x + 1).$$

Table 4.3
Sign of $f'(x) = 12x(x+1)^2$ near $x = -1$

	$x<-1$	$x>-1$
$12x$	$-$	$-$
$(x+1)^2$	$+$	$+$
$f'(x)$	$-$	$-$

We next find all relative maxima and relative minima. Since $f'(x)$ is defined for all x, relative maxima and relative minima occur only when $f'(x) = 0$, that is, when $x = 0$ or $x = -1$. We use the Second Derivative Test to classify these critical values: since $f''(0) = 12$ is positive, $f(x)$ has a relative minimum at $(0, f(0)) = (0, 4)$. Since $f''(-1) = 0$, the Second Derivative Test is inconclusive, so we must use the First Derivative Test. We have $f'(x) = 12x(x + 1)^2$. Table 4.3 shows the determination of the sign of $f'(x)$ on either side of $x = -1$. For $x < -1$, $f'(x)$ is negative, so $f(x)$ is decreasing; for $-1 < x < 0$, $f'(x)$ is still negative, so again $f(x)$ is decreasing. Thus, $f(x)$ has neither a relative maximum nor a relative minimum at $(-1, f(-1))$, even though the tangent line at $x = -1$ is horizontal. See Figure 4.4.

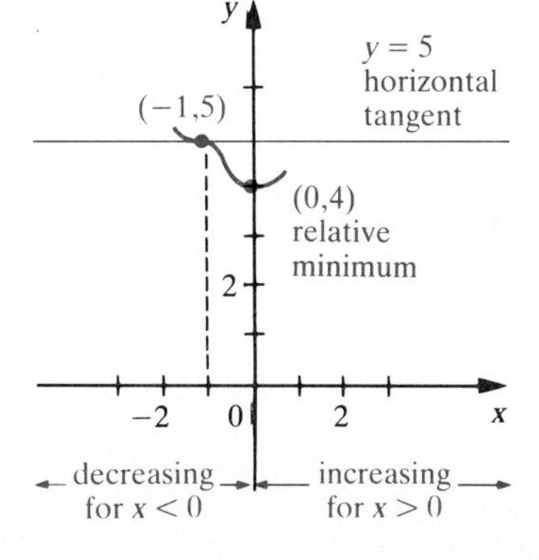

Figure 4.4
Graph of $f(x) = 3x^4 + 8x^3 + 6x^2 + 4$ thus far

Table 4.4
Sign of $f''(x) = 12(3x+1)(x+1)$ near $x = -\frac{1}{3}$ and $x = -1$

	$x<-1$	$-1<x<-\frac{1}{3}$	$x>-\frac{1}{3}$
$3x+1$	$-$	$-$	$+$
$x+1$	$-$	$+$	$+$
$f''(x)$	$+$	$-$	$+$
$f(x)$	concave upward	concave downward	concave upward

Recall that inflection points occur where $f''(x)$ changes sign. To find these points, we must determine where $f''(x)$ equals zero or is not defined. Since $f''(x)$ is defined everywhere, and $f''(x) = 0$ at $x = -1$ and $x = -\frac{1}{3}$, these are the x-values where inflection points may occur. Table 4.4 shows the determination of the sign of $f''(x) = 12(3x + 1)(x + 1)$ for the three regions of the x-axis: $x < -1$, $-1 < x < -\frac{1}{3}$, and $x > -\frac{1}{3}$. Thus $f''(x)$ changes sign when $x = -1$ and when $x = -\frac{1}{3}$; therefore $(-1, f(-1)) = (-1, 5)$ and $(-\frac{1}{3}, f(-\frac{1}{3})) \approx (-\frac{1}{3}, 4.4)$ are inflection points. We have also determined that the concavity of the graph is upward for $x < -1$ and for $x > -\frac{1}{3}$ and downward for $-1 < x < -\frac{1}{3}$. The graph is shown in Figure 4.5. ■

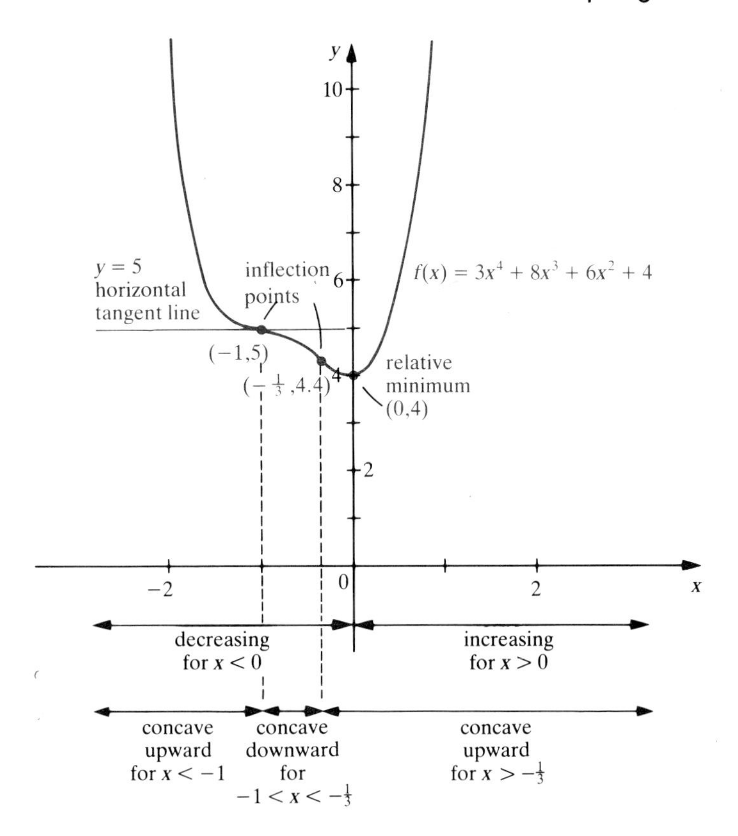

Figure 4.5
Graph of
$\boldsymbol{f(x) = 3x^4 + 8x^3 + 6x^2 + 4}$
Defined for all x
The y-intercept is 4
Horizontal tangent at $(-1, 5)$
Relative minimum at $(0, 4)$
Increasing for $x > 0$, decreasing for $x < 0$
Inflection points at $(-1, 5)$ and $(-1/3, 4.4)$
Concave upward for $x < -1$ and $x > -1/3$
Concave downward for $-1 < x < -1/3$

Example 4.6 Sketch the graph of $f(x) = \dfrac{4x + 1}{x}$.

Solution Since we get a zero denominator when $x = 0$, this function has no y-intercept. In Example 4.3(a) (see Section 4.1), we found that $x = 0$ is a vertical asymptote and that the end arrows are as shown in Figure 4.2(a). We also found that this function has $y = 4$ as a horizontal asymptote.

We calculate $f'(x)$ from the expression $f(x) = 4 + (1/x) = 4 + x^{-1}$:

$$f'(x) = -\frac{1}{x^2}.$$

Since $x^2 > 0$ for all $x \neq 0$, we see that $f'(x) < 0$ for all x in the domain of $f(x)$. Hence, $f'(x)$ is never zero. Even though $f'(x)$ is undefined at $x = 0$, $f(x)$ is also not defined at zero, so $x = 0$ is not a critical value; $f(x)$ has no critical values and no relative maxima or relative minima.

The second derivative is

$$f''(x) = \frac{2}{x^3}$$

and is never zero. It is undefined at $x = 0$, but we have already observed that $x = 0$ is not a point in the domain of this function. Hence, $f(x)$ has no inflection points. For $x < 0$ we have $f''(x) < 0$, so $f(x)$ is concave downward for negative x. For $x > 0$, $f''(x) > 0$, so $f(x)$ is concave upward for positive x. The graph of this function is shown in Figure 4.6. Note that we have plotted three points $(-1, 3)$, $(1, 5)$ and $(3, \frac{13}{3})$ to help place the graph. ■

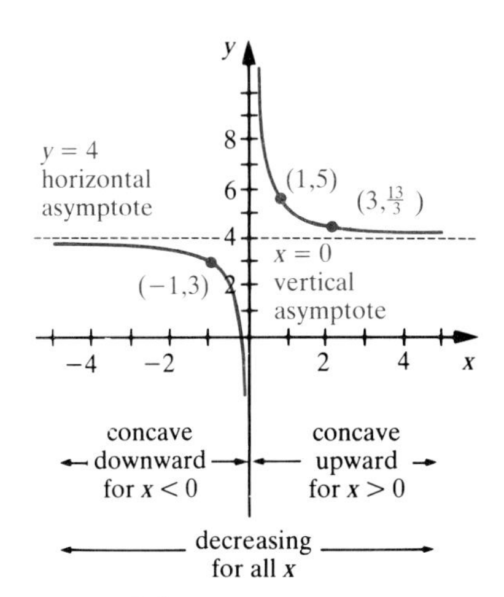

Figure 4.6
Graph of $\boldsymbol{f(x) = \dfrac{4x + 1}{x}}$
No y-intercept
Horizontal asymptote at $y = 4$
Vertical asymptote at $x = 0$
No relative max or min
$f(x)$ is decreasing for all x
No inflection points
Concave upward for $x > 0$
Concave downward for $x < 0$

Example 4.7 Sketch the graph of $f(x)=\dfrac{4x^2+1}{x}$.

Solution This is the function considered in part (c) of Examples 4.2 and 4.3 (in Section 4.1), where we found that $f(x)$ has a vertical asymptote at $x=0$ and no horizontal asymptotes. In fact, $f(x)\to+\infty$ as $x\to+\infty$ and $f(x)\to-\infty$ as $x\to-\infty$. Furthermore, the graph has no y-intercept, since $x=0$ is not in the domain of the function. We can easily compute the derivatives of $f(x)$ by writing the function as

$$f(x)=4x+\frac{1}{x}.$$

Then
$$f'(x)=4-\frac{1}{x^2} \quad\text{and}\quad f''(x)=\frac{2}{x^3}.$$

Hence, $f'(x)=0$ when $1/x^2=4$. This equation yields $x=\pm\frac{1}{2}$ as critical values. We use the Second Derivative Test to determine whether these critical values are relative maxima or relative minima:

$$f''(\tfrac{1}{2})=16>0 \quad\text{and}\quad f''(-\tfrac{1}{2})=-16<0.$$

Consequently, the function has a relative minimum at the point $(\frac{1}{2},4)$ and a relative maximum at the point $(-\frac{1}{2},-4)$. At $x=0$, $f'(x)$ is undefined; since $f(x)$ is also undefined there, there is no critical value at $x=0$.

Finally, $f''(x)=2/x^3$ is positive for $x>0$ and negative for $x<0$, indicating that the graph of $f(x)$ is concave upward for positive x and concave downward for negative x. The value $x=0$ is the only candidate for an inflection point, but the function is not defined there; hence $f(x)$ has no inflection points. Figure 4.7 gives the graph of the function. Compare this graph with that of the function in Figure 4.6. These two functions behave similarly near the vertical asymptote $x=0$, but are vastly different elsewhere. ■

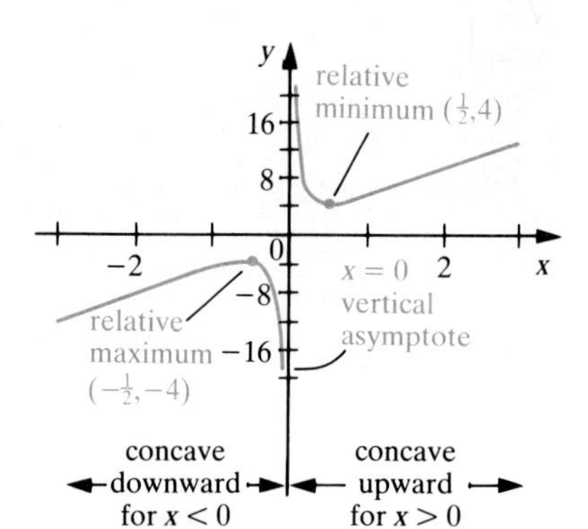

Figure 4.7

Graph of $f(x)=\dfrac{4x^2+1}{x}$

No y-intercept
No horizontal asymptote
Relative max at $(-\frac{1}{2},-4)$
Relative min at $(\frac{1}{2},4)$
No inflection points
Concave upward for $x>0$
Concave downward for $x<0$

Example 4.8 Sketch the graph of the function

$$f(x)=\frac{2x^2-4x+6}{x^2}.$$

Solution Since $x=0$ is not in the domain of the function, the graph has no y-intercept. Since the numerator is not zero at $x=0$, we see that $x=0$ is a vertical asymptote. Furthermore, the denominator x^2 is always positive and the numerator is close to 6 when x is close to zero, so $f(x)>0$ for all x near zero. That is, $f(x)\to+\infty$ as $x\to0^+$, and $f(x)\to+\infty$ as $x\to0^-$. From these observations we can place the end arrows associated with the asymptote $x=0$.

To find a horizontal asymptote, we divide numerator and denominator by x^2, the highest power of x in the denominator, to get

$$f(x)=2-\frac{4}{x}+\frac{6}{x^2}.$$

The second and third terms approach zero as $x\to\pm\infty$, so $f(x)\to2$ and the line $y = 2$ is a horizontal asymptote. This form of $f(x)$ is especially useful for obtaining $f'(x)$ and $f''(x)$. We have

$$f'(x) = \frac{4}{x^2} - \frac{12}{x^3} \quad \text{and} \quad f''(x) = -\frac{8}{x^3} + \frac{36}{x^4}.$$

Combining the two terms in $f'(x)$ gives

$$f'(x) = \frac{4x - 12}{x^3}.$$

We see that $f'(x) = 0$ when $x = 3$ and $f'(x)$ is undefined at $x = 0$. Hence, $x = 3$ is a critical value. Since $f(x)$ is not defined at 0, $x = 0$ is not a critical value. We next determine the sign of $f'(x)$ by analyzing the sign of the factors in $f'(x)$. Since the factor $4x - 12$ changes sign at $x = 3$ and the factor x^3 changes sign at $x = 0$, we consider three intervals: $x < 0$, $0 < x < 3$, and $x > 3$. Table 4.5 shows the results and the behavior of $f(x)$ on each interval. We conclude that $f'(x) < 0$ when $0 < x < 3$, and $f'(x) > 0$ when $x < 0$ and when $x > 3$.

To classify the critical value at $x = 3$, we note that $f''(3) = \frac{4}{27} > 0$, so the Second Derivative Test indicates that the value at $x = 3$ is a relative minimum. The y-coordinate at $x = 3$ is $y = \frac{4}{3}$.

To find any inflection points, we solve $f''(x) = 0$ for x:

$$\begin{aligned} f''(x) &= -\frac{8}{x^3} + \frac{36}{x^4} = \frac{-8x + 36}{x^4} \\ &= \frac{-8(x - 4.5)}{x^4} = 0. \end{aligned}$$

Thus $x = 4.5$ corresponds to the only possible inflection point. Since $x^4 \geq 0$ for all x, the sign of $f''(x)$ is the same as the sign of $-8(x - 4.5)$. Hence, $f''(x) > 0$ for $x < 4.5$ and $f''(x) < 0$ for $x > 4.5$, and we see that $(4.5, f(4.5)) \approx (4.5, 1.4)$ is indeed an inflection point. The graph is concave upward when $x < 4.5$ and concave downward when $x > 4.5$. This information is summarized and a sketch of the graph of the function is given in Figure 4.8. ■

Table 4.5

Sign of $f'(x) = \frac{4x - 12}{x^3}$ near critical values

	$x<0$	$0<x<3$	$x>3$
$4x - 12$	−	−	+
x^3	−	+	+
$f'(x)$	+	−	+
$f(x)$	increasing	decreasing	increasing

Example 4.9 Sketch the graph of $f(x) = \dfrac{x + 1}{\sqrt{x}}$.

Solution We first note that this function is defined only for $x > 0$. We compute

$$\begin{aligned} f'(x) &= \frac{\sqrt{x}\,(1) - (x + 1)\dfrac{1}{2\sqrt{x}}}{x} \\ &= \frac{\dfrac{2\sqrt{x}\sqrt{x}}{2\sqrt{x}} - \dfrac{x + 1}{2\sqrt{x}}}{x} \\ &= \frac{2x - x - 1}{2\sqrt{x}\,x} = \frac{x - 1}{2x^{3/2}} \end{aligned}$$

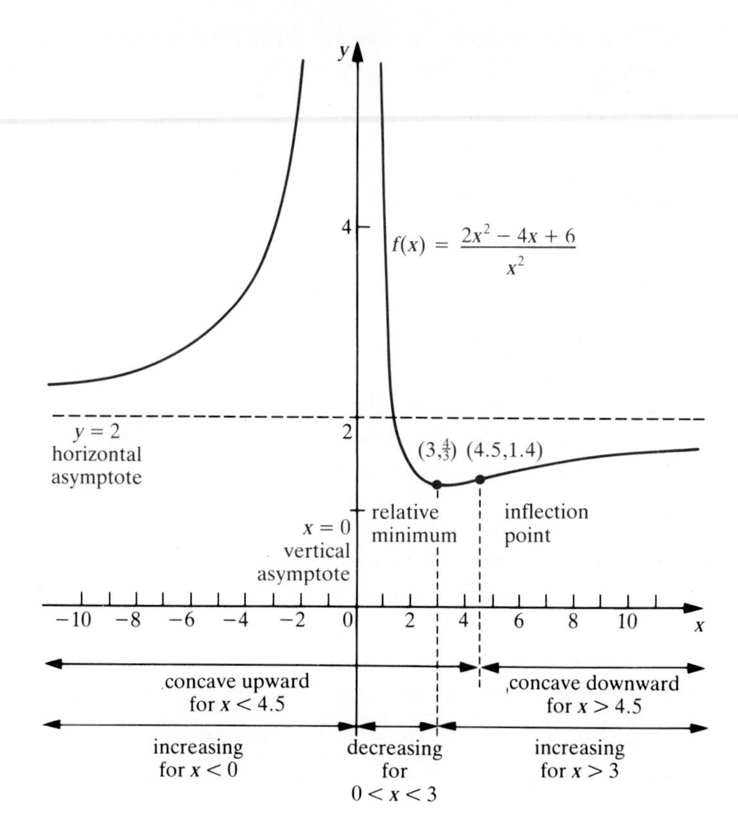

Figure 4.8
Graph of
$$f(x) = \frac{2x^2 - 4x + 6}{x^2} = 2 - \frac{4}{x} + \frac{6}{x^2}$$
No x- or y-intercept
Horizontal asymptote at $y = 2$
Vertical asymptote at $x = 0$
Relative minimum at $(3, 4/3)$
Increasing for $x < 0$ and for $x > 3$
Decreasing for $0 < x < 3$
Inflection point at $(4.5, 1.4)$
Concave upward for $x < 4.5$
Concave downward for $x > 4.5$

and
$$f''(x) = \frac{2x^{3/2}(1) - (x-1)3x^{1/2}}{4x^3} = \frac{x^{1/2}(2x - 3x + 3)}{4x^3}$$
$$= \frac{-x+3}{4x^{5/2}}.$$

Thus, $f'(x) = 0$ when $x = 1$. By the Second Derivative Test, $f''(1) > 0$ implies that the point $(1, f(1)) = (1, 2)$ is a relative minimum. Since $f'(x)$ is defined at all points at which $f(x)$ is defined, this is the only relative maximum or relative minimum.

Setting $f''(x) = 0$ yields $x = 3$. Since the denominator of $f''(x)$ is always positive for $x > 0$ (the domain of $f(x)$), $f''(x)$ is negative for $0 < x < 3$. Thus the graph is concave downward for $0 < x < 3$ and concave upward for $x > 3$. Hence, $(3, 4/\sqrt{3}) \approx (3, 2.3)$ is an inflection point.

Since $x = 0$ produces a zero in the denominator of $f(x)$, $x = 0$ is a vertical asymptote. As $x \to 0^+$, $f(x) \to +\infty$ because both the numerator and denominator are positive. Note that $f(x)$ may be written as $f(x) = \sqrt{x} + (1/\sqrt{x})$. Since $\sqrt{x} \to \infty$ and $1/\sqrt{x} \to 0$ as $x \to \infty$, $f(x)$ also approaches infinity. This indicates that the graph has no horizontal asymptotes.

The graph has no y-intercept because the function is not defined for $x = 0$, and no x-intercept because $\sqrt{x} + 1/\sqrt{x}$ is always positive. We plot the relative maximum and the inflection point and use the other information to sketch the graph of $f(x)$, as shown in Figure 4.9. ∎

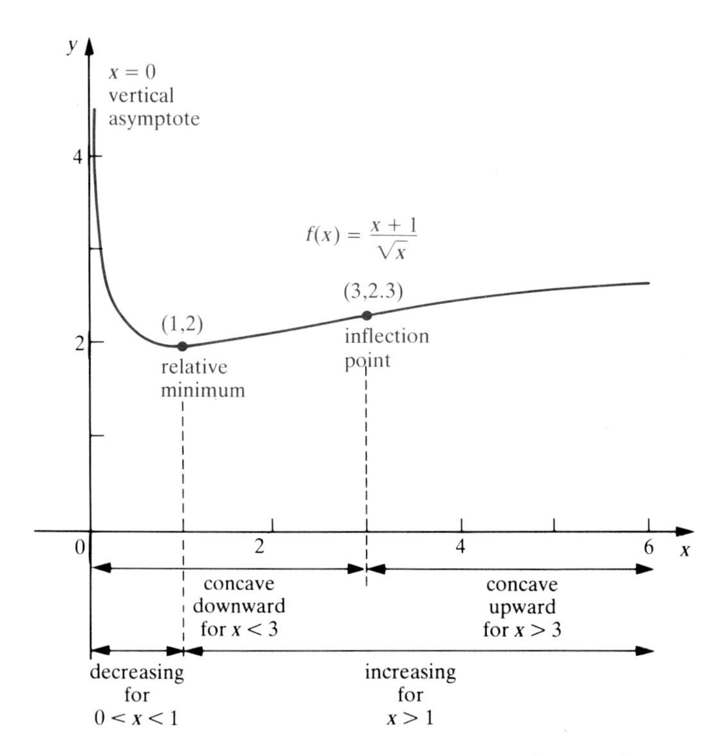

Figure 4.9

Graph of $f(x) = \sqrt{x} + \dfrac{1}{\sqrt{x}} = \dfrac{x+1}{\sqrt{x}}$

Domain $x > 0$
No x- or y-intercept
No horizontal asymptote
Vertical asymptote at $x = 0$
Relative minimum $(1, 2)$
Decreasing for $0 < x < 1$
Increasing for $x > 1$
Point of inflection $(3, 2.3)$
Concave downward for $x < 3$
Concave upward for $x > 3$

Example 4.10 Sketch the graph of $f(x) = \dfrac{4}{x-1} - \dfrac{1}{x-3}$.

Solution When $x = 0$,

$$y = f(0) = \frac{4}{-1} - \frac{1}{-3} = -\frac{11}{3}.$$

The y-intercept is thus $-\frac{11}{3}$. By finding a common denominator, we can rewrite $f(x)$ as

$$f(x) = \frac{3x - 11}{(x-1)(x-3)}.$$

Hence the graph has x-intercept at $x = \frac{11}{3}$, where the numerator is zero. When written in this form, we can see that the denominator of $f(x)$ is zero at $x = 1$ and at $x = 3$, and the numerator is not zero at either of these values. Vertical asymptotes occur at $x = 1$ and $x = 3$. The end arrows for the vertical asymptotes may be determined by evaluating the function near the asymptotes. Table 4.6 shows the results.

Since both terms in the original definition of $f(x)$ approach zero as $x \to \pm\infty$, we see that $y = 0$ is a horizontal asymptote. We may easily compute the derivative of $f(x)$ from the form $f(x) = 4(x-1)^{-1} - (x-3)^{-1}$. We have

$$f'(x) = -4(x-1)^{-2} + (x-3)^{-2}$$

or

$$f'(x) = -\frac{4}{(x-1)^2} + \frac{1}{(x-3)^2}.$$

To find the critical values of $f(x)$, we set this expression equal to zero,

Table 4.6
Data for end arrows for $f(x) = \dfrac{4}{x-1} - \dfrac{1}{x-3}$

x	$f(x)$
0.99	−399.50
1.01	400.5
2.99	102.21
3.01	−98.01

then move the first term to the right side to obtain

$$\frac{1}{(x-3)^2} = \frac{4}{(x-1)^2}.$$

Multiplying both sides of the equation by $(x-1)^2(x-3)^2$ yields

$$(x-1)^2 = 4(x-3)^2.$$

We next take the square root of both sides, including a plus and minus sign on one side:

$$x - 1 = \pm 2(x-3).$$

Choosing the plus sign yields $x - 1 = 2(x-3) = 2x - 6$. Solving for x, we find $x = 5$. Choosing the minus sign gives $x - 1 = -2(x-3)$, whose solution is $x = \frac{7}{3}$. The critical values are $x = \frac{7}{3}$ and $x = 5$. The y-coordinates for these values are $\frac{9}{2}$ and $\frac{1}{2}$, respectively. Figure 4.10 shows the data we have determined thus far.

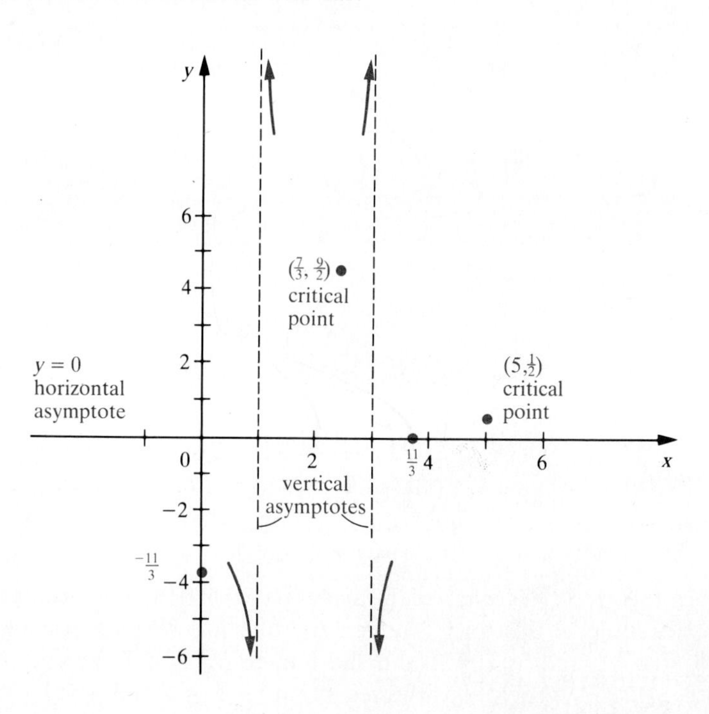

Figure 4.10
End arrows, critical points, and intercepts for $f(x) = \dfrac{4}{x-1} - \dfrac{1}{x-3}$

y-intercept at $-11/3$
x-intercept at $11/3$
Horizontal asymptote at $y = 0$
Vertical asymptotes at $x = 1$ and $x = 3$
Critical points at $(\frac{7}{3}, \frac{9}{2})$ and $(5, \frac{1}{2})$

We have not yet determined if the critical values correspond to relative maxima or relative minima, and we have not found inflection points. However, from the end arrows and the position of the critical points, we see that $(\frac{7}{3}, \frac{9}{2})$ must be a relative minimum and $(5, \frac{1}{2})$ must be a relative maximum. Consequently, the curve is concave downward to the left of $x = 1$ and concave upward between $x = 1$ and $x = 3$. Immediately to the right of $x = 3$ the curve is concave downward, since it rises from a very negative value near $x = 3$ to $\frac{1}{2}$ at $x = 5$ and has a horizontal tangent line at $x = 5$. Beyond $x = 5$, the curve approaches the x-axis asymptotically, which indicates that it must eventually be concave upward, so there is an inflection point somewhere beyond $x = 5$. In

Exercise 33 you are asked to find this inflection point. Figure 4.11 shows the finished graph of the function. ■

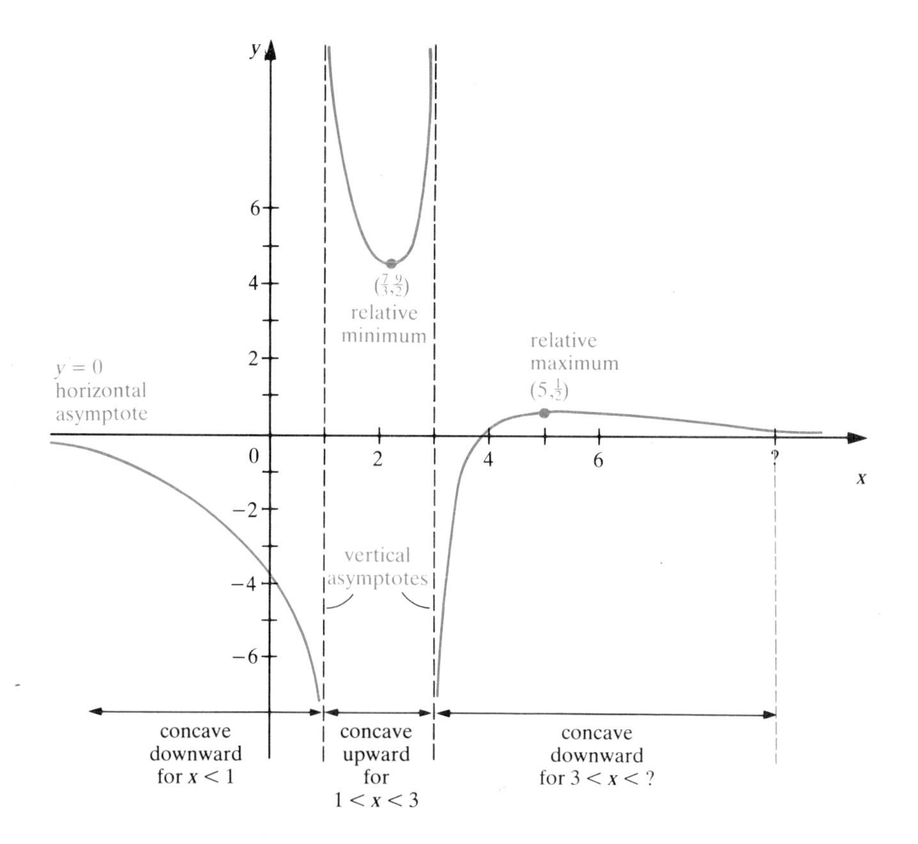

Figure 4.11

Graph of $f(x) = \dfrac{4}{x-1} - \dfrac{1}{x-3}$

y-intercept at $-11/3$
x-intercept at $11/3$
Horizontal asymptote at $y = 0$
Vertical asymptotes at $x = 1$ and $x = 3$
Relative minimum at $(\frac{7}{3}, \frac{11}{3})$
Relative maximum at $(5, \frac{1}{2})$
Concave downward for $x < 1$ and $x > 3$
Concave upward for $1 < x < 3$

Exercises

Exer. 1–32: Graph the function. Record the y-intercept, x-intercepts (if easy to find), all relative maxima, relative minima, inflection points, horizontal asymptotes, and vertical asymptotes. Indicate where the graph is increasing, decreasing, concave upward, and concave downward.

1. $f(x) = x^2 - 1$
2. $f(x) = x^3 - 12x$
3. $f(x) = x^3 - 3x$
4. $f(x) = x^4 - 2x^2$
5. $f(x) = \dfrac{x+3}{x-2}$
6. $f(x) = x^4 - 4x^3$
7. $f(x) = \dfrac{1}{(x-2)^2}$
8. $f(x) = 2x + \dfrac{1}{x^2}$
9. $f(x) = x^4 + 4x^3 - 2$
10. $f(x) = x^{2/3} - x^{5/3}$
11. $f(x) = 12x^{1/5} - x^{6/5}$
12. $f(x) = \dfrac{3x+8}{x+4}$
13. $f(x) = \dfrac{3x^2 - 5x + 10}{x^2}$
14. $f(x) = \dfrac{x^2 - x + 4}{x^2}$
15. $f(x) = \dfrac{x^2}{x^2 - 1}$
16. $f(x) = \dfrac{1}{x^2} + 1$
17. $f(x) = \dfrac{x^2 - 1}{x^2}$
18. $f(x) = x^3 - x^2$
19. $f(x) = x^3 - 6x^2 - 3$
20. $f(x) = \dfrac{x^2 - 1}{x}$
21. $f(x) = x^5 - 80x$
22. $f(x) = \dfrac{1}{x} + x$
23. $f(x) = \sqrt[3]{x} + \dfrac{1}{\sqrt[3]{x}}$
24. $f(x) = 3x^4 - 8x^3 + 6x^2$
25. $f(x) = \dfrac{x}{x^2 + 1}$
26. $f(x) = \dfrac{5 - x}{3x}$
27. $f(x) = \dfrac{x - 2}{(x-1)^2}$
28. $f(x) = \dfrac{2x + 3}{3x^2 + 4x}$
29. $f(x) = \dfrac{x + 2}{x^2 + 4x + 5}$
30. $f(x) = \dfrac{x - 2}{(x+1)^2}$

31. $f(x) = \dfrac{2x - 5}{2x^2 - 8x + 8}$ 32. $f(x) = \dfrac{x}{x^2 - 1}$

33. Refer to Example 4.10, in which we graphed the function

$$f(x) = \frac{4}{x - 1} - \frac{1}{x - 3}.$$

Find the inflection point of the function beyond $x = 5$.

Exer. 34–37: Graph the function and record all pertinent information.

34. $f(x) = \dfrac{2}{x - 2} - \dfrac{2}{x + 2}$

35. $f(x) = \dfrac{2}{(x - 2)^2} - \dfrac{2}{(x + 2)^2}$

36. $f(x) = 1 + \dfrac{1}{(x - 2)^2}$ 37. $f(x) = \dfrac{1}{x^2 - 1} + 2$

4.3 Implicit Differentiation

We have learned a great deal about finding dy/dx when we are given an equation such as $y = 2x^3 + x$, where y is on one side of the equal sign and does not appear in the expression on the other side. In this case, y is said to be given *explicitly* (plainly stated). However, some applications lead to equations where y is given *implicitly* (implied by the equation, but not plainly stated). For example, if $y = 2x^3 + 3$ is written as

$$2x^3 + x - y = 0,$$

it is considered to be an implicit equation (although readily converted to an explicit equation). When y is given implicitly, the equation can sometimes be solved for y, yielding an explicit expression for y. The above equation is an example of this situation. However, solving for y is often not practical or even possible. For example, solving

$$xy^3 + xy - 10 = 0$$

for y is quite difficult. Consequently, if an application requires us to find dy/dx and y is given implicitly, we need a new approach.

To address the problem of finding dy/dx when y is given implicitly, we make the assumption that y is some function of x, even though we do not know how to write the function. Technically, the implicit equation may represent more than one function. For example, in an expression such as $x^2 + y^2 - 25 = 0$, solving for y yields $y = \pm\sqrt{25 - x^2}$, which defines *two* functions, one using the + sign and one using the − sign. Figure 4.12 shows this situation. The function $y = +\sqrt{25 - x^2}$ is the top half of the circle and the function $y = -\sqrt{25 - x^2}$ is the bottom half of the circle. In the implicit form, $x^2 + y^2 - 25 = 0$, the symbol y denotes either of these functions, and the method described below will find the derivative of both functions.

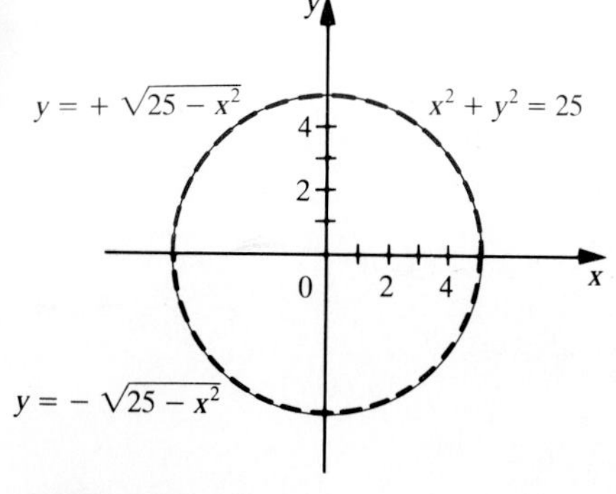

Figure 4.12
The equation $x^2 + y^2 - 25 = 0$ defines two functions

Implicit differentiation is the name given to the technique for finding dy/dx when x and y are related implicitly. If we assume that y is some function of x, the process of finding dy/dx is an application of the chain rule, as the following example illustrates.

Example 4.12 Given $x^2 + y^2 - 25 = 0$,
(a) find y' using the given implicit equation for y;
(b) find y' by solving explicitly for y and then differentiating;
(c) find the slope of the tangent line(s) at $x = 3$.

Solution **(a)** To find y' by implicit differentiation, we first calculate the derivative of each term in the equation with respect to x. We are assuming that y is some function of x, so we emphasize this by writing $y(x)$. By the chain rule

$$\frac{d}{dx}[y(x)]^2 = 2[y(x)]y'(x).$$

Using this formula, we differentiate each term in the equation $x^2 + [y(x)]^2 - 25 = 0$ with respect to x:

$$\frac{d}{dx}\left(x^2 + [y(x)]^2 - 25\right) = 0$$

or

$$2x + 2[y(x)]y' - 0 = 0.$$

We solve this equation for y' in terms of x and $y(x)$, which we now write as y:

$$y' = -\frac{x}{y}.$$

(b) To use explicit differentiation, we write two equations for y

$$y = \sqrt{25 - x^2} \quad \text{and} \quad y = -\sqrt{25 - x^2}$$

and find the derivatives by the chain rule:

$$y' = \tfrac{1}{2}(25 - x^2)^{-1/2}(-2x) \quad \text{and} \quad y' = -\tfrac{1}{2}(25 - x^2)^{-1/2}(-2x)$$

or

$$y' = \frac{-x}{\sqrt{25 - x^2}} \quad \text{and} \quad y' = \frac{-x}{-\sqrt{25 - x^2}}.$$

The denominators in these expressions are the original y, so both may be written as $y' = -x/y$. Thus both differentiation techniques yield the same result.

(c) To find the slope of the tangent lines at $x = 3$, we find the value(s) of y associated with $x = 3$. From $3^2 + y^2 - 25 = 0$ we get $y^2 = 16$, so the associated values of y are $y = 4$ and $y = -4$. The point $(3, 4)$ is a point on the top half of the circle and $(3, -4)$ is a point on the bottom half of the circle. See Figure 4.13. The slope of the tangent line at the point $(3, 4)$ is $-\frac{3}{4}$ and the slope at the point $(3, -4)$ is $-3/(-4) = \frac{3}{4}$. ■

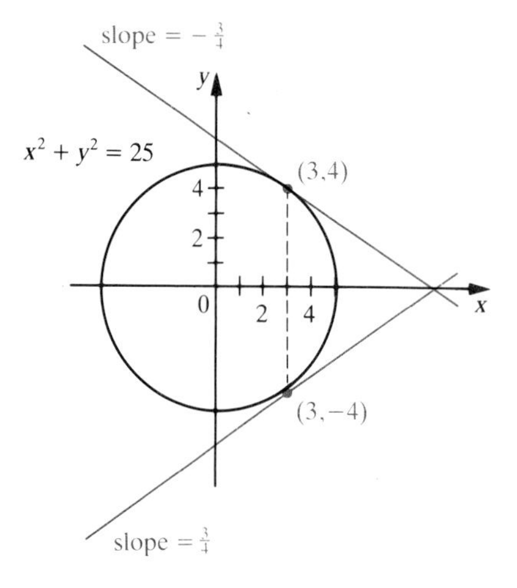

Figure 4.13
Tangent lines to $x^2 + y^2 - 25 = 0$ at $x = 3$

For the implicit differentiation of $[y(x)]^2$ in the previous example, we used

$$\frac{d}{dx}[y(x)]^2 = 2[y(x)]^{2-1}y' = (2y)y'.$$

In general, if $g(y)$ is a function of y where y in turn is a function of x, then the chain rule yields

$$\frac{dg(y)}{dx} = \frac{dg(y)}{dy}\frac{dy}{dx} = g'(y)y'.$$

Example 4.13 Find the slope of the tangent line to the graph of $xy^3 + xy - 10 = 0$ at the point $(1, 2)$.

Solution We differentiate each term in the equation, regarding y as $y(x)$. For emphasis, we rewrite the equation as

$$x[y(x)]^3 + x[y(x)] - 10 = 0$$

and use the product rule and chain rule to find the derivatives. Note that $x[y(x)]^3$ is the product of x and $[y(x)]^3$, both of which are functions of x. Similarly, $x[y(x)]$ is the product of x and $y(x)$. The derivative is

$$\frac{d}{dx}(x[y(x)]^3 + x[y(x)] - 10) = 0$$

or $$x \cdot 3[y(x)]^2 y' + [y(x)]^3(1) + xy' + [y(x)](1) = 0,$$

that is, $$3xy^2y' + y^3 + xy' + y = 0.$$

Collecting the terms involving y' gives

$$y'(3xy^2 + x) + y^3 + y = 0$$

so that $$y' = \frac{-(y^3 + y)}{3xy^2 + x}.$$

To find the value of y' at the point $(1, 2)$ we set $x = 1$ and $y = 2$, and obtain

$$y'|_{(1,2)} = \frac{-(2^3 + 2)}{3(1)(2)^2 + 1} = -\frac{10}{13}.$$

Consequently, the slope of the tangent line at the point $(1, 2)$ is $-1\frac{10}{13}$. ■

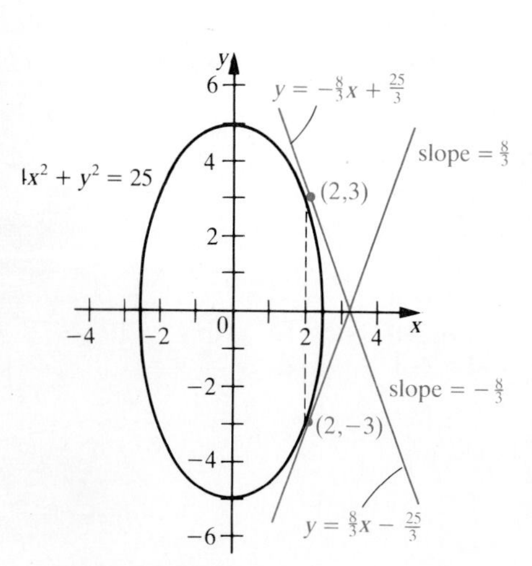

Figure 4.14
Tangent lines to ellipse $4x^2 + y^2 = 25$ at $x = 2$

Example 4.14 Determine the equation of the lines tangent to the curve $4x^2 + y^2 = 25$ when $x = 2$.

Solution Using implicit differentiation to find dy/dx, we obtain

$$\frac{d}{dx}(4x^2 + y^2) = 0$$

$$8x + 2yy' = 0.$$

Solving for y' gives $$y' = -\frac{4x}{y}.$$

To find the slope of the tangent line at $x = 2$, we note that when $x = 2$, $y^2 = 25 - 4(2)^2 = 9$, so $y = \pm 3$. This means that there are two tangent lines: at the points $(2, 3)$ and $(2, -3)$. See Figure 4.14. (The curve in this example is called an *ellipse*.)

For the point $(2, 3)$, the slope of the tangent line is

$$\text{slope} = y'|_{(2,3)} = -\frac{4x}{y}\bigg|_{(2,3)} = -\frac{8}{3}$$

and the equation of the tangent line is

$$y - 3 = -\tfrac{8}{3}(x - 2), \quad \text{or} \quad y = -\tfrac{8}{3}x + \tfrac{25}{3}.$$

In a similar manner we find that the slope at the point $(2, -3)$ is

$-8/(-3) = \frac{8}{3}$, and the equation of this tangent line is

$$y + 3 = \tfrac{8}{3}(x - 2), \quad \text{or} \quad y = \tfrac{8}{3}x - \tfrac{25}{3}.$$

The results of this example can be interpreted by recognizing that the equation $4x^2 + y^2 = 25$ represents two functions: $y = \sqrt{25 - 4x^2}$ (top half of the ellipse) and $y = -\sqrt{25 - 4x^2}$ (bottom half of the ellipse). Implicit differentiation enabled us to find the derivative of both functions in one step. However, since the derivative contained a y term, the slope produced two values, each of which was the slope of a tangent line. ■

Example 4.15 Using implicit differentiation, find d^2y/dx^2 where

$$x^3 + 3x^2y + 17 = 0.$$

Solution Differentiating term by term and using the product rule on the second term, we obtain

$$3x^2 + 3x^2\frac{dy}{dx} + (y)(6x) = 0.$$

When $x \neq 0$, we have

$$\frac{dy}{dx} = \frac{-3x^2 - 6xy}{3x^2} = \frac{-x - 2y}{x}.$$

Therefore, using the quotient rule, we can find the second derivative:

$$\begin{aligned} \frac{d^2y}{dx^2} &= \frac{d}{dx}\left(\frac{-x - 2y}{x}\right) \\ &= \frac{x\left(-1 - 2\dfrac{dy}{dx}\right) - (-x - 2y)(1)}{x^2} \\ &= \frac{-x - 2x\dfrac{dy}{dx} + x + 2y}{x^2} = \frac{2y - 2x\dfrac{dy}{dx}}{x^2}. \end{aligned}$$

Note that the chain rule was used in the second step to find the derivative of $-2y$. Since we already know that $dy/dx = (-x - 2y)/x$, we can write d^2y/dx^2 in terms of x and y by substituting $(-x - 2y)/x$ for dy/dx, which yields

$$\begin{aligned} \frac{d^2y}{dx^2} &= \frac{2y - 2x\left(\dfrac{-x - 2y}{x}\right)}{x^2} \\ &= \frac{2y + 2x + 4y}{x^2} = \frac{2x + 6y}{x^2}. \end{aligned}$$

■

Example 4.16 A nationwide company finds that the demand $D(x)$ (in millions) for one of its products is given by

$$D(x) = \frac{x + A^2}{x^2 + A + 1},$$

where x denotes the price in dollars and A denotes the amount per month spent on advertising (in millions of dollars). Currently, the price is \$2.00 and the company is spending \$3 million per month advertising this product ($A = 3$). The demand is thus $D(2) = 1.375$ million items. The company plans to increase the price to \$2.50. Estimate the amount that should be invested in advertising to maintain the current demand level.

Solution Since the company wants to keep the demand constant at 1.375, we have an equation relating A and x, namely

$$\frac{x + A^2}{x^2 + A + 1} = 1.375. \tag{1}$$

Consequently, we consider A to be an implicit function of x, which we emphasize by writing $A(x)$ temporarily. The change in $A(x)$ caused by a change dx in x is approximately given by the differential

$$\text{change in } A(x) \approx A'(x)\,dx.$$

Since we know the change in x is $dx = 0.50$, we can estimate the required change in $A(x)$ by first evaluating $A'(x)$. We begin by differentiating both sides of equation (1) with respect to x:

$$\frac{d}{dx}\left(\frac{x + [A(x)]^2}{x^2 + A(x) + 1}\right) = 0.$$

By the quotient rule, the derivative can be written as

$$\frac{(x^2 + A(x) + 1)[1 + 2A(x)A'(x)] - (x + [A(x)]^2)[2x + A'(x)]}{(x^2 + A(x) + 1)^2} = 0.$$

We multiply both sides of this equation by the denominator of the left side, expand the products in the numerator, and collect all terms involving $A'(x)$ to get

$$-\big(2x[A(x)]^2 + x^2 - A(x) - 1\big) + A'(x)\big(2A(x)x^2 + [A(x)]^2 + 2A(x) - x\big) = 0.$$

Write A' for $A'(x)$ and A for $A(x)$, and solve for A' to find

$$A' = \frac{2xA^2 + x^2 - A - 1}{2Ax^2 + A^2 + 2A - x}.$$

Evaluating this expression at $x = 2$ and $A = 3$ yields

$$A'\big|_{(2,3)} = \frac{2(2)(9) + 4 - 3 - 1}{2(3)(4) + 9 + 6 - 2} = \frac{36}{37}.$$

The change in advertising cost associated with a change in price of $dx = 0.50$ is estimated by

$$\text{change in } A(x) \approx (A'\big|_{(2,3)})\,dx = \frac{36}{37}(0.50)$$

$$= \frac{18}{37} \approx 0.486.$$

The advertising expenditure should be increased from \$3 million per month to about \$3.486 million per month to keep the product demand constant when the product price is raised from \$2.00 to \$2.50. ■

Exercises

Exer. 1–14: Find dy/dx by implicit differentiation.

1. $yx - 1 = 0$
2. $x^2 + xy = y$
3. $x^2 + x^2y = 14y$
4. $(x - 1)y = 125$
5. $(xy - 2)^2 = 125x$
6. $(xy - 2)(x + y) = 125$
7. $x^2y^3 = 2$
8. $x^2 - 2xy + y^2 = 8$
9. $(x^2y - 3)^2 = 10$
10. $\dfrac{x^2 - xy}{xy} = 10$
11. $x^{1/2} + y^{1/2} = 25$
12. $(xy)^{3/4} = y$
13. $x^{3/4}y^{5/4} = y$
14. $y^2 + y^{-2} = x$

Exer. 14–22: Find the equation of the tangent line to the curve at the point indicated.

15. $xy = 1$; $(1, 1)$
16. $x^2y^3 = 2$; $(\frac{1}{2}, 2)$
17. $x^2y^2 - 3xy^3 + 5y = 3$; $(2, 1)$
18. $x^2 + y^2 + 2xy + 3y = 19$; $(-3, 5)$
19. $(xy - 2)(x + y) = 0$; $(-1, 1)$
20. $\dfrac{x - y}{x + y} = 4$; $(5, -3)$
21. $(x^{-1} + y^{-1})^{1/2} = y/x$; $(2, 2)$
22. $x\sqrt{y^2 - y + 2} = 4$; $(2, 2)$

Exer. 23–32: Find d^2y/dx^2.

23. $x^2 - xy + 25 = 0$
24. $x^2y^3 = 2$
25. $x^2 + y^3 = x$
26. $xy^2 - x^2y = 1$
27. $(x^2 - y^3 - y)^2 = 10$
28. $25xy^3 = 3y^2 - 1$
29. $x^{-1/2} + y^{-1/2} = 4$
30. $\sqrt{x^2 + y^3} = y$
31. $xy^{1/2} = 2x^2$
32. $\sqrt{x^2 + y^3} = x$
33. Determine the value of x where the tangent line to the circle $x^2 + y^2 = 25$ at $(3, 4)$ crosses the x-axis.
34. Find the slope of the tangent line(s) to the circle $x^2 + y^2 = 25$ corresponding to $x = -3$. Write the equation(s) for the tangent line(s).
35. Find the equation of the tangent line to the graph of $5x^2 + 2y^2 = 38$ at $(2, 3)$.
36. Determine where the tangent line to the graph of
$$\frac{x}{y} + \frac{y}{x} = 2xy$$
at $(1, 1)$ intersects the x-axis.

Exer. 37–40: Determine the equation(s) of the line(s) tangent to the given curve at the specified point.

37. $x - 2 = y^2 - 25x + 21$; $x = 4$
38. $y = x^2 - 4xy^3 + 4$; $x = 0$
39. $4x^2 + 9y^2 = 328$; $x = 1$.
40. $x^2 + y^2 - 4x + 3y = 1$; $x = 1$

4.4 Applications to Economics

Economics is one of many fields in which calculus has been used to great advantage. To consider profit in an economic situation, one must consider both cost and revenue. **Revenue** is the term economists use for income and **cost** refers to the cost of production. Both cost and revenue depend on the quantity produced. Suppose the cost in dollars for a weekly production of x units of steel is given by $C(x) = 0.1x^2 + 5x + 2000$. If x units are sold at a price p per unit, then the revenue realized will be $R(x) = px$. The value of the cost function at $x = 0$ has a special significance: it is the *fixed cost* that must be expended even if no item is produced. In our steel example the fixed cost is \$2000 per week.

Throughout this book, cost and revenue will be given explicitly by functions. In actual practice, complicated accounting procedures and mathematical techniques are used to produce a function that approximates certain data. We will consider one of the simplest of these techniques in Chapter 8. For now we will assume that the functions provided have been obtained by a reasonable method.

If $C(x)$ is the cost function, the derivative of $C(x)$ with respect to x is called the **marginal cost**, sometimes denoted by $M_C(x)$. Thus, if the cost in dollars for a weekly production of steel is given by $C(x) = 0.1x^2 + 5x + 2000$, then the marginal cost is $M_C(x) = dC/dx = 0.2x + 5$.

Many elementary economics texts describe marginal cost as the *cost of producing one more item.* Since $f(a+1)$ is the cost of producing $a+1$ items and $f(a)$ is the cost of producing a items, $f(a+1) - f(a)$ is the cost of increasing the production by one unit. Recall that the definition of the derivative of $f(x)$ at $x = a$ is

$$f'(a) = \lim_{h \to 0} \frac{f(a+h) - f(a)}{h}$$

so, for small values of h,

$$f'(a) \approx \frac{f(a+h) - f(a)}{h}.$$

In many practical situations we deal with large values of a so that $h = 1$ is relatively small. Thus, we have $f'(a) \approx f(a+1) - f(a)$ for $h = 1$, which shows that for large a, the marginal cost as we have defined it is approximately equal to the marginal cost given in economics texts which do not use the derivative as a tool.

For linear functions, the mathematics and economics interpretations of marginal cost are identical. Consider, for example, $C = f(x) = 3x + 10$. The cost of producing a items is $f(a) = 3a + 10$ and the cost of producing $a+1$ items is $f(a+1) = 3(a+1) + 10 = 3a + 13$. Thus the cost of producing one more item is $f(a+1) - f(a) = (3a+13) - (3a+10) = 3$. Furthermore, the derivative of $f(x)$ is also 3. The closer the cost function is to being linear, the closer the marginal cost is to the actual cost of producing one more item.

Example 4.17 Suppose that the cost of producing x typewriters per month is given by

$$C(x) = 0.01x^2 + 20x + 5000.$$

(a) If the present production is 9999 typewriters a month, what is the additional cost to increase production to 10,000 per month?
(b) If the company sells each typewriter for \$218, is it advisable to increase production?
(c) What is the marginal cost at 9999 units per month?

Solution **(a)** The amount we seek is the difference in cost between producing 10,000 typewriters and producing 9999 typewriters:

$$C(10{,}000) - C(9999) = 1{,}205{,}000.00 - 1{,}204{,}780.01 = 219.99 \text{ dollars.}$$

(b) Since the cost of producing one more typewriter, \$219.99, exceeds the selling price of the extra typewriter \$218, we advise against increasing production.
(c) To find the marginal cost recall that, by definition, the marginal cost of producing 9999 items is $C'(9999)$. We have $M_C(x) = C'(x) = 0.02x + 20$, so $C'(9999) = 219.98$ dollars. This is indeed a good approximation

to the cost of increasing production by one typewriter per month. This figure would also have warned us not to increase production. ■

Marginal revenue is the derivative of the revenue function with respect to the number of items sold and is denoted by $M_R(x)$. An argument similar to our discussion of marginal cost shows that, when the number of items is sold large, marginal revenue is approximately equal to the additional revenue from selling one more item.

Example 4.18 If the revenue from the sale of a certain product is given by the function $R(x) = x^3 + 5x^2 + 200x$, determine the revenue from the sale of 100 items and the marginal revenue when $x = 100$.

Solution When $x = 100$,

$$R(100) = 100^3 + 5(100)^2 + (200)(100) = 1{,}070{,}000 \text{ dollars.}$$

The marginal revenue is

$$M_R(x) = \frac{dR}{dx} = 3x^2 + 10x + 200.$$

When $x = 100$, the marginal revenue is

$$M_R(100) = 3(100)^2 + 10(100) + 200 = 31{,}200 \text{ dollars.}$$ ■

One measure of the success of a company is **profit**, the difference between revenue and cost. The profit $P(x)$ from manufacturing x items is

$$P(x) = R(x) - C(x).$$

Given the price p of a certain product, the revenue $R(x)$ realized from the sale of x items is

$$R(x) = px,$$

assuming that all items produced are sold. (We will make this assumption throughout this book.) Hence, the price can be determined from the revenue function, and vice versa.

Example 4.19 If the cost and price for a certain product are given by the formulas $C = 50x + 30{,}000$ and $p = 100 - 0.01x$, find the profit function $P(x)$.

Solution We first find the revenue:

$$R(x) = px = (100 - 0.01x)x = 100x - 0.01x^2.$$

Then we can determine profit:

$$\begin{aligned} P(x) &= R(x) - C(x) \\ &= 100x - 0.01x^2 - (50x + 30{,}000) \\ &= -0.01x^2 + 50x - 30{,}000. \end{aligned}$$ ■

To make sound decisions, managers must determine the price at which the maximum profit will be attained. The maximum of the profit function can occur only at the places where the derivative of the profit function is zero. Since $P = R - C$, maximum profit is attained when x is chosen so that

$$\frac{dP}{dx} = \frac{dR}{dx} - \frac{dC}{dx} = 0.$$

Consequently, the optimal production level x is chosen so that marginal revenue equals marginal cost, that is,

$$\frac{dR}{dx} = \frac{dC}{dx}.$$

After we have found the level of production that yields maximum profit, we find the corresponding price by $p = R(x)/x$.

Example 4.20 Suppose that a company's daily cost and revenue functions for manufacturing x motorcycles per day are given by the formulas

$$C = x^2 + 50x + 150{,}000 \quad \text{and} \quad R = -\tfrac{1}{3}x^3 + 50x^2 + 250x.$$

Find the price for which maximum profit will be realized. What is the maximum profit per day?

Solution Our strategy is to find the number x of motorcycles produced that will make $dC/dx = dR/dx$. We then determine the optimal price from $R(x)$, since $p = R(x)/x$. Substituting this value of x into $P(x)$ will give the maximum profit per day. We observe that $dC/dx = dR/dx$ implies

$$2x + 50 = -x^2 + 100x + 250.$$

We move all terms to the left side of the equation to get

$$x^2 - 98x - 200 = 0, \quad \text{or} \quad (x - 100)(x + 2) = 0.$$

We see that $x = 100$ and $x = -2$ are the solutions. However, $x = 100$ is the only reasonable solution, since a company cannot produce -2 motorcycles per day. Using the Second Derivative Test, we may verify that a maximum occurs when $x = 100$. When $x = 100$, the revenue is $R(100) = 191{,}666.67$, so the price that will maximize profit is

$$p = \frac{R(100)}{100} = \frac{191{,}666.67}{100}$$
$$= 1{,}916.67 \text{ dollars.}$$

To calculate the maximum daily profit P, note that at $x = 100$ we have $C = 165{,}000$, so $P = R - C = 191.666.67 - 165{,}000.00 = 26{,}666.67$. Consequently, the motorcycle manufacturing company has a maximum profit of \$26,666.67 per day, and this corresponds to producing 100 motorcycles per day. ■

Example 4.21 Suppose that the cost (in cents) of making x ballpoint pens per minute is

$$C = 0.01x^2 + 12x + 350$$

and the price at which each pen can be sold is $p = 88 - 1.99x$ cents.
(a) Find the production level x for maximum profit and the corresponding price and profit per hour.
(b) Suppose that the government imposes a tax of b cents per pen sold. How does the tax effect the price at which the seller will realize maximum profit;
(c) At what price should the pens be sold if the tax is $b = 3$ cents per pen?

Solution **(a)** Recall that the revenue is $R = px = 88x - 1.99x^2$. We first compute the price that yields maximum profit from

$$\frac{dC}{dx} = 0.02x + 12 \quad \text{and} \quad \frac{dR}{dx} = 88 - 3.98x.$$

Maximum profit occurs when $dC/dx = dR/dx$, that is, when

$$0.02x + 12 = 88 - 3.98x.$$

Solving for x, we obtain $x = 19$. Thus, maximum profit is realized at a production level of 19 pens per minute. The price for maximum profit is $p = 88 - 1.99(19) = 50.19$ cents per pen. Since 19 pens are produced (and sold) per minute, the revenue is $19 \times 50.19 = 953.61$ cents per minute, or \$572.16 per hour. The cost, obtained by evaluating $C(x)$ at $x = 19$ is 581.61 cents per minute, or \$348.97 per hour. Subtracting cost from revenue yields a profit of \$223.19 per hour.
(b) Consider the same market conditions with the tax imposed. The demand function remains the same, but the manufacturer receives only $(p - b)$ cents per pen sold. Thus, revenue is given by $R = (p - b)x = (88 - 1.99x - b)x$. Now the marginal revenue is

$$\frac{dR}{dx} = 88 - 3.98x - b.$$

Marginal cost remains the same as without the tax; $dR/dx = dC/dx$ yields

$$88 - 3.98x - b = 0.02x + 12.$$

Solving for x, we obtain $x = 19 - \frac{1}{4}b$. Thus the maximum profit occurs when the production level is $19 - \frac{1}{4}b$ pens per minute, a smaller value than without the tax. The corresponding price is

$$p = 88 - 1.99\left(19 - \frac{b}{4}\right) = 50.19 + 0.4975b.$$

In other words, the optimal price of each pen with the tax imposed is $0.4975b$ cents more than without the tax. Note that each consumer does not pay b cents more, but only $0.4975b$ cents more than in the no-tax case. Thus, the consumer actually pays slightly less than one-half of the imposed tax.
(c) If a 3-cent tax is imposed on each pen, then $b = 3$, so that the optimal production level is $19 - \frac{3}{4} = 18.25$ pens per minute and the optimal price is $p = 50.19 + 0.4975(3) = 51.68$ cents per pen. To achieve maximum profit with a tax of 3 cents per pen, the manufacturer should

raise the price from 50.19 cents to 51.68 cents per pen. Since an average of 18.25 pens are produced per minute and sold for 51.68 cents each, the revenue is $18.25 \times 51.68 = 943.16$ cents per minute, or \$565.90 per hour. The cost remains \$348.97 per hour, so the profit with a 3-cent tax imposed is $\$565.90 - \$348.97 = \$216.93$. The tax resulted in a loss in profit of \$6.26 per hour.

The graph in Figure 4.15 will help illustrate this discussion. The marginal revenue function $M_R(x)$ is graphed for both the tax and no-tax situations. These functions intersect the marginal cost function $M_C(x)$ at different values of x. ■

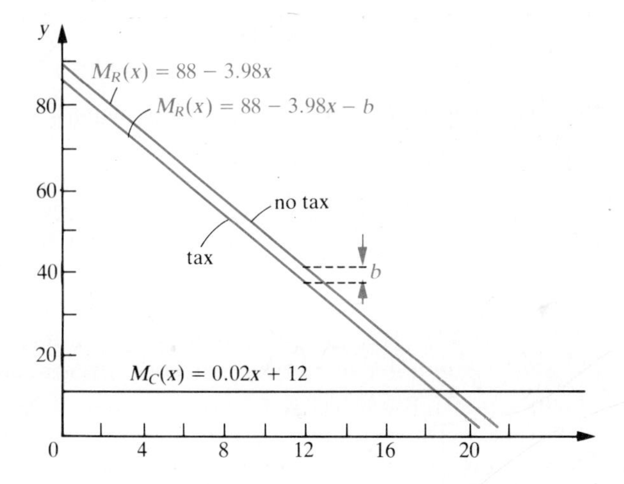

Figure 4.15
Graph of marginal revenue $M_R(x)$ with and without an imposed tax

In many situations we may view revenue as a function of price rather than as a function of the production (sales) level. In this case we use the demand function $D(p)$, the number of items that can be sold at price p. Since revenue is price times amount sold, we have

$$R(p) = pD(p).$$

Example 4.22 As a government-imposed penalty for price-fixing, a small glove manufacturer is required to sell a line of gloves for \$8 per pair for a period of time. Suppose that the demand function for the gloves is $D(p) = 300 - p^2$ pairs per week.
(a) What revenue will the company earn per week when the price is \$8?
(b) What price would yield maximum revenue?
(c) Find the weekly revenue if the price is set at the optimal price.

Solution **(a)** Since $R(p) = pD(p)$, we have $R(p) = p(300 - p^2)$, so

$$R(8) = 8(300 - 8^2)$$
$$= 1888.$$

Therefore, with the price set at \$8, the weekly revenue is \$1888.
(b) To find the price that yields the maximum revenue we first find the price p for which $R'(p) = 0$. Since

$$R(p) = 300p - p^3 \quad \text{we have} \quad R'(p) = 300 - 3p^2.$$

Setting $R'(p) = 0$ and solving for p gives

$$300 - 3p^2 = 0, \quad \text{or} \quad p = \pm 10.$$

The price $p = -10$ makes no sense, so $p = 10$ is the best value of p. Hence, the revenue is greatest when the company charges \$10 per pair of gloves. (Verify this using the Second Derivative Test.)
(c) The revenue at $p = 10$ is $R(10) = 10(300 - 100) = 2000$. Hence, \$2000 per week is the maximum weekly revenue the company could earn. The imposed requirement to sell gloves at \$8 a pair results in a penalty of 2000–1888 = \$112 per week. ■

Exercises

Exer. 1–4: The function represents a cost function for producing x items per month to be sold at the indicated price per item. If 100 items per month are produced now, calculate the cost of producing one more item per month (that is, 101 items). Calculate the marginal cost at $x = 100$. Would you advise the company to increase production to 101 items?

1. $C = 0.01x^2 + 10x + 50$; \$10.00
2. $C = 0.0002x^3 + 0.03x + 100$; \$8.00
3. $C = 0.0004x^3 - 0.02x^2 + 50$; \$7.50
4. $C = 0.005x^3 - 0.2x^2 + 3x + 5000$; \$105.00

Exer. 5–8: The function gives the revenue from the sale of a product. Determine the revenue for the sale of 100 items and the marginal revenue when $x = 100$.

5. $R = 2x^3 + 3x^2 + 50x$
6. $R = 2x^3 - 3x^2 + 25x$
7. $R = 0.01x^4 + 3x^3 + 200x$
8. $R = \dfrac{0.01x^3 - 4x^2 + 20x}{0.01x + 1}$

9. A manufacturer sells television sets for \$250 per set. The cost function for building x sets per month is $C(x) = 0.01x^2 + 20x + 5000$. Determine the lowest monthly production level x such that the company's profit would be greater for x sets than for $x + 1$ sets. Where should the production level be set (that is, when does the marginal cost first exceed \$250)?

Exer. 10–11: For a certain cement manufacturer, the price per ton of x tons of cement is given by the formula $p = 2x^2 - 10x + 1000$, and the company's revenue is given by

$$R(x) = x(2x^2 - 10x + 1000)$$
$$= 2x^3 - 10x^2 + 1000x.$$

10. Find the revenue from the sale of 10 tons of cement.
11. Find the marginal revenue when $x = 10$.

Exer. 12–16: For the given cost function C and revenue function R for producing x items, find the value of x that maximizes the profit, and calculate the maximum profit.

12. $C = 3x^2 + 5$; $R = x^2 + 40x$
13. $C = 5x^2 + 16$; $R = 2x^2 + 24x$
14. $C = \frac{1}{3}x^3 + 2x^2 + 2$; $R = 3x^2 + 3x$
15. $C = 200 + 10x$; $R = 110x - 0.02x^2$
16. $C = 20 + 6x - 0.5x^2$; $R = 10x - x^2$

Exer. 17–21: For the given cost function C and price function p, where x denotes the number of items sold, find the value of x that maximizes profit. Calculate the maximum profit and the price at which the maximum profit is attained.

17. $C = 4x^2 + 7$; $p = 3x + 80$
18. $C = 5x^2 + 5x + 14$; $p = 3x + 73$
19. $C = 8x + x^2 - 1$; $p = 20 - 2x$
20. $C = 10 + 3x - 0.25x^2$; $p = 5 - 0.5x$
21. $C = 100 + 0.1x$; $p = 100 - 0.01x$

22. If the cost of producing x desks is given by $C = 0.02x^2 + 10x + 300$ and the price per desk is given by $p = 90 - x$, find the production level x that yields maximum profit. If a tax of b dollars per desk sold is levied, at what price should the manufacturer sell the desks to maximize profit? Find the price increase if a \$2-tax is imposed.

Exer. 23–25: An importer knows that the cost of importing x snowmobiles is given by $C = 0.1x^2 + 2500x + 800$. The price function indicates that x snowmobiles can be sold at $p = 3000 - 0.1x$ each.

23. How many snowmobiles should be imported to maximize profit?
24. What is the price for each snowmobile at maximum profit?
25. What is the maximum profit?

Exer. 26–28: Refer to Exercises 23–25. Assume that the federal government levies a fee of \$30 per snowmobile imported.

26. How many snowmobiles should now be imported to maximize profit?
27. What will be the price of each snowmobile at maximum profit?
28. Find the maximum profit.

29. A small resort motel with 100 rooms can rent all the rooms at a rate of \$50.00 per night per room. However, for each \$5 increase in the rate, two rooms will be left vacant. How much should be charged per room to maximize revenue? What is the maximum revenue per night? (*Hint:* Let the charge for a room be $50 + 5x$ for $x = 1, 2, \ldots,$ and calculate the number of rooms rented as a function of x; revenue is the product of the number of rooms rented and the rate charged per room.)
30. Suppose that the weekly demand for the hats made by a certain manufacturer is $D(p) = 486 - 2p^2$, where p is the price in dollars per week, and suppose that the government requires that the company sell these hats at \$7.00 each.
(a) Under the government regulations, what is the company's weekly revenue?
(b) What price would yield maximum profit?
(c) Find the company's revenue per week for the price in part (b).
(d) How much potential revenue is lost per week due to government regulation?

4.5 Inventory Problems

A manufacturer attempts to provide an uninterrupted flow of goods while avoiding extra storage costs and, at the same time, taking advantage of mass production. This type of problem—a *production-inventory problem,* a *batching problem,* or a *production-storage problem*—may be solved by a method developed in the following example.

Example 4.23 A manufacturer of citizen's-band radios agrees to supply an auto distributor with 10,000 radios per year, to be delivered in partial shipments at regular intervals throughout the year. Producing all 10,000 radios initially would result in high storage costs before delivery, so the manufacturer will produce a "batch," store the radios until the supply is used up by the regular deliveries, and then produce another batch, which will be stored and shipped, and so on. Suppose the storage cost (including rent, labor, insurance, and so on) is \$1.00 per item per year and the fixed cost is \$20 to start the production of each batch.
(a) Find the batching cost of producing all 10,000 radios in one batch.
(b) Find the batching cost of producing the radios in two batches of 5000 radios.
(c) What batch size will minimize the cost of batching?

Solution We first determine the total cost of batching, that is, the storage cost plus the start-up cost for producing each batch. This is not the *total* production cost, which includes items (such as labor, energy, and materials) that are involved in the actual production but are unaffected by the batching.
(a) If the manufacturer produces all 10,000 radios in one batch, the total start-up cost is \$20. To compute the storage cost, note that at the beginning of the year, all 10,000 radios are on hand (the result of producing only one batch) and this supply dwindles to zero in a regular fashion throughout the year, as shown in Figure 4.16(a). Hence, the average number of radios on hand during the year is 5000, and the storage cost is thus \$5000 (at \$1 per radio stored per year). The cost of

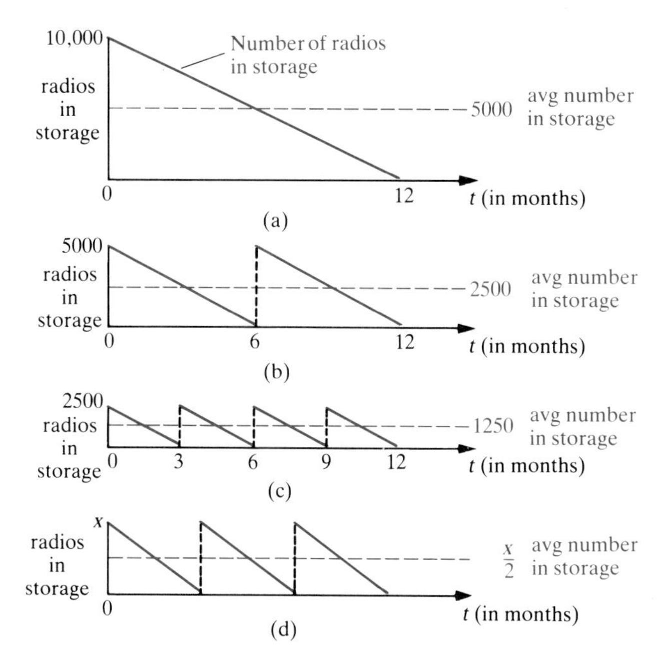

Figure 4.16
Average number of radios stored over a year: (a) one batch of 10,000 radios; (b) two batches of 5,000 radios; (c) four batches of 2,500 radios; (d) (10,000/x) batches of x radios

batching, when all 10,000 radios are manufactured in one batch, is $\$20 + \$5000 = \$5020$.

(b) If the manufacturer produces two batches of 5000 radios each, then the total start-up cost is \$40 (the number of batches times \$20). From Figure 4.16(b) we see that the average number of radios stored throughout the year is 2500. Since the storage cost is \$1 per year per radio, the storage cost incurred is \$2500. Hence, the cost of producing two batches of 5000 radios is $\$40 + \$2500 = \$2540$.

(c) Finally we determine the batch size that yields the minimum cost. Let x represent the number of radios in each batch. Since the number of batches needed to produce the radios in batches of size x is $10{,}000/x$ and the start-up cost for each batch is \$20, the production cost associated with batching is $\$20(10{,}000/x)$. From Figures 4.16(c) and (d) we see that, as in the first two cases, the average number of radios in storage throughout the year is one-half the batch size. Hence, when x is the batch size, $x/2$ is the average number of radios in storage over the year, and the cost for storage is $(x/2) \times 1 = x/2$ dollars. The total cost of batching is

$$f(x) = \frac{x}{2} + \frac{200{,}000}{x}.$$

To determine x such that $f(x)$ is a minimum, we compute $f'(x)$:

$$f'(x) = \frac{1}{2} - \frac{200{,}000}{x^2},$$

and then determine x such that $f'(x) = 0$:

$$\frac{1}{2} - \frac{200{,}000}{x^2} = 0$$

$$x^2 - 400{,}000 = 0 \qquad \text{(multiply by } 2x^2\text{)}$$

Thus $$x = \sqrt{400{,}000} \approx 632.46.$$

Since $f''(x) = 400{,}000/x^3$ is positive at $x = 632.46$, $f(x)$ has a relative minimum there. Consequently, alternating between batches of size 632 and 633 would minimize cost. ■

A variation of the batching problem involving ordering is illustrated in the following example.

Example 4.24 The College Bookstore sells 1200 calculators each semester. The demand for these calculators is constant throughout the semester. Security requirements result in a high storage cost of \$12 per calculator per semester. Analysis of the store's accounting and ordering department indicates that it costs \$20 to place an order for calculators in addition to the actual cost of the calculators. How many orders for calculators should the store place during the semester? How many calculators should be ordered each time?

Solution Let x be the size of each order, so $x \geq 1$. The average number of calculators in storage during the semester is $x/2$, so the storage cost is $(x/2)(12) = 6x$ dollars. To obtain 1200 calculators in orders of size x, the manager should place $1200/x$ orders during the semester. The ordering cost is $(1200/x)(20)$ dollars. Consequently, the total cost of ordering and storage is

$$C(x) = 6x + \frac{24{,}000}{x}.$$

We wish to minimize $C(x)$ for $1 \leq x \leq 1200$. We first compute

$$C'(x) = 6 - \frac{24{,}000}{x^2},$$

and then note that $C'(x) = 0$ when $x^2 = 4000$, that is, when $x = 63.25$. We compare $C(1)$, $C(63.25)$, and $C(1200)$. These costs are \$24,006.00, \$4,219.50, and \$7,220.00, respectively. The minimum occurs at $x = 63.25$. Since $1200/63.25 = 18.97$, the store manager should place 19 orders of size 63 each, which totals 1197 calculators. (To obtain the other three calculators, an extra calculator should be ordered with every sixth order.) ■

The discount offered by a distributor for placing larger orders often influences the size of the order placed. The following example illustrates this factor.

Example 4.25 Refer to Example 4.24. Suppose that the wholesale price is \$40 per calculator for orders of less than 75 and \$39 per calculator for orders of 75 or more. What size order should be placed?

Solution The cost of the calculators themselves is now dependent on the order (or batch) size. Thus the cost function is the sum of the batching cost and the cost of the calculators. The cost of the calculators is $\$40(1200) = \$48{,}000$ if ordered in batches of less than 75 and $\$39(1200) = \$46{,}800$ if ordered in batches of 75 or more. Therefore, the cost function for an order of size x is

$$C(x) = \begin{cases} 6x + \dfrac{24{,}000}{x} + 48{,}000 & \text{for } 1 \le x < 75 \\ 6x + \dfrac{24{,}000}{x} + 46{,}800 & \text{for } 75 \le x \le 1200 \end{cases}$$

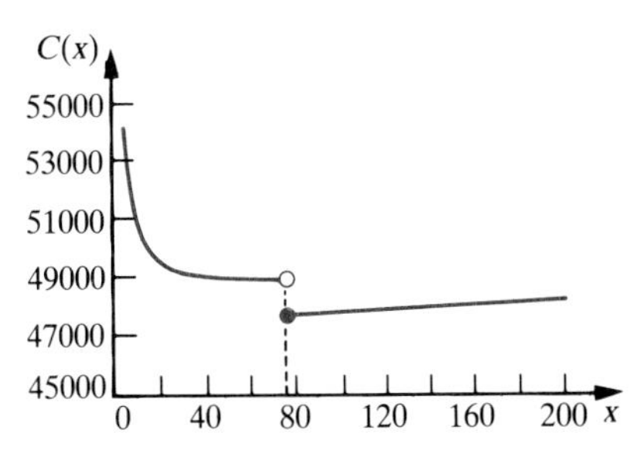

Figure 4.17
Graph of the cost function $C(x)$ in Example 4.25

See Figure 4.17 for a graph of the cost function.

For both of the intervals $(1, 75)$ and $(75, 1200)$, the derivative of $C(x)$ is the same as in the previous example. (However, $C(x)$ has no derivative at $x = 75$ because it is not continuous there.) From the previous example, we know that a relative minimum occurs at $x = 63.25$, and we compute the corresponding cost to be $C(63.25) = 48{,}758.95$. We must also check $x = 75$; since $f'(x)$ does not exist there, $x = 75$ is also a critical value.

Evaluating $C(x)$ at $x = 75$ gives $C(75) = 47{,}570.00$. We want the absolute minimum of $C(x)$ on $[1, 1200]$. While $C(x)$ is not continuous on $[1, 1200]$, it is continuous on each of the intervals $[1, 75)$ and $[75, 1200]$. The latter interval is closed and contains no critical values; therefore, the minimum value on $[75, 1200]$ is the smaller of $C(75) = 47{,}570$ and $C(1200) = 54{,}020$. That is, when orders of size 75 through 1200 are considered the minimum cost is \$47,570 for orders of size 75.

The interval $[1, 75)$ is not closed, but from the definition of $C(x)$ we see that

$$\lim_{x \to 75^-} C(x) = 6(75) + \frac{24{,}000}{75} + 48{,}000 = 48{,}770.$$

Since this value is larger than $C(63.25) = 48{,}758.95$ and $C(1) = 72{,}006$ is also larger than $C(63.25)$, we conclude that the minimum value on the interval $[1, 75)$ is $48{,}758.95$ when $x = 63.25$.

Comparing the minimum cost on the two intervals, we see that the minimum cost for orders of size 1 through 1200 occurs when $x = 75$. Thus, the savings realized by ordering at least 75 calculators each time outweighs the savings realized by minimizing the ordering cost. The manager should therefore order 75 calculators at a time. This strategy results in $1200/75 = 16$ orders per semester. ■

The inventory model in this section provides a framework for the problem of *cash management*. Suppose a firm has a quantity of cash that will be paid out over some period of time. Some of the money could be invested in marketable securities until needed. Thus, having cash on hand costs money in terms of interest foregone on marketable securities. However, each transaction converting securities to cash costs a certain fixed amount, so the company should minimize the number of transactions. The firm needs to determine the optimal amount of cash to maintain, balancing the cost of having the cash on hand (because of lost interest) against the fixed cost of transferring marketable securities to cash.

Suppose a firm with a steady demand for cash over some period of time has invested the cash in marketable securities. The firm sells S

dollars worth of marketable securities whenever the amount of cash on hand reaches zero. (The amount can be greater than zero if a margin is desired or if lead time is necessary for affecting a transaction, but the same principle is involved.) The total cost involved is the sum of the fixed cost of the transfers and the interest foregone by holding cash. We can represent the quantities as follows:

A = total amount of cash needed over the time period,

c = fixed cost of a transaction (assumed independent of the amount),

i = interest rate on marketable securities (assumed constant),

and S = the amount converted to cash in each transaction.

Then $\frac{A}{S}$ = number of transactions needed during the period,

$\frac{S}{2}$ = average cash balance,

$\left(\frac{A}{S}\right)c$ = fixed cost for the period,

and $\left(\frac{S}{2}\right)i$ = interest earnings foregone by holding cash.

Consequently, the total cost is

$$T = \left(\frac{A}{S}\right)c + \left(\frac{S}{2}\right)i.$$

The quantities i, c, and A are typically constant. Hence we regard T as a function of S, and we wish to minimize T.

Example 4.26 Consider a firm that makes cash payments totalling \$6 million over a one-month period, with payments made in a steady flow throughout the month. The firm intends to make these payments from funds that are currently in marketable securities earning 9% interest per year (0.75% per month). If each transaction to sell securities costs \$100, how many dollars worth of securities should be sold per transaction in order to minimize the cost?

Solution Using the notation developed preceding this example, we have $A = 6{,}000{,}000$, $i = 0.0075$, and $c = 100$. Thus the total cost would be

$$T = T(S) = \left(\frac{6{,}000{,}000}{S}\right)100 + \left(\frac{S}{2}\right)0.0075$$

and we want to minimize T:

$$\frac{dT}{dS} = 600{,}000{,}000\left(-\frac{1}{S^2}\right) + 0.00375(1).$$

Setting this derivative to zero, we conclude that the cost is minimized when $S = 400,000$. Consequently, each transaction should convert \$400,000 of securities into cash. The number of transactions needed is $6,000,000/400,000 = 15$ during the one-month period. The average amount of cash on hand is $S/2 = \$200,000$. ■

This cash-management model can also be used when receipts are continuous and are to be invested. The process is then used to determine the optimal size of each transaction to purchase marketable securities. As in the cash-on-hand application, if payments or receipts are not reasonably continuous over the period involved, modifications must be made or another model chosen.

Exercises

1. A supplier of music racks agrees to produce 100,000 racks per year to be delivered regularly to a chain of retail music stores. The storage cost is 75 cents per rack per year and the fixed cost of production is \$33.75 per batch. What batch size will minimize the batching cost?
2. A typewriter store expects to sell 1000 typewriters during the year. Experience has shown that the rate of sale is nearly constant. The cost of storing typewriters is \$8 per typewriter per year. An analysis of the store's accounting procedure shows that the cost of placing an order is \$1. What order size will minimize the cost of ordering and storing the typewriters?
3. Work Exercise 2 using a storage cost of \$2 per typewriter and a cost of \$40 for placing an order.
4. Work Exercise 2 using a storage cost of \$6 per typewriter and a cost of \$30 for placing an order.
5. A skateboard outlet sells 600 skateboards each summer season. Storing each skateboard for the season costs \$2 and \$1.50 per skateboard must be added to the fixed \$6 cost of placing an order. (Thus, the cost of each order of size x will be $6 + 1.5x$.) What order size will minimize ordering and storage cost?
6. Refer to Exercise 5. Suppose the skateboard outlet wants to change to a supplier who charges \$24 plus \$1.20 per skateboard for each order. If the decision is based solely on the annual ordering cost, should the outlet change suppliers?
7. A supply clerk at a pizza outlet estimates the need for flour at 12,000 pounds per year. Storage cost is \$1.50 per 100-pound bag. Ordering cost is a fixed \$10 per order plus \$3 per 100-pound bag. What order size will minimize the cost?
8. An agricultural-equipment retail outlet predicts that 1000 manure spreaders will be sold this year. Storage cost per spreader is \$50 per year and ordering cost is \$250 per order. The wholesaler offers a 10% discount per spreader on orders of 25 or more. The regular price is \$600. What order size will minimize the cost?
9. Work Exercise 8 if, due to a scarcity of spreaders, an extra charge of \$10 per spreader is made for orders of 25 or more?
10. An office estimates that 400 reams of paper will be used in the next year. The storage cost is 80 cents per ream per year and the ordering cost is \$1.20 per order. A paper supplier uses the following pricing scheme:

Reams per order	Price per ream
1–25	\$1.80
26–50	\$1.75
51–100	\$1.70
over 100	\$1.65

What order size should be used?

11. Lillie's Flower Shop stocks spools of ribbon. The annual demand is 8000 spools. Each spool costs \$10, the order cost is \$40 per order, and the carrying cost is \$1 per spool. The owner is currently ordering on an optimum basis. Ribbon Wholesalers, in an effort to reduce its inventory, suggests that the owner place one large order a year. As an inducement, a 4% discount is given if the annual ordering policy is adopted. Evalute this offer and make a recommendation to accept or reject it. Explain your decision. (Ignore interest on any money borrowed to accept the offer.)
12. Refer to Exercise 11. If you reject the offer, make a reasonable counteroffer.
13. The Paul Joseph Concrete & Paint Company introduces a new product called Raincoat with an annual demand of 30,000 cases. The carrying cost per case per year is \$2 and the cost of placing an order is

$12. The cost per case (other than carrying and ordering costs) is $14. Determine the number of orders per year that minimizes the total cost.

14. From prior experience, the MP & C Company has estimated steady cash payments of $600,000 over the next one-month period. The fixed cost of a transaction is $75 and the interest rate on marketable securities is 7.5% annually. Use the inventory model to optimize the cash balance.
15. A subsidiary of Burdett Industries expects $2,500,000 in cash outlays for the next year. The firm must pay a 10% annual interest rate on money borrowed. Each borrowing transaction costs $100. Determine the optimal number of times the firm should borrow money during the year and the amount borrowed each time.
16. Blodget & Sons expects $800,000 in cash outlays for the next year. The company must pay an 8% annual interest rate on money borrowed. Each borrowing transaction costs $80. Determine the optimal number of times the firm should borrow money during the year and the amount borrowed each time.
17. The Lincoln Public School system provides writing booklets for in-class composition assignments. The estimate for the coming school year is that 288,000 booklets (2000 gross) will be used. The cost of the booklets is as follows:

Gross	Price per gross
1–200	$14.40
201–600	$13.68
601–1500	$12.96
1501 or more	$12.60

The handling charge per order is $12.50 plus $0.40 per gross. There is no cost for storage. What order size will minimize cost for the school system?

18. Consider the following variation of Exercise 17. Suppose budget cuts have made it necessary to include a storage charge of $4 per gross. What order size now minimizes cost?
19. Stereo Minneapolis has inadequate storage facilities. The manager estimates that 800 8-inch woofers will be sold during the coming fiscal year with constant demand throughout the year. The storage cost is $5 per speaker per year. The supplier adds $20 per order to cover handling expenses. What order size should be placed to minimize ordering costs?
20. Refer to Exercise 19. Because of a fire, Stereo Minneapolis must acquire new storage for its speakers. The storage cost is now $8 per speaker per year. Due to the lack of an elevator to the storage area, the supplier increases the per-order charge to $50. Find the optimal order size.
21. An office supply company sells about 800 personal copiers per year at a steady rate. The cost of placing an order with the manufacturer is $48 per order. The storage cost is $3 per copier, based on the average number of copiers on hand. What order size minimizes the total copier inventory ordering costs?
22. Refer to Exercise 21. Suppose the manufacturer reduces the order charge to $25 for orders of 200 copiers or more. What order size should the office supply company make now?
23. LaLenses Camera Company sells 1200 35-mm cameras per year. Sales are uniform throughout the year. The supplier charges $18.75 per order. The storage cost is $0.50 per camera per year, based on the average number of cameras in stock. Insurance costs $0.75 per camera per year, based on the maximum number of cameras in stock. What order size will minimize the annual ordering cost?
24. How should the ordering strategy in Exercise 23 be changed if the insurance cost increases to $1.25 per camera per year?
25. Play It Again Sam sells stero turntable equipment at a steady annual rate. Annual sales are 400 units. The supplier charges $20 per order, and storage and insurance costs $2.50 per unit per year, based on the average number of units on hand. What order size minimizes the annual ordering cost?
26. Refer to Exercise 25. Suppose the turntable supplier offers a 0.5% discount on the wholesale price of $35 per unit for orders of 100 or more. What ordering strategy should be used to minimize the annual ordering cost?
27. Through its nationwide chain of service stations, an oil company distributes 16,000 road maps per year. The cost of printing the maps is $100 plus 6 cents per map. Storage and distribution costs 20 cents per map per year (based on the average number of maps produced per batch). Assume that the maps are distributed at a uniform rate throughout the year. How many maps should the company produce in each batch to minimize their cost?
28. A restaurant expects to use 800 cases of cooking oil this year. The oil costs $4 per case, the ordering fee is $10 per order, and storage cost is 20 cents per case per year, based on the maximum number of cases on hand. How many cases should be ordered at one time to minimize the ordering cost? How many shipments will be made per year?
29. A microcomputer manufacturer uses 60,000 cases of transistors per year. The annual cost of storing a case is $0.90, based on the average number of cases in stock. The ordering cost is $30 per order. How many

cases should be ordered each time to minimize the annual ordering cost?

30. Suppose a firm receives goods at a uniform rate throughout the year. The cost of storing the goods is proportional to the amount stored and the ordering cost is fixed, regardless of the size of the order. Show that the minimum ordering cost occurs when the total storage cost equals the total ordering cost.

Relative Change and Elasticity

4.6

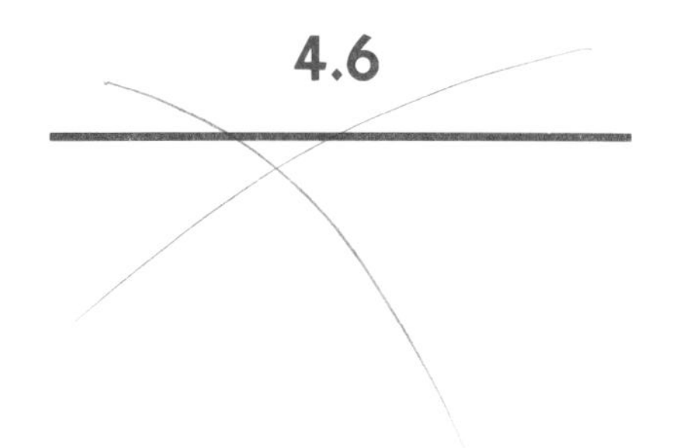

In comparisons of two changing quantities, the rate of change alone may not be meaningful. Knowing that a company is increasing its profits by \$100,000 per year does not give a clear picture of its progress. The significance of \$100,000 is quite different for a company that has an annual profit of \$500,000 than for a company that has an annual profit of \$25,000,000. To obtain a more meaningful measure of change, we use the concept of relative change. If $f(x)$ is a given function of x, then the ratio of the rate of change $f'(x)$ and the original quantity $f(x)$

$$\frac{f'(x)}{f(x)}$$

is called the **relative change** of $f(x)$.

Example 4.27 Suppose that a company's profit (in thousand dollars) at the end of the xth year is given by $P(x) = 2x^2 + 3x + 50$. At what rate is profit rising at the end of the fifth year? At the end of the eighth year? Find the relative change of profit at the end of the fifth and eighth years.

Solution The rate of change at the end of the xth year is $P'(x) = 4x + 3$. Hence, the rate at which the company's profit is rising at the end of the fifth year is $P'(5) = 23$ thousand dollars per year and the rate of change at the end of the eighth year is $P'(8) = 35$ thousand dollars per year. The relative change in the profit at the end of the fifth year is

$$\frac{P'(5)}{P(5)} = \frac{23}{115} = 0.2$$

and the relative change at the end of the eighth year is

$$\frac{P'(8)}{P(8)} = \frac{35}{202} = 0.173.$$

Although the rate of change at the end of the fifth year is less than the rate of change at the end of the eighth year, the relative change is greater. In relation to the total profit, the rate of change is diminishing and so the company's profit growth is slowing. ■

Relative change plays an important role in pricing. Suppose a company that sells a product for p dollars per unit wishes to increase its sales revenue. Recall that the demand function $D(p)$ indicates how many items can be sold at price p. In most cases, an increase in price causes a decrease in demand, so the derivative of the demand function is negative. Since a price increase will generate a decrease in demand, while a price decrease will result in an increase in demand, the company is faced with four possibilities:

1. A price increase results in a small decrease in demand, so the increased price more than offsets the diminished sales. In this case, revenue will increase.
2. A price increase results in many customers choosing another brand, causing a drastic decrease in demand; revenue will decrease.
3. A price decrease results in a substantial increase in demand, with increased sales more than offsetting the lower price, yielding an increase in revenue.
4. A price decrease results in a small increase in demand, which does not offset the lower price. Revenue will decrease.

The proper strategy to ensure increased revenue depends on how the demand is affected by changes in the product price. We need to know the relative change in demand associated with a relative change in the price. The ratio of these relative changes is called the **price elasticity of demand**, which is denoted by $E_D(p)$. Since

$$E_D(p) = \frac{\text{relative change in } D(p)}{\text{relative change in } p} = \frac{\dfrac{D'(p)}{D(p)}}{\dfrac{(p)'}{p}} = \frac{\dfrac{D'(p)}{D(p)}}{\dfrac{1}{p}} = \frac{pD'(p)}{D(p)},$$

we have the following formula.

Price Elasticity of Demand

$$E_D(p) = \frac{pD'(p)}{D(p)}$$

Notice that $E_D(p)$ is never positive because p and $D(p)$ are positive (or possibly zero) and $D'(p)$ is negative.

We can use $E_D(p)$ to find the relative change in demand associated with a relative change in price. In fact,

$$\text{relative change in demand} = E_D(p) \times (\text{relative change in price}).$$

However, $E_D(p)$ can be used directly to answer the question of whether the price should be increased or decreased to ensure an increase in revenue. Since revenue $R(p)$ is given by $R(p) = pD(p)$, the product rule gives

$$R'(p) = \frac{d}{dp} pD(p) = pD'(p) + D(p)(1).$$

If we factor $D(p)$ from both terms, we get

$$R'(p) = D(p)\left(p\frac{D'(p)}{D(p)} + 1\right) = D(p)[E_D(p) + 1].$$

Now $R(p)$ will be increasing when $R'(p) > 0$ and decreasing when $R'(p) < 0$. Since $D(p)$ is always positive (surely the demand for our product will never be zero!), the sign of $R'(p)$ is the same as the sign of $[E_D(p) + 1]$.

> A price increase yields an increase in revenue when $[E_D(p) + 1] > 0$; that is, when $E_D(x) > -1$.
>
> A price increase yields a decrease in revenue when $[E_D(p) + 1] < 0$; that is, when $E_D(x) < -1$.

Example 4.28 The number of transistor radios $D(p)$ (in thousands) that can be sold by an electronics firm at p dollars per radio is given by $D(p) = (2000/p^2) + (400/p) + 10$. Find the price elasticity of demand when the price is \$5 and when the price is \$20.

Solution From $D'(p) = -4000p^{-3} - 400p^{-2}$, we compute $D'(5) = -48$, and since $D(5) = 170$,

$$E_D(5) = 5\left(\frac{-48}{170}\right) = -1.41.$$

Since $E_D(5) + 1 = -0.41$ is negative, a price increase will cause a decrease in revenue. The significance of the value -1.41 is that a price increase of 1% (from \$5.00 to \$5.05) will cause a decrease in demand of 1.41% (from 170 to about 167.4).

When $p = 20$, $D(p) = 35$ and $D'(p) = -1.5$, so

$$E_D(20) = 20\left(\frac{-1.5}{35}\right) = -0.86.$$

At this price, $E_D(p) + 1 = 0.14$ is positive, so increasing the price of the radio will yield an increase in revenue. In both cases a price increase produces a decrease in demand. When the price is low (\$5), the decrease in demand is sufficient to cause revenue (price times demand) to decrease. At a higher price (\$20) the decrease in demand is not large enough to offset the price increase. ■

The sign of $E_D(p) + 1$ depends on whether $E_D(p)$ is larger or smaller than -1. Economists use the following terminology to classify demand (recall that $E_D(p) \le 0$ by definition):

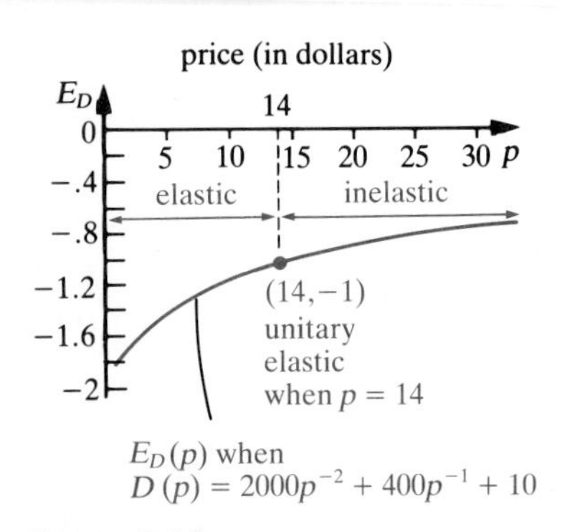

Figure 4.18
Price elasticity in Example 4.28

If $-1 < E_D(p) < 0$, then demand is **inelastic**.	A price increase yields a revenue increase.
If $E_D(p) = -1$, then demand is **unitary elastic**.	A price increase yields no change in revenue.
If $E_D(p) < -1$ then demand is **elastic**.	A price increase yields a revenue decrease.

In Example 4.28, demand is elastic at the price of \$5 but inelastic at the price of \$20. See Figure 4.18.

Example 4.29 A sporting goods store sells a brand of sunglasses for \$5.95 per pair. Based on sales data and data from previous price increases, the owner estimates that the demand for these sunglasses is $D(p) = 1000 - p^3$ for $0 < p < 10$ (pairs per month). Should the owner increase or decrease the price to increase revenue from this product?

Solution We can use the price elasticity of demand to answer this question. Since $D'(p) = -3p^2$, we have

$$E_D(p) = p\frac{D'(p)}{D(p)} = \frac{p(-3p^2)}{1000 - p^3} = \frac{-3p^3}{1000 - p^3}.$$

When $p = 5.95$, we have

$$E_D(5.95) = \frac{-3(5.95)^3}{1000 - (5.95)^3} \approx -0.80.$$

Thus the price elasticity of demand is inelastic, indicating that a price increase will yield a revenue increase. The owner should raise the price of the sunglasses. ■

Example 4.30 A state government currently imposes a tax of 5% on restaurant meals. Economists estimate that the demand for restaurant meals is given by $D(t) = \sqrt[3]{700 - 3t^3}$, where t is the tax rate in percent and $D(t)$ is measured in 10,000 meals per day. The governor proposes to increase the tax to 8% to raise additional funds for education. Is this a viable means to obtain new revenue?

Solution Although this problem does not directly address the price of the product (the meal), it is similar to our previous examples. A low tax will generate a small revenue, while a large tax may discourage people from eating in restaurants, also resulting in a small revenue. We need to determine the relative change in demand associated with the relative change in tax rate, or the *tax elasticity of demand,* which is given by

$$E_D(t) = \frac{tD'(t)}{D(t)},$$

where t is the tax rate imposed on each meal and $D(t)$ is the demand function for meals. If $E_D(t) > -1$, then the demand is inelastic and a tax increase will increase revenue. If $E_D(t) < -1$, the demand is elastic and a tax increase will decrease revenue.

We first determine $E_D(t)$:

$$E_D(t) = \frac{tD'(t)}{D(t)}$$

$$= t\frac{\frac{1}{3}(700 - 3t^3)^{-2/3}(-9t^2)}{(700 - 3t^3)^{1/3}} = \frac{-3t^3}{700 - 3t^3}.$$

When $t = 5$ we calculate $E_D(t) = -375/325 \approx -1.154$. This is an elastic value, indicating that increasing the tax to 8% will result in a decrease in revenue. The governor's plan is not a viable means to raise additional funds. ■

Governments use price elasticity to determine whether or not a particular industry should be controlled. If the demand for a product is elastic at its current price, then the government need not intervene because the company involved would lose revenue by increasing the price. For example, if a company has several competitors that produce a similar product, an increase in the price would cause customers to switch to a competitor's product, resulting in a net loss of revenue.

If the demand is inelastic, the government should impose regulations, since the company can increase its revenue by raising prices. An example of such a situation is the sale of an essential product, such as electric power, by a company that has a monopoly on that product (that is, has no competitor).

Exercises

Exer. 1–14: Calculate the relative change at the indicated value.

1. $f(x) = x^2 + 1;\quad x = 4$
2. $f(x) = \sqrt{x^2 - 4};\quad x = 6$
3. $f(x) = 5x - 3;\quad x = 2$
4. $f(x) = 2x(x^{3/2} - 1);\quad x = 8$
5. $f(x) = 4x + 6;\quad x = -3$
6. $f(x) = x - \frac{2}{x^2};\quad x = 10$
7. $f(x) = 3x^2 + 5x - 1;\quad x = 1$
8. $f(x) = \frac{1}{(4 - x^2)^{2/3}};\quad x = 0$
9. $f(x) = 2x^2 - 12x + 10;\quad x = 3$
10. $f(x) = \frac{4}{x^2} - \frac{5}{x};\quad x = -1$
11. $f(x) = \frac{x - 2}{x};\quad x = 5$
12. $f(x) = \left(x - \frac{1}{x}\right)^{1/2};\quad x = 5$
13. $g(t) = t(3t - 4)^5;\quad t = 2$
14. $w(s) = s^3(5s - 4)^2;\quad s = 2$

Exer. 15–18: A cable-television company has $n(t) = 30t^2 + 7t + 800$ subscribers after t months of an advertising campaign.

15. Find the relative change of the number of subscribers

after 2 months, 6 months, and 9 months of the campaign.

16. Find the relative change of the number of subscribers after 3 months, 8 months, and 12 months of the campaign.
17. Is the campaign gaining or losing steam from the second to the sixth month?
18. Evaluate the success of the campaign from the sixth to the ninth month.

Exer. 19–22: A new president is appointed by an appliance manufacturer. The monthly profit (in thousand dollars) for the company t months after the new president takes charge is $P(t) = t^2 - 20t - 200$.

19. Find the relative change after 2 months and after 6 months.
20. Find the relative change after 5 months and after 8 months.
21. Although the company is losing more in the eighth month than in the fifth month, is the situation improving?
22. Calculate the relative change at the end of one year and interpret your answer.

23. Refer to Example 4.28. Find the elasticity for the given function when the price is \$7 and when it is \$16. Decide in each case whether demand is elastic or inelastic.

Exer. 24–25: A record manufacturer finds that the demand for a new release is $D(p) = -2p^2 - 50p + 700$ thousand copies when the price is p dollars.

24. What is the elasticity of demand when the price is \$5?
25. What is the elasticity when the price is \$7? At what prices is the demand elastic?

26. Refer to Example 4.30. The revenue $R(t)$ from the meals tax may be approximated by $R(t) = tpD(t)$, where t is the tax rate imposed on each meal, p is the average price of a meal, and $D(t)$ is the demand function for meals. Since $D(t)$ is measured in 10,000 meals per day, $R(t)$ is measured in units of \$10,000 per day. Assume that a moderate tax rate change does not affect the average price of a meal. Show that $R'(t)$ is increasing when $E_D(t) > -1$ and $R'(t)$ is decreasing when $E_D(t) < -1$, where $E_D(t)$ is defined by
$$E_D(t) = \frac{tD'(t)}{D(t)}.$$
27. Refer to Example 4.30. The proposal to produce additional revenue by raising the restaurant meal tax rate would actually result in decreased tax revenues. Hence, lowering the tax rate would increase revenue. What tax rate will generate the most revenue?

Exer. 28–29: We discussed supply in Section 1.4. The **price elasticity of supply**, $E_S(p)$ is the relative change in the supply function $S(p)$ divided by the relative change in the price p.

28. Follow the development of $E_D(p)$ in the text to show that
$$E_S(p) = \frac{pS'(p)}{S(p)}.$$
Explain why, under normal conditions, $S'(p) > 0$ and hence $E_S(p) > 0$. Suppose that $0 < E_S(p) < 1$ (supply is *inelastic*). Does an increase in the price raise the revenue supply very little or substantially? What about when $1 < E_S(p)$ (supply is *elastic*)?
29. The number of aluminum cans (in thousands) that will be produced daily if the price of each can is set at p cents is given by the supply function $S(p) = 50p^2 + 70p + 100$. Find the price elasticity of supply when the price is 1 cent and when the price is 5 cents. In each case state whether the supply is elastic or inelastic.

30. When the price of a certain product is p dollars, the demand function for the product is $D(p) = \sqrt{400 - 3p^2}$ for $0 < p < 11$. If the product is currently selling for \$4.50, will a price increase yield a greater or smaller revenue? What price will generate the most revenue?

Summary

The notation $\lim_{x\to\infty} f(x) = b$ indicates that the values of the function $f(x)$ become increasingly close to the number b as x gets larger and larger. The notation $\lim_{x\to-\infty} f(x) = b$ indicates that the values of the function $f(x)$ become increasingly close to the number b as $|x|$ gets larger and larger and $x < 0$.

The line $y = b$ is a **horizontal asymptote** for $f(x)$ if
$$\lim_{x\to\infty} f(x) = b \quad \text{or} \quad \lim_{x\to-\infty} f(x) = b.$$

The line $x = a$ is a **vertical asymptote** for $f(x)$ if $f(x) \to \pm\infty$ as $x \to a^+$ or as $x \to a^-$.

Functions can be graphed systematically using the checklist on page 153.

When y is an *implicit* function of x, the chain rule must be used to determine dy/dx. If $g(y)$ is a function of y, and y is implicitly a function of x, then

$$\frac{d}{dx} g(y) = g'(y) \frac{dy}{dx}.$$

Marginal cost is the derivative of cost with respect to quantity, and **marginal revenue** is the derivative of revenue with respect to quantity. **Profit**, the difference between revenue and cost, achieves a maximum when marginal cost equals marginal revenue.

The **relative change** of $f(x)$ is the ratio of the rate of change $f'(x)$ and $f(x)$, that is,

$$\frac{f'(x)}{f(x)}.$$

Price elasticity of demand $E_D(p)$ is the ratio of the relative change in demand $D(p)$ to the relative change in price p:

$$E_D(p) = \frac{pD'(p)}{D(p)}.$$

When $E_D(p) < -1$, the demand is **elastic** and when $-1 < E_D(p) < 0$, the demand is **inelastic**.

A price increase yields an increase in revenue when $E_D(x) > -1$.

A price increase yields a decrease in revenue when $E_D(x) < -1$.

Review Exercises

Exer. 1–12: Sketch the graph of the function. Give y-intercepts, x-intercepts when possible, asymptotes, relative maxima, relative minima, and inflection points.

1. $f(x) = \dfrac{x^2}{x-1}$
2. $f(x) = \dfrac{x-3}{x+2}$
3. $f(x) = (x^2-1)^3$
4. $f(x) = \dfrac{4}{x-2} - \dfrac{1}{x-5}$
5. $f(x) = \dfrac{x-1}{x^2-x+1}$
6. $f(x) = \dfrac{1}{x^2} + x$
7. $f(x) = \dfrac{x^3}{x^2+1}$
8. $f(x) = x^{1/3} - x^{4/3}$
9. $f(x) = \dfrac{2x}{x-4}$
10. $f(x) = x - x^{4/3}$
11. $f(x) = \dfrac{1}{x^2} - x$
12. $f(x) = \dfrac{x}{\sqrt{x}} - 1$

Exer. 13–20: Find dy/dx by implicit differentiation.

13. $x^3y^3 + \dfrac{x}{y} + 2 = 0$
14. $\dfrac{x+y}{3x-4y} = 1$
15. $x^n + y^{-n} + xy = 1$ (n = a constant)
16. $x^{1/3}y^{1/5} + \dfrac{y}{x} = 1$
17. $x^3 + y^4 + xy - 5x - 6y + 1 = 0$
18. $5x^4 + 5y^4 - 4x^2y^2 + 3 = x + y$
19. $\dfrac{x}{y} + x = \dfrac{y}{x} - y^2$
20. $(x+y)^5 + x - y = 1$
21. Consider the cost function $C = x^2 - 50x + 300$ and the revenue function $R = 30x - x^2$. Find the marginal cost, the marginal revenue, and the value of x that will give maximum profit.
22. The cost of producing x electric toasters per week is given by $C = 0.01x^2 + 5x + 1000$. If the present production is 500 toasters per week, what is the increase in cost to produce 501 toasters per week? What is the marginal cost at 500? If the company sells each toaster for \$15.49, what would you advise about increasing production?
23. The revenue from the sale of electric blankets is given by $R = 0.01x^2 + 5x$, where x is the number of blankets sold. Determine the total revenue from the sale of 100 blankets and the marginal revenue when $x = 100$.
24. A washing-machine manufacturer determines that if x machines are sold, then the price per machine is given by $p = 500 - 0.1x$. Find the total revenue from the sale of 300 washing machines and the marginal revenue when $x = 300$.
25. For a company that manufacturers blue jeans, the cost and revenue functions per month are given by $C = 7000 + 3x - 0.001x^2$ and $R = 25x - 0.01x^2$, where x is the number of pairs of jeans produced during the month. Find the price per pair of jeans that will maximize the profit and find the maximum profit per month.
26. An investor finds that the value of a stock t days after purchase is $V(t)$ dollars, where

$$V(t) = -0.2t^2 + 20t + 3000.$$

What is the rate at which the value is changing 6 days

after purchase? On what day should the stock be sold to realize the greatest return on investment? How much is the stock worth on that day?

27. A pet shop owner knows that the shop will sell 2250 goldfish during the year at a constant rate of sale. If it costs \$30 to place an order for a "batch" of goldfish and \$6 per year to keep each goldfish, how many goldfish should be purchased in each order to minimize the cost?
28. A supplier agrees to produce 100,000 flashlights during the year to be delivered regularly to a chain of retail stores. The storage cost is 5 cents per flashlight and the fixed cost of production is \$5 per batch produced. What size batches will minimize batching costs?
29. A farm-equipment dealer predicts the company will sell 40,000 rolls of barbed wire this year. Storage cost is \$0.50 per roll per year, and ordering cost is \$25 per order. The wholesaler offers a discount of \$1 per roll on orders of 500 rolls or more. The regular price is \$12 per roll. What order size will minimize the cost?
30. An office complex estimates that 50,000 reams of paper will be used this year. Storage cost per ream per year is \$0.75, and ordering cost is \$5 per order. What order size should be used?

Exer. 31–35: Calculate the relative change at the indicated value.

31. $f(x) = 2x^2 - 3x;\ \ x = 2$
32. $f(x) = \sqrt{2x^2 + 1};\ \ x = -\frac{1}{2}$
33. $f(x) = \dfrac{x - 3}{x + 5};\ \ x = 1$
34. $f(x) = x\sqrt[3]{x^2 + x};\ \ x = 1$
35. $f(x) = x^3;\ \ x = 3$

Exer. 36–37: A promotion for a new housecleaning service started June 1. After t weeks, the number of customers enrolled is $c(t) = 20t^2 + 10t + 53$.

36. Find the relative change at $t = 1$, 5, and 9.
37. Is the promotion effectively increasing business between $t = 1$ and $t = 5$? What about between $t = 5$ and $t = 9$?

38. When the price of a certain product is p dollars, the demand function for the product is $D(p) = \sqrt{900 - 3p^3}$ for $0 < p < 6.50$. If the product currently sells for \$4.50, will a price increase yield a greater or smaller revenue? What price will generate the greatest revenue?
39. A manufacturer finds that the demand for a new product is $D(p) = -5p^3 - 30p + 8000$ items when the price is p dollars. What is the price elasticity of demand when the price is \$5? What is the price elasticity at \$7? At what prices is the demand elastic?
40. When the price of a certain product is p dollars, the demand function for the product is $D(p) = \sqrt{1000 - 2p^4}$ for $0 < p < 4.50$. If the product is currently selling for \$2, will a price increase yield a greater or smaller revenue? What price will generate the greatest revenue?

***Facing page*: The age of ancient relics can be determined using carbon dating, which relies on the theory of exponential decay. See Example 5.21.**

Exponential and Logarithmic Functions

· F · I · V · E ·

Many phenomena can be described using functions defined by powers, polynomials, or rational functions; however, some significant situations cannot be described using such functions. For instance, predicting the growth of populations, calculating the accumulation of funds in a continuously compounded bank account, and determining the date of archaeological discoveries by radioactive decay calculations require functions other than powers, polynomials, or rational functions. In this chapter we will study exponential and logarithmic functions, which can be used to analyze these and other applications.

5.1 Exponential Functions

Algebraic manipulation of exponential functions requires the properties of exponents, so we begin our discussion with a brief review of these properties.

Properties of Exponents

If $a > 0$, $b > 0$, p, n are positive integers, and x and y are real numbers, then

(1) $a^x a^y = a^{x+y}$

(2) $\dfrac{a^x}{a^y} = a^{x-y} \quad (a \neq 0)$

(3) $(a^x)^y = a^{xy}$

(4) $(ab)^x = a^x b^x$

(5) $\left(\dfrac{a}{b}\right)^x = \dfrac{a^x}{b^x}$

(6) $a^0 = 1$

(7) $a^{-x} = \dfrac{1}{a^x}$

(8) $\sqrt[n]{a^p} = a^{p/n}$

The following example shows how to use these properties.

Example 5.1 Simplify each expression:

$$\textbf{(a)}\ \frac{2^3 \times 4^2 \times 6^3}{12^2} \qquad \textbf{(b)}\ \frac{3^{2/3} \times 6^{1/3}}{2^{1/6}} \qquad \textbf{(c)}\ \frac{2\sqrt[6]{4}}{3\sqrt{2}}$$

Solution **(a)** Using the properties of exponents, we obtain

$$\begin{aligned}\frac{2^3 \times 4^2 \times 6^3}{12^2} &= \frac{2^3(2^2)^2(2 \times 3)^3}{(2^2 \times 3)^2} \\ &= \frac{(2^3)(2^4)(2^3)(3^3)}{(2^4)(3^2)} && \text{(properties 3, 4)} \\ &= (2^{3+4+3-4})(3^{3-2}) && \text{(properties 1, 2)} \\ &= 2^6 \times 3 = 64 \times 3 = 192.\end{aligned}$$

(b)

$$\frac{3^{2/3} \times 6^{1/3}}{2^{1/6}} = \frac{3^{2/3} \times (3 \times 2)^{1/3}}{2^{1/6}}$$

$$= \frac{3^{2/3}3^{1/3}2^{1/3}}{2^{1/6}} \qquad \text{(property 4)}$$

$$= \frac{3^{(2/3)+(1/3)}2^{1/3}}{2^{1/6}} \qquad \text{(property 1)}$$

$$= \frac{3^1 2^{1/3}}{2^{1/6}}$$

$$= 3(2)^{(1/3)-(1/6)} = 3(2)^{1/6} = 3\sqrt[6]{2} \qquad \text{(property 2)}$$

(c)

$$\frac{2\sqrt[6]{4}}{3\sqrt{2}} = \frac{2 \times 4^{1/6}}{3 \times 2^{1/2}} \qquad \text{(property 8)}$$

$$= \frac{(2)(2^2)^{1/6}(2^{-(1/2)})}{3} \qquad \text{(properties 3 and 2)}$$

$$= \frac{2^{1+(1/3)-(1/2)}}{3} = \frac{2^{5/6}}{3} = \tfrac{1}{3}\sqrt[6]{32}$$

■

Example 5.2 Find the solution to each equation:

(a) $2^x = 32$ **(b)** $3^x = 9\sqrt{3}$ **(c)** $5^{t^2} = \sqrt{5}$

Solution **(a)** Since $32 = 2^5$, the equation $2^x = 32$ is equivalent to $2^x = 2^5$, so $x = 5$.
(b) We rewrite the right side of the equation: $3^x = 9\sqrt{3} = 3^2 3^{1/2} = 3^{5/2}$. Thus $x = \frac{5}{2}$.
(c) Since $\sqrt{5} = 5^{1/2}$, the equation may be written as $5^{t^2} = 5^{1/2}$. Hence, $t^2 = \frac{1}{2}$, so that $t = \pm\sqrt{\frac{1}{2}}$. ■

In many applications, we will need numbers such as 2^x, where x is any real number. For positive integers x, 2^x is defined to be the product of x twos, as in $2^3 = (2)(2)(2)$. For positive rational numbers x, such as $x = \frac{7}{5}$, 2^x is obtained by the definition $2^{7/5} = \sqrt[5]{2^7} = \sqrt[5]{128}$ (a tedious but well-defined process). For negative x, integer or rational, $-x$ is positive and 2^x is defined by $2^x = 1/2^{-x}$. However, since $\sqrt{2}$ is not a rational number, $2^{\sqrt{2}}$ cannot be obtained by any of these definitions. To define what is meant by the expression $2^{\sqrt{2}}$, consider the decimal expansion of $\sqrt{2}$: $\sqrt{2} = 1.414213\ldots$. Each successive number in the list 1, 1.4, 1.41, 1.414, 1.4142, ... is increasingly closer to $\sqrt{2}$, and each of these numbers is rational: 1, $\frac{14}{10}$, $\frac{141}{100}$, $\frac{1414}{1000}$, $\frac{14{,}142}{10{,}000}$, ... and so we can compute 2^1, $2^{1.4}$, $2^{1.41}$, $2^{1.414}$, $2^{1.4142}$, ... Table 5.1 shows the results to seven decimal places. The numbers in the right column are "stabilizing" at about 2.6651... (we say *converging to* 2.6651...). We take this stabilized value as the definition of $2^{\sqrt{2}}$. Every irrational number x can be written as an infinite decimal and the stabilizing effect noted above occurs for any such x. The stabilized value is taken as the definition of 2^x.

For any real number x the calculation $y = 2^x$ yields exactly one number, so the formula $y = 2^x$ defines a function whose domain is all real numbers. This function is an example of an exponential function.

Table 5.1
Approximations to $2^{\sqrt{2}}$

x	2^x
1	2
1.4	2.6390158
1.41	2.6573716
1.414	2.6647496
1.4142	2.6651191
1.41421	2.6651376

Example 5.3 Sketch the graph of $y = 2^x$.

Solution Since we do not yet know how to differentiate the function $y = 2^x$ (its derivative is *not* $x \cdot 2^{x-1}$), we will graph $y = 2^x$ by plotting several points. Table 5.2 shows the results of evaluating 2^x at a variety of x-values. The corresponding points are plotted in Figure 5.1 and connected with a smooth curve. ■

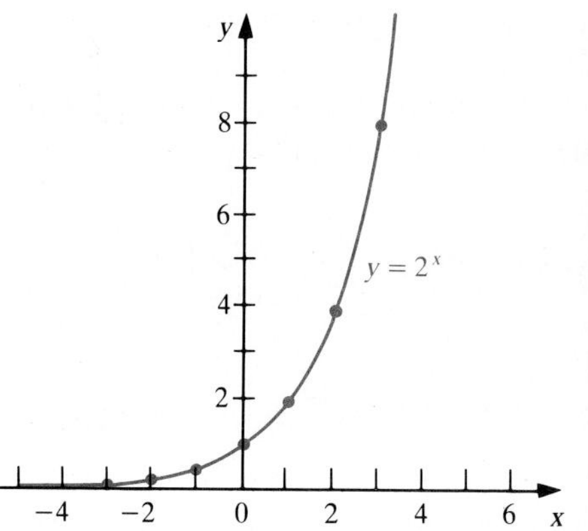

Figure 5.1
Graph of $y = 2^x$

Table 5.2
Values of $y = 2^x$

x	$y = 2^x$
−3	$\frac{1}{8}$
−2	$\frac{1}{4}$
−1	$\frac{1}{2}$
0	1
1	2
2	4
3	8

Let b denote a positive number not equal to 1. A function of the form $y = b^x$ is called an **exponential function** because the independent variable x is in the exponent. The number b is called the **base** of the exponential function. Such a function is defined for all real numbers x; the resulting value of y is always positive. (The case $b = 1$ yields the constant function $y = 1^x = 1$ and is *not* an exponential function.)

Figure 5.2 shows the graphs of $y = f(x) = b^x$ for several values of b. Note that the functions are increasing for $b > 1$ and decreasing for $b < 1$. For $b > 1$, the functions rise more steeply with increasing b. Furthermore, the graph of b^x is the reflection across the y-axis of the graph of $(1/b)^x$. For example, if we think of the y-axis as a mirror, then the graph of 2^x is the reflection of the graph of $(\frac{1}{2})^x$, and vice versa.

An important application of exponential functions is the calculation of the balance in an interest-bearing bank account. Suppose $1000 is deposited in an account that pays 6% interest compounded annually. *Compounded annually* indicates that the bank will add interest to the account once a year. Thus, after the $1000 has been in the bank one year, the bank adds $\$1000 \times 0.06 = \$60.00$ to the account, bringing the balance to $\$1060.00 = \$1000(1 + 0.06) = \$1000(1.06)$. At the end of the second year, the bank pays 6% interest on the new balance. Thus it adds $\$1060.00 \times 0.06 = \$63.60$ to the account, bringing the balance to $\$1123.60 = \$1060(1 + 0.06) = \$1000(1 + 0.06)^2 = \$1000(1.06)^2$. After t years the balance is $B_1(t)$ dollars, where

$$B_1(t) = 1000(1 + 0.06)^t = 1000(1.06)^t. \quad \text{(compounded annually)}$$

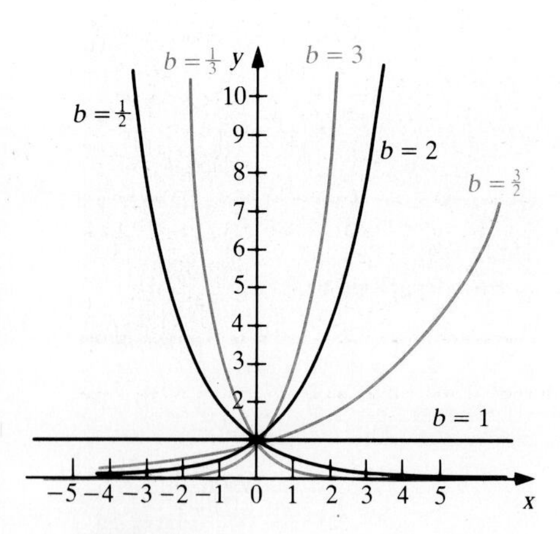

Figure 5.2
Graph of $y = b^x$ for several values of b

(The subscript "1" is a reminder that the interest is compounded *one* time per year.) Therefore, the balance in the account is closely linked to the exponential function $y = 1.06^x$.

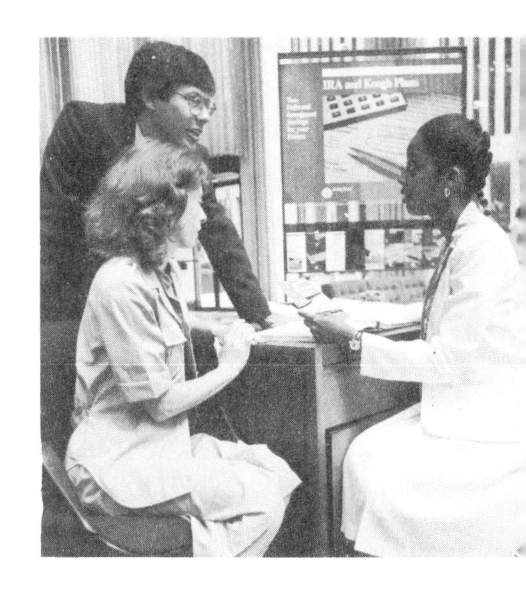

Some banks offer *quarterly compounding,* that is, the bank adds interest to the account four times per year. If \$1000 is deposited in an account paying 6% annual interest compounded quarterly, then after three months the bank adds $\$1000 \times 0.06 \times \frac{1}{4} = \$15.00$ to the account. The factor $\frac{1}{4}$ is included because the 6% rate is an annual rate and the money has been in the account one-fourth of a year. Thus the new balance is $\$1015.00 = \$1000(1 + 0.015) = \$1000(1.015)$. At the end of the next three months, the bank makes the same calculation using the balance of \$1015.00. That is, after six months (two compounding periods) the balance is $\$1030.22 = \$1015.00(1 + 0.015) = \$1000(1 + 0.015)^2 = \$1000(1.015)^2$. After another compounding period has expired (nine months), this balance will be multiplied by $(1 + 0.015)$ to get the new balance. Thus

$$\begin{aligned}\text{balance after 3 compounding periods} &= \$1000(1 + 0.015)^3 \\ &= \$1000(1.015)^3\end{aligned}$$

and

$$\text{balance after one year} = \$1000(1 + 0.015)^4 = \$1000(1.015)^4 = \$1061.36.$$

Consequently, in an account paying 6% annual interest compounded quarterly, \$1000 earns \$1.36 more interest in the first year than in an account paying 6% annual interest compounded annually.

After t years, the balance in a quarterly compounded account is $B_4(t)$ dollars, where

$$B_4(t) = 1000(1 + 0.015)^{4t} = 1000(1.015)^{4t}. \qquad \text{(compounded quarterly)}$$

The exponent $4t$ represents the number of compounding periods in t years. In this case, the balance in the account is closely linked to the exponential function $y = 1.015^x$. For example, after two years the balance in a quarterly compounded account is $\$1000(1.015)^8 = \$1126.49$, a gain of \$2.89 over the annually compounded account.

The bases in these two examples (1.06 and 1.015) are given by $1 + r$ and $1 + \frac{1}{4}r$, respectively, where $r = 0.06$ denotes the annual rate. For an initial deposit of \$1000 and an annual interest rate r compounded n times per year, the balance (in dollars) after t years is given by

$$B_n(t) = 1000\left(1 + \frac{r}{n}\right)^{nt}.$$

Table 5.3 compares the growth of \$1000 after one year at an annual rate of 6% for several compounding periods. Although the differences among these balances seem small, Exercise 61 shows that the effect of daily versus annual compounding can be significant for a bank.

Table 5.3
Growth of \$1000 at 6% annual interest for various compounding periods

Number of compounding periods per year	*Balance after one year*
1 (annually)	1060.00
2 (semiannually)	1060.90
4 (quarterly)	1061.36
12 (monthly)	1061.68
365 (daily)	1061.83

Exercises

Exer. 1–26: Simplify the expression

1. $2^3 2^5$
2. $(x^2y)^3x^{-6}$
3. $\dfrac{4^{3/2}3^4}{9^2}$
4. $(xy)^4x^{-2}$
5. $\dfrac{x^4y^4}{x^{-2}}$
6. $\dfrac{x^3z^{-2}}{(xy)^2}$
7. $\dfrac{3^{-2}4^0}{2^{-4}}$
8. $\dfrac{2^4 4^2}{3^{-2}}$

9. $\dfrac{a^4b^3}{\sqrt{a^6b^4}}$
10. $\dfrac{3^46^3}{2^33^2}$
11. $\dfrac{4^3}{\sqrt{16a^5}}$
12. $\dfrac{x^2y^{-2}}{y^2x^{-2}}$
13. $(a^2)^3$
14. $\dfrac{a^3b^{-5}}{b^3a^{-4}}$
15. $\dfrac{a^3}{a^5}$
16. $\dfrac{5^315^{-2}}{25^29^{-1}}$
17. $(2^{2/3}5^{4/3})^3$
18. $\dfrac{(3^{2/5}2^{3/5})^2}{6^{1/5}}$
19. $\dfrac{2^{3/7}5^{3/14}}{20^{1/28}}$
20. $\dfrac{10^{2/3}14^{1/3}}{35^{5/2}}$
21. $(2^{1/3}3^{-1/4})^8$
22. $\dfrac{(5^{1/5}6^{2/3})^4}{10^{3/5}2^2}$
23. $\dfrac{\sqrt[3]{a}\sqrt[5]{b}}{ab}$
24. $\dfrac{a^{-2/5}}{ab^{-1/5}}$
25. $\dfrac{(2^{1/3}3^{2/3})^6}{a^{-2}b^{-3}}$
26. $\dfrac{(a^{-1/3}b^{-3/4})^2}{(a^2bc)^{-1/2}}$

Exer. 27–42: Solve the equation for x.

27. $2^x = 16$
28. $3^x = 81$
29. $3^x = \frac{1}{9}$
30. $4^{x^2} = 32$
31. $4^x = \sqrt{2}$
32. $4^{2x+1} = \frac{1}{64}$
33. $3^29^{2x} = 27\sqrt{3}$
34. $3^{4x} = \dfrac{1}{9^{3x-2}}$
35. $9^{2x+1} = 27^{x-3}$
36. $2^{3x} = 8^{-x-2}$
37. $\dfrac{1}{3^x} = 3$
38. $2^x3^x5^x = 1$
39. $\sqrt{2}\,3^{2x-1} = \sqrt{6}$
40. $a^x = \dfrac{1}{a^x}$
41. $3(7^x) = 1029$
42. $(3^x)^x = 81$

Exer. 43–56: Sketch a graph for the function. Use a calculator with a y^x key to compute function values when necessary.

43. $y = 4^x$
44. $y = 3^{2x}$
45. $y = 2^{-x}$
46. $y = (\frac{1}{3})^x$
47. $y = 3^x$
48. $y = 1.5^{-x}$
49. $y = (\frac{1}{2})^x$
50. $y = 2^{x+1}$
51. $y = \dfrac{1}{4^x}$
52. $y = 2^{x-1}$
53. $y = 1.5^x$
54. $y = 3^x + 5$
55. $y = \left(\dfrac{1}{1.8}\right)^x$
56. $y = 3(2^x)$
57. Suppose an account that pays interest at an annual rate of 8% compounded quarterly is opened with \$500. Determine the balance after (a) one year, (b) two years, and (c) five years.
58. Suppose an account that pays interest at an annual rate of 8% compounded quarterly is opened with \$1000. Determine the balance after (a) one year, (b) two years, and (c) five years. Compare the results with those from Exercise 57.
59. Suppose an account that pays interest at an annual rate of 6% compounded semiannually is opened with \$1000. Determine the balance after (a) one year, (b) two years, and (c) five years.
60. Suppose an account that pays interest at an annual rate of 6%, compounded monthly is opened with \$1000. Determine the balance after (a) one year, (b) two years, and (c) five years.
61. In the text we computed the increase in balance for an account compounded daily versus an account compounded annually, both at 6% annual interest rate. The increase was very modest on an account with an initial balance of \$1000; however, the total amount of money in a bank's savings accounts may be quite large. Suppose a bank has \$50 million in accounts that are now paying 6% interest compounded annually. Compute the increase in interest payments on this amount, after one year, if the bank decides to compound interest (a) quarterly or (b) monthly.

5.2 The Function e^x

In this section we consider derivatives of exponential functions. This process will lead to a special exponential function, denoted by e^x. To begin, consider the function $y = f(x) = 2^x$ and use the definition of the derivative:

$$f'(x) = \lim_{h\to 0}\frac{f(x+h)-f(x)}{h} = \lim_{h\to 0}\frac{2^{x+h}-2^x}{h}$$

$$= \lim_{h\to 0}\frac{2^x2^h-2^x}{h} = \lim_{h\to 0}\frac{2^x(2^h-1)}{h}. \qquad \text{(factor out } 2^x\text{)}$$

Since 2^x does not depend on h, we may factor it from the limit to obtain

$$f'(x) = 2^x \lim_{h\to 0} \frac{2^h - 1}{h}.$$

The limit portion of this expression does not depend on x at all; the limit is a constant, which we denote by m_2, because the base is 2. Thus

$$m_2 = \lim_{h\to 0} \frac{2^h - 1}{h}.$$

After determining the value of m_2, we will have

$$\frac{d}{dx} 2^x = (m_2)(2^x).$$

Table 5.4 shows the value of the ratio $(2^h - 1)/h$ for several small values of h. From the table, we see that $m_2 \approx 0.693$.

We can do the same calculations with $y = b^x$ for any value of b. For $b = 3$, we find

$$\frac{d}{dx} 3^x = (m_3)(3^x) \quad \text{where } m_3 = \lim_{h\to 0} \frac{3^h - 1}{h} \approx 1.099;$$

for $b = 2.5$,

$$\frac{d}{dx} 2.5^x = (m_{2.5})(2.5^x) \quad \text{where } m_{2.5} = \lim_{h\to 0} \frac{2.5^h - 1}{h} \approx 0.916.$$

These results are summarized in Table 5.5.

Suppose that we could determine the specific value of b such that $m_b = 1$. For that b the derivative of b^x would be extremely simple; in fact, it would be $(1)b^x = b^x$. From Table 5.5 we see that the value of b should be between 2.5 and 3 since $m_{2.5} = 0.916$ and $m_3 = 1.099$. The value of b for which $m_b = 1$ is denoted by the letter e and is approximately $e = 2.7182818284$. For $b = e$,

$$\frac{de^x}{dx} = e^x.$$

We thus have the remarkable result that the derivative of e^x is e^x! The function $f(x) = e^x$ is called *the exponential function* because of its importance and the simplicity of its derivative. Many scientific calculators have an e^x key for computing values of this important function.

We will encounter the function $f(x) = e^x$ often in the remainder of the text. The graph of e^x is shown in Figure 5.3. For positive x, the function $y = e^x$ increases rapidly, and $e^x \to 0$ as $x \to -\infty$. To emphasize this behavior, we have noted several *benchmarks* in Table 5.6. For example, if distances are measured in inches, so that e^1 is approximately 2.72 inches, then the distance from the earth to the sun (about 93 million miles) is approximately $e^{29.4}$ inches; the diameter of a human blood cell is about 3 ten-thousandths of an inch (0.0003 inch), or $e^{-8.1}$ inch.

Table 5.4

Data to estimate $\lim_{h\to 0} \frac{2^h - 1}{h}$

h	$\frac{2^h - 1}{h}$
0.1	0.7177346
0.01	0.6955550
0.001	0.6933875
0.0001	0.6931711
0.00001	0.6931480

Table 5.5

Derivative of $f(x) = b^x$ for three values of b

b	$\frac{d}{dx} b^x$
2.0	$(0.693)2^x$
2.5	$(0.916)2.5^x$
3.0	$(1.099)3^x$

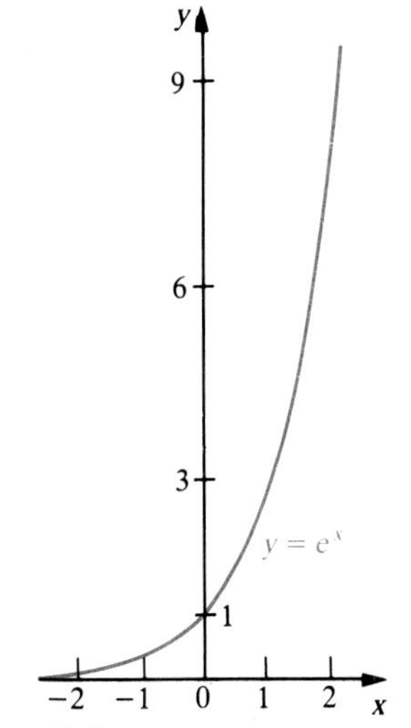

Figure 5.3

Graph of $f(x) = e^x$

Table 5.6
Comparative values for e^x

Description	*Distance or size*	$=e^{\square}$
diameter of an electron	0.55×10^{-13} inch	$e^{-30.5}$ inch
diameter of an oxygen molecule	0.476×10^{-8} inch	$e^{-19.2}$ inch
diameter of a blood cell	0.0003 inch	$e^{-8.1}$ inch
reference unit for comparisons	1 inch	e^{0} inch
height of an average man	6 feet	$e^{4.2}$ inches
length of a football field	100 yards	$e^{8.2}$ inches
one-mile track	1 mile	$e^{11.1}$ inches
distance from New York to Los Angeles	3000 miles	$e^{19.1}$ inches
distance to the moon	240,000 miles	$e^{23.4}$ inches
distance to the sun	93,000,000 miles	$e^{29.4}$ inches
distance from the sun to the nearest star	4 light years = 23 trillion miles	$e^{41.8}$ inches
the 1988–89 federal budget	1 trillion dollars	$e^{27.6}$ dollars

The following properties are obtained from the fact that $(d/dx)e^x = e^x$ and from the graph of $y = e^x$.

Properties of the function $y = e^x$

1. e^x is defined for all x.
2. e^x is continuous for all x.
3. e^x is positive for all x.
4. $e^x \to 0$ as $x \to -\infty$, so $y = 0$ is a horizontal asymptote.
5. e^x has no vertical asymptote.
6. $e^x \to \infty$ as $x \to \infty$.
7. e^x is differentiable for all x.
8. e^x is increasing for all x.
9. e^x is concave upward for all x.
10. Given any number $y > 0$, there is exactly one number x such that $e^x = y$.

Before we give examples involving e^x, we give a geometrical interpretation of the numbers m_b introduced above. Consider the derivative of $y = f(x) = 2^x$ at $x = 0$, that is, the slope of the tangent line to the curve at the point where the curve crosses the y-axis:

$$f'(0) = m_2 2^0 = m_2.$$

Hence m_2 is precisely the slope of the tangent line to $y = f(x) = 2^x$ at $x = 0$. We may make the same interpretation for other exponential functions. In particular, e^x is the exponential function whose graph has slope 1 as it crosses the y-axis. Figure 5.4 shows three exponential functions and their tangent lines at $x = 0$. The function $y = e^x$ is that particular exponential function for which the slope of the tangent line at $x = 0$ is 1.

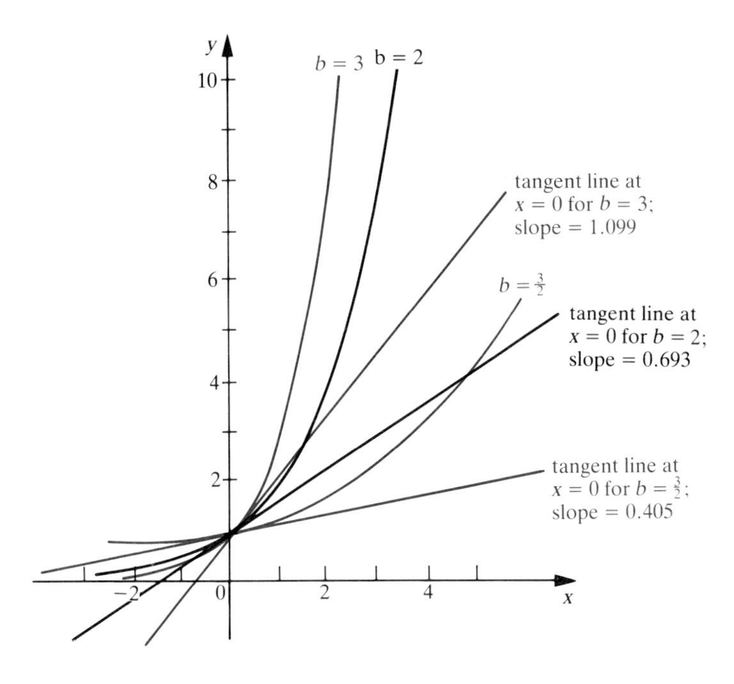

Figure 5.4
Tangent line at $x = 0$ to $y = b^x$ for three values of b

We can use the chain rule (see Section 3.2) with the derivative of e^x to find the derivative of $e^{g(x)}$ for any differentiable function $g(x)$. If we let $y = f(u) = e^u$ and $u = g(x)$, then their composition is $y = f(u) = f(g(x)) = e^{g(x)}$, and the chain rule gives

$$\begin{aligned}\frac{dy}{dx} &= \frac{dy}{du}\frac{du}{dx} \\ &= \frac{de^u}{du}\frac{du}{dx} \\ &= e^u\frac{du}{dx} \\ &= e^{g(x)}g'(x).\end{aligned}$$

Derivative of $e^{g(x)}$

$$\frac{d}{dx}e^{g(x)} = e^{g(x)}\frac{d}{dx}g(x)$$

Example 5.4 Compute the derivative of each function:

(a) $f(x) = e^{2x}$ **(b)** $f(x) = e^{4x+2}$ **(c)** $f(x) = e^{x^2}$ **(d)** $f(x) = e^{e^x}$.

Solution **(a)** The function $g(x)$ in the exponent is $g(x) = 2x$. Hence

$$f'(x) = \frac{d}{dx}e^{2x} = e^{2x}\frac{d}{dx}(2x) = e^{2x}(2) = 2e^{2x}.$$

(b) This case has $g(x) = 4x + 2$, so

$$f'(x) = \frac{d}{dx} e^{4x+2} = e^{4x+2} \frac{d}{dx}(4x+2) = e^{4x+2}(4) = 4e^{4x+2}.$$

(c)
$$f'(x) = \frac{d}{dx} e^{x^2} = e^{x^2} \frac{d}{dx} x^2 = e^{x^2}(2x) = 2xe^{x^2}.$$

(d)
$$f'(x) = \frac{d}{dx} e^{e^x} = e^{e^x} \frac{d}{dx} e^x = e^{e^x}(e^x) = e^x e^{e^x}.$$ ■

Example 5.4 illustrates that we can paraphrase the derivative of $e^{g(x)}$ as *the derivative of e to a power is e to the power times the derivative of the power.* When the exponential function is involved in products and quotients, we use the product and quotient rules in the usual manner.

Example 5.5 Find the derivative of each function:

$$\textbf{(a)}\ f(x) = xe^x \quad \textbf{(b)}\ f(x) = (x^3 + 5x)e^{3x} \quad \textbf{(c)}\ f(x) = \frac{e^{3x}}{x+1}.$$

Solution **(a)** We use the product rule to find $f'(x)$:

$$f'(x) = \frac{d}{dx} xe^x = x\frac{d}{dx} e^x + e^x \frac{d}{dx} x = xe^x + e^x(1)$$

or

$$\frac{d}{dx} xe^x = (x+1)e^x.$$

(b) We again use the product rule:

$$f'(x) = \frac{d}{dx}(x^3 + 5x)e^{3x} = (x^3 + 5x)\frac{d}{dx} e^{3x} + e^{3x} \frac{d}{dx}(x^3 + 5x)$$
$$= (x^3 + 5x)[e^{3x}(3)] + e^{3x}[3x^2 + 5]$$

or, factoring out e^{3x} and combining terms,

$$\frac{d}{dx}(x^3 + 5x)e^{3x} = (3x^3 + 15x + 3x^2 + 5)e^{3x}.$$

(c) We use the quotient rule:

$$f'(x) = \frac{d}{dx}\frac{e^{3x}}{x+1} = \frac{(x+1)\frac{d}{dx} e^{3x} - e^{3x}\frac{d}{dx}(x+1)}{(x+1)^2}$$
$$= \frac{(x+1)e^{3x}(3) - e^{3x}(1)}{(x+1)^2}$$

or

$$\frac{d}{dx}\frac{e^{3x}}{x+1} = \frac{(3x+2)e^{3x}}{(x+1)^2}.$$ ■

The graphing techniques we developed in Chapter 4 also apply when e^x is involved. We have already remarked on the applicability of the product, quotient, and chain rules—the tools necessary for locating critical points and inflection points.

Example 5.6 Let $y = f(x) = xe^{-x}$. Find the intercepts and asymptotes. Locate and classify all critical values and locate all inflection points. Sketch the graph of $f(x)$.

Solution For the intercepts, we note that if $x = 0$, then $y = f(0) = 0e^{-0} = 0$. Conversely, setting $y = 0$ gives $xe^{-x} = 0$. Since the exponential function is never zero, this product will be zero only when $x = 0$. Hence there is exactly one x-intercept and one y-intercept, at $(0, 0)$. This function has no vertical asymptotes. To find horizontal asymptotes, we need to determine the behavior of $f(x) = xe^{-x} = x/e^x$ as x approaches infinity. Table 5.7 supports the conclusion that $x/e^x \to 0$ as $x \to \infty$. This is because as x increases, the denominator increases much faster than the numerator. Since $\lim_{x\to\infty} xe^{-x} = 0$, the line $y = 0$ is a horizontal asymptote. As $x \to -\infty$, xe^{-x} approaches $-\infty$ because the factor x is negative and large in absolute value while the factor e^{-x} is large and positive (see Table 5.8).

Table 5.7
Behavior of xe^{-x} as $x \to \infty$

x	xe^{-x}
5	3.369×10^{-2}
10	4.540×10^{-4}
20	4.122×10^{-8}
50	9.644×10^{-21}
100	3.720×10^{-42}
200	2.768×10^{-85}

Table 5.8
Behavior of xe^{-x} as $x \to -\infty$

x	xe^{-x}
−5	-8.047×10^{0}
−10	-2.303×10^{5}
−20	-9.703×10^{9}
−50	-2.592×10^{23}
−100	-2.688×10^{45}
−200	-1.445×10^{89}

We next determine the critical values of $f(x)$. To find $f'(x)$ we use the product and chain rules:

$$\begin{aligned} f'(x) &= \frac{d}{dx} xe^{-x} = x\frac{d}{dx}e^{-x} + e^{-x}\frac{d}{dx}x \\ &= x[e^{-x}(-1)] + e^{-x}(1) \\ &= (-x+1)e^{-x}. \end{aligned}$$

Since the exponential factor e^{-x} is never zero, $f'(x) = 0$ only when $-x + 1 = 0$, that is, when $x = 1$. The corresponding y-value is $(1)e^{-1} = 1/e \approx 0.368$. The only critical point is $(1, 1/e)$.

We find the second derivative by using the product and chain rules on $f'(x)$. Thus

$$\begin{aligned} f''(x) &= \frac{d}{dx}[(-x+1)e^{-x}] = (-x+1)\frac{d}{dx}e^{-x} + e^{-x}\frac{d}{dx}(-x+1) \\ &= (-x+1)[e^{-x}(-1)] + e^{-x}(-1) \end{aligned}$$

or

$$f''(x) = (x-2)e^{-x}.$$

Now we use the Second Derivative Test to determine the nature of the graph at the critical value $x = 1$. Since $f''(1) = (1-2)e^{-1} = -1/e \approx -0.368$ is negative, the critical value corresponds to a relative maximum at $(1, 1/e) \approx (1, 0.368)$.

We also note that $f''(x)$ is zero when $x = 2$, positive when $x > 2$, and negative when $x < 2$ (recall that the factor e^{-x} is always positive). Hence $(2, 2/e^2) \approx (2, 0.271)$ is an inflection point. Figure 5.5 shows the graph drawn from the data we have accumulated. ■

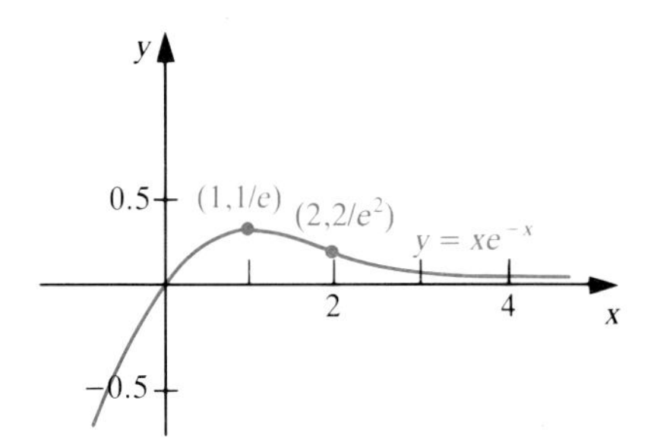

Figure 5.5
Graph of $y = xe^{-x}$
passes through $(0, 0)$
horizontal asymptote at $y = 0$
relative maximum at $(1, 0.368)$
inflection point at $(2, 0.271)$
concave downward for $x < 2$
concave upward for $x > 2$
increasing when $x < 1$
decreasing when $x > 1$

Exercises

Exer. 1–20: Find the derivative of the function.

1. $f(x) = e^{2x}$
2. $f(x) = 3e^x$
3. $f(x) = e^{3x+4}$
4. $f(x) = 5e^{4x-2}$
5. $f(x) = e^{3x^2+5}$
6. $w(s) = 7e^{4s^2-2s+5}$
7. $f(t) = te^{2t}$
8. $f(x) = xe^{x^2}$
9. $f(x) = (3x+2)e^{x^2} + 5$
10. $g(u) = (e^{2u})^{-1/3}$
11. $f(p) = e^{ep}$
12. $f(x) = 2(x^3 + x)e^{4x}$
13. $f(x) = e^{3x}(e^x - x)$
14. $f(t) = \dfrac{t+3}{e^{4t} - t}$
15. $g(t) = \dfrac{e^t + 1}{e^{2t} - 2}$
16. $f(x) = \dfrac{e^{3x} - 9x^2}{xe^x}$
17. $f(x) = x^4 e^{4x}$
18. $f(x) = (e^{4x} + 3x^3 - 8x^2)(x^3 + 5e^{6x})$
19. $f(r) = \dfrac{5r+4}{e^r}$
20. $f(r) = e^{3r+(1/r)}$

Exer. 21–30: Find dy/dx.

21. $y = x^2 - e^x$
22. $y = xe^{-x^2}$
23. $y = e^x - e^{-x}$
24. $y = e^{3x}(3x-1)$
25. $y = x^2 e^{x^2}$
26. $y = (x^3 - 2x)e^{x^2+5}$
27. $y = \sqrt{e^x + x}$
28. $y = (e^x + 3e^{4x} + 5x)^{17}$
29. $y = [e^{x^2 e^x} + x]^4$
30. $y = \sqrt[3]{x^3 e^{x^2} + xe^2}$

Exer. 31–40: Find the equation of the tangent line to the graph of $f(x)$ at the given value of x.

31. $f(x) = e^x;\quad x = 0$
32. $f(x) = e^{3x};\quad x = 0$
33. $f(x) = xe^x;\quad x = 1$
34. $f(x) = \dfrac{e^x + x}{x};\quad x = 1$
35. $f(x) = e^x/x;\quad x = 2$
36. $f(x) = x^e + e^x;\quad x = 1$
37. $f(x) = e^{e^x};\quad x = 0$
38. $f(x) = e^{3x} + ex;\quad x = e$
39. $f(x) = x^2 e^{-x};\quad x = 1$
40. $f(x) = xe^{-x} + x;\quad x = -2$

Exer. 41–46: Find the critical value(s) and inflection point(s) for the function, and sketch its graph.

41. $f(x) = xe^{-2x}$
42. $f(x) = e^x/x$
43. $f(t) = e^{-t^2}$
44. $f(x) = \dfrac{x}{x-1}e^x$
45. $g(x) = x^3 e^{-x}$
46. $h(x) = a - be^{-cx}$, constants $a, b, c > 0$

47. The figure below shows graphs of the functions $y = f(x) = e^x$ and $y = g(x) = mx$ for three values of the constant m. Among the lines whose equations are of the form $y = mx$ there is a value for m such that the line is tangent to the curve $y = e^x$. Find this value of m and the corresponding point of tangency.

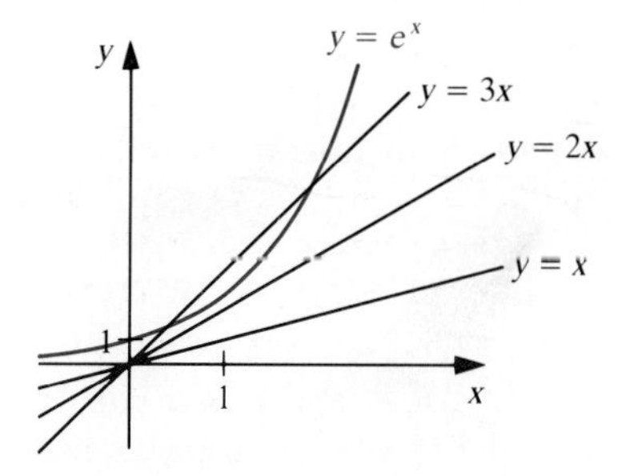

Graphs of $y = e^x$ and $y = mx$ for three values of m

48. Construct a table of values of $x^2 e^{-5x}$ using $x = 5$, 10, 15, 20, and 25 to illustrate the fact that $\lim_{x\to\infty} x^2 e^{-5x} = 0$.

5.3 The Function ln x

In this section we introduce the *logarithm function*, which is denoted $\ln x$. It is closely related to the exponential function e^x. Before we give a definition, consider the following situation. Suppose that on the day of the birth of their grandchild a couple deposits \$1000 in a savings account that pays 6% interest compounded quarterly. The grandparents ask their banker how old the child will be when the balance in the account has grown to \$3000. In Section 5.1 we obtained a formula for the balance in an account such as this. Let $B_4(t)$ denote the balance in the account after t years, where the 4 indicates that interest is compounded *four* times per year. Then $B_4(t)$ is given by

$$B_4(t) = 1000(1.015)^{4t}.$$

To answer the grandparents' question we must find the value of t for which $B_4(t) = 3000$; that is, $1000(1.015)^{4t} = 3000$, or

$$(1.015)^{4t} = 3.$$

We seek the value of t that satisfies this equation. Using the $\boxed{y^x}$ key on a scientific calculator, we can try various "guesses" for t. Suppose we guess that the child will be 15 years old; we try $t = 15$ and obtain

$$(1.015)^{(4)(15)} = (1.015)^{60} \approx 2.44.$$

This is too low, so we consider a higher value for t, say $t = 20$. From $(1.015)^{(4)(20)} = (1.015)^{80} \approx 3.29$, we conclude that $t = 20$ is too large. Since 3.29 is closer to 3 than 2.44 try $t = 18$ (a value between 15 and 20, but closer to 20). Then

$$(1.015)^{(4)(18)} = (1.015)^{72} \approx 2.92.$$

We are getting close. Perhaps the grandparents will be satisfied with the answer "your grandchild will be between 18 and 19 years old." But suppose they ask for a precise answer. We can continuing evaluating $(1.015)^{4t}$ at values of t between 18 and 19 seeking a value of t that yields 3. In fact, we are fortunate to find that $(1.015)^{(4)(18.5)} = (1.015)^{74} \approx 3.009$ is quite close to 3, so a more precise answer is that the grandchild will be about $18\frac{1}{2}$ years old when the account reaches $3000.

We required four evaluations of the exponential function $y = (1.015)^{4t}$ to get an answer roughly accurate to three digits, and additional evaluations were unnecessary only because we happened to get close with our fourth guess. If the grandparents then ask how much sooner the balance will reach $3000 at a competing bank paying 6.25% compounded annually, we must repeat our search to find a solution to $(1.0625)^t = 3$.

This process is certainly tedious, and we would welcome a key on our calculator that gave the answer immediately! While we cannot obtain the answer in a single keystroke, a combination of only a few keystrokes will yield the exact answer to our problem. (The word *exact* here means correct to the accuracy of the calculator being used.)

Suppose that for any positive number k, we could find the number p such that $e^p = k$. With this capability, we can solve

$$(1.015)^{4t} = 3$$

as follows. Find the number p so that $e^p = 1.015$ and the number q such that $e^q = 3$. Then the preceding equation becomes

$$(1.015)^{4t} = (e^p)^{4t} = 3 = e^q$$

or

$$e^{4pt} = e^q.$$

Hence $4pt = q$, or $t = q/4p$. Since we are assuming that we can easily find p and q, this procedure will solve our problem.

The success of this idea lies in the ability to find the number p such that $e^p = k$ for any (given) positive number k. The number p is denoted by $\ln k$; that is,

$$e^p = k \quad \text{is equivalent to} \quad p = \ln k \quad (k > 0).$$

Scientific calculators usually have a key labeled $\ln x$ or ln. Since the solution to $e^p = 1.015$ is $p = \ln 1.015$, we can find p by entering 1.015 into the calculator and pressing the $\ln x$ key. The result is $p = \ln 1.015 \approx 0.014888612$. (This value may vary slightly on different calculators.) Similarly, the solution to $e^q = 3$ is $q = \ln 3 \approx 1.098612289$. Therefore, $t = q/4p \approx 18.44719058$ years. This approach is more precise and does not depend on making good guesses.

Stating the previous equivalence in terms of x and y gives

$$y = e^x \quad \text{is equivalent to} \quad x = \ln y.$$

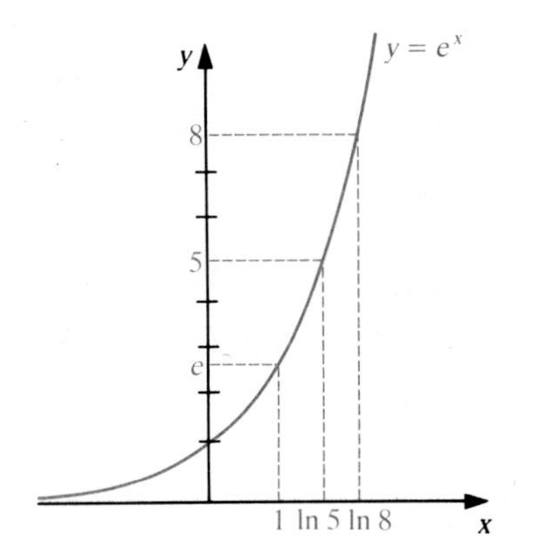

Figure 5.6
To find $\ln a$, locate a on the y-axis and find the value on the x-axis so that $e^x = a$. This is shown for $a = 5$ and $a = 8$

In words, $\ln y$ is the exponent of e for which e to that power equals y. For example, to an accuracy of ten digits, $\ln 5 = 1.609437912$ because $e^{1.609437912} = 5$. Similarly, $\ln e = 1$ because $e^1 = e$. Figure 5.6 shows that $\ln 5$ is the answer to the question "e to what power equals 5?" and $\ln 8$ is the answer to the question "e to what power equals 8?"

Figure 5.6 also shows that given any $y > 0$, there is exactly one value of x such that $e^x = y$. Thus the correspondence between x and y given by $x = \ln y$ defines a function.

When discussing functions our preference is to let y denote the dependent variable and let x be the independent variable, so we write $y = \ln x$ and note that this is equivalent to $e^y = x$. The function $y = \ln x$ is called the **natural logarithm function**. We now consider the construction of its graph and the determination of its properties.

To sketch the graph of $y = \ln x$ we could use a calculator to generate a table of points, plot the points, and connect them with a smooth curve. However, we will use another method that will emphasize the connection between the exponential function e^x and the natural logarithm function $\ln x$.

Let $f(x) = \ln x$ and $g(x) = e^x$. Consider a specific value of x, say $x = 5$. Then $f(5) = \ln 5 = 1.60943\ldots$ (we will abbreviate this as 1.609 throughout this discussion). Hence the point $(5, 1.609)$ is on the graph of $y = f(x) = \ln x$. However, since $\ln 5 = 1.609$ is equivalent to $e^{1.609} = 5$ (that is, $g(1.609) = 5$) the point $(1.609, 5)$ is on the graph of $y = g(x) = e^x$. Consequently, the statement $(5, 1.609)$ *is a point on the graph of* $f(x) = \ln x$ is equivalent to the statement $(1.609, 5)$ *is on the graph of* $g(x) = e^x$.

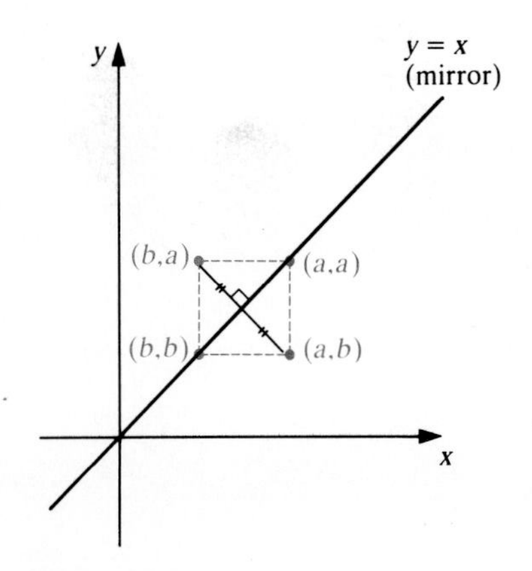

Figure 5.7
Points (a, b) and (b, a) are mirror images using the line $y = x$ as a mirror

Of course there is nothing special about our choice of $x = 5$. The same equivalence could be observed about any point (a, b) with $a > 0$: the statement (a, b) *is a point on the graph of* $f(x) = \ln x$ is equivalent to the statement (b, a) *is a point on the graph of* $g(x) = e^x$. There is an interesting geometric connection between the points (a, b) and (b, a): they are on opposite sides of the line $y = x$ and equidistant from it. See Figure 5.7. If we think of the line $y = x$ as a mirror, then (b, a) is the reflection of (a, b), and vice versa. Because (a, b) is on the graph of $y = \ln x$ whenever (b, a) is on the graph of $y = e^x$, we say that e^x and $\ln x$ are *inverse functions:* $\ln x$ *is the inverse function of* e^x and e^x *is the inverse function of* $\ln x$.

Since (a, b) is on the graph of $f(x) = \ln x$ exactly when (b, a) is on the graph of $g(x) = e^x$, every point (a, b) on the graph of $f(x) = \ln x$ is the reflection of a corresponding point (b, a) on the graph of $g(x) = e^x$.

Figure 5.8 shows the (known) graph of e^x and its reflection across the line $y = x$. This reflection is the graph of $f(x) = \ln x$. The domain of $y = \ln x$ is the set of positive real numbers and the graph of $y = \ln x$ has a vertical asymptote at $x = 0$. (This is our first example of a vertical asymptote that is not produced by a zero denominator.) Thus $\ln x \to -\infty$ as $x \to 0^+$ and $\ln x \to \infty$ as $x \to \infty$.

The function $\ln x$ has many special properties that it inherits from the exponential function.

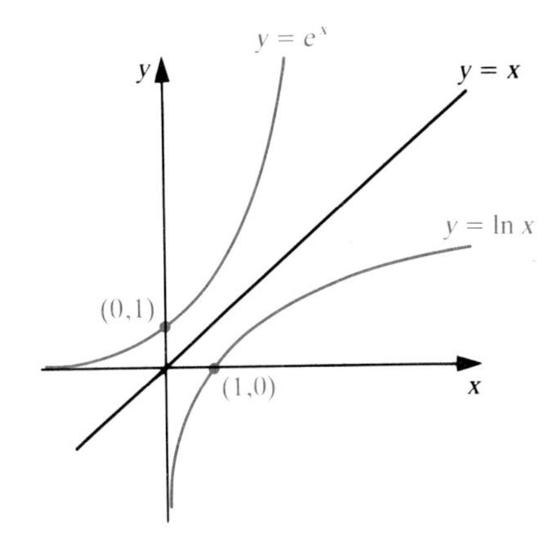

Figure 5.8
Graphs of $g(x) = e^x$ and $f(x) = \ln x$

Properties of the Natural Logarithm

For positive r and s,

(1) $e^{\ln u} = u$ for all $u > 0$
(2) $\ln e^u = u$ for all u
(3) $\ln 1 = 0 \quad \ln e = 1$
(4) $\ln (rs) = \ln r + \ln s$
(5) $\ln \dfrac{r}{s} = \ln r - \ln s$
(6) $\ln r^k = k \ln r$

Properties (1) and (2) are other ways of stating that the solution to $u = e^p$ is $p = \ln u$. For example, to get (1), we replace p in $u = e^p$ with $p = \ln u$ so that $u = e^p = e^{\ln u}$.

Property (3) is a special case of properties (1) and (2) using $u = 0$ in property (2) and $u = 1$ in property (1), respectively. Alternatively, property (3) may be justified as follows. The point (0, 1) is on the graph of $y = e^x$ so its mirror-image point (1, 0) is on the graph of $y = \ln x$. Thus $0 = \ln 1$. Similarly, $(1, e)$ is on the graph of $y = e^x$, so $(e, 1)$ is on the graph of $y = \ln x$; thus $1 = \ln e$.

We will prove property (4) and leave the proof of properties (5) and (6) to the exercises. To prove property (4), let

$$p = \ln r, \quad \text{and} \quad q = \ln s.$$

Then
$$r = e^p \quad \text{and} \quad s = e^q.$$

Consequently, $rs = e^p e^q = e^{p+q}$. Thus $\ln rs = \ln e^{p+q} = p + q$ by property (2). From the definitions of p and q, $\ln rs = p + q = \ln r + \ln s$. This is property (4).

The step from $rs = e^{p+q}$ to $\ln rs = \ln e^{p+q}$ occurs often when using logarithms. We describe this step as *taking the logarithm of both sides* (*of the equation*).

Example 5.7 Use the properties of the natural logarithms to simplify each expression:

$$\textbf{(a)}\ \ln (xe^x) \quad \textbf{(b)}\ \ln \frac{x}{\sqrt{x+1}} \quad \textbf{(c)}\ \ln \frac{x+2}{x^2+1}$$

Solution **(a)** By property (4),

$$\ln(xe^x) = \ln x + \ln e^x;$$

by property (2), $\ln e^x = x$, so

$$\ln(xe^x) = \ln x + x.$$

(b) By property (5)

$$\ln\frac{x}{\sqrt{x+1}} = \ln x - \ln\sqrt{x+1} = \ln x - \ln(x+1)^{1/2}.$$

By property (6) $$\ln(x+1)^{1/2} = \tfrac{1}{2}\ln(x+1).$$

Thus $$\ln\frac{x}{\sqrt{x+1}} = \ln x - \tfrac{1}{2}\ln(x+1).$$

(c) We use property (5) to find

$$\ln\frac{x+2}{x^2+1} = \ln(x+2) - \ln(x^2+1).$$

Note that $\ln(x^2+1)$ cannot be simplified further by the properties of logarithms. For example, $\ln(x^2+1) \neq \ln x^2 + \ln 1$. ∎

Properties (1) and (6) of logarithms are useful in evaluating b^x for any positive number b. Letting $u = b^x$ in property (1) gives

$$e^{\ln b^x} = b^x.$$

By property (6) we may rewrite the left side as

$$e^{\ln b^x} = e^{x\ln b} = e^{(\ln b)x}.$$

Thus $$b^x = e^{(\ln b)x}.$$

For example, $3^x = e^{(\ln 3)x} \approx e^{1.098612x}$; to find the value of $3^{2.7}$ (for instance), evaluate $e^{(1.098612)(2.7)} = e^{2.9662524} \approx 19.41901$. Consequently, if we can compute $\ln b$ and any power of e, then we can compute any power of b. In fact, when x is not an integer, the calculator key $\boxed{y^x}$ computes y^x from $y^x = e^{x\ln y}$.

As illustrated in the next example, the logarithm function may be used to solve an equation where the unknown is in an exponent.

Example 5.8 Solve $(2)3^{x+1} = 5^x$ for x.

Solution Since the unknown is in an exponent, we take the logarithm of both sides of the equation:

$$\ln[(2)3^{x+1}] = \ln 5^x.$$

Then use property (4) to change the product to a sum:

$$\ln 2 + \ln 3^{x+1} = \ln 5^x;$$

now we move the exponents by property (6) to get

$$\ln 2 + (x+1)\ln 3 = x\ln 5$$

or $$\ln 2 + x\ln 3 + \ln 3 = x\ln 5.$$

Subtract $x \ln 3$ from both sides and factor x from the terms on the right side:

$$\ln 2 + \ln 3 = x \ln 5 - x \ln 3 = x(\ln 5 - \ln 3).$$

Finally, divide by the factor $(\ln 5 - \ln 3)$ to obtain

$$x = \frac{\ln 2 + \ln 3}{\ln 5 - \ln 3} \approx 3.5076.$$

■

Property (1) can be used to find the derivative of $y = \ln x$. Since

$$e^{\ln x} = x \quad \text{for all } x > 0,$$

we may differentiate both sides to obtain

$$\frac{d}{dx} e^{\ln x} = \frac{d}{dx} x, \qquad x > 0$$

or, from the chain rule for the exponential function,

$$e^{\ln x} \frac{d}{dx}(\ln x) = 1, \qquad x > 0.$$

Since $e^{\ln x} = x$, this may be written as

$$x \frac{d}{dx}(\ln x) = 1, \qquad x > 0.$$

Divide both sides by x to get the following formula for the derivative.

Derivative of ln *x*

$$\frac{d}{dx}(\ln x) = \frac{1}{x}, \qquad x > 0$$

Since the derivative of $\ln x$ is $1/x$ and $\ln x$ is the inverse function of e^x, we can obtain the following properties of the natural logarithm function.

Properties of $y = \ln x$

1. $\ln x$ is defined for $x > 0$.
2. $\ln x$ is continuous for $x > 0$.
3. $\ln x \to -\infty$ as $x \to 0^+$.
4. $\ln x \to \infty$ as $x \to \infty$.
5. $x = 0$ is a vertical asymptote of $\ln x$.
6. $\ln x$ is negative for $0 < x < 1$ and positive for $x > 1$.
7. $\ln x$ is differentiable for all $x > 0$.
8. $\ln x$ is increasing for all $x > 0$.
9. $\ln x$ is concave downward for all x.

Using the chain rule, we can compute the derivative of $\ln g(x)$ for any positive differentiable function $g(x)$. If $y = f(u) = \ln u$ and $u = g(x)$, then their composition is $y = f(g(x)) = \ln g(x)$, and the chain rule gives

$$\begin{aligned} \frac{dy}{dx} &= \frac{dy}{du}\frac{du}{dx} \\ &= \frac{d \ln u}{du}\frac{du}{dx} \\ &= \frac{1}{u}\frac{du}{dx} \\ &= \frac{1}{g(x)} g'(x) \end{aligned}$$

Derivative of ln $g(x)$

$$\frac{d \ln g(x)}{dx} = \frac{g'(x)}{g(x)} = \frac{\frac{d}{dx} g(x)}{g(x)}$$

Example 5.9 Find the derivative of each function:

(a) $f(x) = \ln(4x^3 + 5)$ **(b)** $f(x) = \ln(x^5 + e^x)$

(c) $f(x) = x \ln(x^2 + 6)$ **(d)** $f(x) = \dfrac{\ln(5x + 2)}{x}$

Solution **(a)** We use the formula for the derivative of $\ln g(x)$ with $g(x) = 4x^3 + 5$:

$$f'(x) = \frac{d}{dx} \ln(4x^3 + 5) = \frac{\frac{d}{dx}(4x^3 + 5)}{4x^3 + 5} = \frac{12x^2}{4x^3 + 5}.$$

(b)

$$f'(x) = \frac{d}{dx} \ln(x^5 + e^x) = \frac{\frac{d}{dx}(x^5 + e^x)}{x^5 + e^x} = \frac{5x^4 + e^x}{x^5 + e^x}.$$

(c) Use the product rule:

$$\begin{aligned} f'(x) = \frac{d}{dx}[x \ln(x^2 + 6)] &= x \frac{d}{dx} \ln(x^2 + 6) + \ln(x^2 + 6) \frac{d}{dx} x \\ &= x \frac{\frac{d}{dx}(x^2 + 6)}{x^2 + 6} + \ln(x^2 + 6)(1) \\ &= \frac{x(2x)}{x^2 + 6} + \ln(x^2 + 6) = \frac{2x^2}{x^2 + 6} + \ln(x^2 + 6). \end{aligned}$$

(d) Use the quotient rule:

$$f'(x) = \frac{d}{dx}\frac{\ln(5x+2)}{x} = \frac{x\dfrac{d}{dx}\ln(5x+2) - \ln(5x+2)\dfrac{d}{dx}x}{x^2}$$

$$= \frac{x\dfrac{5}{5x+2} - [\ln(5x+2)](1)}{x^2}$$

$$= \frac{\dfrac{5x}{5x+2} - \ln(5x+2)}{x^2}. \quad \blacksquare$$

In some cases we may use the properties of logarithms to simplify an expression before we calculate the derivative. The next example illustrates this method.

Example 5.10 Find the derivative of each function:

(a) $f(x) = \ln(2x-1)^4$ **(b)** $f(x) = \ln\dfrac{3x-2}{2x+5}$ **(c)** $f(x) = \ln\sqrt{x^3+5}$

Solution **(a)** Using property (6) for logarithms, we see that $f(x) = \ln(2x-1)^4 = 4\ln(2x-1)$. Hence

$$f'(x) = \frac{d}{dx}4\ln(2x-1) = 4\frac{d}{dx}\ln(2x-1) = 4\frac{2}{2x-1} = \frac{8}{2x-1}.$$

(b) Since $\ln\dfrac{r}{s} = \ln r - \ln s$, we have

$$f(x) = \ln\frac{3x-2}{2x+5} = \ln(3x-2) - \ln(2x+5),$$

so that

$$f'(x) = \frac{d}{dx}[\ln(3x-2) - \ln(2x+5)] = \frac{3}{3x-2} - \frac{2}{2x+5}.$$

(c) By property (6), $\ln\sqrt{x^3+5} = \ln(x^3+5)^{1/2} = \frac{1}{2}\ln(x^3+5)$; thus

$$f'(x) = \frac{d}{dx}\tfrac{1}{2}\ln(x^3+5) = \frac{1}{2}\frac{3x^2}{x^3+5}. \quad \blacksquare$$

Example 5.11 Draw the graph of $f(x) = x\ln x$.

Solution We will find (a) intercepts, (b) asymptotes, (c) critical values, and (d) inflection points.

(a) We obtain the y-intercept by setting $x = 0$; however $x = 0$ is not in the domain of the function because the logarithm function is not defined at zero. Hence there is no y-intercept. Setting $y = f(x) = 0$ gives the x-intercepts. Since $f(x) = 0$ implies that $x\ln x = 0$, and since x cannot be zero, we must have $\ln x = 0$, which implies that $x = e^0 = 1$. Thus $x = 1$ is the only x-intercept.

Table 5.9
Behavior of $f(x)$ near $x = 0^+$

x	$f(x) = x \ln x$
0.1	-2.302×10^{-1}
0.01	-4.605×10^{-2}
0.001	-6.908×10^{-3}
0.0001	-9.210×10^{-4}

(b) The function is not defined for $x < 0$, so we do not consider the behavior of the function as $x \to -\infty$. As $x \to \infty$, both factors in $f(x) = x \ln x$ increase without bound, so $f(x) \to \infty$ also. Therefore no horizontal asymptote is associated with $x \to \infty$, so this function has no horizontal asymptotes.

Since $\ln x$ has a vertical asymptote at $x = 0$, there is the possibility that $f(x)$ also has a vertical asymptote there. As $x \to 0^+$, $\ln x \to -\infty$, but the factor x in $f(x) = x \ln x$ approaches zero. So as $x \to 0^+$ one factor in $f(x)$ is becoming increasingly small and the other is becoming increasingly large (in the negative direction). Table 5.9 indicates that $f(x) \to 0$ as $x \to 0^+$. Thus the function has no vertical asymptote at $x = 0$; $f(x)$ has no vertical asymptotes.

(c) To find the critical values, we find $f'(x)$ using the product rule:

$$f'(x) = \frac{d}{dx}(x \ln x) = x \frac{d}{dx} \ln x + (\ln x) \frac{d}{dx} x$$

$$= x \frac{1}{x} + (\ln x)(1) = 1 + \ln x.$$

Thus $f'(x) = 0$ implies

$$1 + \ln x = 0 \quad \text{or} \quad \ln x = -1.$$

Hence

$$x = e^{-1} = \frac{1}{e}$$

is the only critical value. The y-coordinate associated with this value is $y = e^{-1} \ln e^{-1} = e^{-1}(-1) = -1/e$. Note that $f'(x) = 1 + \ln x \to -\infty$ as $x \to 0^+$ (because $\ln x \to -\infty$ as $x \to 0^+$). We have already determined that $f(x) \to 0$ as $x \to 0^+$. Since the derivative is approaching $-\infty$, the function has a very steep, negative slope close to the origin.

(d) Determination of the inflection points requires the second derivative:

$$f''(x) = \frac{d}{dx}(1 + \ln x) = 0 + \frac{1}{x} = \frac{1}{x},$$

which is never zero. It is undefined only at $x = 0$, a value not in the domain of this function. Thus $f(x)$ has no inflection points. Since $1/x$ is positive for $x > 0$, we conclude that $f(x)$ is always concave upward, so the critical point $(1/e, -1/e)$ is a relative minimum. The graph of $f(x)$ is shown in Figure 5.9. ■

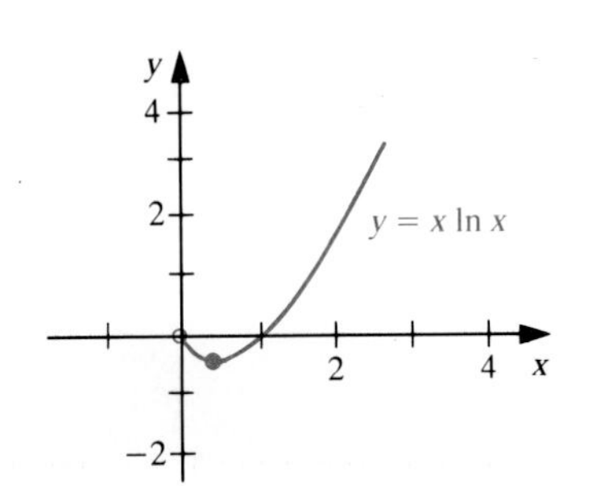

Figure 5.9
Graph of $f(x) = x \ln x$
passes through $(1, 0)$
no y-intercept
no horizontal or vertical asymptote
relative minimum at $(1/e, -1/e)$
always concave upward
no inflection point

Example 5.12 Suppose that the curve of a section of highway is similar to that of the graph of $y = \ln x$, relative to a coordinate system in which one unit represents 100 feet, the y-axis points north, and the x-axis points east (see Figure 5.10). Also suppose that a house is located at $(0, 1)$. Where should a guard rail be located along the highway so that if a westbound vehicle misses the curve it will be deflected away from the house? Assume that any vehicle leaving the highway will travel in a straight line once it is off the road.

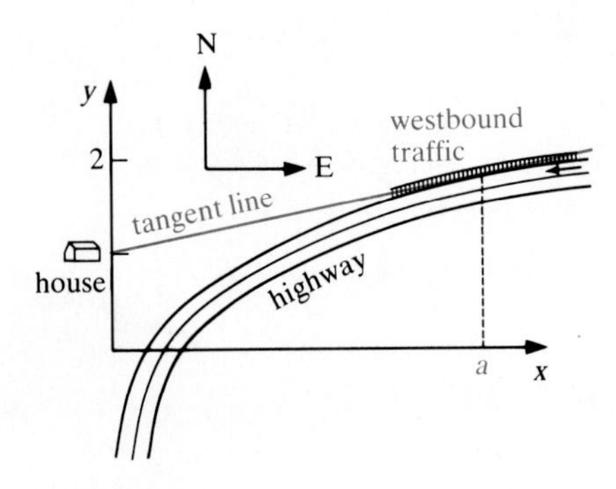

Figure 5.10
The value of a is to be chosen so that the tangent line to $y = \ln x$ at $x = a$ passes through $(0, 1)$

Solution A vehicle traveling west along the highway is traveling in the "direction" of the road at the time it leaves the road; that is, it is traveling in the direction of the tangent line at any moment (the direction of the road). A vehicle that leaves the road travels along the line tangent to the highway at the point where it leaves the road. Thus, the center of the guard rail should be located at the point on the road where the tangent line to the curve (highway) passes through the house. The x-coordinate of this particular point is labeled a in Figure 5.10. We know two things about the tangent line (i) it is a tangent line at the point $(a, \ln a)$; and (ii) it passes through $(0, 1)$. Since the line is tangent to $y = \ln x$ at $x = a$, the slope of the line is $y'(a) = 1/a$. Since the line passes through the points $(0, 1)$ and $(a, \ln a)$, the slope is

$$\text{slope} = \frac{1 - \ln a}{0 - a}.$$

Setting the two expressions for the slope equal to each other yields

$$\frac{1}{a} = \frac{1 - \ln a}{-a} = \frac{-1 + \ln a}{a}.$$

Thus $1 = -1 + \ln a$, so $\ln a = 2$, from which we obtain $a = e^2$. Consequently, $(e^2, \ln e^2) = (e^2, 2)$ is the required point of tangency. The guard rail should be located at $(e^2, 2) \approx (7.39, 2)$; that is, the center of the rail should be 739 feet east of the house and 100 feet north of it. ■

Exercises

Exer. 1–10: Use the properties of the logarithm function to simplify the expression.

1. $\ln x^3$
2. $\ln 3x^4$
3. $\ln x^2 e^{4x}$
4. $\ln s^3(s + 2)^4$
5. $\ln \dfrac{t + 2}{t - 4}$
6. $\ln \dfrac{3x + 4}{x(2x - 5)}$
7. $\ln [(p + 2)(p^3 + p + 2)^3]$
8. $\ln \dfrac{x}{\sqrt{x^4 + 3}}$
9. $\ln \dfrac{x^2 + x + 4}{xe^x}$
10. $\ln \sqrt[3]{3x^2 + x + 1}$

Exer. 11–20: Solve the equation for x.

11. $3^x = 7$
12. $4^x = 3$
13. $(5)4^{x+1} = 9^x$
14. $(8)5^x = 1$
15. $\sqrt{2}\, 3^{2x+1} = 7^{3x-2}$
16. $3^x = 4^{2x}$
17. $(2)4^x = 9$
18. $\ln 2x = 3$
19. $\ln (x^2 - 1) = 2$
20. $\ln (x + 1) = 0$

Exer. 21–44: Find the derivative of the function.

21. $f(x) = 2 \ln x$
22. $f(x) = \ln (3x)$
23. $f(x) = \ln (x + e^x)$
24. $f(x) = \ln (x^3)$
25. $f(t) = \ln (3t^5 + 4t^3 - 7t)$
26. $f(t) = \ln 5t + \ln (t^4 + t)$
27. $f(x) = 4 \ln (2x)$
28. $f(x) = -3 \ln (5x)$
29. $g(x) = \ln (x^2 + 2)$
30. $f(x) = 4 \ln (x^2 + 3)$
31. $f(r) = -2 \ln (r^3 + r - 1)$
32. $g(r) = r \ln (4r)$
33. $f(x) = x^2 \ln (x^2 + 3)$
34. $f(x) = (2x^3 + 1) \ln (5x^2 + 2)$
35. $f(u) = u + \ln (u^2)$
36. $f(x) = \dfrac{x + 3}{\ln (x^3)}$
37. $f(x) = \ln [\ln (2x)]$
38. $f(s) = s \ln [\ln (2s)]$
39. $f(x) = x \ln (x \ln x)$
40. $f(x) = [\ln (2x)]^{2/3}$
41. $f(p) = (\ln p)e^p$
42. $f(x) = (x + \ln x^2)e^{2x}$
43. $f(x) = e^{x^2} \ln x^2$
44. $f(z) = \ln (e^{2z} + 4z)$

Exer. 45–54: Find dy/dx.

45. $y = (\ln x^2)^2$
46. $y = x \ln (10 - x^2)$
47. $y = \ln (2ax - x^2)$
48. $y = (\ln x)^2/(4x)$
49. $y = x^5 \ln x - \frac{1}{5}x^5$
50. $y = \ln (\ln x)$
51. $y = x^2 \ln x^2$
52. $y = \ln \dfrac{e^x}{1 + e^x}$
53. $y = x/\ln x$
54. $y = x + 2 \ln (1 + \sqrt{1 + e^{-x}})$

Exer. 55–62: Find dy/dx. Where possible, use properties of logarithms to simplify the expression before differentiating (see Example 5.10).

55. $y = \ln \dfrac{x}{1 + x^2}$
56. $y = \ln \dfrac{\sqrt[3]{1 + x}}{1 + x}$

57. $y = \ln (2x^2 + 5)^3$

58. $y = \ln \left(\dfrac{e^x - e^{-x}}{e^x + e^{-x}}\right)^{1/2}$

59. $y = \ln (3x^2 + 5x + 1)^{2/3}$

60. $y = \ln (x^2 + 5x - 1)^{-2/3}$

61. $y = \ln \sqrt{5 - 2x + 3x^4}$

62. $y = \ln \dfrac{x + \sqrt{x^2 - 2}}{2}$

Exer. 63–68: In Section 4.6, we defined the relative change of $f(x)$ to be

$$\frac{f'(x)}{f(x)}.$$

Hence the relative change of $f(x)$ can be calculated by taking the derivative of $\ln f(x)$. Find the relative change of the function at the value indicated.

63. $f(x) = 6x - 4;\quad x = 2$
64. $f(x) = x^3 + 2x - 1;\quad x = 4$
65. $f(x) = e^{2x-1};\quad x = \frac{1}{2}$
66. $f(x) = xe^{2x};\quad x = 2$
67. $f(x) = e^{3x}(3x - 1);\quad x = 3$
68. $f(x) = \ln (x^2 + x);\quad x = 2$

Exer. 69–72: Find the critical value(s) and inflection point(s) for the function. Sketch its graph.

69. $f(x) = x + \ln x$

70. $f(x) = \ln \dfrac{x - 4}{x}$

71. $g(x) = \dfrac{\ln x}{x}$

72. $h(x) = \ln \dfrac{x - 4}{x - 2}$

73. What value should be taken for m so that the line $y = mx$ is tangent to the curve $y = x^2 \ln x$? Note: let a denote the point of tangency. Then at $x = a$ the line and curve intersect and have the same slope.
74. Prove property (5) for logarithms: $\ln (r/s) = \ln r - \ln s$, with r, s positive.
75. Prove property (6) for logarithms: $\ln r^k = k \ln r$ with $r > 0$.

5.4 Logarithmic Differentiation

We next turn to the differentiation of exponential functions of the form $y = b^x$ for $b > 0$. Thus far we have differentiated only exponential functions involving $y = e^x$. The differentiation technique presented here uses the chain rule applied to the differentiation of $\ln f(x)$. By the chain rule

$$\frac{d}{dx} \ln f(x) = \frac{f'(x)}{f(x)}.$$

If we let $y = f(x)$, then we can write

$$\frac{d}{dx} \ln y = \frac{y'}{y}.$$

Example 5.13 Find the derivative of $f(x) = 2^x$.

Solution Let $y = f(x) = 2^x$. Then $\ln y = \ln 2^x = x \ln 2$. Differentiating both sides of this equation with respect to x gives

$$\frac{d}{dx}(\ln y) = \frac{d}{dx}(x \ln 2)$$

or

$$\frac{y'}{y} = \ln 2.$$

Note that $\ln 2$ is a constant, so the derivative of $x \ln 2$ is the constant $\ln 2$. Since we wish to find y', we multiply the above equation by $y = 2^x$ to get

$$y' = 2^x \ln 2.$$

■

In Section 5.2 we discovered that the derivative of 2^x is $(m_2)2^x$, where $m_2 \approx 0.693$. We now see that the derivative is $(\ln 2)2^x$, indicating that $m_2 = \ln 2$. Indeed, a calculator will confirm that 0.693 is the correct value of $\ln 2$ to three places.

The procedure shown in Example 5.13 can be carried out for $y = b^x$ for any positive b. The result is

$$y' = \frac{d}{dx} b^x = b^x \ln b.$$

For example, $(d/dx)3^x = 3^x \ln 3$. In Section 5.2 we found that the derivative of this function is $(m_3)3^x$, where $m_3 \approx 1.099$. The exact value of m_3 is $m_3 = \ln 3 \approx 1.098612289$. The value of m_3 we found earlier is correct to three places.

The process of finding $f'(x)$ by taking the logarithm, simplifying, and then differentiating is called **logarithmic differentiation**. It is useful in many situations that involve an independent variable in an exponent.

Example 5.14 Find the derivative of $f(x) = 4^{x^2+x}$.

Solution Let $y = f(x) = 4^{x^2+x}$. Then

$$\ln y = \ln 4^{(x^2+x)} = (x^2 + x) \ln 4.$$

Differentiating both sides with respect to x yields

$$\frac{d}{dx} \ln y = \frac{d}{dx} [(x^2 + x) \ln 4]$$

or

$$\frac{y'}{y} = (2x + 1) \ln 4,$$

because $\ln 4$ is a constant. Multiplying by y gives

$$y' = y(2x + 1) \ln 4 = 4^{(x^2+x)}(2x + 1) \ln 4.$$ ■

Consider the task of finding $f'(x)$ for the function

$$f(x) = \frac{(x^2 + 1)^7(3x - 2)^{15}}{\sqrt{6x + 1}\,(3x + 1)^8}.$$

We can find $f'(x)$ from the quotient, product, and chain rules, but the task would be quite tedious. The following technique simplifies the task. Take the logarithm of both sides of the equation:

$$\ln f(x) = \ln \frac{(x^2 + 1)^7(3x - 2)^{15}}{\sqrt{6x + 1}\,(3x + 1)^8}.$$

The right side may be simplified to

$$\begin{aligned} \ln f(x) &= \ln [(x^2 + 1)^7(3x - 2)^{15}] - \ln [\sqrt{6x + 1}\,(3x + 1)^8] \\ &= \ln (x^2 + 1)^7 + \ln (3x - 2)^{15} - \ln (\sqrt{6x + 1}) - \ln (3x + 1)^8 \\ &= 7 \ln (x^2 + 1) + 15 \ln (3x - 2) - \tfrac{1}{2} \ln (6x + 1) - 8 \ln (3x + 1). \end{aligned}$$

(This last simplification uses $\ln r^k = k \ln r$.) In this form, we can easily differentiate each term on the right side obtaining

$$\frac{f'(x)}{f(x)} = 7\frac{2x}{x^2 + 1} + 15\frac{3}{3x - 2} - \frac{1}{2}\frac{6}{6x + 1} - 8\frac{3}{3x + 1}.$$

Multiplying both sides by $f(x)$ and simplifying the right side gives

$$f'(x) = f(x)\left(\frac{14x}{x^2+1} + \frac{45}{3x \quad 2} - \frac{3}{6x+1} - \frac{24}{3x+1}\right).$$

We have found the derivative with a series of remarkably easy differentiations. We can replace $f(x)$ in this expression with the original definition of $f(x)$ to obtain

$$f'(x) = \frac{(x^2+1)^7(3x-2)^{15}}{\sqrt{6x+1}\,(3x+1)^8}\left(\frac{14x}{x^2+1} + \frac{45}{3x-2} - \frac{3}{6x+1} - \frac{24}{3x+1}\right).$$

Logarithmic differentiation can make the task of finding derivatives easier and, in some cases, may be our only way of finding the derivative. The next example shows such a case.

Example 5.15 Find $f'(x)$ where $f(x) = x^{1-x}$.

Solution This function is neither a power function (x to a *constant* power) nor an exponential function (a *constant* to the x power), so our earlier differentiation rules do not apply. However, if we take the logarithm of both sides and "pull the exponent outside the log," we get

$$\ln f(x) = \ln x^{1-x} = (1-x)\ln x.$$

Now the right-hand side involves a product of functions whose derivatives we know. Differentiating both sides gives

$$\frac{d}{dx}\ln f(x) = \frac{d}{dx}[(1-x)\ln x]$$

$$\frac{f'(x)}{f(x)} = (1-x)\frac{d}{dx}\ln x + \ln x\frac{d}{dx}(1-x)$$

$$= (1-x)\frac{1}{x} + (\ln x)(-1)$$

$$= \frac{1-x}{x} - \ln x.$$

Finally, we multiply both sides by $f(x) = x^{1-x}$ to obtain

$$f'(x) = x^{1-x}\left(\frac{1-x}{x} - \ln x\right).$$

Note that simplification by the properties of the logarithm enabled us to apply our differentiation rules to this problem. ■

We end this section with the calculation of a limit that will be useful in the next section. The limit is

$$\lim_{h\to 0}(1+h)^{1/h}.$$

We might conclude that the value of this limit is 1 because as h approaches zero, $(1+h)$ approaches 1, and 1 raised to any power is 1; alternatively, we might conclude that the limit does not exist, since

$(1+h)>1$ and $1/h \to \infty$, and a number greater than 1 to an infinite power is infinite. Neither conclusion is correct. The correct result is obtained as follows. Let $y=(1+h)^{1/h}$. Then

$$\ln y = \ln (1+h)^{1/h} = \frac{1}{h}\ln(1+h) = \frac{\ln(1+h)-\ln 1}{h}. \tag{1}$$

The term $\ln 1$ (which is zero) has been included so that the last expression suggests the difference quotient we used to find derivatives in Chapter 2. In fact,

$$\lim_{h\to 0}\frac{\ln(1+h)-\ln 1}{h}$$

is the derivative of $f(x)=\ln(1+x)$ at $x=0$ because

$$\begin{aligned} f'(0) &= \lim_{h\to 0}\frac{f(0+h)-f(0)}{h} \\ &= \lim_{h\to 0}\frac{\ln(1+0+h)-\ln(1+0)}{h} \\ &= \lim_{h\to 0}\frac{\ln(1+h)-\ln 1}{h}. \end{aligned}$$

This is precisely the limit we seek. Since $f'(x)=1/(1+x)$, we see that $f'(0)=1$. From equation (1) we now have

$$\lim_{h\to 0}\ln y = \lim_{h\to 0}\frac{\ln(1+h)-\ln 1}{h} = f'(0) = 1,$$

and, since the logarithm function is continuous, $\lim\limits_{h\to 0}\ln y = \ln(\lim\limits_{h\to 0} y) = 1$. Therefore, $\lim\limits_{h\to 0} y = e^1 = e$; that is,

$$\lim_{h\to 0}(1+h)^{1/h} = e,$$

a surprising result! For additional verification, compute the value of $(1+h)^{1/h}$ for several small values of h.

Exercises

Exer. 1–16: Find $f'(x)$.

1. $f(x)=5^x$
2. $f(x)=5^{3x}$
3. $f(x)=7^{2x}$
4. $f(x)=x\,3^x$
5. $f(x)=x\,2^x$
6. $f(x)=2^{\ln x}$
7. $f(x)=4^{6x^2-1}$
8. $f(x)=3^{x^2+2}$
9. $f(x)=3^{4x+1}$
10. $f(x)=(x^2)(2^x)$
11. $f(x)=6^{e^x}$
12. $f(x)=4^x/x$
13. $f(x)=(\frac{1}{2})^{5x}$
14. $f(x)=(\frac{1}{5})^{x^2}$
15. $f(x)=(\frac{1}{3})^{-2x+5}$
16. $f(x)=7^{3x-4}$

Exer. 17–34: Use logarithmic differentiation to find $f'(x)$.

17. $f(x)=\dfrac{(2x+3)^5(4x-1)^3}{\sqrt{2x+1}\,(3x+5)^4}$
18. $f(x)=\dfrac{e^{3x}(9x-5)^4}{x^4(2x+3)^4}$
19. $f(x)=(2x^3+4)^3(1-x^3)^{1/3}\sqrt{4x^4+x}$
20. $f(x)=(3x+1)^4(2x-1)^5(4x+1)^7$
21. $f(x)=x^x$
22. $f(x)=(1+x)^x$
23. $f(x)=(x^2+4)(x^2-4)^x$
24. $f(x)=x^{\ln x}$
25. $f(x)=x^{1/x}$
26. $f(x)=x^{e^x}$
27. $f(x)=(1-x)^{1-1/x}$
28. $f(x)=x^{x^x}$
29. $f(x)=(\ln x)^x$
30. $f(x)=(e^x)(x^x)$
31. $f(x)=x^{x^3}$
32. $f(x)=(2x)^x$
33. $f(x)=x^{1/\ln x}$
34. $f(x)=\left(\dfrac{1}{x}\right)^{1/\ln x}$

5.5 Applications of Exponential and Logarithmic Functions

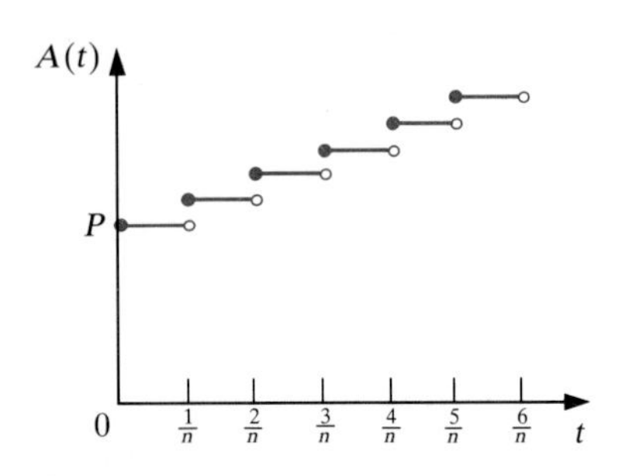

Figure 5.11
Graph of a savings account balance

Consider the accumulation of money in a savings account. Let r denote the annual interest rate (given as a decimal, for example, 0.09 instead of 9%), let n denote the number of compounding periods per year, let P be the initial deposit, and let $A(t)$ denote the amount of money in the account after t years. In Section 5.1 we showed that $A(t)$ is given by

$$A(t) = P\left(1 + \frac{r}{n}\right)^{nt}. \tag{1}$$

Since $A(t)$ increases only when interest is deposited, the graph of $A(t)$ is a "step function" as shown in Figure 5.11.

Example 5.16 Suppose that in 1989 a descendant of Abe Lincoln found a passbook for a savings account opened by Lincoln in 1862. Lincoln opened the account with $100 and made no further deposits or withdrawals. The account paid 3% interest compounded annually.
(a) How much is the account worth in 1989?
(b) How much would it have been worth if the interest rate had been 9%?
(c) How much would it have been worth if the interest had been compounded quarterly at an annual rate of 3%?

Solution **(a)** To use equation (1), we must know the value of P, r, t, and n. According to the passbook, the principal P is $100, the rate r is 0.03, the number n of compounding periods per year is 1, and the number of years t is 127. The value of the account in 1989 is then given by

$$100(1 + 0.03)^{127} = 100(1.03)^{127} = 4269.01 \text{ dollars.}$$

(b) If the interest rate had been 9%, then the value of the account in 1989 would have been the astounding amount of

$$100(1.09)^{127} = 5{,}664{,}547.70 \text{ dollars.}$$

Note that although 9% is three times 3%, the amount of money earned at 9% is a great deal more than three times that earned at 3%.
(c) If the interest rate had been 3%, but compounded quarterly, then we have $P = 100$, $r = 0.03$, $n = 4$, and $t = 127$, yielding an account balance of

$$100\left(1 + \frac{0.03}{4}\right)^{(127)(4)} = 100(1.0075)^{508} = 4451.31 \text{ dollars.}$$

Although the frequency of compounding has quadrupled, the ultimate value of the account has changed very modestly. This example points out that consumers must be cautious when applying common sense to questions about compounding interest. ■

As in the previous example, if a bank pays interest quarterly on an account, then $n = 4$ and the balance in the account after t years is

$$A(t) = P\left(1 + \frac{r}{4}\right)^{4t} = P\left[\left(1 + \frac{r}{4}\right)^{4}\right]^{t}.$$

(Recall that P is the starting balance and r is the annual rate.) If the interest is compounded monthly, then $n = 12$, so the balance is

$$A(t) = P\left(1 + \frac{r}{12}\right)^{12t} = P\left[\left(1 + \frac{r}{12}\right)^{12}\right]^{t}$$

at the end of t years. If the interest is compounded daily, then $n = 365$, and the balance is

$$A(t) = P\left(1 + \frac{r}{365}\right)^{365t} = P\left[\left(1 + \frac{r}{365}\right)^{365}\right]^{t}$$

at the end of t years. Notice that the quantity changing in these expressions for $A(t)$ is

$$\left(1 + \frac{r}{n}\right)^{n}.$$

Only the value of n has changed; n has increased from 4 for quarterly compounding to 365 for daily compounding. If the bank choose to compound the interest every hour, then n would be $24 \times 365 = 8760$. If interest were compounded every minute, then n would be $60 \times 24 \times 365 = 525{,}600$. If we let n approach infinity, then we get "continuous" compounding of interest, which would result in a balance of

$$A(t) = P \lim_{n\to\infty} \left[\left(1 + \frac{r}{n}\right)^{n}\right]^{t}$$

after t years. For convenience we introduce a new symbol h by letting $h = r/n$ (so $n = r/h$). Then $h \to 0$ as $n \to \infty$. Hence

$$\begin{aligned} A(t) &= P \lim_{h\to 0} [(1 + h)^{r/h}]^{t} \\ &= P\left[\lim_{h\to 0} (1 + h)^{1/h}\right]^{rt}. \qquad (2) \end{aligned}$$

In the previous section, we showed that

$$\lim_{h\to 0} (1 + h)^{1/h} = e,$$

so when the interest is compounded continuously the formula for $A(t)$ is given by

$$A(t) = Pe^{rt}.$$

Example 5.17 If \$2000 is invested in an account that pays an 8% ($r = 0.08$) annual rate of interest compounded continuously, how much will be in the account after two years? Compare the balances after two years if the compounding occurs (a) annually, (b) quarterly, (c) monthly, and (d) continuously.

Solution If $A(t)$ is the amount in the account after t years, then

$$A(t) = 2000e^{0.08t}.$$

Hence,
$$A(2) = 2000e^{(0.08)\times 2} = 2347.02 \text{ dollars}.$$

Table 5.10 shows the balance for the various compounding frequencies. ■

Table 5.10
Growth of $2000 in two years at 8% interest

Compounding frequency	*Balance*
annually	2332.80
quarterly	2343.31
monthly	2345.77
continuously	2347.02

We have seen that the growth of money in a continuously compounded account can be described by an exponential function. Exponential functions can also model situations involving both *growth* and *decay*.

Recall that for any constant k, if $y = e^{kx}$, then

$$y' = \frac{d}{dx}e^{kx} = ke^{kx} = ky;$$

that is, the derivative of this function is equal to k times the function itself. The function e^{kx} is almost the only function with this property. The word *almost* is used because the functions $y = 4e^{kx}$ or $y = -7e^{kx}$ also satisfy $y' = ky$. In fact, for any constant C, $y = Ce^{kx}$ has the property that $y' = ky$, and no other function satisfies this equation.

The equation $y' = ky$ is an example of a **differential equation**, which gives a relation between the derivative of the function and the function itself. The key observation to many examples in this section is that $y' = ky$ is equivalent to $y = Ce^{kx}$ for some constant C.

Example 5.18 The population of Italy in 1980 was approximately 57 million. Assume the annual rate of increase of the population is 2% of the population, and develop a formula for the population at any time t. From this formula, predict the population of Italy in the year 2000.

Solution Let $P(t)$ denote Italy's population (in millions) t years after 1980. Then $P(0) = 57$. The rate of change of P is $0.02P$ (that is, 2% of P). Therefore,

$$\frac{dP}{dt} = 0.02P.$$

We can write this equation as $P' = 0.02P$, a differential equation of the type discussed above with $k = 0.02$, so we know that

$$P(t) = Ce^{0.02t}$$

for some constant C. Since $P = 57$ when $t = 0$, we have

$$57 = Ce^0 = C,$$

so
$$P(t) = 57e^{0.02t}$$

is the desired formula for the population at any time t. To find the predicted population in the year 2000 (20 years after 1980), we set $t = 20$ and compute

$$P(20) = 57e^{0.02(20)} \approx 85.0.$$

Hence, Italy's population will be approximately 85 million in the year

2000. Note that this prediction will be valid if the growth rate remains at 2% of the population for the entire 20-year period. ■

In Example 5.18 the rate of change of the population is proportional to the population itself. The function describing such a relationship is always exponential: $f(x) = Ce^{kx}$ for some constants C and k.

In the exponential function $y = Ce^{kt}$ the quantity k is called the **growth constant** when it is positive; if k is negative, then the quantity $\lambda = -k$ is called the **decay constant**. Figure 5.12 shows an exponential growth function and an exponential decay function.

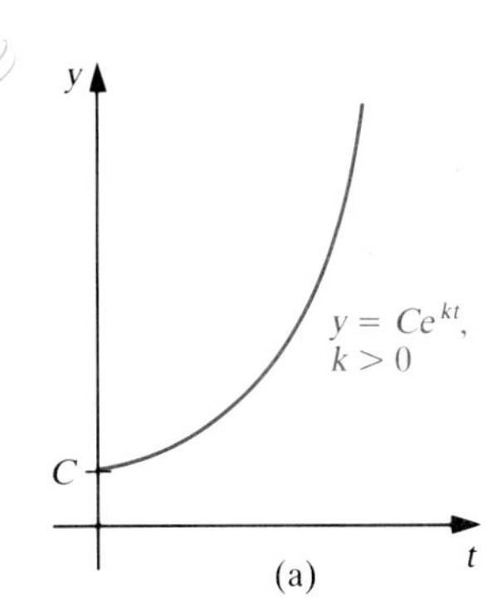

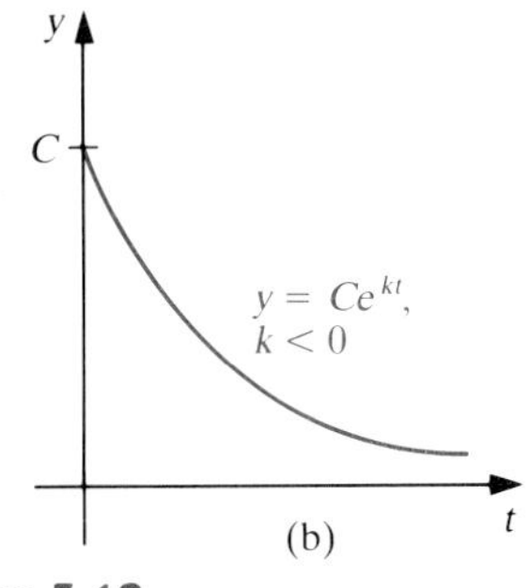

Figure 5.12
(a) Exponential growth function; (b) exponential decay function

Example 5.19 For the radioactive element radium, the rate at which the amount of radium decreases is proportional to the amount of radium present. It takes 1690 years for the amount of radium in a substance to decrease to half of the original amount (its *half-life* is 1690 years). How much of one gram of radium is present after 4000 years?

Solution Let A denote the amount of radium at time t. Since the rate of change of the amount of radium is proportional to the amount present, we have $A' = kA$, where k is a constant of proportionality. Therefore,

$$A = A(t) = A_0 e^{kt}$$

for constants A_0 and k, which may be determined as follows. We know that A is one gram when $t = 0$. Therefore $1 = A_0 e^0 = A_0$. Thus $A = e^{kt}$. Next, $A = 0.5$ gram when $t = 1690$ and so

$$\tfrac{1}{2} = e^{1690k}.$$

Taking the logarithm of both sides gives $\ln \frac{1}{2} = 1690k$, and since $\ln \frac{1}{2} = -\ln 2$,

$$k = \frac{-\ln 2}{1690} \approx -0.00041.$$

Therefore,
$$A = A(t) = e^{-0.00041t}.$$

To find the amount of radium present after 4000 years, we set $t = 4000$ to obtain

$$A(4000) = e^{-0.00041 \times 4000} \approx 0.194.$$

Hence, 0.194 gram of radium is present after 4000 years. ■

Example 5.20 Iodine 131, which is produced by nuclear explosions, has a half-life of 8 days. In the spring of 1986 a nuclear explosion occurred at a nuclear power plant in Chernobyl, near Kiev in the Soviet Union. Suppose that the hay from nearby farms was contaminated by the fallout and could not be fed to dairy cows, since the iodine 131 would contaminate their milk. The hay was found to contain six times the allowable level of iodine 131 shortly after the explosion. How long must the hay be stored to be certain that the amount of iodine 131 is reduced to an acceptable level?

Solution As with all radioactive decay, the rate of change of the amount of iodine 131 is proportional to the amount itself. Let $A = A(t)$ denote the amount of iodine in the hay t days after the explosion. Then $A' = kA$ for some constant of proportionality k. Hence,

$$A = A_0 e^{kt}$$

for some constant A_0. The constant k can be obtained from the half-life. If A_0 is the amount present at time zero, then $0.5A_0$ is the amount present at time $t = 8$. Thus we have

$$0.5A_0 = A_0 e^{8k}.$$

Canceling A_0 and taking the logarithm of both sides gives $\ln 0.5 = 8k$, or $k \approx -0.086643$. Thus

$$A(t) = A_0 e^{-0.086643t}.$$

Although we do not have enough information to find the constant A_0 directly, we can nevertheless find the time required for the hay to be safe. Let L denote the acceptable level of iodine 131. Then at $t = 0$, the amount of iodine 131 is $6L$. We seek the time t when $A(t) = L$, because this is when the amount of iodine 131 present has returned to a safe level. Thus $A(0) = 6L$, so $6L = A_0$, and we seek t so that

$$L = A(t) = 6Le^{-0.086643t}.$$

The quantity L may be canceled; we then divide by 6 and take the logarithm of both sides to get

$$\ln \tfrac{1}{6} = \ln e^{-0.086643t} = -0.086643t.$$

Solving for t and using a calculator or tables from Appendix B gives the desired time of $t = 20.68$ days. Hence, the hay must be stored for three weeks before it can be fed safely to cows. ■

The mathematical description of radioactive decay is an important tool of anthropologists, historians, geologists, and social scientists, since it can be used to date materials accurately. The most common radioactive substance used in this process is carbon 14, a radioactive isotope of carbon. Carbon 14 is produced by the interaction of cosmic rays on nitrogen in the atmosphere. This process has been going on for thousands of years, so that the ratio of carbon 14 to carbon 12 in the atmosphere is relatively constant. The ratio of carbon 14 to carbon 12 found in a plant or an animal matches that in the atmosphere as long as the organism is alive and is absorbing carbon. When that plant or animal dies, the carbon 14 begins to decay; the carbon 12 does not decay. By determining the ratio of carbon 14 to carbon 12 present in a specimen, scientists can determine the date of death.

Example 5.21 The half-life of carbon 14 is 5730 years. How old is a bone uncovered by excavation if it contains one-ninth the expected amount of carbon 14?

Solution The decay of a radioactive substance is an exponential process. Thus, the amount A of carbon 14 present after t years is given by $A = A_0e^{kt}$. The half-life of carbon 14 can be used to determine k. Since $A_0 = A(0)$ and half of the carbon 14 in a sample will decay in 5730 years, we know that $A(5730) = 0.5A_0$. Hence

$$0.5A_0 = A_0e^{5730k}.$$

Taking the logarithm of both sides, we conclude that $k = -(\ln 2)/5730 = -0.000121$. Consequently,

$$A(t) = A_0e^{-0.000121t}.$$

To solve the problem we must determine t such that the fraction of carbon 14 at time t is one-ninth of the amount present at time zero. Let $t = 0$ denote the time of death of the specimen and let A_0 be the amount of carbon 14 at time zero. We know that t years later the amount is $\frac{1}{9}A_0$. Hence, we seek the value of t such that

$$\tfrac{1}{9}A_0 = A_0e^{-0.000121t}.$$

Canceling A_0, taking the logarithm of both sides, and using $\ln \frac{1}{9} = -\ln 9 = -2.1972$, we have $-2.1972 = -0.000121t$, so we conclude that $t = 18,159$. Consequently, the bone is about 18,200 years old. ■

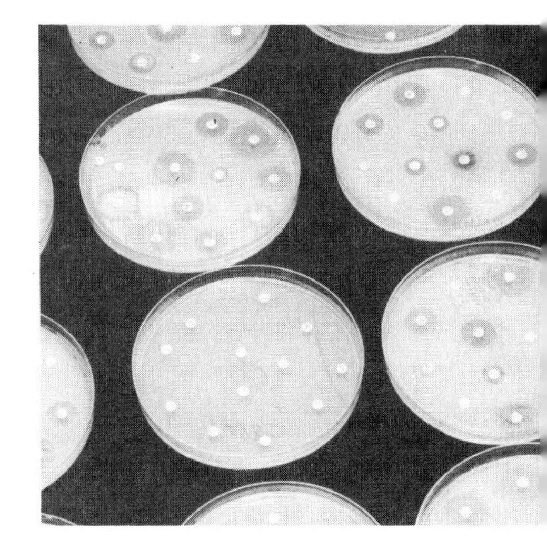

Example 5.22 The population of a certain bacteria culture grows at a rate proportional to its size, doubling every 36 hours. The population weighs 30 grams when an accident kills 90% of the culture. An experiment using the bacteria requires at least 20 grams. How much time is required before the experiment can be performed?

Solution Let $P(t)$ denote the bacteria population (in grams) at time t, with $t = 0$ denoting the moment of the accident. Thus, the population at time zero is $(0.10)(30) = 3$ grams (10% survive). The statement that the rate of growth is proportional to the size may be written as

$$\frac{dP}{dt} = kP$$

for some constant of proportionality k. Therefore, the population is given by

$$P(t) = P_0e^{kt},$$

where P_0 is some constant. In fact, since $P(0) = P_0$, we know that P_0 is the starting population, 3 grams. Since the population doubles every 36 hours, the population at time $t = 36$ is 6 grams. Using this value for $P(t)$ along with $t = 36$ in the preceding formula gives

$$6 = 3e^{36k}.$$

Solving for k by taking logarithms yields $k = \ln \frac{6}{3}/36 \approx 0.019254$. We seek the value of t such that $P(t) = 20$, that is,

$$20 = P(t) = 3e^{0.019254t}.$$

This equation gives $t = \ln(\frac{20}{3})/0.019254 \approx 98.53$ hours, or slightly more than 4 days 2 hours. ■

Suppose that a hospital patient receives medicine intravenously from a bottle suspended above the bed. The medicine is added to the patient's body at the rate of r milligrams per minute, and the body absorbs the medicine at a rate proportional to the amount of medicine present in the bloodstream. Let $A(t)$ be the amount of medicine in the patient's bloodstream t minutes after starting the intravenous injection. Since r milligrams are added each minute and an amount proportional to A is taken out (absorbed) each minute,

$$\frac{dA}{dt} = r - kA$$

for some constant k (called the *absorption constant*). This equation is different from those in previous examples; the r term is new. However, by introducing the quantity $u(t) = r - kA(t)$, we will obtain an equation of the form used in previous examples. Since r and k are constants,

$$\frac{du}{dt} = -k\frac{dA}{dt}.$$

Since $dA/dt = r - kA = u$, we replace dA/dt in this equation with u, obtaining

$$\frac{du}{dt} = -ku.$$

As we have seen before, the solution to this differential equation is $u(t) = Ce^{-kt}$, where $C = u(0)$. Solving $u = r - kA$ for A gives

$$A = \frac{1}{k}(r - u).$$

Since $u = Ce^{-kt}$, the solution to $dA/dt = r - kA$ is

$$A(t) = \frac{1}{k}(r - Ce^{-kt}).$$

Therefore, the amount of medicine in the patient's system when receiving r milligrams per minute for t minutes is given by

$$A(t) = \frac{1}{k}(r - Ce^{-kt})$$

where k and C are constants.

Example 5.23 Suppose a patient receives medicine intravenously at the rate of 50 milligrams per minute and the absorption constant for this drug, given the patient's age, sex, and weight, is 0.0192. Determine the maximum amount of drug in the patient's blood and the time required for the drug level in the blood to reach 75% of the maximum.

Solution Using the notation in the preceding discussion, we have $k = 0.0192$ and $r = 50$. Hence the amount of drug in the blood at time t is given by

$$A(t) = \frac{1}{0.0192}(50 - Ce^{-0.0192t})$$

for some constant C. At time zero (when the intravenous injection begins), the amount of medicine in the blood is zero. Setting $A(0)=0$ gives $C=50$; thus

$$A(t)=\frac{50}{0.0192}(1-e^{-0.0192t})=\frac{50}{0.0192}-\frac{50}{0.0192}e^{-0.0192t}.$$

To find the maximum value of this function, we differentiate with respect to t to get $A'(t)=50e^{-0.0192t}$. Since the exponential function is always positive, $A'(t)>0$ for all t. Consequently, $A(t)$ is an increasing function for all t and $A(t)$ cannot exceed $\lim_{t\to\infty} A(t)$ if this limit exists. This limit does exist because the term containing $e^{-0.0192t}$ goes to zero as $t\to\infty$. Thus the largest amount of drug in the body is $\lim_{t\to\infty} A(t)=50/0.0912\approx 2604.2$ milligrams.

To find the time t when $A(t)=0.75\times 2604.2$, we solve

$$0.75(2604.2)=A(t)=2604.2(1-e^{-0.0192t})$$

or

$$0.75=1-e^{-0.0192t}.$$

Thus $e^{-0.0192t}=0.25$. Taking logarithms and solving for t gives $t=(\ln 4)/0.0912\approx 72.2$. Therefore, in slightly more than 72 minutes the drug will reach 75% of its maximum level. ■

Example 5.24 A monitoring system indicates that the CO_2 level in a movie theater immediately after a movie ends is 4%. The air-conditioning system exchanges 1000 ft^3 of air per minute and the fresh air it pumps into the theater contains 0.02% CO_2. The volume of the theater is 10,000 ft^3. How long will it take for the CO_2 level to drop from 4% to 1%?

Solution Let $A(t)$ denote the cubic feet of CO_2 in the theater t minutes after the patrons have left. Since the theater has a volume of 10,000 ft^3 and 4% of this volume is CO_2, we have $A(0)=0.04(10{,}000)=400\ \text{ft}^3$. The amount of CO_2 is changed by two processes. By exchanging 1000 ft^3 of air per minute, the air-conditioning system (1) adds $0.0002\times 1000=0.2\ \text{ft}^3$ per minute and (2) removes all of the CO_2 contained in the 1000 ft^3 of air removed. Since there are $A(t)$ ft^3 of CO_2 in the entire theater, the 1000 ft^3 removed will contain $(1000/10{,}000)A(t)=0.1A(t)$ ft^3 of CO_2. Consequently, the rate of change of $A(t)$ is given by

$$\frac{dA}{dt}=0.2-0.1A.$$

This is the equation $dA/dt=r-kA$ with $r=0.2$ and $k=0.1$. The solution is

$$A(t)=\frac{1}{0.1}(0.2-Ce^{-0.1t})=2-10Ce^{-0.1t}$$

for some constant C. We can find C by using $A(0)=400$. Thus

$$400=A(0)=2-10C,$$

yielding $C=-39.8$. The amount of CO_2 in the theater at time t is then given by

$$A(t) = 2 + 398e^{-0.1t}.$$

We want to know the value of t when the CO_2 content is 1% of the theater volume, that is, the time when the CO_2 amount is $0.01 \times 10{,}000 = 100 \text{ ft}^3$. We solve

$$100 = 2 + 398e^{-0.1t}$$

for t. This equation is equivalent to $e^{-0.1t} = \frac{98}{398}$. Taking the logarithm of both sides and multiplying by -10 gives $t = 10 \ln\left(\frac{398}{98}\right) \approx 14.01$ minutes; in about 14 minutes the CO_2 content will drop to 1%. ■

Foresters need to predict the average diameter D (in inches) of trees in a mature stand of a particular species. In a given area, D is a function of the number N of trees per acre according to *Reineke's equation*

$$\ln N = k - a \ln D$$

with constants k and a dependent upon the species.

Example 5.25 Solve Reineke's equation for D. For the loblolly pine in a certain region the constants for Reineke's equation are $a = 1.67$ and $k = 9.53$.
(a) Calculate the average diameter of a loblolly pine in a mature stand containing $N = 100$ trees per acre.
(b) Find the average diameter when $N = 500$ trees per acre.
(c) Find the rate of change of D with respect to N when $N = 100$ and when $N = 500$.

Solution **(a)** Isolating D on one side of Reineke's equation, we get

$$\ln D = \frac{k - \ln N}{a}.$$

Hence

$$D = e^{(k - \ln N)/a} = e^{(9.53 - \ln N)/1.67}.$$

When $N = 100$,

$$D = e^{(9.53 - \ln 100)/1.67}$$
$$\approx e^{2.95} \approx 19.09 \text{ inches.}$$

(b) Similarly, if $N = 500$, then $D = 7.28$ inches. Hence the more trees per acre, the smaller the average diameter of the trees.
(c) We find the rate of change dD/dN of D with respect to N by differentiating both sides of

$$\ln D = \frac{k - \ln N}{a} = \frac{k}{a} - \frac{1}{a} \ln N$$

with respect to N. The result is

$$\frac{1}{D}\frac{dD}{dN} = -\frac{1}{a}\frac{1}{N}$$

so that

$$\frac{dD}{dN} = -\frac{D}{aN}.$$

Since $a = 1.67$ and $D = 19.09$ when $N = 100$, we find that

$$\frac{dD}{dN} = -\frac{19.09}{(1.67)(100)} \approx -0.11.$$

Consequently, the *marginal diameter* of the pines is -0.11 inch for 100 trees per acre. Hence when N is near 100, each additional tree planted per acre will result in a decrease of about one-ninth inch in the average diameter of mature trees.

When $N = 500$ $(D = 7.28)$, the rate of change is $dD/dN = -0.01$ inch. The average diameter of mature trees is much smaller for 500 trees per acre than for 100 trees per acre, but the effect on the diameter of mature trees caused by planting an additional tree per acre is much less. In fact, even though five times as many trees have been planted, the effect of each additional tree per acre is 1/11 as much. ■

Exercises

1. The rate of increase of the population of a certain city is proportional to the population. The population increases from 30,000 to 40,000 in 10 years. What will be the population in 30 years?
2. The rate of decay of radium is proportional to the amount present at any time. If 40 milligrams of radium are present now and its half-life is 1690 years, how much radium will be present 75 years from now?
3. A radioactive medicine loses half its radioactivity in three days. Assuming an exponential decay of the radioactivity, what proportion is lost after one day? After three days? After nine days?
4. In a certain culture, the growth rate of bacteria is proportional to the amount present. If 3000 bacteria are present initially and the amount doubles in one hour, how many bacteria will be present in four hours? In six hours?
5. If an amount of money invested doubles in eight years at interest compounded continuously, how long will it take for the original amount to triple? What rate of interest compounded annually is equivalent to 8% compounded continuously over a period of one year?
6. A professor predicts that a typical student will be able to obtain a grade of approximately y percent on the final examination if he or she studies effectively for x hours, where $y = 100 - 75e^{-0.15x}$. How many hours of study will guarantee an A if an 85% grade is required on the final exam?
7. How old is a bone found in an excavation if it has $\frac{1}{10}$ of the expected amount of carbon 14? If it has $\frac{1}{5}$ of it? If it has $\frac{1}{100}$ of it?
8. A sample from an American Indian site has a third of its carbon 14 left. How old is the sample?
9. What percent of the original carbon 14 would be left in a specimen 300,000 years old?
10. A lake contains two bacteria; the population of bacteria A is presently 1000 times that of bacteria B. Furthermore, it is known that the population of bacteria A doubles every 12 days and that of bacteria B triples every 9 days. Determine the time when the two bacteria populations will be equal. (*Hint*: This problem contains two populations $P_A(t)$ and $P_B(t)$ with growth rates k_A and k_B. At $t = 0$, $P_A(0) = 1000P_B(0)$. Determine k_A and k_B and then t when $P_A(t) = P_B(t)$.)

Exer. 11–15: The earth's population is increasing at a rate of 2% per year. Assume there are no limits to growth. The population was 3.5 billion in 1970.

11. When will the population reach 8 billion?
12. When will the population reach 10 billion?
13. When will the population reach 20 billion?
14. How soon will the population double?
15. How soon will the population triple?

Exer. 16–21: If the earth cannot support a population of more than 15 billion people, and the rate of population expansion is proportional to the growth potential (the difference between 15 billion and the present population), then we have $dy/dt = k(15 - y)$, where y is the population (in billions) at time t. Assume that the population in 1975 was 4 billion and that $k = 0.01$.

16. Find the predicted population in 1980.
17. Find the predicted population for 1990.
18. Find the predicted population for 2000.
19. When will the population reach 8 billion?
20. When will the population reach 10 billion?
21. When will the population reach 12 billion?

22. In 1963, Colombia's population was 15 million. In 1980, its population was 28 million. At what

percentage rate is the population increasing annually? What do you project Colombia's population will be in the year 2000?

23. In a psychological experiment, a mouse runs through a maze with food at the end. The mouse requires x seconds to reach the food after running the maze t times, where
$$x = 30 + 20e^{-0.2t}.$$
How many times must the mouse run the maze so that $x = 34$ seconds?
24. How long will it take for an amount invested at 5% interest compounded continuously to double in value?
25. Suppose that the population of a certain town grows at a rate proportional to the population, and that in 1950 the population was 40,000 and in 1960 it was 44,000. Estimate the population in 1990.
26. In 1900 the population of the United States was about 76 million and in 1960 it was about 179 million. Based on these figures, what would the 1970 population have been expected to be?
27. The population of the world in 1930 was about 2 billion and in 1970, about 3 billion. Based on these figures, what is the expected world population in the year 2000?
28. Based on the data given in Exercise 27, when do we expect the world population to be 10 billion?
29. The half-life of strontium 90 is 28 years. Suppose that an island in the Pacific Ocean is contaminated by fallout from an atmospheric nuclear test. The level of strontium 90 is measured at 50 times the level considered safe for humans. How many years will elapse before the island is safe for habitation?
30. Refer to Example 5.23. Suppose the medicine is administered at the rate of 50 milligrams per minute and the absorption constant is $k = 0.03$. Determine the maximum value of drug in the blood and the time needed for the drug concentration to reach 75% of its maximum value.
31. Consider the task of learning random sequences of letters or numbers or a list of nonsense syllables. Experiments show that, within limits, the rate at which this material can be learned is proportional to the amount of material not yet learned. Let $L(t)$ be the percentage of material memorized after t minutes of study. Then $100 - L(t)$ is the material not yet learned, so $L'(t) = k[100 - L(t)]$ for some constant k, and $L(0) = 0$. If 10% of the material is learned in 5 minutes, how long will it take to learn 80%? How long will it take to learn 90%?
32. Work Exercise 31 with the information that 25% of the material is learned in 5 minutes.
33. Refer to Example 5.25. For the slash pine in a certain region the constants of Reineke's equation are $a = 1.97$ and $k = 10.21$. Find the average diameter of slash pines in a mature stand of 50 trees per acre. Find the rate of change of the average diameter with respect to the number of trees per acre, N, when $N = 50$ and when $N = 200$.
34. Work Exercise 33 for the longleaf pine, with Reineke's equation constants $a = 1.56$ and $k = 9.18$.

Summary

A function of the type $y = b^x$, where b is any positive real number (except 1), is an **exponential function**. The rules for computing with exponential functions are as follows. If $a > 0$ and $b > 0$ and x, y are real numbers, and n, p are positive integers, then

1. $b^x b^y = b^{x+y}$
2. $\dfrac{b^x}{b^y} = b^{x-y}$
3. $(b^x)^y = b^{xy}$
4. $a^x b^x = (ab)^x$
5. $\dfrac{a^x}{b^x} = \left(\dfrac{a}{b}\right)^x$
6. $a^0 = 1$
7. $a^{-x} = \dfrac{1}{a^x}$
8. $\sqrt[n]{a^p} = a^{p/n}$

The number e is defined as that particular value of b such that the slope of the curve $y = b^x$ at $x = 0$ is 1. It is given by $e = 2.71828184\ldots.$ Then
$$\frac{d}{dx}e^x = e^x \quad \text{for all } x.$$

The function $y = \ln x$ is defined by the relationship $y = \ln x$ is equivalent to $e^y = x$. Thus, for a given number x, the solution y to the equation $e^y = x$ is $y = \ln x$. The properties of the logarithm function are as follows. For $a, b > 0$,

1. $e^{\ln u} = u$ for all $u > 0$
2. $\ln e^u = u$ for all u
3. $\ln 1 = 0$, $\ln e = 1$
4. $\ln (ab) = \ln a + \ln b$
5. $\ln \dfrac{a}{b} = \ln a - \ln b$
6. $\ln a^k = k \ln a$

The derivative of the function $e^{g(x)}$ is given by

$$\frac{de^{g(x)}}{dx} = e^{g(x)}\frac{dg(x)}{dx} = g'(x)e^{g(x)}$$

for any differentiable function $g(x)$.

The derivative of the logarithm function $\ln g(x)$ is given by

$$\frac{d \ln g(x)}{dx} = \frac{1}{g(x)}\frac{dg(x)}{dx} = \frac{g'(x)}{g(x)}$$

for any positive differentiable function $g(x)$.

If the amount P is invested at an interest rate of r percent compounded continuously, then the total value of the investment after t years is

$$A(t) = Pe^{rt}.$$

The **differential equation** $dy/dt = ky$ (where k is a constant) describes a situation in which the rate of change dy/dt is proportional to the quantity y. The solution to this differential equation is $y = Ce^{kt}$, where C is a constant.

The solution to the differential equation $dA/dt = r - kA$, where k and r are constants, is

$$A = \frac{1}{k}(r - Ce^{-kt}),$$

where C is a constant.

Review Exercises

Exer. 1–4: Sketch a graph of the function.

1. $y = 3^x$
2. $y = (\frac{1}{4})^x$
3. $y = \ln x$
4. $y = e^{2x}$

Exer. 5–10: Solve the equation for x (do not use a calculator). Express the answer as an expression involving logarithms. For example, the solution to $3^x = 5$ is $x = \ln 5/\ln 3$.

5. $8^{2x} = 30$
6. $3^{3x} = 2$
7. $e^{2x+1} = 4$
8. $e^{2x-1} = 5$
9. $5^{3x+2} = 125$
10. $9^x = 3^{x+1}$

Exer. 11–18: Solve the equation for x (use a calculator or the tables in Appendix B).

11. $\ln x = 12$
12. $e^x = 4$
13. $e^x = 7^{5+x}$
14. $300 = 200(1.07)^{2x}$
15. $20 = 4^{5x}$
16. $e^{\ln x^2} = 8$
17. $3^x = 7^{x+2}$
18. $\ln (4x + 2) = 17$

Exer. 19–30: Find $f'(x)$.

19. $f(x) = e^{2x}$
20. $f(x) = xe^{2x}$
21. $f(x) = 4xe^{-5x}$
22. $f(x) = xe^x - xe^{-x}$
23. $f(x) = \ln (2 - x^2)$
24. $f(x) = x^2 \ln (4e^{3x})$
25. $f(x) = \dfrac{e^{3x}}{\ln 3x}$
26. $f(x) = \sqrt[3]{\ln (2x)}$
27. $f(x) = e^{x+\ln x}$
28. $f(x) = x \ln x$
29. $f(x) = x^{2x}$
30. $f(x) = x^{x \ln x}$

31. What is the amount of money after one year in an account with an initial deposit of \$10,000 at 10% interest compounded continuously?
32. How long will it take for \$1,000,000 invested at a 12% interest rate compounded continuously to be worth \$12,000,000?
33. What rate of interest compounded continuously for one year will yield the same amount of money as 8% compounded quarterly for one year?
34. The number (in hundreds) of bacteria in a certain solution is given by $n(t) = 10(1.08)^t$, where t represents the number of hours after the bacteria is added. What is the rate of change of the number of bacteria after 2 hours? After 10 hours?
35. A wooden carving found in an archaeological excavation contains 19% of the expected amount of carbon 14. How old is the carving?
36. The symptoms of an infection caused by a certain bacteria become noticeable when the bacteria population reaches 10,000. The bacteria population grows at a rate proportional to its size and the population triples in size every 2.5 hours. Assume that an average exposure to the bacteria results in the transfer of 100 cells. How long after exposure will the bacteria's presence produce noticeable symptoms?
37. Let $S(t)$ denote the sales per week of a certain product. After a national advertising campaign begins, the rate of change of the sales per week will be given by

$$\frac{dS}{dt} = 0.037S,$$

where t is time (in weeks) with $t = 0$ corresponding to the start of the campaign. At the start of the campaign, sales are 10,000 units per week.
(a) What will the sales per week be after 4 weeks?
(b) After 8 weeks?
(c) When will the sales level reach 16,000 units per week?
38. Suppose the air-conditioning system is switched on when the CO_2 level in a movie theater reaches 4%.

The system exchanges 1000 ft^3 per minute, and the incoming air contains 0.03% CO_2. Suppose further that the patrons contribute 2 ft^3 of CO_2 per minute, and the volume of the theater is 10,000 ft^3. How long must the system run to reduce the CO_2 level to 1%?

39. A savings account with an initial deposit of \$5000 pays interest at a 6% rate compounded monthly. Each month \$75 is withdrawn. How long will the account have a positive balance? (*Hint:* the rate of change of the balance is affected by the withdrawals (-75) and by the interest earned ($(+(0.06/12) \times$ current balance).)

40. Find the values of m and a so that the line $y = mx$ is tangent to $y = x^2e^{-x}$ at $x = a$. Note that at $x = a$ the two curves have the same y-coordinate (so $ma = a^2e^{-a}$) and the same slope (so m is the value at $x = a$ of the derivative of x^2e^{-x}).

***Facing page*: Given the population density function, the population in any region of a city can be calculated using integration. See Example 6.32.**

Integration

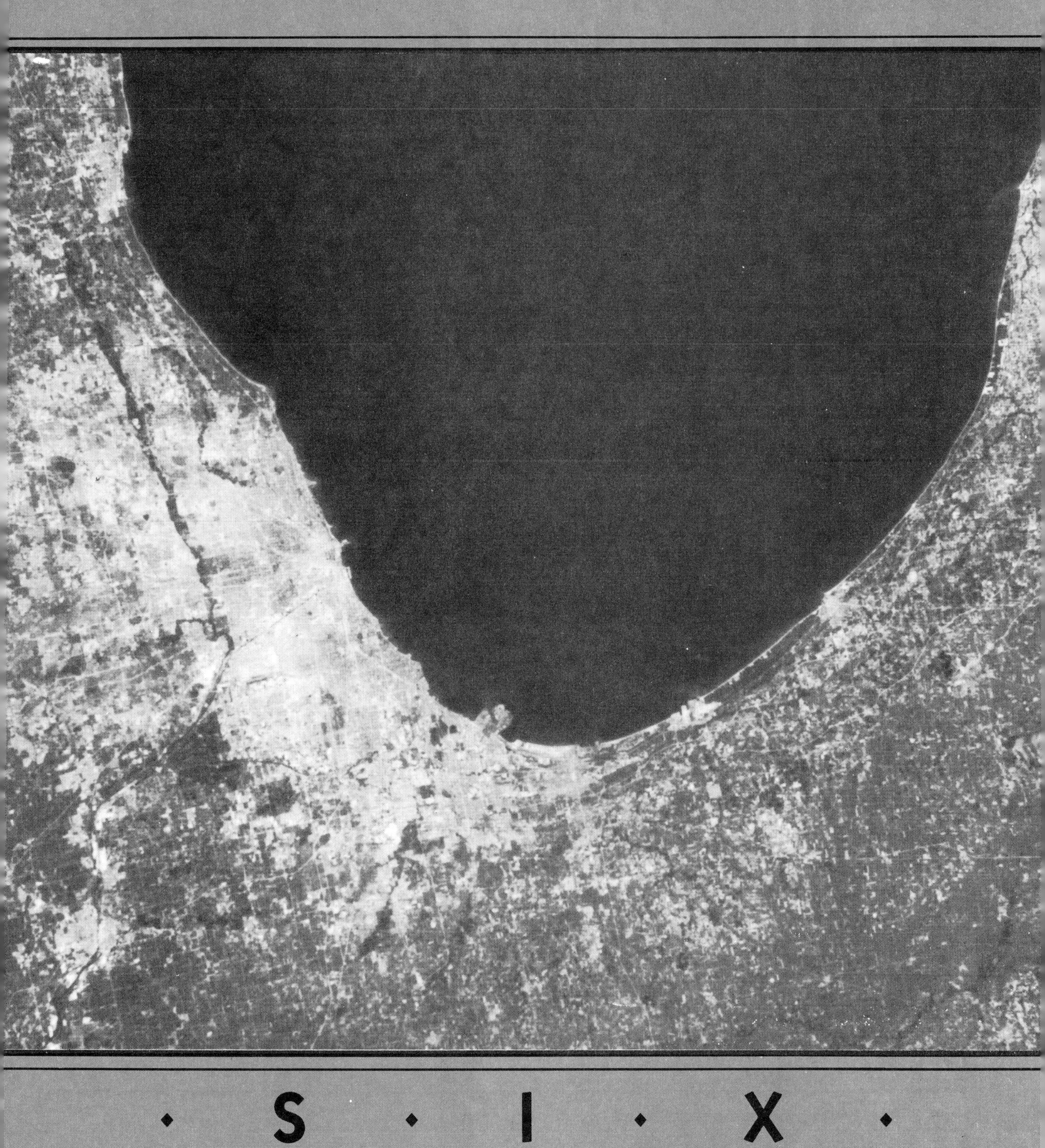

· S · I · X ·

The method of differentiation has enabled us to find the marginal cost of a cost function, the marginal revenue of a revenue function, the velocity of an object given a distance-traveled function, the acceleration of an object given a velocity function, the rate of growth of a quantity, and so forth. By the method of **integration**, which reverses the process of differentiation, we may reverse each of the applications listed above; that is, we can retrieve the cost function from the marginal cost function, the revenue function from the marginal revenue function, the distance-traveled function from the velocity function, the velocity function from the acceleration function, and so on. We can also use integration in many new applications such as the calculation of area and volume.

6.1 Antidifferentiation

In this section we will learn how to reverse the process of differentiation. An **antiderivative** of a function $f(x)$ is a function $F(x)$ such that $F'(x) = f(x)$ for all x.

Example 6.1 **(a)** Three antiderivatives of $f(x) = 2x$ are $F(x) = x^2$, $F(x) = x^2 + 1$ and $F(x) = x^2 - 10$, because $F'(x) = 2x = f(x)$ in each case.
(b) Three antiderivatives of $f(x) = 3x^2$ are $F(x) = x^3$, $F(x) = x^3 + 2$ and $F(x) = x^3 + 10$. (In each case, $F'(x) = 3x^2 = f(x)$).
(c) Three antiderivatives of $f(x) = x^4$ are $F(x) = \frac{1}{5}x^5$, $F(x) = \frac{1}{5}x^5 + 1$ and $F(x) = \frac{1}{5}x^5 + \pi$. ■

Example 6.2 Find an antiderivative of $f(x) = 3x^2 + 2x + 4$.

Solution Since the derivative of a sum is the sum of the derivatives, if we find antiderivatives for $3x^2$, for $2x$, and for 4 separately, then the sum of these three antiderivatives will be an antiderivative of the original function. As we found in the previous example, $F(x) = x^3$ is an antiderivative of $3x^2$ and x^2 is an antiderivative of $2x$. Note that $4x$ is an antiderivative of 4 because $(4x)' = 4$. Hence an antiderivative of $f(x) = 3x^2 + 2x + 4$ is $F(x) = x^3 + x^2 + 4x$. To check this we observe that the derivative of $F(x)$ is $f(x)$, that is, $F'(x) = 3x^2 + 2x + 4 = f(x)$. ■

Suppose that $F(x)$ is one of the antiderivatives of a function $f(x)$ and suppose $f(1) = 4$. What can we say about $F(1)$? Since $F'(1) = f(1) = 4$, the slope of the tangent line to the graph of $F(x)$ at $x = 1$ is 4. We know the graph of $F(x)$ has a slope of 4 at $x = 1$, but we have no idea of the location of the point $(1, F(1))$ in the plane. Consequently, knowing that $F(x)$ is an antiderivative of $f(x)$ tells us the slope of $F(x)$ for any x but gives *no* information about the value $F(1)$.

Example 6.3 Find and graph three antiderivatives of $f(x) = 2x + 3$.

Find an antiderivative of $f(x)$ whose graph passes through the point $(1, 9)$.

Solution Since

$$\frac{d}{dx}(x^2 + 3x) = 2x + 3,$$

we see that $F(x) = x^2 + 3x$ is an antiderivative of $f(x)$. Also $G(x) = x^2 + 3x + 4$ and $H(x) = x^2 + 3x - 2$ are antiderivatives of $f(x)$. These functions are graphed in Figure 6.1. Notice that each of the graphs is a parabola of the same shape (each can be superimposed upon another by moving it up or down). In fact, any parabola of the form $P(x) = x^2 + 3x + C$, where C is a constant, is an antiderivative of $f(x)$. Let $Q(x)$ denote the antiderivative that passes through $(1, 9)$. Then $Q(x) = x^3 + 3x + C$ for some C, and we must choose a value for C that will make $Q(1) = 9$. Since $Q(1) = 1^2 + 3(1) + C$, we have $1 + 3 + C = 9$, so C equals 5. Hence, $Q(x) = x^2 + 3x + 5$ is an antiderivative of $f(x)$ whose graph passes through $(1, 9)$. Note that we can determine a function $F(x)$ completely if we know that *$F'(x) = f(x)$ and a point that $F(x)$ passes through.* ■

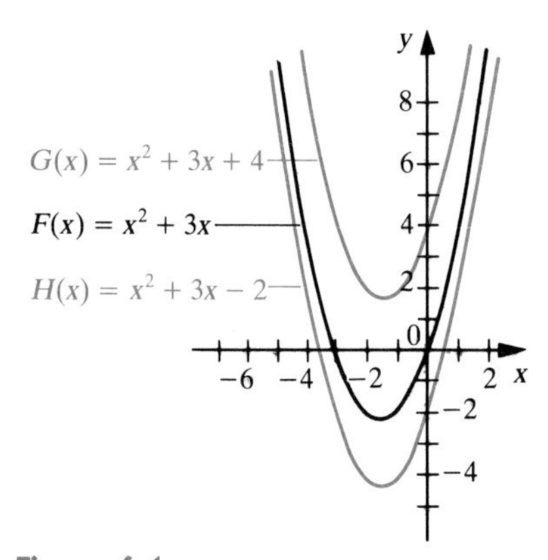

Figure 6.1
Three antiderivatives of $f(x) = 2x + 3$

We have observed that if $F(x)$ is an antiderivative of $f(x)$, then $F(x) + C$ is also an antiderivative of $f(x)$ for any constant C. The reason for this is easy to see: since $F'(x) = f(x)$, then

$$\begin{aligned}[F(x) + C]' &= F'(x) + 0 \\ &= f(x).\end{aligned}$$

Suppose that $F(x)$ and $G(x)$ are any two antiderivatives of $f(x)$. Then, since $F'(x) = G'(x) = f(x)$,

$$\begin{aligned}[G(x) - F(x)]' &= G'(x) - F'(x) \\ &= f(x) - f(x) = 0.\end{aligned}$$

The graph of a function with a zero derivative for all x is neither increasing nor decreasing. In fact, the graph is a horizontal line, so the function is a constant C. Therefore, $G(x) - F(x) = C$ or

$$G(x) = F(x) + C.$$

Consequently, once one antiderivative $F(x)$ for a function $f(x)$ has been found, *all* the antiderivatives of $f(x)$ can be found by adding constants to $F(x)$.

The requirement that the graph of an antiderivative must pass through a given point (a, b) is called an **initial condition**. To find an antiderivative of $f(x)$ that passes through (a, b), we solve $G(a) = F(a) + C = b$ for C, to obtain $C = b - F(a)$. This shows that $G(x) = F(x) + b - F(a)$ is the only antiderivative of $f(x)$ whose graph passes through the given point. In Example 6.3, the initial condition is that an antiderivative of $2x + 3$ must pass through $(1, 9)$. We showed that $Q(x) = x^2 + 3x + 5$ is *an* antiderivative of $2x + 3$ whose graph passes through $(1, 9)$. In fact, it is the *only* such antiderivative of $2x + 3$.

Example 6.4 For a certain product, the marginal cost function for producing x items per day is $M_C(x) = -0.02x + 500$. The cost of producing 20 items per day is $10,600. Find the cost function and determine the cost of producing 100 items per day.

Solution Since the derivative of the cost function is the marginal cost function $M_C(x)$, the cost function is an antiderivative of $M_C(x)$. Every antiderivative of $M_C(x) = -0.02x + 500$ is of the form $-0.01x^2 + 500x + k$, where k is a constant. Hence, the cost function $C(x) = -0.01x^2 + 500x + k$ for some constant k. The initial condition indicates that the graph of $C(x)$ passes through the point $(20, 10600)$. Setting $C(20) = 10{,}600$, we have

$$-0.01(20)^2 + 500(20) + k = 10{,}600$$
$$k = 604.$$

Hence, $C(x) = -0.01x^2 + 500x + 604$. The cost of producing 100 items per day is

$$C(100) = -0.01(100)^2 + 500(100) + 604 = 50{,}504 \text{ dollars.}$$

■

We now turn to the task of finding antiderivatives. Not all functions have antiderivatives, but every continuous function does have antiderivatives. All the functions we will be interested in have antiderivatives everywhere the functions themselves are defined. If $f(x)$ has an antiderivative, then we let the symbol

$$\int f(x)\,dx$$

denote the set of all antiderivatives. We refer to $\int f(x)\,dx$ as the **indefinite integral** of $f(x)$ or just *the integral of* $f(x)$. The function $f(x)$ is called the **integrand**. The dx portion of the symbol indicates that we are reversing the differentiation process with respect to the variable x. We can antidifferentiate with respect to any variable. For example, $\int \cdots dt$ indicates that we are reversing the differentiation process with respect to the variable t. Hence, if $F(x)$ is an antiderivative of $f(x)$, then $\int f(x)\,dx$ can be written as $F(x) + C$, because every antiderivative is of the form $F(x) + C$, where C is a constant and $F(x)$ is any specific antiderivative of $f(x)$.

We have informally reversed the power rule to find antiderivatives, and we now state the power rule for integrals. Since

$$\frac{d}{dx}\left[\frac{1}{n+1}x^{n+1} + C\right] = x^n,$$

we have the following.

Power Rule for Integrals

$$\int x^n\,dx = \frac{1}{n+1}x^{n+1} + C \quad \text{when } n \neq -1.$$

Example 6.5 Find each integral:

(a) $\int x^2\,dx$ **(b)** $\int x^{-4}\,dx$ **(c)** $\int x^{-2/3}\,dx$ **(d)** $\int 1\,dx$

Solution Using the power rule for integrals we have

(a)
$$\int x^2\,dx = \frac{1}{2+1}x^{2+1} + C = \tfrac{1}{3}x^3 + C$$

(b)
$$\int x^{-4}\,dx = \frac{1}{-4+1}x^{-4+1} + C = -\tfrac{1}{3}x^{-3} + C$$

(c)
$$\int x^{-2/3}\,dx = \frac{1}{-\frac{2}{3}+1}x^{-(2/3)+1} + C = 3x^{1/3} + C$$

(d)
$$\int 1\,dx = \int x^0\,dx = \frac{1}{0+1}x^{0+1} + C = x + C$$

We can easily check each case by taking the derivative of the right-hand side of the equality. The result is the original integrand. ■

When $n = -1$ we cannot use the power rule for integrals, since in that case $1/(n+1)$ is undefined. In Section 5.3, we studied the derivative of the natural logarithm function $\ln x$ and found that

$$\frac{d}{dx}\ln x = \frac{1}{x} = x^{-1}.$$

Hence, when $x > 0$, an antiderivative of $1/x$ is $\ln x$ ($\ln x$ is not defined for $x < 0$). If $x < 0$, then $-x > 0$ and

$$\begin{aligned}\frac{d}{dx}\ln(-x) &= \frac{1}{-x}\frac{d(-x)}{dx} \\ &= \frac{1}{-x}(-1) \\ &= \frac{1}{x}.\end{aligned}$$

Hence, when $x < 0$, an antiderivative of $1/x$ is $\ln(-x)$, and when $x > 0$, an antiderivative of $1/x$ is $\ln x$. We may incorporate these two rules into one by writing

$$\int x^{-1}\,dx = \int \frac{1}{x}\,dx = \ln|x| + C \quad \text{for } x \neq 0.$$

In Section 5.2 we found that

$$\frac{d}{dx}e^{kx} = ke^{kx} \quad \text{for a constant } k.$$

Hence, $\int e^{kx}\,dx = (1/k)\,e^{kx} + C$. We may summarize these results as follows.

$$\int x^n\,dx = \frac{1}{n+1}x^{n+1} + C \quad \text{when } n \neq -1.$$

$$\int \frac{1}{x}\,dx = \ln|x| + C \quad \text{when } x \neq 0.$$

$$\int e^{kx}\,dx = \frac{1}{k}e^{kx} + C, \quad \text{for any nonzero constant } k.$$

Suppose that $F(x)$ is an antiderivative of $f(x)$, and $G(x)$ is an antiderivative of $g(x)$. Since $F'(x) = f(x)$ and $G'(x) = g(x)$, we have

$$\begin{aligned}[F(x) + G(x)]' &= F'(x) + G'(x)\\ &= f(x) + g(x).\end{aligned}$$

Thus $F(x) + G(x)$ is an antiderivative of $f(x) + g(x)$. Using integral notation, this fact is written as follows.

The integral of a sum is the sum of the integrals:

$$\int [f(x) + g(x)]\,dx = \int f(x)\,dx + \int g(x)\,dx.$$

Similarly, since $[cF(x)]' = cF'(x)$, if $F(x)$ is an antiderivative of $f(x)$, then $cF(x)$ is an antiderivative of $cf(x)$.

The integral of a constant times a function equals the constant times the integral of the function:

$$\int cf(x)\,dx = c\int f(x)\,dx.$$

These two rules will enable us to use the power rule for integrals to find the integral of many functions, including all polynomials.

Example 6.6 Find $\int (2x^3 - 3x^{1/2} + 5x^{-2/3})\,dx$.

Solution By repeated application of the rules for the integral of a sum and the integral of a constant times a function, and the power rule for integrals, we have

$$\int (2x^3 - 3x^{1/2} + 5x^{-2/3})\,dx = \int 2x^3\,dx + \int (-3)x^{1/2}\,dx + \int 5x^{-2/3}\,dx$$

$$= 2\int x^3\,dx + (-3)\int x^{1/2}\,dx + 5\int x^{-2/3}\,dx.$$

$$= 2(\tfrac{1}{4}x^4 + C_1) + (-3)(\tfrac{2}{3}x^{3/2} + C_2) + 5(3x^{1/3} + C_3)$$

where C_1, C_2, and C_3 are constants. This expression simplifies to

$$\tfrac{1}{2}x^4 - 2x^{3/2} + 15x^{1/3} + 2C_1 - 3C_2 + 5C_3.$$

Since C_1, C_2, and C_3 represent arbitrary constants, we can replace $2C_1 - 3C_2 + 5C_3$ with the single constant C. Therefore,

$$\int (2x^3 - 3x^{1/2} + 5x^{-2/3})\,dx = \tfrac{1}{2}x^4 - 2x^{3/2} + 15x^{1/3} + C.$$ ■

As illustrated in the previous example, when we integrate a sum, it is more efficient to add one constant at the end rather than adding a constant for each term.

Example 6.7 Find all functions $f(x)$ such that the slope of the graph of $f(x)$ at $(x, f(x))$ is $2x - 5$ for all x.

Solution We know that a function $f(x)$ satisfies the requirement if and only if $f'(x) = 2x - 5$. Now $\int (2x - 5)\,dx$ represents all such functions and

$$\int (2x - 5)\,dx = x^2 - 5x + C.$$

Hence, the collection of all functions of the form $f(x) = x^2 - 5x + C$ for some constant C satisfies the stated requirement. ■

Example 6.8 Find

$$\int \left(\frac{1}{\sqrt[3]{x^4}} + \frac{5}{x} + 8e^{4x}\right) dx.$$

Solution Recall that $1/\sqrt[3]{x^4} = x^{-4/3}$. Using our integration rules, we have

$$\int \left(\frac{1}{\sqrt[3]{x^4}} + \frac{5}{x} + 8e^{4x}\right) dx = \int \frac{1}{\sqrt[3]{x^4}}\,dx + \int \frac{5}{x}\,dx + \int 8e^{4x}\,dx$$

$$= \int x^{-4/3}\,dx + 5\int \frac{1}{x}\,dx + 8\int e^{4x}\,dx$$

$$= \frac{x^{-(4/3)+1}}{-4/3+1} + 5\ln|x| + 8\cdot\tfrac{1}{4}e^{4x} + C$$

$$= -3x^{-1/3} + 5\ln|x| + 2e^{4x} + C.$$ ■

Exercises

Exer. 1–10: Verify that $F(x)$ is an antiderivative of $f(x)$ and find an antiderivative of $f(x)$ other than $F(x)$.

1. $f(x)=2x;\quad F(x)=x^2$
2. $f(x)=3x^2+2;\quad F(x)=x^3+2x+3$
3. $f(x)=5x-3;\quad F(x)=\frac{5}{2}x^2-3x+1$
4. $f(x)=-0.04x+600;\quad F(x)=-0.02x^2+600x$
5. $f(x)=3x^2+2x^{-2}+13;\quad F(x)=x^3-2x^{-1}+13x+17$
6. $f(x)=-0.003x^2+2x+250;$ $F(x)=-0.001x^3+x^2+250x+200$
7. $f(x)=\dfrac{-5}{(2x-1)^2};\quad F(x)=\dfrac{3x+1}{2x-1}$
8. $f(x)=\dfrac{2x+2}{x^2+2x};\quad F(x)=\ln(x^2+2x),\quad x>0$
9. $f(x)=\dfrac{1}{x^2-1};\quad F(x)=\frac{1}{2}\ln\left(\dfrac{x-1}{x+1}\right),\quad |x|>1$
10. $f(x)=e^{5x};\quad F(x)=\frac{1}{5}e^{5x}+9$
11. For Exercise 1, sketch the graph of $F(x)$ on the same graph with the antiderivative that you found, and observe their similarities.
12. For Exercise 3, sketch the graph of $F(x)$ on the same graph with the antiderivative that you found, and observe their similarities.
13. Find an antiderivative of $f(x)=3x^2+2$ whose graph passes through the point (2, 22) (see Exercise 2).
14. Find an antiderivative of $f(x)=-5/(2x-1)^2$ whose graph passes through the point (3, −17) (see Exercise 7).

Exer. 15–26: Find an antiderivative of $f(x)$.

15. $f(x)=x^3+x$
16. $f(x)=3x-2x^2$
17. $f(x)=4-2x^2$
18. $f(x)=x^{16}+3x^3$
19. $f(x)=x^{1/2}+3x^2$
20. $f(x)=3x^{3/4}+16$
21. $f(x)=4x^{-1/2}+x^2$
22. $f(x)=x^{-7/4}+2x^{2/3}$
23. $f(x)=3x+\dfrac{4}{x}$
24. $f(x)=\sqrt[4]{x^3}-\dfrac{7}{x}+e^{-x}$
25. $f(x)=4e^{2x}+e^{3x}$
26. $f(x)=6e^{3x}+\dfrac{2}{x}$
27. Find the cost function for the production of x items per day if the marginal cost function for producing x items per day is $M_C(x)=0.04x+600$ and the cost of producing 40 items per day is $24,500. Find the cost of producing 100 items per day.
28. Find the revenue function for selling x items if the marginal revenue function is $M_R(x)=-0.003x^2+0.2x+250$ and the revenue from the sale of 200 items is $30,000.
29. Find the cost function for a manufacturing company if marginal cost is $M_C(x)=5+\frac{1}{50}x$ in thousands of dollars per production run and the fixed cost is $50,000.
30. Find the cost function for a company if its marginal cost is given by $M_C(x)=1+0.2x$, where x is measured in units of 5000 items and $C(x)$ is measured in hundreds of dollars. The fixed cost is $1000.
31. Find the revenue function if the marginal revenue function is $M_R(x)=10+2x-1.2x^2$ in thousands of dollars per production run and the revenue from five production runs is $25,000.

Exer. 32–49: Find the integral and check your answer by differentiating.

32. $\int x\,dx$
33. $\int x^3\,dx$
34. $\int x^{1/2}\,dx$
35. $\int x^{-3}\,dx$
36. $\int x^{-5/2}\,dx$
37. $\int 5\,dx$
38. $\int (3w^2+5w)\,dw$
39. $\int (2t^3-3t^2+5)\,dt$
40. $\int (3r^{-5/2}+2r^{1/2}+7)\,dr$
41. $\int (5u^{-2}+u^{-1/2}+u^{5/3})\,du$
42. $\int \sqrt{x}\,dx$
43. $\int \dfrac{1}{\sqrt[3]{v^2}}\,dv$
44. $\int (4\sqrt{u^3}+\sqrt[3]{u}+2u^3)\,du$
45. $\int \dfrac{7}{x}\,dx$
46. $\int \left(2x^2-\dfrac{4}{x}\right)dx$
47. $\int 3x^{-1}\,dx$
48. $\int 6e^{-2x}\,dx$
49. $\int 10(e^{5t}+t^2)\,dt$

Exer. 50–61: Find the antiderivative of $f(x)$ that passes through the given point.

50. $f(x)=x^2;\quad (3, 12)$
51. $f(x)=x^{-1/3};\quad (8, 6)$
52. $f(x)=2x^3+3x^2+1;\quad (2, 15)$
53. $f(x)=x^{-3/2}+2x^{3/2}+4x^3;\quad (4, 7)$
54. $f(x)=2+(1/x);\quad (e, 2e)$
55. $f(x)=\sqrt[3]{x}+2;\quad (8, 24)$
56. $f(x)=x-x^{-1};\quad (2, 0)$
57. $f(x)=2x-4;\quad (0, 0)$
58. $f(x)=6x^2+x^{-2}\quad (1, 6)$
59. $f(t)=t-2t^2+4t^3;\quad (0, 4)$
60. $f(x)=4e^{2x};\quad (\ln 5, 55)$

61. $f(x) = \frac{1}{x} + e^x$; $(1, 2e)$

Exer. 62–67: Find all functions $f(x)$ such that the slope of the graph of $f(x)$ at $(x, f(x))$ is as indicated.

62. 3
63. $x^2 + 2$
64. $\sqrt{x}$
65. $2 - x$
66. $x^{-1/3}$
67. $-x^3$

Exer. 68–73: Find the integral and check your answer by differentiating. (*Hint:* Simplify the expression before finding the integral.)

68. $\int (3x^2 + 2)x\,dx$
69. $\int x^2(2\sqrt{x} + x)\,dx$
70. $\int (x + 1)^2\,dx$
71. $\int \frac{x^2 + 1}{\sqrt{x}}\,dx$
72. $\int (x^{1/3} + 2)(x^{-1/2} + x^2)\,dx$
73. $\int \frac{3x^2 + 1}{x}\,dx$

74. Suppose that there are specific constants α and β such that

$$\int x^2e^x\,dx = x^2e^x + \alpha xe^x + \beta e^x + C,$$

where C is the constant of integration. Find α and β by differentiation.

Applications of Antiderivatives

6.2

We have seen that we can find a cost or revenue function from the marginal cost or marginal revenue function and an initial condition. We can apply the same procedure to the applications we considered in Section 2.6 on rate of change. We now consider some of these examples in detail.

Example 6.9 Suppose the marginal cost of producing recreational vehicles (RV's) is

$$M_C(x) = 0.06x^2 - 0.4x + 1,$$

where x is the number of RV's (in thousands) and C is the cost (in millions of dollars). Suppose the fixed cost (the cost for producing zero RV's) is \$1 million. Find the cost function $C(x)$ and determine the cost for producing 40,000 items ($x = 40$).

Solution The cost function is an antiderivative of the marginal cost, so for some constant k

$$\begin{aligned} C(x) &= \int (0.06x^2 - 0.4x + 1)\,dx \\ &= 0.06\left(\frac{x^3}{3}\right) - 0.4\left(\frac{x^2}{2}\right) + x + k \\ &= 0.02x^3 - 0.2x^2 + x + k. \end{aligned}$$

The fixed cost $C(0)$ is \$1 million. Since cost is measured in units of \$1 million, we have $C(0) = 1$. Hence

$$1 = C(0) = 0.02(0) - 0.2(0) + (0) + k.$$

Thus $k = 1$, and the cost function is

$$C(x) = 0.02x^3 - 0.2x^2 + x + 1.$$

The cost of producing 40,000 RV's is

$$C(40) = 0.02(40)^3 - 0.2(40)^2 + 40 + 1 = 1001.$$

Hence, the cost of producing 40,000 RV's is \$1,001,000,000. ■

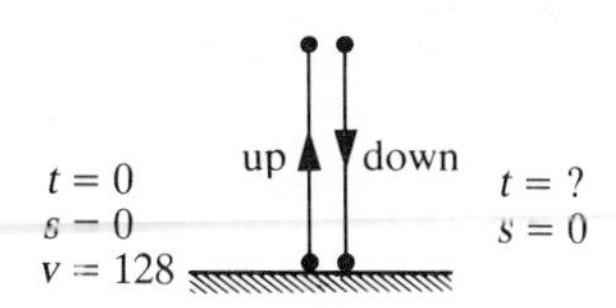

Figure 6.2
Path of a thrown rock

Example 6.10 A rock is thrown upward from the ground with an initial velocity of 128 ft/sec. Figure 6.2 shows the path of the rock during its flight. We know that the acceleration of the rock due to gravity is 32 ft/sec^2 in the downward direction.
(a) Find the vertical velocity function $v(t)$ and the height function $s(t)$.
(b) How high will the rock travel?
(c) How many seconds will it take for the rock to hit the ground?

Solution **(a)** Assume that the rock is thrown at time $t = 0$ and that upward is the positive direction. Then $v(0) = 128$. Since the acceleration is downward, $a(t) = -32$. From $v'(t) = a(t)$, we have

$$\begin{aligned} v(t) &= \int a(t)\,dt \\ &= \int -32\,dt \\ &= -32t + C \end{aligned}$$

for some constant C. Since $v(0) = 128$, we conclude that $C = 128$. Therefore,

$$v(t) = -32t + 128.$$

To find $s(t)$, recall that $s'(t) = v(t)$, so we have

$$\begin{aligned} s(t) = \int v(t)\,dt &= \int (-32t + 128)\,dt \\ &= -16t^2 + 128t + C_1 \end{aligned}$$

for some constant C_1. To evaluate C_1 we use the initial condition that at $t = 0$ the height of the object is zero. Therefore, $s(0) = 0$, so $C_1 = 0$, and

$$s(t) = -16t^2 + 128t$$

is the function that gives the height of the object at any time t.
(b) The highest point of the rock's flight occurs when $v(t) = 0$. Solving $v(t) = -32t + 128 = 0$ for t, we see that $t = 4$, and evaluating the distance function gives $s(4) = 256$ feet as the highest point.
(c) When the object hits the ground, $s(t)$ is zero. Solving $s(t) = -16t^2 + 128t = 16t(-t + 8) = 0$ for t, we have $t = 0$ or $t = 8$. The object was thrown at $t = 0$ and $t = 8$ seconds is when it returns to the ground. The duration of its flight is 8 seconds. The path of the rock is shown in Figure 6.2 and the graph of its distance from the ground is illustrated in Figure 6.3. ■

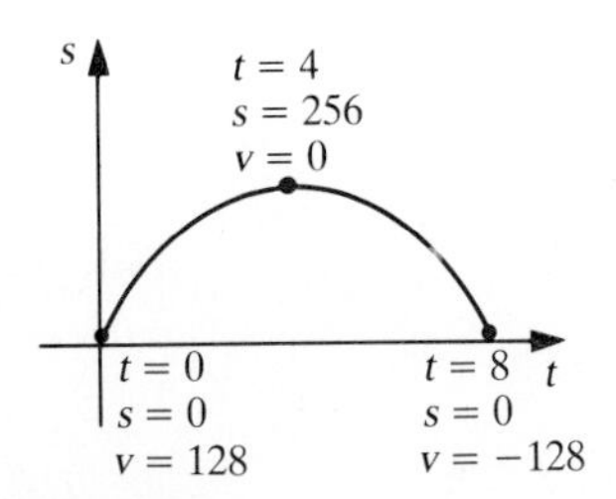

Figure 6.3
The graph of the distance from the rock to the ground at time t

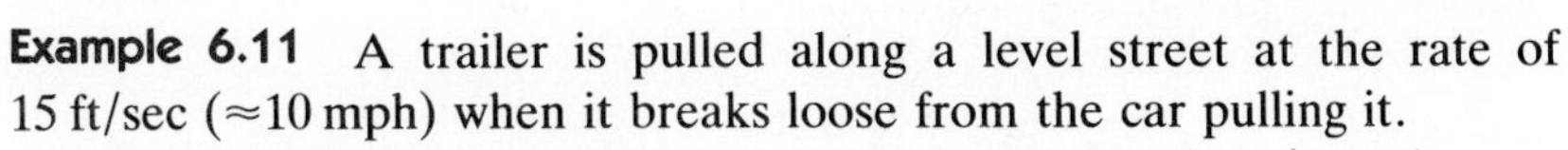

Example 6.11 A trailer is pulled along a level street at the rate of 15 ft/sec ($\approx$10 mph) when it breaks loose from the car pulling it.
(a) If frictional forces produce a constant deceleration (negative acceleration) of the trailer, and it stops after rolling 150 feet, what was the acceleration due to friction?
(b) How long does it take the trailer to stop?

Solution **(a)** Let $t = 0$ be the instant the trailer breaks loose. The initial conditions are $v(0) = 15$ and $s(0) = 0$. We also know that $v(t) = 0$ when $s(t) = 150$. See Figure 6.4.

The acceleration is constant, so $a(t) = k$ for some k. (Since the trailer is decelerating, k is negative.) Thus

$$v(t) = \int a(t)\,dt = \int k\,dt = kt + C.$$

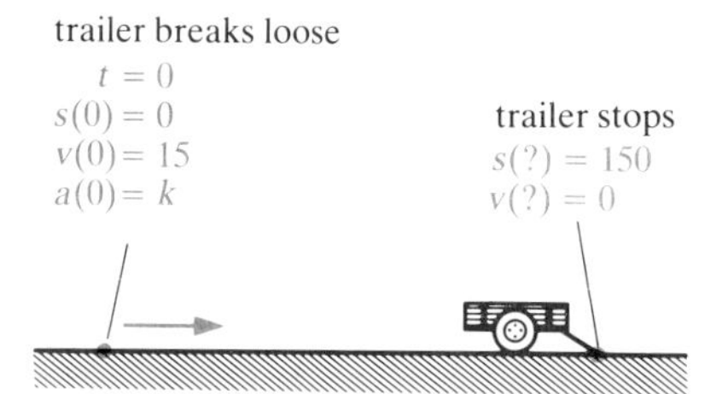

Figure 6.4
A trailer rolling to a stop

Since $v(0) = 15$, we get $C = 15$, so $v(t) = kt + 15$. Hence

$$s(t) = \int v(t)\,dt = \int (kt + 15)\,dt = \tfrac{1}{2}kt^2 + 15t + C_1.$$

From the initial condition $s(0) = 0$ we conclude that $C_1 = 0$. Hence, $s(t) = \frac{1}{2}kt^2 + 15t$. To use the condition that $s(t) = 150$ when $v(t) = 0$, we must find the value of t for which $v(t) = 0$. Solving $v(t) = kt + 15 = 0$ for t, we have $t = -15/k$ (recall that $k < 0$), and so $s(-15/k) = 150$. From $s(t) = \frac{1}{2}kt^2 + 15t$, we have

$$s\left(-\frac{15}{k}\right) = \tfrac{1}{2}k\left(-\frac{15}{k}\right)^2 + 15\left(-\frac{15}{k}\right) = \frac{225}{2k} - \frac{225}{k} = -\frac{225}{2k}.$$

Setting $-(225/2k) = 150$ and solving for k gives $k = -0.75$. Hence, the acceleration due to friction is $a(t) = -0.75\ \text{ft/sec}^2$.
(b) Finally, $v(t) = 0$ when $t = -15/k = 20$, so the trailer stops in 20 seconds. ■

Example 6.12 Suppose that the deceleration of a car when braking is a constant $-10\ \text{ft/sec}^2$. If the car is traveling at a speed v_0 when braking begins, how far will the car travel before coming to a complete stop?

Solution The speed $v(t)$ at time t is

$$v(t) = \int a(t)\,dt = \int -10\,dt = -10t + C.$$

To find the constant C we use the fact that when $t = 0$, $v = v_0$. Thus $C = v_0$ and we have

$$v(t) = v_0 - 10t.$$

When the car has stopped, $v(t) = 0$. Solving $v_0 - 10t = 0$ for t gives $t = v_0/10$ as the time needed to stop completely. The distance-traveled function $s(t)$ is the integral of the velocity function:

$$s(t) = \int v(t)\,dt = \int (v_0 - 10t)\,dt = v_0 t - 5t^2 + C.$$

Furthermore, at time zero the distance traveled is zero, so $s(0) = 0$; setting t equal to 0 gives

$$0 = s(0) = (v_0) \cdot 0 - 5(0)^2 + C = C.$$

Thus $C = 0$ and $s(t) = v_0 t - 5t^2$. Finally, we can get the distance traveled after the brakes were applied by substituting $t = v_0/10$:

$$\text{distance to stop} = v_0\left(\frac{v_0}{10}\right) - 5\left(\frac{v_0}{10}\right)^2 = 0.10v_0^2 - 0.05v_0^2 = 0.05v_0^2.$$

The most important feature of this formula is that the distance needed to stop varies with the *square* of the initial velocity. Table 6.1 shows several initial speeds and the associated stopping distances. Notice that the distance at 40 mph is four times that at 20 mph. ■

Table 6.1
Stopping distance for various speeds

Initial speed (*mph*)	*Initial speed* (*ft/sec*)	*Stopping distance* (*ft*)
10	14.7	11
20	29.3	43
30	44.0	97
40	58.7	172
50	73.3	267
60	88.0	387
80	117.3	688

Exercises

1. A company has a marginal cost function of $0.002x + 4$ dollars and a fixed cost of \$4000. Determine its cost function.
2. A company has a marginal cost function of $-0.04x + 12$ dollars and a fixed cost of \$2000. Determine its cost function.
3. A company has a marginal cost function of $-0.08x + 10$ dollars and a fixed cost of \$6000. Determine its cost function $C(x)$. Calculate the change in cost as the level of production changes from $x = 20$ to $x = 21$, and compare this change with the marginal cost when $x = 20$.
4. A company has a marginal cost function of $-0.20x + 400$ dollars and a fixed cost of \$9000. Determine the cost function $C(x)$. Calculate the change in cost as the level of production changes from $x = 100$ to $x = 101$, and compare this change with the marginal cost when $x = 100$.
5. If the marginal revenue function is $80 - 0.002x$ dollars, find the revenue function. Use the fact that no revenue is earned if no items are sold.
6. If the marginal revenue function is $12 - 0.4x$ dollars, find the revenue function. What revenue will accrue from the sale of 10 items?
7. A company has a marginal revenue function of $-0.18x + 12$ dollars per day on a sales volume of x units per day. Determine the revenue function $R(x)$. Calculate the change in revenue as the sales volume increases from 40 to 41 units per day, and compare this change with the marginal revenue when $x = 40$.
8. A company has a marginal revenue function of $-0.3x + 20$ dollars per day on a sales volume of x units per day. Determine the revenue function $R(x)$. Calculate the change in revenue as the sales volume increases from 20 to 21 units per day, and compare this change with the marginal revenue when $x = 20$.
9. A company has a marginal cost function of $-0.06x + 30$ dollars and a fixed cost of \$9000. Determine the cost function $C(x)$. Calculate the change in cost as the level of production changes from $x = 30$ to $x = 31$, and compare this change with the marginal cost when $x = 30$.
10. Find the cost function for producing x items per day if the marginal cost function is
$$M_C(x) = -0.002x + 12$$
dollars and the cost of producing 20 items is \$360. How much does it cost to produce 150 items per day?
11. A firm has determined that its marginal cost function is given by
$$M_C(x) = x^{-1/2} + 0.001$$
dollars. Suppose the cost of producing 10,000 items is \$1210. What is the cost function?
12. A company finds that it has a marginal cost function $M_C(x) = x^2 + 60x + 904$ dollars when x units are produced per day for $0 \le x \le 30$.
(a) Given that the cost of producing 3 units per day is \$2470 per day, find the cost function.
(b) Evaluate the cost function at $x = 15$, and determine the cost per unit when $x = 3$ and when $x = 15$.
13. A company finds that it has a marginal cost function $M_C(x) = -x^2 + 80x + 20$ dollars when x units are produced per day for $0 \le x \le 30$.

(a) Given that the cost of producing 9 units per day is \$3200 per day, find the cost function.
(b) Evaluate the cost function at $x = 21$, and determine the cost per unit when $x = 9$ and when $x = 21$.
(c) At which level of production in part (b) is the cost per unit the least?

14. A company finds that it has a marginal revenue function $M_R(x) = 0.1x^2 - 10x + 270$ dollars per day when x units are sold per day and $10 \le x \le 40$.
(a) Given that the revenue from the sale of 15 units per day is \$5037.50 per day, find the revenue function.
(b) Evaluate the revenue function at $x = 27$, and determine the cost per unit when $x = 15$ and when $x = 27$.

15. A company finds that it has a marginal revenue function $M_R(x) = -0.2x^2 + 16x + 13$ dollars per day when x units are sold per day and $20 \le x \le 70$.
(a) Given that the revenue from the sale of 40 units per day is \$10,000 per day, find the revenue function.
(b) Evaluate the revenue function at $x = 60$, and determine the cost per unit when $x = 40$ and when $x = 60$.
(c) At which level of sales in part (b) is the revenue per unit the most?

16. A ball is thrown upward with an initial velocity of 64 ft/sec. Carefully set up and describe a suitable frame of reference.
(a) What is the velocity of the ball at the end of 1 second?
(b) What is the ball's velocity at the end of 3 seconds?
(c) How many seconds does it take the ball to reach its highest point?
(d) How many seconds does it take the ball to reach the ground?
(e) How high will the ball travel?
(f) How fast is the ball falling when it hits the ground?

17. Work Exercise 16 for an initial velocity of 96 ft/sec.

18. A block of ice slides down a chute with an acceleration of 18 ft/sec^2.
(a) If the block is sliding 6 ft/sec at a certain instant, how far does it slide in the next 3 seconds?
(b) How far from the first position has the block gone by the time its speed is 72 ft/sec?

19. A ball is thrown up vertically with an initial velocity of 30 m/sec from the edge of the top of a building 50 meters tall. The ball's acceleration due to gravity is -9.7536 m/sec^2.
(a) Find the distance $s(t)$ from the foot of the building up to the ball t seconds later.
(b) How long does it take the ball to hit the ground? (Assume that it misses the building on the way down.)

20. A rock is dropped off a sheer cliff. One second later, a second rock is dropped.
(a) Use the instant at which the second rock is dropped as $t = 0$ to find $s_1(t)$, the distance-fallen function for the first rock, and $s_2(t)$, the distance-fallen function for the second rock.
(b) How far apart are the rocks at $t = 0$?
(c) How far apart are the rocks at $t = 2$?
(d) How far apart are the rocks at $t = 5$?
(e) Explain how the results in parts (b)–(d) can be used to explain why the water near the bottom of a waterfall is more dispersed than the water near the top of the waterfall.

21. A child at the top of a cliff throws a rock straight down, and it hits the ground 189 feet below 2.25 seconds later. With what velocity did the child throw the rock?

22. With what speed (in m/sec) must an arrow be shot upward in order to fall back to its starting point in 8 seconds? How high will it rise?

23. A box slides down an inclined chute 140 feet long in 3.5 seconds and arrives at the bottom with a velocity of 75 ft/sec. Assume constant acceleration, and find the acceleration and the velocity of the box at the top of the chute.

24. A ball is rolling down an incline with an acceleration of 24 ft/sec^2. Find the equation of its motion, given that it has rolled 11 feet after 2 seconds and 120 feet after 3 seconds. (Assume distance is measured down the incline.)

25. A car, traveling 50 ft/sec, starts to slow down with constant (negative) acceleration. In three seconds, its speed is reduced to 20 ft/sec.
(a) Find the acceleration.
(b) Find the time required to stop.
(c) Find the distance traveled while slowing to a stop.

26. A toy car is dropped out of a window 100 feet above the ground. At the same time a toy truck is thrown straight down from a window 200 feet above the ground. Both toys reach the ground at the same time. Find the initial velocity of the toy truck if air resistance is ignored.

27. An airplane drops a bomb from an altitude of 6400 feet at the same time that a shell is shot at it from the ground (sea level). They both reach the ground at the same time. What was the initial velocity of the shell?

28. A rocket is projected upward from the earth with velocity 196 m/sec.
(a) Find the height s as a function of t, taking $t = 0$ and $s = 0$ at the start of the motion with s positive upward.
(b) For how long and how high does the rocket rise?

29. A rock dropped off a cliff hits the ground below T seconds later.
    (a) Obtain a formula for the height of the cliff as a function of T. (Recall that the acceleration of freely falling bodies is a constant -32 ft/sec^2.)
    (b) How high is the cliff if $T = 3$ seconds?
    (c) How high is the cliff if $T = 5$ seconds?
    (d) How high is the cliff if $T = 10$ seconds?
30. The marginal productivity of a team of government researchers working on a project is $M_p(t) = 0.02t - 0.0001t^2 + 0.75$, where t measures the hours per week since the project began. What is their productivity after 20 hours?

6.3 The Definite Integral

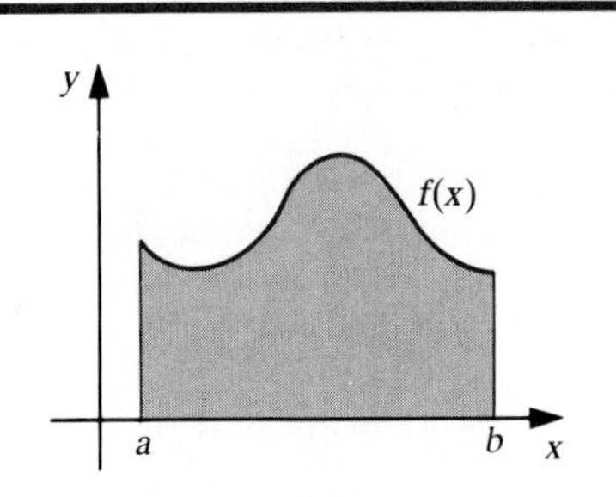

Figure 6.5
The area bounded by the curve $y = f(x)$, the x-axis, and vertical lines at a and b

Let $f(x)$ be a positive continuous function on the interval $[a, b]$ as shown in Figure 6.5. Consider the area bounded by the function, the x-axis, and the vertical lines $x = a$ and $x = b$. This area is shaded in the figure. An essential problem in calculus is the calculation of this area. The following example presents an important interpretation of the area under a curve and illustrates a general procedure for finding area under curves.

Example 6.13 An analysis of the potential effects of an advertising campaign for a bread company shows that t days after the start of the campaign, the company's bread will be selling at a rate of $S(t)$ loaves per day. The function $S(t)$ is graphed in Figure 6.6. Suppose that the bread company's sales follow the predictions of the analysis. Show that the total number of loaves of bread that the company will sell during the first week of the campaign equals the area between the graph of $S(t)$ and the t-axis within the interval $0 \le t \le 7$.

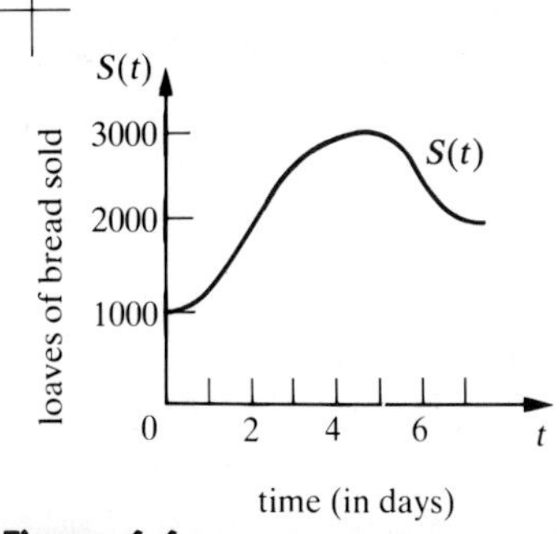

Figure 6.6
Projected rate of sales of loaves of bread during an ad campaign

Solution We see from the graph in Figure 6.6 that at each point in time the company is selling bread at the rate of at least 1000 loaves per day. Hence, if we suppose that the rate is a constant 1000 loaves per day, we will obtain an underestimate of the number of loaves sold for the entire week. This underestimate is shown by the shaded area in Figure 6.7. To calculate this underestimate, we multiply the rate (in this case, 1000 loaves per day) by the length of the time interval, 7 days: the underestimate is 7000 loaves. Notice that the shaded area in Figure 6.7 can also be calculated by multiplying the height of the rectangle (1000) by its width (7), which yields an area of 7000.

From the graph in Figure 6.7 we see that although 1000 is close to the value of $S(t)$ on the first day, it is not a good approximation to $S(t)$ later in the week. We would obtain a more accurate estimate if we consider portions of the week separately. For example, we can divide the week in half. The minimum rate during the first 3.5 days of the week is 1000 loaves per day, and the minimum rate during the second half of the week ($3.5 \le t \le 7$) is 2000 loaves per day. Using these rates, we will again underestimate the number of loaves sold during the week (see Figure 6.8).

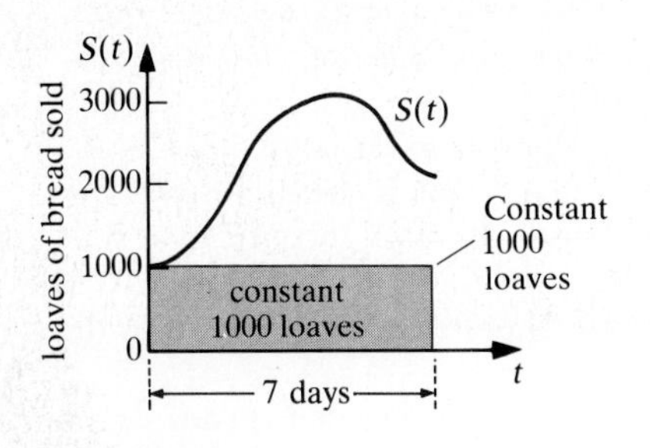

Figure 6.7
First underestimate of loaves sold in first week

To calculate this new underestimate, we consider each subinterval $[0, 3.5]$ and $[3.5, 7]$ separately. Since the rate is 1000 loaves per day for the first subinterval and the length of the first subinterval is 3.5 days, the contribution from this interval is $(1000)(3.5) = 3500$ loaves. Similarly, the contribution from the second subinterval is $(2000)(3.5) = 7000$ loaves. Hence, our new underestimate is $3500 + 7000 = 10{,}500$ loaves. Notice that the shaded area in Figure 6.8 is also calculated as the area of two rectangles $(1000)(3.5) + (2000)(3.5)$, so our new approximation is also represented geometrically by an area. Note that this estimate is better than the first one.

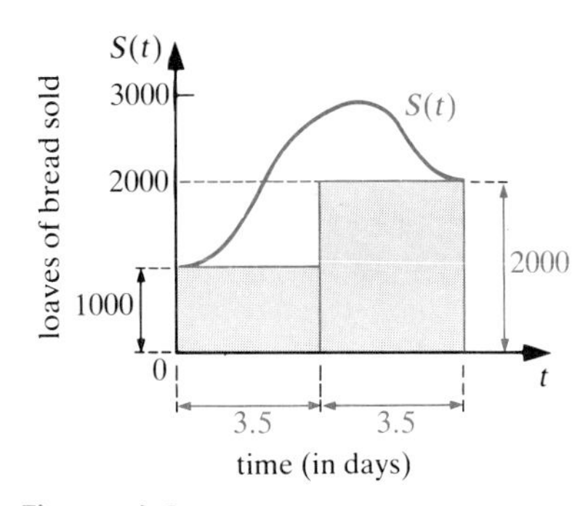

Figure 6.8
Second underestimate of loaves sold in first week

If we divide the interval $[0, 7]$ in more than two pieces (that is, use more than two subintervals), we obtain an even more accurate approximation to the number of loaves sold. In Figure 6.9, the subinterval $[0, 7]$ is broken into seven subintervals, each with width one unit (1 day). To calculate the corresponding underestimate, we use the smallest value of $S(t)$ on each subinterval. We estimate the number of loaves sold in each subinterval by the product of the rate at which the bread is selling and the length of the time interval (1 in this case). Our calculation of total loaves sold is

$$(1000)(1) + (1200)(1) + (2000)(1) + (2600)(1) + (2800)(1) + (2200)(1) + (2000)(1) = 13{,}800 \text{ loaves.}$$

This calculation also yields the shaded area in Figure 6.9.

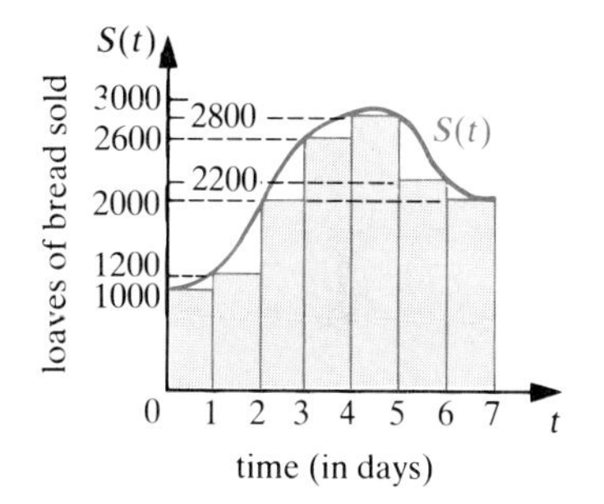

Figure 6.9
Third underestimate of loaves sold in first week

If we continue to refine our approach by breaking the interval into more and more subintervals of smaller and smaller widths, we will obtain approximations that are increasingly closer to the actual number of loaves sold. Geometrically, we would also get values closer and closer to the area under the curve $S(t)$ (and above the t-axis). Consequently, if we could carry out this refinement process indefinitely, we would obtain the exact number of loaves that would be sold as well as the exact area under $S(t)$. Therefore we may interpret the area under the curve as the total number of loaves of bread sold. ■

Riemann Sums

We now return to the problem of determining the area under a positive continuous function $f(x)$. In the previous example, we approximated the area under the graph of $S(t)$ on the interval $[0, 7]$ by dividing the interval into several subintervals, calculating the area of a rectangle that approximates the area under the curve on each subinterval, and then computing the sum of these approximations. Such a sum is called a *Riemann sum*.

To find the area under the graph of $f(x)$ between two vertical lines $x = a$ and $x = b$, we divide the interval $[a, b]$ into n subintervals as shown in Figure 6.10. Any division of $[a, b]$ into subintervals is called a **partition** of $[a, b]$. The endpoints of the first subinterval are denoted by $x_0 = a$ and x_1; the endpoints of the second subinterval are x_1 and x_2; the endpoints of the third subinterval are x_2 and x_3; and so on. We can describe this notation succinctly by saying that the endpoints of the subintervals are x_i, $i = 0, 1, \ldots, n$; $x_0 = a$, $x_n = b$.

To approximate the area under the curve for the ith subinterval, that is, from x_{i-1} to x_i, we calculate the area of an associated rectangle. In the bread loaf example, we were finding *under*estimates of the area,

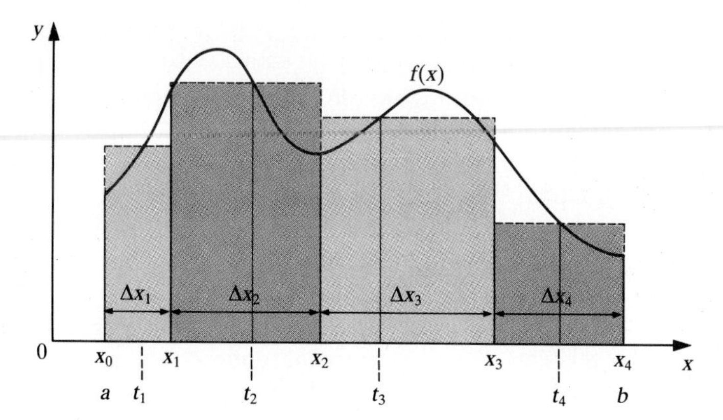

Figure 6.10
A typical Riemann sum partition

so we used the *smallest* value of $f(x)$ on each subinterval for the height of the rectangle. In general, we allow more choice for the height of the rectangle associated with each subinterval. This rectangle has the same width as the ith subinterval and has height $f(t_i)$ for some arbitrarily chosen t_i in the interval $[x_{i-1}, x_i]$, as shown in Figure 6.10; that is, rectangle 1 has width $x_1 - x_0$ (Δx_1 is used to denote this width) and height $f(t_1)$ for some t_1 with $x_0 \leq t_1 \leq x_1$; rectangle 2 has width $\Delta x_2 = x_2 - x_1$ and height $f(t_2)$ of some t_2 with $x_1 \leq t_2 \leq x_2$; and so on. We can summarize this description by saying that the ith rectangle has width $\Delta x_i = x_i - x_{i-1}$ and height $f(t_i)$ for some t_i with $x_{i-1} \leq t_i \leq x_i$. The area of the ith approximating rectangle is $f(t_i)\,\Delta x_i$. The approximation to the area under the curve from $x = a$ to $x = b$ is the sum of the areas of these approximating rectangles:

$$\text{area} \approx f(t_1)\,\Delta x_1 + f(t_2)\,\Delta x_2 + \cdots + f(t_n)\,\Delta x_n.$$

The sum on the right-hand side of this equation is called a **Riemann sum**. The t_i selected from the ith subinterval in the formation of a Riemann sum may be any value in that subinterval. Changing the choice of t_i will, in general, change the value of the sum; however, if the continuous function $f(x)$ is not too erratic and if we divide $[a, b]$ into a large number of small subintervals, then the amount by which the height $f(t_i)$ varies over the ith subinterval will be small. Any Riemann sum formed by such a partition will be a good approximation to the area under the curve.

A function has many Riemann sums on a given interval. The number n of subintervals is arbitrary, the n subintervals may be chosen in any fashion, and within each subinterval each t_i may be chosen randomly. For any particular Riemann sum, we let max Δx_i denote the width of the widest subinterval. Figure 6.11 shows the effect of increasing the number n of subintervals and, simultaneously, decreasing the width of the widest subinterval; each case has n larger and max Δx_i smaller than the preceding case. Also, each case is an increasingly more accurate approximation to the area under the curve. Thus, by using more and more, thinner and thinner subintervals, the corresponding Riemann sums become increasingly accurate approximations to the area under the graph of the function $f(x)$.

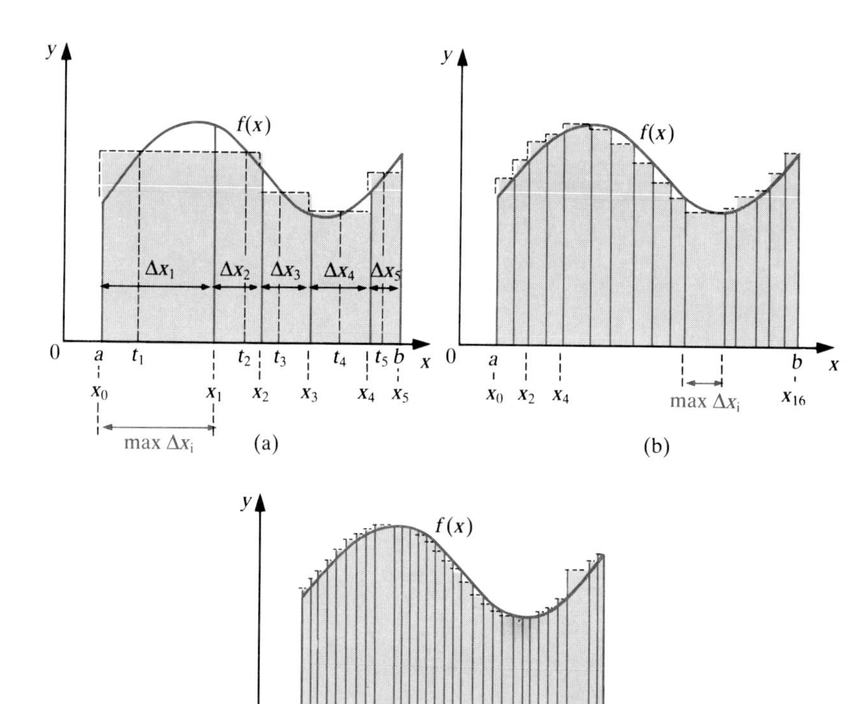

Figure 6.11
Riemann sums with increasing n and decreasing max Δx_i. In cases (b) and (c), $t_i = x_i$.
(a) Area from a Riemann sum with $n = 5$ (max Δx_i is Δx_1);
(b) area from a Riemann sum with $n = 16$ (max Δx_i here is smaller than in case [a]);
(c) area from a Riemann sum with $n = 31$ (max Δx_i here is smaller than in case [b])

The results suggested by Figure 6.11 can be shown to be correct for any continuous function $f(x)$ on a finite closed interval $[a, b]$. For any such function, it can be shown that

$$\lim_{\substack{n \to \infty \\ \max \Delta x_i \to 0}} \left(f(t_1)\,\Delta x_1 + f(t_2)\,\Delta x_2 + \cdots + f(t_n)\,\Delta x_n \right) \tag{1}$$

exists. In the limit, the notation $n \to \infty$ means that we use more and more subintervals and max $\Delta x_i \to 0$ means that we use subintervals such that the widest of them is getting thinner and thinner. The value of this limit is denoted by

$$\int_a^b f(x)\,dx$$

and is called **the definite integral from a to b of $f(x)$**.

Definition of the Definite Integral of $f(x)$

$$\int_a^b f(x)\,dx = \lim_{\substack{n \to \infty \\ \max \Delta x_i \to 0}} \left(f(t_1)\,\Delta x_1 + f(t_2)\,\Delta x_2 + \cdots + f(t_n)\,\Delta x_n \right).$$

If $f(x)$ is nonnegative on $[a, b]$, then the limiting process described calculates the area under the curve so that

$$\int_a^b f(x)\,dx = \text{area under } f(x), \qquad f(x) \geq 0 \text{ on } [a, b].$$

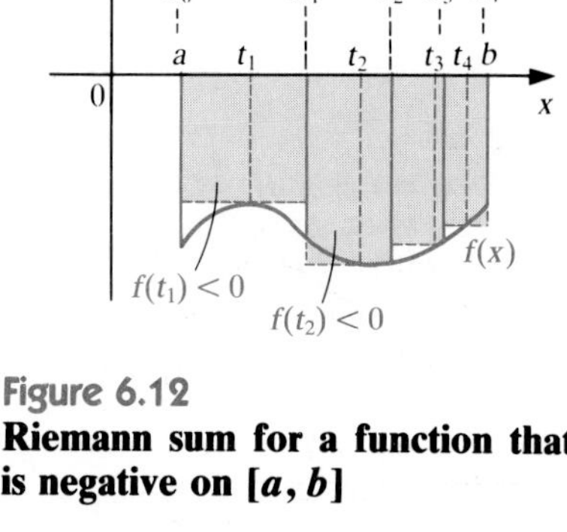

Figure 6.12
Riemann sum for a function that is negative on $[a, b]$

When $f(x)$ is negative on all or on a portion of $[a, b]$, this geometrical interpretation of the definite integral must be modified. First consider the case where $f(x) < 0$ on the entire interval, as shown in Figure 6.12. All the numbers $f(t_i)$ used in expression (1) for the calculation of the area of the individual rectangles are negative. Consequently, all the quantities $f(t_i)\,\Delta x_i$ are negative, and

$$f(t_i)\,\Delta x_i = -(\text{the area of the } i\text{th approximating rectangle}).$$

Adding these "negative areas" will yield an approximation to the *negative* of the area between the curve and the x-axis.

$$\int_a^b f(x)\,dx = -(\text{area between } f(x) \text{ and the } x\text{-axis}),$$

$$f(x) \leq 0 \text{ on } [a, b].$$

Now consider a function that is positive on part of the interval and negative on the rest of the interval, as shown in Figure 6.13. The figure also shows a typical Riemann sum partition. In the regions labeled *I*, the function is positive, so the numbers $f(t_i)\,\Delta x_i$ in expression (1) are positive. The sum of these terms will be positive. Alternatively, in the regions labeled *II* the function is negative, so the numbers $f(t_i)\,\Delta x_i$ are negative, and the sum of these terms will also be negative.

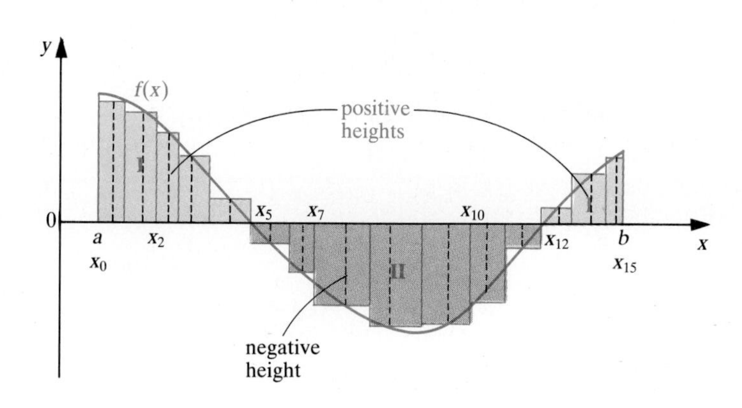

Figure 6.13
Riemann sum partition for a function that is both positive and negative in the interval $[a, b]$

Regions where $f(x)$ is positive will contribute positive area and regions where $f(x)$ is negative will contribute "negative area;" the Riemann sum is (approximately) the *net area* between the curve and the x-axis where

$$\begin{aligned}\text{net area} = &[\text{the area above the } x\text{-axis and below the function}]\\ &-[\text{the area below the } x\text{-axis and above the function}].\end{aligned}$$

Consequently, the definite integral from a to b will (exactly) compute the *net* area between $f(x)$ and the x-axis.

In Section 6.5 we will calculate a few Riemann sums. However, computing area by summing the areas of hundreds or even thousands of rectangles would be tedious, so we now consider the calculation of area from a different point of view.

The Area Function

First suppose that $f(x)$ is nonnegative on $[a, b]$. Let z denote a number between a and b and let $A(z)$ denote the area under the curve $y = f(x)$ from $x = a$ to $x = z$, as shown in Figure 6.14. Since $A(a)$ is the area under $f(x)$ from a to a, and since this area is zero (no width),

$$A(a) = 0.$$

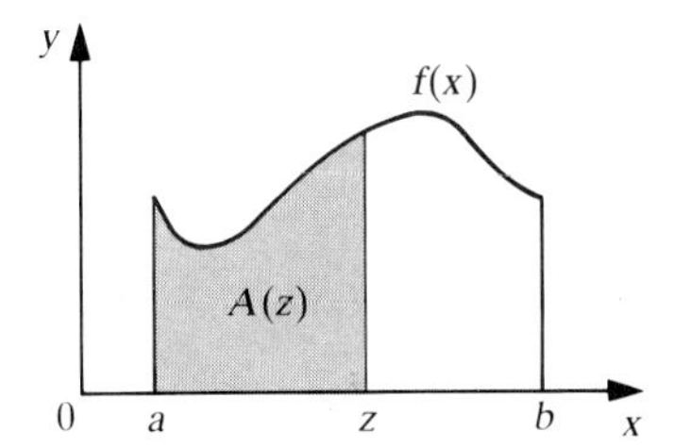

Figure 6.14
$A(z)$ is the area between $f(x)$ and the x-axis from a to z

Also, $A(b)$ is the area under $f(x)$ from a to b, so it is $A(b)$ that we wish to compute. Consider the determination of the derivative $A'(z)$. By the definition of the derivative

$$A'(z) = \lim_{h \to 0} \frac{A(z+h) - A(z)}{h}.$$

Now $A(z+h)$ is the area under the curve from a to $z+h$, and $A(z)$ is the area under the curve from a to z. Consequently, the difference $A(z+h) - A(z)$ is the area under the curve from z to $z+h$. See Figure 6.15. We are assuming $f(x)$ is continuous, so for small values of h, $f(x)$ will not change much on the interval from z to $z+h$. Again referring to the figure, we see that the area under the curve between z and $z+h$ is nearly the area of the rectangle with height $f(z)$ and width h. This area is $hf(z)$, so

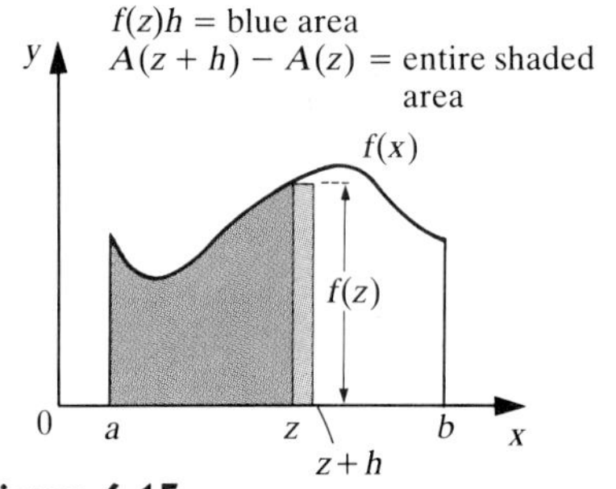

Figure 6.15
A comparison of the area under $f(x)$ from z to $z+h$ to the area of the rectangle of h width and $f(z)$ height

$$A(z+h) - A(z) \approx hf(z),$$

or

$$\frac{A(z+h) - A(z)}{h} \approx f(z).$$

The approximation becomes increasingly better as h gets smaller. Thus

$$\lim_{h \to 0} \frac{A(z+h) - A(z)}{h} = f(z),$$

that is,

$$A'(z) = f(z)$$

for any z in (a, b). Writing x in place of z gives

$$A'(x) = f(x).$$

Thus $A(x)$ is an antiderivative of $f(x)$; in fact, it is the particular antiderivative of $f(x)$ for which $A(a)=0$. We can use this remarkable fact to easily find the area under the graph of many nonnegative continuous functions.

The assumption that $f(x) \geq 0$ on $[a, b]$ played a very minor role in the preceding derivation; it allowed us to rely on the graph in Figure 6.15 to gain insight into the calculations. When we discussed the approximation to the area $A(z+h)-A(z)$ by the area of a rectangle with width h and height $f(z)$, the assumption that $f(x) \geq 0$ made this area positive. If $f(x)$ is negative on part or all of the interval, then we must alter the interpretation of $A(z)$ to be the *net* area between $f(x)$ and the x-axis. The conclusion that $A'(x)=f(x)$ is still correct.

Example 6.14 Find the area under the nonnegative curve $f(x)=x^2$ from 1 to 4.

Solution We know that $A(x)$, the area under the curve from 1 to x, is an antiderivative of $f(x)$. Hence, $A(x)=\frac{1}{3}x^3+C$ for some constant C. Since $A(1)$ is the area under the curve from 1 to 1, we know that $A(1)=0$. Thus $0=\frac{1}{3}(1)^3+C$, so $C=-\frac{1}{3}$ and therefore, $A(x)=\frac{1}{3}x^3-\frac{1}{3}$. The area under the curve from 1 to 4 is $A(4)=\frac{1}{3}(4)^3-\frac{1}{3}=21$ square units. ■

The solution of the previous example shows that to find the area under a nonnegative function $f(x)$ from $x=a$ to $x=b$, we perform the following steps.

1. Find an antiderivative $F(x)$ of $f(x)$.
2. Observe that $A(x)=F(x)+C$ for some constant C.
3. Evaluate the constant C using $A(a)=0$ so that $0=A(a)=F(a)+C$, yielding $C=-F(a)$.
4. The area under the curve from a to b is

$$\text{area} = A(b) = F(b) + C = F(b) - F(a).$$

Note that the quantity $F(b)-F(a)$ does not depend on which antiderivative of $f(x)$ is chosen: if $G(x)$ is any other antiderivative, then $G(x)=F(x)+K$, where K is a constant, so $G(b)-G(a)=F(b)+K-(F(a)+K)=F(b)-F(a)$. Consequently, we usually choose $F(x)$ as the antiderivative with the simplest form, that is, the one for which the constant of integration is zero.

The Fundamental Theorem of Calculus

We have discussed two methods for finding the area under a curve $y=f(x)$: (1) by a limiting process that we approximate by computing the area of many thin rectangles whose heights are calculated from $f(x)$; and (2) by evaluating $F(b)-F(a)$, where $F(x)$ is an antiderivative of $f(x)$. The result by method (1) is denoted by the definite integral

$$\text{area} = \int_a^b f(x)\,dx,$$

and by method (2) this area is $F(b) - F(a)$. Since both of these expressions represent the same area, we have the following.

Fundamental Theorem of Calculus

$$\int_a^b f(x)\,dx = F(b) - F(a), \quad \text{where } F(x) \text{ is any antiderivative of } f(x).$$

Example 6.15 Find the area under $f(x) = 4x^3 + 2x + 1$ from $x = 2$ to $x = 6$.

Solution Note that when $2 \leq x \leq 6$, the terms $4x^3$, $2x$, and 1 are positive and continuous and, hence, so is their sum. Therefore, $f(x) > 0$ on the interval $[2, 6]$. Using the techniques we developed for antiderivatives (in Section 6.1), we see that the function $F(x) = x^4 + x^2 + x$ is an antiderivative of $f(x)$. Hence, the area under $f(x)$ is

$$\begin{aligned} F(6) - F(2) &= (6^4 + 6^2 + 6) - (2^4 + 2^2 + 2) \\ &= 1338 - 22 = 1316. \end{aligned}$$

The area under the curve from $x = 2$ to $x = 6$ is 1316 square units. ∎

We end this section with a comment about the two methods for finding the area under a curve. The first method (Riemann sums) can be used on any continuous function. Although the calculations may be tedious, we can approximate $\int_a^b f(x)\,dx$ to any accuracy by Riemann sums. The success of the second method (antiderivatives) depends on finding an antiderivative of $f(x)$. Unfortunately, not every function has an explicit antiderivative; that is, it is impossible to write a formula for $F(x) = \int f(x)\,dx$ for some functions $f(x)$, such as

$$e^{x^2} \quad \text{and} \quad \frac{e^x}{x}.$$

In the next chapter we will present two important techniques for finding antiderivatives. However, neither method is successful on the integrals

$$\int e^{x^2}\,dx \quad \text{or} \quad \int \frac{e^x}{x}\,dx.$$

Exercises

1. Use Example 6.13 to find *overestimates* for the number of loaves sold per day where (a) the number sold is constant for the whole week, (b) the week is divided in half, and (c) the week is divided into the seven intervals. (*Hint:* Use the maximum number sold in each interval.) In view of your answers and the results of Example 6.13, between what two values does the true number of loaves sold lie? Use Figure 6.9 on page 241 for values of $S(t)$.
2. Suppose that the velocity of an automobile $v(t)$ is as graphed in the figure. Use the approach of Example 6.13 (about the bread loaves) to explain why the area

under the $v(t)$ curve from $t = 0$ to $t = 4$ equals the distance the automobile has traveled during this 4-hour period. (Recall that if the automobile is traveling at a constant speed c over a given time interval, then the distance traveled equals the product of c and the length of the time interval.)

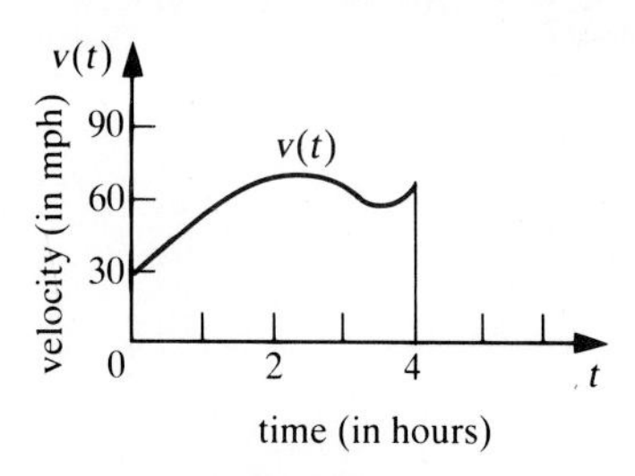

Exercise 2

3. If air friction is ignored, an object falling toward earth has a constant acceleration and, consequently, its velocity is a linear function of t. The graph of the velocity of such an object is shown in the accompanying figure. Calculate the area under the curve to determine how far the object falls in 4 seconds.

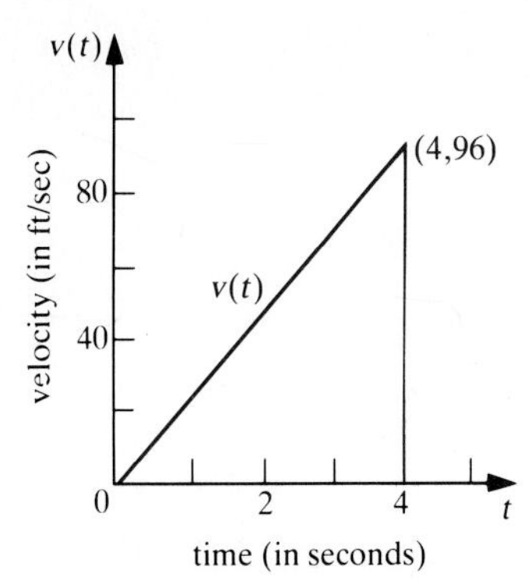

Exercise 3

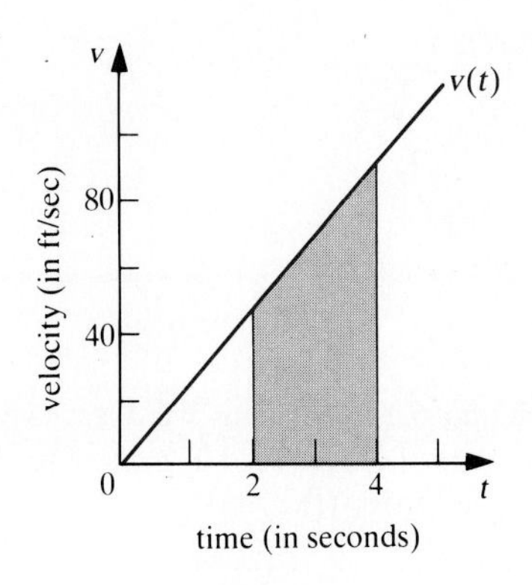

Exercise 4

4. Refer to Exercise 3. The distance the object falls from time $t = 2$ to $t = 4$ is given by the area under the graph of the velocity function between $t = 2$ and $t = 4$, the portion shaded in the preceding figure. Calculate the distance the object falls between the third and fourth seconds by determining the area of the shaded portion.

Exer. 5–16: (a) Find $A(x)$, the area under $f(x)$ from a to x by the method used in Example 6.14. (b) Calculate $A(b)$, the area under $f(x)$ from a to b. (c) Sketch a graph of $f(x)$ and shade the region whose area is being calculated.

5. $f(x) = 3x$; $a = 0$; $b = 5$
6. $f(x) = x^2 + 1$; $a = 0$; $b = 3$
7. $f(x) = 4x + 1$; $a = 2$; $b = 6$
8. $f(x) = 3x^2 + 5$; $a = 1$; $b = 8$
9. $f(x) = 5 - x$; $a = -3$; $b = -1$
10. $f(x) = 5x - x^2$; $a = 0$; $b = 4$
11. $f(x) = 1/x$; $a = 1$; $b = e$
12. $f(x) = 3/x$; $a = 1$; $b = e$
13. $f(x) = e^x$; $a = 0$; $b = 1$
14. $f(x) = 4e^x$; $a = 0$; $b = 1$
15. $f(x) = e^x + x$; $a = 0$; $b = 1$
16. $f(x) = 4x^3$; $a = -1$; $b = 2$

Exer. 17–28: Find the area under $f(x)$ from $x = a$ to $x = b$ using the Fundamental Theorem of Calculus.

17. $f(x) = 2x + 1$; $a = 0$; $b = 4$
18. $f(x) = 3x^2 + 2$; $a = 1$; $b = 5$
19. $f(x) = x^2 + x + 2$; $a = 2$; $b = 4$
20. $f(x) = 2/x$; $a = 1$; $b = 3$
21. $f(x) = 3x^2 + (1/x)$; $a = 1$; $b = e^3$
22. $f(x) = x^{1/2} + 2$; $a = 1$; $b = 4$
23. $f(x) = e^{2x}$; $a = 0$; $b = 1$
24. $f(x) = 3e^{-x}$; $a = 1$; $b = 3$
25. $f(x) = -4 + e^x$; $a = 3$; $b = 5$
26. $f(x) = \sqrt[3]{x}$; $a = 0$; $b = 1$
27. $f(x) = x - (1/x)$; $a = -1$; $b = -\frac{1}{2}$
28. $f(x) = x(x + 1)$; $a = 0$; $b = 1$

29. Suppose that the velocity of an automobile during a 6-hour journey is given by

$$v(t) = 0.1t^3 - 0.5t^2 + 12t + 5$$

for $0 \le t \le 6$. How far did the automobile travel during these 6 hours?

30. Suppose that the rate at which a company is selling loaves of bread at time t for $0 \le t \le 7$ is given by

$$s(t) = -80t^2 + 800t + 1000.$$

How many loaves will the company sell over the 7-day period?

Area in the Plane

6.4

In Section 6.3 we saw that if $f(x)$ is a continuous function with antiderivative $F(x)$, then by the Fundamental Theorem of Calculus, the definite integral of $f(x)$ from $x = a$ to $x = b$ is $F(b) - F(a)$. We also saw that $\int_a^b f(x)\,dx$ equals the net area between the graph of $f(x)$ and the x-axis from a to b.

To facilitate our discussion, we will introduce some notation. The numbers a and b are called the **limits of integration**; a is the **lower limit** and b is the **upper limit**. The interval $[a, b]$ is called the **interval of integration**. The function $f(x)$ is the **integrand.** We will also use the abbreviated notation

$$F(x)\Big|_a^b = F(b) - F(a).$$

For example, $(x^2 + 3x)\,|_2^4 = [4^2 + 3(4)] - [2^2 + 3(2)] = 18.$

The usefulness of this notation will be apparent in the next example, which shows how to evaluate an integral without formally assigning a symbol to the antiderivative.

Example 6.16 Evaluate $\int_4^9 (3x^2 - 2x)\,dx.$

Solution

$$\int_4^9 (3x^2 - 2x)\,dx = (x^3 - x^2)\Big|_4^9$$
$$= [(9)^3 - (9)^2] - [(4)^3 - (4)^2] = 648 - 48 = 600.$$

We have found that the area under $3x^2 - 2x$ from $x = 4$ to $x = 9$ is 600 square units (because the integrand is positive on $[4, 9]$). Using the new notation, the problem was completed without interrupting to name the antiderivative. ■

Example 6.17 Evaluate $\int_e^{e^2} \frac{1}{x}\,dx.$

Solution

$$\int_e^{e^2} \frac{1}{x}\,dx = \ln|x|\Big|_e^{e^2}$$
$$= \ln|e^2| - \ln|e|$$
$$= \ln e^2 - \ln e \qquad (e^2 > 0,\ e > 0)$$
$$= 2\ln e - \ln e = 2 - 1 = 1. \qquad (\ln e = 1)$$

The area under $f(x) = 1/x$ from $x = e$ to $x = e^2$ is one square unit. ■

Properties of the Definite Integral

The definite integral shares many properties with the indefinite integral. The following rules are two important cases.

If $f(x)$ and $g(x)$ are continuous functions on $[a, b]$, then

$$\int_a^b [f(x)+g(x)]\,dx = \int_a^b f(x)\,dx + \int_a^b g(x)\,dx,$$

and

$$\int_a^b kf(x)\,dx = k\int_a^b f(x)\,dx \quad \text{for a constant } k.$$

The definite integral has the following additional property.

If $f(x)$ is a continuous function on $[a, b]$ and $a<c<b$, then

$$\int_a^b f(x)\,dx = \int_a^c f(x)\,dx + \int_c^b f(x)\,dx.$$

We will prove the first rule and leave the second and third as exercises. To prove the first rule, let $F(x)$ and $G(x)$ denote antiderivatives of $f(x)$ and $g(x)$, respectively. Then $F(x)+G(x)$ is an antiderivative of $f(x)+g(x)$, so

$$\begin{aligned}\int_a^b [f(x)+g(x)]\,dx &= [F(x)+G(x)]\Big|_a^b \\ &= [F(b)+G(b)]-[F(a)+G(a)] \\ &= [F(b)-F(a)]+[G(b)-G(a)] \qquad \text{(rearrange terms)} \\ &= F(x)\Big|_a^b + G(x)\Big|_a^b \\ &= \int_a^b f(x)\,dx + \int_a^b g(x)\,dx,\end{aligned}$$

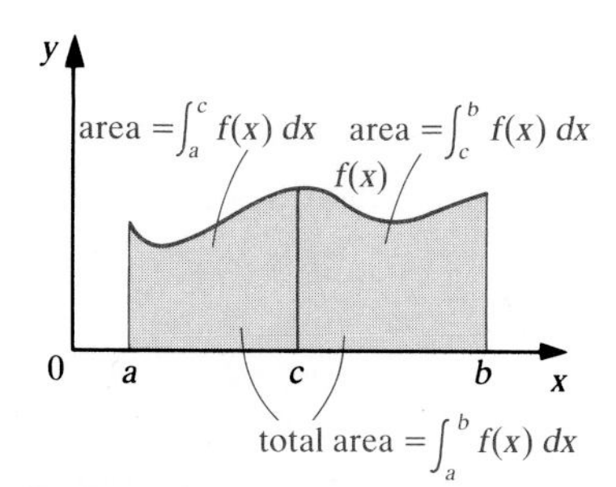

Figure 6.16
The area from a to b is the area from a to c combined with the area from c to b

which proves the first rule. The third rule is illustrated in Figure 6.16 for a nonnegative function.

We have seen that the definite integral computes the *net* area between $f(x)$ and the x-axis. This important property is worth emphasizing. Consider the function $f(x)$ shown in Figure 6.17. This

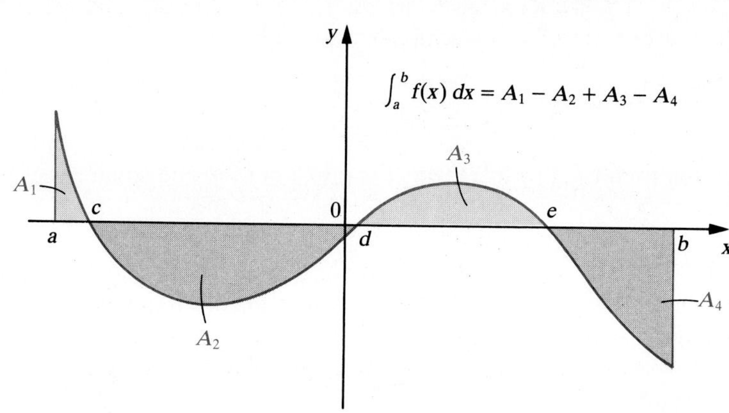

Figure 6.17
Net area for a function that is positive and negative in the interval of integration

function is positive on (a, c), negative on (c, d), positive on (d, e) and negative on (e, b). The symbols A_1, A_2, A_3, and A_4 denote the true (positive) area between the function and the x-axis on the corresponding intervals.

Using the third rule for definite integrals, we have

$$\begin{aligned}\int_a^b f(x)\,dx &= \int_a^c f(x)\,dx + \int_c^b f(x)\,dx \\ &= \int_a^c f(x)\,dx + \int_c^d f(x)\,dx + \int_d^b f(x)\,dx \\ &= \int_a^c f(x)\,dx + \int_c^d f(x)\,dx + \int_d^e f(x)\,dx + \int_e^b f(x)\,dx \\ &= A_1 - A_2 + A_3 - A_4.\end{aligned}$$

The true total area between the function and the x-axis is $A_1 + A_2 + A_3 + A_4$, whereas the definite integral $\int_a^b f(x)\,dx$ represents the *net area* between the x-axis and the function $f(x)$; regions where $f(x)$ is positive contribute positively to the net area and regions where $f(x)$ is negative contribute negatively to the net area.

Example 6.18 Evaluate

$$\int_1^4 (-3x^2 + 12x - 9)\,dx,$$

and interpret the answer in terms of area.

Solution Since $-x^3 + 6x^2 - 9x$ is an antiderivative of $-3x^2 + 12x - 9$, we have

$$\begin{aligned}\int_1^4 (-3x^2 + 12x - 9)\,dx &= (-x^3 + 6x^2 - 9x)\Big|_1^4 \\ &= [-4^3 + 6(4)^2 - 9(4)] - [-1^3 + 6(1)^2 - 9(1)] \\ &= (-4) - (-4) = 0.\end{aligned}$$

The graph of $f(x) = -3x^2 + 12x - 9$ is shown in Figure 6.18. Since the integrand is positive from $x = 1$ to $x = 3$ and negative from $x = 3$ to $x = 4$, and since the definite integral from 1 to 4 is zero, we conclude that the area under the curve from $x = 1$ to $x = 3$ equals the area above the curve from $x = 3$ to $x = 4$. In fact, we have

$$\int_1^3 (-3x^2 + 12x - 9)\,dx = (-x^3 + 6x^2 - 9x)\Big|_1^3 = 4,$$

and

$$\int_3^4 (-3x^2 + 12x - 9)\,dx = (-x^3 + 6x^2 - 9x)\Big|_3^4 = -4.$$

The first integral computes the area bounded by the function and the x-axis between 1 and 3 while the second integral computes the *negative* of the area of the corresponding region from 3 to 4; the two areas are equal, so their sum is zero. Note that the definite integral gives the *net*

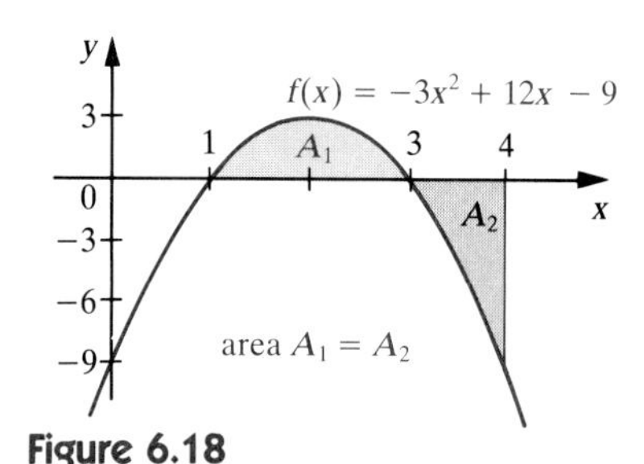

Figure 6.18
Graph of $f(x) = -3x^2 + 12x - 9$; the positive area A_1 and "negative area" A_2 are equal

area of the region bounded by the integrand and the x-axis from 1 to 4, not the actual area, which is $4-(-4)=8$. See Figure 6.18. ∎

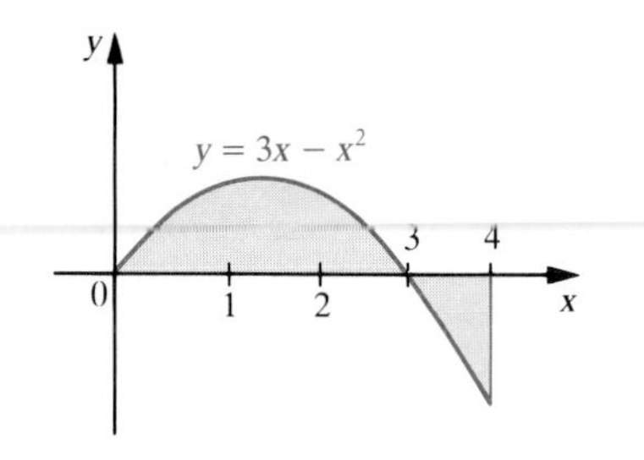

Figure 6.19
Graph of $f(x)=3x-x^2$ showing the area bounded by $f(x)$, the x-axis, and $x=4$

Example 6.19 Find the true area bounded by the curve $f(x)=3x-x^2$, the x-axis, and the vertical lines $x=0$ and $x=4$.

Solution The region described is shown in Figure 6.19. The function $f(x)$ has a zero at $x=3$, within the interval of integration. Hence to find the true area, we must compute the area from 0 to 3 and the area from 3 to 4 separately. From

$$\int_0^3 (3x-x^2)\,dx = \left(\frac{3x^2}{2}-\frac{x^3}{3}\right)\Bigg|_0^3 = \frac{27}{2}-9=\frac{9}{2},$$

we know that the area from 0 to 3 is $\frac{9}{2}$ square units. From

$$\int_3^4 (3x-x^2)\,dx = \left(\frac{3x^2}{2}-\frac{x^3}{3}\right)\Bigg|_3^4 = \left(\frac{3(16)}{2}-\frac{64}{3}\right)-\left(\frac{27}{2}-9\right)=\frac{8}{3}-\frac{9}{2}=-\frac{11}{6},$$

we know that the area from 3 to 4 is $-\frac{11}{6}$ square units. Thus, the true area from 0 to 4 is $\frac{9}{2}+\frac{11}{6}=-\frac{19}{3}$ square units. ∎

For any function $F(x)$, $F(b)$ and $F(a)$ are the values of $F(x)$ at the right and left endpoints of the interval $[a, b]$, so $F(b)-F(a)$ is the **net change** in $F(x)$ as x goes from a to b. Consequently, letting $f(x)$ denote $F'(x)$, we can interpret the Fundamental Theorem of Calculus as follows.

> net change from a to b of any antiderivative $F(x)$ of $f(x)$ $= F(b)-F(a)=\displaystyle\int_a^b f(x)\,dx$

The following examples illustrate interpretations for the definite integral in several situations.

Example 6.20 Suppose that the marginal cost for a given company is $M_C(x)$, where x is the number of items produced per day. What is the significance of $\int_a^b M_C(x)\,dx$?

Solution Let the cost function for the company be $C(x)$ dollars per day. Since marginal cost is the derivative of the cost function, $C'(x)=M_C(x)$, or $C(x)$ is an antiderivative of $M_C(x)$. Thus $\int_a^b M_C(x)\,dx = C(b)-C(a)$ is the net change in the cost function as production changes from $x=a$ to $x=b$ items per day. Therefore, $\int_a^b M_C(x)\,dx$ gives the company's cost of increasing production from a items per day to b items per day. ∎

Example 6.21 Suppose that the marginal cost for a refrigerator company producing x refrigerators per day is $M_C(x)=100x^{-1/2}+150$. Determine the cost of increasing production from 100 refrigerators per day to 400 refrigerators per day.

Solution According to Example 6.20, the cost of increasing production from 100 to 400 items per day would be

$$\begin{aligned}\int_{100}^{400} M_C(x)\,dx &= \int_{100}^{400} (100x^{-1/2} + 150)\,dx \\ &= (200x^{1/2} + 150x)\Big|_{100}^{400} \\ &= [200(20) + 150(400)] - [200(10) + 150(100)] \\ &= \$47{,}000.\end{aligned}$$

■

Example 6.22 Suppose that the velocity of an object at time t is $v(t)$. What is the significance of $\int_a^b v(t)\,dt$?

Solution Let the position of the object at time t be denoted by $s(t)$. In Chapter 2 we saw that $s'(t) = v(t)$. Hence,

$$\int_a^b v(t)\,dt = s(b) - s(a).$$

Now $s(b) - s(a)$ is the net distance between the object's position at time $t = a$ and its position at time $t = b$. For example, if the motion is horizontal with the positive direction to the right and if $\int_a^b v(t)\,dt = 7$, then the object at time $t = b$ is 7 units to the right of its position at time $t = a$. Similarly, if $\int_a^b v(t)\,dt = -4$, then the object at time $t = b$ is 4 units to the left of its position at time $t = a$. ■

Example 6.23 After a wheel is dropped from an airplane, how far does it fall between four and ten seconds, given that its velocity is $v(t) = 32t$ and the positive direction is downward?

Solution According to Example 6.22, the distance between the position of the wheel at time $t = 4$ and at time $t = 10$ is given by

$$\int_4^{10} 32t\,dt = 16t^2\Big|_4^{10} = 1600 - 256 = 1344.$$

Hence, the wheel falls 1334 feet in the interval from 4 to 10 seconds after it is dropped. ■

Example 6.24 A team of biology students undertakes a summer job to perform tests on the 600 largest municipal water supplies in the state. Using results from the first four weeks' work (and a method discussed in Chapter 8), the students compute that the number of tests they will complete in the tth week is $50 + 2t$. Assuming 12 weeks in the summer, how many weeks of summer will be left after the testing has been completed?

Solution The rate of testing if $50 + 2t$, so if the number of tests performed in k weeks is $T(k)$, then

$$T(k) = \int_0^k (50 + 2t)\,dt = 50t + t^2\Big|_0^k = 50k + k^2.$$

Since 600 tests must be completed, we determine the value of k such that $T(k) = 600$:

$$600 = 50k + k^2$$
$$k^2 + 50k - 600 = 0$$
$$(k + 60)(k - 10) = 0$$
$$k = -60 \quad \text{or} \quad k = 10.$$

Certainly $k = 10$ is the only viable solution, so we conclude that the students will test 600 municipal water supplies in ten weeks, leaving them two weeks of summer. ■

The Area Between Two Curves

Consider the region in the xy-plane shown in Figure 6.20(a). The region is bounded above by the function $f(x)$, below by the nonnegative function $g(x)$, and on the sides by the lines $x = a$ and $x = b$. We wish to determine the area of this region. The area of the region can be thought of as the difference between the two areas shown in Figure 6.20(b). One

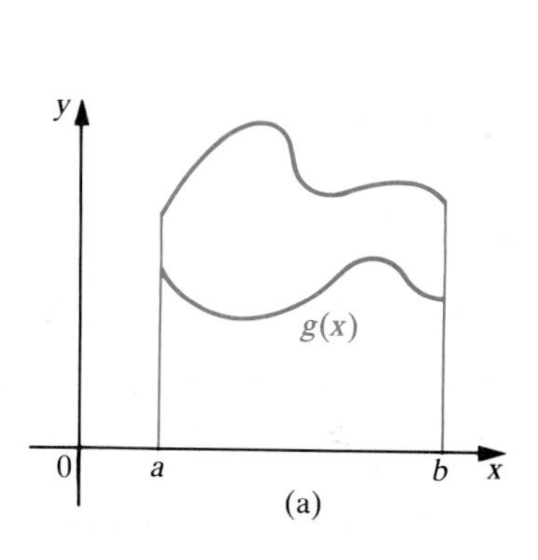

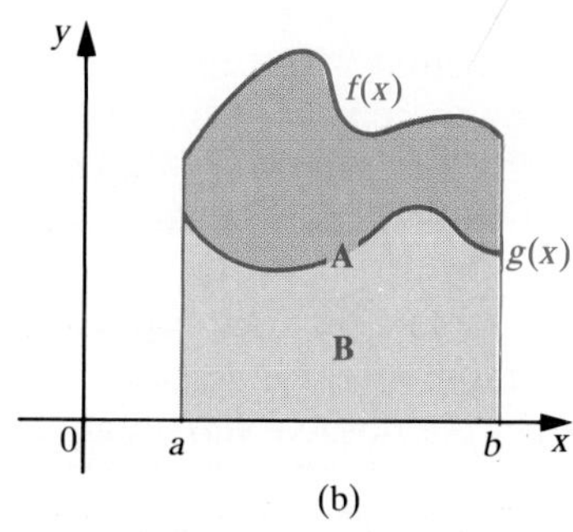

area under $g(x)$ = B = blue area

area under $f(x)$ = A = entire shaded area

Figure 6.20
(a) A region in the xy-plane; (b) the area of the region is the difference of areas A and B

area, A, is the area between the curve $f(x)$ and the x-axis, and the second area, B, is the area between the curve $g(x)$ and the x-axis. If we compute the areas by integration, then the difference of these values is the desired area:

$$\begin{aligned} \text{area between } f(x) \text{ and } g(x) &= \text{area under } f(x) - \text{area under } g(x) \\ &= \int_a^b f(x)\,dx - \int_a^b g(x)\,dx \\ &= \int_a^b [f(x) - g(x)]\,dx \end{aligned}$$

if $f(x) \geq g(x)$ on $[a, b]$. (1)

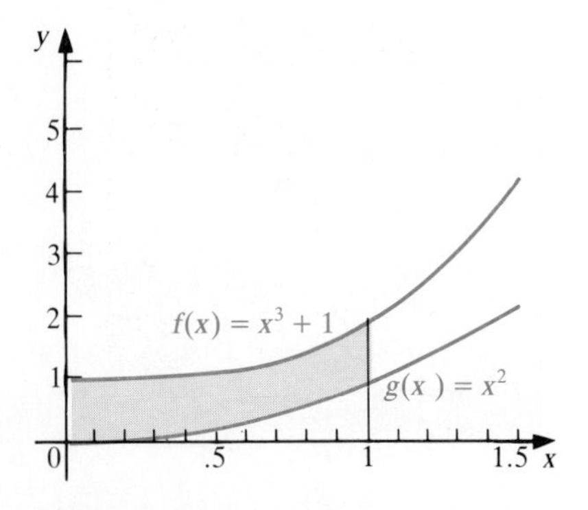

Figure 6.21
Region between $f(x) = x^3 + 1$ and $g(x) = x^2$ from $x = 0$ to $x = 1$

Example 6.25 Find the area between the curves $f(x) = x^3 + 1$ and $g(x) = x^2$ between $x = 0$ and $x = 1$, as shown in Figure 6.21.

Solution Since $f(x) \geq g(x)$ on the interval $[0, 1]$, the area between the curves is given by

$$\text{area} = \int_0^1 [f(x) - g(x)]\,dx = \int_0^1 (x^3 + 1 - x^2)\,dx$$

$$= \left(\tfrac{1}{4}x^4 + x - \tfrac{1}{3}x^3\right)\Big|_0^1 = \left(\tfrac{1}{4} + 1 - \tfrac{1}{3}\right) - (0 + 0 - 0) = \tfrac{11}{12}.$$

The area of the given region is $\frac{11}{12}$ square unit. ■

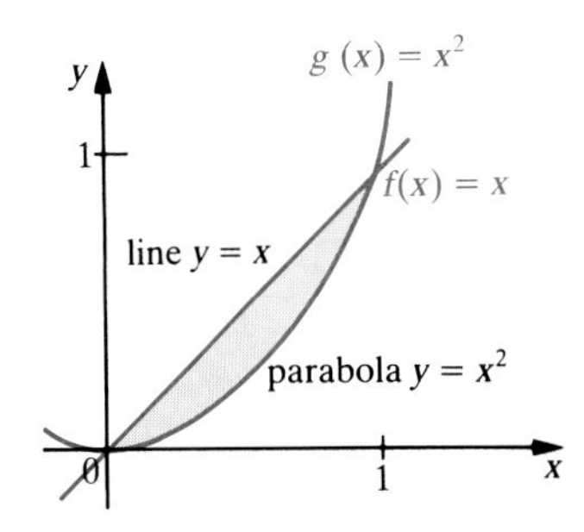

Figure 6.22
Region between the graphs of $f(x) = x$ and $g(x) = x^2$

Example 6.26 Find the area bounded by the curves $y = f(x) = x$ and $y = g(x) = x^2$.

Solution The graphs of the functions are shown in Figure 6.22. We seek the area of the shaded region. To find the points of intersection of the two curves, we set $f(x) = g(x)$ and solve for x:

$$x = x^2$$
$$x^2 - x = 0$$
$$x(x - 1) = 0.$$

Hence $x = 0$ and $x = 1$ are the x-coordinates of intersection. Since $f(x) = x \geq x^2 = g(x)$ on $[0, 1]$, the area between the curves is given by equation (1). The area is

$$\text{area} = \int_0^1 [x - x^2]\,dx = \left(\tfrac{1}{2}x^2 - \tfrac{1}{3}x^3\right)\Big|_0^1 = \tfrac{1}{2} - \tfrac{1}{3} - 0 = \tfrac{1}{6}.$$

The area of the indicated region is $\frac{1}{6}$ square unit. ■

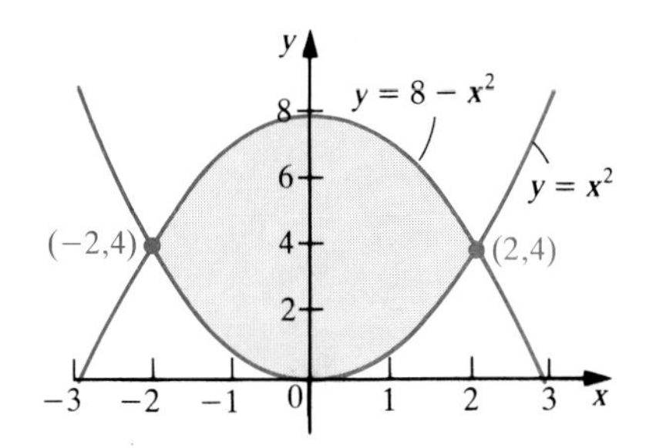

Figure 6.23
Region bounded by $y = x^2$ and $y = 8 - x^2$

Example 6.27 Find the area bounded by the functions $y = x^2$ and $y = 8 - x^2$.

Solution We must determine where the functions intersect and which one is the "top" function. The graphs of both functions are parabolas, the first shaped like a valley and the second like a hill. To find the intersection points, we solve

$$x^2 = 8 - x^2.$$

The roots are $x = 2$ and $x = -2$. Figure 6.23 shows the region; the function $y = 8 - x^2$ is the "top" function and $y = x^2$ is the "bottom" function. Consequently, the area is

$$\text{area} = \int_{-2}^{2} [(8 - x^2) - x^2]\,dx = \int_{-2}^{2} (8 - 2x^2)\,dx$$

$$= \left(8x - \tfrac{2}{3}x^3\right)\Big|_{-2}^{2} = \tfrac{64}{3} \text{ square units.}$$ ■

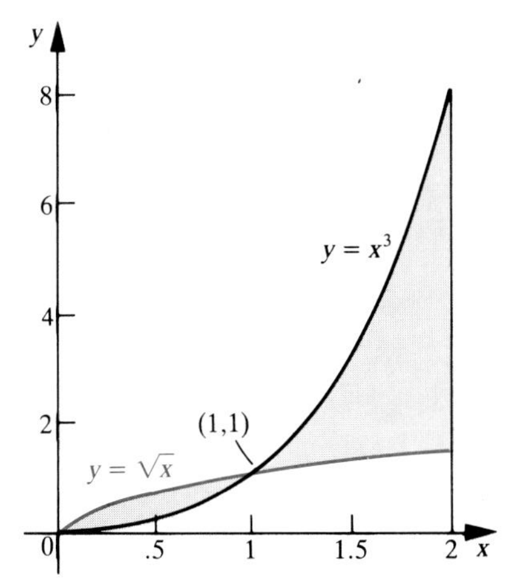

Figure 6.24
Region bounded by $y = x^3$, $y = \sqrt{x}$, and $x = 2$

Example 6.28 Find the area bounded by $y = x^3$, $y = \sqrt{x}$, and $x = 2$.

Solution Figure 6.24 shows the region, which consists of two sections: one for x between 0 and 1 and the other for x between 1 and 2. Notice that the top function is $y = \sqrt{x}$ for $0 \leq x \leq 1$, but it is $y = x^3$ over the interval $[1, 2]$. Since we do not have one function above the other throughout the interval, we *cannot* obtain the area by formula (1) above. This situation is similar to that in Example 6.19, where the function $f(x)$ was above the axis over part of the interval and below it on the rest of the interval. Setting $x^3 = \sqrt{x}$, we find the point of intersection to be $(1, 1)$, so we have

$$\text{area} = \int_0^1 (\sqrt{x} - x^3)\,dx + \int_1^2 (x^3 - \sqrt{x})\,dx$$

$$= \left(\tfrac{2}{3}x^{3/2} - \tfrac{1}{4}x^4\right)\Big|_0^1 + \left(\tfrac{1}{4}x^4 - \tfrac{2}{3}x^{3/2}\right)\Big|_1^2.$$

Completing the evaluation, we find that the area is $\frac{29}{6} - \frac{4}{3}\sqrt{2} \approx 0.9477$ square unit. ■

Exercises

Exer. 1–28: Evaluate the definite integral using the Fundamental Theorem of Calculus.

1. $\int_1^3 2x\,dx$
2. $\int_1^4 (5x + 1)\,dx$
3. $\int_2^4 (x^2 - 2x + 1)\,dx$
4. $\int_{-3}^4 (x^3 + 5x - 2)\,dx$
5. $\int_{-2}^2 x^3\,dx$
6. $\int_1^9 \sqrt{x}\,dx$
7. (a) $\int_{-1}^1 (x^2 - 2)\,dx$ (b) $\int_{-1}^1 (t^2 - 2)\,dt$
8. (a) $\int_{-1}^1 (x^3 - x)\,dx$ (b) $\int_{-1}^1 (t^3 - t)\,dt$
9. (a) $\int_2^3 \sqrt[3]{x}\,dx$ (b) $\int_2^3 \sqrt[3]{u}\,du$
10. (a) $\int_1^4 (3 - x)\,dx$ (b) $\int_1^4 (3 - u)\,du$
11. $\int_1^5 \left(\frac{1}{x} - x^2\right) dx$
12. $\int_{-2}^2 (x^2 + 3e^{x/2})\,dx$
13. $\int_1^e \left(4 - \frac{3}{x}\right) dx$
14. $\int_2^3 \frac{1}{x^2}\,dx$
15. (a) $\int_0^2 e^{3x}\,dx$ (b) $\int_0^2 e^{3t}\,dt$
16. (a) $\int_{-1}^4 e^{-x}\,dx$ (b) $\int_{-1}^4 e^{-t}\,dt$
17. $\int_1^5 (x - x^{-1})\,dx$
18. $\int_1^8 4\left(\frac{2}{x} + e^{x/3}\right) dx$
19. $\int_{-1}^1 (e^x + e^{-x})\,dx$
20. $\int_e^{e^2} \frac{6}{x}\,dx$
21. $\int_{-2}^2 (x + x^3)\,dx$
22. $\int_0^1 (x - 4\sqrt{x})\,dx$
23. $\int_1^4 \frac{1}{\sqrt{x}}\,dx$
24. $\int_1^4 \frac{dx}{x^3}$
25. $\int_{-1}^1 (6 + e^{-x})\,dx$
26. $\int_{-1}^1 \left(6 + \frac{1}{e^{3x}}\right) dx$
27. $\int_1^4 (x^2 + x^{-2})\,dx$
28. $\int_1^3 (3 - e^{-x})\,dx$

Exer. 29–44: Calculate the integral. Sketch a graph of the function being integrated, and shade the area that is equal to the integral.

29. $\int_1^4 x^2\,dx$
30. $\int_2^3 (3x^2 - 2)\,dx$
31. $\int_0^4 (5x + 2)\,dx$
32. $\int_1^5 (8x^3 + 4x^2 + 2x + 1)\,dx$
33. $\int_1^3 \frac{1}{x}\,dx$
34. $\int_{-1}^4 x^3\,dx$
35. $\int_1^3 (x^2 - 1)\,dx$
36. $\int_{-2}^0 (x - x^3 + 2)\,dx$
37. $\int_{-2}^3 (4 - x)\,dx$
38. $\int_0^3 e^x\,dx$
39. $\int_1^4 \frac{1}{x}\,dx$
40. $\int_1^{e^2} \left(\frac{1}{x} + x^2\right) dx$
41. $\int_{-4}^{-1} \frac{1}{x}\,dx$
42. $\int_{-e^2}^{-1} \left(\frac{1}{x} + x^2\right) dx$
43. $\int_{-3}^3 (4 - x^2)\,dx$
44. $\int_1^5 (x^2 - 4)\,dx$

Exer. 45–56: Find the area of the region bounded by the curves and lines; sketch a graph showing the area.

45. $f(x)=x,\quad g(x)=2x,\quad x=2,\quad x=4$
46. $f(x)=x^2,\quad g(x)=x,\quad x=2,\quad x=3$
47. $f(x)=1/x,\quad g(x)=x^2,\quad x=3$
48. $f(x)=1/x^2,\quad g(x)=x,\quad h(x)=8x$ for $x>0$
49. $f(x)=e^{2x},\quad g(x)=e^x,\quad x=2$
50. $f(x)=x+1,\quad g(x)=2x+2,\quad x=2$
51. $f(x)=x+1,\quad g(x)=2x+2,\quad x=-2,\quad x=2$
52. $f(x)=9,\quad g(x)=x^2$
53. $f(x)=x,\quad g(x)=x^3$ for $x\geq 0$
54. $f(x)=2,\quad g(x)=1/x,\quad x=1,\quad x=e^2$
55. $f(x)=2,\quad g(x)=1/x,\quad x=1,\quad x=9$
56. $f(x)=e^{-x},\quad g(x)=1,\quad x=2$

57. A company has found that its marginal revenue when producing x items per day is $M_R(x)$. What is the significance of $\int_a^b M_R(x)\,dx$?
58. Suppose that, in Exercise 57, $M_R(x)=200x^{-3/2}+125$ dollars. How much more revenue will the company generate if production is increased from 64 items per day to 81 items per day?
59. The acceleration of an object in a rectilinear motion problem is $a(t)$ at time t. What is the significance of $\int_b^c a(t)\,dt$?
60. Suppose that, in Exercise 59, $a(t)=4t^2+2t+1$ ft/sec^2. What is the difference between the velocity of the object at time $t=4$ and at time $t=9$?
61. The rate at which a calculator company is selling calculators in hundreds per day is given by $S(t)=30t+1$ for $0\leq t\leq 7$. How many calculators does the company sell during the 7-day period?
62. Under ideal conditions, field mice increase their number at the rate

$$n(t)=3t^2+0.5t^3+40,$$

where t is measured in months. What is the increase in number of field mice between the third and fourth months?
63. The marginal productivity of a team of researchers is given by

$$M_p(t)=0.2t-0.033t^2+0.00001t^3,$$

where t is measured in weeks. What is the productivity of the team between the second and fourth weeks? (Marginal productivity is the derivative of the productivity function.)
64. By analyzing the production of the first 100 pairs of gloves, the amount of labor in a glove factory is found to be $x-\sqrt{x}$ minutes per glove. How much labor is required to produce the next 100 pairs?
65. If the marginal price (in dollars) is $8-0.5x$, what is the change in price from the sale of 20 items to the sale of 30 items? (Marginal price is the derivative of the price function.)
66. The marginal cost (in dollars) of a company is

$$M_C(x)=2+0.02x$$

and the fixed cost is \$200. What is the cost of producing 100 items?
67. A sociologist uses a team of inexperienced interviewers to gather data. The rate at which people are interviewed by the team per week after t weeks is $30+2t$. To complete the survey on time, 1000 people must be interviewed in ten weeks. Will the team complete the survey on time?
68. The egg production rate of a chicken farm after t days is $100+0.25t$ eggs per day. How many eggs are produced in the first 100 days? In the second 100 days?

Exer. 69–70: Prove the given property of definite integrals.

69. $\displaystyle\int_a^b kf(x)\,dx=k\int_a^b f(x)\,dx\quad k$ constant
70. $\displaystyle\int_a^b f(x)\,dx=\int_a^c f(x)\,dx+\int_c^b f(x)\,dx$

Riemann Sums and Applications — 6.5

Consider again the problem of finding the area under the graph of a continuous nonnegative function $f(x)$ between two vertical lines $x=a$ and $x=b$. In Section 6.3 we divided the interval $[a, b]$ into n subintervals as shown in Figure 6.25.

The endpoints of the n subintervals are denoted by x_i, $i=0, 1, \ldots, n$. To approximate the area under the curve for the ith subinterval, from x_{i-1} to x_i, we calculate the area of an associated rectangle, which has the same width as the ith subinterval and has height $f(t_i)$ for some t_i in the interval $[x_{i-1}, x_i]$. We denote the width of the ith subinterval by $\Delta x_i=x_i-x_{i-1}$. Then the area of the ith approximating rectangle is $f(t_i)\,\Delta x_i$. Our approximation to the area

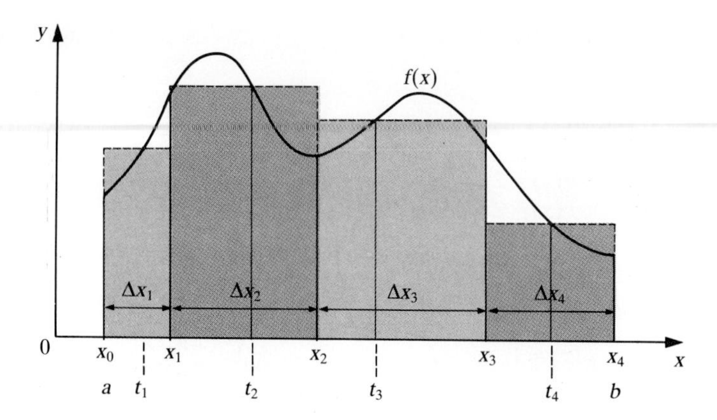

Figure 6.25
A typical Riemann sum partition

under the curve from $x = a$ to $x = b$ is the sum of the areas of these approximating rectangles:

$$\text{area} \approx f(t_1)\,\Delta x_1 + f(t_2)\,\Delta x_2 + f(t_3)\,\Delta x_3 + \cdots + f(t_n)\,\Delta x_n.$$

The Riemann sum on the right side of this approximation is denoted by

$$\sum_{i=1}^{n} f(t_i)\,\Delta x_i,$$

which we read as *the sum from i equal* 1 *to n of* $f(t_i)\,\Delta x_i$; that is,

$$\sum_{i=1}^{n} f(t_i)\,\Delta x_i = f(t_1)\,\Delta x_1 + f(t_2)\,\Delta x_2 + \cdots + f(t_n)\,\Delta x_n.$$

The t_i selected from the ith subinterval in the formation of a Riemann sum may be any value in that subinterval. Changing the choice of t_i will, in general, change the value of the sum; however, if the function $f(x)$ is not too erratic, and if we divide $[a, b]$ into a large number of small subintervals, then the amount by which the height $f(t_i)$ varies over the ith subinterval will be small so that any such Riemann sum will be a good approximation to the area under the curve.

Example 6.29 Let $f(x) = x^2$, $a = 1$, $b = 3$, and $n = 4$. For the endpoints of the four subintervals, take $x_1 = 2$, $x_2 = 2.5$, and $x_3 = 2.8$. We also have $x_0 = a = 1$ and $x_4 = b = 3$. Compute two Riemann sums associated with this information.

Solution Because of the various choices one can make for t_i, the Riemann sum is not uniquely determined by the given information. For the first case, refer to Figure 6.26; we (arbitrarily) take $t_1 = 1.5$, $t_2 = 2.3$, $t_3 = 2.7$, and $t_4 = 3.0$. We calculate the corresponding Δx_i's as 1.0, 0.5, 0.3, and 0.2:

$$\Delta x_1 = x_1 - x_0 = 2.0 - 1.0 = 1.0$$
$$\Delta x_2 = x_2 - x_1 = 2.5 - 1.0 = 0.5$$
$$\Delta x_3 = x_3 - x_2 = 2.8 - 2.5 = 0.3$$
$$\Delta x_4 = x_4 - x_3 = 3.0 - 2.8 = 0.2.$$

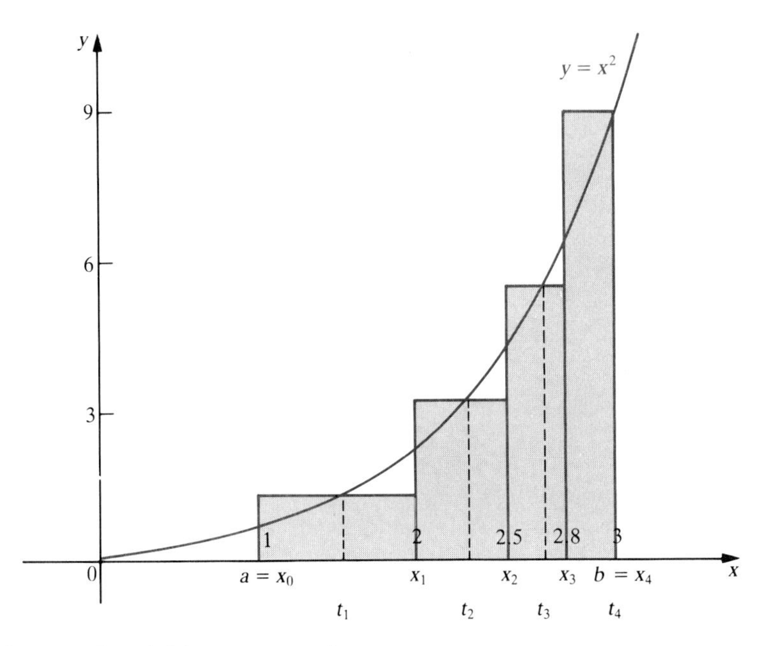

Figure 6.26
First Riemann sum for Example 6.29

The associated Riemann sum is

$$\text{first Riemann sum} = (1.5)^2(1.0) + (2.3)^2(0.5) + (2.7)^2(0.3) + (3.0)^2(0.2) = 8.882.$$

Table 6.2 shows the calculations. Note that $\sum_{i=1}^{n} f(t_i)\,\Delta x_i$ is the total (sum) of the numbers in the last column. The actual area is $\frac{26}{3} \approx 8.667$, so this Riemann sum is close to the correct value even though $n = 4$ is a very small number of subintervals.

Table 6.2

i	x_i (given)	Δx_i (computed)	t_i (chosen)	$f(t_i)$ (computed)	$f(t_i)\,\Delta x_i$ (computed)
0	1.0				
1	2.0	1.0	1.5	2.25	2.250
2	2.5	0.5	2.3	5.29	2.645
3	2.8	0.3	2.7	7.29	2.187
4	3.0	0.2	3.0	9.00	1.800
					$\sum_{i=1}^{4} f(t_i)\,\Delta x_i = 8.882$

For the second sum, we (arbitrarily) take $t_1 = 1.3$, $t_2 = 2.5$, $t_3 = 2.6$, and $t_4 = 2.9$. The Riemann sum for these choices is

$$\text{second Riemann sum} = (1.3)^2(1.0) + (2.5)^2(0.5) + (2.6)^2(0.3) + (2.9)^2(0.2) = 8.525.$$

Table 6.3 shows the calculations. ■

Table 6.3

i	x_i (given)	Δx_i (computed)	t_i (chosen)	$f(t_i)$ (computed)	$f(t_i)\,\Delta x_i$ (computed)
0	1.0				
1	2.0	1.0	1.3	1.69	1.690
2	2.5	0.5	2.5	6.25	3.125
3	2.8	0.3	2.6	6.76	2.038
4	3.0	0.2	2.9	8.41	1.282

$$\sum_{i=1}^{4} f(t_i)\,\Delta x_i = 8.525$$

Clearly there are many Riemann sums corresponding to a given region: the number n of subintervals is arbitrary, the n subintervals may be chosen in any fashion, and within each subinterval, each t_i may be chosen randomly. Nevertheless, for any continuous function $f(x)$,

$$\lim_{\substack{n\to\infty \\ \max \Delta x_i \to 0}} \left(\sum_{i=1}^{n} f(t_i)\,\Delta x_i \right) = \int_a^b f(x)\,dx.$$

(Recall that by $n \to \infty$ in the limit, we mean more and more subintervals are used in the sum. By $\max \Delta x_i \to 0$, we mean that the widest of the subintervals is getting thinner and thinner.) The limit says that by using more and more, thinner and thinner subintervals, the corresponding Riemann sums become increasingly accurate approximations to the area under the graph of the function $f(x)$.

Thinking of integrals in terms of Riemann sums is extremely helpful in setting up integrals to solve problems. Consider the problem of finding the area under $f(x)$, Figure 6.27 shows $f(x)$ and a typical *element* of a Riemann sum for the area. This element is a rectangle of height $f(t_i)$ and width Δx_i. The area of the typical element is

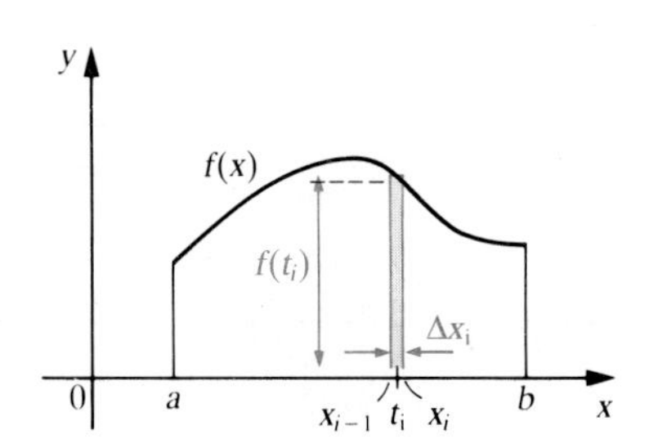

Figure 6.27

A typical element for calculating area by Riemann sums

$$\Delta A_i = f(t_i)\,\Delta x_i.$$

The sum of many such elements is a Riemann sum $\sum_{i=1}^{n} f(t_i)\,\Delta x_i$, and the limit of these sums will be the desired area, $\int_a^b f(x)\,dx$. The important point is the connection between the area of the typical element and the integrand; that is, by computing the desired quantity (area) for the microscopic element and then integrating, we obtain the proper expression for the desired area. The remaining examples in this section use this technique.

Volume of Revolution

Consider the **solid of revolution** obtained by rotating the graph of a function $y = f(x)$ about the x-axis as shown in Figure 6.28. We are interested in finding the volume of such a solid.

Figure 6.29 shows a typical element of a solid of revolution. This element is a disk whose radius is $f(t_i)$ and whose thickness is Δx_i. Hence

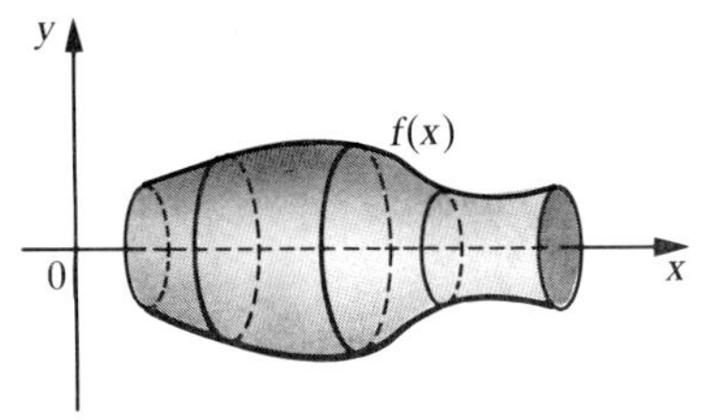

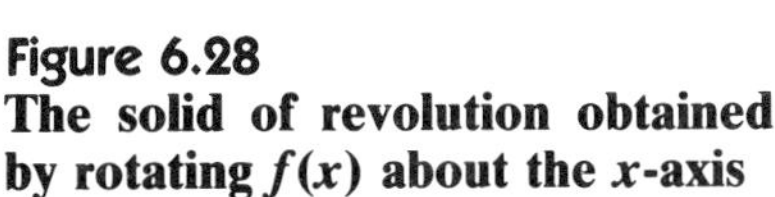

Figure 6.28
The solid of revolution obtained by rotating $f(x)$ about the x-axis

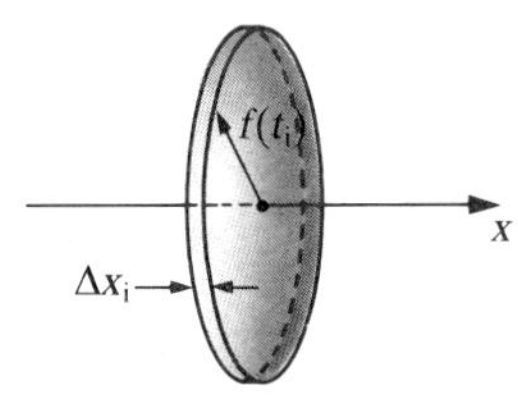

Figure 6.29
Typical element in the solid of revolution shown in Figure 6.28

the volume of the element is [the area of the base of the disk] × [thickness of the disk]. Since the formula for the area of the circular base of the disk is πr^2, where the radius is $r = f(t_i)$, we have

$$\Delta V_i = \pi[f(t_i)]^2\,\Delta x_i.$$

Adding the volumes for all elements gives an approximation to the total volume of the solid:

$$V \approx \sum_{i=1}^{n} \Delta V_i = \sum_{i=1}^{n} \pi[f(t_i)]^2\,\Delta x_i.$$

This is a Riemann sum; the exact volume is given by the corresponding integral:

$$V = \text{volume of revolution} = \int_a^b \pi[f(x)]^2\,dx.$$

Example 6.30 Find the volume of revolution of the region between the x-axis and the curve $y = \sqrt{16 - x^2}$. Figure 6.30 shows the solid of revolution.

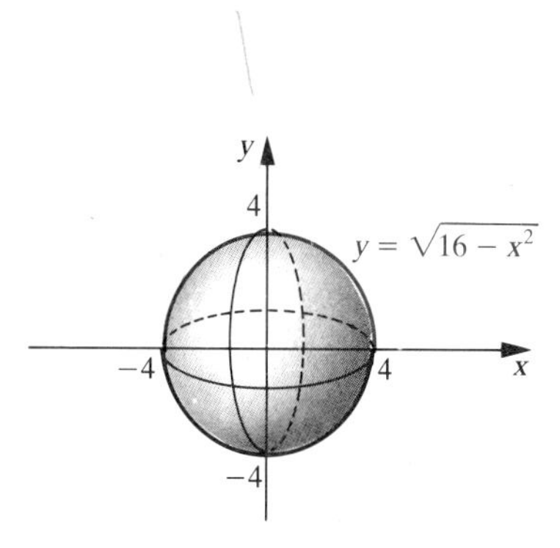

Figure 6.30
The solid of revolution for $f(x) = \sqrt{16 - x^2}$ revolved about the x-axis

Solution The curve is the top half of a circle, so the solid of revolution is a sphere. From our formula for volume, we have

$$\begin{aligned}
\text{volume} &= \int_{-4}^{4} \pi\left(\sqrt{16 - x^2}\right)^2 dx \\
&= \pi \int_{-4}^{4} (16 - x^2)\,dx \\
&= \pi(16x - \tfrac{1}{3}x^3)\Big|_{-4}^{4} \\
&= \pi(64 - \tfrac{64}{3}) - \pi(-64 + \tfrac{64}{3}) \\
&= \frac{256\pi}{3} \approx 268 \text{ cubic units.}
\end{aligned}$$

Since the solid of revolution is a sphere, we can check this result from the formula for the volume of a sphere of radius r: volume $= \frac{4}{3}\pi r^3$. The radius is $r = 4$, so the volume is $\frac{4}{3}\pi(4)^3 = \frac{256\pi}{3}$, as we obtained. ∎

Example 6.31 Find the volume of the solid of revolution created by revolving the graph of $f(x) = x^2$ from $x = 1$ to $x = 3$ about the x-axis.

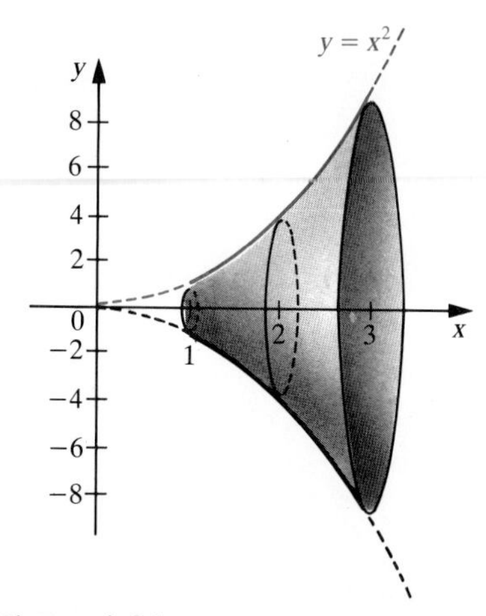

Figure 6.31
The solid of revolution for $f(x) = x^2$ from $x = 1$ to $x = 3$ about the x-axis

Solution The solid is shown in Figure 6.31. Thus

$$\text{volume} = \int_1^3 \pi(x^2)^2\,dx = \pi\frac{x^5}{5}\Big|_1^3 = \frac{242\pi}{5} \approx 152 \text{ cubic units.}$$

Population Density

Consider a large urban area for which the population density is $p(r)$ people per square mile, where r is distance in miles from the city center. Given $p(r)$, how can we find the actual population within a radius of R miles of the center?

Consider the ring formed by two concentric circles shown in Figure 6.32. The inner circle has radius r_i and the outer circle has radius $r_i + \Delta r_i$. If this ring is very narrow, then the change in r is small within the ring, so the population density $p(r)$ will be nearly constant throughout the ring. Let t_i be a number between r_i and $r_i + \Delta r_i$. The population within the ring is approximately $p(t_i)$ times the area of the ring. The area of the ring is $\pi(r_i + \Delta r_i)^2 - \pi r_i^2 = \pi(2r_i\,\Delta r_i + \Delta r_i^2)$. For very small Δr_i, the term Δr_i^2 will be negligible compared to the term $2r_i\,\Delta r_i$. Hence

$$\text{population within the ring} \approx p(t_i)2\pi r_i\,\Delta r_i,$$

and the accuracy of the approximation increases as Δr_i decreases. (The population of the ring is given by this element calculation.) Adding the populations for all such rings, we have

$$\text{population} = \sum_{i=1}^{n}(\text{population in ring } i) \approx \sum_{i=1}^{n} p(t_i)2\pi r_i\,\Delta r_i.$$

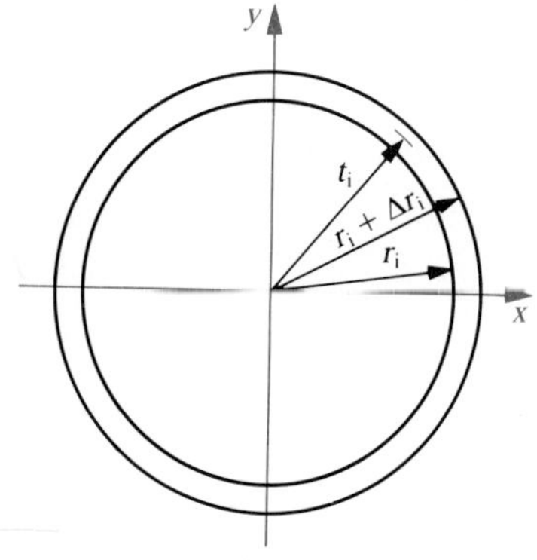

Figure 6.32
Concentric circles form a ring in which the population density may be considered constant

The limit of this Riemann sum is the integral below:

$$\text{population} = \int_0^R 2\pi r p(r)\,dr.$$

Example 6.32 The population density for a certain city is approximately given by the formula

$$p(r) = \frac{4}{r^2 + 20}$$

in units of 100,000 people per square mile.
(a) What is the population within a 2-mile radius of the city center?
(b) How large a radius from the center ensures a population of 500,000 within that radius?

Solution **(a)** The population within a 2-mile radius is

$$\text{population} = P = \int_0^2 2\pi r\frac{4}{r^2 + 20}\,dr = 4\pi\int_0^2 \frac{2r}{r^2 + 20}\,dr.$$

We will discuss a method for determining an antiderivative of $2r/(r^2 + 20)$ in Section 7.1. For now, notice that

$$\frac{d}{dr}\ln(r^2 + 20) = \frac{2r}{r^2 + 20}$$

so
$$\int \frac{2r}{r^2+20}\,dr = \ln(r^2+20) + C.$$

Consequently,
$$\begin{aligned} P &= 4\pi \ln(r^2+20)\Big|_0^2 \\ &= 4\pi \ln 24 - 4\pi \ln 20 \\ &= 4\pi(\ln 24 - \ln 20) \\ &= 4\pi \ln \tfrac{24}{20} \approx 2.291. \end{aligned}$$

Thus, approximately 229,100 people live within 2 miles of the center of the city.

(b) We seek the value of R so that

$$P = 4\pi \int_0^R \frac{2r}{r^2+20}\,dr = 5.$$

The value of the integral is obtained exactly as above in part (a), using R in place of 2. Hence the radius R must satisfy

$$4\pi[\ln(R^2+20) - \ln 20] = 5$$

so
$$4\pi \ln\left(\frac{R^2+20}{20}\right) = 5.$$

Dividing both sides by 4π and using each side as an exponent of e yields

$$\frac{R^2+20}{20} = e^{5/4\pi}.$$

Solving for R gives $R = \sqrt{20e^{5/4\pi} - 20} \approx 3.126$ miles. The population reaches half a million within 3.126 miles of the city center. ■

Average Value of a Function

We can use integration to define the **average value of a function**. Let $f(x)$ be a given continuous function, defined on an interval $[a, b]$. Divide $[a, b]$ into n equally spaced subintervals. The length of each subinterval will be $\Delta x = (b-a)/n$. Within the ith subinterval, we choose a point t_i and compute the average of the numbers $f(t_1), f(t_2), \ldots, f(t_n)$:

$$\begin{aligned} \text{average} &= \frac{f(t_1) + f(t_2) + \cdots + f(t_n)}{n} \\ &= f(t_1)\frac{1}{n} + f(t_2)\frac{1}{n} + \cdots + f(t_n)\frac{1}{n} \\ &= \frac{1}{b-a}\left(f(t_1)\frac{b-a}{n} + f(t_2)\frac{b-a}{n} + \cdots + f(t_n)\frac{b-a}{n}\right) \\ &= \frac{1}{b-a}\Big(f(t_1)\,\Delta x + f(t_2)\,\Delta x + \cdots + f(t_n)\,\Delta x\Big). \end{aligned}$$

The expression in parentheses is a Riemann sum with all $\Delta x_i = \Delta x$. This Riemann sum is approximately equal to $\int_a^b f(x)\,dx$. For this reason, we define

$$\begin{matrix}\text{the average value}\\ \text{of } f(x) \text{ on } [a, b]\end{matrix} = \frac{1}{b-a}\int_a^b f(x)\,dx.$$

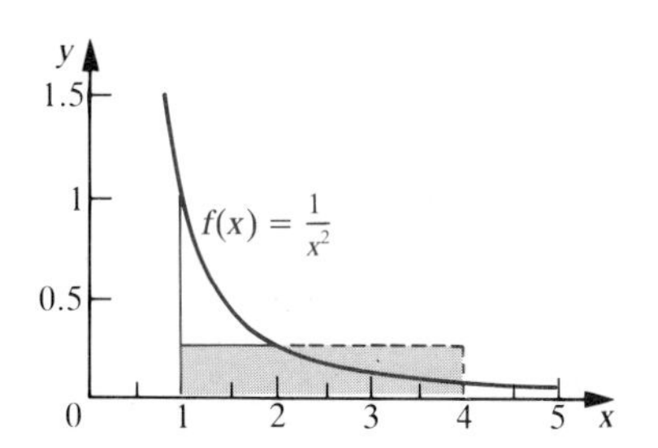

Figure 6.33
The area of the shaded rectangle equals the area under $f(x)$ on [1, 4] because the height of this rectangle is the average value of $f(x)$ on [1, 4]

Example 6.33 Find the average value of $f(x) = 1/x^2$ on [1, 4].

Solution

$$\text{average} = \frac{1}{4-1}\int_1^4 \frac{1}{x^2}\,dx$$
$$= \frac{1}{3}\left.\frac{-1}{x}\right|_1^4$$
$$= \tfrac{1}{3}\left(\tfrac{3}{4}\right) = \tfrac{1}{4}.$$

Thus the average value of $f(x) = 1/x^2$ on the interval [1, 4] is $\frac{1}{4}$. Figure 6.33 shows the function $f(x) = 1/x^2$ on the interval [1, 4] and the average value of $\frac{1}{4}$. The rectangle whose base is the interval [1, 4] and whose height is $\frac{1}{4}$ is shaded in the figure. This rectangle has the same area as the region under the curve $f(x) = 1/x^2$ from $x = 1$ to $x = 4$. ■

Consumer Surplus

Consumer surplus is a measure of the savings that occurs when a product is purchased by consumers at a price below the maximum amount the consumers are willing to pay. Recall that the demand function, discussed in Chapter 1, relates the price consumers are willing to pay for an item to the quantity produced. For this discussion, quantity will be used as the independent variable x and the price p will be the dependent variable: $p = D(x)$.

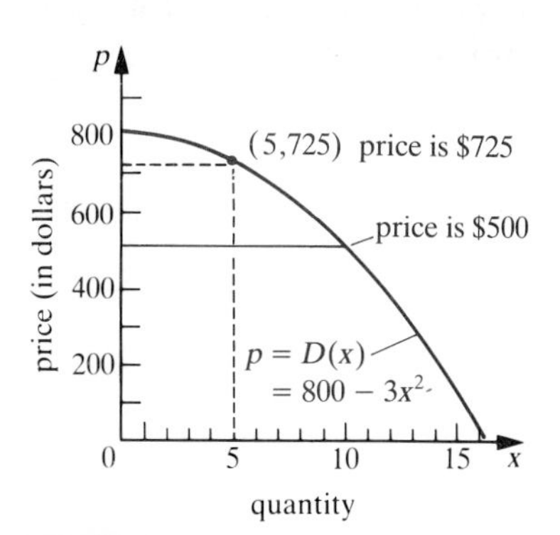

Figure 6.34
Price versus demand curve

Figure 6.34 shows the graph of a demand function $p = D(x) = 800 - 3x^2$. The point (5, 725) on the graph indicates that five items could be sold at a price of \$725 each. If the items were sold at \$500 each, then the savings to the five consumers would be 5(\$725 − \$500) = \$1125. This savings is called a *surplus* for the consumer. Such a surplus is associated with every price larger than the actual selling price and the total of these is what economists call *consumer surplus.*

Consider the situation depicted in Figure 6.35. The demand function $p = D(x)$ gives the price consumers are willing to pay when the available quantity is x. Let the actual selling price be b dollars and let a represent the corresponding demand so that $b = D(a)$. Divide the interval $[0, a]$ into n subintervals and consider one of these intervals, $[x_{i-1}, x_i]$, shown in the figure. Suppose the demand for the product changes from x_{i-1} to x_i, and let Δx_i denote that change: $\Delta x_i = x_i - x_{i-1}$. Now selling Δx_i items at b dollars each will produce a revenue of $b\,\Delta x_i$ dollars.

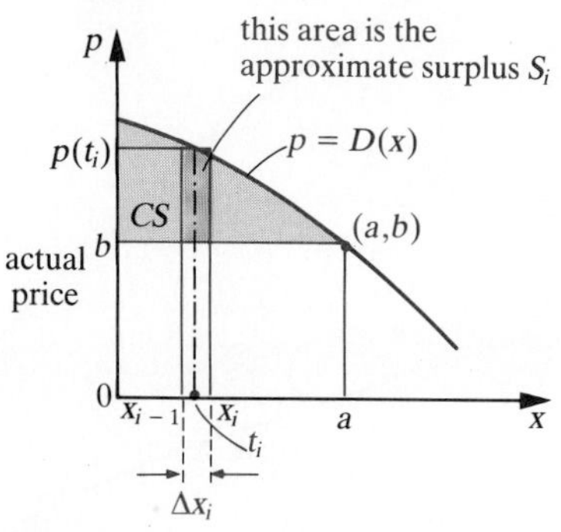

Figure 6.35
Typical demand-price function

If we let t_i be a number between x_{i-1} and x_i, then $p_i = D(t_i)$ is (approximately) the price at which the Δx_i items could have been sold, and this would have yielded a revenue of $p_i\,\Delta x_i = D(t_i)\,\Delta x_i$ dollars. The difference between these revenues is the associated surplus S_i:

$$S_i \approx D(t_i)\,\Delta x_i - b\,\Delta x_i = [D(t_i) - b]\,\Delta x_i.$$

The sum of these S_i is (approximately) the consumer surplus, CS:

$$CS \approx \sum_{i=1}^{n} [D(t_i) - b]\,\Delta x_i.$$

The limit of this Riemann sum is a definite integral that gives the exact consumer surplus, the area shaded in Figure 6.35. Thus

$$CS = \int_0^a [D(x) - b]\,dx, \tag{1}$$

where b is the actual selling price, and $D(a) = b$.

Example 6.34 Compute the consumer surplus if $p = D(x) = 800 - 3x^2$ and the actual selling price is \$500.

Solution This situation is presented in Figure 6.36. Since $D(10) = 500$, the consumer surplus is

$$\begin{aligned} CS &= \int_0^{10} (800 - 3x^2 - 500)\,dx \\ &= \int_0^{10} (300 - 3x^2)\,dx \\ &= (300x - x^3)\,\Big|_0^{10} = 2000. \end{aligned}$$

Thus, for the given demand function, when the price is \$500, consumers as a group pay \$2000 less than what they as individuals are willing to pay. This \$2000 savings is the consumer surplus. ■

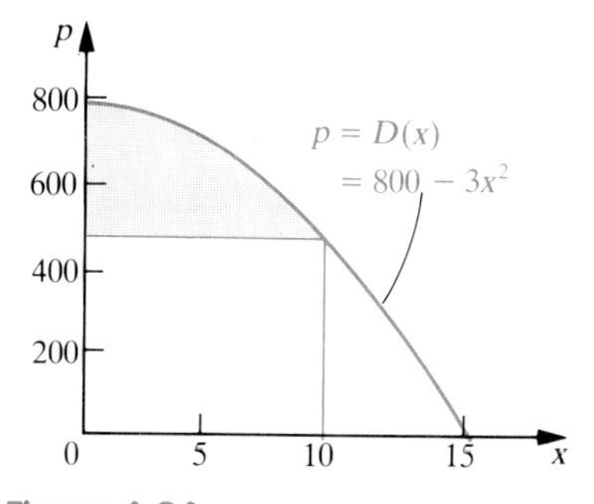

Figure 6.36
The demand curve for $p = D(x) = 800 - 3x^2$

Recall that when the supply and demand curves of a certain product intersect, the point of intersection is the equilibrium point, and the price coordinate of the equilibrium point is the equilibrium price (the price at which consumers will purchase the same quantity of a product that producers wish to sell at that price). The consumer surplus associated with the equilibrium price represents the savings to consumers by a competitive economy.

Example 6.35 The supply and demand functions for a certain product are given by

$$S(x) = 20x + 100 \quad \text{and} \quad D(x) = 400 - x^2,$$

respectively, where x denotes quantity produced in appropriate units. Determine the consumer surplus when the price is the equilibrium price.

Solution We first determine the equilibrium point by solving $D(x) = S(x)$:

$$\begin{aligned} 400 - x^2 &= 20x + 100 \\ x^2 + 20x - 300 &= 0 \\ (x + 30)(x - 10) &= 0 \\ x = -30 \quad &\text{or} \quad x = 10. \end{aligned}$$

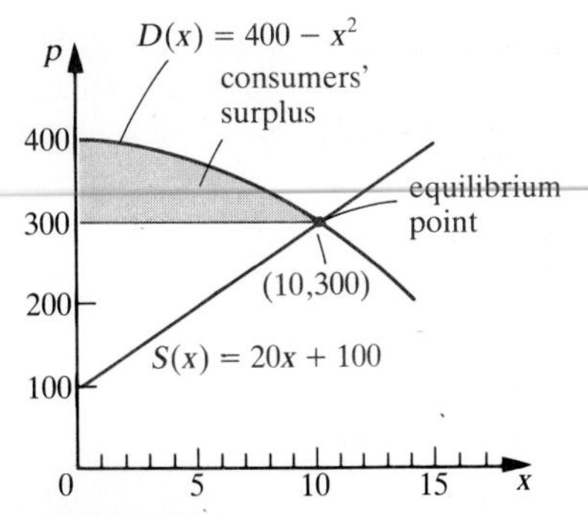

Figure 6.37
Demand and supply functions for Example 6.35

Since $x = -30$ makes no sense, $x = 10$ is the correct value. Since $D(10) = 300$, the equilibrium point is $(10, 300)$ and the equilibrium price is $p = 300$. See Figure 6.37.

The consumer surplus is

$$CS = \int_0^{10} [(400 - x^2) - 300]\, dx$$
$$= \int_0^{10} (100 - x^2)\, dx$$
$$= \left(100x - \frac{x^3}{3}\right)\Big|_0^{10}$$
$$= \$666.67.$$

■

Exercises

1. Let $f(x) = 1/x$, $a = 1$, $b = 4$, $n = 4$, $x_0 = 1$, $x_1 = 1.5$, $x_2 = 2$, $x_3 = 3$, $x_4 = 4$. Write down two Riemann sums associated with these data.
2. Work Exercise 1 with $x_0 = 1$, $x_1 = 2$, $x_2 = 3$, $x_3 = 3.5$, $x_4 = 3.8$.

Exer. 3–9: Find the volume of revolution about the x-axis of the function.

3. $f(x) = 2x, \quad 0 \le x \le 2$
4. $f(x) = x + 4, \quad 0 \le x \le 2$
5. $f(x) = 1/\sqrt{x}, \quad 1 \le x \le 4$
6. $f(x) = 2/x^2, \quad 1 \le x \le 3$
7. $f(x) = \sqrt{x}, \quad 0 \le x \le 4$
8. $f(x) = 5 - x, \quad 0 \le x \le 2$
9. $f(x) = x^2, \quad 0 \le x \le 3$

Exer. 10–12: Use the element method to find the volume of the solid of revolution formed by rotating the region between two functions $f(x)$ and $g(x)$ about the x-axis. This method is appropriately called the *washer method.*

10. $f(x) = 4, \quad g(x) = x, \quad 1 \le x \le 2$
11. $f(x) = x, \quad g(x) = x^2, \quad 0 \le x \le 3$
12. $f(x) = 6 - x, \quad g(x) = x^2, \quad 0 \le x \le 2$

13. Suppose the population density of a city is given by $p(r) = 16 - r$ for $0 \le r \le 16$, measured in 1000 people per square mile. How many people live within 4 miles of the city center? How many people live between 2 and 5 miles from the city center?

Exer. 14–17: Find the average value of $f(x)$.

14. $f(x) = e^{2x}; \quad [1, 3]$
15. $f(x) = x^2 - 4; \quad [0, 2]$
16. $f(x) = x - \sqrt{x}; \quad [0, 1]$
17. $f(x) = x^3 - x; \quad [-2, 3]$

Exer. 18–19: Suppose the temperature in a city over a 24-hour period is given by

$$f(t) = 40 + 5t - 0.2t^2$$

(in °F), where t is measured in hours with $t = 0$ designating the start of the 24-hour period.

18. What is the average temperature for the 24-hour period?
19. What is the average temperature over the first 12 hours?

Exer. 20–23: Suppose that the demand (in dollars) for a commodity is $p = D(x) = 100 - 2x$. Find the consumer surplus for the given price.

20. \$20
21. \$30
22. \$50
23. \$60

Exer. 24–27: Demand and supply functions are given in dollars, with x representing quantity. Assume that price is determined by the intersection of the demand and supply functions. Determine the consumer surplus.

24. $D(x) = 40 - 0.002x; \quad S(x) = 6 + 0.01x$
25. $D(x) = (30 - x)^2; \quad S(x) = x^2 + 2x - 340$
26. $D(x) = 750(x + 5)^{-1}; \quad S(x) = x + 10$
27. $D(x) = 600 - x^2; \quad S(x) = 300 - 5x$

28. Bicycle accessories are sold in units of ten at a price given by $p(x) = D(x) = 120 - 2x$ per unit. The cost per unit is given by $C(x) = x^3 - 26x^2 + 8x + 40$. The seller wants to maximize profit $P(x) = xp(x) - C(x)$. Determine the consumer surplus.

Exer. 29–32: The figure shows the intersection of demand and supply curves. Producers benefit if the price is set at

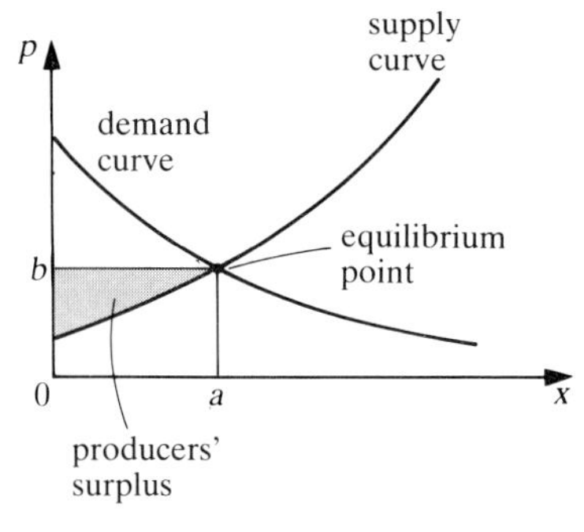

Exercises 29–32

the equilibrium price, since they would be willing to supply the product at prices below the equilibrium price. Their savings, called the *producer surplus*, is represented by the area shaded in the figure. Find a single definite integral expression for the producer surplus, and then compute the producer surplus.

29. $D(x) = 40 - 0.002x$; $S(x) = 4 + 0.01x$
30. $D(x) = (x - 41)^2$; $S(x) = x^2 + 2x + 6$
31. $D(x) = \dfrac{100}{x - 5}$; $S(x) = (0.01)(x + 300)$
32. $D(x) = 600 - x^2$; $S(x) = 300 - 10x$

Summary

An **antiderivative** of $f(x)$ is a function $F(x)$ such that $F'(x) = f(x)$ for all x. The set of all antiderivatives of $f(x)$, denoted by

$$\int f(x)\,dx$$

is the **indefinite integral** of $f(x)$. We may use the following rules to find indefinite integrals:

$$\int x^n\,dx = \frac{1}{n+1}x^{n+1} + C \quad \text{when } n \neq -1$$

$$\int x^{-1}\,dx = \ln|x| + C \quad \text{when } x \neq 0$$

$$\int e^x\,dx = e^x + C$$

$$\int [f(x) + g(x)]\,dx = \int f(x)\,dx + \int g(x)\,dx$$

$$\int cf(x)\,dx = c\int f(x)\,dx$$

Let n be a positive integer, let $[a, b]$ be an interval, and let $a = x_0 < x_1 < x_2 < \cdots < x_n = b$ be points in $[a, b]$. Let t_i be any point in the subinterval $[x_{i-1}, x_i]$. Let $f(x)$ be a function defined on $[a, b]$. Then a **Riemann sum** for $f(x)$ is

$$\sum_{i=1}^{n} f(t_i)\,\Delta x_i = f(t_1)\,\Delta x_1 + f(t_2)\,\Delta x_2 + \cdots + f(t_n)\,\Delta x_n,$$

where $\Delta x_i = x_i - x_{i-1}$. If $f(x)$ is continuous on $[a, b]$, then

$$\lim_{\substack{n\to\infty \\ \max \Delta x_i \to 0}} \sum_{i=1}^{n} f(t_i)\,\Delta x_i = \int_a^b f(x)\,dx.$$

If $A(x)$ is the **area under the graph** of a continuous function $f(x)$ from a to x, then $A'(x) = f(x)$. If $F(x)$ is an antiderivative of $f(x)$, then the area under the graph of $f(x)$ between a and b is $F(b) - F(a)$. The **definite integral** from a to b for any continuous function $f(x)$ on $[a, b]$ is

$$\int_a^b f(x)\,dx = F(b) - F(a)$$

where $F(x)$ is any antiderivative of $f(x)$. The definite integral represents the **net change** of any antiderivative of $f(x)$ from a to b.

The net area between the functions $f(x)$ and $g(x)$ from $x = a$ to $x = b$ is

$$\int_a^b [f(x) - g(x)]\,dx.$$

The **volume of revolution** about the x-axis of a function $f(x)$ from $x = a$ to $x = b$ is

$$\text{volume} = \int_a^b \pi[f(x)]^2\,dx.$$

The **average value** of a function $f(x)$ from $x = a$ to $x = b$ is

$$\text{average value} = \frac{1}{b-a}\int_a^b f(x)\,dx.$$

Consumer surplus is a measure of the saving that results when a product is purchased by consumers at a price below the amount the consumers are willing to pay. If $p = D(x)$ is the demand function, and the price is set at b dollars so that $D(a) = b$, then the consumer surplus is

$$CS = \int_0^a [D(x) - b]\,dx.$$

Review Exercises

1. Verify that $F'(x)$ is an antiderivative of $f(x)$.
   (a) $f(x) = 3x^2 - \frac{2}{x^2} + 1$; $F(x) = x^3 + \frac{2}{x} + x - 4$
   (b) $f(x) = \sqrt{(4x+1)^3}$; $F(x) = \frac{2}{5}(4x+1)^{5/2} - \frac{3}{10}$
   (c) $f(x) = (x^2 - 2) + 2x(x+4)$; $F(x) = (x^2 - 2)(x+4)$
2. Find three antiderivatives of each function.
   (a) $f(x) = \frac{2}{x^3}$ (b) $h(t) = t^2 - t$

Exer. 3–5: Find the antiderivative of $f(x)$ whose graph passes through the given point.

3. $f(x) = x^2 - 2$; $(3, 5)$
4. $f(x) = 1/x^2$; $(1, 1)$
5. $f(x) = 1/x$; $(e, 3)$

Exer. 6–9: Find the integral.

6. $\int (3x^2 - 4)\,dx$
7. $\int \frac{2\,dx}{3x^7}$
8. $\int (4x^{1/3} - 3x^{-1/3})\,dx$
9. $\int (x^{1/2} - x^{-1/2})\,dx$

Exer. 10–17: Evaluate the integral.

10. $\int_{-2}^{2} \frac{1}{2}x\,dx$
11. $\int_{0}^{4} (1 - 2t)\,dt$
12. $\int_{4}^{9} (t - t^{1/2})\,dt$
13. $\int_{1}^{2} x^2(x-2)\,dx$
14. $\int_{-4}^{4} x^3\,dx$
15. $\int_{0}^{3} e^{-2u}\,du$
16. $\int_{2}^{6} \frac{1}{3u}\,du$
17. $\int_{1}^{4} \left(e^{2t} + \frac{1}{2t}\right)dt$
18. The marginal cost function for producing x desks per day is $M_C(x) = 0.04x + 10$. Find the cost function if it costs \$402 to produce 10 desks per day. Find the cost of producing 100 desks per day.
19. As stockman adds water to a livestock watering tank, the rate of change of volume in the tank is $5 - \frac{1}{2}t^{-1/2}$ gallons per minute. The tank contained 1000 gallons when filling began. How many gallons will the tank contain after 3 hours of filling?
20. The marginal profit for sales of berets is $M_p(x) = -0.003x^2 + 0.01x + 0.2$ if x berets are sold per day. When none are sold, the dealer loses \$100 per day. Find the formula for profit when x berets are sold.
21. The rate of growth of bacteria in a culture at time t is given by $r(t) = 3t^{1/2}$ thousands, where t is measured in days. What is the increase in the number of bacteria between the fourth and ninth days?
22. A restaurant owner finds that the rate of increase of customers is given by $r(t) = 3t + 20$, where t is the time measured from the day the restaurant opened. How many customers will the restaurant serve in the second 10 days it is open?
23. A craft shop produces bread boxes at a rate of $30 + 2t - 0.5t^2$ boxes per week. How many boxes will be produced in the first 10 weeks?
24. If the marginal cost of producing x rolls of insulation is $8x^{2/3}$ dollars, find the cost function if the fixed overhead cost is \$20.
25. A spherical balloon of radius 25 feet develops a leak such that the volume is changing at a rate of $-0.008\sqrt{t}$ ft^3 per minute. What is the volume of the balloon 60 hours after the leak occurs?
26. If an object is projected vertically from the earth at a velocity of 144 ft/sec, how high will it rise?
27. Refer to Exercise 26. How long will it take the object to return to earth?
28. A man throws a rock downward from the top of a building 896 feet high. If the rock has an initial velocity of 16 ft/sec, what will its velocity be when it hits the ground?
29. Refer to Exercise 28. If the man throws the rock upward at a rate of 16 ft/sec, what will its velocity be when it hits the ground? (Assume the rock misses the building on its way down.)
30. A block of ice slides down a ramp and stops after 100 meters. If its initial velocity is 10 m/sec, what is its acceleration? (Assume the acceleration is constant.)
31. If a motorcycle accelerates from rest at constant acceleration, find its acceleration if it goes 360 feet in 6 seconds.

Exer. 32–36: Find the distance traveled $s(t)$ in t seconds under the conditions stated.

32. $a(t) = -t$; $v(0) = 100$; $s(1) = 4$
33. $a(t) = t^2 - \sqrt{t}$; $s(0) = 10$; $v(4) = 100$
34. $a(t) = 5t$; $s(0) = 10$; $s(2) = 100$
35. $a(t) = 2 - t^2$; $v(2) = 1$; $v(4) = 20$
36. $a(t) = 1/t^2$; $v(1) = 2$; $s(1) = 1$

37. A hockey puck travels 216 feet before coming to rest. If the acceleration of the puck is a constant -12 feet per second per second, find the velocity of the puck at the time it was struck.
38. Find the area of the region above the x-axis, below the curve $f(x) = 5 - x^2$, and between $x = 0$ and $x = 2$.
39. Find the area of the region above the x-axis, below the curve $f(x) = x - \sqrt{x}$, and between $x = 1$ and $x = 4$.
40. The curve $f(x) = -x^2 + 6x$ is a parabola that intersects the x-axis. Find the area of the region between this

curve and the x-axis. (First find where the curve and axis intersect.)

41. The region between the curve $f(x) = x - 1$, the x-axis, and the line $x = 4$ is a triangle. Compute its area using calculus and using the formula for the area of a triangle.
42. Find the area of the region between the x-axis, the curve $f(x) = x(2 - x)$, $x = -4$, and $x = 0$.
43. The area of the region between the curves $f(x) = x^3 + x$ and $g(x) = x + \sqrt{x}$ and the lines $x = 1$ and $x = 4$ can be considered as the difference between two regions: one under $f(x)$ and the other under $g(x)$. Find the total area of the region.
44. Find the volume of revolution about the x-axis of $y = e^x$ from $x = 0$ to $x = \ln 4$.
45. Find the average value of $y = e^x$ from $x = 0$ to $x = \ln 4$.

Facing page: **The total number of heartbeats during an exercise session can be calculated by monitoring the pulse rate during the session and integrating. See Example 7.14.**

Methods of Integration

· S · E · V · E · N ·

In Chapter 6 we derived the following fundamental integrals:

1. $\int u^n\,du = \frac{1}{n+1}u^{n+1} + C, \quad n \neq -1$

2. $\int \frac{1}{u}\,du = \ln|u| + C$

3. $\int e^{ku}\,du = \frac{1}{k}e^{ku} + C.$

In this chapter we combine these integrals with two important integration techniques to increase the collection of functions that we can integrate. The two techniques are *integration by substitution* and *integration by parts*. The first method is the integration analogue of the chain rule and the second method is the analogue of the product rule.

7.1 Integration by Substitution

The method of substitution is the integration analogue of the chain rule. The following example provides a concrete illustration of the method.

Example 7.1 Find $\int (x^2+8)^{23}2x\,dx$.

Solution If we define the function $u = g(x)$ by $u = x^2 + 8$, then the differential du is given by $du = g'(x)\,dx = 2x\,dx$. In the integrand, we make the substitutions

$$(x^2+8)^{23} \rightarrow u^{23} \quad \text{and} \quad 2x\,dx \rightarrow du$$

to obtain

$$\int (x^2+8)^{23}2x\,dx = \int u^{23}\,du.$$

This integral has the form of fundamental integral (1) and can be integrated easily using the power rule:

$$\int u^{23}\,du = \tfrac{1}{24}u^{24} + C.$$

Recalling that $u = x^2 + 8$, we have

$$\int (x^2+8)^{23}2x\,dx = \tfrac{1}{24}(x^2+8)^{24} + C.$$

We check our answer by differentiation:

$$\frac{d}{dx}\tfrac{1}{24}(x^2+8)^{24} = (x^2+8)^{23}(2x).$$ ■

Note that the verification step at the end of the previous example requires the chain rule. Recall that the chain rule states that if $F(u)$ is a differentiable function of u, and if $u = g(x)$ is a differentiable function of x, then

$$\frac{d}{dx}F(g(x)) = F'(g(x))g'(x).$$

Let $f(x)$ denote $F'(x)$. Then the preceding equation becomes

$$\frac{d}{dx}F(g(x)) = f(g(x))g'(x).$$

We now integrate both sides and write the right-hand side first to get

$$\int f(g(x))g'(x)\,dx = \int \frac{d}{dx}[F(g(x))]\,dx = F(g(x)) + C$$

because $F(g(x))$ is an antiderivative of $\left(\frac{d}{dx}\right)F(g(x))$. Also

$$\int f(u)\,du = \int F'(u)\,du = F(u) + C = F(g(x)) + C.$$

Since the right-hand sides of these equations are equal, we have the following.

Integration by Substitution

$$\int f(g(x))g'(x)\,dx = \int f(u)\,du \quad \text{where } u = g(x).$$

This formula is called **integration by substitution** because it can be obtained formally by substituting $u = g(x)$ and $du = g'(x)\,dx$ in the integral on the left.

In Example 7.1 the function $g(x)$ is $g(x) = x^2 + 8$ and $f(u)$ is $f(u) = u^{23}$. Thus $f(g(x)) = (x^2 + 8)^{23}$ and $g'(x)\,dx = 2x\,dx$. In view of this example and the above formula, we have the following guideline for choosing a substitution $u = g(x)$: a good choice for the substitution $u = g(x)$ is a function $g(x)$ such that both $g(x)$ and its derivative $g'(x)$ appear somewhere in the integrand. *The goal of the substitution is to obtain an integral* $\int f(u)\,du$ *that is easy to compute.*

In the substitution $u = g(x)$, we compute the differential $du = g'(x)\,dx$, or $du = \left(\frac{du}{dx}\right)dx$. This observation is used frequently in the following examples.

Example 7.2 Find $\int \sqrt{x^2 + 3x + 2}\,(2x + 3)\,dx$.

Solution Setting $u = x^2 + 3x + 2$, we have

$$\frac{du}{dx} = 2x + 3,$$

so $du = (2x + 3)\,dx$, and both u and its derivative appear in the integrand. Substituting u and du yields

$$\begin{aligned}\int \sqrt{x^2 + 3x + 2}(2x + 3)\,dx &= \int \sqrt{u}\,du \\ &= \int u^{1/2}\,du \\ &= \tfrac{2}{3}u^{3/2} + C \\ &= \tfrac{2}{3}(x^2 + 3x + 2)^{3/2} + C.\end{aligned}$$ ■

Example 7.3 Find $\int (x^2 + 5)^{12} x\,dx$.

Solution Setting $u = x^2 + 5$, we have $du/dx = 2x$, so $du = 2x\,dx$. Since the integrand contains $x\,dx$ rather than $2x\,dx$, we divide by 2 to get $x\,dx = \frac{1}{2}du$. Then

$$\begin{aligned}\int (x^2 + 5)^{12} x\,dx &= \int u^{12}\tfrac{1}{2}\,du \\ &= \tfrac{1}{2}\int u^{12}\,du \\ &= (\tfrac{1}{2})(\tfrac{1}{13})u^{13} + C = \tfrac{1}{26}(x^2 + 5)^{13} + C.\end{aligned}$$ ■

Example 7.4 Find $\int e^{6x^2+3} x\,dx$.

Solution Setting $u = 6x^2 + 3$, we get $du/dx = 12x$, so $du = 12x\,dx$. Since the integrand contains $x\,dx$, we solve $du = 12x\,dx$ for $x\,dx$, obtaining $x\,dx = \frac{1}{12}du$. Then

$$\begin{aligned}\int e^{6x^2+3} x\,dx &= \int e^u \tfrac{1}{12}\,du \\ &= \tfrac{1}{12}\int e^u\,du \\ &= \tfrac{1}{12}e^u + C \\ &= \tfrac{1}{12}e^{6x^2+3} + C.\end{aligned}$$ ■

The method of substitution may be used to evaluate a definite integral as illustrated in the following example.

Example 7.5 Evaluate $\displaystyle\int_2^5 (x^2 + 3x)^3(2x + 3)\,dx$.

Solution We begin by finding $\int (x^2 + 3x)^3(2x + 3)\,dx$. Since $x^2 + 3x$ and its derivative appear in the integrand, we set $u = x^2 + 3x$. From $du/dx = 2x + 3$ we have $du = (2x + 3)\,dx$, so that

$$\begin{aligned}\int (x^2 + 3x)^3(2x + 3)\,dx &= \int u^3\,du \\ &= \tfrac{1}{4}u^4 = \tfrac{1}{4}(x^2 + 3x)^4.\end{aligned}$$

(Since we are evaluating a definite integral, we do not include a constant of integration.) Evaluating this antiderivative at the given limits yields

$$\begin{aligned}\int_2^5 (x^2+3x)^3(2x+3)\,dx &= \tfrac{1}{4}(x^2+3x)^4\Big|_2^5\\ &= \tfrac{1}{4}[(25+15)^4-(4+6)^4]\\ &= \tfrac{1}{4}(40^4-10^4)=637{,}500.\end{aligned}$$ ■

An alternate method of handling definite integrals when using substitution is useful in many cases. The method changes the limits of integration, which are x-values originally, to u-values. In the previous example, we used the substitution $u=g(x)=x^2+3x$. The original limits of integration are $x=2$ and $x=5$; when $x=2$, $u=g(2)=2^2+3(2)=10$ and when $x=5$, we have $u=g(5)=5^2+3(5)=40$. Thus

$$\begin{aligned}\int_{x=2}^{x=5} (x^2+3x)^3(2x+3)\,dx &= \int_{u=10}^{u=40} u^3\,du\\ &= \tfrac{1}{4}u^4\Big|_{10}^{40}\\ &= \tfrac{1}{4}(40^4-10^4)=637{,}500.\end{aligned}$$

The evaluation step yields the same computation as with the first method, but this method allows us to proceed without changing the substitution variable back to the original variable. This alternate substitution method may be stated as follows.

For the substitution $u=g(x)$,

$$\int_{x=a}^{x=b} f(g(x))g'(x)\,dx = \int_{u=g(a)}^{u=g(b)} f(u)\,du.$$

Example 7.6 Find the value of

$$\int_0^2 \frac{x\,dx}{x^2+1}$$

Solution Let $u=x^2+1$. Then $du=2x\,dx$, so $x\,dx=\frac{1}{2}\,du$. When $x=0$, $u=1$, and when $x=2$, $u=5$. Therefore

$$\begin{aligned}\int_{x=0}^{x=2} \frac{x\,dx}{x^2+1} &= \int_{u=1}^{u=5} \frac{\frac{1}{2}\,du}{u}\\ &= \tfrac{1}{2}\int_{u=1}^{u=5} \frac{du}{u}\\ &= \tfrac{1}{2}\ln|u|\Big|_1^5\\ &= \tfrac{1}{2}(\ln 5-\ln 1)=\tfrac{1}{2}\ln 5\approx 0.80472. \qquad (\ln 1 = 0)\end{aligned}$$ ■

The following example combines substitution and algebraic simplification and shows how we can use a substitution to simplify an integration problem, making it possible to complete the integration.

Example 7.7 Find $\int (x+3)^5(x-4)\,dx$.

Solution The integrand does not appear to be composed of a function and its derivative. Since our goal is to simplify the integrand, and since $(x+3)$ raised to a power is the most complicated feature of the integrand, we make the substitution $u = x+3$. Then $du = dx$. The factor $(x-4)$ has not appeared either in u itself or in its derivative. Nevertheless, if we change the variable from x to u, we must convert all appearances of x to corresponding expressions involving u. Since $u = x+3$, $x = u-3$, so $x-4 = u-3-4 = u-7$. The integral becomes

$$\begin{aligned}\int (x+3)^5(x-4)\,dx &= \int u^5(u-7)\,du \\ &= \int (u^6 - 7u^5)\,du \\ &= \tfrac{1}{7}u^7 - 7\cdot\tfrac{1}{6}u^6 + C \\ &= \tfrac{1}{7}(x+3)^7 - \tfrac{7}{6}(x+3)^6 + C. \qquad (u = x+3)\end{aligned}$$ ■

The integrand must be of the form $f(g(x))g'(x)$ or sums and constant multiples of this type of function for integration by substitution to be useful. The presence of the factor $g'(x)$ is essential. For example,

$$\int \frac{2x}{x^2+1}\,dx$$

is easily evaluated using the substitution $u = g(x) = x^2+1$ (and the result is $\ln(x^2+1)+C$); however, without the factor $g'(x) = 2x$, the integral becomes

$$\int \frac{1}{x^2+1}\,dx,$$

whose evaluation requires methods and functions beyond the scope of this text.

Exercises

Exer. 1–6: Find the integral using the substitution provided.

1. $\int (x^2+3)^{14}(2x)\,dx;\quad u = x^2+3$

2. $\int (5x^2+7)^{13}(10x)\,dx;\quad u = 5x^2+7$

3. $\int (-4x^2+6)^{-5}(-8x)\,dx;\quad u = -4x^2+6$

4. $\int (t^2+4t+6)^{11}(2t+4)\,dt;\quad u = t^2+4t+6$

5. $\int (2t^2+3t-9)^{-4}(4t+3)\,dt;\quad u = 2t^2+3t-9$

6. $\int \frac{1}{x+3}\,dx;\quad u = x+3$

Exer. 7–22: Use an appropriate substitution to find the integral

7. $\int (2x+4)^6\,dx$

8. $\int (3x^2+5)(4x)\,dx$

9. $\int e^{x^2} 2x\,dx$

10. $\int (s^3+7)^{22}(3s^2)\,ds$

11. $\int (5x^3+6x^2+2x-4)^{23}(15x^2+12x+2)\,dx$

12. $\int e^{3x^2+5} 6x\,dx$

13. $\int (-2t^3+5t-6)^{49}(-6t^2+5)\,dt$

14. $\int \frac{3x^2}{x^3+2}\,dx$

15. $\int \frac{x}{3x^2+5}\,dx$

16. $\int (3p-5)^4\,dp$

17. $\int \frac{1}{(3x+1)^4}\,dx$

18. $\int \frac{\ln x}{x}\,dx$

19. $\int \frac{e^w}{3e^w+5}\,dw$

20. $\int (x-3)^7(2x+1)\,dx$

21. $\int (4x+1)e^{2x^2+x}\,dx$

22. $\int \frac{x^2}{4x^3+5}\,dx$

Exer. 23–32: Find the integral.

23. $\int (x^2+7)^{22}x\,dx; \quad u=x^2+7$

24. $\int (3x^2+4)^{19}x\,dx; \quad u=3x^2+4$

25. $\int (4t^2+6t+3)^{33}(4t+3)\,dt$

26. $\int e^{2x^2+3}x\,dx$

27. $\int (3x^2+12x+7)^{-15}(x+2)\,dx$

28. $\int \frac{t^2\,dt}{(4t^3+9)^{12}}$

29. $\int (7x^4+14x+3)^{-35}(2x^3+1)\,dx$

30. $\int e^{4x^3+6x}(2x^2+1)\,dx$

31. $\int (x^2+1)^5 x\,dx$

32. $\int x\sqrt{3x^2+5}\,dx$

Exer. 33–44: Use an appropriate substitution to evaluate the definite integral.

33. $\int_1^4 (2x+4)^2\,2\,dx$

34. $\int_2^5 (3x^2+7)^3\,6x\,dx$

35. $\int_0^2 (2t-5)\,e^{t^2-5t}\,dt$

36. $\int_1^4 (x^2+3)\,2x\,dx$

37. $\int_{-1}^4 (s^2+2s+1)^{3/2}(s+1)\,ds$

38. $\int_0^3 (x^3-19)^{2/3}(4x^2)\,dx$

39. $\int_2^5 (2w^2+12w+18)^{-1/2}(w+3)\,dw$

40. $\int_{-3}^{-1} (x^2-4x+4)^{1/2}(2-x)\,dx$

41. $\int_1^3 \frac{2x}{x^2+5}\,dx$

42. $\int_1^3 (t-1)^4(2t-6)\,dt$

43. $\int_{-1}^2 \frac{x}{x^2+1}\,dx$

44. $\int_0^3 \sqrt{x+1}\,dx$

Exer. 45–53: Find the area under the curve for the given values.

45. $y=4(x+3)$; from $x=1$ to $x=3$
46. $y=e^{2x}$; from $x=0$ to $x=2$
47. $y=x^2-3$; from $x=-1$ to $x=1$
48. $y=(x-3)^2$; from $x=-1$ to $x=1$
49. $y=e^{x-3}$; from $x=1$ to $x=3$
50. $y=e^{x/2}$; from $x=2$ to $x=4$
51. $y=5$; from $x=-2$ to $x=2$
52. $y=\frac{x^2+1}{x^3+3x}$; from $x=1$ to $x=2$
53. $y=x(x-2)^5$; from $x=0$ to $x=2$

54. If the marginal cost of a company is $M_C(x)=4+0.02\sqrt{x}$ and its fixed cost is \$1000, what is the cost of producing 100 items? (Recall that marginal cost is the derivative of the cost function.)

55. A sparrow hawk is flying upward at a velocity of $2\sqrt{t+1}+4$ cm/sec. How high does it rise in the first 8 seconds? (Recall that velocity is the derivative of the distance-traveled function.)

56. An educational psychologist begins to gather data from classroom observations at a rate given by

$$R(t)=75+t\sqrt{t^2+1}.$$

To complete the data-gathering in time for an annual report, 800 observations must be completed in 8 weeks. Will the data be gathered in time?

57. The marginal revenue of a small firm is given by

$$M_R(x)=125+200(x+1)^{-3/2}.$$

How much will revenue increase if production is increased from 99 items per day to 143 items per day?

58. The acceleration (in ft/sec^2) of an object moving along a line is given by

$$a(t)=4(t-3)^2+2t-5.$$

Find the change in velocity from $t = 4$ to $t = 8$. (Recall that acceleration is the derivative of velocity with respect to time.)

59. Use integration by substitution to show that for any constant $k \neq 0$,

$$\int e^{kx}\,dx = \frac{1}{k}e^{kx} + C,$$

where C is a constant of integration. What is the result when $k = 0$?

7.2 Integration by Parts

Since integration is the opposite of differentiation, the rules of differentiation can be reversed to become rules of integration: the substitution method of integration is essentially the reversal of the chain rule for differentiation. In this section we study the reversal of the product rule, which is called **integration by parts**.

The product rule for differentiation may be written as:

$$\frac{d}{dx}f(x)g(x) = f(x)g'(x) + g(x)f'(x).$$

If we integrate both sides, we have

$$\int \frac{d}{dx}f(x)g(x)\,dx = \int f(x)g'(x)\,dx + \int g(x)f'(x)\,dx.$$

The term on the left-hand side is $f(x)g(x)$, since $f(x)g(x)$ is an antiderivative of $(d/dx)f(x)g(x)$. Rearranging terms yields the following formula.

Integration by Parts Formula

$$\int f(x)g'(x)\,dx = f(x)g(x) - \int g(x)f'(x)\,dx$$

The integration by parts formula is frequently written in a form somewhat different from the preceding version. Suppose we write u for $f(x)$, du for $f'(x)\,dx$, v for $g(x)$, and dv for $g'(x)\,dx$. Then the formula becomes

$$\int u\,dv = uv - \int v\,du.$$

This formula is also called the **integration by parts formula**.

Integration by Parts Formula

$$\int u\,dv = uv - \int v\,du$$

Notice that the integral on the right-hand side has $v\,du$ as its integrand, and the left-hand side has $u\,dv$ as integrand. For integration by parts to be successful, the integral on the right must be *easier* to integrate than the integral on the left. However, as with integration by substitution, this method is not successful on all integrals. (Making good choices for u and v is not always easy and some experience is required to become proficient.) Its use is best illustrated by examples.

Example 7.8 Find $\int xe^{2x}\,dx$.

Solution The first step is to choose the functions u and dv that appear in the integral on the left-hand side of the integration by parts formula. We take $u = x$ and $dv = e^{2x}\,dx$. We get du by differentiation:

$$\frac{du}{dx} = 1, \quad \text{so} \quad du = dx.$$

We get v by integration:

$$v = \int dv = \int e^{2x}\,dx = \tfrac{1}{2}e^{2x}.$$

(No constant of integration has been included, since any antiderivative will do.) To summarize:

$$\begin{aligned} u &= x & dv &= e^{2x}\,dx \\ du &= dx & v &= \tfrac{1}{2}e^{2x} \end{aligned}$$

We use these in the integration by parts formula to give

$$\begin{aligned} \int xe^{2x}\,dx &= uv - \int v\,du \\ &= x(\tfrac{1}{2}e^{2x}) - \int \tfrac{1}{2}e^{2x}\,dx \\ &= \tfrac{1}{2}xe^{2x} - \tfrac{1}{2}\int e^{2x}\,dx. \end{aligned}$$

Notice that the remaining integral is easier to determine than the original integral. In fact, it is exactly in the form of the integration formula $\int e^{kx}\,dx = e^{kx}/k + C$, so we have achieved the goal of integration by parts, converting a difficult integral into an easier one. To finish the problem, we have

$$\int xe^{2x}\,dx = \tfrac{1}{2}xe^{2x} - \tfrac{1}{2}\int e^{2x}\,dx = \tfrac{1}{2}xe^{2x} - \tfrac{1}{4}e^{2x} + C.$$

Note that we supply the constant of integration C at the end. As verification, we use the product and chain rules to compute

$$\frac{d}{dx}(\tfrac{1}{2}xe^{2x} - \tfrac{1}{4}e^{2x} + C) = \tfrac{1}{2}x(2e^{2x}) + e^{2x}(\tfrac{1}{2}) - (\tfrac{1}{4})(2e^{2x}) = xe^{2x}.$$

■

Example 7.9 Find $\int x^3 \ln x\,dx$.

Solution In this case, we let $u = \ln x$ and $dv = x^3\,dx$. (The choice

$u = x^3$ and $dv = \ln x\,dx$ is not useful because we cannot yet antidifferentiate $\ln x$ to find v.) Then $du = dx/x$ and $v = \frac{1}{4}x^4$. In summary,

$$u = \ln x \qquad dv = x^3\,dx$$

$$du = \frac{1}{x}\,dx \qquad v = \tfrac{1}{4}x^4$$

Consequently,

$$\begin{aligned}\int x^3 \ln x\,dx &= uv - \int v\,du \\ &= (\ln x)(\tfrac{1}{4}x^4) - \int (\tfrac{1}{4}x^4)\left(\frac{1}{x}\right) dx \\ &= \tfrac{1}{4}x^4 \ln x - \tfrac{1}{4}\int x^3\,dx \\ &= \tfrac{1}{4}x^4 \ln x - \tfrac{1}{16}x^4 + C.\end{aligned}$$

■

In the preceding example we used the integration by parts method to "trade" a difficult integral for an easier one. The next example illustrates that an easy integral may not be obtained in a single step. This example also illustrates a slightly shortened form of the integration by parts process.

Example 7.10 Find $\int x^3 e^{2x}\,dx$.

Solution We begin by choosing u and dv to be $u = x^3$ and $dv = e^{2x}\,dx$. Then

$$u = x^3 \qquad dv = e^{2x}\,dx$$

$$du = 3x^2\,dx \qquad v = \tfrac{1}{2}e^{2x}$$

The integration by parts formula gives

$$\int x^3 e^{2x}\,dx = \tfrac{1}{2}x^3 e^{2x} - \int \tfrac{3}{2}x^2 e^{2x}\,dx.$$

This expression still contains a difficult integral, but we have made some improvement. The power of x in the integrand is diminished by one from the original integral. We use integration by parts again on the last term in the preceding equation. Taking

$$u = 3x^2 \qquad dv = \tfrac{1}{2}e^{2x}\,dx$$

$$du = 6x\,dx \qquad v = \tfrac{1}{4}e^{2x}$$

we have

$$\int \tfrac{3}{2}x^2 e^{2x}\,dx = \tfrac{3}{4}x^2 e^{2x} - \tfrac{6}{4}\int x e^{2x}\,dx.$$

The power is again diminished by one. Furthermore, the remaining integral is the one we evaluated in Example 7.8. Combining the two uses of integration by parts gives

$$\int x^3 e^{2x}\,dx = \tfrac{1}{2}x^3 e^{2x} - \tfrac{3}{4}x^2 e^{2x} + \tfrac{6}{4}\int x e^{2x}\,dx.$$

Had we not already done Example 7.8, we would use integration by parts a third time. From Example 7.8 we have

$$\int x^3e^{2x}\,dx = \tfrac{1}{2}x^3e^{2x} - \tfrac{3}{4}x^2e^{2x} + \tfrac{3}{4}xe^{2x} - \tfrac{3}{8}e^{2x} + C.$$

This process may be organized more efficiently by proceeding as follows. First, as above, choose the part of the integrand that will be differentiated (called u above) and the part that will be integrated (called dv above). Write these in two columns, with the differentiation candidate on the left:

$$x^3 \qquad\qquad e^{2x}$$

(the dx won't be needed; we know the right-hand side is to be integrated). Next, we differentiate the left-hand column, and integrate the right-hand column:

differentiate	***integrate***
x^3	e^{2x}
$3x^2$	$\frac{1}{2}e^{2x}$

Now we ask if we can (easily) integrate the product of the functions in the bottom row. Since we cannot, we repeat the differentiation/integration process. The table expands to:

differentiate	***integrate***
x^3	e^{2x}
$3x^2$	$\frac{1}{2}e^{2x}$
$6x$	$\frac{1}{4}e^{2x}$

Again, we determine if we can integrate the product of the functions in the last row. The answer is still no, so we repeat the differentiation/integration process, obtaining

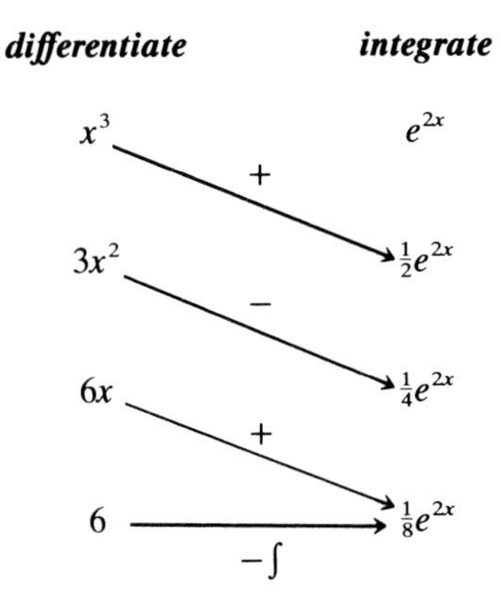

(We will explain the arrows and signs in this table momentarily.) Since

the product of the functions in the last row is easy to integrate, we stop the differentiation/integration process. Due to the minus sign in the integration by parts formula, repeated uses of integration by parts cause the terms to alternate in sign. As a reminder of this, and to indicate which products to form, we draw diagonal arrows in our table and label them alternately +, −, +, −, and so on. The arrows indicate products of functions, except for the last arrow, which has an associated integral sign to remind us that the product must be integrated. Furthermore, this product is not of diagonal functions, but of functions on the same row. From this chart, we can write down the result of repeated use of integration by parts:

$$\int x^3 e^{2x}\,dx = \tfrac{1}{2}x^3e^{2x} - \tfrac{3}{4}x^2e^{2x} + \tfrac{6}{8}xe^{2x} - \int \tfrac{6}{8}e^{2x}\,dx$$

or

$$\int x^3 e^{2x}\,dx = \tfrac{1}{2}x^3e^{2x} - \tfrac{3}{4}x^2e^{2x} + \tfrac{3}{4}xe^{2x} - \tfrac{3}{8}e^{2x} + C.$$

Of course, this is our earlier result. (Compare this method to the procedure given at the beginning of this example.) ■

The second method illustrated in the previous example is still integration by parts; it is just a different layout of the work. In fact, if we use this layout on the integral $\int f(x)g'(x)\,dx$, we have

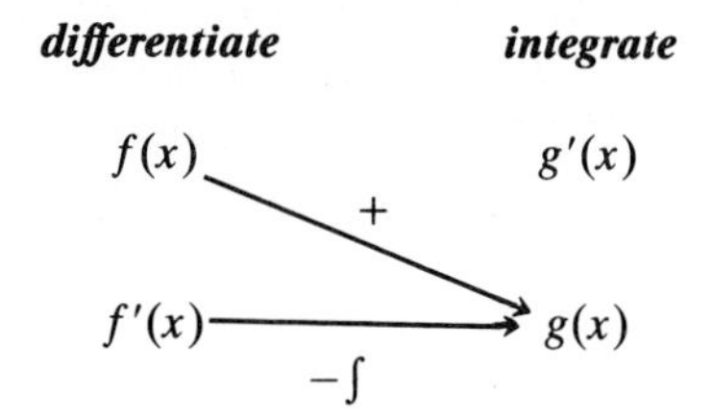

and the operations indicated by the arrows yield

$$\int f(x)g'(x)\,dx = f(x)g(x) - \int g(x)f'(x)\,dx.$$

This equation is the first version of the integration by parts formula we obtained at the beginning of this section. The current version of integration by parts is helpful in deciding what portion of the integrand to include in the function to be differentiated and what portion to put in the integration column. We must be careful that we can integrate the function placed in the integration column.

Example 7.11 Find $\int x^6(3x-4)^2\,dx$.

Solution We will differentiate $(3x-4)^2$ and integrate x^6 because x^6 is easier to integrate than $(3x-4)^2$ and the table ends sooner than if these functions are reversed. We display the integration by parts steps with the arrows already in place:

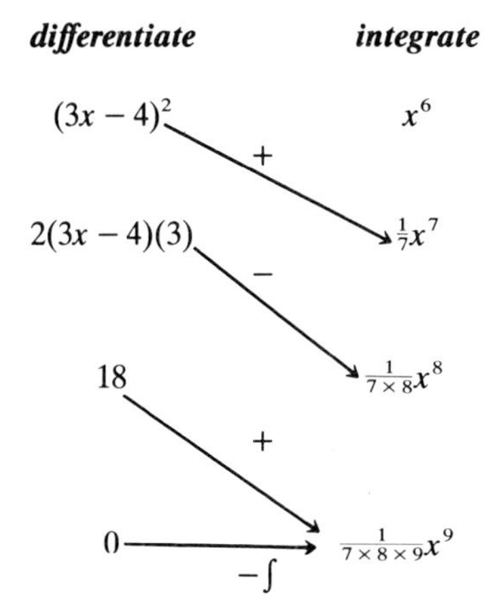

Note that we have carried the process one line farther than necessary. Even though we can integrate the product of the functions on the third line, the product of the functions on the last line is zero, which is even easier to integrate! The result is

$$\int x^6(3x-4)^2\,dx = \tfrac{1}{7}x^7(3x-4)^2 - \tfrac{3}{28}x^8(3x-4) + \tfrac{1}{28}x^9 + C. \quad \blacksquare$$

Most of the preceding examples have had powers of x that are decreasing to zero as the differentiation step is repeated, but the following example shows that this is not always the case. Remember that the criterion for stopping the differentiation/integration process is that the product of the last two functions is easy to integrate. (Occasionally, as in the previous example, continuing one more step may be helpful.)

Example 7.12 Find $\int \ln x\,dx$.

Solution Integration by parts does not seem to apply because the integrand contains only one function. However, if we interpret the integrand to be the product of $\ln x$ and 1, we can use integration by parts:

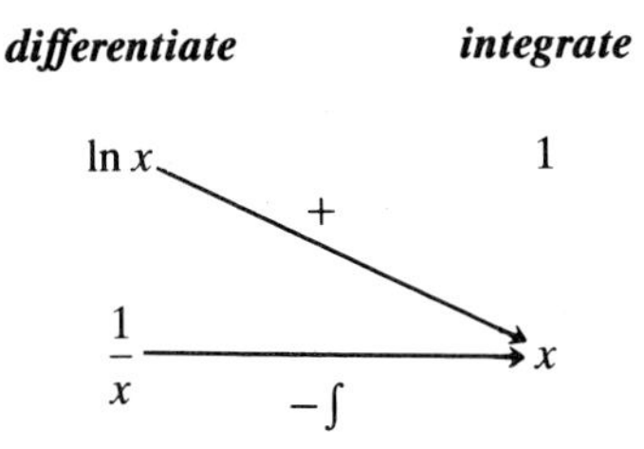

The product of the functions in the last row is 1, which is easy to integrate. The result is

$$\int \ln x\,dx = x\ln x - \int 1\,dx = x\ln x - x + C. \quad \blacksquare$$

Before discussing definite integrals, we make two observations: (1) The headings *differentiate* and *integrate* are supplied only as a reminder. In the actual use of this method these headings need not be written. (2) Even for a problem that "stops" after only one step, the table method is slightly shorter than the $u\,dv$ approach. Furthermore, the table method provides a graphic structure to recall the integration by parts formula. Integration by parts can be used to evaluate definite integrals as illustrated in the following example.

Example 7.13 Find $\int_0^3 (x+1)^{2/3}x\,dx$.

Solution

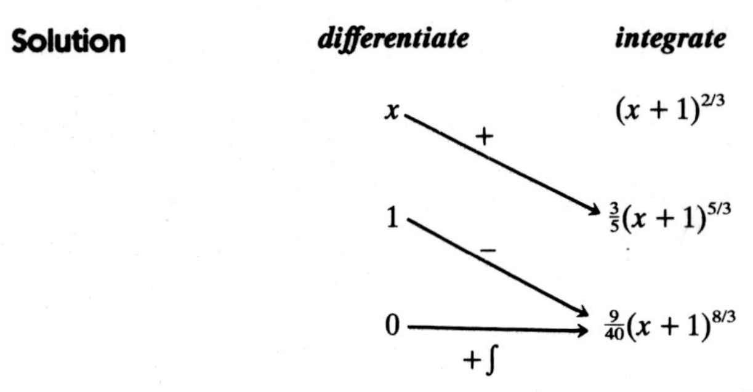

Consequently,

$$\int_0^3 (x+1)^{2/3}x\,dx = [\tfrac{3}{5}x(x+1)^{5/3} - \tfrac{9}{40}(x+1)^{8/3}]\Big|_0^3$$

$$= \tfrac{9}{5}(4)^{5/3} - \tfrac{9}{40}(4)^{8/3} - (-\tfrac{9}{40}) \approx 9.2964.$$ ■

Example 7.14 During a workout, an athlete has a pulse rate of $p(t)$ beats per minute where, t minutes after the start of the workout,

$$p(t) = 58 + 10.62t^2e^{-0.25t}.$$

During the first 20 minutes of the workout (from $t = 0$ to $t = 20$), how many times does the athlete's heart beat?

Solution Note that if $h(t)$ denotes the number of times the athlete's heart beats from time 0 until time t, then $h'(t)$ is the rate of change of the number of heartbeats; that is, $h'(t)$ is the pulse rate $p(t)$. The number of heartbeats from $t = 0$ to $t = 20$ is the net change in $h(t)$ from $t = 0$ to $t = 20$, that is, $h(20) - h(0)$. Since $h'(t) = p(t)$,

$$h(20) - h(0) = \int_0^{20} p(t)\,dt.$$

Substituting the given formula for $p(t)$ into the integrand, we obtain

$$\int_0^{20} p(t)\,dt = \int_0^{20} (58 + 10.62t^2e^{-0.25t})\,dt$$

$$= \int_0^{20} 58\,dt + 10.62\int_0^{20} t^2e^{-0.25t}\,dt$$

$$= 58t\Big|_0^{20} + 10.62\int_0^{20} t^2e^{-0.25t}\,dt$$

To integrate the remaining integral, we use integration by parts:

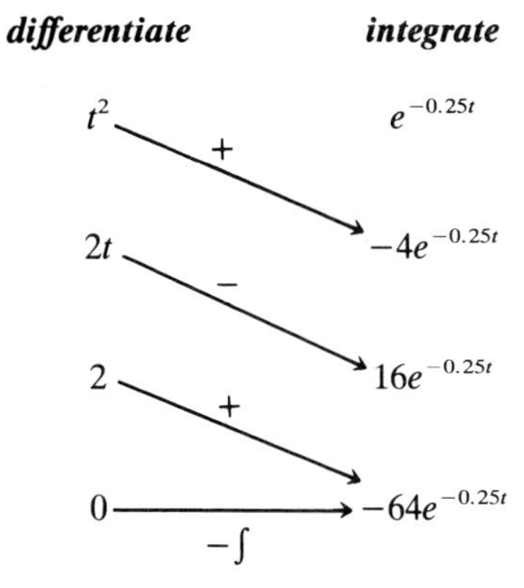

$$\text{Hence } \int_0^{20} t^2 e^{-0.25t}\,dt = -4t^2e^{-0.25t} - 32te^{-0.25t} - 128e^{-0.25t}\Big|_0^{20}$$

$$\approx (-10.78 - 4.31 - 0.86) - (-128) = 112.05.$$

$$\text{Therefore, } \int_0^{20} p(t)\,dt \approx 58(20) + 10.62(112.05) = 2349.97,$$

that is, the athlete's heart beats about 2350 times during the 20-minute period. ■

Exercises

Exer. 1–40: Find the indefinite integral using integration by parts.

1. $\int xe^x\,dx$
2. $\int (4x+1)(2x+3)^3\,dx$
3. $\int x(x-3)^4\,dx$
4. $\int (x+2)(3x-2)^5\,dx$
5. $\int (5x-4)^3e^x\,dx$
6. $\int (3-2t)(5t-1)^{-3/4}\,dt$
7. $\int (4-3x)(2-3x)^5\,dx$
8. $\int (5-4x)^{3/2}(4+3x)\,dx$
9. $\int x^2e^{-x}\,dx$
10. $\int x^2(8x+7)^{3/2}\,dx$
11. $\int (2t+1)^2(4t-2)^{3/2}\,dt$
12. $\int s(5s+1)^{-1/2}\,ds$
13. $\int (2x^2+3x-4)(6x+5)^{5/2}\,dx$
14. $\int (3x^2-5x+2)(2x-3)^{-7/2}\,dx$
15. $\int s^2(s-1)^{3/2}\,ds$
16. $\int x^2(x+1)^3\,dx$
17. $\int xe^{2x}\,dx$
18. $\int te^{4t}\,dt$
19. $\int te^{-2t}\,dt$
20. $\int (2x+3)e^{3x}\,dx$
21. $\int (3-4x)e^{-4x}\,dx$
22. $\int (2w+5)e^{-6w}\,dw$
23. $\int (3x-4)e^{2x+3}\,dx$
24. $\int (5-2x)e^{7+3x}\,dx$
25. $\int t^2e^{3t}\,dt$
26. $\int x^2e^{-4x}\,dx$
27. $\int (2x^2+5x-3)e^{2x}\,dx$
28. $\int (4t^2-3t+1)e^{-t/4}\,dt$
29. $\int (3u^2+5u-2)e^{3u+2}\,du$
30. $\int x^3e^{4x}\,dx$
31. $\int x^3e^{x/2}\,dx$
32. $\int (2x+3)^3e^{x/3}\,dx$
33. $\int (2t^2+3t+1)e^{2t}\,dt$
34. $\int (3s^3-4s+2)e^{s/2}\,ds$
35. $\int x^4\ln x\,dx$
36. $\int x^{-3}\ln x\,dx$

37. $\int (2x+3)^4 \ln(2x+3)\,dx$

38. $\int (5t-1)^3 \ln(5t-1)\,dt$

39. $\int 4p^{-3} \ln 5p\,dp$

40. $\int 5x^{-4} \ln 3x\,dx$

Exer. 41–54: Evaluate the definite integral using integration by parts.

41. $\int_2^4 x(3x+4)^2\,dx$

42. $\int_0^1 (3x+2)(x+3)^{-1}\,dx$

43. $\int_1^3 t(2t+3)^{1/2}\,dt$

44. $\int_2^3 (x-2)^2x^3\,dx$

45. $\int_0^1 x^2(3x+1)^3\,dx$

46. $\int_1^3 s^3(s-1)^4\,ds$

47. $\int_1^3 se^s\,ds$

48. $\int_{-1}^3 (2x+1)e^{x+3}\,dx$

49. $\int_2^4 x^2e^{3x}\,dx$

50. $\int_1^3 (3p-1)^2e^{5p+1}\,dp$

51. $\int_1^4 w\ln(w^2)\,dw$

52. $\int_0^3 (2x+1)^2\ln(2x+1)\,dx$

53. $\int_0^1 x^4e^{-x}\,dx$

54. $\int_{-2}^0 t^3e^{3-2t}\,dt$

Exer. 55–62: Find the area of the region of the plane bounded by the given function and vertical lines.

55. $y = x^3(x-1)^3$; from $x=1$ to $x=2$
56. $y=(x-2)^3e^x$; from $x=0$ to $x=2$
57. $y=(x+1)^3\ln(x+1)$; from $x=1$ to $x=4$
58. $y=x^3e^{x^2}$; from $x=0$ to $x=1$
59. $y=(1-x)^3x^4$; from $x=0$ to $x=1$
60. $y=(2x-1)^3e^{3x}$; from $x=0$ to $x=3$
61. $y=(2x-1)^3\ln(2x-1)$; from $x=1$ to $x=3$
62. $y=x^3e^{-x^2}$; from $x=-1$ to $x=1$

63. Refer to Example 7.14. Suppose that the athlete has a pulse rate of $p(t)=49+2.4t^2e^{-0.1t}$. How many heart beats will occur from 10 minutes into the workout to 30 minutes into the workout?
64. Refer to Exercise 63. Suppose that $p(t)=51+3.6t^2e^{-0.13t}$, where $p(t)$ is the pulse rate of the athlete. How many heartbeats occur in the first 25 minutes of the workout?
65. Suppose that the rate of change of sales (in 100,000 items per week) of a certain item t weeks after the end of a sales campaign is given by
$$dS/dt = te^{-0.2t}.$$
At the end of the campaign ($t=0$), sales were 250,000 items per week. How many items were sold in the first 4 weeks following the end of the campaign?
66. The region bounded by the graphs of $y=0$, $x=e$, and $y=\ln x$ is revolved around the y-axis. Compute the volume of the resulting solid.
67. The velocity of a point moving along the x-axis is given by $v(t)=6te^{-2t}$ m/sec. If the point is at the origin when $t=0$, where is it when $t=2$?
68. Find the area of the region of the plane bounded by the graphs of $y=x^2\ln x$ and $y=0$.
69. Consider the graph of $y=xe^{-x}$ from $x=1$ to $x=N$ for a fixed large number N. Compute the area between the graph and the x-axis.

7.3 Improper Integrals

Suppose that an object travels in a straight line so that its velocity $v(t)$ at time t is $v(t)=1/t^2$ ft/sec and that its journey began at time $t=1$. Since $v(t)>0$ for all $t\geq 1$, the object is always moving forward, so its distance-traveled function is an increasing function. If the object travels in this manner forever, how far will it travel?

If we rely on our intuition, it seems as though the object will travel an infinite distance, since its velocity is always positive. However, as we saw in our discussion of horizontal asymptotes, some increasing functions do not increase without bound. To analyze the distance the object travels, we let $s(x)$ denote the distance traveled by time $t=x$. Then, for the velocity function given above,

$$s(x)=\int_1^x v(t)\,dt=\int_1^x \frac{1}{t^2}\,dt$$
$$=\int_1^x t^{-2}\,dt=-t^{-1}\Big|_1^x=-x^{-1}-(-1)=1-\frac{1}{x}.$$

Hence, when $x = 10$ seconds, the object will have traveled a distance of $1 - \frac{1}{10} = 0.9$ feet; when the time $x = 100$ seconds, the distance traveled is 0.99 feet; when $x = 1000$ seconds, the distance traveled is 0.999 feet. As x gets larger and larger, the distance traveled by the object gets closer and closer to 1 foot, but never exceeds 1 foot. Therefore, if the object travels forever in the manner described, it would travel only 1 foot!

The process in the preceding calculation can be used to define any integral of the form

$$\int_a^\infty f(t)\,dt.$$

In general, we define this integral to be the value that $\int_a^x f(t)\,dt$ approaches as x gets larger and larger, if such a value exists; that is,

$$\int_a^\infty f(t)\,dt = \lim_{x\to\infty} \int_a^x f(t)\,dt$$

if the limit exists. Integrals of this type are called **improper integrals**. We evaluate such integrals by determining if $\int_a^x f(t)\,dt$ approaches some number N as x increases without bound. If the integral does approach a number N as x approaches infinity, then the improper integral is said to **converge** to N, and we call N the **value of the integral**. If no such number N exists, then the improper integral is said to **diverge**.

Example 7.15 Determine if $\int_1^\infty \frac{4}{t^4}\,dt$ converges or diverges. If it converges, find its value.

Solution We begin by replacing the ∞ with x and calculating the integral:

$$\begin{aligned}\int_1^x \frac{4}{t^4}\,dt &= -\tfrac{4}{3}t^{-3}\Big|_1^x \\ &= -\tfrac{4}{3}x^{-3} - (-\tfrac{4}{3})(1^{-3}) \\ &= \tfrac{4}{3} - \tfrac{4}{3}x^{-3} = \frac{4}{3} - \frac{4}{3x^3}.\end{aligned}$$

Since $\lim_{x\to\infty} \frac{4}{3x^3} = 0$, the function $\frac{4}{3} - \frac{4}{3x^3}$ approaches $\frac{4}{3}$ as x approaches infinity. Thus, the improper integral converges and

$$\int_1^\infty \frac{4}{t^4}\,dt = \frac{4}{3}.$$

■

Example 7.16 Determine if $\int_2^\infty \frac{dx}{\sqrt{x}}$ converges or diverges. If it converges, find the value of the integral.

Solution We begin by replacing ∞ with q (since x is already used in the integral) and calculating the integral as follows:

$$\int_2^\infty \frac{dx}{\sqrt{x}} = \lim_{q\to\infty} \int_2^q x^{-1/2}\,dx = \lim_{q\to\infty} 2x^{1/2}\Big|_2^q = \lim_{q\to\infty} [2q^{1/2} - 2(2)^{1/2}].$$

As q gets larger and larger, this function does not approach any number (in fact, it gets increasingly large without bound), so the improper integral diverges. ■

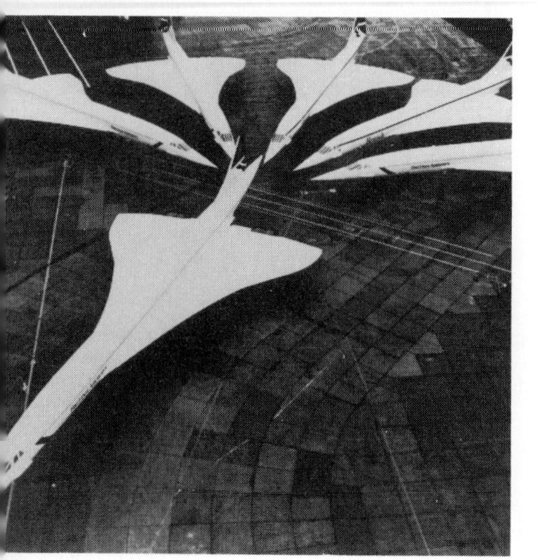

Example 7.17 A manufacturing company built a special fleet of supersonic transport planes and then shut down production. The company has promised to provide its customers a lifetime supply of a special lubricant for these planes. The rate at which this fleet of planes will use the special lubricant (in gallons per year) after one year in service is given by

$$r(t) = \frac{300}{t^{3/2}},$$

where t is the number of years since the planes were put into service (so $t \geq 1$). The company wants to produce all of the lubricant needed after the first year in one batch and then dispense it as needed. How many gallons should be produced?

Solution Since $r(t)$ is the rate at which the lubricant is used, $\int_1^x r(t)\,dt$ is the amount used between time $t = 1$ and time $t = x$. Therefore, the amount needed for all time is $\int_1^\infty r(t)\,dt$. To evaluate this integral, we proceed as follows:

$$\begin{aligned}\int_1^\infty r(t)\,dt &= \lim_{x\to\infty} \int_1^x \frac{300}{t^{3/2}}\,dt = \lim_{x\to\infty} \int_1^x 300t^{-3/2}\,dt \\ &= \lim_{x\to\infty} 300(-2t^{-1/2})\Big|_1^x = \lim_{x\to\infty} [-600x^{-1/2} - (-600)] \\ &= \lim_{x\to\infty} \left(600 - \frac{600}{\sqrt{x}}\right) = 600,\end{aligned}$$

because $600/\sqrt{x}$ approaches 0 as x gets larger and larger. Thus we conclude that 600 gallons will provide a lifetime supply of lubricant. ■

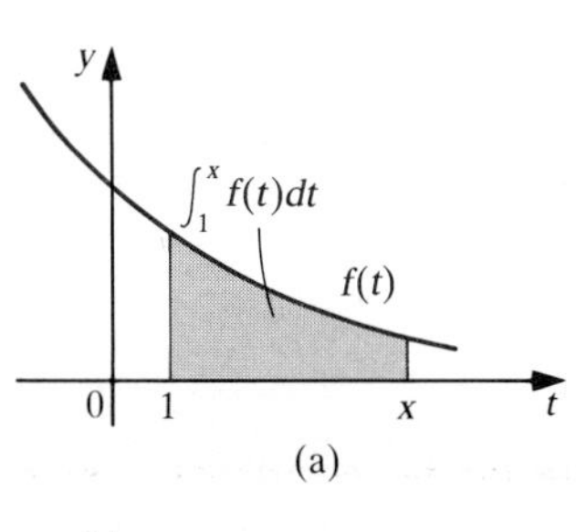

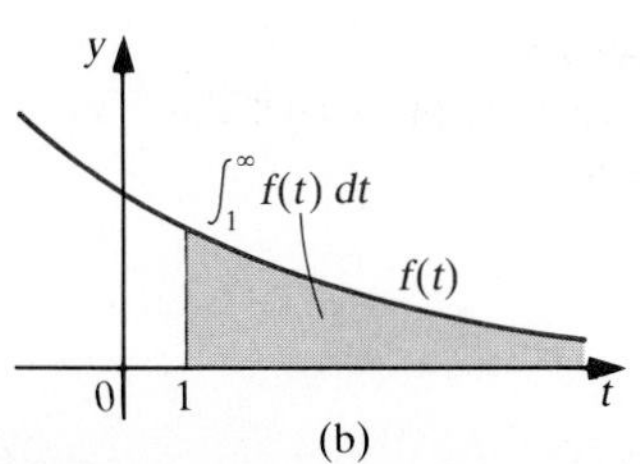

Figure 7.1
(a) Area under $f(t)$ from $t=1$ to $t=x$; (b) area under $f(t)$ from $t=1$ to $x=\infty$

Geometrically, the improper integral has an interesting interpretation. When $f(t)$ is positive for all $t \geq 1$, the integral $\int_1^x f(t)\,dt$ is the area under $f(t)$ on the interval $[1, x]$ as shown in Figure 7.1(a). Consequently, $\int_1^\infty f(t)\,dt$ is the area under $f(t)$ for all $t \geq 1$, as shown in Figure 7.1(b). For instance, from Example 7.15, the total area under the curve $f(t) = 4/t^4$ to the right $t = 1$ is $\frac{4}{3}$; that is, we have a geometric region with an infinitely long side (the horizontal t-axis, $t \geq 1$) with a finite area!

Example 7.18 Determine if $\int_1^\infty x^2e^{-x}\,dx$ converges or diverges. If it converges, find its value.

Solution We have

$$\int_1^\infty x^2e^{-x}\,dx = \lim_{q\to\infty} \int_1^q x^2e^{-x}\,dx.$$

The indefinite integral $\int x^2e^{-x}\,dx$ can be found by integration by parts:

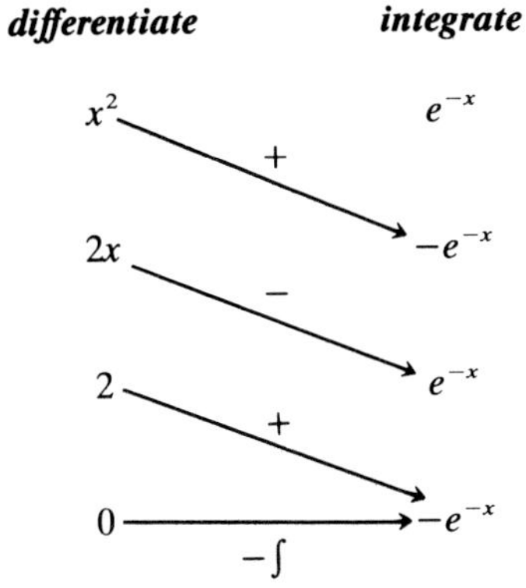

Consequently, $\displaystyle\int x^2e^{-x}\,dx = -x^2e^{-x} - 2xe^{-x} - 2e^{-x}.$

(We won't need a constant of integration.) If we factor out $-e^{-x}$, then

$$\int_1^\infty x^2e^{-x}\,dx = \lim_{q\to\infty} \left[-(x^2+2x+2)e^{-x}\right]\Big|_1^q$$

$$= \lim_{q\to\infty} \left[-(q^2+2q+2)e^{-q} + 5e^{-1}\right].$$

Consider the behavior of the function $(q^2+2q+2)e^{-q}$ for large q. The factor (q^2+2q+2) is increasing as $q\to\infty$, and the factor e^{-q} is decreasing as $q\to\infty$. However, from Table 7.1, we see that the latter factor decreases much more rapidly than the first factor increases, and hence the product goes to zero; that is,

$$\lim_{q\to\infty} (q^2+2q+2)e^{-q} = 0,$$

and $$\int_1^\infty x^2e^{-x}\,dx = 5e^{-1} = \frac{5}{e}.$$ ■

q	q^2+2q+2	e^{-q}	$(q^2+2q+2)e^{-q}$
5	37	6.74×10^{-3}	2.49×10^{-1}
10	122	4.54×10^{-5}	5.54×10^{-3}
15	257	3.06×10^{-7}	7.86×10^{-5}
20	442	2.06×10^{-9}	9.11×10^{-7}
50	2602	1.93×10^{-22}	5.02×10^{-19}

Table 7.1
Behavior of $(q^2+2q+2)e^{-q}$ as $q\to\infty$

We can also consider integrals of the form $\int_{-\infty}^a f(t)\,dt$ by calculating $\int_x^a f(t)\,dt$ and then letting x approach negative infinity.

Example 7.19 Determine if $\int_{-\infty}^{-1} \frac{1}{t^6}\,dt$ converges or diverges. If it converges, find its value.

Solution We begin by calculating the integral

$$\int_x^{-1} \frac{1}{t^6}\,dt = \int_x^{-1} t^{-6}\,dt = -\tfrac{1}{5}t^{-5}\Big|_x^{-1}$$

$$= -\tfrac{1}{5}(-1)^{-5} - (-\tfrac{1}{5}x^{-5}) = \frac{1}{5} + \frac{1}{5x^5}.$$

As x approaches $-\infty$, the term $1/(5x^5)$ approaches zero, so the improper integral converges and

$$\int_{-\infty}^{-1} \frac{1}{t^6}\,dt = \frac{1}{5}.$$

■

Finally, we can also handle improper integrals of the form $\int_{-\infty}^{\infty} f(t)\,dt$ by breaking them up as the sum of two improper integrals:

$$\int_{-\infty}^{\infty} f(t)\,dt = \int_{-\infty}^{0} f(t)\,dt + \int_{0}^{\infty} f(t)\,dt.$$

This operation is permissible provided that both of the improper integrals on the right converge.

Example 7.20 Evaluate $\int_{-\infty}^{\infty} e^{-|t|}\,dt.$

Solution We begin by breaking the interval of integration into two intervals, as described above. Thus

$$\int_{-\infty}^{\infty} e^{-|t|}\,dt = \int_{-\infty}^{0} e^{-|t|}\,dt + \int_{0}^{\infty} e^{-|t|}\,dt.$$

To evaluate the integrals on the right-hand side, note that in the first integral, $|t| = -t$, because $t \le 0$ on the interval of integration. In the second integral on the right, $|t| = t$, since $t \ge 0$ for this integral. Hence

$$\int_{-\infty}^{0} e^{-|t|}\,dt = \lim_{q\to-\infty} \int_{q}^{0} e^{t}\,dt = \lim_{q\to-\infty} \left(e^{t}\Big|_q^0\right)$$

$$= \lim_{q\to-\infty} (1 - e^{q}) = 1.$$

Similarly,

$$\int_{0}^{\infty} e^{-|t|}\,dt = \lim_{q\to\infty} \int_{0}^{q} e^{-t}\,dt = \lim_{q\to\infty} \left(-e^{-t}\Big|_0^q\right)$$

$$= \lim_{q\to\infty} (-e^{-q} + 1) = 1.$$

Combining these results, we obtain

$$\int_{-\infty}^{\infty} e^{-|t|}\,dt = 1 + 1 = 2.$$

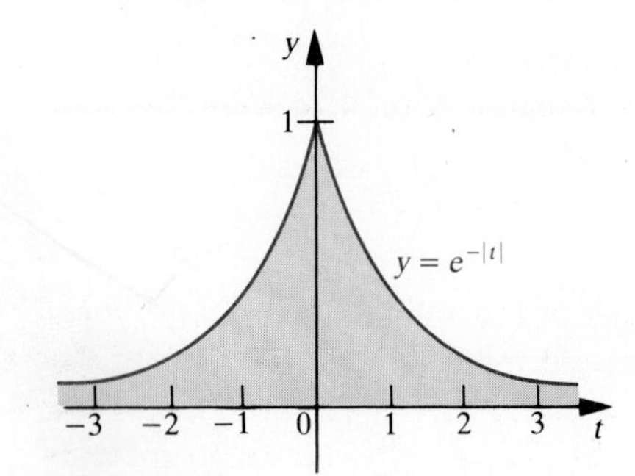

Figure 7.2
Graph of $e^{-|t|}$ and the area computed by $\int_{-\infty}^{\infty} e^{-|t|}\,dt$

Figure 7.2 shows the graph of $e^{-|t|}$ and the area computed by this integral.

■

Exercises

Exer. 1–20: Determine if the improper integral converges or diverges. If it converges, find its value.

1. $\int_1^\infty \frac{1}{t^3}\,dt$
2. $\int_4^\infty \frac{1}{t^2}\,dt$
3. $\int_2^\infty \frac{1}{t^{2/3}}\,dt$
4. $\int_1^\infty \frac{6}{t^{3/2}}\,dt$
5. $\int_2^\infty \frac{1}{t^{1.01}}\,dt$
6. $\int_1^\infty \frac{1}{t^{0.99}}\,dt$
7. $\int_0^\infty e^{-x}\,dx$
8. $\int_1^\infty x^{-3}\,dx$
9. $\int_1^\infty xe^{-x^2}\,dx$
10. $\int_0^\infty (x+4)^{-3}\,dx$
11. $\int_1^\infty \frac{\ln x}{x}\,dx$
12. $\int_{-1}^\infty xe^{-x}\,dx$
13. $\int_2^\infty \frac{x}{x^2-1}\,dx$
14. $\int_5^\infty \frac{x^2}{(x-3)^4}\,dx$
15. $\int_1^\infty \frac{\ln x}{x^5}\,dx$
16. $\int_1^\infty (2x+1)^2 x^{-5}\,dx$
17. $\int_2^\infty te^{-t}\,dt$
18. $\int_{-2}^\infty t^2 e^{-0.1t}\,dt$
19. $\int_3^\infty \frac{t\,dt}{(t^2+8)^3}$
20. $\int_3^\infty \frac{x}{(x^2+8)^3}\,dx$

Exer. 21–30: Determine if the improper integral converges or diverges. If it converges, find its value.

21. $\int_{-\infty}^{8} \frac{dt}{t^{5/3}}$
22. $\int_{-\infty}^{4} \frac{6\,dt}{(t-6)^3}$
23. $\int_{-\infty}^{3} \frac{dt}{t^5}$
24. $\int_{-\infty}^{-1} \frac{t^2\,dt}{(t^3-1)^2}$
25. $\int_{-\infty}^{-3} \frac{dt}{t^{1/3}}$
26. $\int_{-\infty}^{5} \frac{t\,dt}{(t^2+9)^{1/2}}$
27. $\int_{-\infty}^{0} te^t\,dt$
28. $\int_{-\infty}^{2} t^2 e^{t^3}\,dt$
29. $\int_{-\infty}^{3} \frac{t^2}{(t^3+2)^4}\,dt$
30. $\int_{-\infty}^{-2} x^{-3}\ln(-x)\,dx$

Exer. 31–34: Evaluate the improper integral.

31. $\int_{-\infty}^{\infty} \frac{|t|\,dt}{(t^2+7)^3}$
32. $\int_{-\infty}^{\infty} \frac{6t\,dt}{(t^2+3)^4}$
33. $\int_{-\infty}^{\infty} \frac{(t+1)\,dt}{(t^2+2t+5)^2}$
34. $\int_{-\infty}^{\infty} |t|\,e^{-|t|}\,dt$ (*Hint:* See Example 7.20.)

35. (a) Refer to Example 7.17. If the manufacturing company finds that the rate at which the lubricant will be used is $r(t) = 500/t^2$ gallons per year, how many gallons should be made to ensure a lifetime supply?
(b) What is the lifetime supply if $r(t) = 400/t^{1/2}$? Would you recommend that the company honor its promise to supply a lifetime supply?
36. If an object is traveling in a straight line at a velocity of $1/(t+2)^{3/2}$ mph, and if its journey began at $t = 0$, how far will the object travel, assuming it travels forever? (Use the substitution $u = t + 2$.)
37. A pest is destroying the Douglas firs in a national forest; it is predicted that t years after the beginning of a pest-control program the firs will be destroyed at a rate of $400t/(t^2+3)^2$ acres per year. Assuming that the program continues indefinitely, how many acres will eventually be destroyed by the pest after the program is implemented?

Exer. 38–39: For what numbers p does the improper integral converge?

38. $\int_2^\infty \frac{dt}{t^p}$
39. $\int_2^\infty \frac{\ln t\,dt}{t^p}$

7.4 The Trapezoidal Rule and Simpson's Rule

We can use the definite integral to solve many applied problems, but its power is limited by our ability to find antiderivatives. Finding an antiderivative of a simple function, such as $f(x) = 1/(x^3+1)$, is very difficult and for some functions, such as $f(x) = e^{-x^2}$, it is impossible to find an antiderivative. Consequently, we need a method for approximating the value of a definite integral for which an antiderivative cannot be found. In this section we present two methods: the Trapezoidal Rule and Simpson's Rule. Both methods are relatively easy to use and give good results in most cases.

The Trapezoidal Rule

Suppose we wish to estimate the area under the graph of a nonnegative function $f(x)$ from $x = a$ to $x = b$. Let n be a positive integer, and divide the interval $[a, b]$ into n subintervals of *equal* length. Let h denote the length of each subinterval so that $h = (b - a)/n$. The endpoints of these subintervals are $a, a + h, a + 2h, \ldots, a + nh = b$. It will be convenient to have a notation for these values, so we let $x_i = a + ih$, $i = 0, 1, 2, \ldots, n$.

We next draw a line segment connecting the points $(x_0, f(x_0))$ and $(x_1, f(x_1))$, another segment connecting $(x_1, f(x_1))$ and $(x_2, f(x_2))$, another connecting $(x_2, f(x_2))$ and $(x_3, f(x_3))$, and so on. This process forms a collection of trapezoids (see Figure 7.3).

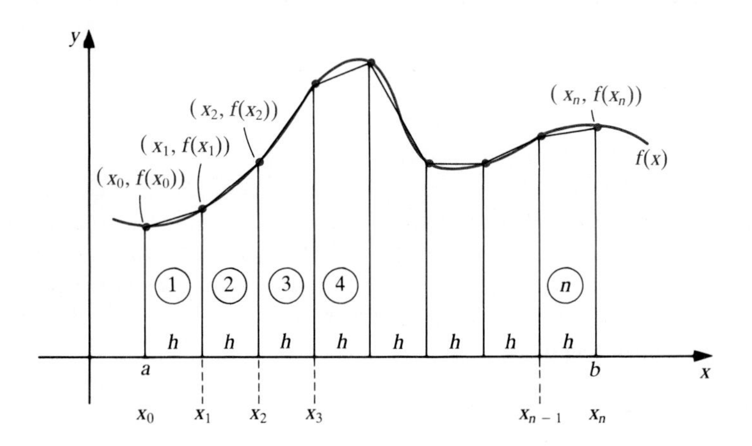

Figure 7.3
Line segments forming the trapezoids in the Trapezoidal Rule

The Trapezoidal Rule estimates the area under the curve by computing the combined area of all the trapezoids. Recall that the formula for the area of a trapezoid with bases b_1 and b_2 (the parallel sides) and height h is $\frac{1}{2}h(b_1 + b_2)$. Referring to Figure 7.3, trapezoid 1 has bases $f(x_0)$ and $f(x_1)$; trapezoid 2 has bases $f(x_1)$ and $f(x_2)$; trapezoid 3 has bases $f(x_2)$ and $f(x_3)$, and so on. Thus

$$\text{area of trapezoid } 1 = \tfrac{1}{2}h[f(x_0) + f(x_1)],$$
$$\text{area of trapezoid } 2 = \tfrac{1}{2}h[f(x_1) + f(x_2)],$$
$$\text{area of trapezoid } 3 = \tfrac{1}{2}h[f(x_2) + f(x_3)],$$
$$\vdots$$
$$\text{area of trapezoid } n = \tfrac{1}{2}h[f(x_{n-1}) + f(x_n)].$$

We add these individual areas to get an approximation to the entire area under the curve from a to b, and we denote this approximation by T_n. Note that $f(x_1)$ appears in the formula for the area of trapezoid 1 and for trapezoid 2; $f(x_2)$ appears in the formula for the area of trapezoid 2 and trapezoid 3; and so forth. Combining these common functional values gives the **Trapezoidal Rule with n subintervals**, as follows.

> **Trapezoidal Rule**
>
> $$\int_a^b f(x)\,dx \approx T_n$$
>
> where
>
> $$T_n = \tfrac{1}{2}h[f(x_0) + 2f(x_1) + 2f(x_2) + \cdots + 2f(x_{n-1}) + f(x_n)],$$
>
> $$h = \frac{b-a}{n} \quad \text{and} \quad x_i = a + ih; \quad i = 0, 1, \ldots, n.$$

For our first example of the Trapezoidal Rule, we choose an integral for which we can find an antiderivative, so we can compare the results with the exact value.

Example 7.21 Use the Trapezoidal Rule to estimate the value of $\int_1^5 x^2\,dx$ using (a) $n = 4$ and (b) $n = 8$. Compare the results with the exact value.

Solution (a) Since $n = 4$ and $b - a = 5 - 1 = 4$, we have $h = (b-a)/n = 1$. Hence the endpoints of the four intervals are $x_0 = 1$, $x_1 = 2$, $x_2 = 3$, $x_3 = 4$, and $x_4 = 5$ (see Figure 7.4). The corresponding values of $f(x)$ are the squares of these because $f(x) = x^2$. Table 7.2 shows our computations.

Table 7.2

Trapezoidal Rule on $\int_1^5 x^2\,dx$ using $n = 4$

x	$f(x) = x^2$	*Weights*	*Product*
1	1.0	1	1.0
2	4.0	2	8.0
3	9.0	2	18.0
4	16.0	2	32.0
5	25.0	1	25.0
			84.0

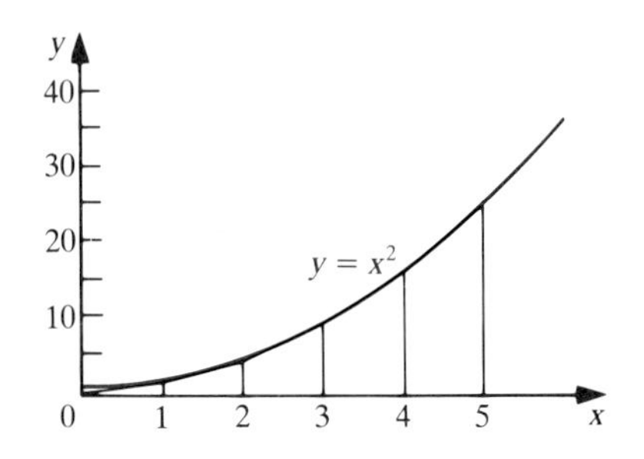

Figure 7.4
Graph of $f(x) = x^2$ with subdivisions corresponding to $n = 4$ for the Trapezoidal Rule on the interval [1, 5]

The *x column* of Table 7.2 is obtained by starting with the lower limit of integration (1 in this case) and adding h (1 in this case) repeatedly until we reach the upper limit of integration (5 in this case). The $f(x)$ column is the integrand evaluated at the corresponding x.

The *weights column* contains the coefficients in the Trapezoidal Rule: 1, 2, 2, 2, . . . , 2, 1. The *product column* is the product of the $f(x)$ column and the weights column. The last entry in the *product* column is the total (or sum) of that column. This total equals the quantity in brackets in the Trapezoidal Rule formula. Therefore (recall $h = 1$)

$$T_4 = \tfrac{1}{2}(1)(84.0) = 42.0$$

The exact value is $\displaystyle\int_1^5 x^2\,dx = \tfrac{1}{3}x^3\Big|_1^5 = \tfrac{124}{3} \approx 41.33.$

Consequently, T_4 is incorrect by $\frac{2}{3}$ unit out of about 41.33, or approximately 1.6% error.

(b) Using $n = 8$ gives $h = (b - a)/n = \frac{4}{8} = 0.5$. Table 7.3 shows our calculations. Therefore (recall $h = \frac{1}{2}$)

$$T_8 = \tfrac{1}{2}(\tfrac{1}{2})(167.00) = 41.75,$$

an error of only 1%. ■

Table 7.3

Trapezoidal Rule on $\int_1^5 x^2\,dx$ using $n = 8$

x	$f(x) = x^2$	*Weights*	*Product*
1.0	1.00	1	1.00
1.5	2.25	2	4.50
2.0	4.00	2	8.00
2.5	6.25	2	12.50
3.0	9.00	2	18.00
3.5	12.25	2	24.50
4.0	16.00	2	32.00
4.5	20.25	2	40.50
5.0	25.00	1	25.00
			167.00

Example 7.22 Approximate $\displaystyle\int_0^6 \sqrt{1 + x^2}\,dx$ using the Trapezoidal Rule with $n = 6$.

Solution Since $n = 6$ and $b - a = 6$, we have $h = 1$. Hence the endpoints of the six subintervals are 0, 1, 2, 3, 4, 5 and 6 (see Figure 7.5). The corresponding functional values are

$$\sqrt{1 + 0^2} = \sqrt{1}, \qquad \sqrt{1 + 1^2} = \sqrt{2},$$
$$\sqrt{1 + 2^2} = \sqrt{5}, \qquad \sqrt{1 + 3^2} = \sqrt{10},$$
$$\sqrt{1 + 4^2} = \sqrt{17}, \qquad \sqrt{1 + 5^2} = \sqrt{26},$$

and $\sqrt{1 + 6^2} = \sqrt{37}$.

Our calculations are shown in Table 7.4.

Table 7.4

Trapezoidal Rule on $\int_0^6 \sqrt{1+x^2}\,dx$ using $n = 6$

x	$f(x) = \sqrt{1+x^2}$	*Weights*	*Product*
0	1.0000	1	1.0000
1	1.4142	2	2.8284
2	2.2361	2	4.4722
3	3.1623	2	6.3246
4	4.1231	2	8.2462
5	5.0990	2	10.1980
6	6.0828	1	6.0828
			39.1522

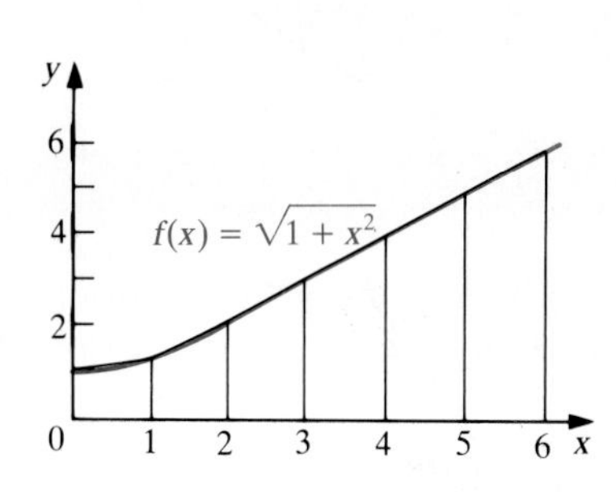

Figure 7.5
Graph of $f(x) = \sqrt{1 + x^2}$ and subdivisions for the Trapezoidal Rule on the interval [0, 6] corresponding to $n = 6$

Hence $T_6 = \frac{1}{2}(1)(39.1522) = 19.5761.$ ■

In general, the larger the value of n, the more accurate the approximation will be. Table 7.5 shows the integral of Example 7.22 approximated by the Trapezoidal Rule for several values of n. The exact value of the integral to eight places is 19.49417752.

Table 7.5
Results of the Trapezoidal Rule on $\int_0^6 \sqrt{1+x^2}\,dx$

n	T_n
6	19.576066
10	19.523768
20	19.501575
50	19.495361
100	19.494473
1000	19.494180
10000	19.494178

Example 7.23 Suppose that when a company produces x items per day its marginal revenue is given by

$$M_R(x) = \frac{x^2(x^2+1)}{x^3+100}$$

in hundreds of dollars. Find the change in revenue that the company would realize if production increases from 10 items per day to 20 items per day.

Solution The required change in revenue is given by

$$\int_{10}^{20} M_R(x)\,dx = \int_{10}^{20} \frac{x^2(x^2+1)}{x^3+100}\,dx.$$

Evaluation of this difficult integral cannot be done by the methods presented in this book. However, we can approximate the integral by the Trapezoidal Rule. We take $n = 10$. Table 7.6 shows our computations.

Table 7.6
Trapezoidal Rule on $\int_{10}^{20} \frac{x^2(x^2+1)}{x^3+100}\,dx$ **using** $n = 10$

x	$f(x) = \frac{x^2(x^2+1)}{x^3+100}$	*Weights*	*Product*
10	9.1818	1	9.1818
11	10.3159	2	20.6318
12	11.4223	2	22.8446
13	12.5076	2	25.0152
14	13.5767	2	27.1534
15	14.6331	2	29.2662
16	15.6797	2	31.3594
17	16.7185	2	33.4370
18	17.7512	2	35.5024
19	18.7788	2	37.5576
20	19.8025	1	19.8025
			291.7519

We approximate the integral by multiplying the sum of the product column by $\frac{1}{2}h$. Since $h = 1$, we have

$$\int_{10}^{20} \frac{x^2(x^2+1)}{x^3+100}\,dx \approx \tfrac{1}{2}(1)(291.7519) = 145.8759.$$

Thus the revenue increases by approximately \$14,587.59. ■

Example 7.24 Suppose that $a(t) = \sqrt{t^3 + 4}$ ft/sec^2 is the acceleration of an object along a straight line. What is the net change in the velocity of the object between $t = 2$ and $t = 6$ seconds? If the object is traveling at a velocity of 80 ft/sec at $t = 2$, what is its velocity at $t = 6$?

Solution Since we cannot evaluate

$$\int_2^6 a(t)\,dt = \int_2^6 \sqrt{t^3 + 4}\,dt$$

by the methods in this text, we approximate the integral by the Trapezoidal Rule. Taking $n = 8$, we get the results in Table 7.7.

Table 7.7
Results of the Trapezoidal Rule on $\int_2^6 \sqrt{t^3 + 4}\,dt$ using $n = 8$

t	$f(t) = \sqrt{t^3+4}$	*Weights*	*Product*
2.0	3.4641	1	3.4641
2.5	4.4300	2	8.8600
3.0	5.5678	2	11.1356
3.5	6.8465	2	13.6930
4.0	8.2462	2	16.4924
4.5	9.7532	2	19.5064
5.0	11.3578	2	22.7156
5.5	13.0528	2	26.1056
6.0	14.8324	1	14.8324
			136.8051

The integral is then approximately $\frac{1}{2}h$ times the sum of the product column. Hence the change in velocity is approximately $\frac{1}{2}(\frac{1}{2})(136.8051) = 34.2013$. The velocity after 6 seconds is about $80 + 34.2 = 114.2$ ft/sec. ■

Simpson's Rule

We use the Trapezoidal Rule to estimate the area under a curve $y = f(x)$ by approximating the curve by line segments and then computing the area under each approximating segment. Sometimes a more accurate estimate of the area is obtained by approximating the curve by parabolic segments and then computing the area under each segment. See Figure 7.6. This idea leads to Simpson's Rule for approximating integrals.

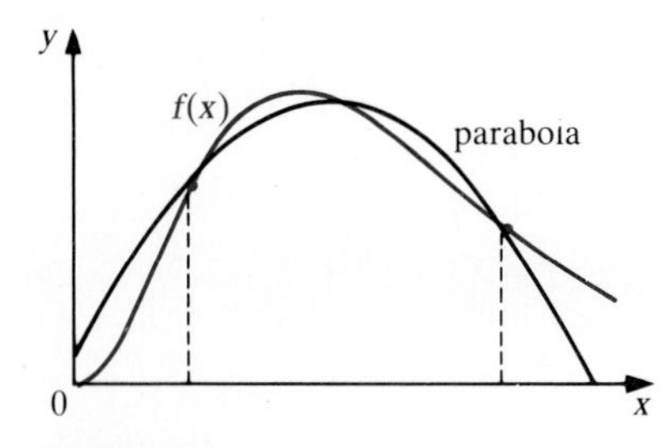

Figure 7.6
Approximating a portion of a curve by a parabola

While two points are sufficient to determine the equation of a straight line, we need three points to find the equation of a parabola. For convenience, let $y_i = f(x_i)$, where the values of x_i are obtained as in the Trapezoidal Rule; that is, $x_i = a + ih$, $i = 0, 1, 2, \ldots, n$ for a given value of n. The first parabola passes through the three points (x_0, y_0), (x_1, y_1) and (x_2, y_2); the second parabola passes through the points (x_2, y_2), (x_3, y_3) and (x_4, y_4); and so on. By determining the equation of the parabola through the points (x_{i-1}, y_{i-1}), (x_i, y_i) and (x_{i+1}, y_{i+1}) and integrating from x_{i-1} to x_{i+1}, we find that

$$\begin{array}{l}\text{the area under parabolic} \\ \text{segment from } x_{i-1} \text{ to } x_{i+1}\end{array} = \tfrac{1}{3}h(y_{i-1} + 4y_i + y_{i+1}).$$

If the original interval $[a, b]$ is divided into n equal subintervals with endpoints $x_i = a + ih$, then the approximate area is the sum of the area under each parabolic segment. Thus

$$\begin{array}{l}\text{total area} \\ \text{under curve}\end{array} \approx \begin{array}{c}\text{area under} \\ \text{parabola} \\ \text{on } [x_0, x_2]\end{array} + \begin{array}{c}\text{area under} \\ \text{parabola} \\ \text{on } [x_2, x_4]\end{array} + \begin{array}{c}\text{area under} \\ \text{parabola} \\ \text{on } [x_4, x_6]\end{array} + \cdots$$

If we let S_n denote the approximate area using n subintervals, then

$$S_n = \tfrac{1}{3}h[(y_0 + 4y_1 + y_2) + (y_2 + 4y_3 + y_4) + (y_4 + 4y_5 + y_6) + \cdots]$$

Combining terms and writing $f(x_i)$ for y_i gives **Simpson's Rule**.

Simpson's Rule

$$\int_a^b f(x)\,dx \approx S_n$$

where

$$S_n = \tfrac{1}{3}h[f(x_0) + 4f(x_1) + 2f(x_2) + 4f(x_3) + 2f(x_4) + \cdots + 4f(x_{n-1}) + f(x_n)],$$

where $h = \dfrac{b-a}{n}$ and $x_i = a + ih$, $i = 0, 1, \ldots, n$.

Because of the way the formula for Simpson's Rule is derived, n must be an *even* integer. The first and last coefficients are 1, and the other coefficients alternate between 4 and 2, *beginning and ending with* 4.

Simpson's Rule can be arranged in table form similar to the Trapezoidal Rule. In fact, the only differences in the two rules are the numbers in the *weights column* and the factor $\frac{1}{3}h$ versus $\frac{1}{2}h$ used to multiply the total of the *product column.*

Example 7.25 Use Simpson's Rule with $n = 6$ to estimate $\int_0^6 \sqrt{1 + x^2}\,dx$.

Solution As with the Trapezoidal Rule, we compute $h = (6 - 0)/6 = 1$. Our computations are shown in Table 7.8. The first number in the *x column* is the lower limit 0, and each succeeding x is found by adding h to the former x until we reach the upper limit. The *$f(x)$ column* is the integrand evaluated at the corresponding x. The *weights column* contains the Simpson's Rule weights 1 4 2 4 2 4 2 4 $\cdots$ 2 4 1. The *product column* is the product of the $f(x)$ column and the weights column.

Table 7.8
Simpson's Rule on $\int_0^6 \sqrt{1+x^2}\,dx$ using $n=6$

x	$f(x)=\sqrt{1+x^2}$	*Weights*	*Product*
0	1.0000	1	1.0000
1	1.4142	4	5.6568
2	2.2361	2	4.4722
3	3.1623	4	12.6492
4	4.1231	2	8.2462
5	5.0990	4	20.3960
6	6.0828	1	6.0828
			58.5032

Then $$S_6 = \tfrac{1}{3}(1)(58.5032) = 19.5011.$$ ■

Recall from Example 7.22 that the Trapezoidal Rule with $n=6$ gives $T_6 = 19.5761$ for this integral, and the exact value rounded to four decimal places is 19.4942. Consequently, T_6 has a percentage error of $100(19.5761 - 19.4942)/19.4942 = 0.42\%$, while Simpson's Rule has a percentage error of only 0.035%. As we hoped, Simpson's Rule is more accurate.

Example 7.26 Refer to Example 7.23. Use Simpson's Rule with $n = 10$ to estimate the integral

$$\int_{10}^{20} \frac{x^2(x^2+1)}{x^3+100}\,dx.$$

Solution We have $h = 1$; our calculations are shown in Table 7.9.

Table 7.9
Simpson's Rule on $\int_0^6 \frac{x^2(x^2+1)}{x^3+100}\,dx$ with $n=10$

x	$f(x)=\frac{x^2(x^2+1)}{x^3+100}$	*Weights*	*Product*
10	9.1818	1	9.1818
11	10.3159	4	41.2636
12	11.4223	2	22.8446
13	12.5076	4	50.0304
14	13.5767	2	27.1534
15	14.6331	4	58.5324
16	15.6797	2	31.3594
17	16.7185	4	66.8740
18	17.7512	2	35.5024
19	18.7788	4	75.1152
20	19.8025	1	19.8025
			437.6597

The value of the integral is approximately

$$\int_{10}^{20} \frac{x^2(x^2+1)}{x^3+100}\, dx \approx \tfrac{1}{3}(1)(437.6597) = 145.8866.$$

According to this computation, the increase in revenue attained by increasing production from 10 items to 20 items is \$14,588.66. This is \$1.07 more than the estimate obtained by the Trapezoidal Rule. ■

Both the Trapezoidal Rule and Simpson's Rule can be used to estimate the value of a definite integral when a formula for the function is not available, but individual values of the function are available. The following example illustrates this situation.

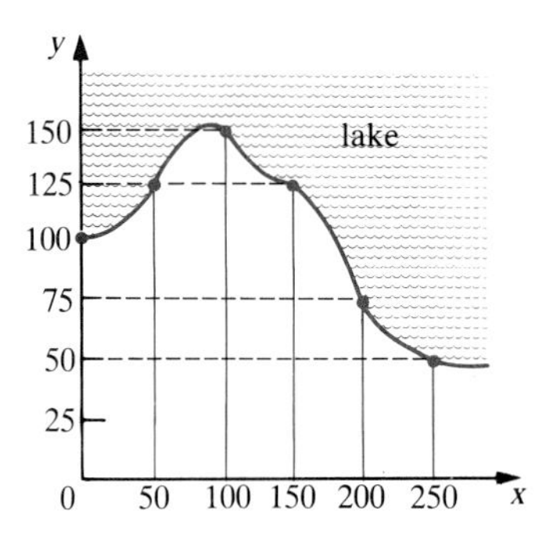

Figure 7.7
A piece of lakefront property

Example 7.27 A piece of property 250 feet wide has three straight sides and an irregular side bordering a lake as shown in Figure 7.7. Distances from the lake shoreline to the opposite side of the property are measured at equal intervals of 50 feet. Determine the approximate size of the property in acres (1 acre = 43,560 ft^2).

Solution If we let $f(x)$ denote the function whose graph is the irregular side formed by the lake shoreline, and let the origin be located at the lower left corner of the property, then the area we wish to find is

$$\text{area} = \int_0^{250} f(x)\, dx.$$

Since the measurements are spaced 50 feet apart, we take $h = 50$ (so $n = 5$). Since n is odd, we use the Trapezoidal Rule. Our computations are shown in Table 7.10. The area is approximately $(\frac{1}{2})(50)(1100) = 27{,}500$ ft$^2 \approx 0.63$ acre. ■

Table 7.10
Trapezoidal Rule for the lakefront property

x	$f(x)$	Weights	Product
0	100	1	100
50	125	2	250
100	150	2	300
150	125	2	250
200	75	2	150
250	50	1	50
			1100

Suppose that for a function $f(x)$ we want to approximate $\int_a^b f(x)\, dx$ to a given accuracy. For most functions Simpson's Rule will produce an acceptable answer with a smaller value of n than the Trapezoidal Rule, and the Trapezoidal rule will require a smaller value of n than a Riemann sum (which uses rectangles to approximate area). In practice, Simpson's Rule is preferred over the Trapezoidal Rule because it is usually much more accurate without being more difficult to apply. Also, either method may be used with a computer, so that the value of n may be increased until two approximations agree to a desired accuracy.

For a comparison of the Trapezoidal Rule, Simpson's Rule, and Riemann sums, consider the evaluation $\int_1^3 \ln x\, dx$. To five places, the correct answer is 1.29584. Table 7.11 shows the smallest value of n that will yield this value (1.29584) using Simpson's Rule, the Trapezoidal Rule and *right* Riemann sums, that is, Riemann sums where the right-hand endpoint of each subinterval is chosen as the point at which to evaluate $f(x)$ within that subinterval. Note that Simpson's Rule outperforms the other methods by a wide margin, and the Trapezoidal Rule is far better than this application of Riemann sums. Generally, if n subintervals are needed to obtain the desired accuracy using Simpson's Rule, then n^2 subintervals will be required for the Trapezoidal Rule and n^4 subintervals for right Riemann sums (the data in Table 7.11 support this remark).

Table 7.11
Minimum number of subintervals needed to approximate $\int_1^3 \ln x\, dx$ to five decimal places (1.29584)

Method	*Smallest n*
Simpson's Rule	18
Trapezoidal Rule	345
"Right" Riemann sums	135,065

Exercises

Exer. 1–12: Approximate the integral using the Trapezoidal Rule with the value of n given.

1. $\int_1^4 2x\,dx;\quad n=6$
2. $\int_0^4 x^2\,dx;\quad n=8$
3. $\int_1^4 \sqrt{x^2+2}\,dx;\quad n=6$
4. $\int_1^6 \sqrt{2x+1}\,dx;\quad n=5$
5. $\int_2^8 \sqrt{x^2-2x}\,dx;\quad n=6$
6. $\int_1^3 \sqrt{2x^2+8}\,dx;\quad n=2,\quad n=4,\quad n=6$
7. $\int_0^{10} \sqrt{x^3+3}\,dx;\quad n=5,\quad n=10$
8. $\int_{-2}^2 \frac{1}{x^2+1}\,dx;\quad n=8$
9. $\int_{-4}^4 \frac{1}{\sqrt{2x+10}}\,dx;\quad n=8$
10. $\int_1^5 \ln(x^2+2)\,dx;\ n=4$
11. $\int_0^6 \frac{\sqrt{x}}{\sqrt{x+10}}\,dx;\quad n=6$
12. $\int_{-2}^2 e^{x+2}\,dx;\quad n=4$
13. The rate at which cases of a French champagne are selling per day during a special sale is given by $s(t)=\sqrt{t^2+3t+2}$ for $0\le t\le 7$. Approximately how many cases will be sold during the 7-day period? Use $n=7$.
14. A forest is managed so that the rate at which lumber can be harvested for the next six years is $L(t)$ hundred thousand board feet per year, where $L(t)=(t^3+t^2+20)^{3/2}$ for $0\le t\le 6$. Approximately how many board feet will be produced during this 6-year period? Use $n=6$.
15. A sociologist finds that the rate of burglaries in a given city during a 5-day police strike is $b(t)$ burglaries per day, where $b(t)=(t^3+6t^2+12t+8)^{2/3}$ for $0\le t\le 5$ with t measured in days. According to this formula, approximately how many burglaries were committed during the 5-day strike? Use $n=5$.
16. The Trapezoidal Rule gave an exact answer for the integral in Exercise 1. Explain why this is true.
17. Use Simpson's Rule on $\int_1^4 x^3\,dx$ with $n=4$.

18–29. Work Exercises 1–12 by Simpson's Rule. In cases where $n=5$, use $n=6$.

30. For the piece of property shown in the following figure, use the Trapezoidal Rule and Simpson's Rule to estimate the area of the property (in square feet).

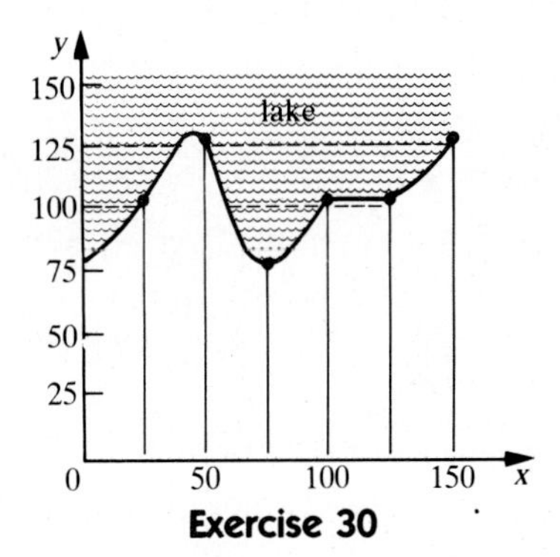

Exercise 30

7.5 Integration by the Use of Tables

The integrals we have encountered thus far that could be integrated rather than only approximated have been reduced to one of the fundamental integrals by using substitution or integration by parts. An alternative method for determining many integrals is to use a table of integrals.

A **table of integrals** is a list of commonly occurring functions and their antiderivatives. If an integral we seek is in the table, we can determine its antiderivative immediately. Usually a table of integrals does not contain an integral as specific as

$$\int \frac{1}{3x+4}\,dx.$$

Instead the table contains an entry such as

$$\int \frac{1}{ax+b}\,dx = \frac{1}{a}\ln|ax+b|.$$

(No constants of integration are included in the table.) When using the table to find the antiderivative of a particular integral, we find the entry that will match that integral when we assign specific values to the constants a, b, and so on.

Table 7.12 gives several common integrals. Most of these integrals can be determined by substitution or by integration by parts.

Table 7.12
A short table of integrals

1. $\int \frac{1}{ax+b}\,dx = \frac{1}{a}\ln|ax+b|$
2. $\int \frac{x}{(ax^2+b)^n}\,dx = \frac{1}{2a(n+1)}\frac{1}{(ax^2+b)^{n+1}}, \quad n \neq -1$
3. $\int \frac{x}{ax+b}\,dx = \frac{1}{a^2}(ax+b-b\ln|ax+b|)$
4. $\int \frac{x}{(ax+b)^2}\,dx = \frac{1}{a^2}\left(\frac{b}{ax+b}+\ln|ax+b|\right)$
5. $\int \frac{1}{\sqrt{ax+b}}\,dx = \frac{2}{a}\sqrt{ax+b}$

6. $\int \sqrt{x^2+a^2}\,dx = \frac{x}{2}\sqrt{x^2+a^2}+\frac{a^2}{2}\ln\left|x+\sqrt{x^2+a^2}\right|$
7. $\int \frac{dx}{\sqrt{x^2+a^2}} = \ln\left|x+\sqrt{x^2+a^2}\right|$
8. $\int \frac{dx}{x\sqrt{x^2+a^2}} = \frac{1}{a}\ln\left|\frac{\sqrt{x^2+a^2}+a}{x}\right|$

9. $\int \sqrt{x^2-a^2}\,dx = \frac{x}{2}\sqrt{x^2-a^2}-\frac{a^2}{2}\ln\left|x+\sqrt{x^2-a^2}\right|$
10. $\int x\sqrt{x^2-a^2}\,dx = \frac{1}{3}(x^2-a^2)^{3/2}$
11. $\int x^2\sqrt{x^2-a^2}\,dx = \frac{x}{8}(2x^2-a^2)\sqrt{x^2-a^2}-\frac{a^4}{8}\ln\left|x+\sqrt{x^2-a^2}\right|$
12. $\int \frac{dx}{\sqrt{x^2-a^2}} = \ln\left|x+\sqrt{x^2-a^2}\right|$

13. $\int xe^{ax}\,dx = \frac{1}{a^2}(ax-1)e^{ax}$
14. $\int \frac{dx}{1+be^{ax}} = x-\frac{1}{a}\ln(1+be^{ax})$
15. $\int x^n\ln x\,dx = \frac{x^{n+1}}{(n+1)^2}[(n+1)\ln x-1], \quad n \neq -1$
16. $\int \frac{dx}{x\ln x} = \ln|\ln x|$

Example 7.28 Find $\int \frac{5x}{(3x+4)^2}\,dx$.

Solution Factoring 5 out of the integral yields

$$\int \frac{5x}{(3x+4)^2}\,dx = 5\int \frac{x}{(3x+4)^2}\,dx.$$

Comparing this integral with those in Table 7.12, we find that integral 4 matches this integral if $a = 3$ and $b = 4$. Hence

$$\int \frac{5x}{(3x+4)^2}\,dx = (5)\frac{1}{3^2}\left(\frac{4}{3x+4} + \ln|3x+4|\right) + C. \qquad \blacksquare$$

Example 7.29 Find $\int \sqrt{2x^2+50}\,dx$.

Solution Factoring $\sqrt{2}$ out of the radical and out of the integral gives

$$\int \sqrt{2x^2+50}\,dx = \sqrt{2}\int \sqrt{x^2+25}\,dx$$

This integral matches integral 6 in Table 7.12 if $a = 5$. Consequently,

$$\int \sqrt{2x^2+50}\,dx = \sqrt{2}\left(\tfrac{1}{2}x\sqrt{x^2+25} + \tfrac{25}{2}\ln|x+\sqrt{x^2+25}|\right) + C. \qquad \blacksquare$$

Example 7.30 Find $\int \dfrac{dx}{1-7e^{0.5x}}$.

Solution This integral has the same form as integral 14 in Table 7.12 with $b = -7$ and $a = 0.5$. Thus

$$\int \frac{dx}{1-7e^{0.5x}} = x - \frac{1}{0.5}\ln|1-7e^{0.5x}| + C$$
$$= x - 2\ln|1-7e^{0.5x}| + C. \qquad \blacksquare$$

Exercises

Exer. 1–22: Use Table 7.12 to find the integral.

1. $\int \dfrac{1}{4x-9}\,dx$
2. $\int \dfrac{1}{\sqrt{5x^2+4}}\,dx$
3. $\int \dfrac{x}{(-3x^2+2)^4}\,dx$
4. $\int \dfrac{dx}{x\sqrt{20+x^2}}$
5. $\int \dfrac{1}{\sqrt{-2x+1}}\,dx$
6. $\int \sqrt{3x^2-12}\,dx$
7. $\int \sqrt{4t^2+12}\,dt$
8. $\int x^2\sqrt{x^2-7}\,dx$
9. $\int \dfrac{x}{x+6}\,dx$
10. $\int \dfrac{dt}{1+4e^{3t}}$
11. $\int x^2\sqrt{x^2-9}\,dx$
12. $\int x^2\sqrt{x^2-3}\,dx$
13. $\int \dfrac{1}{\sqrt{8t+9}}\,dt$
14. $\int x\sqrt{5x^2-15}\,dx$
15. $\int \dfrac{dx}{\sqrt{x^2-1}}$
16. $\int \dfrac{dt}{t\sqrt{2t^2+8}}$
17. $\int \dfrac{x}{(3x-2)^3}\,dx$
18. $\int \dfrac{dx}{1-3e^{-2x}}\,dx$
19. $\int x^2\sqrt{x^2-4}\,dx$
20. $\int \dfrac{1}{\sqrt{4x-1}}\,dx$
21. $\int w^{17}\ln w\,dw$
22. $\int \dfrac{3x}{(3-x)^2}\,dx$

Exer. 23–28: Use Table 7.12 to find the value of the integral.

23. $\int_0^{\sqrt{3}} \sqrt{x^2+1}\,dx$
24. $\int_1^2 x^2\sqrt{x^2-1}\,dx$
25. $\int_0^{\ln 2} \dfrac{4}{1+2e^x}\,dx$
26. $\int_1^2 \sqrt{x^2-1}\,dx$

27. $\int_2^4 \frac{dx}{\sqrt{x^2-3}}$

28. $\int_1^4 \frac{4}{\sqrt{2x-1}}\,dx$

29. Use the substitution $u = ax + b$ to derive integral 4 in Table 7.12.

30. Refer to Table 7.12. By differentiating x and integrating $x\sqrt{x^2-1}$ (by substitution), show that one step of integration by parts converts integral 11 into integral 9 (along with some extra terms).

Summary

The method of **integration by substitution** is based on the formula

$$\int f(g(x))g'(x)\,dx = \int f(u)\,du,$$

where $f(u)$ is a function of u and $u = g(x)$ is a function of x.

Integration by parts is a method of integration based on the formula

$$\int u\,dv = uv - \int v\,du,$$

where u and v represent differentiable functions of x.

The **improper integral** $\int_a^\infty f(t)\,dt$ indicates the value of $\lim_{x\to\infty} \int_a^x f(t)\,dt$, provided this limit exists. If the limit exists, we say that the improper integral **converges**. If the limit does not exist, the integral **diverges**.

The approximation of the integral $\int_a^b f(x)\,dx$ by the **Trapezoidal Rule** is

$$T_n = \tfrac{1}{2}h[f(x_0) + 2f(x_1) + \cdots + 2f(x_{n-1}) + f(x_n)],$$

where $h = \frac{b-a}{n}$ and $x_i = a + ih, \quad i = 0, 1, 2, \ldots, n$

and where n is the number of subintervals used in the approximation.

With the same notation, **Simpson's Rule** for the approximation of $\int_a^b f(x)\,dx$ is

$$S_n = \tfrac{1}{3}h\,[f(x_0) + 4f(x_1) + 2f(x_2) + \cdots + 4f(x_{n-1}) + f(x_n)],$$

where n must be an even integer.

Review Exercises

Exer. 1–20: Perform the integration.

1. $\int (x+3)^5\,dx$

2. $\int (2x-5)^4\,dx$

3. $\int 2x\sqrt{x^2-4}\,dx$

4. $\int \frac{x\,dx}{(x^2+10)^4}$

5. $\int \frac{(2t-3)\,dt}{\sqrt{t^2-3t+1}}$

6. $\int 6x(3x^2+2)^{3/2}\,dx$

7. $\int \frac{e^x\,dx}{e^x+1}$

8. $\int e^{-2t}\,dt$

9. $\int xe^{x^2+1}\,dx$

10. $\int x^2 \ln x\,dx$

11. $\int x\sqrt{3x+5}\,dx$

12. $\int x^2e^{4x}\,dx$

13. $\int (2t+3)(t+2)^{-1}\,dt$

14. $\int \frac{e^t\,dt}{(e^t+1)^4}$

15. $\int \frac{2\,dx}{2x-5}$

16. $\int \frac{x\,dx}{3x^2-1}$

17. $\int x\sqrt{x^2+1}\,dx$

18. $\int e^{2x^3+3x}(2x^2+1)\,dx$

19. $\int \frac{\ln s}{s}\,ds$

20. $\int (x+1)(x^2+2x)^5 + xe^{-x}\,dx$

Exer. 21–26: Evaluate the definite integral.

21. $\int_0^1 (x+1)^2\,dx$

22. $\int_{-3}^{-2} (5+2x)^5\,dx$

23. $\int_0^1 x^4e^{-x}\,dx$

24. $\int_1^3 \frac{(x+2)\,dx}{\sqrt{x^2+4x+4}}$

25. $\int_3^4 \frac{dx}{x-2}$

26. $\int_0^{\ln 2} \frac{3\,dx}{e^x}$

Exer. 27–30: Determine if the improper integral converges or diverges. If it converges, determine the value of the integral.

27. $\int_1^\infty \frac{dt}{2t^2}$

28. $\int_2^\infty \sqrt{3t-2}\,dt$

29. $\int_{-\infty}^{-2} \frac{dt}{t^{1/5}}$

30. $\int_{-\infty}^{\infty} \frac{|t|}{(t^2+3)^2}\,dt$

Exer. 31–32: Use the Trapezoidal Rule with the given value of n to approximate the definite integral.

31. $\int_1^5 \frac{dx}{x+1}$; $n = 4$

32. $\int_{-2}^{2} \frac{dx}{x^2+1}$; $n = 8$

Exer. 33–34: Use Simpson's Rule with the given value of n to approximate the definite integral.

33. $\int_1^5 \frac{dx}{x+1}$; $n = 4$

34. $\int_{-2}^{2} \frac{dx}{x^2+1}$; $n = 8$

Exer. 35–38: Find the area under the curve for the given interval.

35. $y = x^2 \ln x$; from $x = 1$ to $x = e$

36. $y = \frac{\ln x^2}{x}$; from $x = 1$ to $x = e^2$

37. $y = \frac{x}{3x+5}$; from $x = 0$ to $x = 1$

38. $y = \frac{x}{3x^2+5}$; from $x = 0$ to $x = 1$

Exer. 39–40: Use Table 7.12 to evaluate the integral.

39. $\int \frac{dx}{\sqrt{x^2-16}}$

40. $\int_3^4 x\sqrt{4x^2-36}\,dx$

41. The marginal revenue of a company is given by $M_R(x) = 110 + 300(2x+1)^{-1/2}$ in dollars when production is at the level x items per day. How much will the revenue change if production is increased from 24 items per day to 40 items per day?

42. An investment guarantees a return of \$50 per month. Suppose that over the last year, the rate of return $r(t)$ at time t (in months) is given by $r(t) = 50 + t^2e^{t/3}$. Find the total amount paid by this investment for the full year.

Functions of More Than One Variable

· E · I · G · H · T ·

So far we have limited our discussion to functions of one independent variable. However, many interesting and important concepts in the managerial, social, and life sciences depend on more than one variable. Consider the following examples:

- The cost function of a shoe factory that makes two types of shoes depends on the number of leather shoes produced and the number of synthetic shoes produced.
- The amount of pollution in a lake used by a nuclear power plant and a chemical manufacturer depends on the quantity of heat entering the lake from the power plant and the amount of chemical waste pouring into the lake.
- The number of bushels per acre of a certain crop depends on the number of plants per square foot, the number of hours of labor applied, weather conditions, the amount of fertilizer per acre, and so on.
- The number of bacteria in a certain culture is a function of time and temperature.
- The reaction time of a treatment for a certain illness is a function of the amounts of two distinct drugs administered.
- A company's sales depend on the amount of newspaper advertising and the amount of television advertising.

As we did with functions of one variable, we will use the concept of the derivative in our study of the behavior of functions of two or more variables. In this chapter we develop appropriate methods of differentiation and integration for functions of more than one variable.

8.1 Functions of Several Variables

A function of two variables x and y is a rule that assigns a unique value z to each pair of real numbers x and y. This relationship is usually denoted by $z=f(x, y)$. In particular, the value assigned to the pair $(3, 4)$ is denoted by $f(3, 4)$.

Example 8.1 Find $f(1, 2)$, $f(2, 1)$, and $f(-1, 0)$ if $z=f(x, y)$ is defined by the equation

$$f(x, y) = 3x + 4y^2 + 3.$$

Solution The evaluation of functions of two variables is similar to that of one variable. To evaluate $f(1, 2)$, we substitute 1 for x and 2 for y in $z=f(x, y)$, obtaining

$$f(1, 2) = 3(1) + 4(2)^2 + 3 = 22.$$

Similarly,

$$f(2, 1) = 3(2) + 4(1)^2 + 3 = 13,$$

and

$$f(-1, 0) = 3(-1) + 4(0)^2 + 3 = 0.$$

■

Example 8.2 A firm produces two models of lawn statuettes. It takes 6 hours of labor to produce each unit of one model and 11 hours to produce each unit of the other. Thus, the total hours of labor needed to produce x of the first model and y of the second model is given by $h = h(x, y) = 6x + 11y$. How many hours are needed to produce 4 units of the first model and 7 units of the second?

Solution The number of hours needed is $h(4, 7) = 6(4) + 11(7) = 101$ hours of labor. ■

Example 8.3 Through experimentation, the yield (in bushels) of rutabagas has been determined to be given by

$$B = f(x, y, z) = \frac{240xyz - 2500x^2 - 36y^2z}{x + y + 15z},$$

where y is the number of acres of land used, $100x$ is the number of hours of labor employed, and z is the number of bags of fertilizer used. What is the yeld if 10 acres are cultivated, 400 hours of labor are used, and 100 bags of fertilizer are applied?

Solution From the statement of the problem, $y = 10$, $z = 100$, and since $100x = 400$, we have $x = 4$. Thus, the yield is

$$f(4, 10, 100) = \frac{240(4)(10)(100) - 2500(4)^2 - 36(10)^2(100)}{4 + 10 + 15(100)}$$

$$= \frac{960000 - 40000 - 360000}{1514} \approx 369.9$$

So 370 bushels of rutabagas are produced. ■

To sketch the graph of functions of two variables, we need three axes rather than the two we have used to graph functions of one variable. Traditionally, the positive z-axis is shown rising vertically above the plane determined by the x-axis and y-axis. See Figure 8.1. This system is often referred to as a *three-dimensional coordinate system.*

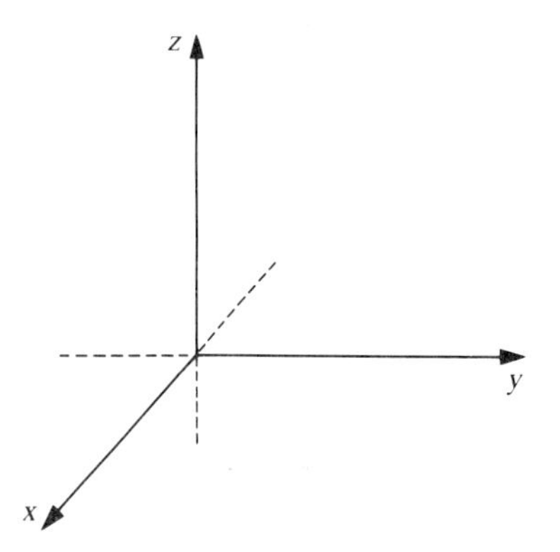

Figure 8.1
A three-dimensional coordinate system

A point is determined by a **triple** (x_0, y_0, z_0). For example, the triple $(2, 3, 4)$ is a point that is reached by moving two units in the x-direction, followed by 3 units in the y-direction, then 4 units in the z-direction (up). Figure 8.2 illustrates how this and other points are plotted in three-dimensional space.

Table 8.1

Point	*Location of point*
(0, 0, 0)	Origin
(2, 0, 0)	on x-axis
(0, 3, 0)	on y-axis
(0, 0, 4)	on z-axis
(2, 3, 0)	in the xy-plane
(2, 0, 4)	in the xz-plane
(0, 3, 4)	in the yz-plane

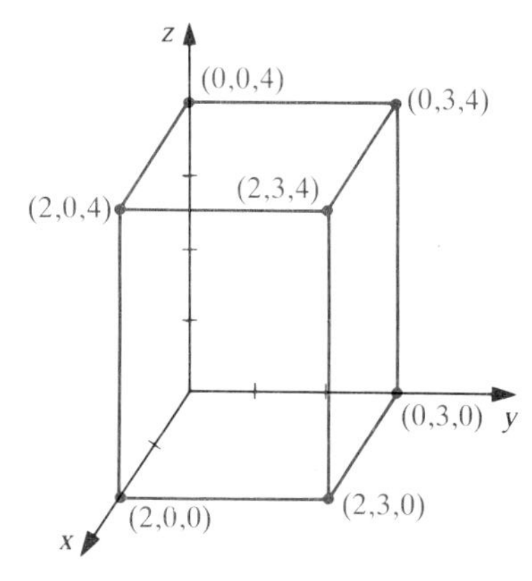

Figure 8.2
Several points in a three-dimensional coordinate system

To visualize a three-dimensional coordinate system, imagine being in a room with the origin in the front corner to the left and on the floor. The positive x-axis is along the left wall at the floor, pointing toward the back of the room; the positive y-axis is along the front wall at the floor, pointing to the right, and the positive z-axis is in the corner where the left and front walls meet, pointing up. The xy-plane is the floor of the room; it is the set of all points with z-coordinate equal to zero. The xz-plane is the left wall; it is the set of points with y-coordinate zero. The yz-plane is the front wall; it is the set of points with x-coordinate zero. If the ceiling is 10 feet high, then the ceiling is the plane described by $z = 10$; it is the set of all points with z-coordinate 10. Note that the plane determined by the x- and y-axes is the plane $z = 0$; this is important! The x- and z-axes determine the plane $y = 0$; the y- and z-axes determine the plane $x = 0$.

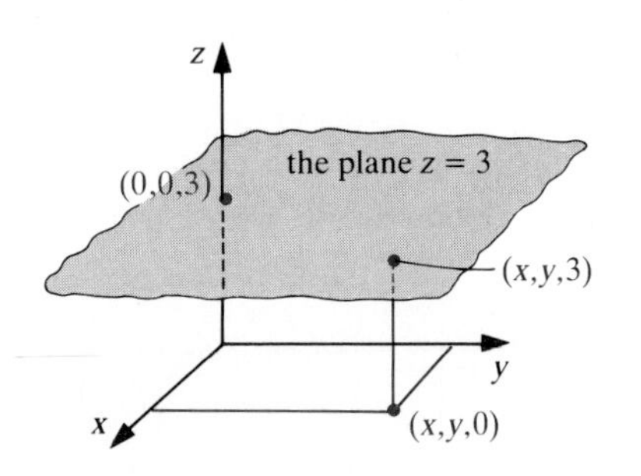

Figure 8.3
Graph of all points of the form $(x, y, 3)$ is the plane $z = 3$

Example 8.4 Locate all points of the form $(x, y, 3)$.

Solution Since z has the fixed value 3 for all x and y, the points of the form $(x, y, 3)$ must be 3 units above the xy-plane. Thus, the graph of all points $(x, y, 3)$ is the plane parallel to the xy-plane and 3 units above it. Figure 8.3 shows this plane. The equation describing the plane is $z = 3$. ∎

In two dimensions a linear equation such as $2x + 3y = 6$ has a straight line as its graph. However, in three dimensions a linear equation such as $2x + 3y + 4z = 12$ is a plane. Since any three non-colinear points determine a plane, we can graph a plane by finding three points that satisfying the equation and sketching the plane that passes through the three points. The points $(0, 0, 3)$, $(6, 0, 0)$, and $(0, 4, 0)$ satisfy the equation $2x + 3y + 4z = 12$, and Figure 8.4 shows a graph of the plane that passes through these points.

Graphs of nonlinear functions are somewhat more difficult to construct in three dimensions than in two dimensions. Familiarity with some graphs will be helpful in visualizing the topics discussed in this chapter. We have included the graphs of three nonlinear functions in Figures 8.5–8.7.

The equation $z = f(x, y)$ expresses z in terms of x and y. The variables x and y are *independent variables*, and z is the *dependent variable*. The **domain** of the function $f(x, y)$ is the set of all x and y for which $f(x, y)$ is defined and is some region R of the xy-plane. For each point (x, y) in R, the function provides a way to compute an *altitude* z above (or below) the xy-plane. The collection of these points for all (x, y) in R forms a *surface*. Three surfaces are shown in Figures 8.5–8.7.

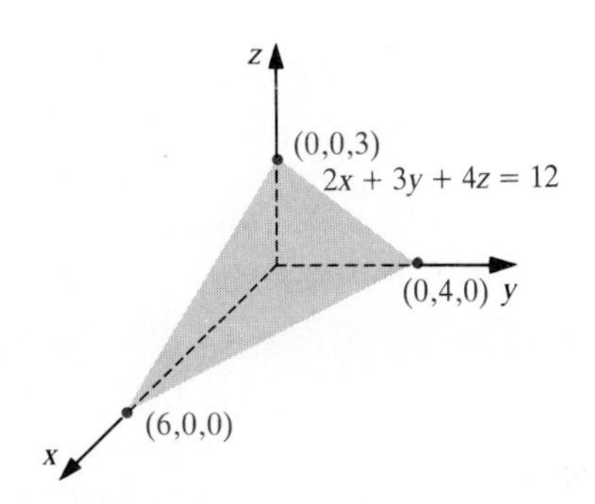

Figure 8.4
The plane $2x + 3y + 4z = 12$

The graphs shown in Figures 8.5–8.7 are two-dimensional representations of three-dimensional surfaces. For example, the graph in Figure 8.7 looks like a hollow hill that goes down forever. It is difficult to imagine what this hill would look like from the side or from the top. To help visualize these shapes we can view *slices* of the graph. For example, if we look at how the yz-plane slices through the surface we will obtain a curve in that plane. Let $z = f(x, y)$ denote the surface. The

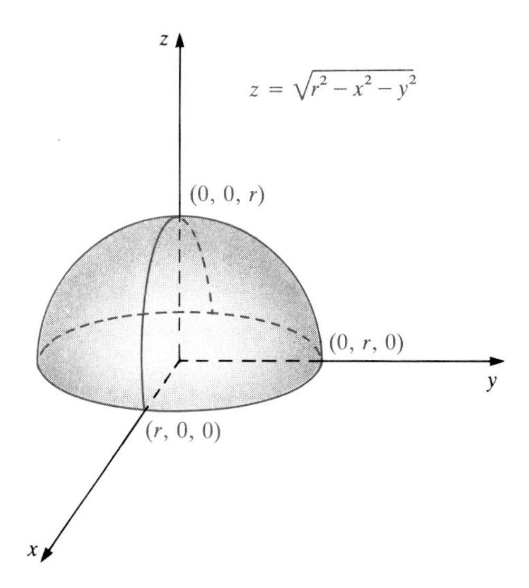

Figure 8.5
Graph of $x^2 + y^2 + z^2 = r^2$ is a sphere of radius r

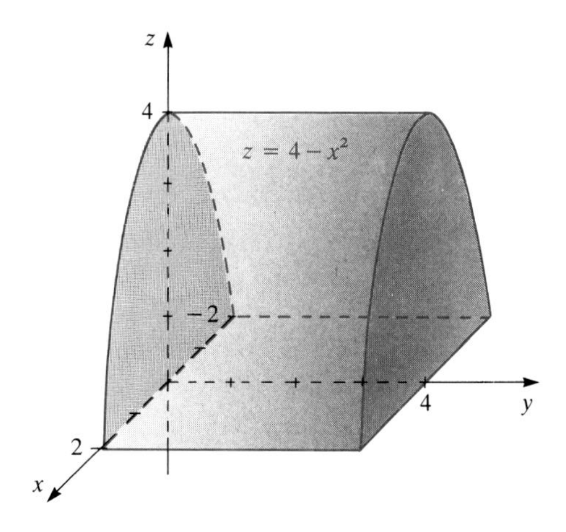

Figure 8.6
Graph of $z = 4 - x^2$, a parabolic Quonset-hut shape

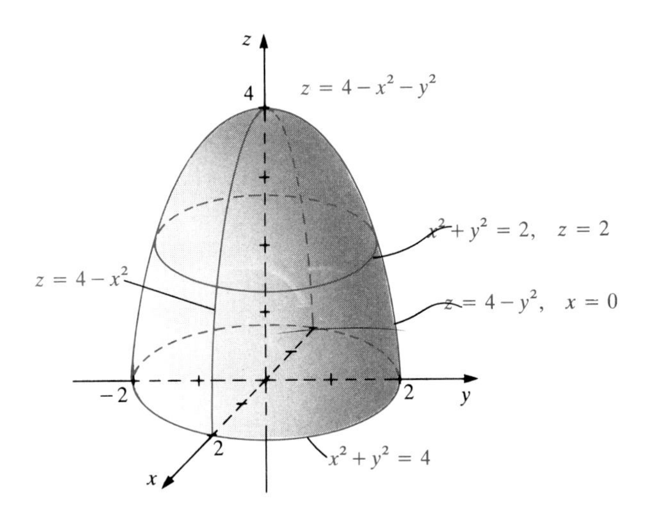

Figure 8.7
Graph of $z = 4 - x^2 - y^2$, a parabolic hill

equation of the yz-plane is $x=0$, so $z=f(0, y)$ is the equation of the slice of the surface lying in the yz-plane.

For the surface in Figure 8.7, the slice in the yz-plane has the equation $z=f(0, y)=4-0^2-y^2=4-y^2$. The parabola $z=4-y^2$ is graphed in Figure 8.8. Similarly, for any surface $z=f(x, y)$ the slice in the xz-plane ($y=0$) has equation $z=f(x, 0)$. For the surface in Figure 8.7 the equation of the slice in the xz-plane is $z=4-x^2$; it is graphed in Figure 8.9. The slice in the xy-plane ($z=0$) for $z=f(x, y)$ is given by the equation $0=f(x, y)$. For the surface in Figure 8.7, the equation of this slice is $0=4-x^2-y^2$, and its graph is a circle with radius 2 centered at the origin, as shown in Figure 8.10.

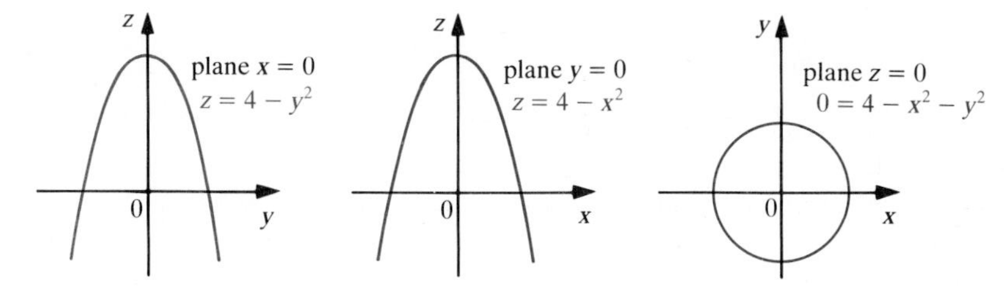

Figure 8.8
Intersection of $z=4-x^2-y^2$ with the yz-plane

Figure 8.9
Intersection of $z=4-x^2-y^2$ with the xz-plane

Figure 8.10
Intersection of $z=4-x^2-y^2$ with the xy-plane

The slicing technique not only gives us a better understanding of the shape of a three-dimensional surface but, more importantly, it enables us to apply our two-dimensional definitions of slope and derivative to the three-dimensional case. To accomplish this goal, we will slice a surface with planes parallel to the yz-plane and with planes parallel to the xz-plane. A plane parallel to the yz-plane passing through the x-axis at $x=a$ has the equation $x=a$; hence, the slice of $z=f(x, y)$ that lies in the plane $x=a$ has the equation $z=f(a, y)$. See Figure 8.11. Similarly, the slice lying in the plane parallel to the xz-plane passing through $y=b$ has equation $z=f(x, b)$ (see Figure 8.12).

Slicing the function $z=f(x, y)$ with a plane $z=c$ parallel to the xy-plane produces the equation $c=f(x, y)$. This equation relating x and y describes a curve in the xy-plane, called a **level curve**. Slicing the function $z=4-x^2-y^2$ with the plane $z=2$ produces the equation $2=4-x^2-y^2$, or $x^2+y^2=2$, which describes a circle of radius $\sqrt{2}$ with

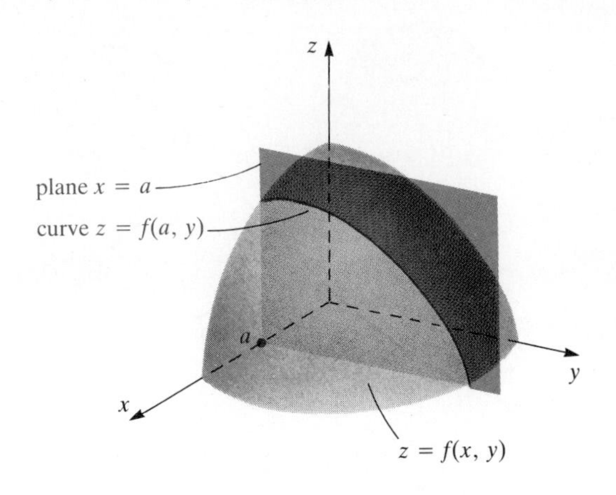

Figure 8.11
A slice through a surface at $x=a$

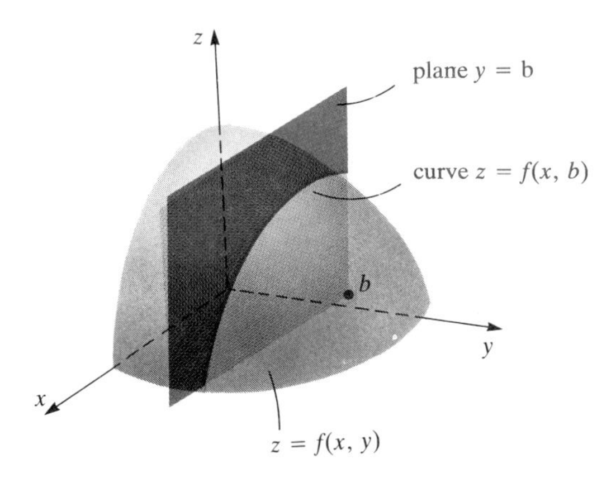

Figure 8.12
A slice through a surface at $y = b$

center at the origin. Since $z = 2$, the circle is located 2 units above the xy-plane. Likewise, the level curve for $z = -3$ is $x^2 + y^2 = 7$, a circle of radius $\sqrt{7}$ with center at the origin and located 3 units below the xy-plane. Figure 8.13 shows the graph of several level curves for the function $z = 4 - x^2 - y^2$. We are looking down from the "top of the hill" as we proceed from $z = 4$ (the top) through smaller values of z.

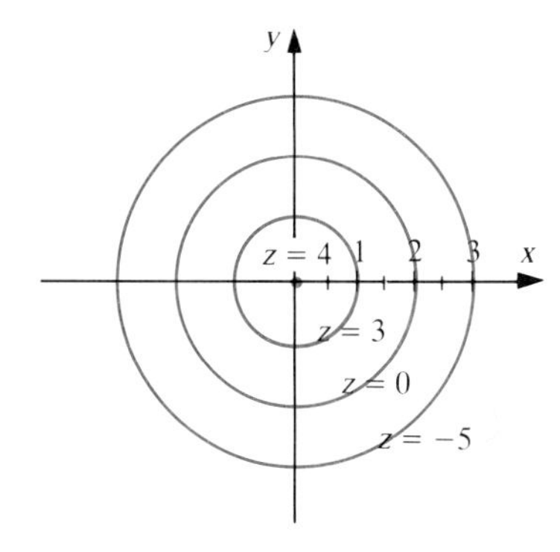

Figure 8.13
Level curves for $z = 4 - x^2 - y^2$

Level curves are used to show curves of constant functional values in other situations. For example, topographical maps of mountainous terrain show curves of constant altitude—level curves for the "function" describing the mountains. On such a map level curves may be shown for values of altitude in 1000-foot increments, with curves indicating where the altitude is 6000 feet, 7000 feet, and so on. Several level curves close to one another indicate very steep terrain.

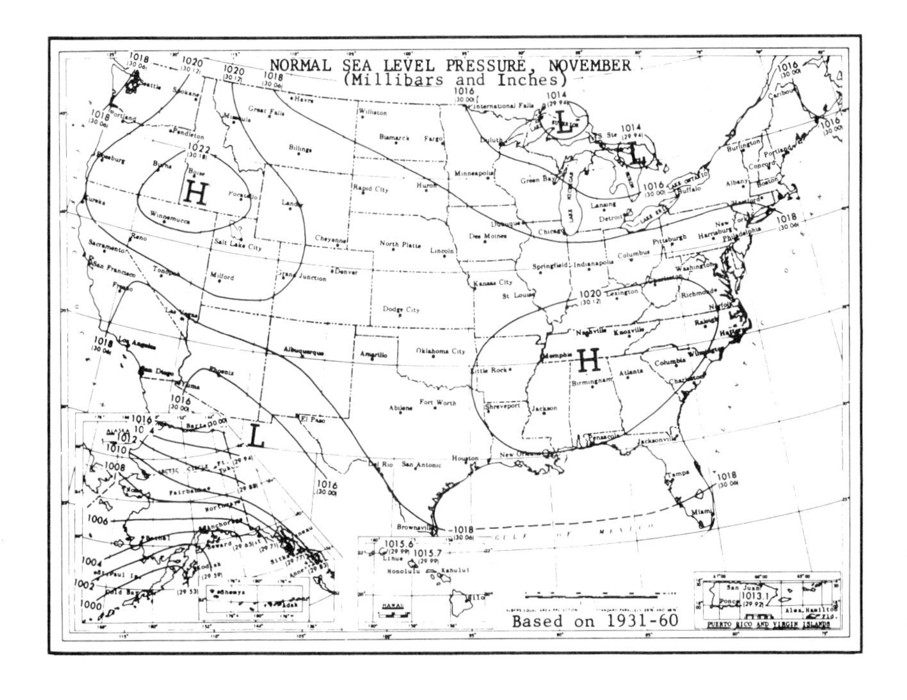

Figure 8.14
A portion of a U.S. weather map showing (a) isobars (curves of constant pressure); (b) isotherms (curves of constant temperature) (shown at top of following page)

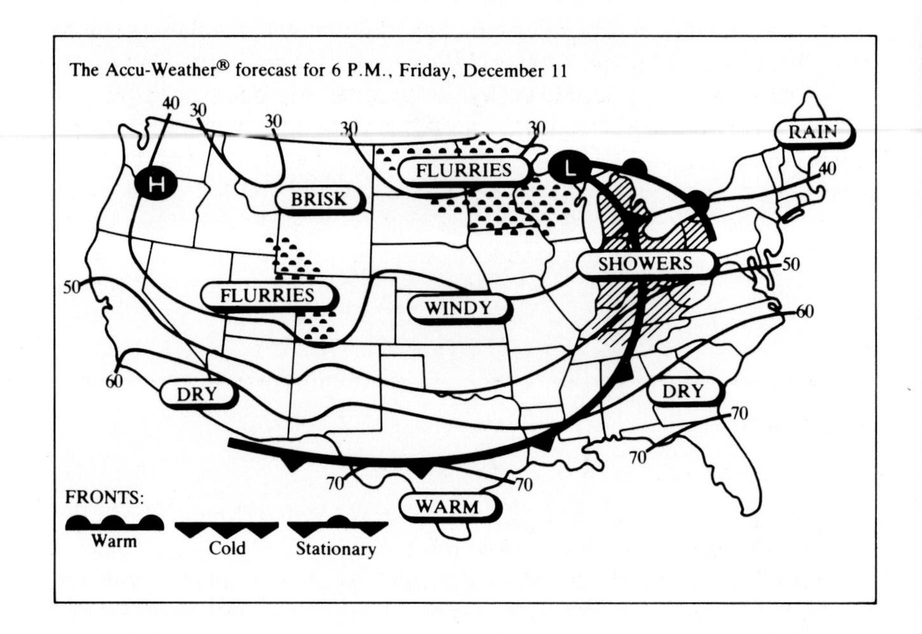

Level curves are used on weather maps to show regions of constant conditions, such as pressure or temperature. If we think of atmospheric pressure as a function of position (x, y) over the region, then *isobars* are level curves for atmospheric pressure. See Figure 8.14(a). Closely spaced isobars indicate a rapid change in pressure over a small region, a condition that produces strong winds. Similarly, *isotherms*, which are level curves of the temperature function, show curves of constant temperature. See Figure 8.14(b).

The two-dimensional analogue of an interval such as $[a, b]$ is a *rectangular region* such as $a \leq x \leq b$, $c \leq y \leq d$. In a one-dimensional domain, the phrase *some interval about a* designates an interval such as $[a - h, a + h]$ for some number h. For a two-dimensional domain, *some rectangular region about* (a, b) refers to the set of all points (x, y) such that $a - h \leq x \leq a + h$, $b - k \leq y \leq b + k$ for some numbers h and k. See Figure 8.15.

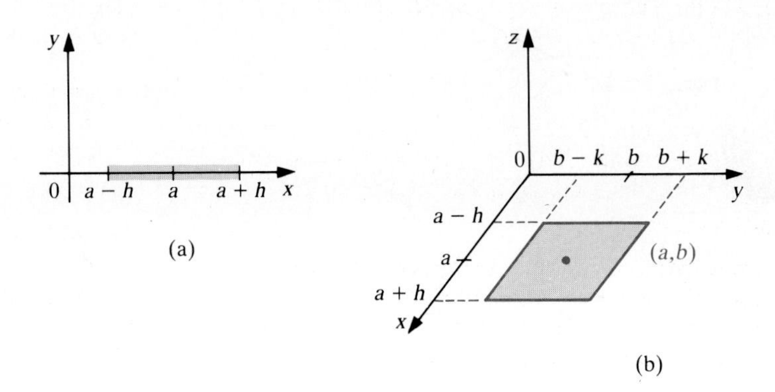

Figure 8.15
(a) One-dimensional domain: an interval about a; (b) two-dimensional domain: a rectangular region about (a, b)

Most of our investigations in this chapter will involve functions that have two independent variables, such as $z = f(x, y)$. However, some situations are described by a function of many variables. In Example 8.3 the yield of a rutabaga harvest was dependent on acreage, labor, fertilizer, and, perhaps, also on rainfall. The air temperature at a given location depends on at least four variables: latitude, longitude, altitude, and time. A company's sales S may depend on the size of the sales force s, the time of year t, printed advertising α, television advertising τ, and price p, and may be given by a function $S = f(s, t, \alpha, \tau, p)$.

Exercises

Exer. 1–6: Evaluate the function at the indicated points.

1. $f(x, y) = x^2 + y^2 - 7xy$; $(2, 3)$, $(0, 4)$, $(3, 0)$
2. $f(x, y) = 3x^2y - 4x + 7y^2$; $(2, 3)$, $(-1, 3)$, $(12, 9)$
3. $f(x, y) = 4(x + y)^2 - 17(x - 2y)^3$; $(2, 3)$, $(4, 1)$
4. $f(x, y) = \dfrac{\sqrt{x^2 + y^2}}{x^2 - y^2 + 3}$; $(2, 0)$, $(-2, 2)$, $(0, 0)$
5. $f(s, t, u) = su^2 + \dfrac{t^2 - 2su}{s^2 - t^2 + u^2}$; $(3, 1, 1)$, $(2, 0, -1)$, $(2, -1, 0)$
6. $f(x, y, z) = \dfrac{xz^2 - 2yz^3}{14xy^2} + 30x^2$; $(4, 4, 4)$, $(1, 7, 2)$

Exer. 7–10: The cost of producing two types of textbooks is \$7.50 for each hardbound volume and \$3.45 for each paperback.

7. Write an equation expressing the cost of producing the two types of books, with x representing the number of hardbound copies produced and y representing the number of paperback copies produced.
8. What is the cost of producing 3000 hardbound and 5000 paperback books?
9. What is the cost of producing 10,000 hardbound and 40,000 paperback books?
10. What is the cost of producing 1000 of each kind?

Exer. 11–13: The cost of replacement of parts for an item that is subject to two inspections is

$$C(x, y) = 2x^2 + 17xy^2 + 140y + 12x + 200,$$

where x represents the number of deficient parts found during the first inspection and y represents the number of deficient parts found during the second. Find the cost for the given condition.

11. Three parts are found to be deficient in the first inspection and none in the second.
12. Forty parts are found to be deficient in the first inspection and 100 in the second.
13. Two parts are found to be deficient in the first inspection and 3 in the second.

Exer. 14–16: A candy factory produces two kinds of lemon drops in quantities of x kilograms of the first kind and y kilograms of the second. The total cost is $C(x, y) = 4x^2 + 80x + 60y^2 + 100y - 40xy$ and the total revenue is $R(x, y) = 240x + 300y + 40xy$. Find the total cost and total revenue for the given production quantities.

14. 100 kilograms of the first kind and 80 kilograms of the second
15. 10 kilograms of the first kind and 10 kilograms of the second
16. 100 kilograms of the first kind and 100 kilograms of the second

Exer. 17–20: Plot these points on a three-dimensional coordinate system.

17. $(2, 1, 4)$
18. $(-3, 0, 4)$
19. $(-1, -2, 3)$
20. $(4, -1, 10)$

21. Sketch the graph of all points of the form $(x, 2, z)$, and describe it with an equation.
22. Sketch the graph of all points of the form $(1, y, z)$, and describe it with an equation.

Exer. 23–30: Describe and sketch the graph of all points (x, y, z) under the given conditions.

23. $x = 0$, $y = 0$
24. $x = 1$, $y = 0$
25. $x = 2$, $y = 3$
26. $x = 0$, $z = 1$
27. $x = y = z$
28. $x = 3$
29. $y = 1$
30. $z = 4$

Exer. 31–36: Sketch a graph of the plane.

31. $x + y + z = 8$
32. $x - y + 2z = 6$
33. $x + y = 4$
34. $y - z = 2$
35. $z = 4$
36. $x = -2$

Exer. 37–42: Sketch the graph of the indicated slice(s).

37. $z = f(x, y) = x + y$; slice $z = f(1, y)$
38. $z = f(x, y) = x^2$; slice $z = f(x, 1)$
39. $z = f(x, y) = x^2 + y^2$; slices $z = f(x, 2)$ and $z = f(1, y)$
40. $z = f(x, y) = xy$; slices $z = f(x, 2)$ and $z = f(2, y)$
41. $z = f(x, y) = 2 - x^2 - y^2$; slices $z = f(x, 1)$ and $z = f(1, y)$
42. $z = f(x, y) = 6 + xy$; slices $z = f(x, 1)$ and $z = f(3, y)$

Exer. 43–46: Draw the indicated level curves.

43. $z = 2x + y$; $z = 1, 2, 3$
44. $z = y - x^2$; $z = 0, 1, 2$ z = 1
45. $z = \frac{y}{e^x}$; $z = 0, 1, 2$
46. $z = x^2 + y^2$; $z = 1, 2, 3$

8.2 Partial Derivatives

Consider a function of two independent variables, for example, $z = f(x, y) = x^2 + 3xy + y^3$. The graph of this function is a surface above (or below) some region of the xy-plane. If we substitute a particular value for y, say $y = 4$, then we obtain a function of x:

$$f(x, 4) = x^2 + 3x(4) + 4^3 = x^2 + 12x + 64.$$

We can calculate the derivative of this function easily: $2x + 12$. If we substitute another value for y, say $y = -2$, then we get another function of x:

$$f(x, -2) = x^2 - 6x - 8,$$

which has the derivative $2x - 6$. If we substitute any constant b for y we obtain the function of x:

$$g(x) = f(x, b) = x^2 + 3xb + b^3.$$

The derivative of this function is $2x + 3b$. Since $y = b$, we can write this derivative as $2x + 3y$.

We found this derivative by holding the variable y constant and viewing the function as a function of the remaining variable. A derivative obtained in this way is called a **partial derivative** of the original function $f(x, y)$. Of course, we can hold x constant and compute the derivative with y as the variable. For example, if $x = 7$ then

$$f(7, y) = 49 + 3(7)y + y^3,$$

and its derivative with respect to y is $3(7) + 3y^2$ or since $x = 7$, this expression is $3x + 3y^2$. Hence for any x, the derivative with x constant and y the variable is $3x + 3y^2$. The two derivatives with respect to x and with respect to y are different, so the notation for them must indicate whether we consider the *variable* to be x or y. The most common partial derivative notation is

$$\frac{\partial f}{\partial x} \text{ and } \frac{\partial f}{\partial y}.$$

(The *rounded d* symbol, ∂, indicates that f depends on at least two variables.) These notations are read as *the partial derivative of f with respect to x* and *the partial derivative of f with respect to y*, respectively.

In the first case, y is considered constant, so the *variable* is x; in the second case, x is considered constant, so y is the variable.

Since the derivative represents a rate of change, the partial derivative of $z = f(x, y)$ with respect to x is the *rate of change of* $z = f(x, y)$ *in the x-direction,* that is, the rate of change of f with respect to x when y is held fixed. For $z = f(x, y)$, the symbols for the partial derivative of $f(x, y)$ with respect to x are

$$\frac{\partial f}{\partial x} \quad \text{or} \quad \frac{\partial z}{\partial x} \quad \text{or} \quad f_x(x, y).$$

These symbols are interchangeable. To find the partial derivative of $z = f(x, y)$ with respect to x, we treat y as a constant and differentiate with respect to x in the usual way. All rules for differentiating apply. The *value* of the partial derivative of $z = f(x, y)$ with respect to x at the point $(1, 2, f(1, 2))$ is denoted by any of these expressions:

$$\left.\frac{\partial f}{\partial x}\right|_{(1,2)} \quad \text{or} \quad \left.\frac{\partial z}{\partial x}\right|_{(1,2)} \quad \text{or} \quad f_x(1, 2).$$

Example 8.5 Find $\left.\dfrac{\partial z}{\partial x}\right|_{(2,1)}$ if $z = f(x, y) = 3x^2 - 4xy + 5y^3$.

Solution First we compute $\partial z/\partial x$. The partial derivative of $3x^2$ with respect to x is $6x$. Since we treat y as a constant, the partial derivative of $-4xy$ with respect to x is $-4y$, and the partial derivative of $5y^3$ with respect to x is 0. Thus

$$\frac{\partial z}{\partial x} = 6x - 4y.$$

To find $(\partial z/\partial x)\Big|_{(2,1)}$, we evaluate $\partial z/\partial x$ at $x = 2$ and $y = 1$:

$$\left.\frac{\partial z}{\partial x}\right|_{(2,1)} = 6(2) - 4(1) = 8. \qquad ■$$

The *partial derivative of* $z = f(x, y)$ *with respect to* y is the rate of change of $z = f(x, y)$ in the y-direction with x held fixed. We denote it by one of these expressions:

$$\frac{\partial f}{\partial y} \quad \text{or} \quad \frac{\partial z}{\partial y} \quad \text{or} \quad f_y(x, y)$$

and compute it by treating x as a constant and differentiating with respect to y. The *value* of the partial derivative with respect to y when $x = a$ and $y = b$ is denoted by the symbols

$$\left.\frac{\partial f}{\partial y}\right|_{(a,b)} \quad \text{or} \quad \left.\frac{\partial z}{\partial y}\right|_{(a,b)} \quad \text{or} \quad f_y(a, b).$$

The following discussion of the geometric interpretation of partial derivatives is based on our description of "slicing" a surface with planes in the previous section. The graph of the function $z = f(x, y)$ is a surface in a three-dimensional coordinate system. Imagine this surface as rolling terrain; we are standing at the origin facing south, the z-axis pointing toward the sky, the y-axis pointing east, and the x-axis pointing south as in Figure 8.16(a). Suppose a and b are positive, so that the point (a, b) is south and east of our position. The plane $y = b$ will cut vertically through the rolling-terrain surface from north to south. This plane is parallel to the xz-plane, and its intersection with the surface forms a curve, as shown in Figure 8.16(a). (Imagine a north-to-south slice of the Rocky Mountains.) Note that north to south is the direction of the x-axis. The slope of this north-to-south curve at $x = a$ and $y = b$ is

$$\left.\frac{\partial z}{\partial x}\right|_{(a,b)}.$$

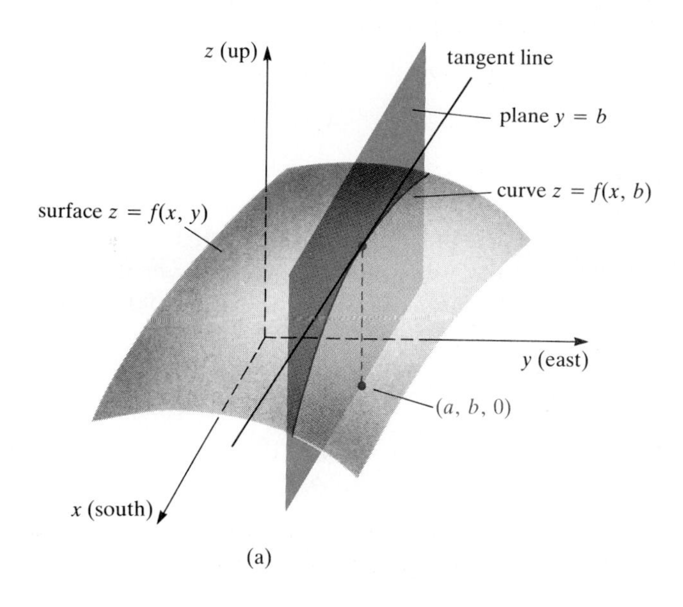

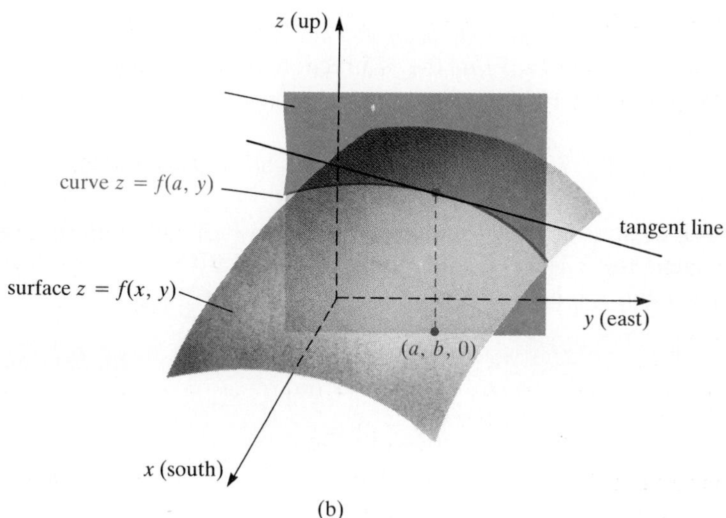

Figure 8.16
Geometric interpretations of partial derivatives: (a) $y = b$ slice; (b) $x = a$ slice

Next consider the plane $x = a$ parallel to the yz-plane and cutting through the rolling-terrain surface from west to east. This forms another curve, as shown in Figure 8.16(b). (Imagine a west-to-east slice of the Rocky Mountains.) Note that west to east is the direction of the y-axis. The slope of this west-to-east curve at $x = a$ and $y = b$ is

$$\left.\frac{\partial z}{\partial y}\right|_{(a,b)}.$$

Example 8.6 Determine the slope in the y-direction at the point $(-2, 1)$ of the surface defined by

$$z = f(x, y) = x^2 + y^3 + 5y - 6.$$

Solution The slope in the y-direction at $(-2, 1)$ is $f_y(-2, 1)$ or, equivalently, $(\partial f/\partial y)|_{(-2,1)}$. To find $f_y(x, y)$, we treat x as a constant and differentiate with respect to y:

$$f_y(x, y) = 0 + 3y^2 + 5 - 0 = 3y^2 + 5.$$

Hence, $$f_y(-2, 1) = 3(1)^2 + 5 = 8.$$ ■

Example 8.7 Find the slope of $f(s, t) = s^3t^2 + 4s^2t - 3s - 2t^2s$ in both the t-direction and the s-direction at $(2, 1)$.

Solution Since $f_s(s, t) = 3s^2t^2 + 8st - 3 - 2t^2$, we have

$$f_s(s, t)\,|_{(2,1)} = 3(2)^2(1)^2 + 8(2)(1) - 3 - 2(1)^2 = 23.$$

Since $f_t(s, t) = 2s^3t + 4s^2 - 4ts$, we have

$$f_t(s, t)\,|_{(2,1)} = 2(2)^3(1) + 4(2)^2 - 4(1)(2) = 24.$$

The slope in the s-direction is 23, and the slope in the t-direction is 24. ■

The following example shows how we use the rules of differentiation to compute partial derivatives.

Example 8.8 If $z = 3x^2(x^2 + xy)^3$, find $\partial z/\partial x$ and $\partial z/\partial y$.

Solution First we find $\partial z/\partial x$. The function is a product of $3x^2$ and $(x^2 + xy)^3$, so by the product rule we have

$$\frac{\partial z}{\partial x} = (3x^2)3(x^2 + xy)^2(2x + y) + (x^2 + xy)^3(6x).$$

The chain rule was used in the first term, since $(x^2 + xy)^3$ is a power of a function of x. Similarly,

$$\frac{\partial z}{\partial y} = (3x^2)3(x^2 + xy)^2x.$$

Here we did not use the product rule since $3x^2$ is a constant when differentiating with respect to y. ■

Example 8.9 Given $z = x^2 + y^2 - xy + x + 3$, find all points (x, y) in the xy-plane where both $\partial z/\partial x$ and $\partial z/\partial y$ equal zero.

Solution We first find the partial derivatives:

$$\frac{\partial z}{\partial x} = 2x - y + 1 \quad \text{and} \quad \frac{\partial z}{\partial y} = 2y - x.$$

To determine when both derivatives are zero, we solve the system of equations

$$2x - y + 1 = 0 \quad \text{and} \quad -x + 2y = 0.$$

The second equation says that $x = 2y$. Using this relation in the first equation gives $2(2y) - y + 1 = 0$, or $3y + 1 = 0$, so $y = -\frac{1}{3}$. Then $x = 2y = -\frac{2}{3}$. There is only one point (x, y) where both partial derivatives are zero, namely, $(-\frac{2}{3}, -\frac{1}{3})$. ■

A warning about terminology may be in order here. We cannot speak of *the partial derivative of z* without the modifier *with respect to,* as in *the partial derivative of z with respect to x*. Occasionally, we may speak of *the partial derivatives of a function* when we mean the collection of all partial derivatives with respect to each independent variable.

Example 8.10 An earth-moving project has produced a hill, the surface of which can be described by the equation $z = 3000 - 30x^2 - 20y^2$, where z is the vertical coordinate. Figure 8.17 shows a portion of the hill. Is the slope at $(5, 10, 250)$ steeper in the x-direction or in the y-direction?

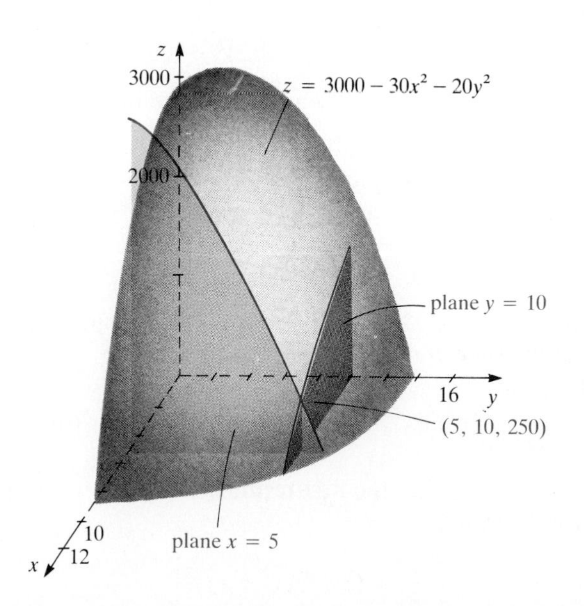

Figure 8.17
A portion of the hill in Example 8.10

Solution The slope in the x-direction is given by $\partial z/\partial x = -60x$; hence,

$$\left.\frac{\partial z}{\partial x}\right|_{(5,10)} = -300.$$

Since $\partial z/\partial y = -40y$, the slope in the y-direction at (5, 10) is given by

$$\left.\frac{\partial z}{\partial y}\right|_{(5,10)} = -40(10) = -400.$$

Consequently, the decline in the y-direction is greater (steeper) than in the x-direction. ■

Applications

Effective use of partial derivatives depends upon our ability to interpret their meanings in various situations. Consider the following applications, which illustrate the significance of partial derivatives in nongeometric settings.

Suppose that the productivity P of a manufacturing plant is a function of the number L of units of labor and the amount K of capital invested. (Capital includes the cost of buildings, equipment, materials, and so on.) Then $\partial P/\partial L$ is called the **marginal productivity of labor** since for $L=a$ and $K=b$ the productivity of the company changes approximately

$$\left.\frac{\partial P}{\partial L}\right|_{(a,b)}$$

units per unit increase in labor. Similarly, $\partial P/\partial K$ is called the **marginal productivity of capital** since for $L=a$ and $K=b$ the productivity of the company changes approximately

$$\left.\frac{\partial P}{\partial K}\right|_{(a,b)}$$

units per unit increase in capital. Economists have found that production functions are frequently of the form $P(L, K) = AL^rK^{1-r}$, where A and r are constants and $0<r<1$. Such a production function is known as a *Cobb–Douglas production function.*

Example 8.11 Suppose $P(L, K) = 300L^{2/3}K^{1/3}$. Determine the marginal productivity of labor and the marginal productivity of capital when 8 units of labor and 125 units of capital are being used.

Solution We compute

$$\frac{\partial P}{\partial L} = 300(\tfrac{2}{3})L^{-1/3}K^{1/3} = \frac{200K^{1/3}}{L^{1/3}}$$

and

$$\frac{\partial P}{\partial K} = 300(\tfrac{1}{3})L^{2/3}K^{-2/3} = \frac{100L^{2/3}}{K^{2/3}}.$$

If $L=8$ and $K=125$, adding one unit of labor will increase production approximately

$$\left.\frac{\partial P}{\partial L}\right|_{(8,125)} = \frac{200(125)^{1/3}}{8^{1/3}} = 500 \text{ units}$$

and adding one unit of capital will increase production approximately

$$\left.\frac{\partial P}{\partial K}\right|_{(8,125)} = \frac{100(8)^{2/3}}{125^{2/3}} = 16 \text{ units.}$$

■

Example 8.12 The air pollution index P in a certain city is determined by two factors: the amount x of solid waste (soot, dust and so forth) in the air and the amount y of noxious gases (sulphur dioxide, carbon monoxide and so forth) in the air. The index is described by

$$P = x^2 + 2xy + 4xy^2.$$

What is the significance of $\partial P/\partial x$ and $\partial P/\partial y$ at a point (a, b)? Compute these quantities at $(10, 5)$. Use partial derivatives to estimate which changes P more, a 10% increase in x or a 10% increase in y.

Solution The notation $(\partial P/\partial x)|_{(a,b)}$ means that if y is held constant at the value b and x is allowed to vary, then the pollution P changes approximately $(\partial P/\partial x)|_{(a,b)}$ units for a unit change in x when x has the value a. A similar interpretation applies for $(\partial P/\partial y)|_{(a,b)}$. For the given function,

$$\frac{\partial P}{\partial x} = 2x + 2y + 4y^2 \quad \text{and} \quad \frac{\partial P}{\partial y} = 2x + 8xy.$$

Therefore,

$$\left.\frac{\partial P}{\partial x}\right|_{(10,5)} = 20 + 10 + 100 = 130.$$

If y is held constant at $y = 5$, then 130 is the rate of change of P with respect to x when $x = 10$. Since a 10% increase changes x from 10 to 11, P will increase by about $(130)(1) = 130$ units. (In fact, $P(10, 5) = 1200$, and $P(11, 5) = 1331$, an increase of 131.)

Similarly,

$$\left.\frac{\partial P}{\partial y}\right|_{(10,5)} = 20 + 400 = 420.$$

If x is held constant at $x = 10$, then 420 is the per unit rate of change of P with respect to y when $y = 5$. Since a 10% increase in y (from 5 to 5.5) is an increase of 0.5 unit, the increase in P is about $420(0.5) = 210$ units. Thus, the pollution index is much more sensitive to a 10% increase in y (gases) than in x (solids). ■

Example 8.13 Empirical data shows that the number n of bushels per acre of a certain crop on a large farm is given by

$$n = n(p, L) = 300\left(6p - \frac{p^2}{30L}\right),$$

where p is the number of plants per acre and L is the number of labor-hours per acre used in cultivating the crop. If 100 labor-hours per acre are used, how many plants per acre will produce the maximum yield?

Solution We are asked to maximize n when $L = 100$. Since

$$n(p, 100) = 300\left(6p - \frac{p^2}{3000}\right) = 1800p - \tfrac{1}{10}p^2,$$

we want p such that

$$\frac{\partial n}{\partial p} = 0$$

or

$$1800 - \tfrac{1}{5}p = 0.$$

Thus $p = 9000$. So when 100 labor-hours per acre are used, the maximum yield will be produced by 9000 plants per acre. ∎

Second Partial Derivatives

For functions of one variable, we found it useful to differentiate the derivative of f to obtain the second derivative of f. Since partial derivatives themselves are functions of two (or more) variables, we can calculate the partial derivatives of $\partial z/\partial x$ and $\partial z/\partial y$, just as we calculated the partial derivatives of z. Thus, the function $z = f(x, y)$ has four **second partial derivatives** (with a variety of notations):

$$\frac{\partial}{\partial x}\left(\frac{\partial z}{\partial x}\right) = \frac{\partial^2 z}{\partial x^2} = \frac{\partial^2 f}{\partial x^2} = f_{xx}$$

$$\frac{\partial}{\partial y}\left(\frac{\partial z}{\partial x}\right) = \frac{\partial^2 z}{\partial y\,\partial x} = \frac{\partial^2 f}{\partial y\,\partial x} = f_{xy}$$

$$\frac{\partial}{\partial x}\left(\frac{\partial z}{\partial y}\right) = \frac{\partial^2 z}{\partial x\,\partial y} = \frac{\partial^2 f}{\partial x\,\partial y} = f_{yx}$$

$$\frac{\partial}{\partial y}\left(\frac{\partial z}{\partial y}\right) = \frac{\partial^2 z}{\partial y^2} = \frac{\partial^2 f}{\partial y^2} = f_{yy}.$$

The derivatives f_{xx} and f_{yy} are called **second partial derivatives with respect to x and to y**, respectively, and the derivatives f_{xy} and f_{yx} are called **mixed partial derivatives**. The mixed partial derivatives f_{xy} and f_{yx} are distinguished by the order in which f is differentiated successively with respect to x and to y; that is, in f_{xy} the partial derivative with respect to x is taken first, so f_{xy} denotes the partial derivative of f_x with respect to y, or

$$f_{xy} = \frac{\partial}{\partial y}(f_x) = \frac{\partial}{\partial y}\left(\frac{\partial f}{\partial x}\right) = \frac{\partial^2 f}{\partial y\,\partial x}.$$

Note that the order of x and y in the equivalent notations is reversed:

$$f_{xy} = \frac{\partial^2 f}{\partial y\,\partial x}.$$

Similarly, for the partial derivative of f_y with respect to x,

$$f_{yx} = \frac{\partial^2 f}{\partial x\,\partial y}.$$

In certain special cases $f_{xy} \neq f_{yx}$. However, for all functions $f(x, y)$ in this text, $f_{yx} = f_{xy}$, provided that f_{xy} and f_{yx} exist.

Example 8.14 Let $f(x, y) = e^{xy} + xy^2$. Find f_x, f_y, f_{xy}, and f_{yx}.

Solution First, we find

$$f_x = e^{xy}(y) + y^2 = ye^{xy} + y^2$$

and

$$f_y = e^{xy}(x) + 2xy = xe^{xy} + 2xy.$$

Then

$$f_{xy} = \frac{\partial}{\partial y} f_x = \frac{\partial}{\partial y}(ye^{xy} + y^2) = xye^{xy} + e^{xy} + 2y$$

and

$$f_{yx} = \frac{\partial}{\partial x} f_y = \frac{\partial}{\partial x}(xe^{xy} + 2xy) = xye^{xy} + e^{xy} + 2y.$$

■

Example 8.15 If $f(x, y) = \ln(xy + y + 3)$, find $f_{xy}(2, 1)$.

Solution We first compute f_x:

$$f_x = \frac{\partial}{\partial x} \ln(xy + y + 3) = \frac{y}{xy + y + 3}$$

and then use the quotient rule to differentiate this expression with respect to y, obtaining

$$\begin{aligned} f_{xy} &= \frac{\partial}{\partial y}\left(\frac{y}{xy + y + 3}\right) \\ &= \frac{(xy + y + 3)(1) - (y)(x + 1)}{(xy + y + 3)^2} \\ &= \frac{3}{(xy + y + 3)^2}. \end{aligned}$$

Therefore, $$f_{xy}(2, 1) = \left.\frac{\partial^2 f}{\partial y\, \partial x}\right|_{(2,1)} = \frac{3}{[(2)(1) + 1 + 3]^2} = \frac{1}{12} \approx 0.08333.$$ ■

Exercises

Exer. 1–11: For the given function, find the indicated partial derivatives or values:

1. $z = x^2 + y^2 - xy$; $\frac{\partial z}{\partial x}$, $\left.\frac{\partial z}{\partial x}\right|_{(2,2)}$, $\frac{\partial z}{\partial y}$
2. $z = 3x^2 - 2x^2y + y^3 + 4x$; $\frac{\partial z}{\partial y}$, $\left.\frac{\partial z}{\partial y}\right|_{(2,-1)}$
3. $z = (x^2 + y^2)^4$; $\frac{\partial z}{\partial x}$, $\left.\frac{\partial z}{\partial x}\right|_{(3,4)}$
4. $z = (x^2 - y^2)^3$; $\frac{\partial z}{\partial x}$, $\frac{\partial z}{\partial y}$
5. $z = (3x + 2y - 4)^5$; $\frac{\partial z}{\partial x}$, $\left.\frac{\partial z}{\partial y}\right|_{(3,2)}$
6. $F(s, t) = s^2 + 3s^2t - 14st - 6t$; $F_s(s, t)$, $F_t(2, -2)$
7. $z = f(u, v) = u^2 - v^2 + uv - 14u$; $\frac{\partial z}{\partial u}$, $f_u(u, v)$, $f_v(u, v)$, $f_v(2, 3)$
8. $z = \frac{(x^2 + y^2)^2}{x + y}$; $\frac{\partial z}{\partial x}$, $\left.\frac{\partial z}{\partial x}\right|_{(2,2)}$
9. $z = x\sqrt{x^2 + 4y}$; $\frac{\partial z}{\partial x}$, $\frac{\partial z}{\partial y}$, $\left.\frac{\partial z}{\partial y}\right|_{(2,1)}$
10. $F(s, t) = (s^2 + t^2)^{1/2}(s^2 - t^2)^{3/2}$; $F_s(s, t)$ $F_t(s, t)$

11. $u = G(x, y, z) = 4x^2z + 6y^2(x^2z^2 + 3xyz)^2$; $\dfrac{\partial u}{\partial x}$, $\dfrac{\partial u}{\partial z}$

12. Find the slope of $z = \dfrac{x^2 + y^2}{x - y}$ in the x-direction and in the y-direction at $x = 2$, $y = 4$.
13. Find the slope of $f(s, t) = s\sqrt{s^2 - t^2}$ in the s-direction and in the t-direction at $s = 5$, $t = 3$.
14. Find the slope of $u = F(s, t) = (s^2 + t^2)^{3/2}$ in the t-direction when $s = 2$, $t = 2$.
15. Find the slope of $z = 3xy - x^2 + y^2$ in the x-direction and in the y-direction when $x = -1$, $y = -2$.
16. Find $f_s(s, t)$ and $f_t(s, t)$ when $f(s, t) = \sqrt{s^2 - t^2} + st - t^2$.
17. If $u = f(s, t) = \dfrac{s^2}{t} - \dfrac{t^2}{4 - s}$, find $\dfrac{\partial u}{\partial s}$, $\dfrac{\partial u}{\partial t}$, and $\left.\dfrac{\partial u}{\partial t}\right|_{(2,1)}$.

Exer. 18–24: For the given function, find the indicated partial derivative or value:

18. $z = e^{xy}$; $\dfrac{\partial z}{\partial x}$, $\left.\dfrac{\partial z}{\partial x}\right|_{(2,1)}$
19. $z = xe^{x^2+y^2}$; $\dfrac{\partial z}{\partial x}$, $\dfrac{\partial z}{\partial y}$
20. $z = x^2 + e^{x+y}$; $\dfrac{\partial z}{\partial x}$, $\dfrac{\partial z}{\partial y}$, $\left.\dfrac{\partial z}{\partial y}\right|_{(0,1)}$
21. $z = \ln(x^2 + y^2)$; $\dfrac{\partial z}{\partial x}$, $\left.\dfrac{\partial z}{\partial x}\right|_{(2,3)}$
22. $z = (x^2 + y^3)\ln(xy)$; $\dfrac{\partial z}{\partial x}$, $\dfrac{\partial z}{\partial y}$
23. $z = \ln(xe^{y^3})$; $\dfrac{\partial z}{\partial x}$, $\dfrac{\partial z}{\partial y}$, $\left.\dfrac{\partial z}{\partial y}\right|_{(4,2)}$
24. $z = e^{x^2+y^3}\ln(x^2 - y^2)$; $\dfrac{\partial z}{\partial x}$, $\dfrac{\partial z}{\partial y}$

25. A manufacturing plant uses x units of one type of fuel and y units of another type. The amount of air pollutants exhausted by the plant is given by

$$p = 0.008x^2 + 0.03xy + 0.002y^2.$$

If the company uses 10 units of x and 18 units of y, find the rate of change with respect to x and the rate of change with respect to y.
26. If the cost of producing an item is given by the function $C = s - 0.02s^2 + 9t + 0.01t^3 + 500$, find the rate of change with respect to s and the rate of change with respect to t when $s = 11$ and $t = 9$.
27. Find the slope in the x-direction and the slope in the y-direction of $f(x, y) = x^2 + (x/y) + y^{-2}$ at $(3, 3)$.
28. Find the slopes in the x-direction and y-direction of $f(x, y) = xy\sqrt{xy}$ at $(2, 3)$.
29. Find the slopes in the s-direction and t-direction of $f(s, t) = \sqrt{s - t}$ at $(3, 2)$.

Exer. 30–34: Find all the points (x, y) in the xy-plane where f_x and f_y are both zero.

30. $z = 2x^2 - xy + y$
31. $z = x^3 - 3x + y^2 + y - 3$
32. $z = x^3 + x^2 - x - y^2 + y^3 - y$
33. $z = 3x^3 - 9x + 4 + 2y^3 - 12y$
34. $z = x^2 + xy + y^2 - 3y + 4$

35. A company finds that its productivity P is a function of the number L of units of labor and the amount K of capital invested according to the rule $P = 3L + 2K^2 + 2LK$.
    (a) Find $\partial P/\partial L$ and $\partial P/\partial K$ (the marginal productivity of labor and of capital).
    (b) Calculate P, $\partial P/\partial L$, and $\partial P/\partial K$ when $L = 10$ and $K = 12$.
    (c) Using the concept of marginal productivity, and the value $P(10, 12)$, approximate P when $L = 11$, $K = 12$ and when $L = 10$, $K = 13$.
    (d) Evaluate P at the two points given in part (c), and compare the answer to your approximations.
36. If the number of bacteria in thousands is a function of time h and temperature t according to $f(h, t) = 3h^2 - 4h + \frac{1}{10}t$, what is the rate of change with respect to time and the rate of change with respect to temperature when $h = 4$ and $t = 80$?
37. The demand for a certain product is a function of its price x and the price y of its largest competitor. If the demand function varies according to $f(x, y) = 500 - 3x + x^2 - xy + 36 - y^2$, find the marginal demand of this product with respect to x when $x = 10$ and $y = 15$.
38. Suppose a nuclear power plant and a chemical company are situated on a large lake. The pollution index P of the lake is a function of x and y, where x measures the quantity of heat entering the lake in power-plant waste water and y measures the amount of chemical waste being poured into the lake. Interpret

$$\left.\frac{\partial P}{\partial x}\right|_{(a,b)} \quad \text{and} \quad \left.\frac{\partial P}{\partial y}\right|_{(a,b)}$$

for some point (a, b).

Exer. 39–41: A toothpaste company finds that daily sales are a function of the number of times x that its commercial is shown on television and the number y of people (in millions) that see the commercial, according to $f(x, y) = 50xy + 1000$.

39. Find f_x, f_y.
40. Interpret f_x when $y = 5$.
41. Interpret f_y when $x = 2$.

42. The rate R at which a chemical substance can be absorbed into a bacterium and distributed throughout

its entire volume is given by $R = a(S/V)$, where S is the surface area, V is the volume of the bacterium, and a is a constant. If the bacterium is cylindrical with radius r and length l, then $S = 2\pi rl + 2\pi r^2$ and $V = \pi r^2 l$. Compute R_r and R_l, and interpret their meanings. How does an increase in length affect R? How does an increase in radius affect R?

43. Refer to Exercise 42. Suppose the bacterium is cylindrical with length l and radius r, and with two hemispherical caps. Then $S = 2\pi rl + 4\pi r^2$ and $V = \pi r^2 l + \frac{4}{3}\pi r^3$. In this case, how does an increase in length affect the rate R and how does an increase in radius affect R?

Exer. 44–45: The total yield from y acres of a certain crop is found to be z tons where

$$z = \frac{240xy + 25x^2 - 3.6y^2}{x + y}$$

and $100x$ is the number of labor-hours of labor employed to produce the crop.

44. For a labor force of 200 labor-hours, how many acres will produce a maximum yield?
45. Suppose 800 labor-hours of labor and 4 acres are devoted to the crop. Describe quantitatively the effects of (a) a change in labor time with fixed acreage and (b) a small increase in acreage with fixed labor time.

46. Two drugs are used simultaneously as a treatment for a certain disease. The reaction R (measured in appropriate units) to x units of the first drug and y units of the second drug is $R = x^2y^2(a - x)(b - y)$. For a fixed amount x of the first drug, what amount y of the second drug produces the maximum reaction?

Exer. 47–54: Find f_{xx}, f_{xy}, and evaluate each at the given point.

47. $f(x, y) = x^2 + x^3y^2 + 2y^4$; $(1, 3)$
48. $f(x, y) = x^5y + y^2x^3 - 3xy^4$; $(2, -1)$
49. $f(x, y) = x^2y^4 - 3xy$; $(1, 4)$
50. $f(x, y) = (x^2y - 2x)^{1/2}$; $(1, 3)$
51. $f(x, y) = 3x^2y^2(y^3 - x^3)$; $(3, -1)$
52. $f(x, y) = e^{x^2+y^2}$; $(1, -1)$
53. $f(x, y) = x \ln(2x^2y)$; $(1, 2)$
54. $f(x, y) = \dfrac{xy}{x^2 + y + 3}$; $(2, -2)$

Exer. 55–68: Find $\dfrac{\partial^2 z}{\partial x\, \partial y}$.

55. $z = x^4y^2 - x^3y^5$
56. $z = \dfrac{y^4}{x} + \dfrac{x^3}{y^2}$
57. $z = x^2y^2 - \sqrt{x^3 + y^3}$
58. $z = x^{y^3+2} - 7x^3y^3$
59. $z = (x^2 - y^2)(x^2 + y^2)$
60. $z = \sqrt{x^2y^3 + 3}$
61. $z = x^2 - y^2x^2$
62. $z = 4 - y^2 + x^2$
63. $z = 4xy^3 - 4x^3y$
64. $z = (x^2 - 3xy)^3$
65. $z = (y - 3x + xy^2)^{2/3}$
66. $z = 4xye^{2xy^2}$
67. $z = 4 \ln(2x + e^{xy})$
68. $z = \ln(x^2 + y^{-2})$

Exer. 69–74. If $z = f(x, y) = (x^3 + y^3)/2xy$, find the partial derivative or value.

69. $\dfrac{\partial^2 z}{\partial x^2}$
70. f_{xy}
71. f_{xx}
72. $\left.\dfrac{\partial^2 z}{\partial x\, \partial y}\right|_{(2,3)}$
73. $f_{yy}(1, 2)$
74. $f_{xy}(-2, 2)$

8.3 Optimization Problems

Consider a point $(a, b, f(a, b))$ on a surface $z = f(x, y)$. Assuming that both $\partial z/\partial x$ and $\partial z/\partial y$ exist at (a, b), we can compute

$$\left.\frac{\partial z}{\partial x}\right|_{(a,b)} \quad \text{and} \quad \left.\frac{\partial z}{\partial y}\right|_{(a,b)},$$

the slopes of two lines that are *tangent* to the surface. The plane that contains these two lines is tangent to the surface at $(a, b, f(a, b))$. See Figure 8.18.

Finding an explicit formula for a tangent plane is beyond the scope of this text. However, as we will see, tangent planes play a similar role in optimization problems in two variables as tangent lines do in one variable. We will also find that information about two lines in a tangent plane is sufficient to solve these optimization problems.

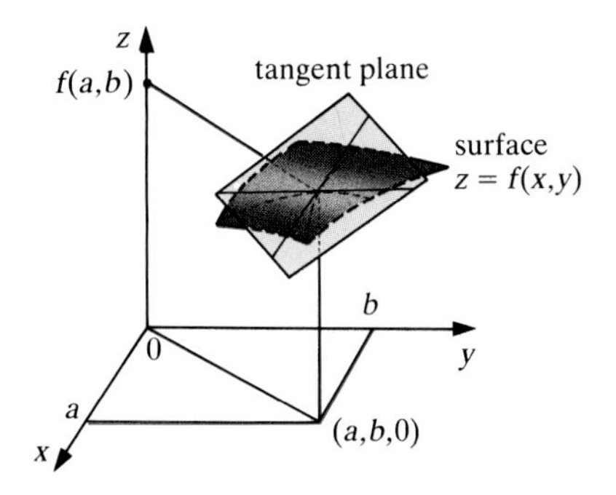

Figure 8.18
A tangent plane to the surface $z = f(x, y)$ at $(a, b, f(a, b))$

A function $z = f(x, y)$ has a **relative maximum** at a point $(a, b, f(a, b))$ if $f(a, b) \geq f(x, y)$ for all (x, y) in some rectangular region about (a, b), that is, the graph of $f(x, y)$ has the shape of a hill that peaks at (a, b). The surface in Figure 8.19 has several relative maxima. We say that $z = f(x, y)$ has a **relative minimum** at $(a, b, f(a, b))$ if $f(a, b) \leq f(x, y)$ for all (x, y) in some rectangular region about (a, b); that is, its graph has the shape of a valley or bowl, with its lowest point at (a, b) (see Figure 8.19).

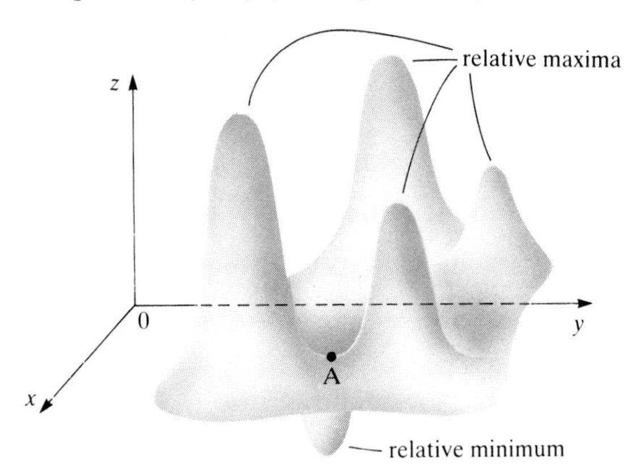

Figure 8.19
A surface with several relative maxima and a relative minimum

In Figure 8.19, notice that at a relative maximum (the top of the hill), the tangent plane is level, or parallel to the xy-plane. Thus, the two partial derivatives are zero at that point since the slopes in both the x-direction and y-direction are zero. We conclude that if a surface $z = f(x, y)$ has a relative maximum at a point $(a, b, f(a, b))$ and if both partial derivatives exist, then these partial derivatives are zero there. The same conclusion holds for a relative minimum because both the x-slope and the y-slope are zero at the bottom of a bowl-shaped graph. Thus, if both $\partial f/\partial x$ and $\partial f/\partial y$ exist, and if the function has a relative maximum or relative minimum at (a, b), then it must be the case that

$$\left.\frac{\partial f}{\partial x}\right|_{(a,b)} = 0 \quad \text{and} \quad \left.\frac{\partial f}{\partial y}\right|_{(a,b)} = 0.$$

Any point (a, b) in the domain of f where both $\partial f/\partial x$ and $\partial f/\partial y$ are zero is called a **critical point of f**. However, not every critical point is a relative maximum or relative minimum. Again referring to Figure 8.19, the point labeled A is called a **saddle point**. It is a critical point, since both partial derivatives are zero there; however, it is neither a maximum nor a minimum.

We have seen (in Section 2.4) that a function of one variable can have a relative maximum or relative minimum at points for which $f'(x)$ is undefined. Similarly, functions of two or more variables can have a relative maximum or minimum at points for which either $\partial f/\partial x$ or $\partial f/\partial y$ is not defined. For the problems given in this text, if $z = f(x, y)$ has a relative maximum or minimum at (a, b), then $\left.\frac{\partial f}{\partial x}\right|_{(a,b)}$ and $\left.\frac{\partial f}{\partial y}\right|_{(a,b)}$ will always be defined there and will be zero.

Example 8.16 Locate all points that could be relative maxima or relative minima of the surface $z = f(x, y) = 5x^2 + 3y^2 + 2xy + 500$; that is, find all critical points of f.

Solution First we find the partial derivatives:

$$\frac{\partial z}{\partial x} = 10x + 2y \quad \text{and} \quad \frac{\partial z}{\partial y} = 6y + 2x.$$

Both partial derivatives must be zero for a relative maximum or relative minimum to occur; so in order to locate these points, we solve the system of equations

$$10x + 2y = 0 \quad \text{and} \quad 2x + 6y = 0.$$

The only solution is $x = 0,\ y = 0$. Thus, the only point on the surface at which a relative maximum or relative minimum could occur is $(0, 0, 500)$. ■

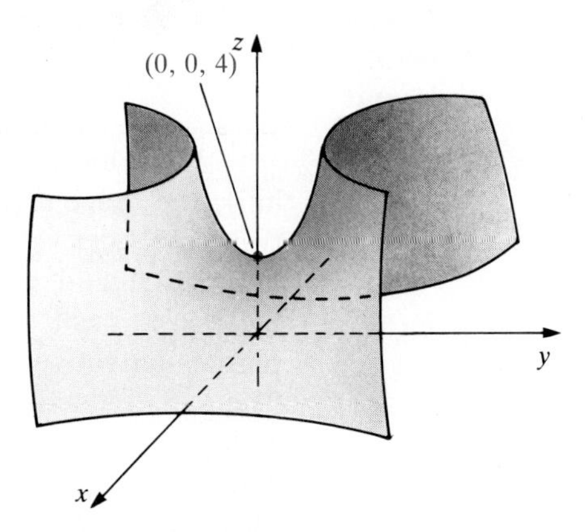

Figure 8.20
The function $f(x, y) = 4 - x^2 + y^2$ has saddle point at (0, 0, 4)

The saddle-shaped surface in Figure 8.20 is called a *hyperbolic paraboloid.* Its equation is $z = 4 - x^2 + y^2$, and both partial derivatives are zero at $(0, 0, 4)$. As we can see by inspecting the graph, there is no relative maximum or relative minimum at $(0, 0, 4)$. In fact, $(0, 0, 4)$ is a saddle point. In the xz-plane $(y = 0)$, the point $(0, 0, 4)$ is a relative maximum, and in the yz-plane $(x = 0)$, the point $(0, 0, 4)$ is a relative minimum. This behavior is typical of a saddle point.

The determination of relative maxima and relative minima in three dimensions $(z = f(x, y))$ is similar to that for two dimensions. In two dimensions $(y = f(x))$, a point where the first derivative is zero is a possible relative maximum or relative minimum. More information is obtained from either the First Derivative Test or the Second Derivative Test to classify the critical point. In three dimensions, a point where both partial derivatives are simultaneously zero may be either a relative maximum or relative minimum or a saddle point. To distinguish among relative maxima, relative minima, and saddle points, we use a three-dimensional analogue of the Second Derivative Test.

Second Derivative Test for Two Variables

Suppose that $z = f(x, y)$ has partial derivatives at all points near a point (a, b) and that (a, b) is a critical point of $f(x, y)$, so that

$$f_x(a, b) = 0 \quad \text{and} \quad f_y(a, b) = 0.$$

Let

$$A = f_{xx}(a, b)$$
$$B = f_{xy}(a, b)$$
$$C = f_{yy}(a, b).$$

Rule 1: $f(a, b)$ is a relative maximum if $AC - B^2 > 0$ and $A < 0$;
Rule 2: $f(a, b)$ is a relative minimum if $AC - B^2 > 0$ and $A > 0$;
Rule 3: $(a, b, f(a, b))$ is a saddle point if $AC - B^2 < 0$;
Rule 4: The test gives no information about the type of critical point if $AC - B^2 = 0$.

Note that the values of A, B, and C depend upon the point (a, b) and must be determined independently for each critical point.

Example 8.17 Find all relative maxima, relative minima, and saddle points of $z = f(x, y) = x^2 + y^4 - 2y^2 - 1$.

Solution We first compute the first partial derivatives:

$$f_x(x, y) = 2x \quad \text{and} \quad f_y(x, y) = 4y^3 - 4y.$$

The points (a, b) where critical points are possible are the simultaneous solutions to

$$2x = 0 \quad \text{and} \quad 4y^3 - 4y = 0.$$

Thus $x = 0$ and, noting that $4y^3 - 4y = 4y(y^2 - 1) = 4y(y - 1)(y + 1)$, we conclude that $(0, 0)$ $(0, 1)$, and $(0, -1)$ provide the only solutions to both equations.

To use the Second Derivative Test, we must evaluate the three second derivatives at each of these points. The results are displayed in Table 8.2.

For the points $(0, 1)$ and $(0, -1)$, $AC - B^2$ is positive, so we must also consider the sign of A to distinguish between a relative maximum and a relative minimum. Since A is positive for both points, these critical points are relative minima. The z-coordinates for the critical points may be computed from the formula for $f(x, y)$; they are $f(0, 0) = -1$, $f(0, 1) = -2$, and $f(0, -1) = -2$. ■

Table 8.2
Computations for classifying possible extreme points of $z = f(x, y) = x^2 + y^4 - 2y^2 - 1$

Critical point (x, y)	$(0, 0)$	$(0, 1)$	$(0, -1)$
$A = f_{xx} = 2$	2	2	2
$B = f_{xy} = 0$	0	0	0
$C = f_{yy} = 12y^2 - 4$	-4	8	8
$AC - B^2$	$-8 < 0$	$16 > 0$	$16 > 0$
sign of A	+	+	+
type of critical point	saddle point	relative minimum	relative minimum

Example 8.18 A botanist finds that the growth rate of a particular plant is a function of the temperature and the humidity of the greenhouse. After a great deal of experimentation the botanist concludes that the growth rate $g(x, y)$ in inches per week, when the greenhouse has a temperature of x °F and humidity of y percent, is given by

$$g(x, y) = -2x^2 + 196x - 3y^2 + 242y + 2xy - 16{,}360$$

where $75 \le x \le 85$ and $50 \le y \le 75$. At what temperature and humidity should the greenhouse be kept to maximize the growth rate of the plant? How fast will the plant grow under these optimal conditions?

Solution We begin by finding the first partial derivatives:

$$g_x(x, y) = -4x + 196 + 2y, \qquad g_y(x, y) = -6y + 242 + 2x.$$

We must find values of x and y such that

$$-4x + 2y + 196 = 0 \quad \text{and} \quad 2x - 6y + 242 = 0.$$

Solving these equations simultaneously, we find that $x = 83$ and $y = 68$ is the only solution. Since

$$A = g_{xx}(83, 68) = -4,$$
$$C = g_{yy}(83, 68) = -6,$$
and
$$B = g_{xy}(83, 68) = 2,$$

we have $AC - B^2 > 0$ with $A < 0$ and so, from the Second Derivative Test, we conclude that $x = 83$ and $y = 68$ yield a relative maximum value of $g(x, y)$. Therefore, the botanist should keep the greenhouse at a temperature of 83 °F and at 68% humidity. Under these optimal conditions, the growth rate of the plant is

$$g(83, 68) = -2(83)^2 + 196(83) - 3(68)^2 + 242(68) + 2(83)(68) - 16{,}360$$
$$= 2 \text{ inches per week.}$$

■

Example 8.19 A store's profit is dependent on the number x of salespersons and the amount y of inventory on hand (in thousands of dollars). If the profit is given by the function

$$P = 2600 - x^2 + 24x - y^2 + 80y,$$

determine the number of salespersons and amount of inventory that will produce the maximum profit.

Solution First find the first partial derivatives:

$$P_x = -2x + 24 \quad \text{and} \quad P_y = -2y + 80.$$

Setting these equal to zero and solving yields that the only point at which a relative maximum, relative minimum, or saddle point can occur is $(12, 40)$. Since $A = C = -2 < 0$ and $B = 0$, $AC - B^2 = 4 > 0$ implies that P has a relative maximum at $(12, 40)$. The relative maximum profit is

$$P(12, 40) = 2600 - 12^2 + 24(12) - 40^2 + 80(40) = 4344,$$

which occurs when 12 salespersons are used and inventory worth $40,000 is on hand. ■

Example 8.20 Find and classify all critical points of

$$f(x, y) = 2x^2y^2 + 12xy + 4x^2 + y^2.$$

Solution We compute the first partial derivatives and set them equal to zero:

$$f_x = 4xy^2 + 12y + 8x = 0,$$
$$f_y = 4x^2y + 12x + 2y = 0.$$

This system of nonlinear equations can be solved by multiplying the first equation by x and the second equation by y, and subtracting to remove the xy^2 and x^2y terms:

$$(4x^2y^2 + 12xy + 8x^2) - (4x^2y^2 + 12xy + 2y^2) = 0$$
$$8x^2 - 2y^2 = 0.$$

Consequently, $4x^2 = y^2$, or $y = \pm 2x$. If we use $y = 2x$ in the equation $f_x = 0$, we obtain

$$16x^3 + 32x = 16x(x^2 + 2) = 0.$$

Since $x^2 + 2$ is never zero, $x = 0$ is the only solution. Associated with $y = 2x$ we have one critical point $(0, 0)$.

Substituting $y = -2x$ into the equation $f_x = 0$ yields

$$16x^3 - 16x = 16x(x^2 - 1) = 0.$$

The roots are $x = 0$, 1, and -1. Since $y = -2x$, we have two more critical points: $(1, -2)$ and $(-1, 2)$. Table 8.3 shows the calculations and results of the Second Derivative Test. ■

Table 8.3
Computations for classifying critical points of $z = f(x, y) = 2x^2y^2 + 12xy + 4x^2 + y^2$

Critical point (x, y)	$(0, 0)$	$(1, -2)$	$(-1, 2)$
$A = f_{xx} = 4y^2 + 8$	8	24	24
$B = f_{xy} = 8xy + 12$	12	−4	−4
$C = f_{yy} = 4x^2 + 2$	2	6	6
$AC - B^2$	$16 - 144 < 0$	$24 \times 6 - 16 > 0$	$24 \times 6 - 16 > 0$
sign of A	+	+	+
type of critical point	saddle point	relative minimum	relative minimum

Exercises

Exer. 1–10: Find all relative extrema and saddle points for the function.

1. $f(x, y) = x^2 - y^2$
2. $f(x, y) = x^2 - y^2 + 2$
3. $f(x, y) = x^2 - xy$
4. $f(x, y) = x^2 + y^2 - yx + 9y + 3$
5. $f(x, y) = 2x^4 - x^2 - 3y^2$
6. $f(x, y) = xy(x - y - 5)$
7. $f(x, y) = xy(4 - x - y)$
8. $f(x, y) = x^3 - 4y^2$
9. $f(x, y) = x^2 + \frac{1}{4}xy + y^2 + 4$
10. $f(x, y) = x^4 - x^2y^2 + 3y^2$
11. Suppose that a publisher's profit on the sales of a novel is given by

$$P(x, y) = \frac{200}{x^2}(1 - x^2)(y^2 - 10y + 9) - 100x - 1000$$

where x is the amount spent on promotion (in thousands of dollars) and y is the publisher's wholesale price (in dollars). Determine the wholesale price and promotion budget that provides maximum profit.

12. The revenue obtained by a firm is $f(x, y) = 4x - 0.07x^2 + 5y - 100y^2 + 2xy$, where x and y are the amounts (in thousands of dollars) spent on newspaper and radio advertising, respectively. For what values of x and y is $f(x, y)$ a maximum?
13. A sociological analysis indicates that a parameter measuring the crime rate R depends on the amount spent on welfare w (in hundred of millions of dollars) and on the amount spent on prisons p (in hundreds of millions of dollars) according to the formula

$$R = w^3 + 6p^2 - 12p - 6pw + 30.$$

How should funds be allocated to reduce the crime rate as much as possible?

14. A pharmaceutical company wants to formulate its new cold medication to minimize the duration of symptoms due both to the cold and to the side effects of the drug. If the medication is to consist of x units of one chemical and y units of another chemical, and if the duration $d(x, y)$ of the symptoms for the average patient is

$$d(x, y) = x^2 - 20x + 2y^2 - 26y + 2xy + 113$$

(in days), how should x and y be chosen to minimize the duration of the symptoms? How long will the symptoms last for the average person taking this medication?

15. A farmer finds that the yield of corn is a function of the amounts of fertilizer and insecticide used. If the number of bushels of corn $N(x, y)$ per acre when x barrels of liquid fertilizer and y barrels of liquid insecticide are applied is

$$N(x, y) = -3x^2 + 66x - 4y^2 + 66y - 2xy - 456,$$

how many barrels of each should be applied to maximize the yield?

16. A rectangular box with a volume of 48 ft^3 is constructed from three different materials. The cost of the bottom is 40¢ per ft^2, the sides cost 30¢ per ft^2, and the top costs 20¢ per ft^2. Find the dimensions of the most economical box. (*Hint:* If x, y, and z represent the dimensions of the depth, width, and height, respectively, of the box, then volume $= xyz = 48$; hence, $z = 48/xy$. Substitute this into the cost function to obtain a function of two variables.)
17. Work Exercise 16 if the bottom costs 50¢ per ft^2, and the sides and top each costs 25¢ per ft^2.

Exer. 18–20: Two competing companies produce cameras. The Flash Camera Company makes x cameras and the

Snap Camera Company makes y cameras. Flash's profit is $f(x, y) = 120x - x^2 - y^2$ and Snap's profit is $s(x, y) = 150y - 2x^2 - \frac{1}{2}y^2$.

18. How many cameras should Flash make to maximize profit?
19. How many cameras should Snap make to maximize profit?
20. Suppose the companies cooperate in an effort to maximize $f(x, y) + s(x, y)$. Find the optimal values of x and y. Compare the profit of each company with the shared profit.

21. A rectangular tank is open at the top and holds 108 ft^3. Find its dimensions so that the surface area of the tank is a minimum.
22. Find the dimensions of a rectangular box of maximum volume, if the box has no top and has a surface area of 64 in^2.
23. A psychologist conducting motivational research has found that when all else is held constant, a customer's receptivity is a function of both the temperature of the store and the volume level of the music in the store. If the temperature is x °F and the volume level is y decibels, the psychologist finds the receptivity $R(x, y)$ is given by

$$R(x, y) = -2x^2 + 234x - 3y^2 + 229y + xy - 14{,}024.$$

What values of x and y will cause the customer's receptivity to be the greatest?
24. Refer to Example 8.18. Suppose that the growth rate of another plant is $h(x, y)$ inches per week when the greenhouse is at x °F and at y percent humidity, where

$$h(x, y) = -5x^2 + 548x - 36y^2 - 8y + 4xy - 15{,}330.$$

At what temperature and humidity should the greenhouse be kept to maximize the growth of this plant? Under the optimal conditions, how fast will the plant grow?
25. A company finds that its profit (in hundreds of dollars) varies according to the formula $P = 20x + 20y - 5x^2y$, where x is the number of units produced by the night shift and y the number produced by the day shift. What production goals should be set for each shift to maximize profit?
26. A farm's profit is given by $P = 100x + 80y + 2xy - x^2 - 2y^2 - 5000$, where x is the number of hogs produced and y is the number of beef cattle produced. How many of each should be produced to maximize the profits?
27. A manufacturer finds that the profit from two sizes of a product is given by $P = 24x + 30y + 3xy - 3x^2 - 6y^2 - 500$, where x is the number of the smaller size made and y is the number of the larger size made. How many of each size should be produced to maximize the profit?
28. A manufacturer produces safety razors and blades at a cost of 50¢ per razor and 15¢ per dozen blades. If the company charges x cents per razor and y cents per dozen blades, daily sales will be $(19494)10^4/(x^2y)$ razors and $(7776)10^4/(xy^2)$ dozen blades. How should the company set prices to maximize profit?
29. A rectangular enclosure is to be built with fiberglass paneling covering the top, two ends, and the back. The volume enclosed is to be 3456 ft^3. What is the least possible square footage of paneling needed? (Assume that there is an absolute minimum area.)
30. In a certain economic model, the cost z of a given item is related to two factors, measured as x and y, by the equation $z = x^2 + y^2 - \sqrt{xy}$. Determine the values of x and y that correspond to a minimum value of z.
31. The profit P from the sales of a commodity is related to cost, production factors, advertising costs, and so on, by the equation

$$P = -4a^2 - 2b^2 - ab + 18b + 20a,$$

where a and b are variables describing actual costs (in thousands of dollars). Determine how much should be invested in each factor to maximize profit.

Constrained Optimization Problems and Lagrange Multipliers

8.4

In many applications, we must find the largest or smallest values of a function $f(x, y)$ when certain restrictions are placed on the variables x and y. Such restrictions are called *constraints*. For instance, a plant manager wishes to minimize costs but must produce at a certain level. A company that provides monitoring services to regulate pollution wishes to minimize the cost for inspections while doing the required number of inspections. An economist wishes to maximize profit while restricting expenditures to a budgeted amount.

A constraint can usually be expressed in the form $g(x, y) = 0$. We wish to determine values of x and y such that $f(x, y)$ is a relative maximum (or relative minimum) for those points (x, y) satisfying $g(x, y) = 0$; this is called a **constrained optimization problem**. The function $f(x, y)$ is called the **objective function** and $g(x, y)$ is the **constraint function**. We now introduce two methods for solving such problems: **algebraic substitution** and **Lagrange multipliers**.

Algebraic Substitution

Example 8.21 If x hours of labor at \$8 per hour and y acres of land at \$200 per acre per year are used in the production of a certain agricultural product, what amounts of land and labor should be utilized to obtain maximum production on a budget of \$400,000 if the production function is $P(x, y) = 20x\sqrt{y}$?

Solution The constraint is the budget, so our choices of x and y must satisfy $8x + 200y = 400{,}000$. We set $g(x, y) = 400{,}000 - 8x - 200y$ to write the constraint in the form $g(x, y) = 0$. In other words, we are considering only those values of x and y that use the entire budget. If we solve $g(x, y) = 0$ for one of the variables, say x, we get $x = 50{,}000 - 25y$. We then substitute this into the production function to obtain

$$\begin{aligned} P(x, y) &= P(50000 - 25y, y) \\ &= 20(50000 - 25y)\sqrt{y} \\ &= 1{,}000{,}000y^{1/2} - 500y^{3/2}. \end{aligned}$$

Since this is a function of y only, we can use the techniques of Chapter 3 to maximize production. Differentiation yields

$$\frac{dP}{dy} = 500{,}000y^{-1/2} - 750y^{1/2}.$$

We next solve $dP/dy = 0$ by multiplying by $y^{1/2}$ to get $500{,}000 - 750y = 0$; hence, production is maximized when $y = \frac{500{,}000}{750} = \frac{2000}{3} \approx 667$. (The Second Derivative Test for one variable may be used to affirm that this is a relative maximum.)

To find the value of x that maximizes production, we use

$$g(x, \tfrac{2000}{3}) = 400{,}000 - 8x - 200(\tfrac{2000}{3}) = 0$$

to get $x = \frac{100{,}000}{3} \approx 33{,}333$. Thus, production is maximized on a budget of \$400,000 by using approximately 33,333 hours of labor and 667 acres of land. ■

When an application involves three independent variables, the constraint is of the form $g(x, y, z) = 0$, where the function to be maximized (or minimized) is a function of three variables x, y, and z. The algebraic substitution method can be used to eliminate one of the variables and thus reduce an optimization problem with constraints involving three variables to the two-variable case.

Example 8.22 A firm uses x hours of labor at \$10 per hour, rents y acres of land at \$40 per acre per month, and buys z units of raw materials at \$100 per unit. The firm's production is given by $P = 0.001xyz$. The total budget is \$30,000. How should labor, land, and raw materials be allocated to maximize production?

Solution The constraint is the fixed budget. The \$30,000 pays for labor, land, and raw materials, so

$$30{,}000 = 10x + 40y + 100z.$$

If $g(x, y, z) = 30{,}000 - 10x - 40y - 100z$, then the constraint is $g(x, y, z) = 0$. We solve this equation for one of the variables, say x, to get $x = 3000 - 4y - 10z$. Substitute this into the production equation to produce

$$\begin{aligned} P(y, z) &= 0.001(3000 - 4y - 10z)yz \\ &= 0.001(3000yz - 4y^2z - 10yz^2). \end{aligned}$$

We find the maximum of $P(y, z)$ by the methods discussed in Section 8.3:

$$\frac{\partial P}{\partial y} = 0.001(3000z - 8yz - 10z^2)$$

$$\frac{\partial P}{\partial z} = 0.001(3000y - 4y^2 - 20yz).$$

To obtain possible extrema of $P(y, z)$ we set these equations equal to zero, divide by 0.001, and factor z from the first equation and y from the second. This yields

$$(3000 - 8y - 10z)z = 0 \qquad (3000 - 4y - 20z)y = 0$$

If $z = 0$ then the first equation is just $0 = 0$ and the second equation is $(3000 - 4y)y = 0$, which has solutions $y = 0$ and $y = 750$. So $(0, 0)$ and $(750, 0)$ are critical points. Similarly, if $y = 0$ then the first equation becomes $(3000 - 10z)z = 0$, yielding $z = 0$ and $z = 300$, while the second equation becomes $0 = 0$. Thus $(0, 0)$ and $(0, 300)$ are critical points. If $y \neq 0$ and $z \neq 0$, then we can cancel z in the first equation and y in the second equation to get

$$3000 - 8y - 10z = 0 \qquad 3000 - 4y - 20z = 0.$$

Substracting two times the second equation from the first gives $-3000 + 30z = 0$, or $z = 100$. The corresponding value of y is $y = 250$ (by substituting $z = 100$ in either equation). Hence $(250, 100)$ is a critical point.

There are four critical points: $(0, 0)$, $(0, 300)$, $(750, 0)$, and $(250, 100)$. Using the Second Derivative Test for two variables, we see that $(0, 0)$, $(0, 300)$, and $(750, 0)$ are saddle points and a relative maximum occurs at $y = 250$, $z = 100$. See Table 8.4. To determine the value of x that maximizes production, we calculate x from the expression used in the substitution, namely, $x = 3000 - 4y - 10z$; thus, $x = 1000$. We conclude that production is maximized by 1000 hours of labor, 250 acres of land, and 100 units of raw materials. ■

Table 8.4

Second Derivative Test for Example 8.22

critical point (y, z)	**(0, 0)**	**(0, 300)**	**(750, 0)**	**(250, 100)**
$A = p_{yy} = -8z$	0	-8×300	0	-800
$B = p_{yz} = 3000 - 20z$	3000	-3000	3000	1000
$C = p_{zz} = -20y$	0	0	-20×750	-5000
$AC - B^2$	$-3000^2 < 0$	$-(-2000)^2 < 0$	$-3000^2 < 0$	$3{,}000{,}000 > 0$
Sign of A	none	$-$	none	$+$
type of critical point	saddle point	saddle point	saddle point	relative maximum

Lagrange Multipliers

To use algebraic substitution, we must solve for one variable in terms of the others in the constraint equation. Sometimes this is difficult or impossible. In these situations we use the method of **Lagrange multipliers**, named after the French mathematician Joseph Louis Lagrange (1736–1813). Lagrange's method is more general than algebraic substitution and may be used on any constrained optimization problem, including those that may be solved by algebraic substitution. We state the results for objective functions of three variables $w = f(x, y, z)$ and a constraint of $g(x, y, z) = 0$.

The **Lagrange function** is the following combination of the objective function $f(x, y, z)$ and the constraint function $g(x, y, z)$:

$$F(x, y, z, \lambda) = f(x, y, z) + \lambda g(x, y, z).$$

The variable λ (the Greek letter *lambda*) is called the *Lagrange multiplier*. The connection between the Lagrange function and the maxima (minima) of $f(x, y, z)$ is given by the following theorem, which we present without proof.

Lagrange Multiplier Theorem

The relative maxima and relative minima of $f(x, y, z)$, subject to the constraint $g(x, y, z) = 0$, are among those points (x_0, y_0, z_0) for which $(x_0, y_0, z_0, \lambda_0)$ is a solution to the system

$$F_x(x, y, z, \lambda) = 0$$
$$F_y(x, y, z, \lambda) = 0$$
$$F_z(x, y, z, \lambda) = 0$$
$$F_\lambda(x, y, z, \lambda) = 0$$

where $\quad F(x, y, z, \lambda) = f(x, y, z) + \lambda g(x, y, z),$

provided all partial derivatives exist.

To find relative maxima and relative minima of $f(x, y, z)$ constrained by the condition $g(x, y, z) = 0$, we find all $(x_0, y_0, z_0, \lambda_0)$ where all of the partial derivatives of $F(x, y, z, \lambda)$ are zero. Among these points lie the relative maxima and relative minima of $f(x, y, z)$. As a first illustration of the Lagrange multiplier method, we will rework Example 8.22.

Example 8.23 Rework Example 8.22 by the Lagrange multiplier method. Recall that the objective function is $P(x, y, z) = 0.001xyz$ and the constraint is $g(x, y, z) = 30000 - 10x - 40y - 100z = 0$.

Solution The Lagrange function is

$$F(x, y, z, \lambda) = 0.001xyz + \lambda(30000 - 10x - 40y - 100z).$$

Setting the partial derivatives of $F(x, y, z, \lambda)$ equal to zero yields

$$\frac{\partial F}{\partial x} = 0.001yz - 10\lambda = 0,$$

$$\frac{\partial F}{\partial y} = 0.001xz - 40\lambda = 0,$$

$$\frac{\partial F}{\partial z} = 0.001xy - 100\lambda = 0,$$

$$\frac{\partial F}{\partial \lambda} = 30000 - 10x - 40y - 100z = 0.$$

Solving the first three equations for λ gives

$$\lambda = \frac{0.001yz}{10}$$

$$\lambda = \frac{0.001xz}{40}$$

$$\lambda = \frac{0.001xy}{100}.$$

Equating the first two expressions for λ, we have

$$\frac{yz}{10} = \frac{xz}{40}$$

and equating the first and third expression for λ results in

$$\frac{yz}{10} = \frac{xy}{100}.$$

The equation $yz/10 = xz/40$ is satisfied if $z = 0$ or $y = \frac{1}{4}x$ and the equation $yz/10 = xy/100$ implies $y = 0$ or $z = \frac{1}{10}x$. The possibilities $y = 0$ and $z = 0$ yield a production level of zero and hence do not correspond to a maximum production. If we substitute $y = \frac{1}{4}x$ and $z = \frac{1}{10}x$ into the constraint equation, we obtain

$$30000 = 10x + 40(\tfrac{1}{4}x) + 100(\tfrac{1}{10}x) = 30x.$$

Thus, $x = 1000$ and hence $y = \frac{1}{4}x = 250$ and $z = \frac{1}{10}x = 100$ yield the maximum production. These values are the same as those obtained by algebraic substitution.

In this problem λ has a special meaning. Since

$$\lambda = \frac{0.001yz}{10},$$

and $y = 250$, $z = 100$ at the optimal production level,

$$\lambda = \frac{0.001(250)(100)}{10} = 2.5.$$

Since the constraint is the amount of *money* available, λ represents *marginal productivity of money*: if an additional dollar is available for production and x, y, and z are kept at their optimal values, then production will increase by about $\lambda = 2.5$ units. (A derivation of this fact is beyond the scope of this text.) ■

The procedure we used in the previous example is a common approach when using the Lagrange multiplier method. By solving each "partial derivative equation" for λ, we obtain relationships among the variables (such as $y = x/4$ and $z = x/10$). The constraint equation is then used to find numeric values for the variables.)

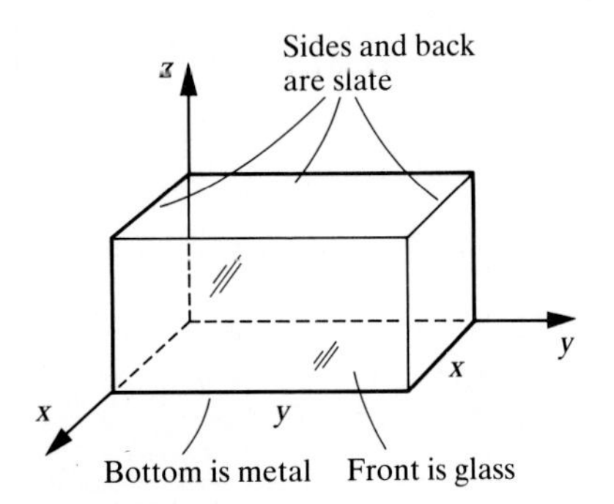

Figure 8.21
An aquarium

Example 8.24 An aquarium is to be built in the form of a rectangular box with an open top. The base is metal, costing 5.4¢ per in^2. The sides and back are slate, costing 4.5¢ per in^2. The front is glass and costs 1.5¢ per in^2. Find the dimensions of the cheapest aquarium that can be constructed to contain 10,800 in^3. What is the cost of the materials needed to build it? See Figure 8.21.

Solution Let x be the width, y the length, and z the height of the aquarium. Then the area of the base is xy and costs $5.4xy$ cents = $0.054xy$ dollars; the area of the two sides are xz and cost $0.045xz$ dollars each; the area of the front and back are yz, so the front costs $0.015yz$ dollars and the back costs $0.045yz$ dollars. Consequently, the cost in dollars of building the aquarium is

$$\begin{aligned} C = f(x, y, z) &= 0.054xy + 2(0.045xz) + 0.015yz + 0.045yz \\ &= 0.054xy + 0.09xz + 0.06yz. \end{aligned}$$

The constraint is volume $= xyz = 10{,}800$, so we write

$$F(x, y, z, \lambda) = 0.054xy + 0.09xz + 0.06yz + \lambda(10800 - xyz)$$

and set each partial derivative equal to zero:

$$\frac{\partial F}{\partial x} = 0.054y + 0.09z - \lambda yz = 0,$$

$$\frac{\partial F}{\partial y} = 0.054x + 0.06z - \lambda xz = 0,$$

$$\frac{\partial F}{\partial z} = 0.09x + 0.06y - \lambda xy = 0,$$

$$\frac{\partial F}{\partial \lambda} = 10800 - xyz = 0.$$

Since none of the dimensions x, y, or z can be zero, we can perform the necessary divisions to solve the first three equations for λ:

$$\lambda = \frac{0.054}{z} + \frac{0.09}{y}, \qquad \lambda = \frac{0.054}{z} + \frac{0.06}{x}, \qquad \lambda = \frac{0.09}{y} + \frac{0.06}{x}.$$

Equating the first two expressions for λ gives

$$\frac{0.054}{z} + \frac{0.09}{y} = \frac{0.054}{z} + \frac{0.06}{x}$$

so that $0.09/y = 0.06/x$, or $x = \frac{2}{3}y$. Likewise, equating the last two expressions for λ and solving for z yields $z = \frac{54}{90}y = \frac{3}{5}y$. The equation $\partial F/\partial \lambda = 0$ is the constraint $xyz = 10800$; substituting $x = \frac{2}{3}y$ and $z = \frac{3}{5}y$ gives

$$(\tfrac{2}{3}y)y(\tfrac{3}{5}y) = 10800,$$

or

$$y^3 = 27000 = (30)^3.$$

Hence, $y = 30$, $x = \frac{2}{3}y = 20$, and $z = \frac{3}{5}y = 18$. The cheapest aquarium will have width 20 inches, length 30 inches, and height 18 inches. Its cost will be \$97.20:

$$C = 0.054(20)(30) + 0.09(20)(18) + 0.06(30)(18) = 97.20.$$

Note that the choice of expressing x and z in terms of y is completely arbitrary. We could have chosen x or z as the *final* variable. Also, from $xyz = 10800$, we wrote the constraint equation as $10800 - xyz = 0$; the choice $xyz - 10800 = 0$ is equally acceptable, and we would obtain the same result.

The variable λ may be interpreted as follows: suppose the volume allowed is increased by one unit to 10,801 in^3. The corresponding cost increase is approximately given by λ. Since $\lambda = 0.054/z + 0.09/y = 0.054/18 + 0.09/30 = 0.006$, the extra cubic inch of volume will add about \$0.006 = 0.6¢ to the cost. In this problem, λ represents the *marginal cost of volume*. ■

Example 8.25 Find the point on the surface of the function $z = \sqrt{x^2 + y^2}$ that is closest to the point $(4, 3, 10)$.

Solution Writing the relation between x, y, and z in the form

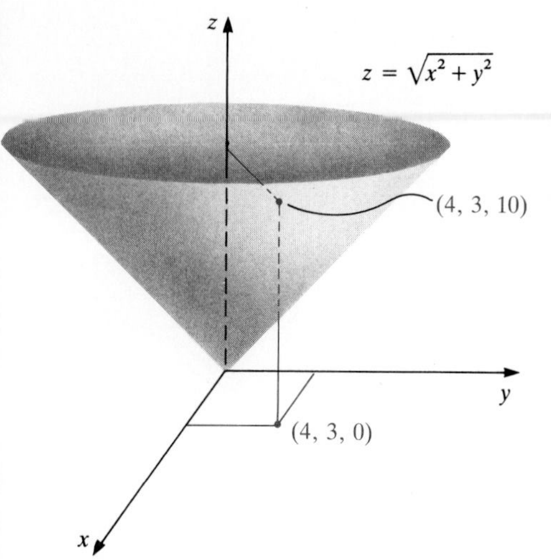

Figure 8.22
The cone $z = \sqrt{x^2 + y^2}$

$z^2 = x^2 + y^2$, $z > 0$ helps determine the graph of the function. For $z = c$, we get the level curve $x^2 + y^2 = c^2$, which is a circle of radius c. For example, the level curve for $z = 1$ generates a circle of radius 1. If $z = 2$, the level curve is a circle of radius 2. In fact, the larger the value of z, the larger the level-curve circle. The graph of the function is a cone as shown in Figure 8.22.

By analogy to the distance formula between two points in the xy-plane, the distance from the point $(4, 3, 10)$ to a point (x, y, z) on the cone is $\sqrt{(x-4)^2 + (y-3)^2 + (z-10)^2}$. If we find the point on the cone that minimizes the *square* of the distance, that same point will minimize the distance itself. Hence, we can avoid using the square-root function. Consequently, we seek x, y, z such that $(x-4)^2 + (y-3)^2 + (z-10)^2$ is minimized, subject to the constraint $z^2 - x^2 - y^2 = 0$. The Lagrange function is

$$F(x, y, z, \lambda) = (x-4)^2 + (y-3)^2 + (z-10)^2 + \lambda(z^2 - x^2 - y^2).$$

Finding the partial derivatives and setting them equal to zero yields

$$F_x = 2(x-4) - 2\lambda x = 0 \qquad F_y = 2(y-3) - 2\lambda y = 0$$
$$F_z = 2(z-10) + 2\lambda z = 0 \qquad F_\lambda = z^2 - x^2 - y^2 = 0.$$

We solve each equation for λ, by dividing by x, y, or z. If the minimum or maximum occurs where any of these is zero, then we would miss the extremum, since the division requires a nonzero divisor. However, it is easy to see that neither $x = 0$ nor $y = 0$ nor $z = 0$ is a possibility. Consider $x = 0$, for example. Substituting $x = 0$ into the $F_x = 0$ equation yields $-8 = 0$. The possibilities $y = 0$ and $z = 0$ yield similarly absurd results. Hence we may divide by x, y, or z without risking the loss of a critical point. Solving the equations for λ gives

$$\lambda = \frac{x-4}{x}, \qquad \lambda = \frac{y-3}{y}, \qquad \lambda = -\frac{z-10}{z}.$$

These three expressions for λ must be equal to one another; setting the first two equal implies

$$\frac{x-4}{x} = \frac{y-3}{y}.$$

We multiply both sides of the equation by xy to get $xy - 4y = xy - 3x$, and then cancel the xy term to get $y = \frac{3}{4}x$. Likewise, equating the first and last expression for λ gives

$$\frac{x-4}{x} = -\frac{z-10}{z};$$

multiplying both sides of the equation by xz, we get $xz - 4z = -xz + 10x$. Thus $2xz - 4z = 10x$. We factor z from the left-hand side and divide by the remaining factor, which yields

$$z = \frac{10x}{2x-4} = \frac{5x}{x-2}.$$

We have found relationships among the variables. We now use the constraint equation to find values for the variables; since $z^2 = x^2 + y^2$,

$$\left(\frac{5x}{x-2}\right)^2 = x^2 + \left(\frac{3x}{4}\right)^2,$$

or

$$\frac{25x^2}{(x-2)^2} = x^2 + \frac{9x^2}{16}.$$

Dividing both sides by x^2 makes the right-hand side $1 + \frac{9}{16} = \frac{25}{16}$. Hence

$$\frac{25}{(x-2)^2} = \frac{25}{16}$$

so that $(x-2)^2 = 16$. Thus $x - 2 = \pm 4$. We therefore have two critical points: one for which $x = 6$ and one for which $x = -2$. From $y = \frac{3}{4}x$ and $z = 5x/(x-2)$, we can compute the other coordinates of the critical points; they are $y = \frac{9}{2}$ and $z = \frac{15}{2}$ for $x = 6$ and $y = -\frac{3}{2}$ and $z = \frac{5}{2}$ for $x = -2$.

We have two candidates for the point on the cone that is closest to $(4, 3, 10)$: the points $(6, \frac{9}{2}, \frac{15}{2})$ and $(-2, -\frac{3}{2}, \frac{5}{2})$. The distance from $(6, \frac{9}{2}, \frac{15}{2})$ to $(4, 3, 10)$ is 3.54, and the distance from $(-2, -\frac{3}{2}, \frac{5}{2})$ to $(4, 3, 10)$ is 10.06. So the minimum distance of 3.54 units occurs at the point $(6, \frac{9}{2}, \frac{15}{2})$ on the cone. (While we have not developed the necessary procedure, it may be shown that $(-2, -\frac{3}{2}, \frac{5}{2})$ is a saddle point.) ■

Exercises

Exer. 1–5: Use both algebraic substitution and the Lagrange multiplier method to find the indicated relative extrema of the function.

1. The maximum of $f(x, y) = x(y + 6)$ when $x + y = 2$
2. The maximum of $f(x, t) = (3x + 1)(2t - 3)$ when $3x + 6t = 14$
3. The maximum of $f(x, y) = x^2 - y^2 - y$ subject to $x^2 + y^2 = 1$
4. The maximum of $f(x, y) = x^2 + y^2$ subject to $2x - y = 4$
5. The minimum of $f(s, t) = (5s - 3)(4t + 1)$ subject to $5s = 4t + 18$

6. Use the Lagrange multiplier method to find the maximum and minimum, if they exist, of $f(x, y) = y^2 - 4xy + 4x^2$ when $x^2 + y^2 = 1$.
7. Minimize $x^2 + y^2$ subject to $2x = 4 - 3y$.
8. Find the minimum of $21s + 14t$ when $st = 600$ and both s and t are positive.
9. If $2x + y + z = 20$, what values of x, y, and z make $f(x, y, z) = x^2 + y^2 + z^2 - 3x - 5y - z$ as large as possible?
10. Maximize $f(x, y, z) = x + 5z$ if $2x^2 + 3y^2 + 5z^2 = 198$.

Exer. 11–16: Use Lagrange multipliers to determine where the maximum and minimum of the function occurs.

11. $G(x, y, z) = 5 - 6x - 3y - 2z$ subject to $4x^2 + 2y^2 + z^2 = 70$.
12. $H(x, y) = xy$ subject to $\frac{1}{6}x^2 + \frac{1}{4}y^2 = 1$

13. $f(x, y, z) = 6x + 3y + 2z - 5$ with the requirement that the solution be on the surface $4x^2 + 2y^2 + z^2 = 70$.
14. $f(x, y) = 2x + y$ subject to $\frac{1}{18}x^2 + \frac{1}{9}y^2 = 1$.
15. $f(x, y) = (x + 1)^2 + y^2 - 1$ subject to $3x^2 + 2y^2 = 48$.
16. $M(x, y) = x^2y^2$ and the solution must be on the curve $x^2 + 4y^2 = 24$.

17. In a large calculus course, an index of student achievement is given by

$$f(x, y) = 6x^2 + 60xy - 10y^2,$$

where x is the number of hours devoted to lectures and y is the number of hours devoted to recitations. A five-semester-credit-hour course includes 80 hours of instruction. How should the course be scheduled to maximize the index of student achievement?

18. Output in a factory is given by $-3x^2 + 10xy - 3y^2$, where x is the amount of labor used at \$1100 per unit and y is the amount of land used at \$3000 per acre. The production level must be kept at 80,000 units. How can this be accomplished and minimize cost?
19. A diet for a group of animals in a biological study supplies h calories per individual per day from x units of food A, which costs \$2 per unit, and y units of food B, which costs \$5 per unit. It has been found that $h = 2x^2 + 8xy + 17y^2$. If $h = 612$, find x and y that will minimize the total cost of the diet.
20. An electric generator uses two kinds of fuel. The number of tons of air pollutants exhausted by the plant is given by $p(x, y) = 0.06x + 0.03y + 0.0001x^2y$, where x and y are the number of tons of the two fuels consumed. What amount of each fuel should be used to minimize pollutants emitted when 1,000,000 tons of fuel are used to generate the required amount of electricity?
21. Suppose that each laborer works x hours at one task and y at a second task during an 8-hour work day. An industrial psychologist indicates that a worker's job satisfaction can be indexed by $s(x, y) = 2\sqrt{x} + \sqrt{y}$. How should each laborer be assigned to maximize job satisfaction?
22. A firm has $z = xy^{1/2}$ as its production function, with labor x costing \$10 per hour and land y costing \$3 per acre per month. If the firm has a total budget of \$25,000 for labor and land, what amounts of these inputs should be used for maximum output?
23. A firm has $z = 8xy - x^2 - 2y^2$ as its production function. The cost function for the inputs x and y is $C(x, y) = 2x + 4y$ (in thousands of dollars). What is the maximum production on a total budget of \$12,000?
24. A small company with a total budget of \$8500 makes two models of the same item, making profits of \$8 and \$20, respectively. The cost of producing x of the cheaper model and y of the other is given by $C(x, y) = 500 + (x^2/200) + 4y$. What is the maximum profit?
25. A family has \$2400 to spend while on vacation. They have no time limit, and they spend \$200 per week when camping and \$400 per week when staying at motels. If x is the number of weeks spent camping and y is the number of weeks in motels, and index of their satisfaction is given by

$$s(x, y) = \sqrt{x} + \sqrt{y}.$$

Find x and y such that $s(x, y)$ is maximized.
26. Maximize output for a firm with production function $z = xyw$ if labor x costs \$6 per hour, land y costs \$3 per acre per month, and raw materials w cost \$12 per unit. The total budget is \$7200.
27. A firm utilizes x units of labor, y units of land, and s units of raw material. The production function is $z = x^{1/2}ys^{2/3}$ and labor costs \$10 per hour, land rental is \$4 per acre per month, and raw materials cost \$10 per unit. The total budget is \$13,000. Maximize production.
28. The U.S. postal service limits the size of packages by requiring that the length plus girth of a package may not exceed 108 inches. (The girth is the measurement around the package.) Find the dimensions of the package with largest volume that may be mailed.
29. Use Lagrange multipliers to find the position on the parabola $y = x^2$ that is closest to the point $(16, 0.5)$.
30. Suppose that x units of labor and y units of capital are needed to produce $P(x, y) = 500x^{1/4}y^{3/4}$ units of a product. The amount of money available to invest is \$20,000, each unit of labor costs \$200, and each unit of capital costs \$100. How many units of labor and capital should be used to maximize production? If labor and capital are kept at optimal values and the available money is increased by one dollar, what is the approximate increase in production level?
31. Refer to Exercise 30. Compute $\partial P/\partial x$ and $\partial P/\partial y$. Evaluate these at the optimal values found in Exercise 30. Note that the ratio of these partial derivatives at the optimal point is the same as the ratio of the cost of a unit of labor to the cost of a unit of capital; that is, if a is the optimal amount to spend on labor and b is the optimal amount to invest in capital, then

$$\frac{\left.\dfrac{\partial P}{\partial x}\right|_{(a,b)}}{\left.\dfrac{\partial P}{\partial y}\right|_{(a,b)}} = \frac{\text{cost per unit of labor}}{\text{cost per unit of capital}}.$$

This result is a law of economics: *if labor and capital are at optimal values, then the ratio of their marginal productivity is equal to the ratio of their unit cost.*
32. Find the dimensions of a circular can (cylinder) for which the surface area is minimal for a given (fixed) volume.
33. Suppose that a mix of x pounds of one pesticide and y pounds of another is sprayed on a crop using an airplane that can carry 500 pounds. Agricultural scientists have found that the fraction of pests killed by these amounts of pesticide is given by

$$f(x, y) = 1 - \tfrac{1}{2}e^{-x/20} - \tfrac{1}{2}e^{-y/100}.$$

How much of each pesticide should be mixed to maximize the fraction of pests killed by the spraying?

The Total Differential

8.5

Although the definition of partial derivatives allows only one variable to change, these derivatives may be used to approximate the effect of changing both independent variables at once. Consider the rolling-terrain image of the function $z = f(x, y)$ with the x-axis pointing south and the y-axis pointing east. Suppose we are at the position (a, b), and we want to move to a nearby position $(a + h, b + k)$. To simplify matters, suppose both h and k are positive. One way to get to the new position is to move h units south (parallel to the x-axis), reaching the point $(a + h, b)$ and then move k units east (parallel to the y-axis) to the desired position. See Figure 8.23.

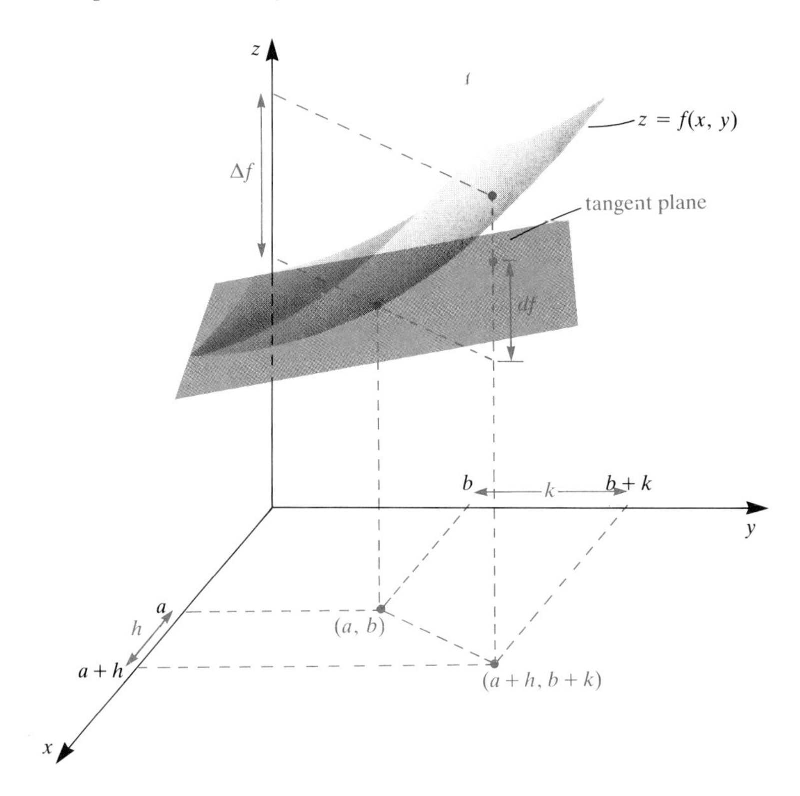

Figure 8.23
The total differential of $f(x, y)$

During the first step of the move, the change in altitude is approximately the slope of the surface times the distance moved, or $[f_x(a, b)]h$. During the second step, the change in altitude is again approximately the slope of the surface times the distance moved, or $[f_y(a, b)]k$. If we let df denote the total change in altitude caused by the two moves combined, then

$$df = [f_x(a, b)]h + [f_y(a, b)]k.$$

We call df the **total differential** of $f(x, y)$. Of course, the exact change in altitude when moving from (a, b) to $(a + h, b + k)$ is the difference in function values at the two points: $f(a+h, b+k) - f(a, b)$. We thus have

$$f(a+h, b+k) - f(a, b) \approx f_x(a, b)h + f_y(a, b)k = df.$$

The accuracy of this approximation increases as h and k become smaller.

An important application of the total differential is the assessment of the impact that an error in measurement may have on the computed value of a function. As an illustration, consider the computation of the area of a rectangle from measured values of its length and width.

Example 8.26 Suppose the length x of a rectangle is measured to be 24 inches with a possible error of 0.5 inch and the width y is measured to be 10.6 inches with a possible error of 0.05 inch. Estimate the largest error in the computed area of the rectangle.

Solution The area function is $A(x, y) = xy$. The total differential is

$$dA = \frac{\partial A}{\partial x}h + \frac{\partial A}{\partial y}k,$$

where h and k represent the errors in measuring x and y, respectively. Since $\partial A/\partial x = y$ and $\partial A/\partial y = x$ are both positive at (24, 10.6), the largest error will occur when h and k are each as large as possible. Thus we take $h = 0.5$ and $k = 0.05$. The total differential $yh + xk$ at (24, 10.6) is

$$dA\,|_{(24,\,10.6)} = (10.6)(0.5) + (24)(0.05) = 6.5.$$

Thus our estimate of the largest error in the measurement of the area of the rectangle is 6.5 in^2. Since the measured area is $24 \times 10.6 = 254.4$ in^2, we say that the area is 254.4 ± 6.5 in^2. ■

Another application of the total differential is the approximation of functions involving roots.

Example 8.27 Use the total differential to estimate $\sqrt[3]{26}\,\sqrt[4]{79}$.

Solution Observe that if

$$f(x, y) = \sqrt[3]{x}\,\sqrt[4]{y},$$

then $f(26, 79) = \sqrt[3]{26}\,\sqrt[4]{79}$. Since $\sqrt[3]{27} = 3$ and $\sqrt[4]{81} = 3$, we can easily evaluate $f(27, 81) = 3(3) = 9$. The total differential will give an estimate of the change in $f(x, y)$ from (27, 81) to (26, 79). If we let $a = 27$ and $b = 81$, then for $a + h$ to equal 26 we must take $h = -1$; for $b + k$ to equal 79 we must take $k = -2$. Thus, the total differential is

$$\begin{aligned}
df &= [f_x(a, b)]h + [f_y(a, b)]k.\\
&= \left[\tfrac{1}{3}x^{-2/3}y^{1/4}\Big|_{(27,81)}\right](-1) + \left[\tfrac{1}{4}x^{1/3}y^{-3/4}\Big|_{(27,81)}\right](-2)\\
&= \left[\frac{1}{3}\frac{1}{(27)^{2/3}}(81)^{1/4}\right](-1) + \left[\frac{1}{4}\frac{1}{(81)^{3/4}}(27)^{1/3}\right](-2)\\
&= (\tfrac{1}{3})(\tfrac{1}{9})(3)(-1) + (\tfrac{1}{4})(\tfrac{1}{27})(3)(-2)\\
&= -\tfrac{1}{9} - \tfrac{1}{18} = -\tfrac{1}{6} \approx -0.167.
\end{aligned}$$

Since $f(27, 81) = 9$, we have $f(26, 79) \approx f(27, 81) + df$,

or
$$\sqrt[3]{26}\,\sqrt[4]{79} \approx 9 - 0.167 = 8.833.$$

Rounded to three places after the decimal point, the exact value of $\sqrt[3]{26}\,\sqrt[4]{79}$ is 8.832, so our estimate is quite good, and it does not require the use of a calculator. ■

Example 8.28 Refer to Example 8.12. The air pollution index P in a city is determined by two factors: the amount x of solid waste (soot, dust and so forth) in the air and the amount y of noxious gases (sulphur dioxide, carbon monoxide and so forth) in the air. The index is described by

$$P = x^2 + 2xy + 4xy^2.$$

At the values $x = 10$, $y = 5$, the pollution index is 1200 (measured in some appropriate unit); that is, $P(10, 5) = 1200$. Suppose the value of x changes to 10.2 and the value of y changes to 4.7. Estimate the change in the air pollution index.

Solution The total differential dP may be used to approximate the change in the pollution index from $(10, 5)$ to $(10.2, 4.7)$ as follows. Letting $a = 10$ and $b = 5$, we have $10.2 = a + h$ and $4.7 = b + k$ if we let $h = 0.2$ and $k = -0.3$. Since $P_x(10, 5) = (2x + 2y + 4y^2)\,|_{(10,5)} = 130$ and $P_y(10, 5) = (2x + 8xy)\,|_{(10,5)} = 420$, we find

$$dP = [P_x(a, b)]h + [P_y(a, b)]k = (130)(0.2) + (420)(-0.3) = -100.$$

This amount is an approximation of the *change* in P; since $P = 1200$ at the original point $(10, 5)$, we estimate that $P(10.2, 4.7) \approx 1100$. (Using the formula for P we get $P(10.2, 4.7) = 1101.192$; the relative error of our estimate is about 0.11%.) ■

Consider a function of three independent variables $w = f(x, y, z)$. The approximate change in the value of $f(x, y, z)$ as (x, y, z) moves from (a, b, c) to $(a + h, b + k, c + m)$ is

$$df\,|_{(a,b,c)} = \left[\frac{\partial f}{\partial x}(a, b, c)\right]h + \left[\frac{\partial f}{\partial y}(a, b, c)\right]k + \left[\frac{\partial f}{\partial z}(a, b, c)\right]m.$$

Example 8.29 One of the functions of the kidneys is to remove urea (a byproduct of protein decomposition) from the blood. A common measure of the health of the kidneys is the *standard urea clearance function* $C(u, v, s)$ given by

$$C(u, v, s) = \frac{u\sqrt{v}}{s},$$

where u denotes the concentration of urea in the urine (in mg/100 ml), v denotes volume of urine excreted per minute (in ml/min) and s denotes the concentration of urea in the blood (in mg/100 ml). Suppose a clinical test yields $u = 30$, $v = 1$, and $s = 0.5$ with an error of 1% in the measurements. Then $C(30, 1, 0.5) = 30$. Estimate the maximum change in C and the maximum percent change in C.

Solution If we let h be the actual error in u, let k be the error in v, and let m denote the error in s, then an error of no more than 1% in u, v, and s indicates that $-0.3 \le h \le 0.3$, $-0.01 \le k \le 0.01$, and $-0.005 \le m \le 0.005$. The total differential is

$$dC = \left[\frac{\sqrt{v}}{s}\bigg|_{(30,1,0.5)}\right]h + \left[\frac{u}{2s\sqrt{v}}\bigg|_{(30,1,0.5)}\right]k + \left[-\frac{u\sqrt{v}}{s^2}\bigg|_{(30,1,0.5)}\right]m$$

or

$$dC = 2h + 30k - 120m.$$

We wish to estimate the largest possible error in C. Since h and k have positive coefficients in dC and m has a negative coefficient, we use $h = 0.3$, $k = 0.01$, and $m = -0.005$ to get

$$dC_{\text{max}} = 2(0.3) + 30(0.01) - 120(-0.005) = 1.5.$$

Since $C(30, 1, 0.5) = 30$, the largest percent error is

$$\frac{(100)dC_{\text{max}}}{C} = \frac{100(1.5)}{30} = 5\%.$$

Consequently, an error in the measured data of only 1% results in an error of 5% in the computed quantity. Therefore it is *incorrect* to assume that, if the input data are accurate to within 1%, the kidney health indicator C will also be accurate to within 1%. ■

Example 8.30 Estimate $7\sqrt{24}\sqrt[4]{18}\sqrt[3]{30}$.

Solution We consider the function $f(x, y, z) = 7\sqrt{x}\sqrt[4]{y}\sqrt[3]{z}$ and take $a = 25$, $b = 16$, $c = 27$, because these are values near $x = 24$, $y = 18$, $z = 30$ for which we can easily evaluate the function. Thus we have $h = 24 - 25 = -1$, $k = 18 - 16 = 2$, and $m = 30 - 27 = 3$. The total differential is

$$\begin{aligned}
df &= \left(\frac{\partial f}{\partial x}\right)(-1) + \left(\frac{\partial f}{\partial y}\right)(2) + \left(\frac{\partial f}{\partial z}\right)(3) \\
&= [7(\tfrac{1}{2})x^{-1/2}y^{1/4}z^{1/3}(-1) + 7(\tfrac{1}{4})x^{1/2}y^{-3/4}z^{1/3}(2) \\
&\quad + 7(\tfrac{1}{3})x^{1/2}y^{1/4}z^{-2/3}(3)]_{(25,16,27)} \\
&= (\tfrac{7}{2})(\tfrac{1}{5})(2)(3)(-1) + (\tfrac{7}{4})(5)(\tfrac{1}{8})(3)(2) + (\tfrac{7}{3})(5)(2)(\tfrac{1}{9})(3) \\
&= -\tfrac{21}{5} + \tfrac{105}{16} + \tfrac{70}{9} \approx 10.14.
\end{aligned}$$

Since $f(25, 16, 27) = 7(5)(2)(3) = 210$, we have

$$f(24, 18, 30) \approx f(25, 16, 27) + df \approx 210 + 10.14 = 220.14.$$

This result is about 0.3% too large. ■

Example 8.31 At the current level of operation, a certain company has a marginal productivity of labor of 150 items produced per unit of labor and a marginal productivity of capital of 70 items produced per unit of capital. Estimate the increase in production obtained by adding 0.5 unit of labor and 1 unit of capital.

Solution If we let $P(L, K)$ denote the production function, then the total differential of P is

$$dP = \frac{\partial P}{\partial L}\,dL + \frac{\partial P}{\partial K}\,dK.$$

The marginal productivity of labor is $\partial P/\partial L = 150$ and the marginal productivity of capital is $\partial P/\partial K = 70$. The contribution of labor is increased by $dL = 0.5$ and the contribution of capital is increased by $dK = 1$. Hence, the increase in production is approximately

$$dP = (150)(0.5) + (70)(1) = 145 \text{ units.}$$

■

Exercises

Exer. 1–9: Find the total differential.

1. $f(x, y) = x^2 + xy + 3x$ at $(1, -2)$ when $h = 0.1$, $k = 0.5$.
2. $f(x, y) = x^2y - xy^2$ at $(2, 2)$ with $h = k = 0.1$.
3. $f(x, y) = xe^y - ye^x$ at $(1, -1)$ with $h = 0.1$, $k = -0.2$.
4. $f(x, y, z) = xy + yz + xz$ at $(1, 2, 3)$ with $3h = 2k = m = 0.3$.
5. $f(x, y, z) = x^2ye^{2z}$ at $(4, 4, 1)$ with $h = k = m = 0.1$.
6. $f(x, y) = \ln(x + y)$ at $(2, 3)$ with $h = 0.5$, $k = -0.25$.
7. $f(x, y) = (xy)/(x + y)$ at $(2, 5)$ with $h = 1$, $k = 0.5$.
8. $f(x, y, z) = e^{xyz}$ at $(1, 2, 3)$ with $h = -1$, $k = 1$, $m = 0.5$.
9. $f(x, y) = x/y$ at $(-2, 1)$ with $h = -1$, $k = 1$.

Exer. 10–19: Use the total differential to approximate the following.

10. $\sqrt{125}\,\sqrt[4]{17}$
11. $\sqrt[3]{210}\,\sqrt[4]{250}$
12. $\sqrt{8}\,\sqrt[3]{28}\,\sqrt[4]{84}$
13. $\sqrt[6]{60}\,\sqrt[3]{25}\,\sqrt[4]{80}$
14. $72\,\sqrt[3]{340}\,\sqrt[5]{30}$
15. $\sqrt{17}\,\sqrt[4]{620}$
16. $2.95e^{1.01}$
17. $\ln(1.2^2 + 0.9^2)$
18. $6.1 \ln 2.7$
19. $\dfrac{e^{1.1}}{\sqrt{4.1}}$

20. Consider the function $z = f(x, y) = (x - y)/(x + y)$. Suppose x changes from 20 to 20.5 and y changes from 40 to 40.5. Estimate the change in z.
21. Estimate the change in volume of a right circular cylinder ($V = \pi r^2 h$) if the height changes from 9 to 9.2 and the radius changes from 4 to 4.1.
22. Find the percent change in the volume of a right circular cone ($V = \frac{1}{3}\pi r^2 h$) if the radius is increased by 2% and the height is decreased by 5%.
23. Estimate the error in computing
$$F(x, y, z) = \frac{x^2 - y^2}{x^2 + z^2}$$
at $(3, 1, 4)$ if the error in determining x is 0.01, the error in determining y is 0.001, and the error in determining z is 0.005.
24. Use the total differential to estimate the error that can arise from calculating
$$H(x, y, z, w) = x^2y^3 + yzw^4$$
if $x = 5 \pm 0.01$, $y = -2 \pm 0.05$, $z = 10 \pm 0.1$, and $w = 3 \pm 0.005$.

Exer. 25–28: Suppose a nuclear power plant and a chemical company are situated on a large lake. Let the pollution index P of the lake be a function of x and y, where x measures the quantity of heat entering the lake due to the power plant waste water and y measures the amount of chemical waste entering the lake. Suppose the pollution index P for the lake is such that $\partial P/\partial x = 0.06x^2y + 2x$ and $\partial P/\partial y = 0.02x^3 + 10y$. Also suppose $P = 3000$ when $x = 25$ and $y = 12$. Use the total differential to estimate the value of P for the following values:

25. $x = 26$, $y = 12$
26. $x = 25$, $y = 13$
27. $x = 24$, $y = 13$
28. $x = 26$, $y = 11$

29. Referring to Example 8.29, suppose $u = 100$, $v = 1.44$, and $s = 3$. Compute the maximum change in $C = u\sqrt{v}/s$. Compute the maximum percent change in C.
30. Suppose a driver slams on the car brakes and skids to a stop. Let v denote the speed of the car when the brakes were applied, let L denote the length of the

skid marks, and let w be the weight of the car. Then v is related to L and w by

$$v = \frac{c\sqrt{L}}{\sqrt{w}}$$

for some constant c. Suppose a car is involved in an accident: L is measured to be 400 feet, plus or minus 30 feet; and the weight of the car is estimated to be 2500 pounds, give or take 500 pounds. From the measured data it is computed that the car's speed when the brakes were applied was 50 mph. Estimate the accuracy of this result in view of the possible error in the data.

31. A certain company has a marginal productivity of labor of 90 items produced per unit of labor and a marginal productivity of capital of 60 items produced per unit of capital. Estimate the increase in production obtained by removing one unit of labor and adding two units of capital.

8.6 The Method of Least Squares

Thus far we have provided functions describing cost, revenue, productivity, population, distance, velocity, and so on. In practice, an analyst or scientist must determine these functions from data before calculus can be used to analyze them. In this section we discuss one method for the determination of particular functions from gathered data.

Two types of considerations are involved in determining a function that describes a phenomenon: *theoretical* and *empirical.* Theoretical considerations are based on previous knowledge of similar situations and usually predict the "shape" and general behavior of the function sought. For instance, an economist knows that the function describing the total production P of a commodity in terms of the amount spent on labor L and on capital K would be one of the forms

$$P = \alpha L^{\beta} K^{1-\beta} \quad \text{or} \quad P = (\alpha L^{\beta} + \gamma K^{\beta})^{1/\beta}.$$

where α, β, and γ are constants. We have seen that population functions are of an exponential form $y = Ce^{at}$. The function is linear, $y = mx + b$, if a change in one variable always creates a proportional change in the other variable. In these situations, the constants α, β, γ, C, a, m, b must be determined from the available data to produce a function describing a specific situation.

Finding constants requires empirical data obtained from measurements or other experimental means. For instance, in a population problem, the data is the set of points (t_i, p_i), where t_i is the ith time at which the population was measured and p_i is the corresponding population.

Unfortunately, a function of the general form satisfying the theoretical conditions and passing through the given data points does not exist for every situation. Consequently, we must determine the function that best satisfies both the theoretical and empirical considerations. We will discuss what we mean by *best* in a moment. The area of mathematics that deals with the determination of functions that best satisfy certain criteria is *curve fitting* or *approximation theory.*

The examples and exercises in this section are presented with a small amount of empirical data to allow reasonable computation by hand. We begin by considering phenomena for which theoretical considerations indicate that the function is linear.

The form of the function we seek is $y = mx + b$. The data used to determine m and b are points (x_i, y_i) for $i = 1, 2, \ldots, n$. If n were 2, we could find m and b easily, because two points determine a straight line uniquely. However, n is typically much larger than 2, and all the measured data points (x_i, y_i) will seldom be collinear. Many possible sources may contribute to the lack of collinearity: instruments recording the data may fluctuate slightly, measured quantities may be slightly affected by other variables, or the researcher may make an error. Consequently, we are usually unable to find values for m and b such that $y_i = mx_i + b$ for all i.

We therefore seek values of m and b such that the line $y = mx + b$ is the "best approximation" to all the data (x_i, y_i). At $x = x_i$ the y-coordinate of this line is $mx_i + b$, so the difference between the y-coordinate of the "best" line and the measured y-coordinate is

$$E_i = mx_i + b - y_i.$$

We are interested in the "best" line for all the data, not just the error at one point. Naively, we might conclude that by adding the errors at every point, we would obtain a total error

$$E_1 + E_2 + E_3 + \cdots + E_n.$$

However, we must reject this as an accurate measure of total error because each E_i may be positive or negative, so their sum could be small (even zero) although some of the individual errors may be quite large. For example, $E_1 = 100$, $E_2 = -100$, $E_3 = -1000$ and $E_4 = 1000$ gives a total error of zero with large individual errors. Summing the absolute values of the individual errors would give

$$|E_1| + |E_2| + |E_3| + \cdots + |E_n|$$

as a measure of the total error. This measure is acceptable except that we will need to differentiate this sum to calculate a minimum total error. The absolute value function has no derivative at the origin. Consequently, we reject this measure of total error. Another way to avoid the cancellation of errors is to square the errors. Since the square function *is* differentiable everywhere, we choose m and b so that the sum of the squares of the E_i is as small as possible; that is, we define the "best" line to be the line that minimizes

$$S = E_1^2 + E_2^2 + \cdots + E_n^2,$$

or $$S = (mx_1 + b - y_1)^2 + (mx_2 + b - y_2)^2 + \cdots + (mx_n + b - y_n)^2.$$

Figure 8.24 shows the "best" line for some given data points and their E_i. Since m and b are chosen to make this sum of squares have the least value, the method is called the **method of least squares**. It is a very common measure of total error.

Since S is a function of only b and m, to minimize S we set $\partial S/\partial b$

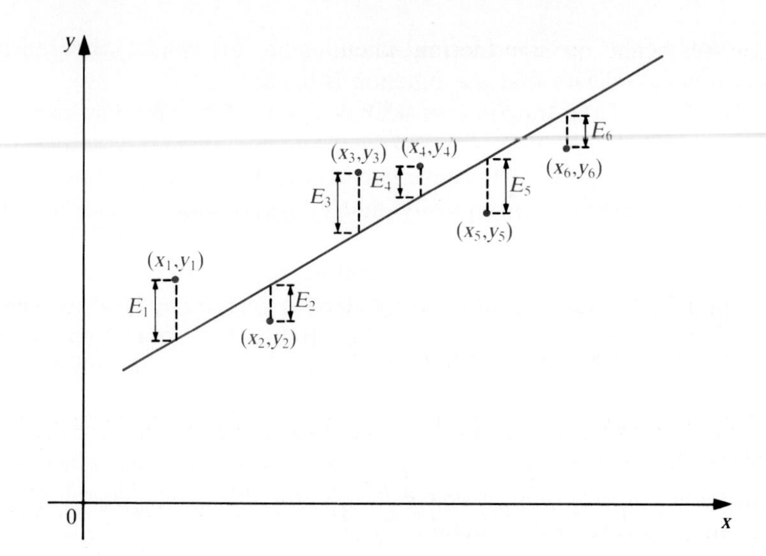

Figure 8.24
Some data points and a line passing "among" them

and $\partial S/\partial m$ equal to zero and solve for m and b. We have

$$\frac{\partial S}{\partial b} = 2(mx_1 + b - y_1) + 2(mx_2 + b - y_2) + \cdots + 2(mx_n + b - y_n) = 0$$

and

$$\frac{\partial S}{\partial m} = 2(mx_1 + b - y_1)x_1 + 2(mx_2 + b - y_2)x_2 + \cdots + 2(mx_n + b - y_n)x_n = 0.$$

We divide both sides of each equation by 2 and collect all m and b terms on the left-hand sides of the equations, other terms on the right-hand sides. The results are

$$m(x_1 + x_2 + \cdots + x_n) + b(1 + 1 + \cdots + 1) = y_1 + y_2 + \cdots + y_n$$

and

$$m(x_1^2 + x_2^2 + \cdots + x_n^2) + b(x_1 + x_2 + \cdots + x_n) = x_1y_1 + x_2y_2 + \cdots + x_ny_n.$$

These two equations, which we call the **normal equations**, may be solved for m and b to give the required line $y = mx + b$. The line obtained in this fashion is called the **least squares line of regression.**

Recall that a common notation for a sum of terms such as $x_1 + x_2 + \cdots + x_n$ is

$$\sum_{i=1}^{n} x_i.$$

Using this notation, we have

$$x_1 + x_2 + \cdots + x_n = \sum_{i=1}^{n} x_i$$

$$x_1^2 + x_2^2 + \cdots + x_n^2 = \sum_{i=1}^{n} x_i^2$$

$$1 + 1 + \cdots + 1 = \sum_{i=1}^{n} 1.$$

The normal equations may then be written as

$$m \sum_{i=1}^{n} x_i + b \sum_{i=1}^{n} 1 = \sum_{i=1}^{n} y_i$$

and

$$m \sum_{i=1}^{n} x_i^2 + b \sum_{i=1}^{n} x_i = \sum_{i=1}^{n} x_i y_i.$$

$\left(\text{Note that } \sum_{i=1}^{n} 1 = n.\right)$ We may use the Second Derivative Test to verify that we have indeed found a minimum.

Example 8.32 Find the least squares line for the data given in the accompanying table.

x_i	y_i
1	0
2	1
4	2
7	5

Solution To find the sum of x_i^2 and $x_i y_i$, we rewrite and extend the given table:

x_i	y_i	x_i^2	$x_i y_i$
1	0	1	0
2	1	4	2
4	2	16	8
7	5	49	35
14	8	70	45

We obtain the last row of the table by summing the numbers in each column. These sums are (from left to right):

$$\sum_{i=1}^{4} x_i = 14, \qquad \sum_{i=1}^{4} y_i = 8, \qquad \sum_{i=1}^{4} x_i^2 = 70, \qquad \sum_{i=1}^{4} x_i y_i = 45.$$

Also note that $\sum_{i=1}^{4} 1 = 4$. The normal equations are therefore

$$14m + 4b = 8$$
$$70m + 14b = 45.$$

We multiply the first equation by 5 and subtract from the second equation; the result is $-6b = 5$. Hence, $b = -5/6$. Substituting this value in the first equation gives $m = \frac{17}{21}$, so the least squares line is

$$y = \tfrac{17}{21}x - \tfrac{5}{6}.$$

Note that none of the data points are on the line. See Figure 8.25. ■

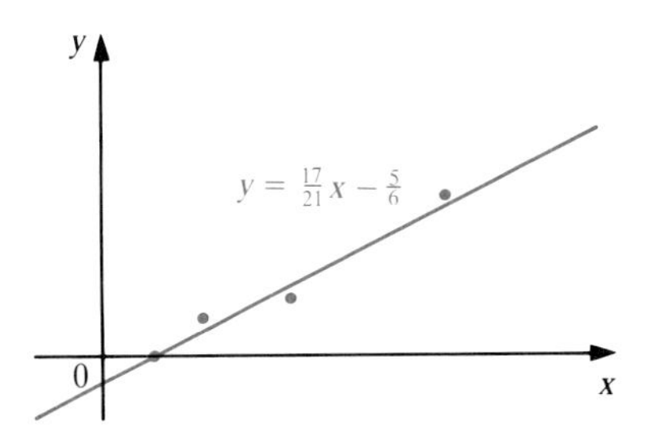

Figure 8.25
Least squares line for data in Example 8.32

Example 8.33 A sociologist is studying the relationship between the number of rooms in a family's living quarters and the number of quarrels per week for families of a given ethnic background and size. The data in the first two columns of Table 8.5 are obtained from surveys and interviews. If five-room living quarters could be found for every family of this size, how many quarrels per week would be expected?

Table 8.5
Data for size of living quarters and number of quarrels

Number of rooms	*Quarrels per week*		
N_i	Q_i	N_i^2	N_iQ_i
2	6	4	12
2	5	4	10
2	8	4	16
2	3	4	6
2	4	4	8
3	5	9	15
3	2	9	6
3	3	9	9
4	2	16	8
4	3	16	12
4	1	16	4
31	42	95	106

Solution We assume a linear relationship between the number of rooms N and the number of quarrels Q:

$$Q = mN + b.$$

The normal equations are

$$31m + 11b = 42$$
$$95m + 31b = 106.$$

We multiply the first equation by 31 and the second by 11 and subtract to get $-84m = 136$. Hence $m = -\frac{136}{84} \approx -1.62$ and $b \approx 8.4$. The relation between Q and N is

$$Q = -1.62N + 8.4.$$

When $N = 5$, then $Q = 0.3$. Thus, approximately three quarrels every ten weeks would be expected if all families of the category under study have five rooms for living quarters. ■

Example 8.34 The first column in Table 8.6 gives the percentage of employed citizens without a college education in a certain community. The second column gives the difference between the median incomes of those with a college education and those without. Find the line of regression. Predict the difference between the median incomes when the percentage of those without a college education is 8%.

Table 8.6
Least squares table for differences in income versus percent without college education

Percentage without a college education x_i	*Difference in income* y_i	x_i^2	$x_i y_i$
2.13	809	4.5369	1723.17
2.52	763	6.3504	1922.76
11.86	612	140.6596	7258.32
2.55	492	6.5025	1254.60
2.87	679	8.2369	1948.73
4.23	635	17.8929	2686.05
4.62	859	21.3444	3968.58
5.19	228	26.9361	1183.32
6.43	897	41.3449	5767.71
6.70	867	44.8900	5808.90
1.53	513	2.3409	784.89
1.87	335	3.4969	626.45
10.38	868	107.7444	9009.84
62.88	8557	432.2768	43943.32

Solution From Table 8.6 we find that the normal equations are

$$62.88m + 13b = 8557$$
$$432.2768m + 62.88b = 43{,}943.32.$$

Solving as usual, we obtain $m = 19.931$ and $b = 561.83$. Hence, the linear relationship between the quantities is given by

$$y = 19.931x + 561.83.$$

Our prediction is made by substituting 8 for x:

$$y = 19.931(8) + 561.83 = 721.278.$$

We predict that the difference in median income of those with a college education and those without would be $721.28 when the precentage of non-college-educated citizens is 8%. ■

Table 8.7
Population data for a bacteria population

Time	t_i	P_i
8 A.M.	0	500
9 A.M.	1	800
10 A.M.	2	1400
11 A.M.	3	2300
12 noon	4	4200

Example 8.35 The population of a bacteria culture is expected to behave according to the exponential curve $P = Ce^{kt}$ for constants C and k. The data in Table 8.7 show measurements of the population P_i at various times t_i. Determine the best values for C and k. What will the population be at 2 P.M.?

Solution Let t denote time (in hours) with $t = 0$ being 8 A.M. The method of least squares assumes a *linear* dependence between the variables, but this population model is based on the exponential relationship

$$P = Ce^{kt}.$$

However, if we take the logarithm of both sides and use the properties of the logarithm, we get

$$\ln P = \ln Ce^{kt} = \ln C + \ln e^{kt} = \ln C + kt.$$

Letting $y = \ln P$ and $b = \ln C$, we may write this as

$$y = kt + b,$$

which is a linear relation between y and t. After converting the data for P to the new variable y, we can apply the method of least squares. Table 8.8 shows our calculations.

Table 8.8
Data for the linear version of the bacteria population model

t_i	$y_i = \ln P_i$	t_i^2	$t_i y_i$
0	6.215	0	0
1	6.685	1	6.685
2	7.244	4	14.488
3	7.741	9	23.223
4	8.343	16	33.372
10	36.228	30	77.768

The normal equations are

$$10k + 5b = 36.228$$
$$30k + 10b = 77.768.$$

We subtract two times the first equation from the second to get $10k = 5.312$. Thus, $k = 0.5312$; then $b = 6.183$. Now k was part of the original function $P = Ce^{kt}$, but b was not; b and C are related by $b = \ln C$, so $C = e^b = e^{6.183} \approx 484$. Thus

$$P = 484e^{0.5312t}.$$

At 2 P.M., $t = 6$, and the population will be approximately $P = 484e^{0.5312 \times 6} \approx 11{,}723$. ■

Suppose that a quantity y is related to a quantity x according to a function of the form

$$y = kx^m$$

for some real numbers k and m. We wish to determine the best values of k and m from data for x and y. Taking the natural logarithm of both sides, we obtain

$$\ln y = \ln kx^m = \ln k + \ln x^m = \ln k + m \ln x.$$

In other words, if x and y are related by $y = kx^m$, then $\ln y$ is a linear function of $\ln x$ (with slope m and y-intercept $\ln k$). Many important relationships have this form. The following zoological example involves the mouse-to-elephant function, which accurately relates weight to metabolism for a variety of animals.

Example 8.36 The metabolism of an animal is measured by the number of calories of heat the animal produces during a 24-hour period (its metabolic heat production). Table 8.9 gives weight W and metabolic heat production H for a sample of animals. Find the function of the form $H = kW^m$ that best fits the data given in the table. Given that the weight of a pony is 253 kg and its metabolic heat production is 4588 calories, what is the percentage error that occurs if we use our formula to calculate a pony's metabolic heat production from its weight?

Table 8.9
Weight and metabolic heat production for a variety of animals

Animal	*Weight W (in kg)*	*Metabolic heat production H (in calories)*
Canary	0.016	4.9
Albino mouse	0.021	3.6
Dove	0.15	17.2
Pigeon	0.278	28.3
Guinea pig	0.41	35.1
Rabbit	2.6	117
Goose	5.0	272
Dog	14.0	485
Sheep	45.0	1160
Sow	122.4	2400
Elephant	3672.0	49,000

Solution Taking the logarithm of both sides of $H = kW^m$, we obtain

$$\ln H = \ln k + m \ln W.$$

Letting $y = \ln H$, $x = \ln W$, and $b = \ln k$, we have the linear equation

$$y = mx + b.$$

We will determine the best values for $b = \ln k$ and m by the least squares method. We must first translate the information in Table 8.9 into the new notation and find the sums needed to construct the normal equations. See Table 8.10.

The normal equations are

$$9.959m + 11b = 54.807$$
$$153.493m + 9.959b = 158.445.$$

Solving this system, we get $m = 0.753$ and $b = 4.301$, so $\ln k = b = 4.301$. Thus $k = e^{4.301} = 73.77$. Putting these values of m and k into the

x_i	y_i	x_i^2	$x_i y_i$
−4.135	1.589	17.098	−6.571
−3.863	1.281	14.923	−4.949
−1.897	2.845	3.599	−5.397
−1.280	3.343	1.638	−4.279
−0.892	3.558	0.796	−3.174
0.956	4.762	0.914	4.552
1.609	5.606	2.589	9.020
2.639	6.184	6.964	16.320
3.807	7.056	14.493	26.862
4.807	7.783	23.107	37.413
8.208	10.800	67.371	88.646
9.959	54.807	153.493	158.445

Table 8.10

Transformed data for the mouse-to-elephant function in Example 8.36

original equation, we have

$$H = 73.77W^{0.753}.$$

If we apply this equation to the pony that weighs 253 kg, its metabolic heat production should be

$$H_{\text{pony}} = 73.77(253)^{0.753} = 4758.$$

Since the pony's actual metabolic heat production is given to be 4588, the percent error is

$$\frac{4758 - 4588}{4588} 100 = 3.7\%,$$

which is a small error for approximations of this type. Studies taking many more animals into account have found that the function

$$H = 70.5W^{0.734}$$

fits the larger set of data best; this is reasonably close to the function we found with our small sample of animals. ■

A common relationship used in economics for the production level P of a commodity when labor cost is L dollars and capital expenditure is K dollars is the Cobb–Douglas production function

$$P = \alpha L^m K^{1-m}$$

for constants m and α. Although this relationship is nonlinear and contains three variables, we can change it to a linear form involving two variables. Since $P = \alpha L^m K^{1-m}$,

$$\begin{aligned}
\ln P &= \ln(\alpha L^m K^{1-m}) \\
&= \ln \alpha + \ln L^m + \ln K^{1-m} \\
&= \ln \alpha + m \ln L + (1-m)\ln K \\
&= \ln \alpha + \ln K + m(\ln L - \ln K)
\end{aligned}$$

so that
$$\ln P - \ln K = \ln \alpha + m(\ln L - \ln K),$$
or
$$\ln \frac{P}{K} = \ln \alpha + m \ln \frac{L}{K}.$$

Now we let $y = \ln(P/K)$, $x = \ln(L/K)$, and $b = \ln \alpha$. We then have the linear relationship
$$y = mx + b.$$

Example 8.37 Over a period of several years a company collects the data in Table 8.11. Assuming that productivity P, labor cost L, and capital cost K are related by $P = \alpha L^m K^{1-m}$ for constants α and m, find the values of α and m that best describe the company's performance record.

Table 8.11
Productivity versus labor and capital costs

Productivity P_i (in millions)	*Labor cost L_i (in thousands)*	*Capital cost K_i (in thousands)*
2.0	100	100
2.3	150	75
2.5	200	60
2.6	150	100
2.2	100	125
1.7	75	100
1.8	75	125

Solution By letting $y = \ln(P/K)$, $x = \ln(L/K)$, and $b = \ln \alpha$ we can transform the relationship into a linear form, for which the method of least squares is applicable. Table 8.12 shows the usual least squares data. We obtain these entries (for example, row 1) by computing

$$x_1 = \ln \frac{L_1}{K_1} = \ln \frac{100}{100} = 0 \quad \text{and} \quad y_1 = \ln \frac{P_1}{K_1} = \ln \frac{2.0}{100} = -3.912.$$

Table 8.12
Data for the linear version of the productivity-labor-capital model

x_i	y_i	x_i^2	$x_i y_i$
0.000	−3.912	0.000000	0.000000
0.693	−3.485	0.480249	−2.415105
1.204	−3.178	1.449616	−3.826312
0.405	−3.650	0.164025	−1.478250
−0.223	−4.040	0.049729	0.900920
−0.288	−4.075	0.082944	1.173600
−0.511	−4.241	0.261121	2.167151
1.280	−26.581	2.487684	−3.477996

The normal equations are

$$1.280m + 7b = -26.58$$
$$2.488m + 1.280b = -3.478,$$

rounded to four significant digits. Solving these equations gives $m = 0.6135$ and $b = -3.9095$. Then $\alpha = e^b = 0.0200$. Consequently, the productivity of the company is best described by

$$P = 0.02L^{0.6135}K^{0.3865}.$$

For example, to determine the company's productivity level if it invested \$150,000 in labor cost and \$130,000 in capital cost, we use $L = 150$ and $K = 130$ in our formula to get

$$P = 0.02(150)^{0.6135}(130)^{0.3865} = 2.84.$$

The company would produce approximately 2,840,000 units with these labor and capital costs. ■

We again point out that one should be wary of making judgments based on small samples, such as those in this section. Our intention was to introduce the method of least squares, using small amounts of data to keep the arithmetic simple. Statisticians use the phrase *to predict statistically* to indicate that predictions based on such procedures are not expected to predict the outcome of a given event with absolute certainty. Interested students are urged to study the field of statistics to become familiar with the validity of predictions.

Exercises

Exer. 1–10: Find the equation of the least squares line of regression for the given data.

1. $(0, 2)$, $(1, 2)$, $(3, 5)$, $(2, 3)$
2. $(0, 5)$, $(2, 1)$, $(-1, 3)$
3. $(1, 2)$, $(3, 5)$, $(4, 3)$
4. $(-1, 6)$, $(0, 2)$, $(2, 3)$, $(1, 2)$, $(3, 5)$
5. $(0, 1)$, $(1, 1)$, $(2, 2)$, $(3, 1)$
6. $(-2, -3)$, $(-1, 1)$, $(1, 4)$, $(3, 7)$, $(5, 12)$
7. $(-50, 10)$, $(-40, 5)$, $(-30, 0)$, $(-20, 7)$, $(-10, 11)$
8. $(100, 10)$, $(200, 5)$, $(300, -1)$ $(400, -6)$, $(500, -13)$
9. $(2, 2)$, $(4, 5)$, $(6, 11)$, $(8, 23)$
10. $(2, 7)$, $(4, 5)$, $(4, 8)$, $(5, 10)$, $(5, 11)$

11. For four years an executive has recorded the total amount y of sales (in millions of dollars) of the company per year and the total amount x (in millions of dollars) spent on advertising per year:

advertising x	1.0	0.5	3.0	4.5
sales y	8.0	3.0	16.5	30.0

The executive asks a financial analyst to determine the line of regression and to predict the amount that should be spent on advertising to raise sales to \$37 million next year. Find that amount.

12. Four graduate students in economics obtain the following Graduate Record Exam (GRE) percentile scores and achievement scores:

Achievement x	170	250	160	200
GRE (%) y	85	98	82	90

Find the regression line. If this study is predictive, what achievement score would you expect for a student who scores 95 on the GRE?

13. Find the line of regression for the scores of a student group:

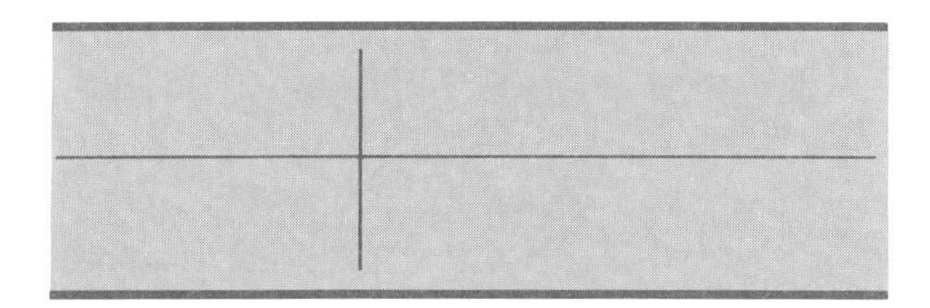

14. Find the regression line for the data points on this graph:

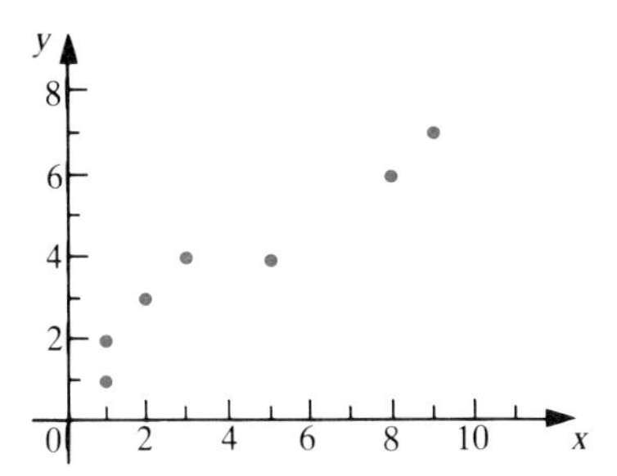

15. The scores of workers on a test administered at the time of initial employment are given in the table along with efficiency scores obtained later:

Worker	***Test score***	***Efficiency score***
A	3	9
B	15	45
C	13	51
D	10	33
E	7	27
F	12	51

(a) Find the line of regression.
(b) Based on these data, what efficiency score is predicted for a worker whose test score is 8?

Exer. 16–18: Suppose a sociologist studying the extent of community cooperation in eight cities determines the data shown in the table:

City	***Community cooperation index***	***Heterogeneity index***	***Mobility index***
A	16.4	20.4	15.6
B	16.7	31.2	19.2
C	12.3	19.7	20.4
D	19.0	28.3	30.6
E	9.8	40.1	28.7
F	13.0	25.5	22.2
G	11.1	28.7	32.6
H	12.0	36.0	19.2

The community cooperation index is the percentage of the population participating in community projects. Heterogeneity is measured in terms of the relative number of minorities in the population. The mobility index measures the relative number of persons moving in and out of the cities. (a) Plot the data relating the given indices and (b) determine and graph the line of regression for the data in part (a).

16. Community cooperation and heterogeneity
17. Mobility and heterogeneity
18. Community cooperation and mobility

19. A study of six youths in a foster-care center ranked the individuals in terms of participation in group discussions and popularity, as measured by friendship choices. Determine the line of regression for the data given in this table:

	Popularity rank	***Participation rank***
Bob	1	2
Tom	5	3
Sally	2.5	2
Betty	2.5	1
Mike	4	6
Paul	3	5

Exer. 20–23: Using the data given in this table, find the line of regression for the specified relationships:

Average age	*Male income ($)*	*Female income ($)*	*Male educational level (years)*	*Female educational level (years)*
20	13,000	8000	10.1	8.5
25	14,000	7500	11.2	8.5
30	16,300	8300	11.6	8.9
35	17,000	7100	12.1	9.1
40	16,400	6400	12.0	9.6
45	18,000	6100	11.8	9.0
50	17,600	5800	10.9	8.4

20. Average age and male educational level
21. Average age and female educational level
22. Male income and male educational level
23. Female income and female educational level

Exer. 24–30: Determine C and k so that the least squares fit of the form $y = Ce^{kt}$ is obtained for the given data.

24. $(1, 1)$, $(2, 2)$, $(3, 5)$, $(4, 12)$
25. $(0, 3)$, $(1, 2)$, $(3, 0.65)$, $(5, 0.25)$, $(7, 0.01)$
26. $(1, 405)$, $(2, 1650)$, $(3, 6700)$, $(4, 27000)$
27. $(0, 2)$, $(1, 1.21)$, $(2, 0.74)$, $(4, 0.45)$
28. $(0, -5)$, $(0.5, -2.25)$, $(1.0, -1)$, $(1.5, -0.45)$, $(2.0, -0.2)$
29. $(-1, 3.0)$, $(0, 6)$, $(2, 20.0)$, $(3, 36.0)$, $(5, 120.0)$
30. $(1, 30)$, $(2, 90)$, $(3, 280)$, $(4, 850)$

31. Based on theoretical considerations, a relationship of the form $y = ax^2 + b$ is expected between two quantities x and y. Using the data in this table, find the values of a and b that give the best approximation. (*Hint:* Let $t = x^2$.)

x_i	0	1	3	4	5
y_i	−5	−3	14	27	44

32. A relationship of the form $y = a \ln x + b$ is expected between two quantities x and y. Using the data in this table, find the values of a and b that give the best approximation:

x_i	1	2	3	4	5
y_i	5.0	7.1	8.3	9.1	9.8

33. A company has collected the data shown:

Productivity (in thousands)	*Labor Cost (in hundreds)*	*Capital Cost (in hundreds)*
3.0	100	120
3.3	125	100
3.7	200	80
4.0	100	100
3.2	50	175
2.1	75	140

Assuming productivity P, labor cost L, and capital cost K are related by $P = \alpha L^m K^{1-m}$ for constants α and m, find the values of α and m that best describe the company's performance record.

8.7 Double Integrals

In Chapter 6 we found that a geometric interpretation of the definite integral of a function of one variable is the area under a curve. In this section we discuss the double integral, the two-dimensional analogue of the definite integral. Its geometrical interpretation is the volume under a surface.

The definite integral $\int_a^b f(x)\,dx$ depends on the integrand $f(x)$ and on the interval $[a, b]$ containing the integration variable x. In the two-dimensional version $f(x, y)$ is the integrand and the integration variables x and y belong to some region R of the xy-plane. Suppose that we wish to find the volume of the solid shown in Figure 8.26. The base of the solid is some portion R of the xy-plane, the top of the solid is part of a surface $z = f(x, y)$, and the sides are parallel to the z-axis.

Consider Figure 8.27. The region R is given by two curves $y = g(x)$ and $y = h(x)$ and by two lines $x = a$ and $x = b$; that is $a \le x \le b$ and $g(x) \le y \le h(x)$. See Figure 8.28(a). At a position s_i along the x-axis between a and b, we "cut" a slice from the solid. The thickness of the slice is Δx_i. If this thickness is very small, then the volume of the slice will be approximately the area of the slice times its thickness. The area of the slice is the area under the curve $z = f(s_i, y)$ from $y = g(s_i)$ to $y = h(s_i)$. See Figure 8.28(b). If we denote this area by $A(s_i)$, then

$$A(s_i) = \int_{y=g(s_i)}^{y=h(s_i)} f(s_i, y)\, dy.$$

In the limits of integration we have added "$y =$" to emphasize that y is the variable of integration.

Figure 8.26 (LEFT)
A solid formed by a region R in the xy-plane and a surface $z = f(x, y)$

Figure 8.27 (RIGHT)
Slicing a solid parallel to the x-axis

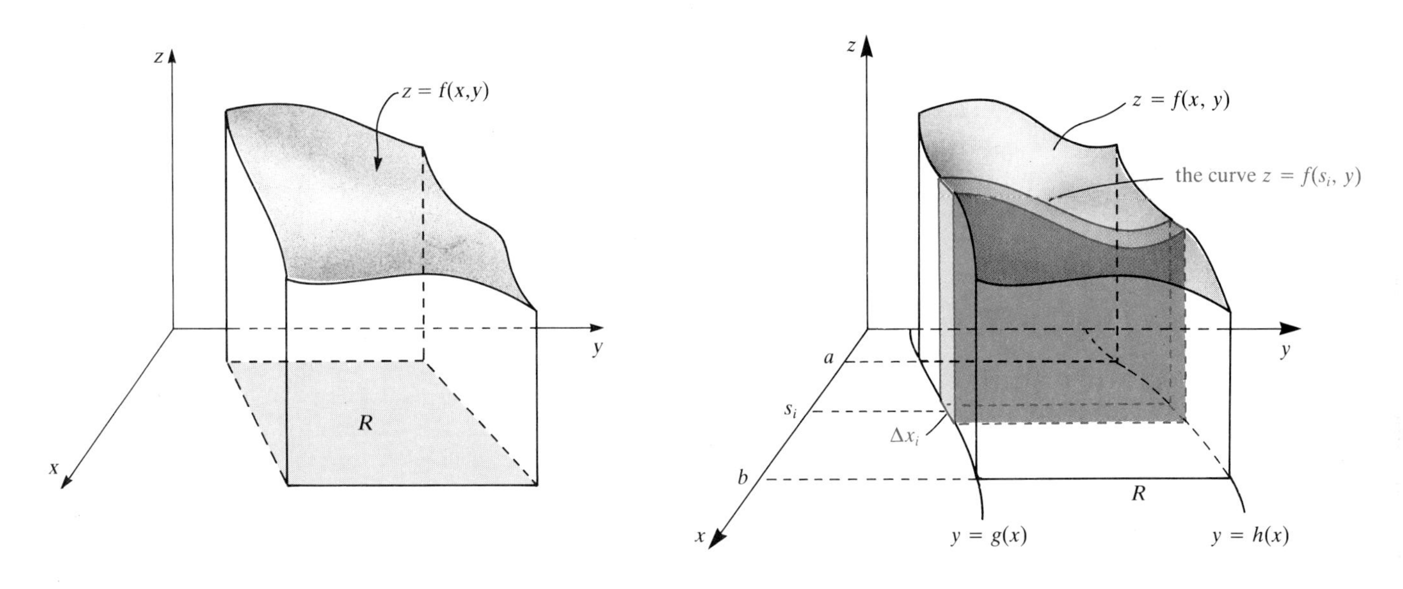

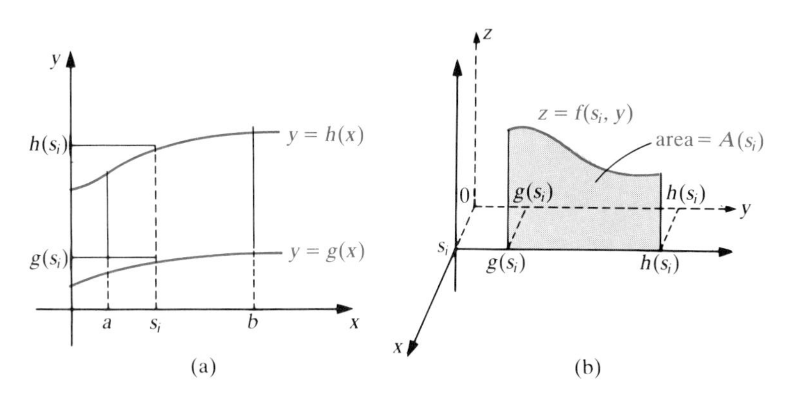

Figure 8.28
(a) The region R in the xy-plane; (b) the area of a typical slice

Letting ΔV_i denote the volume of the slice, we have

$$\Delta V_i \approx A(s_i)\, \Delta x_i.$$

The volume of the solid is approximately the sum of the volumes of these slices:

$$V \approx \sum_{i=1}^{n} A(s_i)\,\Delta x_i.$$

This is a Riemann sum approximation for the volume. We find the exact volume by integration:

$$\text{volume} = V = \int_{x=a}^{x=b} A(x)\,dx.$$

From the definition of A given above,

$$A(x) = \int_{y=g(x)}^{y=h(x)} f(x, y)\,dy.$$

Substituting this expression into the expression for volume, we obtain

$$V = \int_{x=a}^{x=b} \left(\int_{y=g(x)}^{y=h(x)} f(x, y)\,dy \right) dx.$$

This combination of two integrals is called an **iterated integral**, or a **double integral** and is evaluated by computing the innermost integral first and then the outer integral, as shown in the following example.

Example 8.38 Evaluate $\int_{x=0}^{x=2} \left(\int_{y=1}^{y=3} x^3y^2\,dy \right) dx$.

Solution We first evaluate the innermost integral. The dy associated with this integral indicates that y is the variable of integration, so x is regarded as a constant for this step only:

$$\int_{y=1}^{y=3} x^3y^2\,dy = x^3 \int_1^3 y^2\,dy$$

$$= x^3(\tfrac{1}{3}y^3)\Big|_1^3 = x^3(\tfrac{27}{3} - \tfrac{1}{3}) = \tfrac{26}{3}x^3.$$

We complete the evaluation of the double integral by integrating this result ($\tfrac{26}{3}x^3$) from 0 to 2, obtaining

$$\int_{x=0}^{x=2} \left(\int_{y=1}^{y=3} x^3y^2\,dy \right) dx = \int_0^2 \tfrac{26}{3}x^3\,dx$$

$$= \tfrac{26}{12}x^4\Big|_0^2 = \frac{26 \times 16}{12} = \frac{104}{3}.$$

Geometrically, this double integral gives the volume of the solid formed by the portion of the surface $z = x^3y^2$ lying above the region bounded by the lines $x = 0$, $x = 2$, $y = 1$, and $y = 3$ in the xy-plane. ■

The roles of x and y in the calculation of volume by an iterated integral may be interchanged; that is, we may slice the solid parallel to the xz-plane rather than the yz-plane. In this case, the region R in the xy-plane is described as $c \le y \le d$, $p(y) \le x \le q(y)$. See Figure 8.29. We then have

$$\text{volume} = \int_{y=c}^{y=d} \left(\int_{x=p(y)}^{x=q(y)} f(x, y)\, dx \right) dy.$$

To simplify this notation we write

$$\int_a^b \int_{g(x)}^{h(x)} f(x, y)\, dy\, dx \quad \text{for} \quad \int_{x=a}^{x=b} \left(\int_{y=g(x)}^{y=h(x)} f(x, y)\, dy \right) dx.$$

The order of the dy and dx indicate which variable is associated with the innermost integral: when dy appears first, from left to right, y is the variable for the innermost integral and, consequently, the limits on the innermost integral are for the variable y.

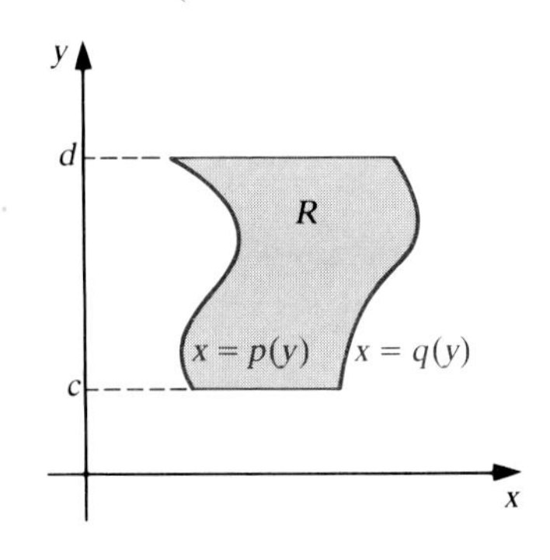

Figure 8.29
A region described by $c \le y \le d$, $p(y) \le x \le q(y)$

Example 8.39 Evaluate (a) $\int_0^1 \int_{x^2}^{x} 2x^3 y\, dy\, dx$ (b) $\int_0^1 \int_{y^2}^{y} 2x^3 y\, dx\, dy$

Solution (a) The innermost integral has y as the variable of integration; x is constant for this integration. Thus

$$\int_0^1 \int_{x^2}^{x} 2x^3 y\, dy\, dx = \int_0^1 x^3 y^2 \Big|_{y=x^2}^{y=x} dx$$

$$= \int_0^1 [x^3(x)^2 - x^3(x^2)^2]\, dx$$

$$= \int_0^1 (x^5 - x^7)\, dx$$

$$= (\tfrac{1}{6}x^6 - \tfrac{1}{8}x^8)\Big|_0^1 = \frac{1^6}{6} - \frac{1^8}{8} = \frac{1}{24}.$$

(b) The innermost integral has x as the variable, so y is constant during the first integration. Thus

$$\int_0^1 \int_{y^2}^{y} 2x^3 y\, dx\, dy = \int_0^1 \frac{x^4 y}{2} \Big|_{x=y^2}^{x=y} dy$$

$$= \int_0^1 \left[\frac{(y)^4 y}{2} - \frac{(y^2)^4 y}{2} \right] dy$$

$$= \frac{1}{2} \int_0^1 (y^5 - y^9)\, dy$$

$$= \tfrac{1}{2} (\tfrac{1}{6}y^6 - \tfrac{1}{10}y^{10}) \Big|_0^1 = \tfrac{1}{2}(\tfrac{1}{6} - \tfrac{1}{10}) = \tfrac{1}{30}. \quad ■$$

An iterated integral may be denoted by

$$\iint_R f(x, y)\, dA$$

where R denotes the region of the xy-plane over which the integration is performed.

The order of the integration (first y, then x, or vice versa) is not specified by this notation. Either order may be chosen to evaluate the integral. Thus dA represents either $dx\,dy$ or $dy\,dx$.

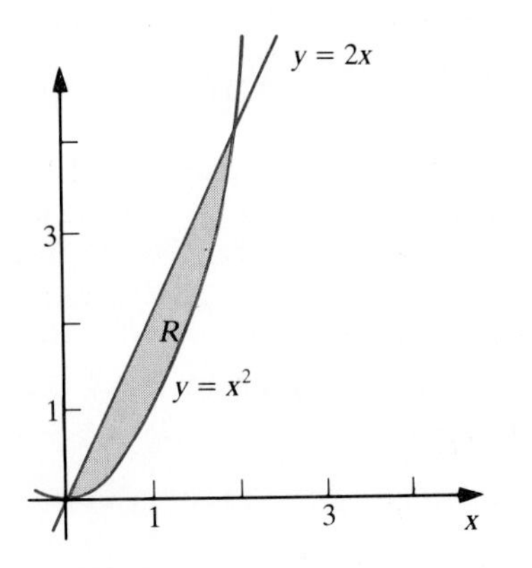

Figure 8.30
The base of the solid in Example 8.40

Example 8.40 Find the volume of the solid bounded by the region R in the xy-plane, shown in Figure 8.30, and by the surface $z = 4xy$.

Solution The region R can be described by $x^2 \le y \le 2x$ when $0 \le x \le 2$. Since the limits on x are constants, we integrate with respect to x last. Thus

$$\text{volume} = \iint_R 4xy\,dA = \int_0^2 \int_{x^2}^{2x} 4xy\,dy\,dx.$$

$$= \int_0^2 2xy^2 \Big|_{y=x^2}^{y=2x} dx = \int_0^2 [2x(2x)^2 - 2x(x^2)^2]\,dx$$

$$= \int_0^2 (8x^3 - 2x^5)\,dx = (2x^4 - \tfrac{1}{3}x^6)\Big|_0^2 = 32 - \tfrac{64}{3} = \tfrac{32}{3}.$$

Thus the volume of the solid is $\frac{32}{3}$ cubic units. ■

Exercises

Exer. 1–18: Evaluate the iterated integral.

1. $\int_0^2 \int_2^5 xy\,dy\,dx$
2. $\int_1^2 \int_2^3 \frac{2x}{y}\,dx\,dy$
3. $\int_1^2 \int_0^1 e^x e^y\,dx\,dy$
4. $\int_0^1 \int_y^{e^y} xy\,dx\,dy$
5. $\int_{-1}^1 \int_x^{x^3} (x+y)\,dy\,dx$
6. $\int_{-1}^2 \int_0^4 xe^{xy}\,dy\,dx$
7. $\int_0^1 \int_{x^2}^{3x^2} x(y+1)^5\,dy\,dx$
8. $\int_0^1 \int_{y^3}^{y^5} \frac{x}{y}\,dx\,dy$
9. $\int_0^1 \int_x^{\sqrt{x}} xye^{-y^2}\,dy\,dx$
10. $\int_0^1 \int_1^{y^2} xy^3(x^2-1)\,dx\,dy$
11. $\int_0^2 \int_0^x x^2y^2(y^3+1)^3\,dy\,dx$
12. $\int_0^1 \int_0^{y/2} e^{2x-y}\,dx\,dy$
13. $\int_0^4 \int_1^4 (y^2 \ln x + ye^{xy})\,dx\,dy$
14. $\int_0^4 \int_0^4 (\tfrac{1}{2}x + \tfrac{1}{2}y)\,dy\,dx$
15. $\int_0^4 \int_0^x y(x^2+y^2)\,dy\,dx$
16. $\int_0^4 \int_1^{\sqrt{y}} x\sqrt{y+x^2}\,dx\,dy$
17. $\int_{-1}^1 \int_{-y}^y (x^2+y^4)\,dx\,dy$
18. $\int_0^4 \int_x^4 (x+\sqrt{y})\,dy\,dx$

Exer. 19–28: Find the volume of the solid above the region R and below the surface $z = f(x, y)$.

19. R: $0 \le x \le 2$, $1 \le y \le 3$; $f(x, y) = x^2 + y^2$
20. R: $-2 \le y \le 2$, $1 \le x \le 4$; $f(x, y) = y/x$
21. R: $y \le x \le 2y$, $1 \le y \le 2$; $f(x, y) = x^2y$
22. R: $0 \le x \le \ln 2$, $0 \le y \le x$; $f(x, y) = xe^y$
23. R: $1 \le x \le 2$, $0 \le y \le 2x$; $f(x, y) = e^{y-x}$
24. R: $0 \le y \le 1$, $0 \le x \le e^y$; $f(x, y) = x$
25. R: $0 \le y \le 1$, $-1 \le x \le y$; $f(x, y) = xy - y^3$
26. R: area bounded by $x + y = 4$ and $\sqrt{x} + \sqrt{y} = 2$; $f(x, y) = 1$
27. R: $0 \le y \le 1$, $y^2 \le x \le y$; $f(x, y) = \sqrt{xy}$
28. R: area bounded by $y^2 = 2x$ and $y^2 = 8 - 2x$; $f(x, y) = 4 - y^2$
29. Find the volume of the solid whose base is the region between the curves $y = x^2$ and $y = x^3$ and whose height is $z = f(x, y) = x^2 + y^2$.
30. Find the volume of the region whose base is the region between the curves $y = x$, $y = 3x$, and $x = 2$ and whose height is $z = xe^y$.

Summary

The **partial derivative of $z = f(x, y)$ with respect to x** is the rate of change of $z = f(x, y)$ when y is held constant. The following are equivalent notations for the partial derivative with respect to x:

$$\frac{\partial f}{\partial x}, \quad \frac{\partial z}{\partial x}, \quad f_x(x, y).$$

The **partial derivative of $z = f(x, y)$ with respect to y**, the rate of change of $z = f(x, y)$ when x is held fixed, is denoted by

$$\frac{\partial f}{\partial y}, \quad \frac{\partial z}{\partial y}, \quad f_y(x, y).$$

The **slope in the x-direction** of the graph of $z = f(x, y)$ is $\partial z/\partial x$ and the **slope in the y-direction** is $\partial z/\partial y$. The **second partial derivatives** are denoted by

$$\frac{\partial}{\partial x}\left(\frac{\partial f}{\partial x}\right) = \frac{\partial^2 f}{\partial x^2} = f_{xx},$$

$$\frac{\partial}{\partial y}\left(\frac{\partial f}{\partial x}\right) = \frac{\partial^2 f}{\partial y\, \partial x} = f_{yx}, \qquad \frac{\partial}{\partial y}\left(\frac{\partial f}{\partial y}\right) = \frac{\partial}{\partial x}\left(\frac{\partial f}{\partial y}\right) = f_{yx}$$

For the function $z = f(x, y)$, if $\partial z/\partial x$ and $\partial z/\partial y$ both exist and if $z = f(x, y)$ has a relative maximum or relative minimum at (a, b), then

$$\left.\frac{\partial z}{\partial x}\right|_{(a,b)} = 0 \quad \text{and} \quad \left.\frac{\partial z}{\partial y}\right|_{(a,b)} = 0.$$

If $A = f_{xx}(a, b)$, $B = f_{xy}(a, b)$, and $C = f_{yy}(a, b)$, then the following rules can be used to determine if the surface has a relative maximum or relative minimum at (a, b):

1. If $AC - B^2 > 0$ and $A < 0$, then $f(a, b)$ is a relative maximum.
2. If $AC - B^2 > 0$ and $A > 0$, then $f(a, b)$ is a relative minimum.
3. If $AC - B^2 < 0$, then $(a, b, f(a, b))$ is a saddle point.
4. If $AC - B^2 = 0$, further work is needed before conclusions can be made.

Algebraic substitution or Lagrange multipliers may be used to solve an **optimization problem with a constraint**. The method of Lagrange multipliers is based on the fact that the values of x, y, and z that yield the relative extrema of the function $f(x, y, z)$, subject to the constraint $g(x, y, z) = 0$, are among the values of x, y, and z that satisfy the equations

$$f_x(x, y, z) + \lambda g_x(x, y, z) = 0$$
$$f_y(x, y, z) + \lambda g_y(x, y, z) = 0$$
$$f_z(x, y, z) + \lambda g_z(x, y, z) = 0$$
$$g(x, y) = 0,$$

provided that all indicated partial derivatives exist.

The quantity

$$df = [f_x(a, b)]h + [f_y(a, b)]k$$

is the **total differential** of $f(x, y)$. We have the approximation

$$f(a + h, b + k) - f(a, b) \approx df = f_x(a, b)h + f_y(a, b)k.$$

Given the data (x_1, y_1), (x_2, y_2), . . . , (x_n, y_n), the **method of least squares** determines the best approximating linear function $y = mx + b$ for the data by solving the **normal equations**

$$m \sum_{i=1}^{n} x_i + b \sum_{i=1}^{n} 1 = \sum_{i=1}^{n} y_i \quad \text{and} \quad m \sum_{i=1}^{n} x_i^2 + b \sum_{i=1}^{n} x_i = \sum_{i=1}^{n} x_i y_i.$$

The volume of a solid whose base R is given by $a \le x \le b$, $g(x) \le y \le h(x)$ and whose height at (x, y) is $z = f(x, y)$ is given by the **iterated integral**

$$\text{volume} = \int_a^b \int_{g(x)}^{h(x)} f(x, y)\, dy\, dx.$$

Review Exercises

Exer. 1–2: Evaluate the function at the given points.

1. $f(x, y) = x^2 - y^2 - 2xy$; $(4, 2)$, $(2, -1)$
2. $f(x, y) = \dfrac{x - y}{x^2}$; $(2, 2)$, $(-4, -1)$
3. Plot the points $(1, 2, 3)$, $(-4, 1, 2)$, and $(1, -3, -2)$ on a three-dimensional coordinate system.

Exer. 4–7: Describe and sketch the graph of the surface on a three-dimensional coordinate system.

4. $x = 3$
5. $z = -2$
6. $x - 2y + 4z = 8$
7. $z = x^2$

Exer. 8–13: Find $\frac{\partial z}{\partial x}$, $\frac{\partial z}{\partial y}$, and $\left.\frac{\partial z}{\partial x}\right|_{(2,-2)}$

8. $z = x^2y^2 - 3x - 2y$
9. $z = (x - y)^{3/2}$
10. $z = e^{x^2+y^2}$
11. $z = \ln \frac{x^2 + y}{x + y + 1}$
12. $z = -\frac{x + y}{x - y}$
13. $z = x^y$
14. Find the slope of $z = x/y$ in the x-direction and in the y-direction at the point where $x = 4$, $y = 2$.

Exer. 15–20: Find $f_{xx}(x, y)$, $f_{xy}(x, y)$ and $f_{xy}(-1, 2)$.

15. $f(x, y) = x^3y + x^5y^2$
16. $f(x, y) = \frac{3x}{2y + 1}$
17. $f(x, y) = e^{2x-3y}$
18. $f(x, y) = \ln(x/y)$
19. $f(x, y) = \frac{y}{x}$
20. $f(x, y) = \frac{x - y}{x + y}$

Exer. 21–26: Find all relative maxima, relative minima, and saddle points.

21. $z = 1 - x^2 - y^2$
22. $z = 2x^2 + y^2 - 2xy - 4x + 3$
23. $z = \frac{xy}{4} - \frac{1}{2x} - \frac{1}{y} + 7$
24. $z = 3xy + 4x^2 - 2y^2 - 5x - 7y + 3$
25. $z = x^3 + y^2 - 3x + 6y$
26. $z = x^3 + y^2 - 3x - 8y + 5$
27. A company produces x units of item A and y units of item B, and its profit varies according to the formula
$$p(x, y) = -2x^2 - 4y^2 + 2xy - 2x + 36y.$$
How many units of each item should be produced to maximize the profit?
28. A rectangular dumpster, open at the top, is to have a capacity of 4000 ft³. Find the dimensions that will minimize the amount of material needed for the dumpster.
29. If a company spends x thousand dollars on newspaper advertising and y thousand dollars on television advertising, its income (in thousands of dollars) due to the two types of advertising is given by the function $f(x, y) = 2xy - x^2 - 2y^2 + 3x + 4$. How much should be spent on each type of advertising to maximize income?
30. A company finds that the number of hours of labor needed to produce x items of type A and y items of type B is given by $F(x, y) = 4x^2 - xy + 300y + 800$. If 5000 items must be produced, how many of each type should be produced to minimize the number of hours of labor used? (Use algebraic substitution.)
31. The cost of shipping x items from warehouse A is given by $9x^2 + 600$ and the cost of shipping y items from warehouse B is given by $16y^2 + 400$. What is the largest number of items that can be shipped for $15,400? (Use Lagrange multipliers.)
32. A farmer plans to use the edge of a river as one boundary of the rectangular holding pen for cows and sheep shown in the figure. Fencing for the outside of the area will cost $4 per foot. To separate the cows and sheep, a fence will divide the area into two rectangles. This interior fence will cost $2.50 per foot. Find the dimensions x and y of the rectangle that will give an area of 1200 ft² for the least cost of fencing.

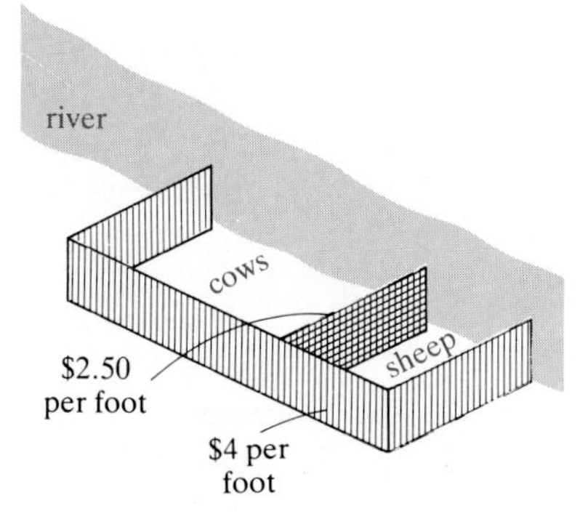

33. Use the total differential to estimate the value of $\sqrt{38}\,\sqrt[3]{62}$.
34. Find the line of regression of y versus x for the points $(-1, 1)$, $(0, 2)$, $(2, 2)$, $(4, 3)$, $(4, 5)$.
35. Find the least squares line for the data $(1, 3)$, $(2, 6)$, $(4, 13)$, $(6, 24)$, $(7, 40)$.
36. Height and weight for six males on their nineteenth birthday are as follows:

Height x (in inches)	70	63	72	60	66	70
Weight y (in pounds)	155	150	180	135	156	168

Find the line of regression of weight as a function of height. If a male is 68 inches tall on his nineteenth birthday, make a guess at his weight using the line of regression.

Exer. 37–40: Evaluate the integral.

37. $\int_0^1 \int_2^4 (x^2y + xy)\, dy\, dx$
38. $\int_{-1}^1 \int_0^{y^2} e^{x/y}\, dx\, dy$
39. $\int_0^2 \int_{x^2}^x xy\, dy\, dx$
40. $\int_1^e \int_x^{2x} \ln xy\, dy\, dx$
41. Let R be the region of the xy-plane bounded by the curves $y = \sqrt{x}$ and $y = x$. Find the volume of the solid above the region R and below the surface $z = x^2y^3$.

Review of Algebra

Appendix A

A.1 Exponents and Fractions

Success in simplifying algebraic expressions often depends upon adeptness at using exponents. In the definition of exponents and their properties that we examine in this section, we assume that the expressions given obey two fundamental properties of real numbers: no division by zero and no taking even roots of negative numbers (for instance, the square root of -4 is not defined).

Laws of Exponents

For all positive real numbers a and positive integers m, n, and k,

$$a^m = a \cdot a \cdot \cdots \cdot a \quad (m \text{ factors of } a)$$

$$a^{1/n} = n\text{th root of } a \quad (\text{also denoted by } \sqrt[n]{a})$$

$$(a^m)^{1/n} = \sqrt[n]{a^m} = (\sqrt[n]{a})^m$$

$$a^{-k} = \left(\frac{1}{a}\right)^k$$

$$a^0 = 1$$

$$(-1)^n = \begin{cases} -1 & \text{if } n \text{ is even} \\ 1 & \text{if } n \text{ is odd} \end{cases}$$

$$(-a)^n = (-1)^n a^n$$

$(-a)^{1/n} = -(a)^{1/n}$ for n an odd integer. It is not a real number for n an even integer.

The following examples demonstrate how to apply the laws.

Example A.1

$$2^3 = 2 \cdot 2 \cdot 2 = 8$$

$$(-3)^4 = (-1)^4 3^4 = 81$$

$$(4)^{1/2} = \sqrt{4} = 2$$

$$(27)^{1/3} = \sqrt[3]{27} = 3$$

$$(-8)^{2/3} = ((-8)^{1/3})^2 = (\sqrt[3]{-8})^2 = (-2)^2 = 4$$

$$(16)^{3/4} = (16^{1/4})^3 = (\sqrt[4]{16})^3 = 2^3 = 8$$

$$2^{-2} = \frac{1}{2^2} = \frac{1}{4}$$

$$(-3)^{-5} = (-1)^{-5} 3^{-5} = (-1)\frac{1}{3^5} = \frac{-1}{243}$$

$$(32)^{-2/5} = \frac{1}{(32)^{2/5}} = \frac{1}{(\sqrt[5]{32})^2} = \frac{1}{2^2} = \frac{1}{4}$$

$$(14)^0 = 1$$

$$(-2)^0 = 1$$

■

Some useful properties of exponents are listed below.

Properties of Exponents

For all positive real numbers a and rational numbers r and s,

$$a^r \times a^s = a^{r+s} \qquad (a^r)^s = a^{rs}$$

$$\frac{a^r}{a^s} = a^{r-s} \qquad a^r \times b^r = (ab)^r$$

The next example shows how these properties are used.

Example A.2

$$2^2 \times 2^3 = 2^{2+3} = 2^5$$

$$3^4 \times 3^{-2} = 3^{4-2} = 3^2$$

$$(8^2)^4 = 8^{2\times 4} = 8^8$$

$$((\tfrac{1}{2})^3)^{-2} = (\tfrac{1}{2})^{3\times(-2)} = (\tfrac{1}{2})^{-6} = 2^6 = 64$$

$$\frac{3^7}{3^2} = 3^{7-2} = 3^5$$

$$2^3 \times 3^3 = (2 \times 3)^3 = 6^3$$

■

Rules for Computing with Fractions

For all integers a, b, c, and d,

$$\frac{a}{b} \times \frac{c}{d} = \frac{ac}{bd}; \quad (b, d \neq 0)$$

$$\frac{ac}{bc} = \frac{a}{b}; \quad (b, c \neq 0)$$

$$\frac{a}{b} + \frac{c}{d} = \frac{ad}{bd} + \frac{bc}{bd} = \frac{ad + bc}{bd}; \quad (b, d \neq 0).$$

These rules are familiar when numbers are involved. They may also be applied when a, b, c, and d represent variables, as illustrated in the next example.

Example A.3

(a) $\dfrac{2}{x-2} \times \dfrac{3x}{4} = \dfrac{2(3x)}{(x-2)4} = \dfrac{6x}{4(x-2)}$

(b) $\dfrac{2}{x-2} + \dfrac{3x}{4} = \dfrac{2(4)}{(x-2)4} + \dfrac{3x(x-2)}{4(x-2)}$

$$= \frac{8+3x(x-2)}{4(x-2)} = \frac{8+3x^2-6x}{4(x-2)}$$

(c) $\dfrac{3x}{x-2} + 6x = \dfrac{3x}{x-2} + \dfrac{6x(x-2)}{x-2}$

$$= \frac{3x+6x(x-2)}{x-2}$$

$$= \frac{3x+6x^2-12x}{x-2} = \frac{6x^2-9x}{x-2}$$

(d) $\dfrac{1}{x} + \dfrac{1}{x+2} = \dfrac{(x+2)}{x(x+2)} + \dfrac{x}{x(x+2)}$

$$= \frac{x+2+x}{x(x+2)} = \frac{2x+2}{x(x+2)}$$

■

Exercises

Exer. 1–26: Simplify, using properties of exponents.

1. 5^3
2. $(-4)^2$
3. $16^{3/2}$
4. $4^{-1/2}$
5. $(\frac{1}{8})^{1/3}$
6. $16^{-3/4}$
7. $(\frac{1}{2})^0$
8. $(\frac{1}{8})^{2/3}$
9. $(\frac{1}{6})^2$
10. $25^{-5/2}$
11. $(\frac{1}{4})^{-3}$
12. $32^{3/5}$
13. $4^3 \times 4^2$
14. $(-2)^2 \times (-2)^4$
15. $6^2 \times 6^{-4}$
16. $(\frac{1}{2})^2 \times (\frac{1}{2})^4$
17. $\dfrac{3^5}{3^4}$
18. $(a^{1/2}b^2)^3$
19. $(b^2a^2)^3$
20. $\dfrac{3^5}{3^7}$
21. $4^3 \times 5^3$
22. $(\frac{1}{3})^4 \times (\frac{1}{3})^2$
23. $(2^4)^3$
24. $(6^{-2})^3$
25. $(a^2b^{1/2})^3(a^{-1/2}b^3)^{-2}$
26. $(a^2)^0(b^{-2}a)^{1/3}$

Exer. 27–34: Perform the indicated operation.

27. $\dfrac{x}{x-3} \times \dfrac{2}{1-x}$
28. $\dfrac{x-2}{x} \times \dfrac{3x}{5}$
29. $\dfrac{2x-4}{x+3} + \dfrac{6}{x}$
30. $\dfrac{x-2}{x} + \dfrac{x}{x-2}$
31. $\dfrac{x}{x+2} + (x-1)$
32. $\dfrac{x+h}{h} + 3x$
33. $\dfrac{x+h}{x^2} \times \dfrac{3x}{x+h}$
34. $\dfrac{1}{x} + x^2$

A.2 Products and Factoring

Some rules used for finding products and decomposing products into factors are listed as follows.

Rules for Products

For all real numbers a, b, and x,

$$a(x+b) = ax + ab$$
$$(x+a)^2 = x^2 + 2ax + a^2$$
$$(x+a)(x-a) = x^2 - a^2$$
$$(x+a)(x+b) = x^2 + (a+b)x + ab$$
$$(x+a)^3 = x^3 + 3ax^2 + 3a^2x + a^3$$
$$(x-a)^3 = x^3 - 3ax^2 + 3a^2x - a^3$$

These rules are demonstrated in the next example.

Example A.4

$$3(x+2) = 3x + (3)(2) = 3x + 6$$
$$(t+3)^2 = t^2 + 2(3)t + (3)^2 = t^2 + 6t + 9$$
$$(p+4)(p-4) = p^2 - 4^2 = p^2 - 16$$
$$(u+1)(u+5) = u^2 + (1+5)u + (1)(5) = u^2 + 6u + 5$$
$$(x+5)^3 = x^3 + 3(5)x^2 + 3(5)^2x + (5)^3 = x^3 + 15x^2 + 75x + 125$$
$$(y-2)^3 = y^3 - 3(2)y^2 + 3(2)^2y - (2)^3 = y^3 - 6y^2 + 12y - 8$$ ■

A root of a quadratic polynomial $ax^2 + bx + c$ is any number that satisfies the equation $ax^2 + bx + x = 0$. Sometimes the roots can be found by factoring. When that fails, the quadratic formula can be used to find the roots (if they exist).

Quadratic Formula

The solutions to the equation $ax^2 + bx + c = 0$ with $a \neq 0$ (that is, the roots of $ax^2 + bx + c$) are

$$x = \frac{-b \pm \sqrt{b^2 - 4ac}}{2a}.$$

One solution uses the plus sign before the square root and the other uses the minus sign. If $b^2 - 4ac$ is negative, the square root cannot be taken; in this case, there are no real solutions and, hence, no real roots for the polynomial. Examples A.5–A.7 illustrate applications of the quadratic formula.

Example A.5 Find the solutions of the equation $2x^2 - 5x + 2 = 0$.

Solution Here $a = 2$, $b = -5$, and $c = 2$. So the quadratic formula yields

$$x = \frac{-b \pm \sqrt{b^2 - 4ac}}{2a} = \frac{-(-5) \pm \sqrt{(-5)^2 - 4(2)(2)}}{2(2)}$$
$$= \frac{5 \pm \sqrt{25 - 16}}{4} = \frac{5 \pm 3}{4}.$$

Thus the roots are

$$x = \frac{5+3}{4} = 2 \quad \text{and} \quad x = \frac{5-3}{4} = \frac{1}{2}.$$

It is also possible to solve this quadratic equation by factoring,

$$0 = 2x^2 - 5x + 2 = (2x-1)(x-2),$$

which yields $x = \frac{1}{2}$ and $x = 2$. ■

Example A.6 Find the solutions to $2x^2 + 3x - 3 = 0$.

Solution We have $a = 2$, $b = 3$, and $c = -3$. Thus

$$x = \frac{-3 \pm \sqrt{(3)^2 - 4(2)(-3)}}{2(2)} = \frac{-3 \pm \sqrt{33}}{4}.$$

Thus the solutions are $x = \dfrac{-3 + \sqrt{33}}{4}$ and $x = \dfrac{-3 - \sqrt{33}}{4}$. ■

Example A.7 Find the solutions to $3x^2 + 2x + 4 = 0$.

Solution Here $a = 3$, $b = 2$, $c = 4$. Since $b^2 - 4ac = (2)^2 - 4(3)(4) = 4 - 48 = -44 < 0$, we cannot take the square root in the quadratic formula. We conclude that this equation has no real roots. ■

Sometimes a quadratic polynomial can be factored by inspection by using the product rule

$$x^2 + (p+q)x + pq = (x+p)(x+q).$$

Example A.8 Factor each quadratic polynomial by inspection:

(a) $x^2 + 6x + 8$ (b) $x^2 - 2x - 15$ (c) $2x^2 + 10x + 8$

Solution **(a)** We want two numbers p and q whose product pq is 8 (the constant term) and whose sum $p + q$ is 6 (the coefficient of x). The choice $p = 2$ and $q = 4$, or vice versa, satisfies these conditions. Thus

$$x^2 + 6x + 8 = (x+2)(x+4).$$

(b) The numbers 3 and -5 have sum -2 and product -15 as needed. Thus

$$x^2 - 2x - 15 = (x+3)(x-5).$$

(c) We factor out a 2 and then factor by inspection:

$$2x^2 + 10x + 8 = 2(x^2 + 5x + 4) = 2(x+1)(x+4).$$ ■

If a polynomial is factored as $(x+p)(x+q)$, then $x=-p$ and $x=-q$ are roots of the polynomial, since they satisfy the equation $(x+p)(x+q)=0$. Therefore, if it is difficult to factor a polynomial by inspection, we can factor it by first determining the roots of the quadratic polynomial from the quadratic formula. If we find the roots to be $x=r_1$ and $x=r_2$, then

$$ax^2+bx+c=a(x-r_1)(x-r_2).$$

Note that a has been factored out. In the next example, $a=1$, so we do not have to factor it out. Example A.10 shows that this step is needed when $a \neq 1$.

Example A.9 Factor $x^2+8x+12$ by finding the roots from the quadratic formula.

Solution By the quadratic formula

$$x=\frac{-8\pm\sqrt{64-4(12)}}{2}=\frac{-8\pm 4}{2}$$

$$=-6 \quad \text{or} \quad -2.$$

Thus $x^2+8x+12=[x-(-6)][x-(-2)]=(x+6)(x+2)$, ■

Example A.10 Factor $2x^2-6x+1$ by finding the roots from the quadratic formula.

Solution By the quadratic formula

$$x=\frac{6\pm\sqrt{36-4(2)}}{2(2)}=\frac{6\pm 2\sqrt{7}}{4}$$

$$=\frac{3+\sqrt{7}}{2} \quad \text{or} \quad \frac{3-\sqrt{7}}{2}.$$

Let $r_1=\dfrac{3+\sqrt{7}}{2}$ and $r_2=\dfrac{3-\sqrt{7}}{2}$. Then

$$(x-r_1)(x-r_2)=x^2-(r_1+r_2)x+r_1r_2$$

$$=x^2-\left(\frac{3+\sqrt{7}}{2}+\frac{3-\sqrt{7}}{2}\right)x+\left(\frac{3+\sqrt{7}}{2}\right)\left(\frac{3-\sqrt{7}}{2}\right)$$

$$=x^2-\frac{6}{2}x+\frac{9-7}{4}$$

$$=x^2-3x+\tfrac{1}{2}$$

or $\quad (x-r_1)(x-r_2)=\tfrac{1}{2}(2x^2-6x+1)$.

Multiplying both sides by 2, writing the right side on the left, and using the values for r_1 and r_2 gives

$$2x^2-6x+1=2\left(x-\frac{3+\sqrt{7}}{2}\right)\left(x-\frac{3-\sqrt{7}}{2}\right).$$

Note that the leading coefficient has been factored out. This is always necessary when $a \neq 1$. ■

Example A.11 Factor $3x^2 + 6x + 5$ by using the quadratic formula.

Solution Using the quadratic formula to find the roots, we get

$$x = \frac{-6 \pm \sqrt{36 - 4(3)(5)}}{2(3)} = \frac{-6 \pm \sqrt{-24}}{6}$$

and so there are no real roots. Hence $3x^2 + 6x + 5$ cannot be factored. ■

Exercises

Exer. 1–14: Find the product.

1. $2(x - 1)$
2. $-3(x + 2)$
3. $(x + 1)^2$
4. $(x + \frac{1}{2})^2$
5. $(x - 1)(x + 1)$
6. $(x - 1)(x + 3)$
7. $(x + 4)(x - 2)$
8. $(x + 4)^3$
9. $(x + 4)(x - 1)$
10. $(x - 3)(x + 4)$
11. $(x - 3)(x + 3)$
12. $(x - 2)^3$
13. $(x - \frac{1}{2})^2$
14. $(x + 2)^3$

Exer. 15–24: Find the solutions to the equation.

15. $x^2 - 3x + 2 = 0$
16. $9x^2 + 12x + 4 = 0$
17. $x^2 - 7x + 12 = 0$
18. $x^2 - 5 = 0$
19. $2x^2 - 9x + 4 = 0$
20. $x^2 - x - 12 = 0$
21. $3x^2 + 2x + 1 = 0$
22. $x^2 - 6x = 0$
23. $6x^2 + 13x - 5 = 0$
24. $x^2 + x + 1 = 0$

Exer. 25–32: Factor the quadratic polynomial by inspection.

25. $x^2 + 10x + 16 = 0$
26. $x^2 - 6x + 8 = 0$
27. $x^2 + 4x - 32 = 0$
28. $2x^2 - 14x + 12 = 0$
29. $3x^2 - 15x + 18 = 0$
30. $-3x^2 + 18x - 27 = 0$
31. $-2x^2 - 8x - 6 = 0$
32. $2x^2 - \frac{1}{2} = 0$

Exer. 33–38: Factor the quadratic polynomial by finding the roots.

33. $x^2 - 5x - 50$
34. $x^2 - 12x - 28$
35. $3x^2 + 5x + 2$
36. $2x^2 - 6x + 4$
37. $4x^2 + 10x - 6$
38. $6x^2 - x - 1$

A.3 Inequalities and Intervals

Inequalities show relationships between quantities and are very important in calculus.

Definition of Inequality Symbols

For all real numbers a and b,

$a < b$ means a is less than b

$a \leq b$ means a is less than or equal to b

$a > b$ means a is greater than b

$a \geq b$ means a is greater than or equal to b.

Example A.12

$3 < 5$ is true	$-2 < -4$ is false
$3 \leq 3$ is true	$-6 < -2$ is true
$3 < 3$ is false	$6 \geq 4$ is true
$-2 < 4$ is true	$6 \geq 6$ is true

■

Properties of Inequalities

1. If $a < b$, then $a + c < b + c$ for all c.
2. If $a < b$ and $c > 0$, then $ac < bc$.
3. If $a < b$ and $c < 0$, then $ac > bc$.

Properties 2 and 3 say that when both sides of an inequality are multiplied (or divided) by a positive number, the direction of the inequality is preserved; when multiplied (or divided) by a negative number, the direction of the inequality is reversed.

Example A.13 The following are true for any real numbers a and b:

If $a < b$ then $a + 4 < b + 4$.

If $a < b$ then $a - 6 < b - 6$.

If $a < b$ then $\frac{a}{2} < \frac{b}{2}$.

If $a < b$ then $-2a > -2b$.

■

The **absolute value** of a real number a, denoted by $|a|$, is defined by

$$|a| = \begin{cases} a & \text{for } a \geq 0 \\ -a & \text{for } a < 0 \end{cases}$$

Example A.14 Find $|3|$, $|-4|$, and $|0|$.

Solution Since $3 > 0$, we use the rule $|a| = a$ to get $|3| = 3$. Since $-4 < 0$, we use the rule $|a| = -a$ to get $|-4| = -(-4) = 4$. Similarly, $|0| = 0$. ■

One interpretation of the absolute value is that the absolute value of a number represents its distance from zero (in either direction) on the x-axis. See Figure A.1.

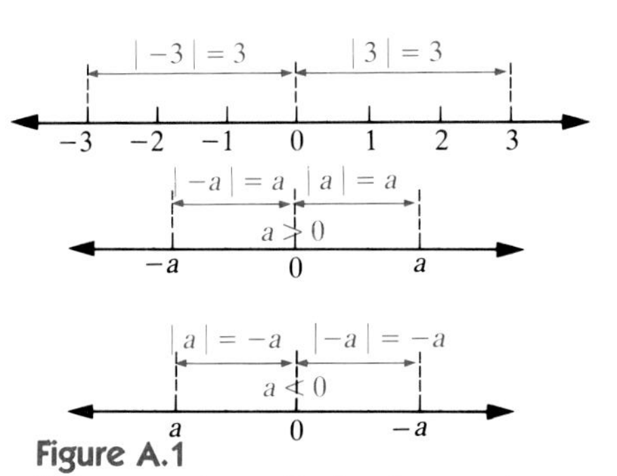

Figure A.1
Interpretation of absolute value as distance from the origin

Let b be a fixed positive number. If x is such that $-b < x < b$, then the distance between x and zero is less than the distance between b and zero (see Figure A.2); that is, $|x| < b$ *whenever* $-b < x < b$. Conversely, if $|x| < b$, then x lies closer to zero than b (or $-b$) do, so it must lie between $-b$ and b. Therefore, $-b < x < b$ implies that $|x| < b$. We have now established the important fact

$$|x| < b \quad \text{is equivalent to} \quad -b < x < b.$$

The inequality $|x| > b$ specifies two intervals: x is in $(-\infty, -b)$ or in (b, ∞); that is, $x < -b$ or $x > b$.

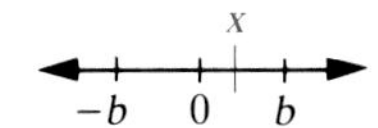

Figure A.2
When $-b < x < b$, x is closer to 0 than either b or $-b$ and so $|x| < |b|$

Example A.15 Indicate on the x-axis the set of points that satisfy $|x - 2| < 3$.

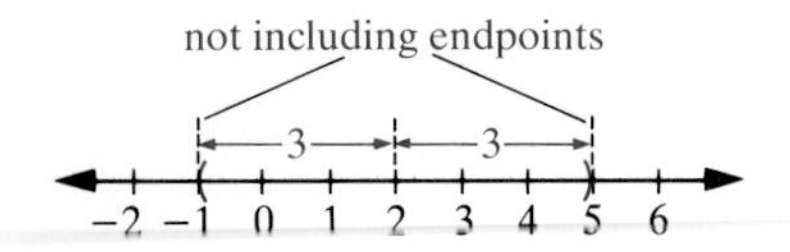

Figure A.3
Solutions to $|x-2|<3$

Solution We use the preceding property of inequalities, with $b=3$ and x replaced by $x-2$, to obtain

$$|x-2|<3 \quad \text{is equivalent to} \quad -3<x-2<3.$$

We now add 2 to all terms to get $-1<x<5$. We show these points in Figure A.3. ■

Observe that the points in Example A.14 are those whose distance from 2 is less than 3. We can generalize this example and conclude the following: The points x satisfying $|x-a|<b$ are those whose distance from a is less than b. See Figure A.4.

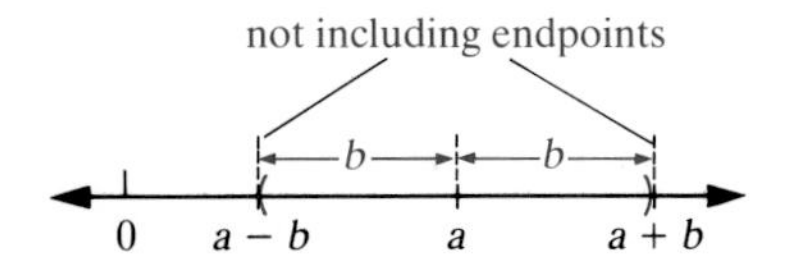

Figure A.4
Set of values of x such that $|x-a|<b$

We conclude this section with notation for describing intervals. The interval $0 \le x \le 150$ includes both endpoints 0 and 150; it is called a **closed interval** and is denoted by [0, 150], where the square brackets indicate that the endpoints 0 and 150 are included. The region $150<x<190$ includes neither of the endpoints and is called an **open interval**; it is denoted by (150, 190), where the parentheses indicate that the endpoints are excluded. We can also denote regions where only one endpoint is included; for example, (20, 35] denotes the region $20<x\le 35$. Such an interval is called a **half-open interval**. Table A.1 lists the notations for intervals and their meanings.

Table A.1
Interval notation

Notation	*Meaning*	*Graphic Representation*
$[a, b]$	All x such that $a \le x \le b$	—[——]—
(a, b)	All x such that $a<x<b$	—(——)—
$[a, b)$	All x such that $a \le x<b$	—[——)—
$(a, b]$	All x such that $a<x\le b$	—(——]—

Example A.16 Indicate on the x-axis the interval denoted by each notation:

$$[-2, 3],\ (2, 4),\ [3, 5),\ \text{and}\ (-1, 4].$$

Solution See Figure A.5. ■

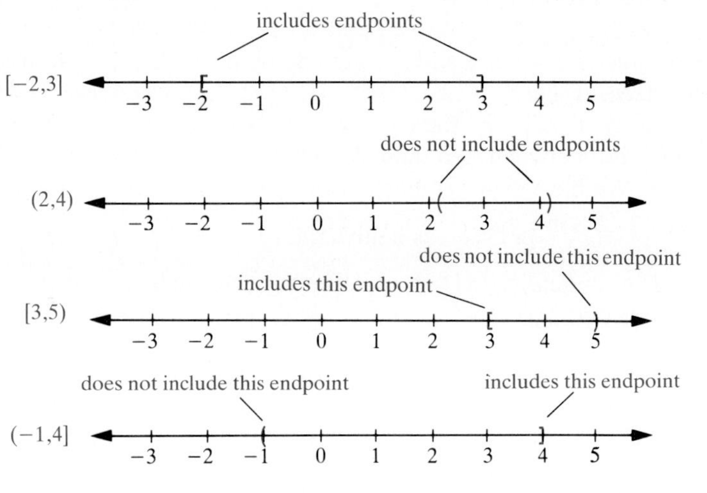

Figure A.5
The intervals from Example A.15

There is a useful connection between absolute value and open or closed intervals. For example, we showed that $|x-a|<b$ is equivalent to $a-b<x<a+b$; this is the open interval $(a-b, a+b)$. Similarly, the set of all x such that $|x-a| \leq b$ is the set of all x whose distance from a is less than or equal to b. Hence, this set is the closed interval $[a-b, a+b]$.

Example A.17 Express the interval $(-2, 6)$ as an inequality using absolute value.

Solution The center of the interval is

$$\frac{6+(-2)}{2}=2$$

and the distance between 2 and the endpoints is 4 (since $6-2=4$ and $2-(-2)=4$). Hence the required expression is

$$|x-2|<4.$$

We use $<$ since we are dealing with an open interval. ■

Exercises

Exer. 1–8: Is each inequality true or false?

1. $3>2$
2. $1 \leq 2$
3. $2 \leq -3$
4. $-2<-2$
5. $-2<-3$
6. $-4>2$
7. $6>7$
8. $13 \leq 13$

Exer. 9–18: Supply $<$ or $>$.

9. If $a<b$ and $c=4$, then $a+c$ ____ $b+c$.
10. If $a<b$ and $c=-\frac{1}{2}$, then $a+c$ ____ $b+c$.
11. If $a<b$ and $c=2$, then $a-c$ ____ $b-c$.
12. If $a<b$ and $c=\frac{1}{4}$ then ac ____ bc.
13. If $a<b$ and $c=-3$, then ac ____ bc.
14. If $a<b$, then $-\frac{a}{6}$ ____ $-\frac{b}{6}$.
15. If $a<b$, then $3a-6$ ____ $3b-6$.
16. If $a<b$ then $\frac{1}{a}$ ____ $\frac{1}{b}$.
17. If $a<-a$, then a ____ 0.
18. If $a<b$, then $a-b$ ____ 0.

Exer. 19–28: Give the value of the expression.

19. $|-4|$
20. $|2|-|-2|$
21. $|\frac{1}{2}|$
22. $||3|-|-11||$
23. $|3|-3$
24. $|-\frac{1}{2}|$
25. $|-4|+(-4)$
26. $\frac{|-11|}{|-2|}$
27. $|2-3|$
28. $|2|-|3|$

Exer. 29–38: Sketch the interval on the real line.

29. $(2, 4)$
30. $[4, 8]$
31. $(3, 6]$
32. $(4, 8]$
33. $[-1, 0]$
34. $[-2, 2]$
35. $[-6, -2]$
36. $(-6, -4)$
37. $[2, 6]$
38. $(-8, 4]$

Exer. 39–46: Find the interval described by the inequality, and graph the interval on the x-axis.

39. $|x-1|<3$
40. $|2-x|<4$
41. $|2x-3| \leq 2$
42. $|2x+5| \leq 3$
43. $|2x-1| \leq -2$
44. $|2x+3| \leq 5$
45. $|x-3|<3$
46. $|3+x| \leq 1$

Exer. 47–52: Express the interval as an inequality using absolute value.

47. $[2, 6]$
48. $[1, 9]$
49. $(-2, 8)$
50. $(-3, 5)$
51. $[2, 7]$
52. $(-3, 10)$

Tables

· A · P · P · E · N · D · I · X · · B ·

B.1 Using the Table of Natural Logarithms

To use Table 1 on natural logarithms (pp. 544–547), the following properties of logarithms are helpful. These are discussed in Chapter 5:

1. $\ln(xy) = \ln x + \ln y$
2. $\ln\left(\frac{x}{y}\right) = \ln x - \ln y$
3. $\ln x^c = c \ln x$

The following examples show how to use Table 1.

Example B.1 Find the natural logarithm of each number:

(a) 1.03 (b) 3.14 (c) 2230
(d) 0.0214 (e) 33.36

Solution **(a)** $\ln 1.03 = 0.0296$. Look under 0.03 in the row of 1.0:

N	.00	.01	.02	.03	.04	.05	.06	.07	.08	.09
1.0	0.0000	0.0100	0.0198	0.0296	0.0392	0.0488	0.0583	0.0677	0.0770	0.0862

(b) $\ln 3.14 = 1.1442$. Look under 0.04 in the row of 3.1:

N	.00	.01	.02	.03	.04	.05	.06	.07	.08	.09
1.0	0.0000	0.0100	0.0198	0.0296	0.0392	0.0488	0.0583	0.0677	0.0770	0.0862
1.1	0.0953	0.1044	0.1133	0.1222	0.1310	0.1398	0.1484	0.1570	0.1655	0.1740
1.2	0.1823	0.1906	0.1989	0.2070	0.2151	0.2231	0.2311	0.2390	0.2469	0.2546
3.0	1.0986	1.1019	1.1053	1.1086	1.1119	1.1151	1.1184	1.1217	1.1249	1.1282
3.1	1.1314	1.1346	1.1378	1.1410	1.1442	1.1474	1.1506	1.1537	1.1569	1.1600
3.2	1.1632	1.1663	1.1694	1.1725	1.1756	1.1787	1.1817	1.1848	1.1878	1.1909
3.3	1.1939	1.1969	1.2000	1.2030	1.2060	1.2090	1.2119	1.2149	1.2179	1.2208

(c) Note that $2230 = 2.23 \times 1000$. We use property 1 to write

$$\begin{aligned}\ln 2230 &= \ln(2.23 \times 1000)\\ &= \ln 2.23 + \ln 1000\end{aligned}$$

Since $\ln 1000 = 6.9078$ (see the notes to Table 1), and since $\ln 2.23 = 0.8020$, we have

$$\begin{aligned}\ln 2230 &= \ln 2.23 + \ln 1000\\ &= 0.8020 + 6.9078 = 7.7098\end{aligned}$$

(d) Although 0.0214 is not in the table note that $0.0214 = 2.14 \times 0.01$. From the notes to Table 1, we read that $\ln 0.01 = 0.3948 - 5$. We leave the expression in this form and do not perform the subtraction. Using property 1 of logarithms, we get

$$\begin{aligned}\ln 0.0214 &= \ln 2.14 + \ln 0.01 \\ &= 0.7608 + (0.3948 - 5) \\ &= 1.1556 - 5 \\ &= -3.8444\end{aligned}$$

(e) Although 33.36 is not in the table, we observe that $33.36 = 3.336 \times 10$. To determine the natural logarithm of a number such as 3.336, which is "between" two numbers in Table 1 (in this case, between 3.33 and 3.34), we use *linear interpolation*, proceeding as follows:

N	.00	.01	.02	.03	.04	.05	.06	.07	.08	.09
3.0	1.0986	1.1019	1.1053	1.1086	1.1119	1.1151	1.1184	1.1217	1.1249	1.1282
3.1	1.1314	1.1346	1.1378	1.1410	1.1442	1.1474	1.1506	1.1537	1.1569	1.1600
3.2	1.1632	1.1663	1.1694	1.1725	1.1756	1.1787	1.1817	1.1848	1.1878	1.1909
3.3	1.1939	1.1969	1.2000	1.2030	1.2060	1.2090	1.2119	1.2149	1.2179	1.2208
3.4	1.2238	1.2267	1.2296	1.2326	1.2355	1.2384	1.2413	1.2442	1.2470	1.2499

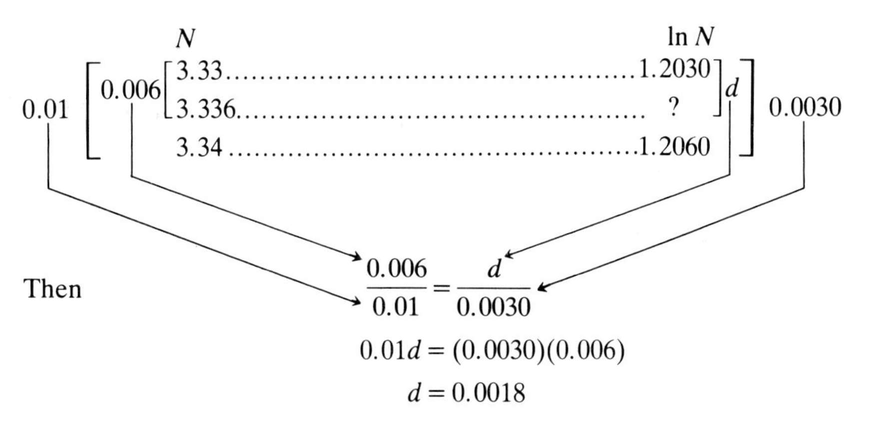

Then

$$\frac{0.006}{0.01} = \frac{d}{0.0030}$$

$$0.01d = (0.0030)(0.006)$$

$$d = 0.0018$$

We add the amount d to 1.2030 to obtain $\ln 3.336$; thus, $\ln 3.336 = 1.2048$. Now we have

$$\begin{aligned}\ln 33.36 &= \ln 3.336 + \ln 10 \\ &\approx 1.2048 + 2.3026 = 3.5074\end{aligned}$$

■

Table 1
Natural logarithms

N	.00	.01	.02	.03	.04	.05	.06	.07	.08	.09
1.0	0.0000	0.0100	0.0198	0.0296	0.0392	0.0488	0.0583	0.0677	0.0770	0.0862
1.1	0.0953	0.1044	0.1133	0.1222	0.1310	0.1398	0.1484	0.1570	0.1655	0.1740
1.2	0.1823	0.1906	0.1989	0.2070	0.2151	0.2231	0.2311	0.2390	0.2469	0.2546
1.3	0.2624	0.2700	0.2776	0.2852	0.2927	0.3001	0.3075	0.3148	0.3221	0.3293
1.4	0.3365	0.3436	0.3507	0.3577	0.3646	0.3716	0.3784	0.3853	0.3920	0.3988
1.5	0.4055	0.4121	0.4187	0.4253	0.4318	0.4383	0.4447	0.4511	0.4574	0.4637
1.6	0.4700	0.4762	0.4824	0.4886	0.4947	0.5008	0.5068	0.5128	0.5188	0.5247
1.7	0.5306	0.5365	0.5423	0.5481	0.5539	0.5596	0.5653	0.5710	0.5766	0.5822
1.8	0.5878	0.5933	0.5988	0.6043	0.6098	0.6152	0.6206	0.6259	0.6313	0.6366
1.9	0.6419	0.6471	0.6523	0.6575	0.6627	0.6678	0.6729	0.6780	0.6831	0.6881
2.0	0.6931	0.6981	0.7031	0.7080	0.7129	0.7178	0.7227	0.7275	0.7324	0.7372
2.1	0.7419	0.7467	0.7514	0.7561	0.7608	0.7655	0.7701	0.7747	0.7793	0.7839
2.2	0.7885	0.7930	0.7975	0.8020	0.8065	0.8109	0.8154	0.8198	0.8242	0.8286
2.3	0.8329	0.8372	0.8416	0.8459	0.8502	0.8544	0.8587	0.8629	0.8671	0.8713
2.4	0.8755	0.8796	0.8838	0.8879	0.8920	0.8961	0.9002	0.9042	0.9083	0.9123
2.5	0.9163	0.9203	0.9243	0.9282	0.9322	0.9361	0.9400	0.9439	0.9478	0.9517
2.6	0.9555	0.9594	0.9632	0.9670	0.9708	0.9746	0.9783	0.9821	0.9858	0.9895
2.7	0.9933	0.9969	1.0006	1.0043	1.0080	1.0116	1.0152	1.0188	1.0225	1.0260
2.8	1.0296	1.0332	1.0367	1.0403	1.0438	1.0473	1.0508	1.0543	1.0578	1.0613
2.9	1.0647	1.0682	1.0716	1.0750	1.0784	1.0818	1.0852	1.0886	1.0919	1.0953
3.0	1.0986	1.1019	1.1053	1.1086	1.1119	1.1151	1.1184	1.1217	1.1249	1.1282
3.1	1.1314	1.1346	1.1378	1.1410	1.1442	1.1474	1.1506	1.1537	1.1569	1.1600
3.2	1.1632	1.1663	1.1694	1.1725	1.1756	1.1787	1.1817	1.1848	1.1878	1.1909
3.3	1.1939	1.1969	1.2000	1.2030	1.2060	1.2090	1.2119	1.2149	1.2179	1.2208
3.4	1.2238	1.2267	1.2296	1.2326	1.2355	1.2384	1.2413	1.2442	1.2470	1.2499
3.5	1.2528	1.2556	1.2585	1.2613	1.2641	1.2669	1.2698	1.2726	1.2754	1.2782
3.6	1.2809	1.2837	1.2865	1.2892	1.2920	1.2947	1.2975	1.3002	1.3029	1.3056
3.7	1.3083	1.3110	1.3137	1.3164	1.3191	1.3218	1.3244	1.3271	1.3297	1.3324
3.8	1.3350	1.3376	1.3403	1.3429	1.3455	1.3481	1.3507	1.3533	1.3558	1.3584
3.9	1.3610	1.3635	1.3661	1.3686	1.3712	1.3737	1.3762	1.3788	1.3813	1.3838
4.0	1.3863	1.3888	1.3913	1.3938	1.3962	1.3987	1.4012	1.4036	1.4061	1.4085
4.1	1.4110	1.4134	1.4159	1.4183	1.4207	1.4231	1.4255	1.4279	1.4303	1.4327
4.2	1.4351	1.4375	1.4398	1.4422	1.4446	1.4469	1.4493	1.4516	1.4540	1.4563
4.3	1.4586	1.4609	1.4633	1.4656	1.4679	1.4702	1.4725	1.4748	1.4770	1.4793
4.4	1.4816	1.4839	1.4861	1.4884	1.4907	1.4929	1.4951	1.4974	1.4996	1.5019
4.5	1.5041	1.5063	1.5085	1.5107	1.5129	1.5151	1.5173	1.5195	1.5217	1.5239
4.6	1.5261	1.5282	1.5304	1.5326	1.5347	1.5369	1.5390	1.5412	1.5433	1.5454
4.7	1.5476	1.5497	1.5518	1.5539	1.5560	1.5581	1.5602	1.5623	1.5644	1.5665
4.8	1.5686	1.5707	1.5728	1.5748	1.5769	1.5790	1.5810	1.5831	1.5851	1.5872
4.9	1.5892	1.5913	1.5933	1.5953	1.5974	1.5994	1.6014	1.6034	1.6054	1.6074
5.0	1.6094	1.6114	1.6134	1.6154	1.6174	1.6194	1.6214	1.6233	1.6253	1.6273
5.1	1.6292	1.6312	1.6332	1.6351	1.6371	1.6390	1.6409	1.6429	1.6448	1.6467
5.2	1.6487	1.6506	1.6525	1.6544	1.6563	1.6582	1.6601	1.6620	1.6639	1.6658
5.3	1.6677	1.6696	1.6715	1.6734	1.6752	1.6771	1.6790	1.6808	1.6827	1.6845
5.4	1.6864	1.6882	1.6901	1.6919	1.6938	1.6956	1.6974	1.6993	1.7011	1.7029

ln 100 = 4.6052*
ln 1000 = 6.9078
ln 10,000 = 9.2103
ln 100,000 = 11.5129
ln 1,000,000 = 13.8155
ln 10,000,000 = 16.1181

ln .1 = .6974 − 3
ln .01 = .3948 − 5
ln .001 = .0922 − 7
ln .0001 = .7896 − 10
ln .00001 = .4871 − 12
ln .000 001 = .1845 − 14

* In general, $\ln 10^k = k \ln 10 = k(2.3026)$.

Table 1 (continued)

N	.00	.01	.02	.03	.04	.05	.06	.07	.08	.09
5.5	1.7047	1.7066	1.7084	1.7102	1.7120	1.7138	1.7156	1.7174	1.7192	1.7210
5.6	1.7228	1.7246	1.7263	1.7281	1.7299	1.7317	1.7334	1.7352	1.7370	1.7387
5.7	1.7405	1.7422	1.7440	1.7457	1.7475	1.7492	1.7509	1.7527	1.7544	1.7561
5.8	1.7579	1.7596	1.7613	1.7630	1.7647	1.7664	1.7681	1.7699	1.7716	1.7733
5.9	1.7750	1.7766	1.7783	1.7800	1.7817	1.7834	1.7851	1.7867	1.7884	1.7901
6.0	1.7918	1.7934	1.7951	1.7967	1.7984	1.8001	1.8017	1.8034	1.8050	1.8066
6.1	1.8083	1.8099	1.8116	1.8132	1.8148	1.8165	1.8181	1.8197	1.8213	1.8229
6.2	1.8245	1.8262	1.8278	1.8294	1.8310	1.8326	1.8342	1.8358	1.8374	1.8390
6.3	1.8405	1.8421	1.8437	1.8453	1.8469	1.8485	1.8500	1.8516	1.8532	1.8547
6.4	1.8563	1.8579	1.8594	1.8610	1.8625	1.8641	1.8656	1.8672	1.8687	1.8703
6.5	1.8718	1.8733	1.8749	1.8764	1.8779	1.8795	1.8810	1.8825	1.8840	1.8856
6.6	1.8871	1.8886	1.8901	1.8916	1.8931	1.8946	1.8961	1.8976	1.8991	1.9006
6.7	1.9021	1.9036	1.9051	1.9066	1.9081	1.9095	1.9110	1.9125	1.9140	1.9155
6.8	1.9169	1.9184	1.9199	1.9213	1.9228	1.9242	1.9257	1.9272	1.9286	1.9301
6.9	1.9315	1.9330	1.9344	1.9359	1.9373	1.9387	1.9402	1.9416	1.9430	1.9445
7.0	1.9459	1.9473	1.9488	1.9502	1.9516	1.9530	1.9544	1.9559	1.9573	1.9587
7.1	1.9601	1.9615	1.9629	1.9643	1.9657	1.9671	1.9685	1.9699	1.9713	1.9727
7.2	1.9741	1.9755	1.9769	1.9782	1.9796	1.9810	1.9824	1.9838	1.9851	1.9865
7.3	1.9879	1.9892	1.9906	1.9920	1.9933	1.9947	1.9961	1.9974	1.9988	2.0001
7.4	2.0015	2.0028	2.0042	2.0055	2.0069	2.0082	2.0096	2.0109	2.0122	2.0136
7.5	2.0149	2.0162	2.0176	2.0189	2.0202	2.0215	2.0229	2.0242	2.0255	2.0268
7.6	2.0281	2.0295	2.0308	2.0321	2.0334	2.0347	2.0360	2.0373	2.0386	2.0399
7.7	2.0412	2.0425	2.0438	2.0451	2.0464	2.0477	2.0490	2.0503	2.0516	2.0528
7.8	2.0541	2.0554	2.0567	2.0580	2.0592	2.0605	2.0618	2.0631	2.0643	2.0656
7.9	2.0669	2.0681	2.0694	2.0707	2.0719	2.0732	2.0744	2.0757	2.0769	2.0782
8.0	2.0794	2.0807	2.0819	2.0832	2.0844	2.0857	2.0869	2.0882	2.0894	2.0906
8.1	2.0919	2.0931	2.0943	2.0956	2.0968	2.0980	2.0992	2.1005	2.1017	2.1029
8.2	2.1041	2.1054	2.1066	2.1078	2.1090	2.1102	2.1114	2.1126	2.1138	2.1150
8.3	2.1163	2.1175	2.1187	2.1199	2.1211	2.1223	2.1235	2.1247	2.1258	2.1270
8.4	2.1282	2.1294	2.1306	2.1318	2.1330	2.1342	2.1353	2.1365	2.1377	2.1389
8.5	2.1401	2.1412	2.1424	2.1436	2.1448	2.1459	2.1471	2.1483	2.1494	2.1506
8.6	2.1518	2.1529	2.1541	2.1552	2.1564	2.1576	2.1587	2.1599	2.1610	2.1622
8.7	2.1633	2.1645	2.1656	2.1668	2.1679	2.1691	2.1702	2.1713	2.1725	2.1736
8.8	2.1748	2.1759	2.1770	2.1782	2.1793	2.1804	2.1815	2.1827	2.1838	2.1849
8.9	2.1861	2.1872	2.1883	2.1894	2.1905	2.1917	2.1928	2.1939	2.1950	2.1961
9.0	2.1972	2.1983	2.1994	2.2006	2.2017	2.2028	2.2039	2.2050	2.2061	2.2072
9.1	2.2083	2.2094	2.2105	2.2116	2.2127	2.2138	2.2148	2.2159	2.2170	2.2181
9.2	2.2192	2.2203	2.2214	2.2225	2.2235	2.2246	2.2257	2.2268	2.2279	2.2289
9.3	2.2300	2.2311	2.2322	2.2332	2.2343	2.2354	2.2364	2.2375	2.2386	2.2396
9.4	2.2407	2.2418	2.2428	2.2439	2.2450	2.2460	2.2471	2.2481	2.2492	2.2502
9.5	2.2513	2.2523	2.2534	2.2544	2.2555	2.2565	2.2576	2.2586	2.2597	2.2607
9.6	2.2618	2.2628	2.2638	2.2649	2.2659	2.2670	2.2680	2.2690	2.2701	2.2711
9.7	2.2712	2.2732	2.2742	2.2752	2.2762	2.2773	2.2783	2.2793	2.2803	2.2814
9.8	2.2824	2.2834	2.2844	2.2854	2.2865	2.2875	2.2885	2.2895	2.2905	2.2915
9.9	2.2925	2.2935	2.2946	2.2956	2.2966	2.2976	2.2986	2.2996	2.3006	2.3016

ln 100 = 4.6052
ln 1000 = 6.9078
ln 10,000 = 9.2103
ln 100,000 = 11.5129
ln 1,000,000 = 13.8155
ln 10,000,000 = 16.1181

ln .1 = .6974 − 3
ln .01 = .3948 − 5
ln .001 = .0922 − 7
ln .0001 = .7896 − 10
ln .00001 = .4871 − 12
ln .000 001 = .1845 − 14

Table 1 (continued)

N	.0	.1	.2	.3	.4	.5	.6	.7	.8	.9
10	2.3026	2.3125	2.3224	2.3321	2.3418	2.3514	2.3609	2.3702	2.3795	2.3888
11	2.3979	2.4069	2.4159	2.4248	2.4336	2.4423	2.4510	2.4596	2.4681	2.4765
12	2.4849	2.4932	2.5014	2.5096	2.5177	2.5257	2.5337	2.5416	2.5494	2.5572
13	2.5649	2.5726	2.5802	2.5878	2.5953	2.6027	2.6101	2.6174	2.6247	2.6319
14	2.6391	2.6462	2.6532	2.6603	2.6672	2.6741	2.6810	2.6878	2.6946	2.7014
15	2.7081	2.7147	2.7213	2.7279	2.7344	2.7408	2.7473	2.7537	2.7600	2.7663
16	2.7726	2.7788	2.7850	2.7912	2.7973	2.8034	2.8094	2.8154	2.8214	2.8273
17	2.8332	2.8391	2.8449	2.8507	2.8565	2.8622	2.8679	2.8736	2.8792	2.8848
18	2.8904	2.8959	2.9014	2.9069	2.9124	2.9178	2.9232	2.9285	2.9339	2.9392
19	2.9444	2.9497	2.9549	2.9601	2.9653	2.9704	2.9755	2.9806	2.9857	2.9907
20	2.9957	3.0007	3.0057	3.0106	3.0155	3.0204	3.0253	3.0301	3.0350	3.0397
21	3.0445	3.0493	3.0540	3.0587	3.0634	3.0681	3.0727	3.0773	3.0819	3.0865
22	3.0910	3.0956	3.1001	3.1046	3.1091	3.1135	3.1179	3.1224	3.1268	3.1311
23	3.1355	3.1398	3.1442	3.1485	3.1527	3.1570	3.1612	3.1655	3.1697	3.1739
24	3.1781	3.1822	3.1864	3.1905	3.1946	3.1987	3.2027	3.2068	3.2108	3.2149
25	3.2189	3.2229	3.2268	3.2308	3.2347	3.2387	3.2426	3.2465	3.2504	3.2542
26	3.2581	3.2619	3.2658	3.2696	3.2734	3.2771	3.2809	3.2847	3.2884	3.2921
27	3.2958	3.2995	3.3032	3.3069	3.3105	3.3142	3.3178	3.3214	3.3250	3.3286
28	3.3322	3.3358	3.3393	3.3429	3.3464	3.3499	3.3534	3.3569	3.3604	3.3638
29	3.3673	3.3707	3.3742	3.3776	3.3810	3.3844	3.3878	3.3911	3.3945	3.3979
30	3.4012	3.4045	3.4078	3.4111	3.4144	3.4177	3.4210	3.4243	3.4275	3.4308
31	3.4340	3.4372	3.4404	3.4436	3.4468	3.4500	3.4532	3.4563	3.4595	3.4626
32	3.4657	3.4689	3.4720	3.4751	3.4782	3.4812	3.4843	3.4874	3.4904	3.4935
33	3.4965	3.4995	3.5025	3.5056	3.5086	3.5115	3.5145	3.5175	3.5205	3.5234
34	3.5264	3.5293	3.5322	3.5351	3.5381	3.5410	3.5439	3.5467	3.5496	3.5525
35	3.5553	3.5582	3.5610	3.5639	3.5667	3.5695	3.5723	3.5752	3.5779	3.5807
36	3.5835	3.5863	3.5891	3.5918	3.5946	3.5973	3.6000	3.6028	3.6055	3.6082
37	3.6109	3.6136	3.6163	3.6190	3.6217	3.6243	3.6270	3.6297	3.6323	3.6350
38	3.6376	3.6402	3.6428	3.6454	3.6481	3.6507	3.6533	3.6558	3.6584	3.6610
39	3.6636	3.6661	3.6687	3.6712	3.6738	3.6763	3.6788	3.6814	3.6839	3.6864
40	3.6889	3.6914	3.6939	3.6964	3.6988	3.7013	3.7038	3.7062	3.7087	3.7111
41	3.7136	3.7160	3.7184	3.7209	3.7233	3.7257	3.7281	3.7305	3.7329	3.7353
42	3.7377	3.7400	3.7424	3.7448	3.7471	3.7495	3.7519	3.7542	3.7565	3.7589
43	3.7612	3.7635	3.7658	3.7682	3.7705	3.7728	3.7751	3.7773	3.7796	3.7819
44	3.7842	3.7865	3.7887	3.7910	3.7932	3.7995	3.7977	3.8000	3.8022	3.8044
45	3.8067	3.8089	3.8111	3.8133	3.8155	3.8177	3.8199	3.8221	3.8243	3.8265
46	3.8286	3.8308	3.8330	3.8351	3.8373	3.8395	3.8416	3.8437	3.8459	3.8480
47	3.8501	3.8523	3.8544	3.8565	3.8586	3.8607	3.8628	3.8649	3.8670	3.8691
48	3.8712	3.8733	3.8754	3.8774	3.8795	3.8816	3.8836	3.8857	3.8877	3.8898
49	3.8918	3.8939	3.8959	3.8979	3.9000	3.9020	3.9040	3.9060	3.9080	3.9100
50	3.9120	3.9140	3.9160	3.9180	3.9200	3.9220	3.9240	3.9259	3.9279	3.9299
51	3.9318	3.9338	3.9357	3.9377	3.9396	3.9416	3.9435	3.9455	3.9474	3.9493
52	3.9512	3.9532	3.9551	3.9570	3.9589	3.9608	3.9627	3.9646	3.9665	3.9684
53	3.9703	3.9722	3.9741	3.9759	3.9778	3.9797	3.9815	3.9834	3.9853	3.9871
54	3.9890	3.9908	3.9927	3.9945	3.9964	3.9982	4.0000	4.0019	4.0037	4.0055

ln 100 = 4.6052
ln 1000 = 6.9078
ln 10,000 = 9.2103
ln 100,000 = 11.5129
ln 1,000,000 = 13.8155
ln 10,000,000 = 16.1181

ln .1 = .6974 − 3
ln .01 = .3948 − 5
ln .001 = .0922 − 7
ln .0001 = .7896 − 10
ln .00001 = .4871 − 12
ln .000 001 = .1845 − 14

Table 1 (continued)

N	.0	.1	.2	.3	.4	.5	.6	.7	.8	.9
55	4.0073	4.0091	4.0110	4.0128	4.0146	4.0164	4.0182	4.0200	4.0218	4.0236
56	4.0254	4.0271	4.0289	4.0307	4.0325	4.0342	4.0360	4.0378	4.0395	4.0413
57	4.0431	4.0448	4.0466	4.0483	4.0500	4.0518	4.0535	4.0553	4.0570	4.0587
58	4.0604	4.0622	4.0639	4.0656	4.0673	4.0690	4.0707	4.0724	4.0741	4.0758
59	4.0775	4.0792	4.0809	4.0826	4.0843	4.0860	4.0877	4.0893	4.0910	4.0927
60	4.0943	4.0960	4.0977	4.0993	4.1010	4.1026	4.1043	4.1059	4.1076	4.1092
61	4.1109	4.1125	4.1141	4.1158	4.1174	4.1190	4.1207	4.1223	4.1239	4.1255
62	4.1271	4.1287	4.1304	4.1320	4.1336	4.1352	4.1368	4.1384	4.1400	4.1415
63	4.1431	4.1447	4.1463	4.1479	4.1495	4.1510	4.1526	4.1542	4.1558	4.1573
64	4.1589	4.1604	4.1620	4.1636	4.1651	4.1667	4.1682	4.1698	4.1713	4.1728
65	4.1744	4.1759	4.1775	4.1790	4.1805	4.1821	4.1836	4.1851	4.1866	4.1881
66	4.1897	4.1912	4.1927	4.1942	4.1957	4.1972	4.1987	4.2002	4.2017	4.2032
67	4.2047	4.2062	4.2077	4.2092	4.2106	4.2121	4.2136	4.2151	4.2166	4.2180
68	4.2195	4.2210	4.2224	4.2239	4.2254	4.2268	4.2283	4.2297	4.2312	4.2327
69	4.2341	4.2356	4.2370	4.2384	4.2399	4.2413	4.2428	4.2442	4.2456	4.2471
70	4.2485	4.2499	4.2513	4.2528	4.2542	4.2556	4.2570	4.2584	4.2599	4.2613
71	4.2627	4.2641	4.2655	4.2669	4.2683	4.2697	4.2711	4.2725	4.2739	4.2753
72	4.2767	4.2781	4.2794	4.2808	4.2822	4.2836	4.2850	4.2863	4.2877	4.2891
73	4.2905	4.2918	4.2932	4.2946	4.2959	4.2973	4.2986	4.3000	4.3014	4.3027
74	4.3041	4.3054	4.3068	4.3081	4.3095	4.3108	4.3121	4.3135	4.3148	4.3162
75	4.3175	4.3188	4.3202	4.3215	4.3228	4.3241	4.3255	4.3268	4.3281	4.3294
76	4.3307	4.3320	4.3334	4.3347	4.3360	4.3373	4.3386	4.3399	4.3412	4.3425
77	4.3438	4.3451	4.3464	4.3477	4.3490	4.3503	4.3516	4.3529	4.3541	4.3554
78	4.3567	4.3580	4.3593	4.3605	4.3618	4.3631	4.3644	4.3656	4.3669	4.3682
79	4.3694	4.3707	4.3720	4.3732	4.3745	4.3758	4.3770	4.3783	4.3795	4.3808
80	4.3820	4.3833	4.3845	4.3858	4.3870	4.3883	4.3895	4.3907	4.3920	4.3932
81	4.3944	4.3957	4.3969	4.3981	4.3994	4.4006	4.4018	4.4031	4.4043	4.4055
82	4.4067	4.4079	4.4092	4.4104	4.4116	4.4128	4.4140	4.4152	4.4164	4.4176
83	4.4188	4.4200	4.4212	4.4224	4.4236	4.4248	4.4260	4.4272	4.4284	4.4296
84	4.4308	4.4320	4.4332	4.4344	4.4356	4.4368	4.4379	4.4391	4.4403	4.4415
85	4.4427	4.4438	4.4450	4.4462	4.4473	4.4485	4.4497	4.4509	4.4520	4.4532
86	4.4543	4.4555	4.4567	4.4578	4.4590	4.4601	4.4613	4.4625	4.4636	4.4648
87	4.4659	4.4671	4.4682	4.4694	4.4705	4.4716	4.4728	4.4739	4.4751	4.4762
88	4.4773	4.4785	4.4796	4.4807	4.4819	4.4830	4.4841	4.4853	4.4864	4.4875
89	4.4886	4.4898	4.4909	4.4920	4.4931	4.4942	4.4954	4.4965	4.4976	4.4987
90	4.4998	4.5009	4.5020	4.5031	4.5042	4.5053	4.5065	4.5076	4.5087	4.5098
91	4.5109	4.5120	4.5131	4.5142	4.5152	4.5163	4.5174	4.5185	4.5196	4.5207
92	4.5218	4.5229	4.5240	4.5250	4.5261	4.5272	4.5283	4.5294	4.5304	4.5315
93	4.5326	4.5337	4.5347	4.5358	4.5369	4.5380	4.5390	4.5401	4.5412	4.5422
94	4.5433	4.5444	4.5454	4.5465	4.5475	4.5486	4.5497	4.5507	4.5518	4.5528
95	4.5539	4.5549	4.5560	4.5570	4.5581	4.5591	4.5602	4.5612	4.5623	4.5633
96	4.5643	4.5654	4.5664	4.5675	4.5685	4.5695	4.5706	4.5716	4.5726	4.5737
97	4.5747	4.5757	4.5768	4.5778	4.5788	4.5799	4.5809	4.5819	4.5829	4.5839
98	4.5850	4.5860	4.5870	4.5880	4.5890	4.5901	4.5911	4.5921	4.5931	4.5941
99	4.5951	4.5961	4.5971	4.5981	4.5992	4.6002	4.6012	4.6022	4.6032	4.6042

ln 100 = 4.6052
ln 1000 = 6.9078
ln 10,000 = 9.2103
ln 100,000 = 11.5129
ln 1,000,000 = 13.8155
ln 10,000,000 = 16.1181

ln .1 = .6974 − 3
ln .01 = .3948 − 5
ln .001 = .0922 − 7
ln .0001 = .7896 − 10
ln .00001 = .4871 − 12
ln .000 001 = .1845 − 14

Table 2
The Values of Exponential Functions

x	e^x	e^{-x}	x	e^x	e^{-x}	x	e^x	e^{-x}
0.00	1.0000	1.00 000	**0.50**	1.6487	.60 653	**1.00**	2.7183	.36 788
0.01	1.0101	0.99 005	0.51	1.6653	.60 050	1.01	2.7456	.36 422
0.02	1.0202	.98 020	0.52	1.6820	.59 452	1.02	2.7732	.36 059
0.03	1.0305	.97 045	0.53	1.6989	.58 860	1.03	2.8011	.35 701
0.04	1.0408	.96 079	0.54	1.7160	.58 275	1.04	2.8292	.35 345
0.05	1.0513	.95 123	**0.55**	1.7333	.57 695	**1.05**	2.8577	.34 994
0.06	1.0618	.94 176	0.56	1.7507	.57 121	1.06	2.8864	.34 646
0.07	1.0725	.93 239	0.57	1.7683	.56 553	1.07	2.9154	.34 301
0.08	1.0833	.92 312	0.58	1.7860	.55 990	1.08	2.9447	.33 960
0.09	1.0942	.91 393	0.59	1.8040	.55 433	1.09	2.9743	.33 622
0.10	1.1052	.90 484	**0.60**	1.8221	.54 881	**1.10**	3.0042	.33 287
0.11	1.1163	.89 583	0.61	1.8404	.54 335	1.11	3.0344	.32 956
0.12	1.1275	.88 692	0.62	1.8589	.53 794	1.12	3.0649	.32 628
0.13	1.1388	.87 810	0.63	1.8776	.53 259	1.13	3.0957	.32 303
0.14	1.1503	.86 936	0.64	1.8965	.52 729	1.14	3.1268	.31 982
0.15	1.1618	.86 071	**0.65**	1.9155	.52 205	**1.15**	3.1582	.31 664
0.16	1.1735	.85 214	0.66	1.9348	.51 685	1.16	3.1899	.31 349
0.17	1.1853	.84 366	0.67	1.9542	.51 171	1.17	3.2220	.31 037
0.18	1.1972	.83 527	0.68	1.9739	.50 662	1.18	3.2544	.30 728
0.19	1.2092	.82 696	0.69	1.9937	.50 158	1.19	3.2871	.30 422
0.20	1.2214	.81 873	**0.70**	2.0138	.49 659	**1.20**	3.3201	.30 119
0.21	1.2337	.81 058	0.71	2.0340	.49 164	1.21	3.3535	.29 820
0.22	1.2461	.80 252	0.72	2.0544	.48 675	1.22	3.3872	.29 523
0.23	1.2586	.79 453	0.73	2.0751	.48 191	1.23	3.4212	.29 229
0.24	1.2712	.78 663	0.74	2.0959	.47 711	1.24	3.4556	.28 938
0.25	1.2840	.77 880	**0.75**	2.1170	.47 237	**1.25**	3.4903	.28 650
0.26	1.2969	.77 105	0.76	2.1383	.46 767	1.26	3.5254	.28 365
0.27	1.3100	.76 338	0.77	2.1598	.46 301	1.27	3.5609	.28 083
0.28	1.3231	.75 578	0.78	2.1815	.45 841	1.28	3.5966	.27 804
0.29	1.3364	.74 826	0.79	2.2034	.45 384	1.29	3.6328	.27 527
0.30	1.3499	.74 082	**0.80**	2.2255	.44 933	**1.30**	3.6693	.27 253
0.31	1.3634	.73 345	0.81	2.2479	.44 486	1.31	3.7062	.26 982
0.32	1.3771	.72 615	0.82	2.2705	.44 043	1.32	3.7434	.26 714
0.33	1.3910	.71 892	0.83	2.2933	.43 605	1.33	3.7810	.26 448
0.34	1.4049	.71 177	0.84	2.3164	.43 171	1.34	3.8190	.26 185
0.35	1.4191	.70 469	**0.85**	2.3396	.42 741	**1.35**	3.8574	.25 924
0.36	1.4333	.69 768	0.86	2.3632	.42 316	1.36	3.8962	.25 666
0.37	1.4477	.69 073	0.87	2.3869	.41 895	1.37	3.9354	.25 411
0.38	1.4623	.68 386	0.88	2.4109	.41 478	1.38	3.9749	.25 158
0.39	1.4770	.67 706	0.89	2.4351	.41 066	1.39	4.0149	.24 908
0.40	1.4918	.67 032	**0.90**	2.4596	.40 657	**1.40**	4.0552	.24 660
0.41	1.5068	.66 365	0.91	2.4843	.40 252	1.41	4.0960	.24 414
0.42	1.5220	.65 705	0.92	2.5093	.39 852	1.42	4.1371	.24 171
0.43	1.5373	.65 051	0.93	2.5345	.39 455	1.43	4.1787	.23 931
0.44	1.5527	.64 404	0.94	2.5600	.39 063	1.44	4.2207	.23 693
0.45	1.5683	.63 763	**0.95**	2.5857	.38 674	**1.45**	4.2631	.23 457
0.46	1.5841	.63 128	0.96	2.6117	.38 289	1.46	4.3060	.23 224
0.47	1.6000	.62 500	0.97	2.6379	.37 908	1.47	4.3492	.22 993
0.48	1.6161	.61 878	0.98	2.6645	.37 531	1.48	4.3929	.22 764
0.49	1.6323	.61 263	0.99	2.6912	.37 158	1.49	4.4371	.22 537
0.50	1.6487	.60 653	**1.00**	2.7183	.36 788	**1.50**	4.4817	.22 313

Table 2 (continued)

x	e^x	e^{-x}	x	e^x	e^{-x}	x	e^x	e^{-x}
1.50	4.4817	.22 313	**2.00**	7.3891	.13 534	**2.50**	12.182	.082 085
1.51	4.5267	.22 091	2.01	7.4633	.13 399	2.51	12.305	.081 268
1.52	4.5722	.21 871	2.02	7.5383	.13 266	2.52	12.429	.080 460
1.53	4.6182	.21 654	2.03	7.6141	.13 134	2.53	12.554	.079 659
1.54	4.6646	.21 438	2.04	7.6906	.13 003	2.54	12.680	.078 866
1.55	4.7115	.21 225	**2.05**	7.7679	.12 873	**2.55**	12.807	.078 082
1.56	4.7588	.21 014	2.06	7.8460	.12 745	2.56	12.936	.077 305
1.57	4.8066	.20 805	2.07	7.9248	.12 619	2.57	13.066	.076 536
1.58	4.8550	.20 598	2.08	8.0045	.12 493	2.58	13.197	.075 774
1.59	4.9037	.20 393	2.09	8.0849	.12 369	2.59	13.330	.075 020
1.60	4.9530	.20 190	**2.10**	8.1662	.12 246	**2.60**	13.464	.074 274
1.61	5.0028	.19 989	2.11	8.2482	.12 124	2.61	13.599	.073 535
1.62	5.0531	.19 790	2.12	8.3311	.12 003	2.62	13.736	.072 803
1.63	5.1039	.19 593	2.13	8.4149	.11 884	2.63	13.874	.072 078
1.64	5.1552	.19 398	2.14	8.4994	.11 765	2.64	14.013	.071 361
1.65	5.2070	.19 205	**2.15**	8.5849	.11 648	**2.65**	14.154	.070 651
1.66	5.2953	.19 014	2.16	8.6711	.11 533	2.66	14.296	.069 948
1.67	5.3122	.18 825	2.17	8.7583	.11 418	2.67	14.440	.069 252
1.68	5.3656	.18 637	2.18	8.8463	.11 304	2.68	14.585	.068 563
1.69	5.4195	.18 452	2.19	8.9352	.11 192	2.69	14.732	.067 881
1.70	5.4739	.18 268	**2.20**	9.0250	.11 080	**2.70**	14.880	.067 206
1.71	5.5290	.18 087	2.21	9.1157	.10 970	2.71	15.029	.066 537
1.72	5.5845	.17 907	2.22	9.2073	.10 861	2.72	15.180	.065 875
1.73	5.6407	.17 728	2.23	9.2999	.10 753	2.73	15.333	.065 219
1.74	5.6973	.17 552	2.24	9.3933	.10 646	2.74	15.487	.064 570
1.75	5.7546	.17 377	**2.25**	9.4877	.10 540	2.75	15.643	.063 928
1.76	5.8124	.17 204	2.26	9.5831	.10 435	2.76	15.800	.063 292
1.77	5.8709	.17 033	2.27	9.6794	.10 331	2.77	15.959	.062 662
1.78	5.9299	.16 864	2.28	9.7767	.10 228	2.78	16.119	.062 039
1.79	5.9895	.16 696	2.29	9.8749	.10 127	2.79	16.281	.061 421
1.80	6.0496	.16 530	**2.30**	9.9742	.10 026	**2.80**	16.445	.060 810
1.81	6.1104	.16 365	2.31	10.074	.09 9261	2.81	16.610	.060 205
1.82	6.1719	.16 203	2.32	10.176	.09 8274	2.82	16.777	.059 606
1.83	6.2339	.16 041	2.33	10.278	.09 7296	2.83	16.945	.059 013
1.84	6.2965	.15 882	2.34	10.381	.09 6328	2.84	17.116	.058 426
1.85	6.3598	.15 724	**2.35**	10.486	.09 5369	**2.85**	17.288	.057 844
1.86	6.4237	.15 567	2.36	10.591	.09 4420	2.86	17.462	.057 269
1.87	6.4883	.15 412	2.37	10.697	.09 3481	2.87	17.637	.056 699
1.88	6.5535	.15 259	2.38	10.805	.09 2551	2.88	17.814	.056 135
1.89	6.6194	.15 107	2.39	10.913	.09 1630	2.89	17.993	.055 576
1.90	6.6859	.14 957	**2.40**	11.023	.09 0718	**2.90**	18.174	.055 023
1.91	6.7531	.14 808	2.41	11.134	.08 9815	2.91	18.357	.054 476
1.92	6.8210	.14 661	2.42	11.246	.08 8922	2.92	18.541	.053 934
1.93	6.8895	.14 515	2.43	11.359	.08 8037	2.93	18.728	.053 397
1.94	6.9588	.14 370	2.44	11.473	.08 7161	2.94	18.916	.052 866
1.95	7.0287	.14 227	**2.45**	11.588	.08 6294	**2.95**	19.106	.052 340
1.96	7.0993	.14 086	2.46	11.705	.08 5435	2.96	19.298	.051 819
1.97	7.1707	.13 946	2.47	11.822	.08 4585	2.97	19.492	.051 303
1.98	7.2427	.13 807	2.48	11.941	.08 3743	2.98	19.688	.050 793
1.99	7.3155	.13 670	2.49	12.061	.08 2910	2.99	19.886	.050 287
2.00	7.3891	.13 534	**2.50**	12.182	.08 2085	**3.00**	20.086	.049 787

Table 2 (continued)

x	e^x	e^{-x}	x	e^x	e^{-x}	x	e^x	e^{-x}
3.00	20.086	.04 9787	**3.50**	33.115	.030 197	**4.00**	54.598	.01 8316
3.01	20.287	.04 9292	3.51	33.448	.029 897	4.01	55.147	.01 8133
3.02	20.491	.04 8801	3.52	33.784	.029 599	4.02	55.701	.01 7953
3.03	20.697	.04 8316	3.53	34.124	.029 305	4.03	56.261	.01 7774
3.04	20.905	.04 7835	3.54	34.467	.029 013	4.04	56.826	.01 7597
3.05	21.115	.04 7359	**3.55**	34.813	.028 725	**4.05**	57.397	.01 7422
3.06	21.328	.04 6888	3.56	35.163	.028 439	4.06	57.974	.01 7249
3.07	21.542	.04 6421	3.57	35.517	.028 156	4.07	58.557	.01 7077
3.08	21.758	.04 5959	3.58	35.874	.027 876	4.08	59.145	.01 6907
3.09	21.977	.04 5502	3.59	36.234	.027 598	4.09	59.740	.01 6739
3.10	22.198	.04 5049	**3.60**	36.598	.027 324	**4.10**	60.340	.01 6573
3.11	22.421	.04 4601	3.61	36.966	.027 052	4.11	60.947	.01 6408
3.12	22.646	.04 4157	3.62	37.338	.026 783	4.12	61.559	.01 6245
3.13	22.874	.04 3718	3.63	37.713	.026 516	4.13	62.178	.01 6083
3.14	23.104	.04 3283	3.64	38.092	.026 252	4.14	62.803	.01 5923
3.15	23.336	.04 2852	**3.65**	38.475	.025 991	**4.15**	63.434	.01 5764
3.16	23.571	.04 2426	3.66	38.861	.025 733	4.16	64.072	.01 5608
3.17	23.807	.04 2004	3.67	39.252	.025 476	4.17	64.715	.01 5452
3.18	24.047	.04 1586	3.68	39.646	.025 223	4.18	65.366	.01 5299
3.19	24.288	.04 1172	3.69	40.045	.024 972	4.19	66.023	.01 5146
3.20	24.533	.04 0762	**3.70**	40.447	.024 724	**4.20**	66.686	.01 4996
3.21	24.779	.04 0357	3.71	40.854	.024 478	4.21	67.357	.01 4846
3.22	25.028	.03 9955	3.72	41.264	.024 234	4.22	68.033	.01 4699
3.23	25.280	.03 9557	3.73	41.679	.023 993	4.23	68.717	.01 4552
3.24	25.534	.03 9164	3.74	42.098	.023 754	4.24	69.408	.01 4408
3.25	25.790	.03 8774	**3.75**	42.521	.023 518	**4.25**	70.105	.01 4264
3.26	26.050	.03 8388	3.76	42.948	.023 284	4.26	70.810	.01 4122
3.27	26.311	.03 8006	3.77	43.380	.023 052	4.27	71.522	.01 3982
3.28	26.576	.03 7628	3.78	43.816	.022 823	4.28	72.240	.01 3843
3.29	26.843	.03 7254	3.79	44.256	.022 596	4.29	72.966	.01 3705
3.30	27.113	.03 6883	**3.80**	44.701	.022 371	**4.30**	73.700	.01 3569
3.31	27.385	.03 6516	3.81	45.150	.022 148	4.31	74.440	.01 3434
3.32	27.660	.03 6153	3.82	45.604	.021 928	4.32	75.189	.01 3300
3.33	27.938	.03 5793	3.83	46.063	.021 710	4.33	75.944	.01 3168
3.34	28.219	.03 5437	3.84	46.525	.021 494	4.34	76.708	.01 3037
3.35	28.503	.03 5084	**3.85**	46.993	.021 280	**4.35**	77.478	.01 2907
3.36	28.789	.03 4735	3.86	47.465	.021 068	4.36	78.257	.01 2778
3.37	29.079	.03 4390	3.87	47.942	.020 858	4.37	79.044	.01 2651
3.38	29.371	.03 4047	3.88	48.424	.020 651	4.38	79.838	.01 2525
3.39	29.666	.03 3709	3.89	48.911	.020 445	4.39	80.640	.01 2401
3.40	29.964	.03 3373	**3.90**	49.402	.020 242	**4.40**	81.451	.01 2277
3.41	30.265	.03 3041	3.91	49.899	.020 041	4.41	82.269	.01 2155
3.42	30.569	.03 2712	3.92	50.400	.019 841	4.42	83.096	.01 2034
3.43	30.877	.03 2387	3.93	50.907	.019 644	4.43	83.931	.01 1914
3.44	31.187	.03 2065	3.94	51.419	.019 448	4.44	84.775	.01 1796
3.45	31.500	.03 1746	**3.95**	51.935	.019 255	**4.45**	85.627	.01 1679
3.46	31.817	.03 1430	3.96	52.457	.019 063	4.46	86.488	.01 1562
3.47	32.137	.03 1117	3.97	52.985	.018 873	4.47	87.357	.01 1447
3.48	32.460	.03 0807	3.98	53.517	.018 686	4.48	88.235	.01 1333
4.49	32.786	.03 0501	3.99	54.055	.018 500	4.49	89.121	.01 1221
3.50	33.115	.03 0197	**4.00**	54.598	.018 316	**4.50**	90.017	.01 1109

Table 2 (continued)

x	e^x	e^{-x}	x	e^x	e^{-x}	x	e^x	e^{-x}
4.50	90.017	.011 109	**5.00**	148.41	.00 67379	**7.50**	1 808.0	.000 5531
4.51	90.922	.010 998	5.05	156.02	.00 64093	7.55	1 900.7	.000 5261
4.52	91.836	.010 889	5.10	164.02	.00 60967	7.60	1 998.2	.000 5005
4.53	92.759	.010 781	5.15	172.43	.00 57994	7.65	2 100.6	.000 4760
4.54	93.691	.010 673	5.20	181.27	.00 55166	7.70	2 208.3	.000 4528
4.55	94.632	.010 567	**5.25**	190.57	.00 52475	**7.75**	2 321.6	.000 4307
4.56	95.583	.010 462	5.30	200.34	.00 49916	7.80	2 440.6	.000 4097
4.57	96.544	.010 358	5.35	210.61	.00 47482	7.85	2 565.7	.000 3898
4.58	97.514	.010 255	5.40	221.41	.00 45166	7.90	2 697.3	.000 3707
4.59	98.494	.010 153	5.45	232.76	.00 42963	7.95	2 835.6	.000 3527
4.60	99.484	.010 052	**5.50**	244.69	.00 40868	**8.00**	2 981.0	.000 3355
4.61	100.48	.009 9518	5.55	257.24	.00 38875	8.05	3 133.8	.000 3191
4.62	101.49	.009 8528	5.60	270.43	.00 36979	8.10	3 294.5	.000 3035
4.63	102.51	.009 7548	5.65	284.29	.00 35175	8.15	3 463.4	.000 2887
4.64	103.54	.009 6577	5.70	298.87	.00 33460	8.20	3 641.0	.000 2747
4.65	104.58	.009 5616	**5.75**	314.19	.00 31828	**8.25**	3 827.6	.000 2613
4.66	105.64	.009 4665	5.80	330.30	.00 30276	8.30	4 023.9	.000 2485
4.67	106.70	.009 3723	5.85	347.23	.00 28799	8.35	4 230.2	.000 2364
4.68	107.77	.009 2790	5.90	365.04	.00 27394	8.40	4 447.1	.000 2249
4.69	108.85	.009 1867	5.95	383.75	.00 26058	8.45	4.675.1	.000 2139
4.70	109.95	.009 0953	**6.00**	403.43	.00 24788	**8.50**	4 914.8	.000 2035
4.71	111.05	.009 0048	6.05	424.11	.00 23579	8.55	5 166.8	.000 1935
4.72	112.17	.008 9152	6.10	445.86	.00 22429	8.60	5 431.7	.000 1841
4.73	113.30	.008 8265	6.15	468.72	.00 21335	8.65	5 710.1	.000 1751
4.74	114.43	.008 7386	6.20	492.75	.00 20294	8.70	6.002.9	.000 1666
4.75	115.58	.008 6517	**6.25**	518.01	.00 19305	**8.75**	6 310.7	.000 1585
4.76	116.75	.008 5656	6.30	544.57	.00 18363	8.80	6 634.2	.000 1507
4.77	117.92	.008 4804	6.35	572.49	.00 17467	8.85	6 974.4	.000 1434
4.78	119.10	.008 3960	6.40	601.85	.00 16616	8.90	7 332.0	.000 1364
4.79	120.30	.008 3125	6.45	632.70	.00 15805	8.95	7 707.9	.000 1297
4.80	121.51	.008 2297	**6.50**	665.14	.00 15034	**9.00**	8 103.1	.000 1234
4.81	122.73	.008 1479	6.55	699.24	.00 14301	9.05	8 518.5	.000 1174
4.82	123.97	.008 0668	6.60	735.10	.00 13604	9.10	8 955.3	.000 1117
4.83	125.21	.007 9865	6.65	772.78	.00 12940	9.15	9 414.4	.000 1062
4.84	126.47	.007 9071	6.70	812.41	.00 12309	9.20	9 897.1	.000 1010
4.85	127.74	.007 8284	**6.75**	854.06	.00 11709	**9.25**	10 405	.000 0961
4.86	129.02	.007 7505	6.80	897.85	.00 11138	9.30	10 938	.000 0914
4.87	130.32	.007 6734	6.85	943.88	.00 10595	9.35	11 499	.000 0870
4.88	131.63	.007 5970	6.90	992.27	.00 10078	9.40	12 088	.000 0827
4.89	132.95	.007 5214	6.95	1 043.1	.00 09586	9.45	12 708	.000 0787
4.90	134.29	.007 4466	**7.00**	1 096.6	.00 09119	**9.50**	13 360	.000 0749
4.91	135.64	.007 3725	7.05	1 152.9	.00 08674	9.55	14 045	.000 0712
4.92	137.00	.007 2991	7.10	1 212.0	.00 08251	9.60	14 765	.000 0677
4.93	138.38	.007 2265	7.15	1 274.1	.00 07849	9.65	15 522	.000 0644
4.94	139.77	.007 1546	7.20	1 339.4	.00 07466	9.70	16 318	.000 0613
4.95	141.17	.007 0834	**7.25**	1 408.1	.00 07102	**9.75**	17 154	.000 0583
4.96	142.59	.007 0129	7.30	1 480.3	.00 06755	9.80	18 034	.000 0555
4.97	144.03	.006 9431	7.35	1 556.2	.00 06426	9.85	18 958	.000 0527
4.98	145.47	.006 8741	7.40	1 636.0	.00 06113	9.90	19 930	.000 0502
4.99	146.94	.006 8057	7.45	1 719.9	.00 05814	9.95	20 952	.000 0477
5.00	148.41	.006 7379	**7.50**	1 808.0	.00 05531	**10.00**	22 026	.000 0454

BIBLIOGRAPHY

Applications in Life Sciences

Folk, George E., Jr. *Textbook of Environmental Physiology*. Philadelphia: Lee and Febiger, 1974.
COMMENTS: Many graphs and functions are used to describe the relationship between various physiological aspects of organisms and the environment of those organisms. Scatter diagrams are presented with the corresponding lines of linear regression.

Hall, Charles A.S., and John W. Day, Jr. *Ecosystem Modeling in Theory and Practice: An Introduction with Case Histories*. New York: John Wiley and Sons, 1977.
COMMENTS: Models that are constructed to describe our ecosystem are discussed and case histories are presented to show the power of these models. Logistic curves, linear regression, differential equations, consumer surplus, and other ideas that are pertinent to calculus are discussed at a level that can be understood by students in this course.

Harte, John. *Consider a Spherical Cow, A Course in Environmental Problem Solving*. Los Altos, Calif.: William Kaufmann, Inc., 1985.
COMMENTS: Methods are given for constructing mathematical models for a variety of environmental problems. This delightful book entertains as well as teaches. Students in this calculus course have the mathematical background needed to understand nearly all the applications presented.

Kanwisher, J.W., and S. Ridgeway. "The Physiological Ecology of Whales and Porpoises." *Scientific American* 248, no. 6 (June 1983): 110–20.
COMMENTS: An investigation of the metabolism of whales and porpoises indicates that mass-to-metabolism data for these marine mammals do not lie along the mouse-to-elephant curve (see Example 8.36). This observation leads to some interesting conclusions about the evolution of these mammals.

Klein, Richard M. and Deana T. Klein. *Research Methods In Plant Science*. Garden City, N.Y.: The Natural History Press, 1970.
COMMENTS: Graphs are used throughout. Exponential rates of growth are discussed (p. 252). Diffusion is used to discuss seed germination (p. 641), and the derivative is used to describe the rate of change of the volume of a seed under certain conditions (p. 642).

Livingstone, Frank B. *Abnormal Hemoglobins in Human Population*. Chicago: Aldine, 1967.
COMMENTS: Both differentiation and integration are used extensively in the study of human genetics presented in this book. In particular, a very nice discussion of this type is given in Chapter IV, entitled "The Population Genetics of the Red Cell Defects" (pp. 22–36).

Olson, Theodore A. and Frederick J. Burgess, eds. *Pollution and Marine Ecology*. New York: Interscience Publishers, 1967.
COMMENTS: Many of the essays in this book make use of graphs, functions, and calculus. In the essay by E.A. Pearson, P.N. Stone, and R.E. Selleck, entitled "Some Physical Parameters and Their Significance in Marine Waste

Disposal" (pp. 297–315), some interesting diffusion problems are discussed using models for diffusion that are somewhat more complicated than the one discussed in this book.

Shaffer, P.L. and H.J. Gold, "A Simulation Model of Population Dynamics of the Codling Moth, *cydia pomonella.*" *Ecological Modelling* 30 (1985): 247–74.
COMMENTS: Numerous aspects of the life cycle of the codling moth are modelled, using several types of models, including quadratic and logistic. Examples 1.36 and 3.32 are based on data extracted from this article. The journal itself is a wonderful source of models applying the concepts in this calculus course to the life sciences.

Sissons, C.J., M. Cross, and S. Robertson. "A New Approach to the Mathematical Modelling of Biodegradation Processes." *Appl. Math. Modelling* 10 (1986): 33–40.
COMMENTS: Several models are presented for the growth rate of biological organisms under certain conditions, and the article shows how models are adjusted to reflect increasingly more of the physical realities in an application. Functions similar to logistic functions are used. The journal is a source of interesting applications related to the material studied in calculus.

Tannebaum, Steven R., and Daniel I.C. Wang, eds. *Single-Cell Protein II.* Cambridge, Mass.: MIT Press, 1975.
COMMENTS: Many graphs and functions are used to illustrate the ideas discussed. Derivatives are used to describe the productivity of single-cell protein from methane (pp. 359–362). A horizontal asymptote is used to discover a point of diminishing return for the expenditure of energy to increase the production of single-cell protein (p. 430). Many graphs employ logarithmic scales (for example, a graph describing the death of yeast during heat treatment, pp. 92–93).

Zeide, B. "Tolerance and Self-tolerance of Trees." *Forest Ecology and Management* 13 (1985): 149–66.
COMMENTS: The affect of the density of trees in a stand on the mature size of individual trees in the stand is modelled by exponential and logarithmic functions. Example 5.23 is based on this paper. The journal contains other papers that show interesting models of forestry problems that are accessible to students using this calculus text.

Applications in Social and Behavioral Sciences

Bower, Gordon H., ed. *The Psychology of Learning and Motivation.* Vol. 10. New York: Academic Press, 1976.
COMMENTS: The essay by Howard Rachlin, "Economic Demand Theory and Psychological Studies of Choice," shows how economic models similar to those discussed in this calculus text can be used to describe psychological phenomena. Many graphs and their tangents are used.

Cooper, L.A. and R.N. Shepard. "Turning something over in the mind." *Scientific American* 251, 6 (December 1984): 106–15.
COMMENTS: Spatial thinking by making objective observations is studied by analyzing the length of time subjects require to determine if two objects in different orientations are identical. Linear equations are used extensively. (See Example 1.22.)

Keyfitz, Nathan, and Wilhelm Flieger. *Population—Facts and Methods of Demography.* San Francisco: W.H. Freeman, 1971.
COMMENTS: Types of functions used by demographers are shown and the concepts of differentiation, integration, exponential functions, and im-

proper integration are discussed. Of particular interest is Chapter 11, entitled "Inferences from Incomplete Data."

Luce, Robert D.; Bush, Robert R.; and Galanter, Eugene, eds. *Handbook of Mathematical Psychology* (3 vols.). New York: John Wiley & Sons, 1963.
COMMENTS: These volumes assume a good background in calculus. An excellent sampling of the mathematics involved in research psychology is presented. Ideas that have been developed for use in this area are introduced in this calculus text.

Strasser, Gabor, and Eugene M. Simons, eds. *Science and Technology Policies.* Cambridge, Mass.: Ballinger, 1973.
COMMENTS: Of particular interest is Chapter 18, entitled "Benefit-Cost Considerations in National Planning." Reasonable approaches to analyzing many different problems facing sociologists are given. Many graphs, some using logarithmic scales, are shown. When dealing with functions, the emphasis is upon interpretation instead of mathematical techniques. Diffusion problems are considered and exponential functions are used.

Applications in Business and Economics

Alpert, Mark I. *Pricing Decisions.* Glenview, Ill.: Scott, Foresman, 1971.
COMMENTS: Topics studied in this calculus text are utilized with very little added sophistication. Derivatives, demand curves, max-mins, scatter diagrams, linear regression, and marginal analysis are all used. An excellent book to explain the inclusion of topics in this text.

Childress, Robert L. *Mathematics For Managerial Decisions.* Englewood Cliffs, N. J.: Prentice-Hall, 1974.
COMMENTS: The basic mathematical tools needed for economics, finance, production, and marketing are presented. Topics include matrix theory, linear programming, calculus, linear regression, and many types of models.

Davis, S. *Zero-Coupon Bonds,* Consortium: The Newsletter of the Consortium for Mathematics and its Applications, no. 20 (November 1986). [The Hi-Map pullout section (no page numbers)]
COMMENTS: The concept of the present value of money is applied to such applications as zero-coupon bonds and the real value of salaries (such as to athletes) that are paid out over a long period of time. The discussion is lively, interesting, and uses only techniques developed in this calculus text. This series of newsletters is a source of interesting current applications.

Day, Ralph L. *Marketing Models: Quantitative and Behavioral.* Scranton, Pa.: International Textbook, 1964.
COMMENTS: Good discussions using calculus techniques to make marketing decisions. Derivatives and integrals are both used.

Donnelly, James H., and John M. Ivancevich. *Analysis for Marketing Decisions.* Homewood, Ill.: Richard D. Irwin, 1970.
COMMENTS: Instead of using calculus as such, this book uses a "small change in x" approach (somewhat like dealing with differentials). Good reading for students beginning a calculus course since it presents the rationale for using calculus to analyze marketing problems without being burdened with the complication of calculus.

Forsythe, Robert E. and David A. Walker. *Mathematics for Economic and Business Analysis.* Pacific Palisades, Calif.: Goodyear Publishing, 1976.
COMMENTS: Topics of interest to business and economics majors, such as linear programming, matrix theory, and probability, as well as calculus are presented. Many interesting models and examples are discussed.

Frank, Ronald E., and William F. Massy. *An Econometric Approach to a Marketing Decision Model.* Cambridge, Mass.: MIT Press, 1971.

COMMENTS: This book is hard reading but demonstrates what is involved in constructing an elaborate marketing model. Uses a great deal of calculus as well as topics from linear algebra. Illustrations of the kind of mathematics that lies ahead for students interested in this area.

Maxwell, David W. *Price Theory and Applications in Business Administration.* Pacific Palisades, Calif.: Goodyear Publishing, 1970.

COMMENTS: Extensive use is made of differentiation to develop the major concepts involved in price theory.

Shane, Harold D. *Mathematics for Business Applications.* Columbus, Ohio: Charles E. Merrill, 1976.

COMMENTS: Mathematical topics such as linear programming, matrix theory, probability, and calculus are given. Strong emphasis is placed on applications in business and finance.

Teichroew, Daniel. *An Introduction to Management Science: Deterministic Models.* New York: John Wiley & Sons, 1964.

COMMENTS: An excellent text making frequent use of calculus to solve management problems. The concepts of differentiation, integration, max-mins, differentials, diffusion, functions of several variables, and fitting curves to scatter diagrams are all applied.

Answers

Section 1.1

1. $f(2) = 3(2) + 11 = 17$
   $f(0) = 3(0) + 11 = 11$
   $f(-1) = 3(-1) + 11 = 8$
   $f(-4) = 3(-4) + 11 = -1$
   $f(h) = 3(h) + 11 = 3h + 11$
3. $f(0) = 2(0)^2 - 3(0) = 0$
   $f(2) = 2(2)^2 - 3(2) = 2$
   $f(-3) = 2(-3)^2 - 3(-3) = 27$
   $f(h) = 2(h)^2 - 3(h) = 2h^2 - 3h$
5. $h(2) = \sqrt{(2)^2 - 3} = 1$
   $h(3) = \sqrt{(3)^2 - 3} = \sqrt{6}$
   $h(6) = \sqrt{(6)^2 - 3} = \sqrt{33}$
   $h(1) = \sqrt{(1)^2 - 3} = \sqrt{-2}$ not real
   $h(x + a) = \sqrt{(x + a)^2 - 3}$
   $h(x) + h(a) = \sqrt{x^2 - 3} + \sqrt{a^2 - 3}$

7. $f(0) = -17$
   $f(-1) = -17$
   $f(5) = -17$
   $f(x + h) = -17$

9. $f(0) = 0$
   $f(2) = \frac{4}{5}$
   $f(-2) = -\frac{4}{5}$
   $f(x + h) = \dfrac{2x + 2h}{x^2 + 2xh + h^2 + 1}$

11. $f(2) = 4; f(5) = 25; f(7) = 49; f(2) + f(5) = 29$
13. $f(2) = \sqrt{5} \approx 2.236; f(5) = \sqrt{8} \approx 2.828;$
    $f(7) = \sqrt{10} \approx 3.162; f(2) + f(5) \approx 5.064$
15. $f(2) = 6; f(5) = 120; f(7) = 336; f(2) + f(5) = 126$

17. yes 19. yes 21. yes 23. yes
25. yes 27. no

29. $C(10) = 80 + 45(10) = 530$
    $C(100) = 80 + 45(100) = 4580$
    $C(200) = 80 + 45(200) = 9080$
31. $C(-10) = 80 + 45(-10) = -370$ (not meaningful)
33. $D(0) = 200 - 0.5(0) = 200$ (they could give away 200 trailers)
35. $D(5000) = 200 - 0.5(5000) = -2300$ (impossible, no trailers should be produced)
37. $f(40) = 300 - 0.1(40)^2 = 140$
39. $x \neq 0$; denominator $\neq 0$
41. $x \neq 1, 2$; denominator $\neq 0$
43. $x \leq 3$; no negative radicals
45. All t

47. $\dfrac{(2x - 3) - 1}{x - 2} = 2$
49. $\dfrac{(0.5 + x) - (-0.5)}{x - (-1)} = 1$
51. $\dfrac{(x^2 + x) - 6}{x - 2} = x + 3$
53. $\dfrac{6 - 6}{x - 3} = 0$
55. $\dfrac{\sqrt{x} - \sqrt{2}}{x - 2} = \dfrac{\sqrt{x} - \sqrt{2}}{(\sqrt{x} - \sqrt{2})(\sqrt{x} + \sqrt{2})} = \dfrac{1}{\sqrt{x} + \sqrt{2}}$
57. $x \neq 2$
59. $x \neq 2$
61. $x \geq 2$; no negative radicals
63. $x \neq -1, 3$
65. $x \neq 1, -3$

Section 1.2

1, 3, 5.

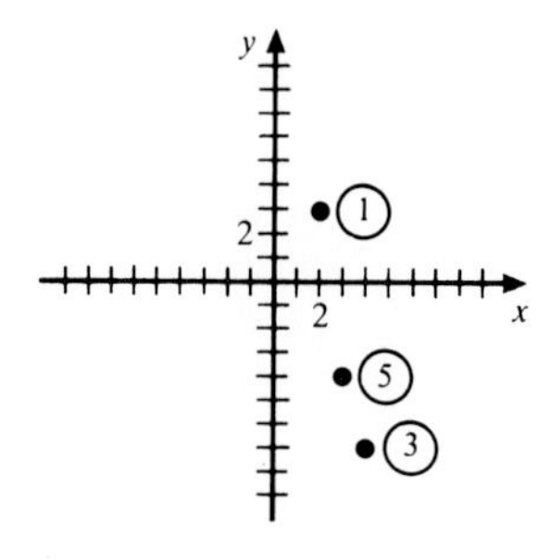

7, 9, 11, 13.

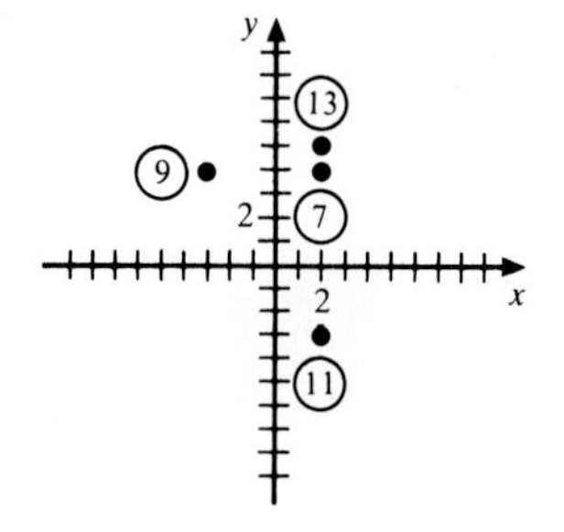

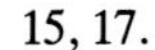
15, 17.

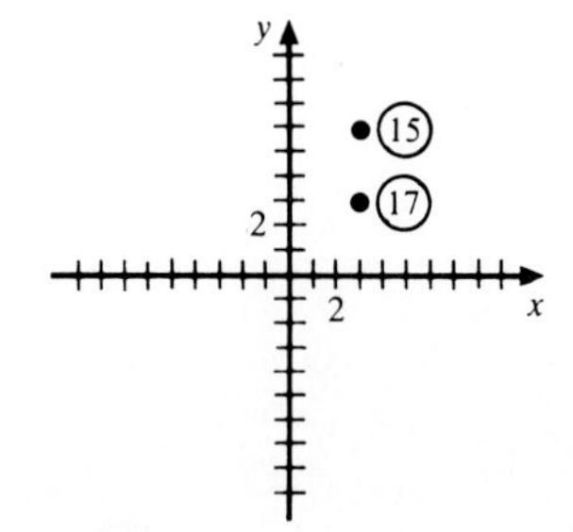

19.

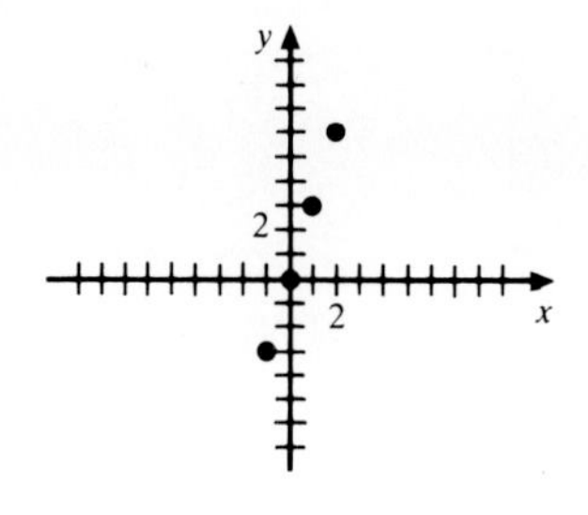

21.

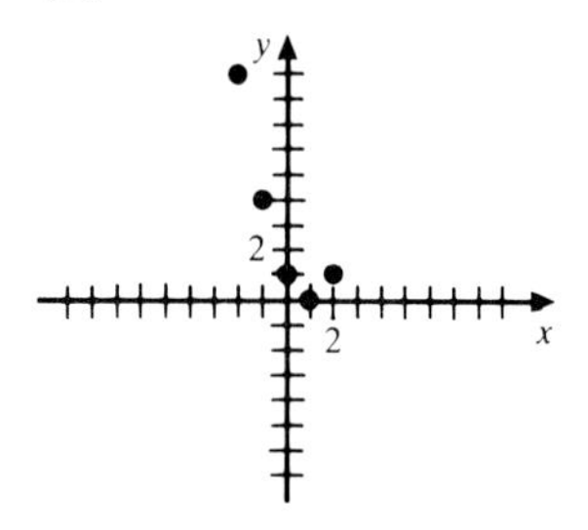

23. One of the points is (c, b).

25.

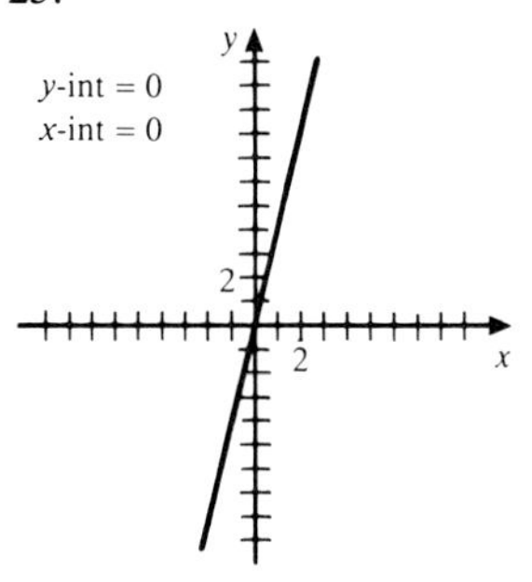

27.

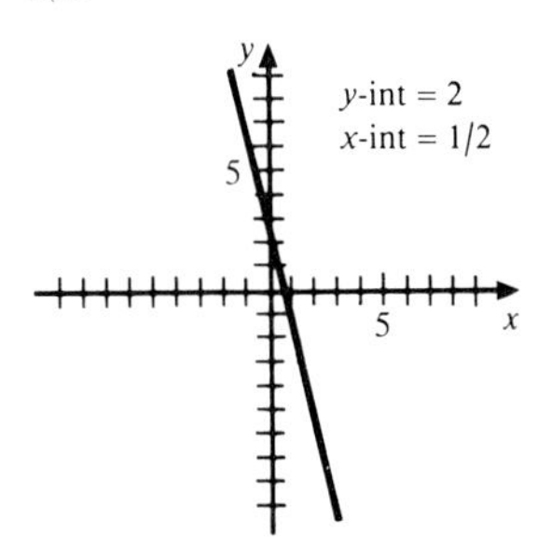

29.

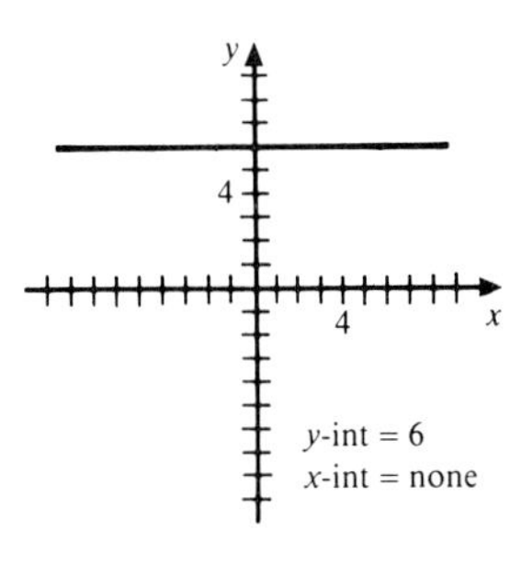

31.

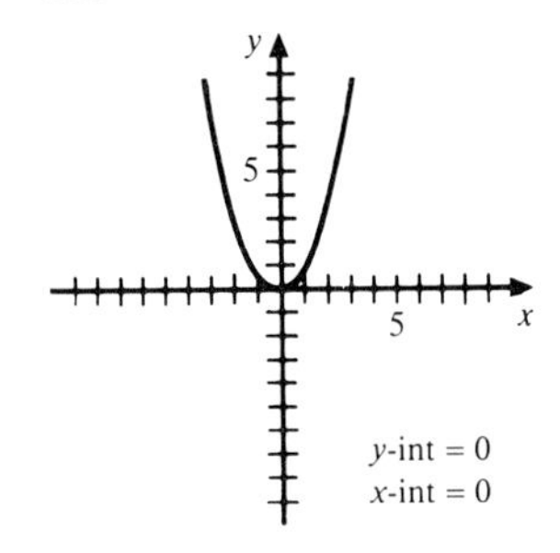

33.

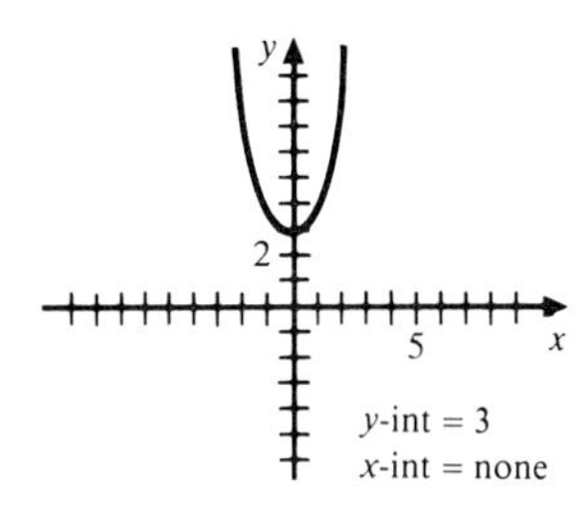

35.

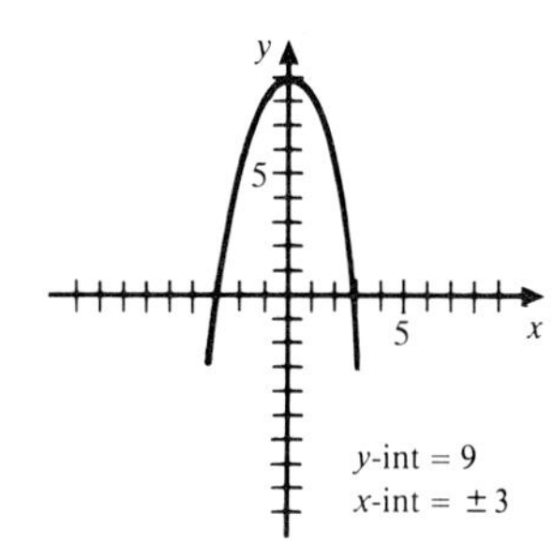

37.

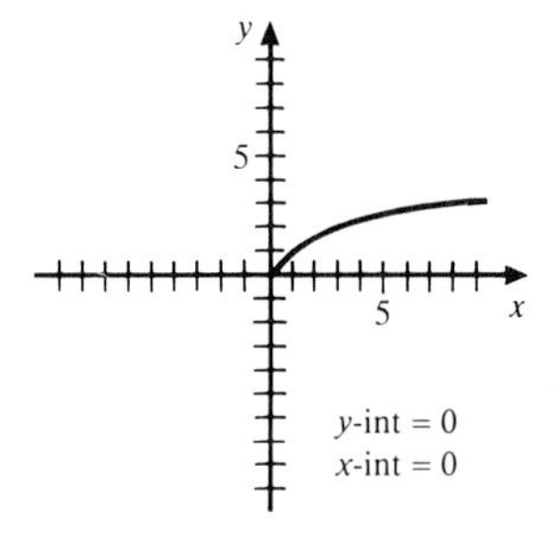

39.

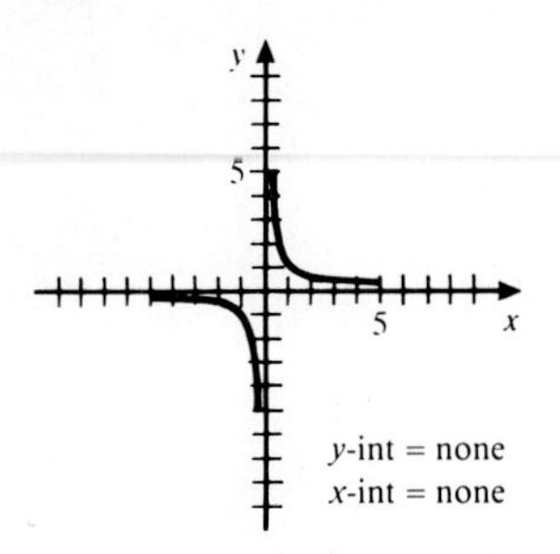

41.

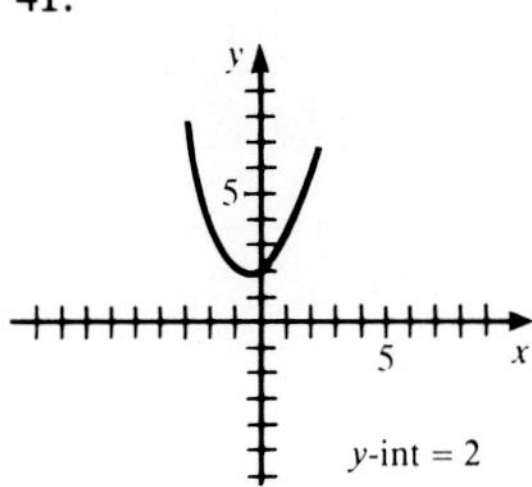

43.

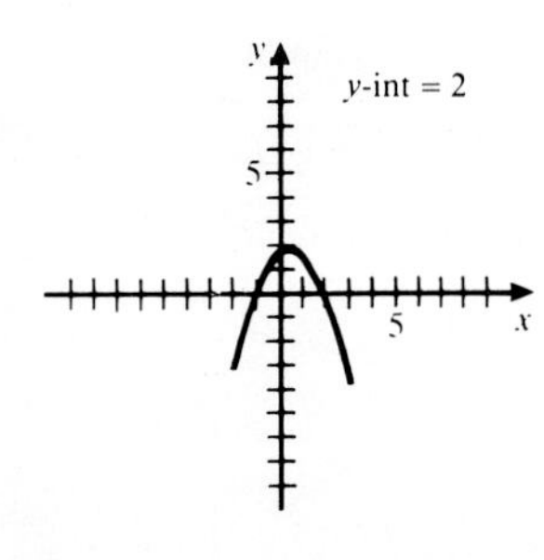

45. yes 47. no 49. no

51.

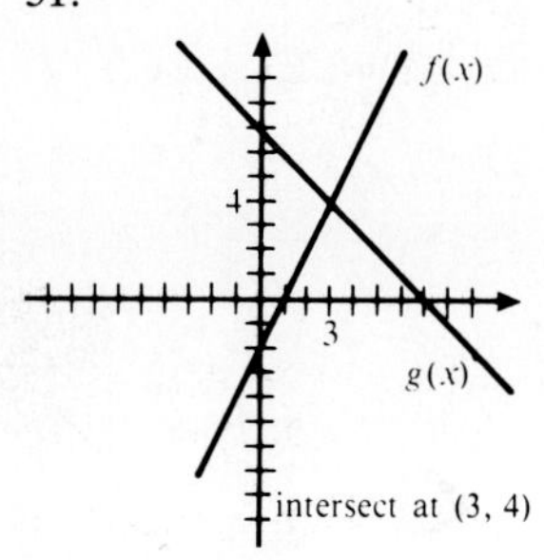

53.

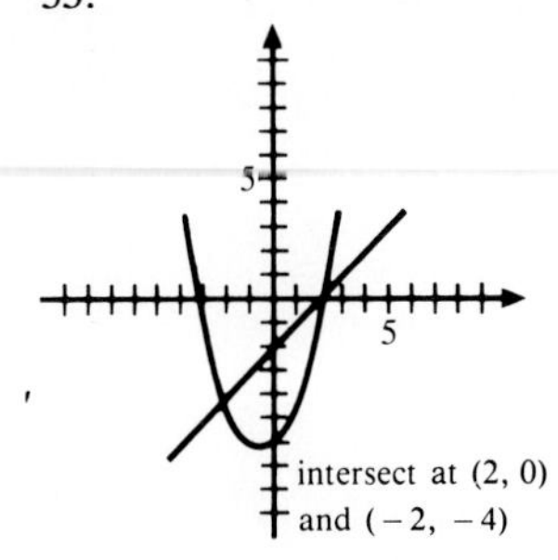

Section 1.3

1.

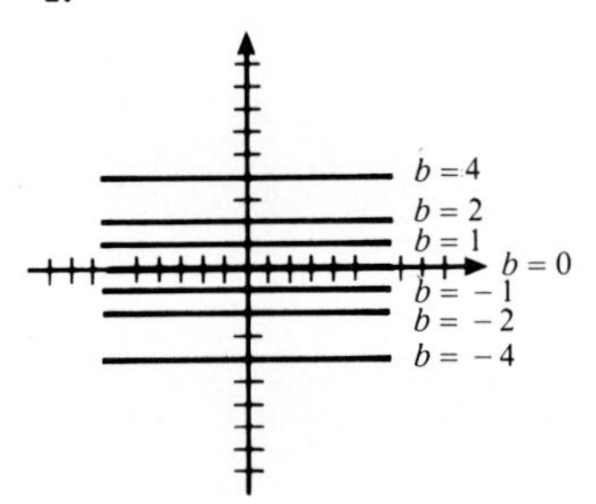

3.

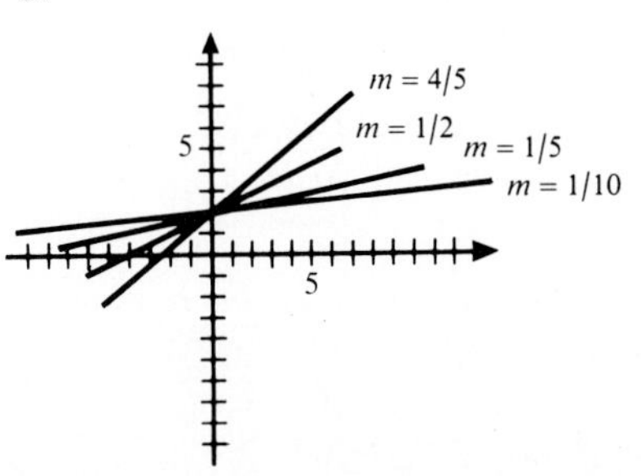

5.

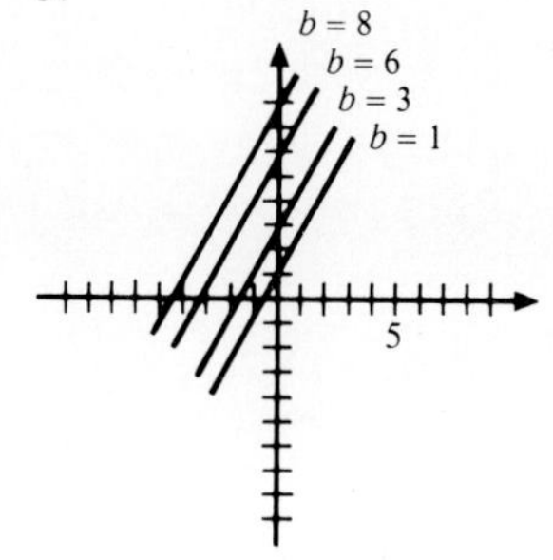

7.

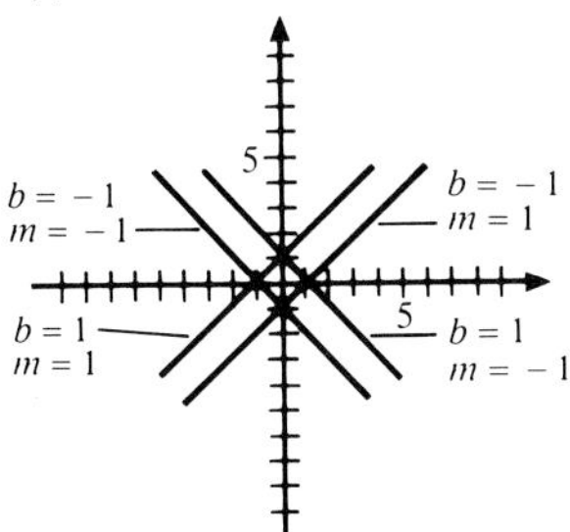

9.

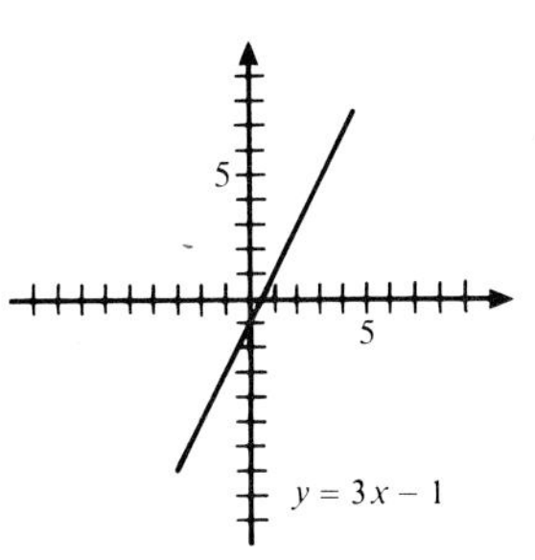

11.

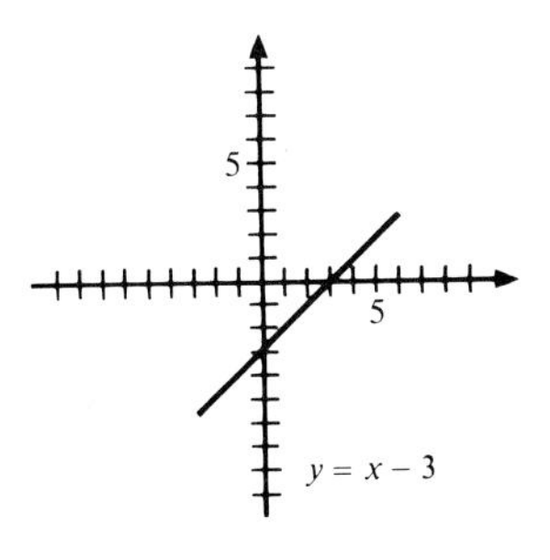

13.

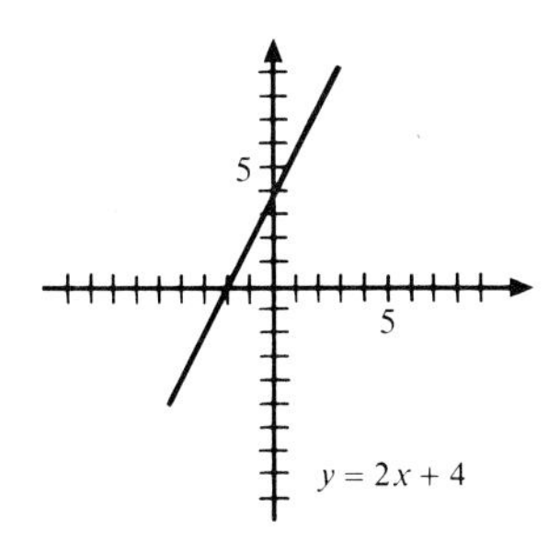

15.

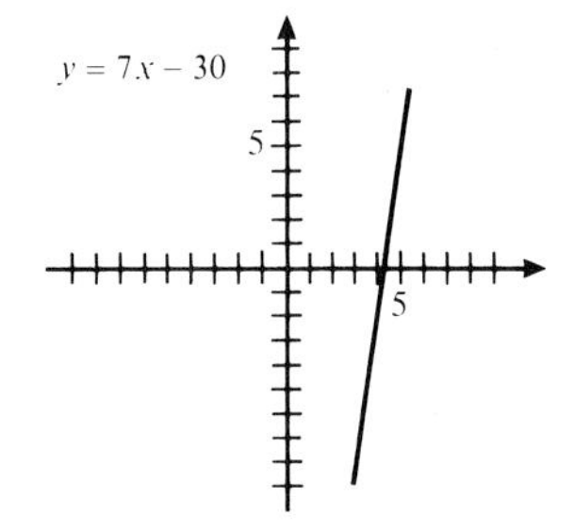

17.

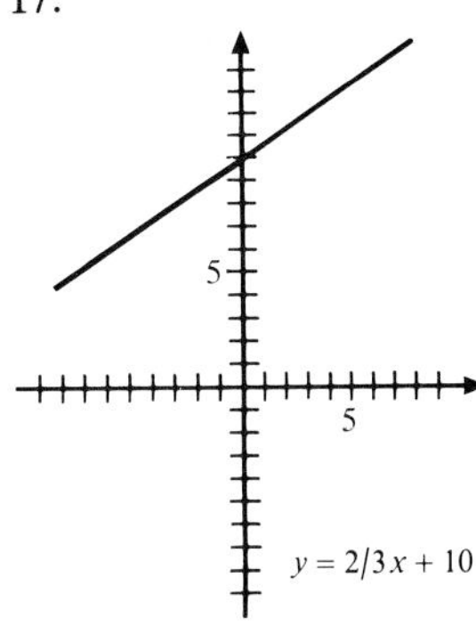

19.

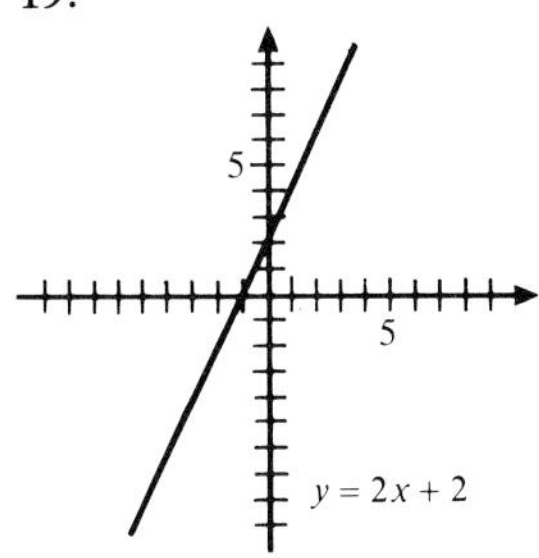

21. $y - 1 = \frac{1}{2}(x - 2)$
23. $y - 4 = 1(x - 4)$
25. $y - 2 = -2(x - 0)$
27. $y - 2 = -2(x - 1)$
29. $y - 0 = 3(x - 1)$
31. $y - 0 = 4(x - 3)$
33. $y - 6 = -3(x - 0)$
35. $y + 3 = \frac{-3}{2}(x - 0)$
37. Linear
39. Linear
41. Not linear. It is quadratic.
43. Linear
45. Linear
47. $f(0) = 1.5$ seconds
49. $33\frac{1}{3}$ degrees/second
51. $y = -2x + 12$
53. $y = 1.2x + 1.8$
55. $y = 4x + 23$
57. $y = -2x$
59. $y = \frac{1}{5}x + \frac{33}{5}$

61. $y = -2x$
63. (3, 0)
65. (3, −2)
67. (3, 1)
69. (3, 3)
71. $\sqrt{58}$
73. $2\sqrt{17}$
75. $4\sqrt{2}$
77. $\sqrt{13}$

79. There will never be enough money in the college fund.

Section 1.4

1. (a) $F = \frac{9}{5}C + 32$
   $F(60) = \frac{9}{5}(50) + 32 = 122$ degrees
   (b) Same as (a).
3. $m = 0.6215k$
   $k = 1.609m$
   You are not speeding.
5. $C(x) = 10{,}000 + \frac{1000}{11}(x - 90)$
   $C(300) = \$29{,}090.90$
   $C(0) = \$1818.18$
7. $S(2) = 49.4$ and $S(10) = 55$
9. $C(t) = 3 + 1.5t$
   $C(12) = 21$ dollars
11. $C(t) = 30 + 16t$
    8 minutes
13. The cost function is $C(x) = 22.5x + 10{,}000$.
    $C(6500) = 156{,}250$.
15. $p(9) = 10.125$
17. (65, 405)
19. (140, 320)
21. (3.91, 30.725)
23. (5.91, 46.36)
25. The demand will drop by 20 pies.
27. Revenue decreases from \$75 to zero.
29. Demand decreases by 145 items.

Section 1.5

1. y-intercept: 0
   x-intercept: 0

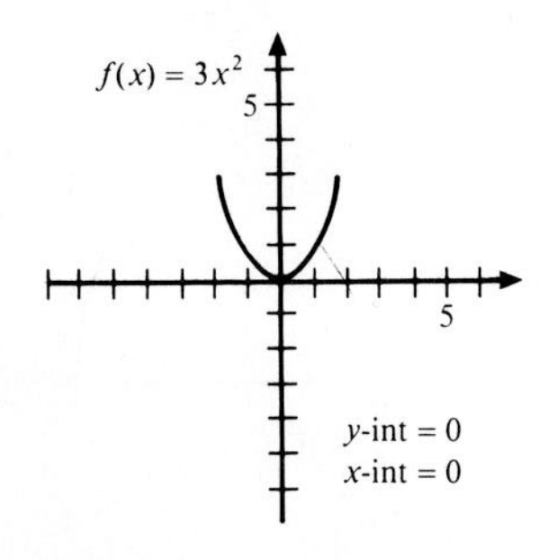

3. y-intercept: 5
   x-intercepts: $-1, -5$

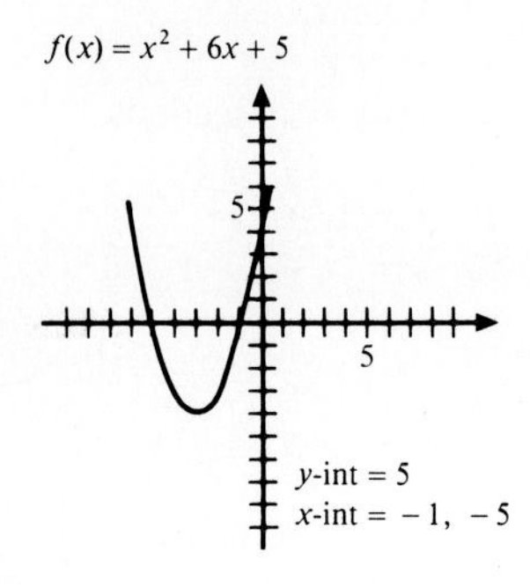

5. y-intercept: -8
   x-intercepts: $-1 + \sqrt{5}, -1 - \sqrt{5}$

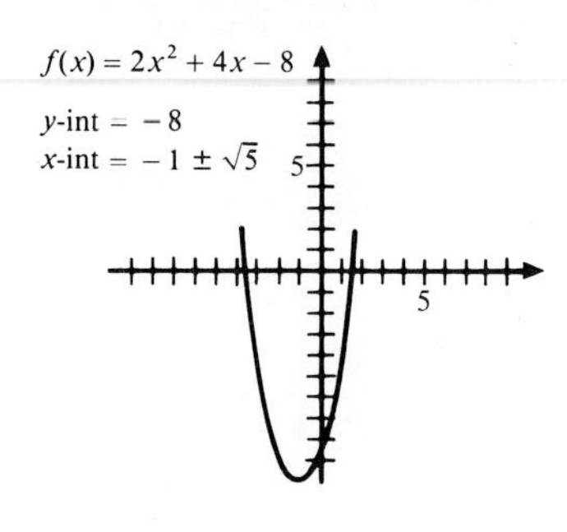

7. y-intercept: 5
   x-intercepts: $-2 + \frac{1}{3}\sqrt{21}, -2 - \frac{1}{3}\sqrt{21}$

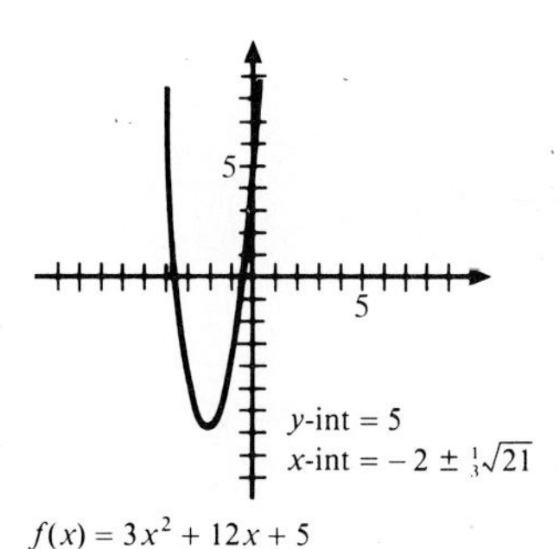

9. y-intercept: 12
   x-intercepts: $-2 + \sqrt{10}, -2 - \sqrt{10}$

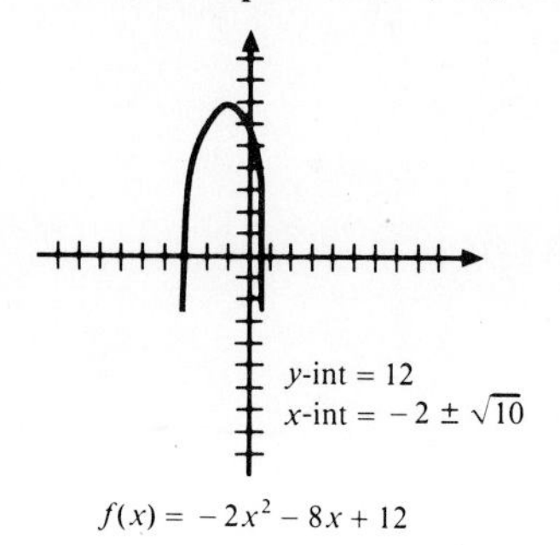

11. y-intercept: -3
x-intercepts: $3 + \frac{1}{2}\sqrt{30}$, $3 - \frac{1}{2}\sqrt{30}$

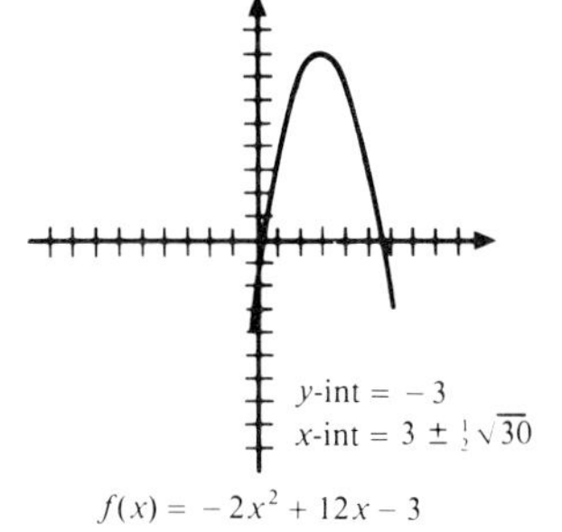

13. y-intercept: 8
x-intercepts: none

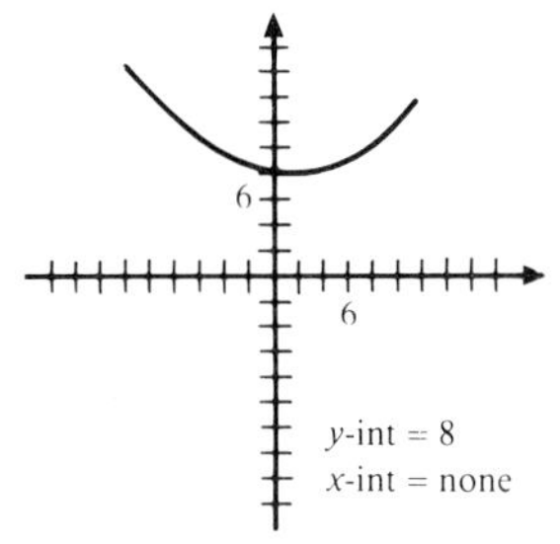

15. $(-1, 0)$ and $(3, 16)$

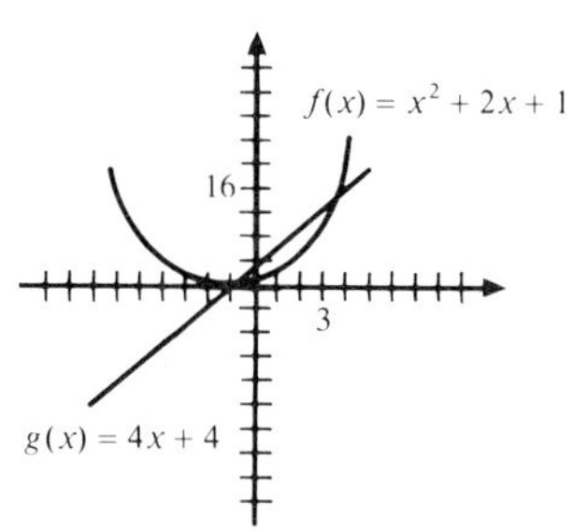

17. $(-2, -5)$ and $(-\frac{1}{2}, -\frac{1}{2})$

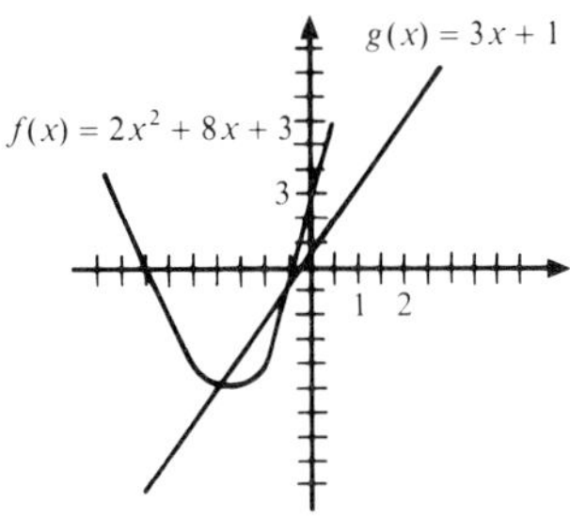

19. $f(4) = 10.6$ $\quad g(4) = 4$
$f(10) = 20$ $\quad g(10) = 10$
$f(30) = 0$ $\quad g(30) = 30$
$f(x) = g(x)$ at 20 months.
$f(12) = 21.6$ $\quad g(12) = 12$
$f(x)$ should be adopted if the program will last 12 months.
$g(x)$ should be adopted if the program will last indefinitely.

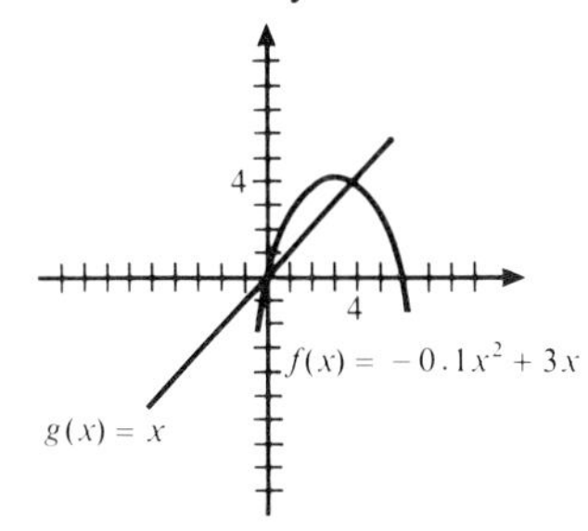

21. $N(15) = 46.75\%$
At 30 degrees Celsius 64% survive.

23. 2 $\qquad$ 25. 0 $\qquad$ 27. 9 $\qquad$ 29. 2

31. $|2 + (-4)| = |(-2)| = 2$

33. $f(x) = \begin{cases} x - 1, & x \geq 1 \\ 1 - x, & x < 1 \end{cases}$

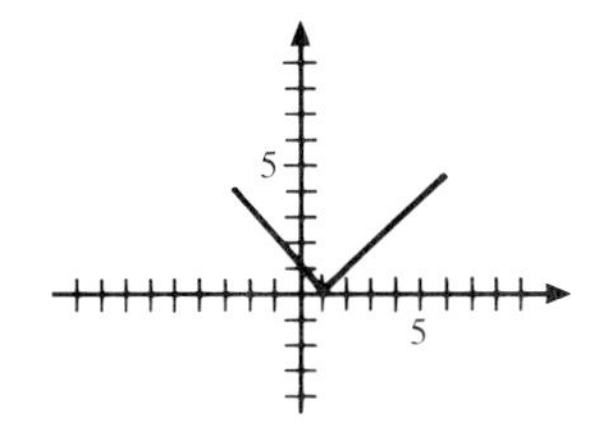

35. $f(x) = \begin{cases} -x, & x \leq 0 \\ x, & x > 0 \end{cases}$

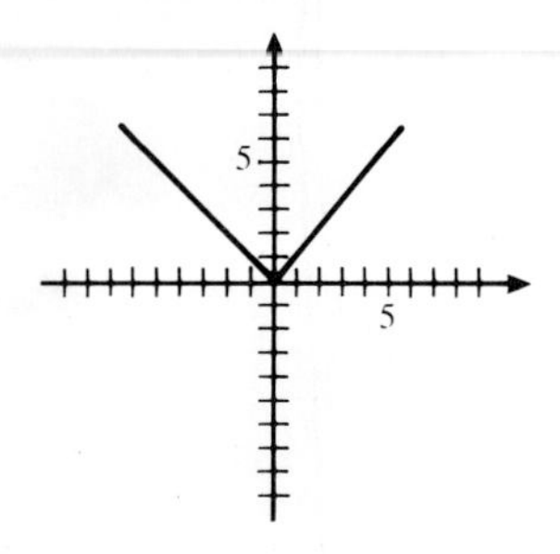

37. $f(x) = \begin{cases} 2x + 3, & x \geq -\frac{3}{2} \\ -2x - 3, & x < -\frac{3}{2} \end{cases}$

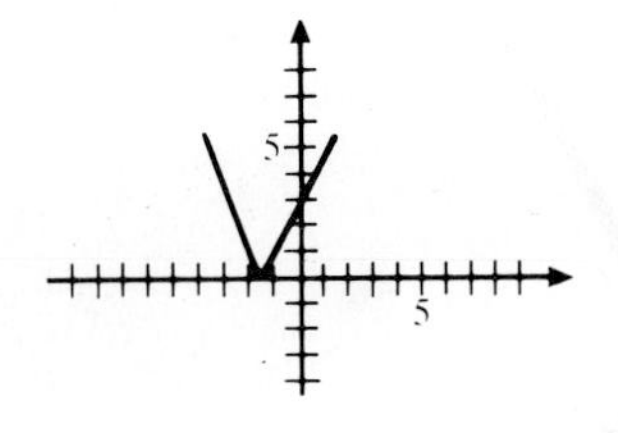

Section 1.6

1. 4
3. 10
5. 2
7. 0.59
9. 27
11. 51
13. −5
15. 22,050
17. 6350
19. −12
21. −3
23. 9.615
25. 6.24
27. \$25.08
29. \$1.97
31. \$2.00
33. −3200
35. 10 million per year
    −8 million per year

Chapter 1 Review Exercises

1. $f(2) = 5$
   $f(a-3) = -a^2 + 6a$
3. $x \geq -3$ and $x \neq -2$
5. $y = -5x + 3$
7. $y = 2x + 5$
9. $y = 4$
11. $y + 2 = -2(x - 1)$
13. $y = -2x$
15. $I(s) = 0.15s + 150$
    $I(800) = \$270$
17. y-intercept: -2
    x-intercept: $\frac{2}{3}$
19. y-intercept: -1
    x-intercepts: $-1, \frac{1}{2}$
21. $(2, 9)$
23. $(2, 5)$ and $(-6, 13)$

25.

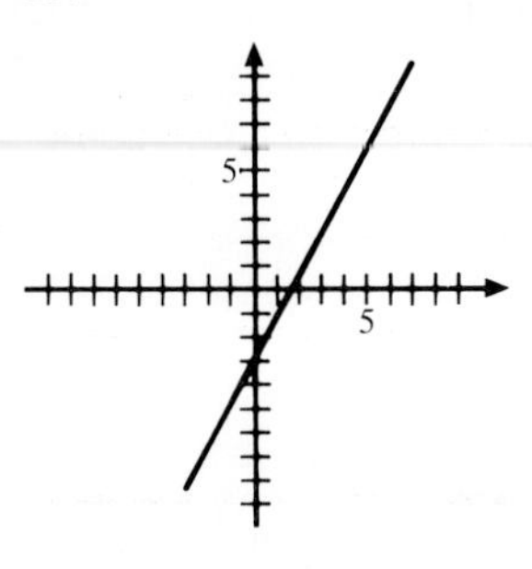

27.

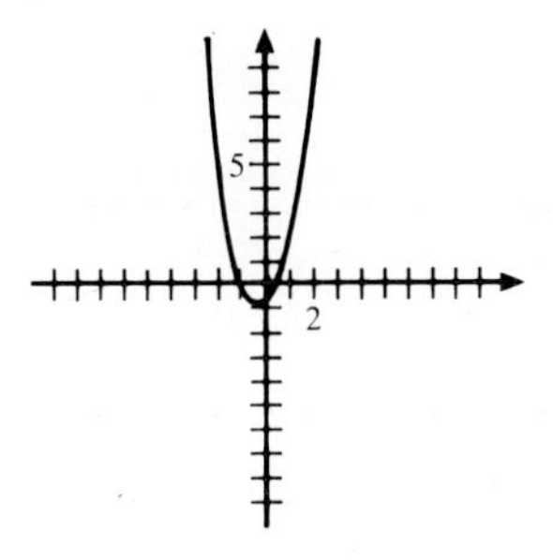

29.

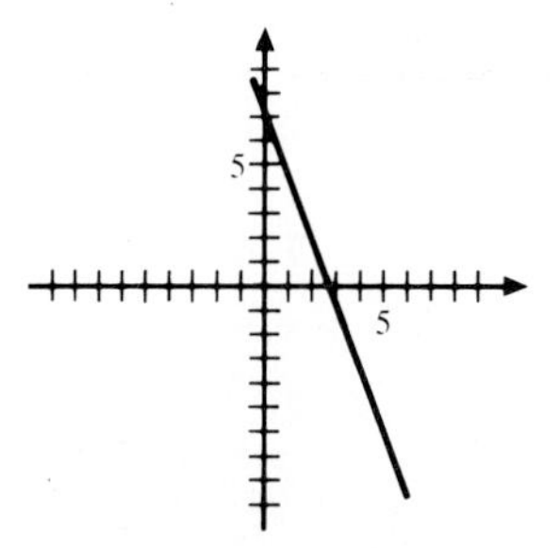

31. The first would raise more in a five-week campaign. The second would raise more in a twelve-week campaign.
    $f(x) = g(x)$ at $x = 10$ weeks.
33. Increase by 50 radios
35. 3
37. 5
39. 4
41. \$2.01

Section 2.1

1. 9; 7.5; 7.25; 7.125; 7.025; 7.0025; approaches 7
3. −6.5
5. −1.22
7. −0.06005
9. 0
11. 1.755
13. 0.05995
15. 0
17. −0.45
19. 0
21. 212.5
23. 4.495
25. 0.04
27. The tangent line must be horizontal.

Section 2.2

1. $\lim_{x\to2}(x+2)=4$

x	1	1.5	1.9	1.99	2.01	2.1	2.5	3
$f(x)$	3	3.5	3.9	3.99	4.01	4.1	4.5	5

3. $\lim_{x\to3}(2x+5)=11$

x	2	2.5	2.9	2.99	3.01	3.1	3.5	4
$f(x)$	9	10	10.8	10.98	11.02	11.2	12	13

5. $\lim_{x\to4}\frac{1}{x-4}$ does not exist since the left-hand and right-hand limits do not exist.

x	3	3.5	3.9	3.99	4.01	4.1	4.5	5
$f(x)$	−1	−2	−10	−100	100	10	2	1

7. $\lim_{x\to3}\frac{x}{(x-3)^2}$ does not exist.

x	2	2.5	2.9	2.99	3.01	3.1	3.5	4
$f(x)$	2	10	290	29,900	30,100	310	14	4

9. 0 11. 56 13. 4
15. The limit does not exist.
17. 0 19. $\frac{4}{3}$ 21. 13 23. −2
25. Does not exist. 27. −28 29. −33
31. 20 33. 14 35. 2 37. 8
39. $-\frac{1}{9}$ 41. −12 43. 12 45. 10,944 ft
47. 13 49. Undefined. 51. −2
53. 21 55. 0

Section 2.3

1. 2 3. −2 5. $\frac{\sqrt{2}}{4}$
7. 0 9. $-\frac{1}{4}$ 11. $f(x)=x^2-3;\ a=4$
13. $f(x)=x^2-1$, $a=2$ 15. $f(x)=x^2+2$, $a=2$
17. 7 19. 1 21. $\frac{-1}{x^2}$
23. 2 25. $10x-3$ 27. $\frac{-1}{x^2}$
29. $8x+4$ 31. $\frac{1}{(x+1)^2}$ 33. $3x^2$
35. $20x^4$ 37. $-\frac{3}{4}x^{-7/4}$ 39. $4x^{-1/3}+4$
41. $6x+2$ 43. $4x^3+9x^2-3$ 45. $6x-\frac{14}{3}x^{-5/3}$
47. 16 49. $2x+84x^5$ 51. 0
53. $-4x^{-3}$ 55. $\frac{4}{3}x^{-2/3}+6$ 57. $6\frac{1}{3}$
59. $\frac{4}{3}x^{-2/3}+6$ 61. $-9x^2$ 63. Does not exist.
65. $-6x^{-3}-\frac{2}{3}x$ 67. $-6x^{-3}-\frac{2}{3}x$ 69. −9
71. $\frac{25}{12}$ 73. 1 75. None
77. $x=0$ 79. $x=0, \frac{32}{3}$ 81. $x=0, 3$

Section 2.4

1. $y=-1$ 3. $y=-12x+20$
5. $y=7x-6$ 7. $y=\frac{1}{3}x+\frac{16}{3}$
9. $y=x-3$ 11. $y=-\frac{9}{64}x+\frac{15}{16}$
13. $y=x-1$ 15. $y=\frac{1}{8}x+6$
17. $x=3$ 19. $x=\frac{-2\sqrt{3}}{3}, \frac{2\sqrt{3}}{3}$
21. $x=3, 0$ 23. $x=1$
25. There are no values of x for which the slope of the tangent line is −4.
27. If $x=-1$ then $m=\frac{3}{2}$.
If $x=0$ then $m=1$.
If $x=2$ then $m=0$.
If $x=4$ then $m=-1$.
If $x=5$ then $m=-\frac{3}{2}$.

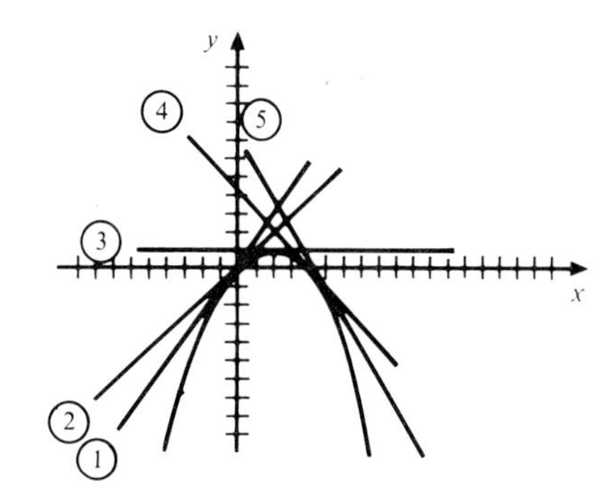

29. $x=-2$ 31. There is no x.
33. $x=2$ 35. $x=0, 3$
37. At (3, 12), $y=4x$. At (−3, 24), $y=-8x$.
39. 78.87 cubic ft/acre
77.25 cubic ft/acre
26.05 years
41. 3 43. 1.4
45. Cost is \$27.10. dC at $x=10$ is 27.

Section 2.5

1. $f(x)$ is increasing for $x>-3$ and decreasing for $x<-3$.
3. Critical value at $x=0$; $f(x)$ decreasing for $x<0$ and increasing for $x>0$.

5. No critical values, always increasing.
7. No critical values, $f(x)$ is always increasing.
9. Critical values at $x = -1, 2$; $f(x)$ increasing for $x < -1$ and $x > 2$, decreasing for $-1 < x < 2$.
11. No critical values, always increasing.
13. $f(x)$ is increasing when $x > \frac{3}{2}$ and decreasing when $x < \frac{3}{2}$.

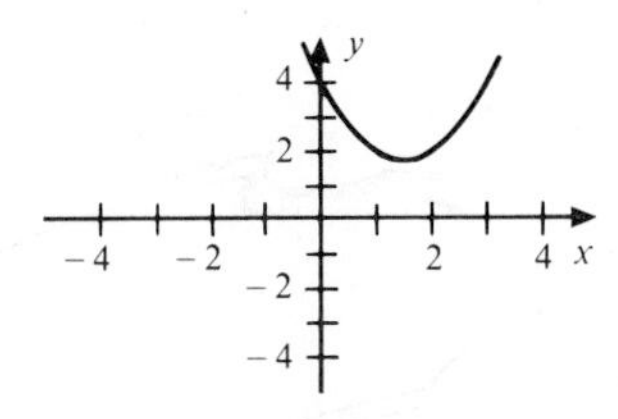

15. Critical values at $x = -1, 1$; $f(x)$ increasing for $x < -1$ and $x > 1$ and decreasing for $-1 < x < 1$.

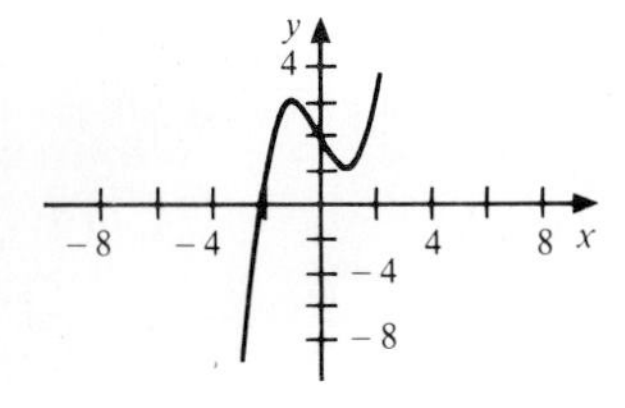

17. Critical values at $x = -1, 0$; $f(x)$ increasing for $-1 < x < 0$ and decreasing for $x < -1$ and $x > 0$.

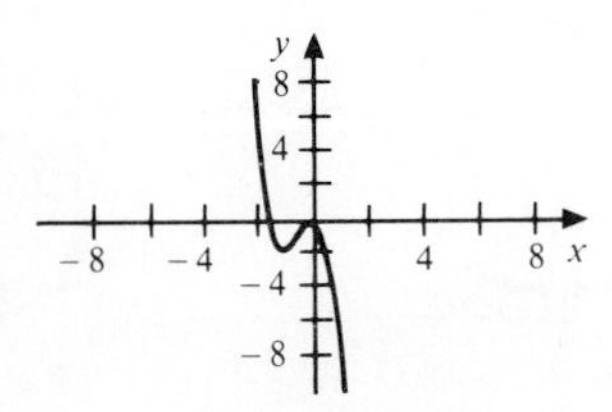

19. Critical values at $x = 0, 1$; $f(x)$ decreasing for $x < 1$ and decreasing for $x > 1$.

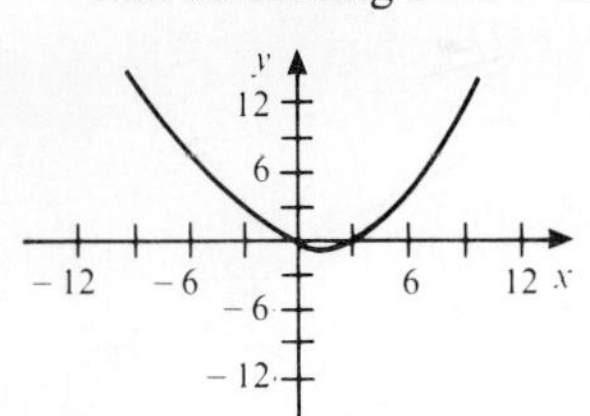

21. Critical value at $x = \frac{1}{4}$; $f(x)$ increasing for $x > \frac{1}{4}$ and decreasing for $0 < x < \frac{1}{4}$.

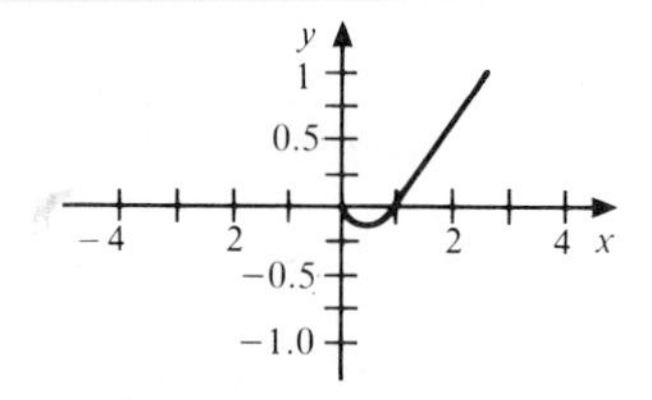

23. $0 \le x < 3$
25. (a) $R(x) = 3x^2$
    (b) $R'(x) = 6x$. Revenue is always increasing.
27. (a) 0 to 22 minutes
    (b) 22 to 52.7 minutes
    (c) at 22 minutes
    (d) yes
29. (a) $x > 80$
    (b) Decrease
    (c) Increase

Section 2.6

1. $f'(2) = 20$
3. 105
5. $v(3) = 117$
   $a(8) = 40$
7. $v(2) = 45$
   $a(4) = 336$
9. (a) 0
   (b) 40.96
   (c) 3.2 seconds
   (d) -32
11. (a) 0
    (b) 252
    (c) 6
    (d) -32
13. The ceiling must be 64 feet high. Pat's muzzle velocity is 64 ft/sec.
15. (a) -1990
    (b) -9990
    (c) $-19{,}990$
17. (a) 10
    (b) 28.54
    (c) 42.43
19. The manufacturer should increase spending on advertising.
21. 1
23. (a) 1000 ft/sec
    (b) 981.6 ft/sec
    (c) -1.53 ft/sec^2
25. 39 feet deep
27. (a) 92.32 cubic ft/year
    (b) 21.9 cubic ft/year
29. 63 units/hr
31. -15 °/hr

Section 2.7

1. 11,000
3. Continuous, value is 13
5. Continuous, value is $\frac{7}{6}$
7. Discontinuous
9. Discontinuous
11. Discontinuous
13. Discontinuous
15. Continuous, value is 0
17. Not differentiable but is continuous.
19. Continuous at $x = 3$
    Not differentiable at $x = 3$
21. Differentiable
23. Not differentiable but is continuous.

25. Not continuous, not differentiable at $x = 2$.
27. Not differentiable, not continuous
29. Not differentiable but is continuous.

Chapter 2 Review Exercises

1. 0
3. 3
5. 1
7. −3
9. 4
11. 2
13. $-\frac{4}{25}$
15. $5x^{2/3}$
17. $-2x^{-5/3} + 10x^{2/3}$
19. $-\dfrac{3}{x^2}$
21. −13
23. $x = -2$
25. $a = \pm\sqrt{3}$. Two tangent lines: $y = (5 - 4\sqrt{3})x$ and $y = (5 + 4\sqrt{3})x$.
27. Increasing for $x < 2$; decreasing for $x > 2$; Critical Point at (2, 18)
29. $x = 2$
31. $t = -3$
33. $x = -3, 1$
35. Increasing
37. Discontinuous at $x = 2$
39. Continuous but not differentiable
41. Continuous and differentiable

Section 3.1

1. $2x(5x^3 - 8)$
3. $x^2(10x^2 - 3)$
5. $3t^2 + 2t + 1$
7. $6x + \dfrac{1}{3} - \dfrac{1}{3x^2}$
9. $\dfrac{2}{5}u^{1/5}(13u^4 + 9)$
11. $-48r^3 + 12r^2 - 12r + 2$
13. $8x(2x^2 - 1)$
15. $\dfrac{1}{(x+1)^2}$
17. $\dfrac{-3(2x+3)}{x^2(x+3)^2}$
19. $\dfrac{38t}{(1-2t^2)^2}$
21. $6\left(1 + \dfrac{x(x-1)}{(2x-1)^2}\right)$
23. $\dfrac{14(r^2 + 3r + 1)}{(2r+3)^2}$
25. $\dfrac{x\sqrt{x}}{(x-3)(2+x^2)}\left[\dfrac{2-3x^2}{2(2+x^2)} + \dfrac{(x-6)}{(x-3)}\right]$
27. $\dfrac{3u^{2/3}(u^2 + 28u + 3)}{(3-u^2)^2} + \dfrac{2(u+14)}{u^{1/3}(3-u^2)^2}$
29. $-\frac{5}{3}$
31. $-\frac{5119}{12}$
33. −416
35. $\frac{5}{64}$
37. $-\frac{5}{9}$
39. $\frac{9}{4}$
41. $y = -\frac{5}{48}x + \frac{4}{3}$
43. $y = 2x - 2$
45. $y = -\frac{7}{2}x + \frac{3}{2}$
47. $(2x+5)(4x+1) + 2(x-3)(4x+1) + 4(x-3)(2x+5)$
49. $2x(3x-5)(x^3-x) + (x^2-1)(3)(x^3-x) + (x^2-1)(3x-5)(3x^2-1)$
53. 0.9978

Section 3.2

1. $g(x) = x^{-3}$
$h(x) = 3x^2 - 4$
3. $g(x) = \sqrt{x}$
$h(x) = x^2 + 4$
5. $g(t) = 2t^{1/3} + 5$
$h(t) = t^2 - 4$
7. $g(s) = \sqrt{s} + 1.03$
$h(s) = 4s^2 + 1$
9. $g(x) = \dfrac{14}{x^2}$
$h(x) = 3x^2 - 4$
11. $f(g(x)) = 9x^2 + 3x$
$g(f(x)) = 3x^2 + 3x$
13. $f(g(x)) = \dfrac{x^2}{x^2+3}$
$g(f(x)) = \left(\dfrac{x}{x+3}\right)^2$
15. $f(g(x)) = \dfrac{2\sqrt{x}+3}{\sqrt{x}-2}$
$g(f(x)) = \sqrt{\dfrac{2x+3}{x-2}}$
17. $15(3x-2)^4$
19. $f'(x) = 4x(x^2 - 4)$
21. $f'(x) = \dfrac{3}{4}(x-3)^{-1/4}$
23. $f'(x) = \dfrac{3}{2}x(x^2-3)^{-1/4}$
25. $f'(x) = -\dfrac{4.3}{3}(4 - 2.15t)^{-1/3}$
27. $f'(x) = -(x^2 - 3x + 1)^{-3/2}(2x - 3)$
29. $\dfrac{3(1-4\sqrt{r})(\sqrt{r}-2r)^2}{2\sqrt{r}}$
31. $f'(x) = 2x(x^2+2)(3x^2+14)$
33. $\dfrac{-3(3t^2 - 4t + 6)}{2\sqrt{(t^2+6)(6-3t)}}$
35. $2x(3-x)(3-2x)$
37. $y' = \dfrac{(r-3)3(3r+2)^2(3) + (3r+2)^3}{2\sqrt{(r-3)(3r+2)^3}}$
$= \dfrac{(12r-25)\sqrt{3r+2}}{2\sqrt{r-3}}$
39. $y = 6x - 5$
41. $y = -\frac{2}{3}u + \frac{7}{3}$
43. $y = 54x - 81$
45. $R'(p) = 6000 - \dfrac{64\left(\dfrac{p^2}{50} + 250\right)}{\sqrt{\dfrac{p^2}{100} + 250}}$
47. 288π cubic inches/sec
49. 3 points per year
51. $\dfrac{dA}{dt} = 20\text{ m}^2\text{/hr}$; $\dfrac{dr}{dt} = \dfrac{20}{\pi r}$ m/hr
53. It is moving away from the wall at a rate of $\frac{9}{4}$ ft/min.
55. $\dfrac{dA}{dt} = (20\pi)(2) = 40\pi\text{ cm}^2\text{/min}$
61. $\dfrac{3x^2+1}{x^3+x}$
63. 36
65. 80

Section 3.3

1. $f''(x) = 12x^2 - 18x$; $f''(2) = 12$
3. $f''(t) = 6t - 4$
$f''(3) = 14$
5. $f''(x) = -\dfrac{1}{4}x^{-3/2} - 24x$
$f''(4) = -\dfrac{3073}{32}$
7. $y'' = 30x(x^3 - 62)^3(7x^3 - 62)$
$y''(4) = 370{,}560$
9. $-\frac{3}{32}$

11. $\dfrac{d^2y}{dx^2} = -\dfrac{3}{4}x^{-3/2} - \dfrac{10}{9}x^{-4/3}$
$\left.\dfrac{d^2y}{dx^2}\right|_{x=1} = -\dfrac{67}{36}$

13. $y'' = 72x^2 - 6$
$y''(3) = 642$

15. $y''' = -\dfrac{504}{125}x^{-17/5} - 30$
$y'''(-1) = -\dfrac{4254}{125}$

17. $f^{(4)}(x) = 1680x^3 + 120$
$f^{(7)}(x) = 10{,}080$

19. $y^{(4)} = 360s + \dfrac{15}{8}s^{-7/2}$
$y^{(5)} = 360 - \dfrac{105}{16}s^{-9/2}$
$y^{(5)}(4) = 359.99$

21. Inflection Point: $(1, 1)$
Concave Down: $x < 1$
Concave Up: $x > 1$

23. Inflection Point: $(0, 6)$
Concave Down: $t < 0$
Concave Up: $t > 0$

25. Inflection Point: none
Concave Down: nowhere
Concave Up: everywhere

27. Inflection Points: $\left(1 + \dfrac{\sqrt{3}}{3}, -\dfrac{50}{9}\right), \left(1 - \dfrac{\sqrt{3}}{3}, -\dfrac{50}{9}\right)$
Concave Down: $1 - \dfrac{\sqrt{3}}{3} < x < 1 + \dfrac{\sqrt{3}}{3}$
Concave Up: $x < 1 - \dfrac{\sqrt{3}}{3}$ and $1 + \dfrac{\sqrt{3}}{3} < x$

29. Inflection Points: $(1, 47)$, $(3, 63)$
Concave Down: $1 < r < 3$
Concave Up: $r < 1$ and $3 < r$

31. Inflection Points: $(0, 0)$, $(\frac{2}{3}, -\frac{16}{27})$
Concave Down: $0 < x < \frac{2}{3}$
Concave Up: $x < 0$ and $x > \frac{2}{3}$

33. Inflection Points: $(0, 0)$, $\left(-\sqrt{3}, -\dfrac{\sqrt{3}}{4}\right), \left(\sqrt{3}, \dfrac{\sqrt{3}}{4}\right)$
Concave Down: $x < -\sqrt{3}$ and $0 < x < \sqrt{3}$
Concave Up: $-\sqrt{3} < x < 0$ and $\sqrt{3} < x$

35. Inflection Points: $\left(-\dfrac{2\sqrt{3}}{3}, \dfrac{28}{9}\right), \left(\dfrac{2\sqrt{3}}{3}, \dfrac{28}{9}\right)$
Concave Down: $-\dfrac{2\sqrt{3}}{3} < x < \dfrac{2\sqrt{3}}{3}$
Concave Up: $x < -\dfrac{2\sqrt{3}}{3}$ and $\dfrac{2\sqrt{3}}{3} < x$

37. Inflection Points: $(0, 15)$, $(-\sqrt{2}, 15 + 28\sqrt{2})$, $(\sqrt{2}, 15 - 28\sqrt{2})$
Concave Down: $x < -\sqrt{2}$ and $0 < x < \sqrt{2}$
Concave Up: $-\sqrt{2} < x < 0$ and $\sqrt{2} < x$

41.

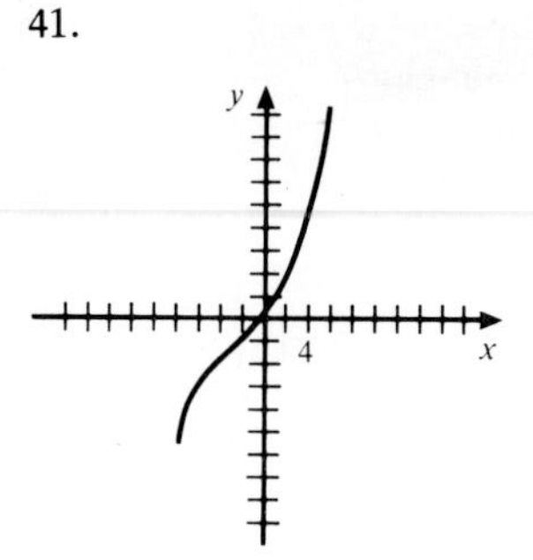

43.

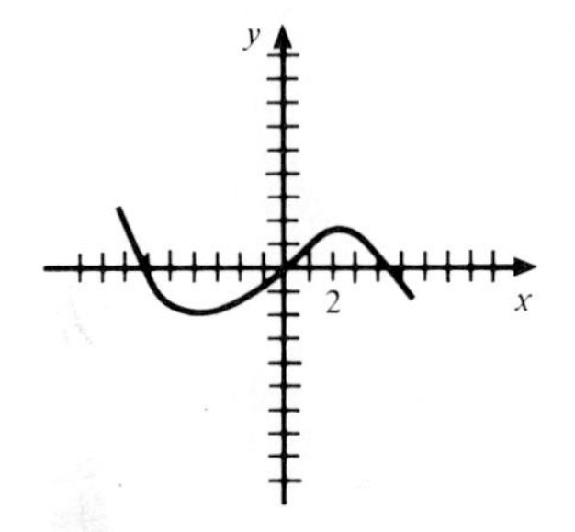

45.

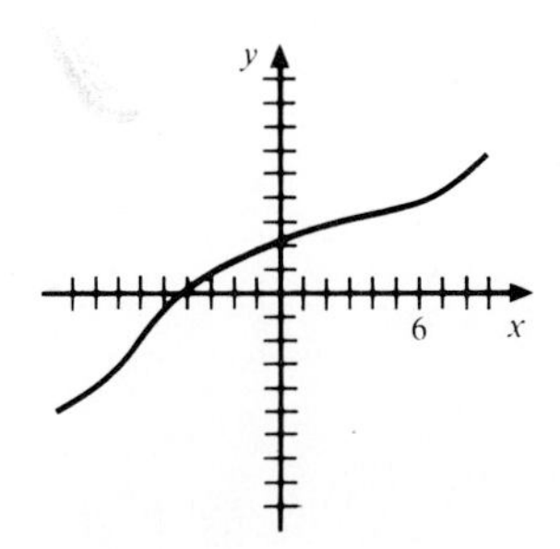

47. 62 ft/sec^2

49. **(a)** The government should anticipate that the program will last 14 months.
(b) It will take 37.66 months before the number of unemployed people begins to decrease.
(c) 25,776,000 people will be unemployed.
(d) 52,279,870 is the maximum number of people who will be unemployed.

Section 3.4

1. Relative Minimum: $x = 1$

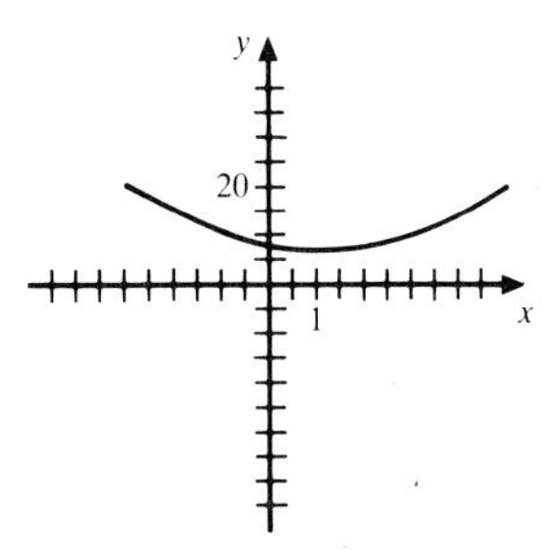

3. Relative Maximum: $x = 3$

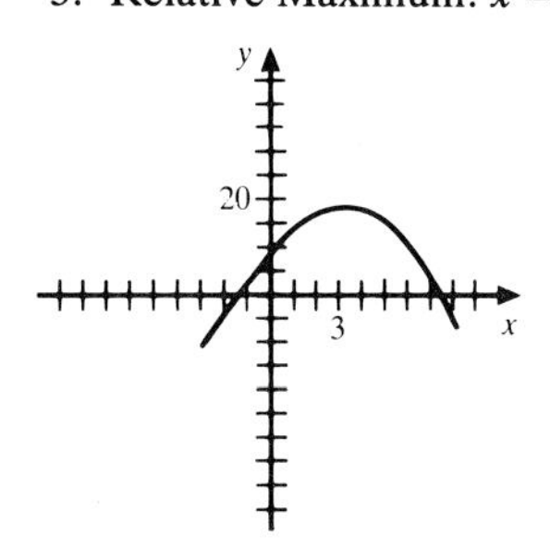

5. Relative Maximum: $t = \frac{1}{4}$

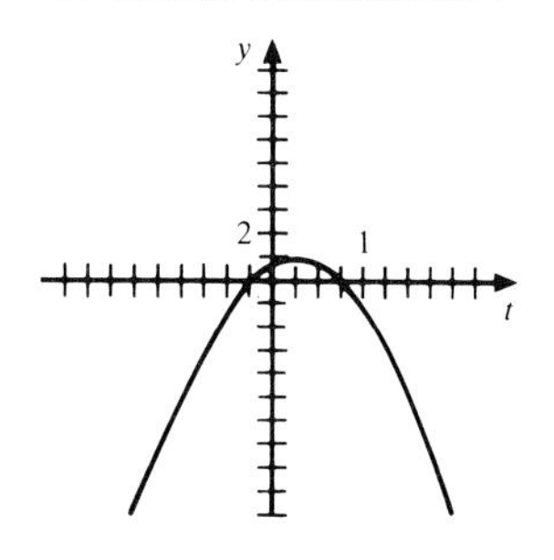

7. Relative Maximum: $x = 1$
   Relative Minimum: $x = 3$

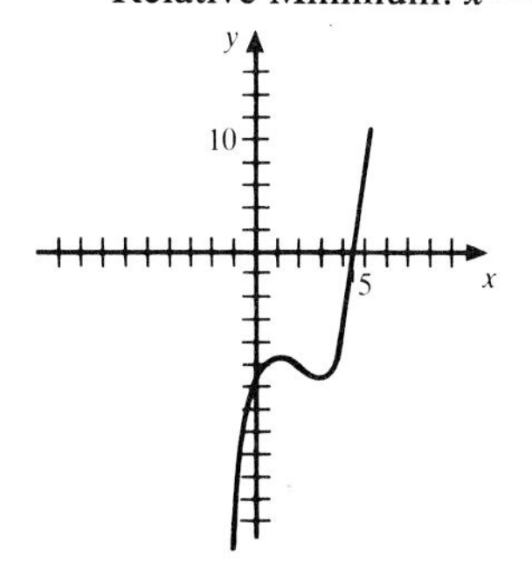

9. Relative Maximum: $u = -\sqrt{3}$
   Relative Minimum: $u = \sqrt{3}$

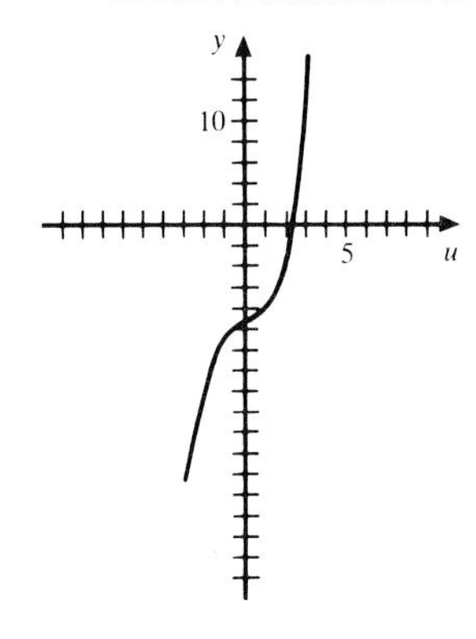

11. Critical Value: none

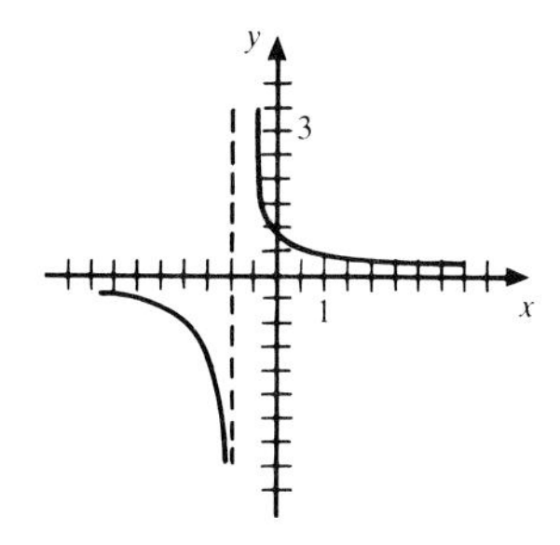

13. Relative Maximum: $x = 0$

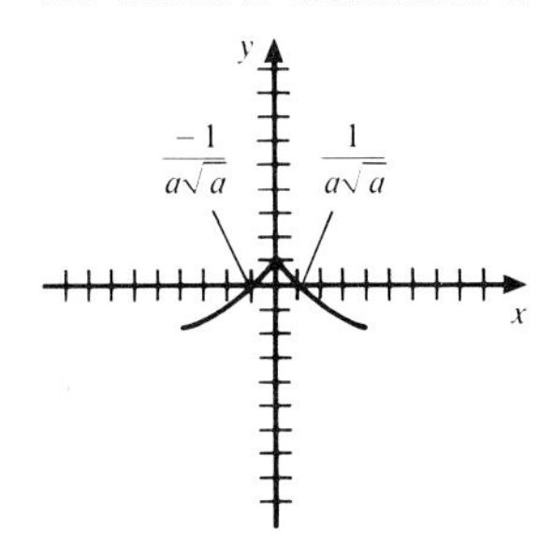

15. Relative Maximum: $s = 0$

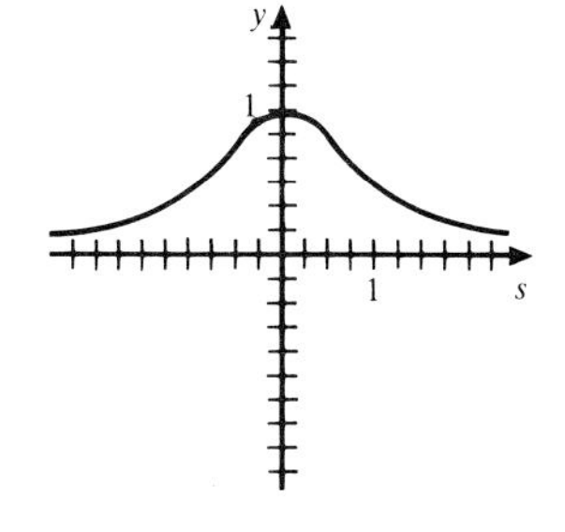

17. Relative Maximum: $x = 3$

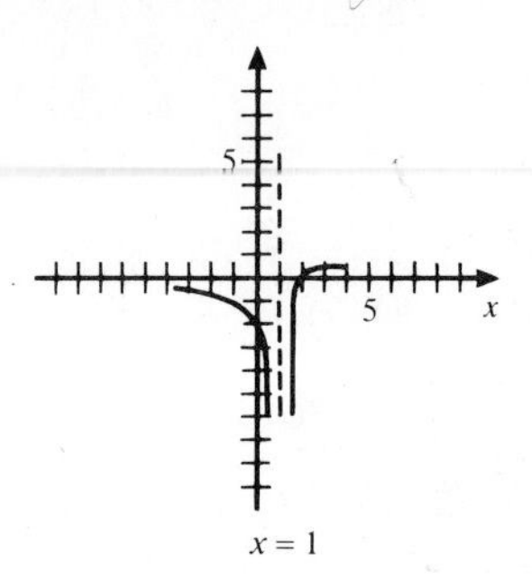

19. Relative Minimum: $p = -3$
Relative Maximum: $p = -2$

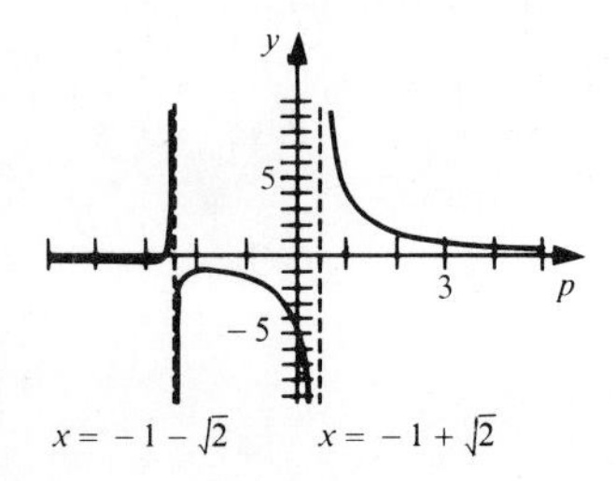

21. Relative Minimum: $x = 1$
23. Relative Maximum: $x = -\frac{5}{4}$
25. Relative Maximum: $t = \frac{1}{3}$
27. Relative Maximum: $x = 2$
Relative Minimum: $x = -2$
29. Relative Maximum: $s = \sqrt{1/3}$
Relative Minimum: $s = -\sqrt{1/3}$
31. Relative Minimum: $x = 2 + \sqrt{2}$
Relative Maximum: $x = 2 - \sqrt{2}$
33. Relative Maximum: $x = -\sqrt{2}$
Relative Minimum: $x = \sqrt{2}$
35. Relative Minimum at $x = 32$
37. No relative extreme points
39. None
41. Relative Minima: $x = 1, -1$
43. Relative Minimum: $x = 1$
45. Relative Minimum: $x = \frac{3}{5}\sqrt{\frac{3}{5}}$
Relative Maximum: $x = -\frac{3}{5}\sqrt{\frac{3}{5}}$
47. Relative Maximum: $x = 0$
Relative Minima: $x = \frac{16\sqrt{6}}{9}$ and $x = -\frac{16\sqrt{6}}{9}$
49. Relative Maximum: $x = -1$
Relative Minimum: $x = 1$

Section 3.5

1. Absolute Minimum at $f(-1) = -2$
Absolute Maximum at $f(2) = 7$
3. Absolute Maximum at $f(2) = 13$
Absolute Minimum at $f(5) = -5$
5. Absolute Maximum at $f(-1) = 27$
Absolute Minimum at $f(-3) = -13$
7. Absolute Minimum at $f(3) = -7$
Absolute Maximum at $f(4 - \sqrt{14}) = 8 - 2\sqrt{14} \approx 0.52$
9. Absolute Maximum at $g(6) = \frac{1}{8}$
Absolute Minimum at $g(0) = -\frac{5}{2}$
11. Absolute Maximum at $f(0) = -0.84$
Absolute Minimum at $f(2) = -6.35$
13. Absolute Minimum at $Q(1) = -2$
Absolute Maximum at $Q(-1) = 2$
15. Absolute Maximum at $f(2) = \frac{1}{3}$
Absolute Minimum at $f(0) = -1$
17. Absolute Maximum at $f(0) = -\frac{1}{4}$
Absolute Minimum at $f(3) = -10$
19. Absolute Minimum at $f(1) = 0$
Absolute Maximum at $f(4) = 4\sqrt{15}$
21. Absolute Maximum at $f\left(\sqrt{\frac{1}{3}}\right) = \frac{2\sqrt{3}}{9}$
Absolute Minimum at $f(0) = f(1) = 0$
23. Absolute Maximum at $f(2 - \sqrt{2}) = 3 - 2\sqrt{2} \approx 0.17$
Absolute Minimum at $f(-1) = -\frac{2}{3}$
25. Absolute Minimum at $w(3) = 3\sqrt{5}$
Absolute Maximum at $w(4) = 8\sqrt{3}$
27. Absolute Maximum at $f(1) = 4$
29. Absolute Minimum at $f(1) = 3$
31. Absolute Minimum at $g(0) = -2$
Absolute Maximum at $g(5) = \frac{1}{12}$
33. Absolute Maximum at $f(4) = \frac{4}{17}$
35. Absolute Maximum at $f(\sqrt[3]{1/3}) = \frac{3}{4\sqrt[3]{3}} \approx 0.57$
Absolute Minimum at $f(-\sqrt[3]{1/3}) = \frac{-3}{4\sqrt[3]{3}} \approx -0.57$
37. Absolute Minimum at $f(\sqrt{1/3}) = \frac{-2}{3\sqrt{3}} \approx -0.38$
No Absolute Maximum
39. Absolute Minimum at $Q(1) = 2$
No Absolute Maximum

Section 3.6

1. $x = 10$ ft and $h = 5$ ft
3. \$11.00 per day making 100 spools
5. 200 refrigerators for a \$10,000 weekly profit
7. 10 and 10
9. 20 and 20
11. 2 ft wide, 3 ft high
13. $r = 1.366$ inches and $h = 2.729$ inches
15. 40,000 ft

17. A \$14 billion decrease in spending will yield 47.8% of the electorate; not enough to win.
19. The fifth attempt will yield the highest level of proficiency where 7 errors will occur.
21. (a)

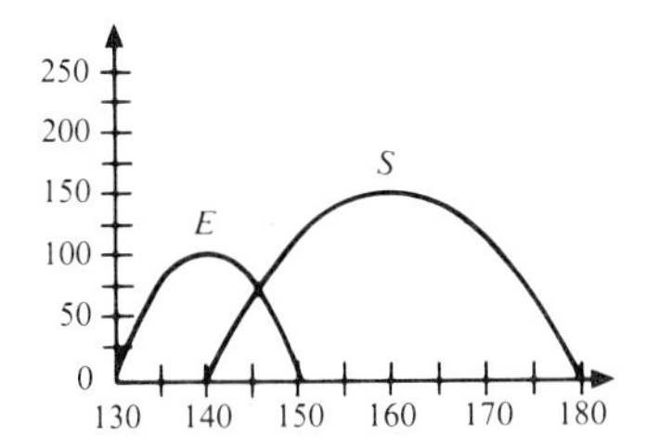

(b) 140 lb
(c) 160 lb
(d)

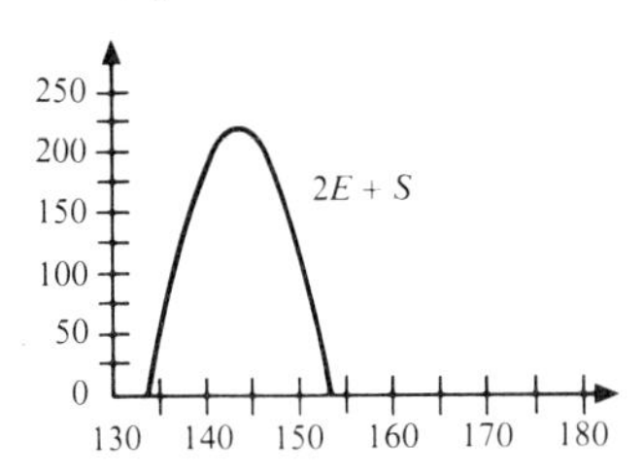

(e) 143.3 lb
23. $d = 0$
25. 3 barrels of pesticide yield 19 bushels per acre.
27. 8.67 trees
29. Maximum profit of \$9200 at 2000 books. Maximum profit of \$11,450 at 3500 books.

Chapter 3 Review Exercises

1. $f'(x) = 42x^6 - 60x^5$
3. $\dfrac{-6(x^2+5)}{(x^2-5)^2}$
5. $2 - \dfrac{x+4}{2x\sqrt{x}}$
7. $\dfrac{x^2-1}{2x^2\sqrt{x+\dfrac{1}{x}}}$
9. $\dfrac{1}{3}(x^4+5x+3)^{-2/3}(4x^3+5)$
11. $-\dfrac{5t^2+10t+1}{(3t+5)^{2/3}(t^2-1)^2}$
13. $\dfrac{-4s^6+10s^5-36s^4+84s^3-15s^2+45}{(s^4+3s)^4}$
15. $\dfrac{(x^3+1)\left(\dfrac{1}{2\sqrt{x}}+\dfrac{1}{3\sqrt[3]{x^2}}\right)-(\sqrt{x}+\sqrt[3]{x})3x^2}{(x^3+1)^2}$

17. 91 items
19. 0.03 units/year
21. Relative Minimum at $f(3) = -4$
23. Relative Maximum at $f(1) = 4$
Relative Minimum at $f(3) = 0$
25. Relative Minimum at $f(-3) = -13$
Relative Maximum at $f(3) = 23$
27. Relative Maximum at $f(0) = \frac{3}{5}$
29. Relative Maximum at $Q(0) = 0$
Relative Minimum at $Q(\sqrt[3]{2/17}) = -\dfrac{15}{17}\left(\dfrac{2}{17}\right)^{2/3}$
31. Relative Minimum at $f(64) = -32$
Relative Maximum at $f(0) = 0$
33. Absolute Maximum at $f(5) = 12$
Absolute Minimum at $f(\frac{3}{2}) = -\frac{1}{4}$
35. Absolute Maximum at $f(3) = 20$
Absolute Minimum at $f(1) = 0$
37. Absolute Maximum at $f(0) = \frac{3}{5}$
Absolute Minimum at $f(2) = \frac{1}{3}$
39. Absolute Maximum at $f(\sqrt[3]{1/13}) = \sqrt[3]{1/13}\left(\dfrac{12}{13}\right)^4$
Absolute Minimum at $f(-1) = -16$
41. Absolute Maximum at $g(1) = \frac{1}{3}$
Absolute Minimum at $g(0) = 0$
43. $\dfrac{x^2+x}{x^2+x+3}(2x+1)$
45. $r = 1$ inch
$h = \sqrt{2}$ inches

Section 4.1

1. $y = 2$
3. $y = -5$
5. $y = \frac{2}{3}$
7. $y = -\frac{3}{2}$
9. $y = -1$
11. $y = -1$
13. $x = 0$

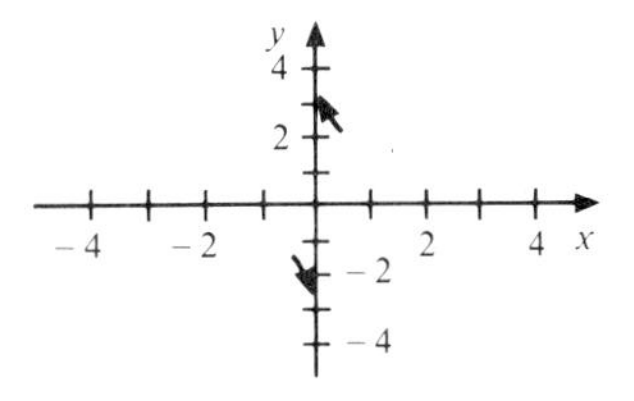

15. $x = 2$

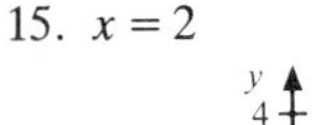

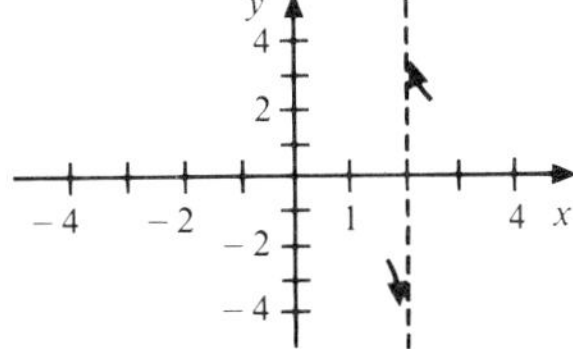

17. $x = -3$

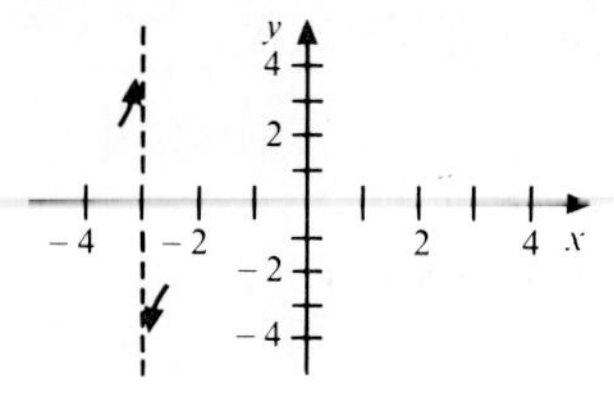

19. No vertical asymptotes

21. $x = 2, x = -2$

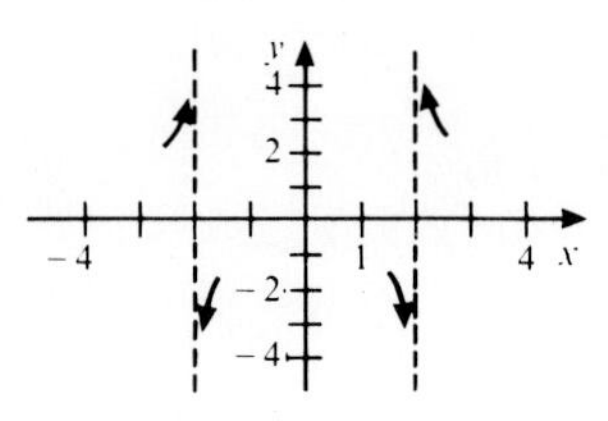

23. $x = 1, x = 4$

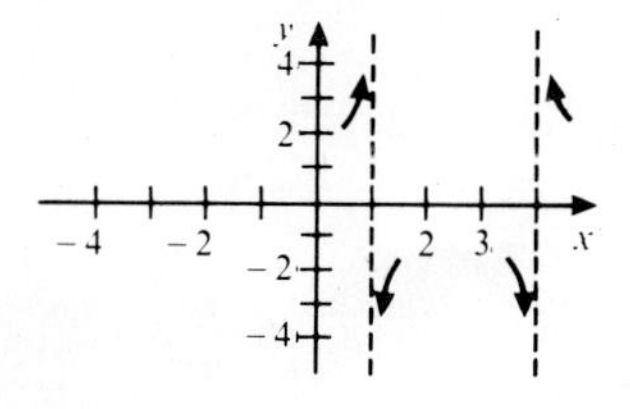

25. $x = 0, x = 4$

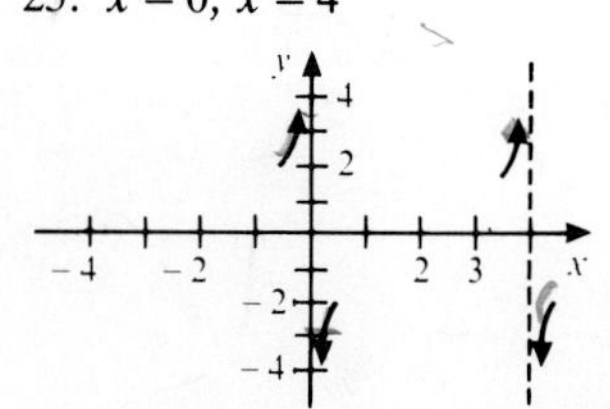

27. $x = 0$

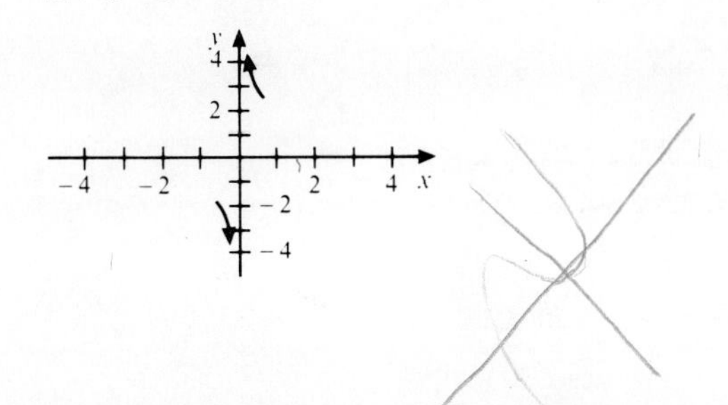

29. $x = 2, x = 6$

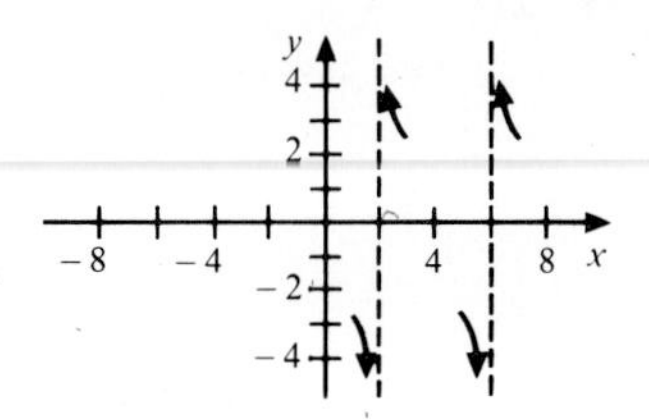

Section 4.2

1. y-intercept: $(0, -1)$
   x-intercepts: $(-1, 0)$, $(1,0)$
   Relative Minimum: $(0, -1)$
   Increasing: $x > 0$
   Decreasing: $x < 0$
   Concave Upward: everywhere

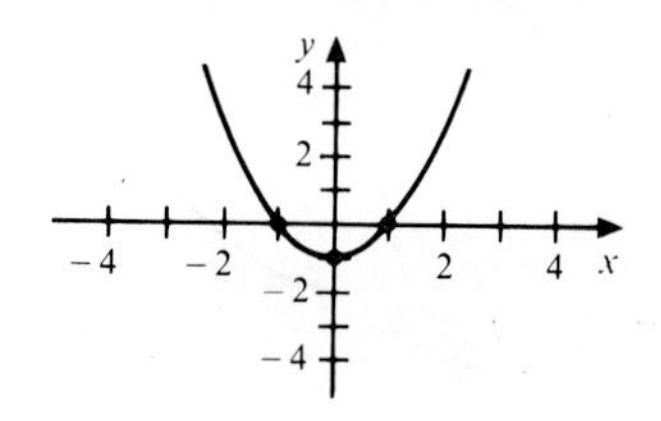

3. y-intercept: $(0, 0)$
   x-intercepts: $(-1, 0)$, $(0, 0)$, and $(1, 0)$
   Relative Maximum: $(-1, 2)$
   Relative Minimum: $(1, -2)$
   Inflection Point: $(0, 0)$
   Increasing: $x < -1$ and $x > 1$
   Decreasing: $-1 < x < 1$
   Concave Upward: $x > 0$
   Concave Downward: $x < 0$

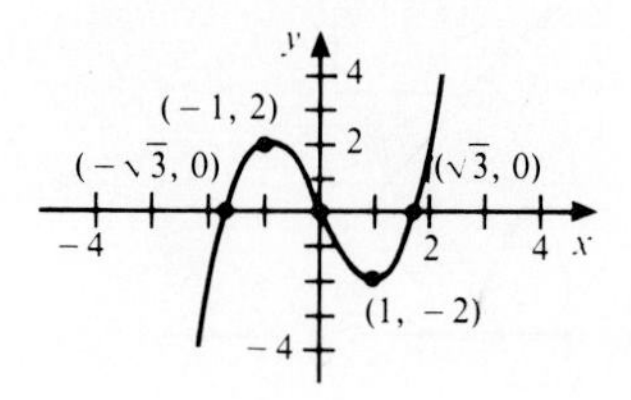

5. y-intercept: $(0, -\frac{3}{2})$
   x-intercept: $(-3, 0)$
   Horizontal Asymptote: $y = 1$
   Vertical Asymptote: $x = 2$
   Decreasing: everywhere

Concave Upward: $x > 2$
Concave Downward: $x < 2$

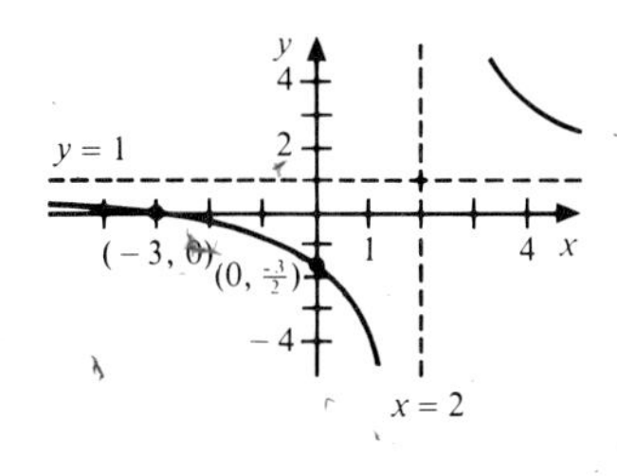

7. y-intercept: $(0, \frac{1}{4})$
   Horizontal Asymptote: $y = 0$
   Vertical Asymptote: $x = 2$
   Increasing: $x < 2$
   Decreasing: $x > 2$
   Concave Upward: everywhere

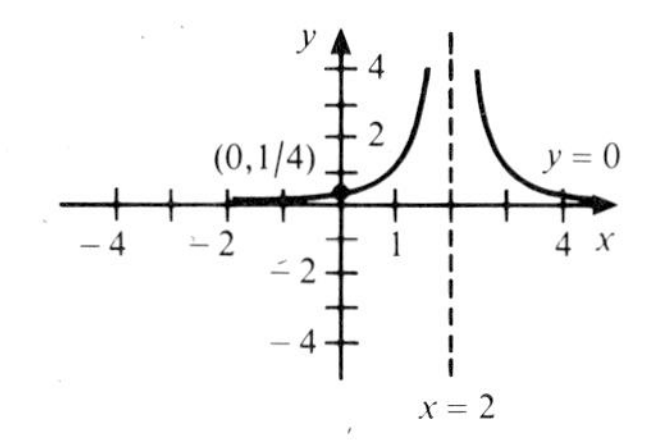

9. y-intercept: $(0, -2)$
   Relative Minimum: $(-3, -29)$
   Inflection Points: $(-2, -18)$, $(0, -2)$
   Increasing: $x > -3$
   Decreasing: $x < -3$
   Concave Upward: $x < -2$ and $x > 0$
   Concave Downward: $-2 < x < 0$

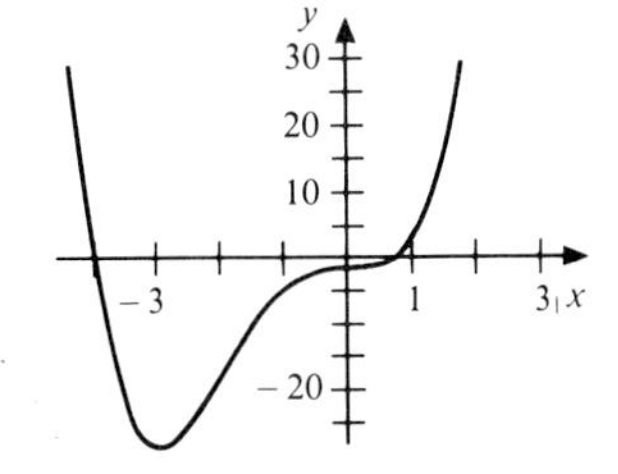

11. y-intercept: $(0, 0)$
    x-intercepts: $(0, 0)$, $(12, 0)$

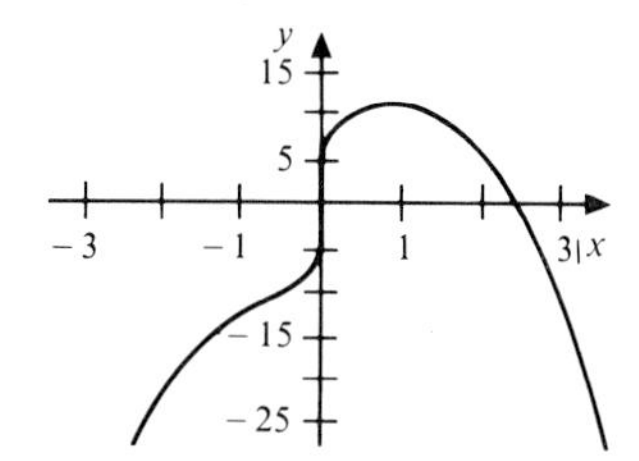

Relative Maximum: $(2, 11.5)$
Inflection Points: $(-8, -30.3)$, $(0, 0)$
Increasing: $x < 2$
Decreasing: $x > 2$
Concave Upward: $-8 < x < 0$
Concave Downward: $x < -8$ and $x > 0$

13. Relative Minimum: $(4, \frac{19}{8})$
    Inflection Point: $(6, \frac{22}{9})$
    Horizontal Asymptote: $y = 3$
    Vertical Asymptote: $x = 0$
    Increasing: $x < 0$ and $x > 4$
    Decreasing: $0 < x < 4$
    Concave Upward: $x < 6$
    Concave Downward: $x > 6$

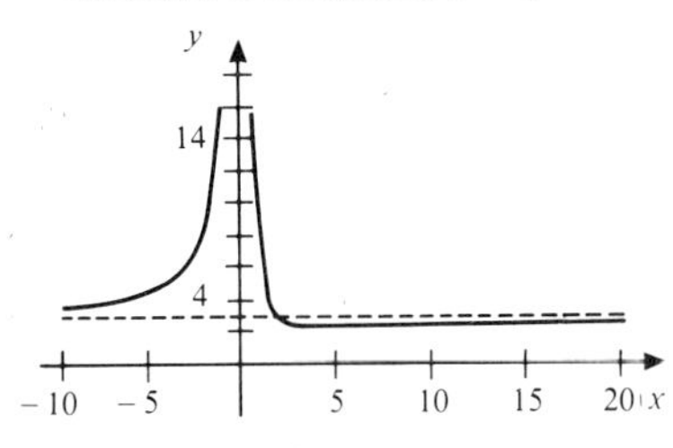

15. y-intercept: $(0, 0)$
    x-intercept: $(0, 0)$
    Relative Maximum: $(0, 0)$
    Horizontal Asymptote: $y = 1$
    Vertical Asymptotes: $x = -1$ and $x = 1$
    Increasing: $x < 0$
    Decreasing: $x > 0$
    Concave Upward: $x < -1$ and $x > 1$
    Concave Downward: $-1 < x < 1$

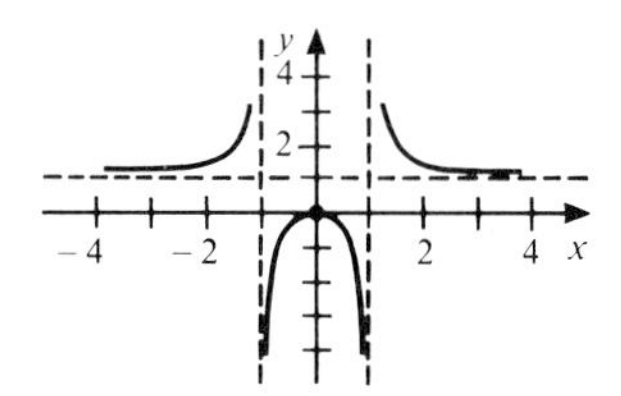

17. x-intercepts: $(-1, 0)$, $(1, 0)$
    Horizontal Asymptote: $y = 1$
    Vertical Asymptote: $x = 0$
    Increasing: $x > 0$
    Decreasing: $x < 0$
    Concave Downward: everywhere

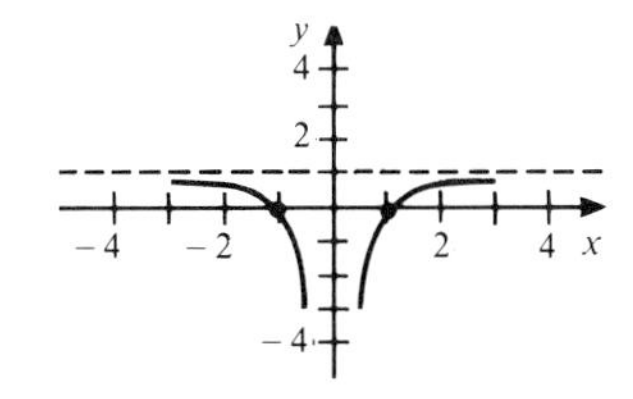

19. y-intercept: $(0, -3)$
Relative Maximum: $(0, -3)$
Relative Minimum: $(4, -35)$
Inflection Point: $(2, -19)$
Increasing: $x < 0$ and $x > 4$
Decreasing: $0 < x < 4$
Concave Upward: $x > 2$
Concave Downward: $x < 2$

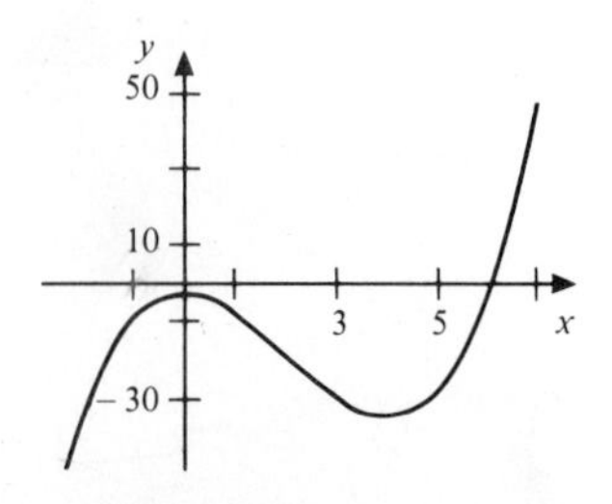

21. y-intercept: $(0, 0)$
x-intercepts: $(0, 0)$, $(2.99, 0)$, $(-2.99, 0)$
Relative Maximum: $(-2, 128)$
Relative Minimum: $(2, -128)$
Inflection Point: $(0, 0)$
Increasing: $x < -2$ and $x > 2$
Decreasing: $-2 < x < 2$
Concave Upward: $x > 0$
Concave Downward: $x < 0$

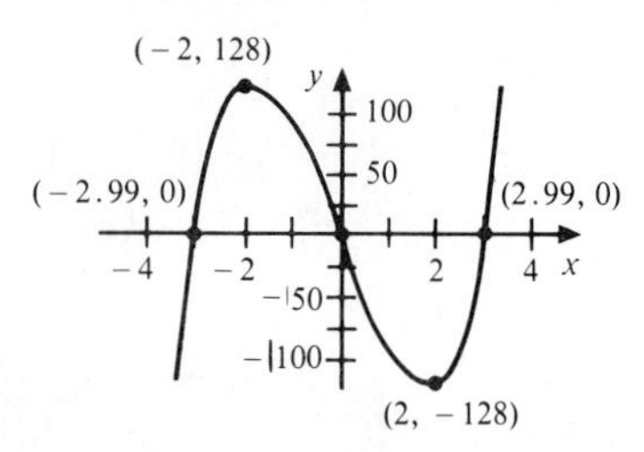

23. Relative Maximum: $(-1, -2)$
Relative Minimum: $(1, 2)$
Inflection Points: $(2\sqrt{2}, 2.12)$ and $(-2\sqrt{2}, -2.12)$
Vertical Asymptote: $x = 0$
Increasing: $x < -1$ and $x > 1$
Decreasing: $-1 < x < 0$ and $0 < x < 1$
Concave Upward: $x < -2\sqrt{2}$ and $0 < x < 2\sqrt{2}$
Concave Downward: $-2\sqrt{2} < x < 0$ and $2\sqrt{2} < x$

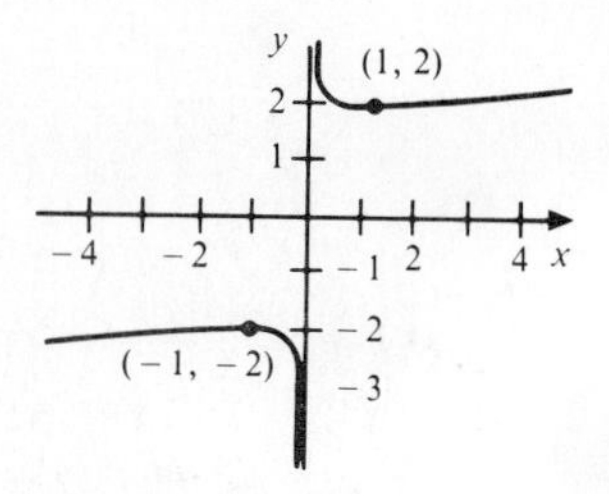

25. y-intercept: $(0, 0)$
x-intercept: $(0, 0)$
Relative Maximum: $(1, \frac{1}{2})$
Relative Minimum: $(-1, -\frac{1}{2})$
Inflection Points: $(0, 0)$, $(-\sqrt{3}, -0.433)$, $(\sqrt{3}, 0.433)$
Horizontal Asymptote: $y = 0$
Increasing: $-1 < x < 1$
Decreasing: $x < -1$ and $x > 1$
Concave Upward: $-\sqrt{3} < x < 0$ and $x > \sqrt{3}$
Concave Downward: $x < -\sqrt{3}$ and $0 < x < \sqrt{3}$

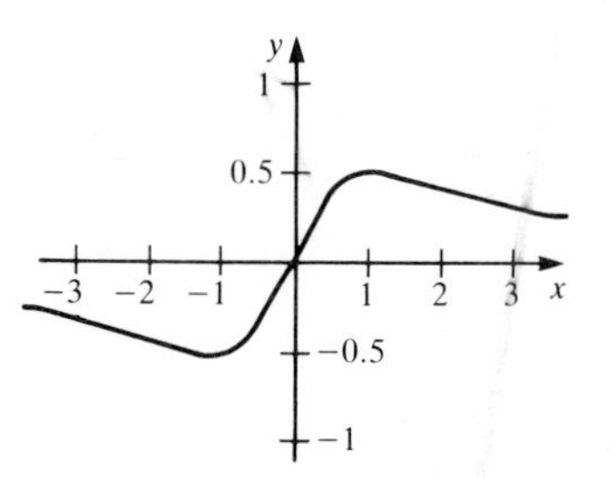

27. y-intercept: $(0, -2)$
x-intercept: $(2, 0)$
Relative Maximum: $(3, \frac{1}{4})$
Inflection Points: $(4, \frac{2}{9})$
Horizontal Asymptote: $y = 0$
Vertical Asymptote: $x = 1$
Increasing: $1 < x < 3$
Decreasing: $x > 3$
Concave Upward: $x > 4$
Concave Downward: $x < 4$

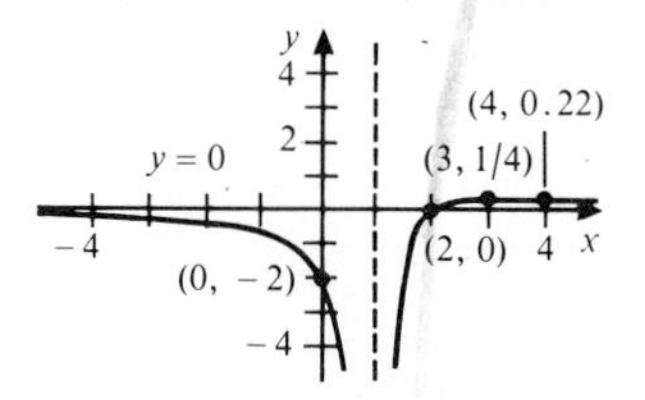

29. y-intercept: $(0, \frac{2}{5})$
x-intercept: $(-2, 0)$
Relative Maximum: $(-1, 0.5)$
Relative Minimum: $(-3, -0.5)$
Inflection Points: $(-3.73, -0.433)$, $(-2, 0)$, $(-0.268, 0.433)$

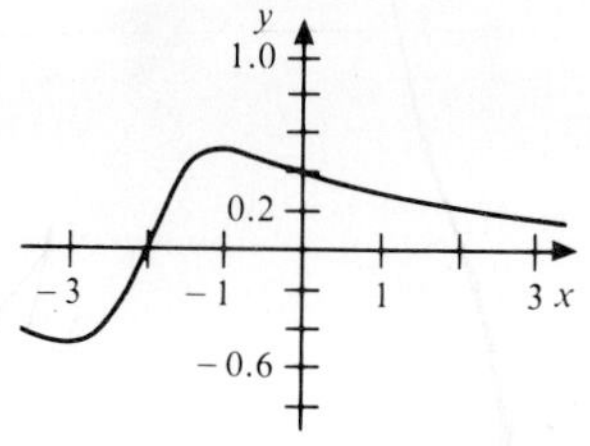

Horizontal Asymptote: $y = 0$
Increasing: $-3 < x < -1$
Decreasing: $x < -3$ and $x > -1$
Concave Upward: $-3.73 < x < -2$ and $x > -0.268$
Concave downward: $x < -3.73$ and $-2 < x < -0.268$

31. y-intercept: $(0, -\frac{5}{8})$
x-intercept: $(\frac{5}{2}, 0)$
Relative Maximum: $(3, \frac{1}{2})$
Inflection Point: $(\frac{7}{2}, \frac{4}{9})$
Horizontal Asymptote: $y = 0$
Vertical Asymptote: $x = 2$
Increasing: $2 < x < 3$
Decreasing: $x < 2$ and $x > 3$
Concave Upward: $x > \frac{7}{2}$
Concave Downward: $x < \frac{7}{2}$

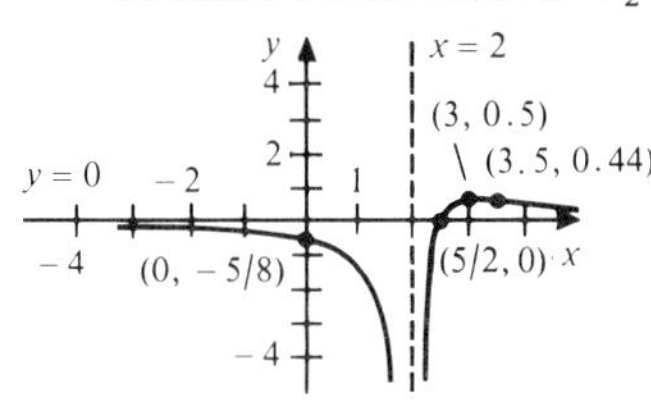

33. $(6.4, 0.45)$
35. y-intercept: $(0, 0)$
x-intercept: $(0, 0)$
Inflection Point: $(0, 0)$
Horizontal Asymptote: $y = 0$
Vertical Asymptotes: $x = -2$ and $x = 2$
Increasing: $-2 < x < 2$
Decreasing: $x < -2$ and $x > 2$
Concave Upward: $x > 0$
Concave Downward: $x < 0$

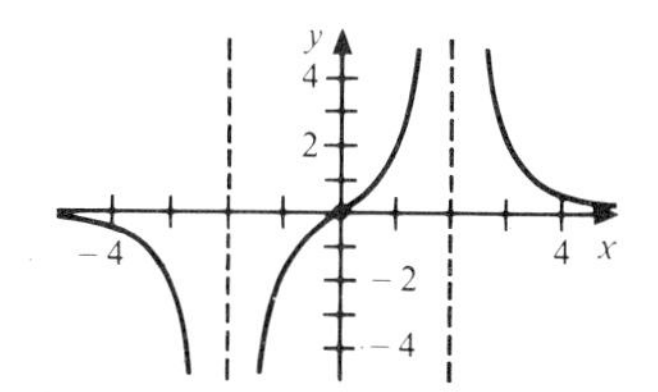

37. y-intercept: $(0, 1)$
x-intercepts: $\left(\frac{-1}{\sqrt{2}}, 0\right), \left(\frac{1}{\sqrt{2}}, 0\right)$

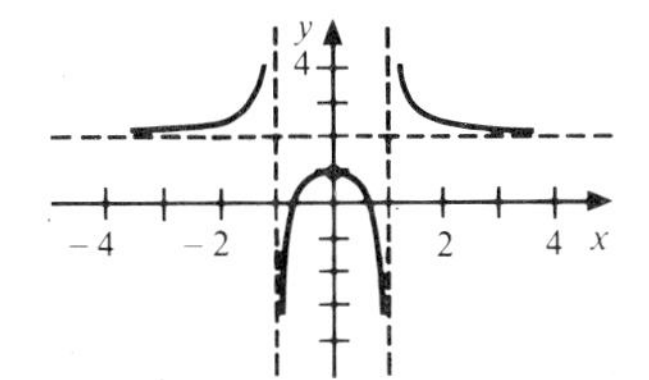

Relative Maximum: $(0, 0)$
Horizontal Asymptote: $y = 2$
Vertical Asymptotes: $x = -1, x = 1$
Increasing: $x < 0$
Decreasing: $x > 0$
Concave Upward: $x < -1$ and $x > 1$
Concave Downward: $-1 < x < 1$

Section 4.3

1. $\frac{dy}{dx} = -\frac{y}{x}$
3. $\frac{2x + 2xy}{14 - x^2}$
5. $\frac{125 - 2y(xy - 2)}{2x(xy - 2)}$
7. $-\frac{2y}{3x}$
9. $-\frac{2y}{x}$
11. $-\frac{\sqrt{y}}{\sqrt{x}}$
13. $-\frac{3x^{-1/4}y^{5/4}}{5x^{3/4}y^{1/4} - 4}$
15. $y = -x + 2$
17. $y = \frac{1}{5}x + \frac{3}{5}$
19. $y = -x$
21. $y = \frac{3}{5}x + \frac{4}{5}$
23. $\frac{2y - 2x}{x^2}$
25. $\frac{d^2y}{dx^2} = -\frac{6y^3 - 2(1 - 2x)^2}{9y^5}$
27. $\frac{2[(3y^2 + 1)^2 - 12x^2y]}{(3y^2 + 1)^3}$
29. $\frac{3(y^2 - y\sqrt{y}\sqrt{x})}{2x^3}$
31. $\frac{d^2y}{dx^2} = 32 - \frac{24\sqrt{y}}{x} + \frac{6y}{x^2}$
33. 6
35. $y = -\frac{5}{3}x + \frac{19}{3}$
37. $y = \frac{13}{9}x + \frac{29}{9}$
$y = -\frac{13}{9}x - \frac{29}{9}$
39. $y = -\frac{2}{27}x + \frac{164}{27}$
$y = \frac{2}{27}x - \frac{164}{27}$

Section 4.4

1. $f(101) - f(100) = \$12.01$
$M_C(100) = \$12.00$
Do not step up production.
3. \$8.10
$M_C(100) = \$8.00$
Do not step up production.
5. $f(100) = \$2{,}035{,}000$
$M_R(100) = \$60{,}650$
7. Revenue at $x = 100$ is $f(100) = \$4{,}020{,}000$
$M_R(100) = f'(100) = \$130{,}200$
9. 11,500 sets per month
11. \$400
13. $P(4) = \$32.00$
15. $x = 2500$
$P(2500) = \$124{,}800$
17. $x = 40$
$P(40) = \$1593$
$p(40) = \$200$
19. $P(2) = \$13.00$
$p(2) = \$16$
21. $x = 4995$
$P(4995) = \$249{,}400.25$
$p(4995) = \$50.05$

23. $x = 1250$
25. $P(1250) = \$311,700$
27. \$2882.50
29. \$150
    \$9000

Section 4.5

1. 3000
3. 200 typewriters per batch
5. 60 skateboards per order
7. 40 100-pound bags
9. 20 spreaders per order
11. Presently the shop's minimal cost for the 8000 rolls is \$80,800. Under the new conditions the minimal cost would be \$80,840. Therefore, reject the offer.
13. The order size should be 600.
15. 50 transactions of \$50,000 each
17. 253 gross per order
19. The order size should be 80.
21. 160 copiers per order
23. 150 cameras per order
25. The order size should be 80.
27. 4000 maps per batch
29. 2000 cases per order

Section 4.6

1. $\frac{8}{17}$
3. $\frac{5}{7}$
5. $-\frac{2}{3}$
7. $\frac{11}{7}$
9. 0
11. $\frac{2}{15}$
13. 8
15. At two months, the relative change is 0.1317; at six months, 0.1909; at nine months, 0.1661.
17. Gaining
19. 0.0678
21. No
23. $E_D(7) = -1.285$ Demand is elastic.
    $E_D(16) = -0.949$ Demand is inelastic.
25. -2.167
    When $p > \$5.55$ the demand is elastic.
27. $4\frac{3}{4}\%$
29. At $p = 1$ the supply is inelastic.
    At $p = 5$ the supply is elastic.

Chapter 4 Review Exercises

1. y-intercept: $(0, 0)$
   x-intercept: $(0, 0)$
   Relative Maximum: $(0, 0)$
   Relative Minimum: $(2, 4)$
   Vertical Asymptote: $x = 1$

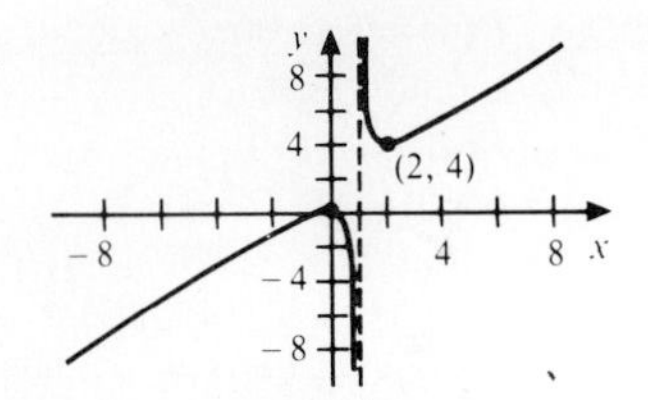

3. y-intercept: $(0, -1)$
   x-intercepts: $(1, 0)$, $(-1, 0)$
   Relative Minimum: $(0, -1)$
   Inflection Points: $(-1, 0)$, $\left(\frac{1}{\sqrt{5}}, \frac{64}{125}\right)$, $\left(\frac{1}{\sqrt{5}}, \frac{-64}{125}\right)$, $(1, 0)$

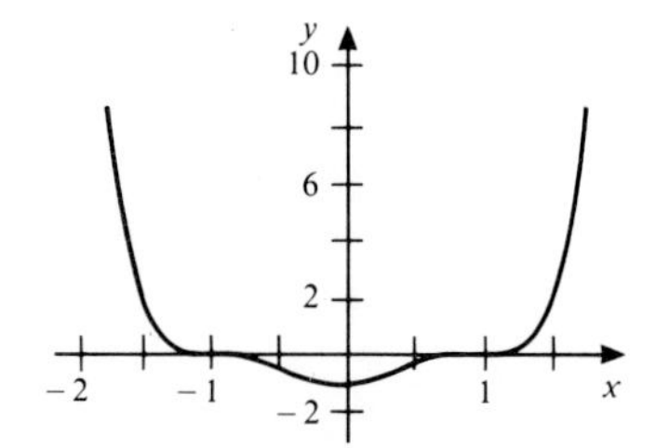

5. y-intercept: $(0, -1)$
   x-intercept: $(1, 0)$
   Horizontal Asymptote: $y = 0$
   Relative Maximum: $(2, \frac{1}{3})$
   Relative Minimum: $(0, -1)$

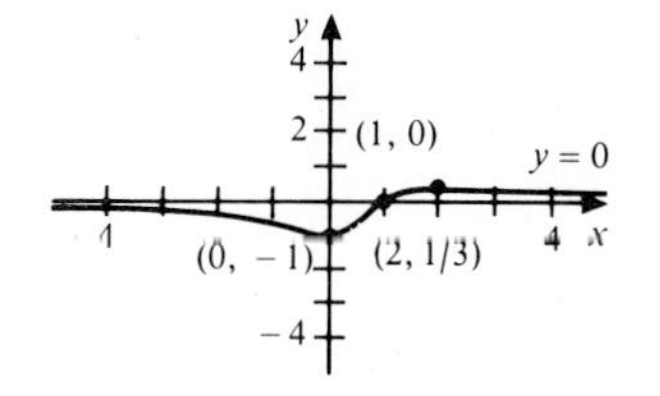

7. y-intercept: $(0, 0)$
   x-intercept: $(0, 0)$
   Inflection Points: $(-\sqrt{3}, -0.75\sqrt{3})$, $(0, 0)$,

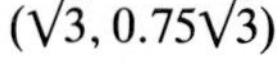

$(\sqrt{3}, 0.75\sqrt{3})$

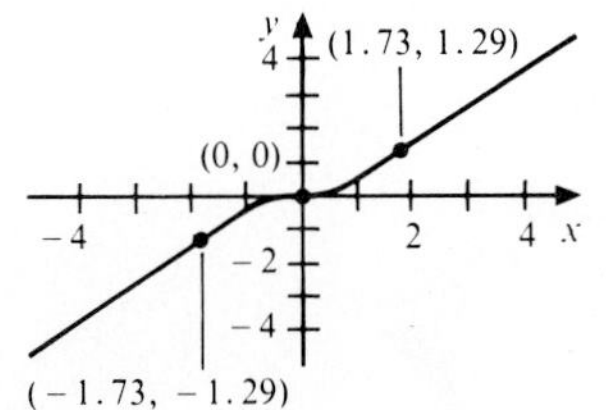

9. y-intercept: $(0, 0)$
   x-intercept: $(0, 0)$

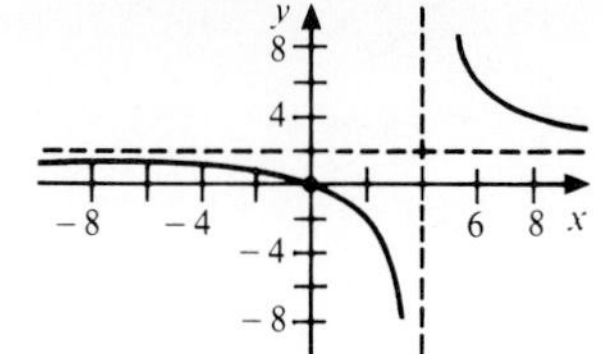

Vertical Asymptote: $x = 4$
Horizontal Asymptote: $y = 2$

11. x-intercept: $(1, 0)$
Relative Minimum: $(-\sqrt[3]{2}, 1.5\sqrt[3]{2})$
Vertical Asymptote: $x = 0$

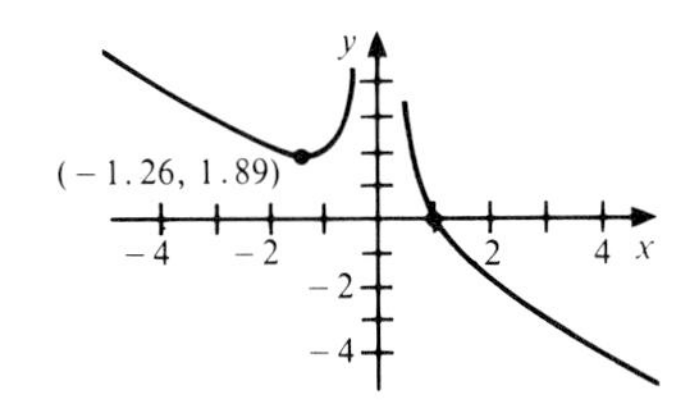

13. $\dfrac{dy}{dx} = \dfrac{3x^2y^3 - \dfrac{1}{y}}{xy^{-2} - 3x^3y^2}$

15. $\dfrac{nx^{n-1} + y}{ny^{-n-1} - x}$

17. $-\dfrac{3x^2 + y - 5}{4y^3 + x - 6}$

19. $\dfrac{dy}{dx} = \dfrac{1 + \dfrac{1}{y} + \dfrac{y}{x^2}}{\dfrac{1}{x} + \dfrac{x}{y^2} - 2y}$

21. $M_C(x) = 2x - 50$
$M_R(x) = 30 - 2x$
$x = 20$

23. $600
$7

25. $15.00

27. 150

29. 2000

31. $\frac{5}{2}$

33. $-\frac{2}{3}$

35. 1

37. yes
no

39. Larger revenue, $4.93.

43.

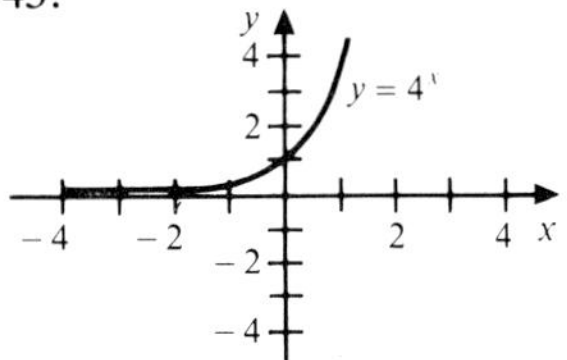

45.

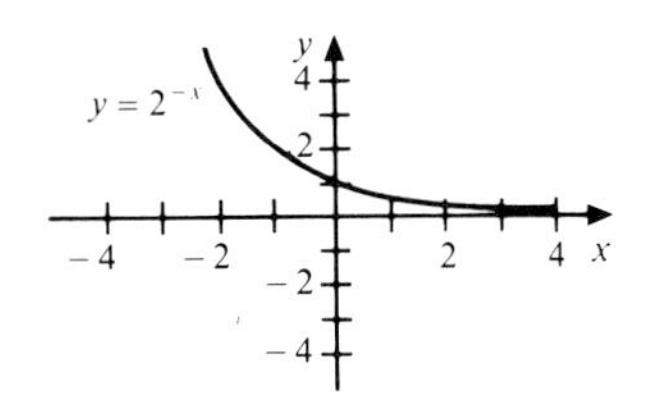

47.

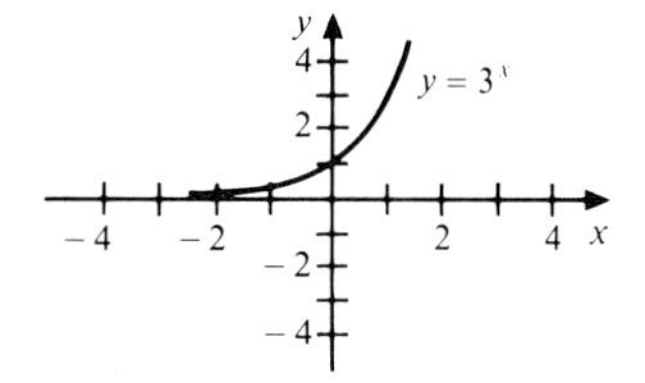

Section 5.1

1. $2^3 2^5 = 2^8 = 256$

3. 8

5. $x^6 y^4$

7. $\dfrac{3^{-2}4^0}{2^{-4}} = \dfrac{2^4 \cdot 1}{3^2} = \dfrac{16}{9}$

9. ab

11. $\dfrac{16}{a^{5/2}}$

13. $(a^2)^3 = a^{2 \cdot 3} = a^6$

15. a^{-2}

17. 2500

19. $\dfrac{2^{3/7}5^{3/14}}{20^{1/28}} = \dfrac{2^{3/7}5^{3/14}}{(2^2 \cdot 5^1)^{1/28}} = 20^{5/28}$

21. $\dfrac{4\sqrt[3]{4}}{\sqrt{3}}$

23. $a^{-2/3}b^{-4/5}$

25. $\dfrac{(2^{\frac{1}{3}}3^{\frac{2}{3}})^6}{a^{-2}b^{-3}} = \dfrac{2^{\frac{1}{3}\cdot 6}3^{\frac{2}{3}\cdot 6}}{a^{-2}b^{-3}} = 324a^2b^2$

27. $x = 4$

29. $x = -2$

31. $4^x = 2^{2x} = \sqrt{2}$
So $x = \frac{1}{4}$

33. $\frac{3}{8}$

35. $x = -11$

37. $\dfrac{1}{3^x} = 3^{-x} = 3;\ x = -1$

39. $x = \frac{3}{4}$

41. $x = 3$

49.

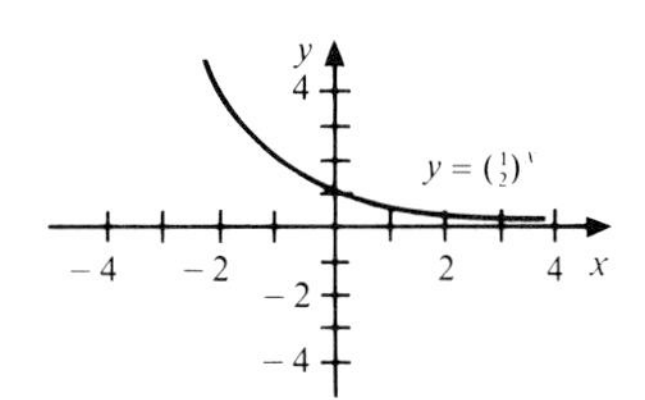

51.

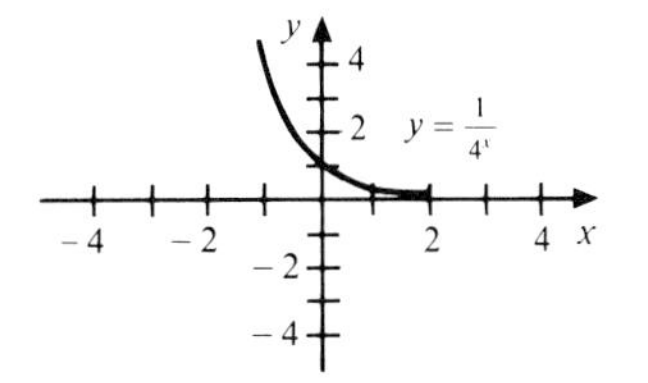

53.

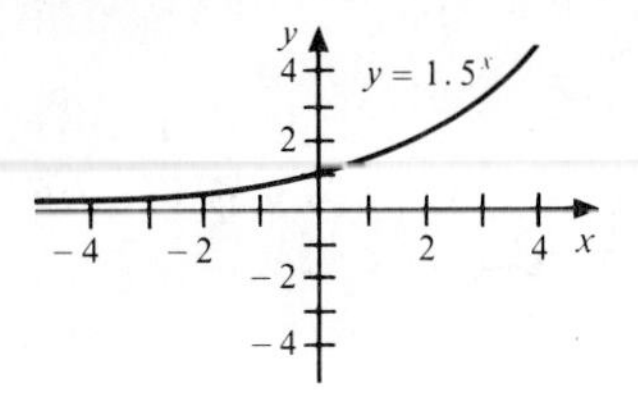

55.

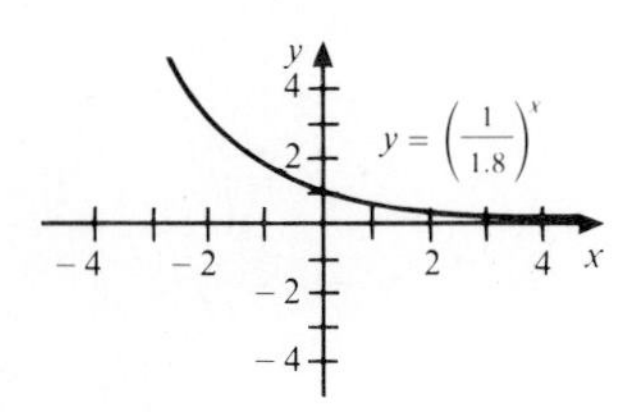

57. (a) \$541.21
(b) \$585.82
(c) \$742.97

59. (a) \$1060.90
(b) \$1125.50
(c) \$1343.91

61. quarterly: \$68,177.53; monthly: \$83,890.59

Section 5.2

1. The derivative of $2x$ is 2; therefore $f'(x) = 2e^{2x}$.
3. $3e^{3x+4}$
5. $6xe^{3x^2+5}$
7. Use the product rule: $f'(t) = t[2e^{2t}] + e^{2t}[1] = 2te^{2t} + e^{2t}$
9. $(6x^2 + 4x + 3)e^{x^2}$
11. $e^p e^{e^p}$
13. Use the product rule: $f'(x) = [e^{3x}][e^x - 1] + [e^x - x][3e^{3x}] = 4e^{4x} - e^{3x}(1 + 3x)$
15. $-\dfrac{e^{3t} + 2e^{2t} + 2e^t}{(e^{2t} - 2)^2}$
17. $4x^3(x + 1)4e^{4x}$
19. Use the quotient rule: $\dfrac{[e^r][5] - [5r + 4][e^r]}{[e^r]^2} = \dfrac{1 - 5r}{e^r}$
21. $2x - e^x$
23. $e^x + e^{-x}$
25. Use the product rule: $\dfrac{dy}{dx} = [x^2][2xe^{x^2}] + [e^{x^2}][2x]$
$= 2x(x^2 + 1)e^{x^2}$
27. $\dfrac{e^x + 1}{2\sqrt{e^x + x}}$
29. $4[e^{x^2e^x} + x]^3[(x^2e^x + e^x 2x)e^{x^2e^x} + 1]$
31. $f(0) = 1; f'(x) = e^x; f'(0) = 1$; the equation of the tangent line is $y = x + 1$.
33. $y = 2ex - e$
35. $y = \frac{1}{4}e^2 x$
37. $f(0) = e; f'(x) = e^x e^{e^x}$; the equation of the tangent line is $y = ex + e$.
39. $y = \dfrac{x}{e}$
41. Relative Maximum at $\left(\dfrac{1}{2}, \dfrac{1}{2e}\right)$

Inflection Point at $\left(1, \dfrac{1}{e^2}\right)$
43. $f'(t) = -2te^{-t^2} = 0$ when $t = 0$; $f''(t) = -4t^2e^{-t^2} - 2e^{-t^2}$; (0, 1) is a relative maximum; $\left(-\dfrac{1}{\sqrt{2}}, \dfrac{1}{\sqrt{e}}\right)$ and $\left(\dfrac{1}{\sqrt{2}}, \dfrac{1}{\sqrt{e}}\right)$ are inflection points.
45. Relative Maximum at $\left(3, \dfrac{27}{e^3}\right)$

Inflection Points at (0, 0), $(3 + \sqrt{3}, 0.93)$, $(3 - \sqrt{3}, 0.57)$
47. Let (a, e^a) be the point of tangency. The line passes through $(a, f(a))$ and (0, 0); $a = 1$. The point of tangency is $(1, e)$; $m = e$.

Section 5.3

1. Use $\ln x^c = c \ln x$. So $\ln x^3 = 3 \ln x$.
3. $4x + 2 \ln x$
5. $\ln (t + 2) - \ln (t - 4)$
7. $\ln (p + 2) + 3 \ln (p^2 + p + 2)$
9. Use $\ln \dfrac{a}{b} = \ln a - \ln b$; $\ln ab = \ln a + \ln b$ and $\ln e^x = x$.
$\ln \dfrac{x^2 + x + 4}{xe^x} = \ln (x^2 + x + 4) - \ln x - x.$
11. $\dfrac{\ln 7}{\ln 3}$
13. $(5)4^{x+1} = 9^x$ implies $20 = \left(\dfrac{9}{4}\right)^x$; $x \approx 3.6942$
15. 1.4660
17. 1.085
19. $\ln (x^2 - 1) = 2$; $e^{\ln (x^2-1)} = e^2$; $x = \pm\sqrt{e^2 + 1}$
21. $f'(x) = 2/x$
23. $\dfrac{1 + e^x}{x + e^x}$
25. $\dfrac{15t^4 + 12t^2 - 7}{3t^5 + 4t^3 - 7t}$
27. The derivative of $\ln 2x$ is $\dfrac{1}{x}$; $f'(x) = \dfrac{4}{x}$
29. $\dfrac{2x}{x^2 + 2}$
31. $\dfrac{-6r^2 - 2}{r^3 + r - 1}$

33. Use the product rule: $f'(x) = [x^2]\left[\frac{2x}{x^2+3}\right] + [\ln(x^2+3)][2x] = \frac{2x^3}{x^2+3} + 2x\ln(x^2+3)$

35. $1 + \frac{2}{u}$

37. $\frac{1}{x\ln(2x)}$

39. Use the product and chain rules:
$f'(x) = [x]\left[\frac{1}{x\ln x}\left(x\cdot\frac{1}{x} + \ln x\right)\right] + \ln(x\ln x)$
$= \frac{1+\ln x}{\ln x} + \ln(x\ln x)$

41. $(\ln p)e^p + \frac{e^p}{p}$

43. $\frac{2e^{x^2}}{x} + 2xe^{x^2}\ln x$

45. Use the power and chain rules: $f'(x) = 2(\ln x^2)\left(\frac{1}{x^2}\right)(2x) = \frac{8\ln x}{x}$

47. $\frac{2a-2x}{2ax-x^2}$

49. $x^4(5\ln x + \frac{2}{5})$

51. Use the product and chain rules: $f'(x) = [x^2]\left[\frac{1}{x^2}(2x)\right] + [\ln x^2][2x] = 2x(1+\ln x^2)$

53. $\frac{\ln x - 1}{(\ln x)^2}$

55. $\frac{1}{x} - \frac{2x}{1+x^2}$

57. $f(x) = 3\ln(2x^2+5);\ f'(x) = 3\left[\frac{1}{2x^2+5}\right][4x] = \frac{12x}{2x^2+5}$

59. $\frac{12x+10}{9x^2+15x+3}$

61. $\frac{6x^3-1}{3x^4-2x+5}$

63. $\frac{3}{4}$

65. Relative change $= \frac{f'(x)}{f(x)} = (\ln f(x))' = 2$; the relative change is 2.

67. $\frac{27}{8}$

69. No critical values; no inflection points; vertical asymptote at $x = 0$.

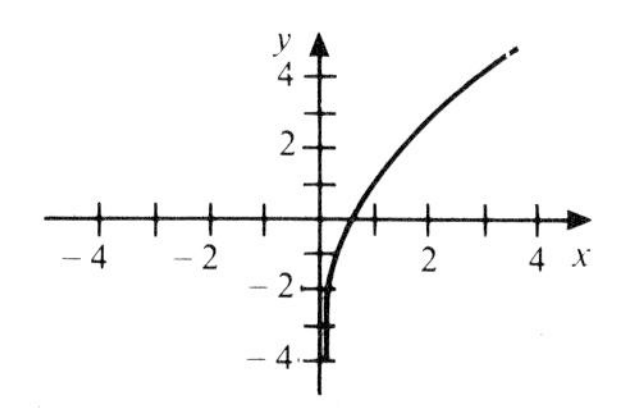

71. $g'(x) = \frac{1-\ln x}{x^2} = 0$ when $x = e$. $(e, e^{-1}) \approx (2.72, 0.37)$ is a relative maximum; $g''(x) = \frac{-3+2\ln x}{x}$; $(e^{3/2}, 3/(2e^{3/2}))$ is an inflection point; vertical asymptote at $x = 0$; horizontal asymptote at $y = 0$.

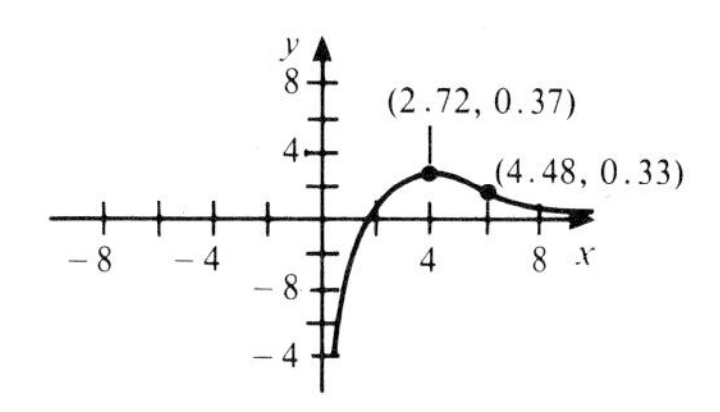

73. $m = \frac{-1}{e}$

Section 5.4

1. $f(x) = 5^x = e^{\ln 5^x} = e^{x\ln 5}$
$f'(x) = 5^x \ln 5$

3. $7^{2x}(2\ln 7)$

5. $2^x(1 + x\ln 2)$

7. $\ln f(x) = (6x^2-1)\ln 4;\ f'(x)/f(x) = (12x\ln 4);$
$f'(x) = (12x\ln 4)4^{6x^2-1}$

9. $(4\ln 3)3^{4x+1}$

11. $(e^x\ln 6)6^{e^x}$

13. $\ln f(x) = 5x\ln\left(\frac{1}{2}\right);\ \frac{f'(x)}{f(x)} = 5\ln\left(\frac{1}{2}\right);$
$f'(x) = 5\ln\left(\frac{1}{2}\right)\left(\frac{1}{2}\right)^{5x}$

15. $(\ln 9)\left(\frac{1}{3}\right)^{-2x+5}$

17. $f'(x) = f(x)\left[\frac{10}{2x+3} + \frac{12}{4x-1} - \frac{1}{2x+1} - \frac{12}{3x+5}\right]$

19. $\ln f(x) = 3\ln(2x^3+4) + \frac{1}{3}\ln(1-x^3) + \frac{1}{2}\ln(4x^4+x)$
$f'(x) = f(x)\left[\frac{18x^2}{2x^3+4} - \frac{x^2}{1-x^3} + \frac{8x^3+\frac{1}{2}}{4x^4+x}\right]$

21. $f'(x) = x^x(1+\ln x)$

23. $f'(x) = f(x)\left[\frac{2x}{x^2+4} + \frac{2x^2}{x^2-4} + \ln(x^2-4)\right]$

25. $\ln f(x) = \frac{\ln x}{x}$
$f'(x) = x^{1/x}\left[\frac{1-\ln x}{x^2}\right]$

27. $f'(x) = f(x)\left[\frac{\ln(1-x)}{x^2} + \frac{1}{x}\right]$

29. $f'(x) = (\ln x)^x\left[\frac{1}{\ln x} + \ln\ln x\right]$

31. $\ln f(x) = x^3\ln x$; use the product rule on the right side:
$f'(x) = x^{x^3}[x^2 + 3x^2\ln x]$

33. 0

Section 5.5

1. Let $P(t)$ represent the population at time t. $P = Ce^{kt}$; $C = 30{,}000$; $k = 0.02877$; $P(30) = 71{,}111$

3. 79.4%
 50%
 12.5%

5. 16 years
 8.329%

7. (a) The amount of carbon 14 left after t years is given by $A(t) = A_0e^{-0.000121t}$, where A_0 is the initial amount present. $A(t) = \frac{1}{10}A_0 = A_0e^{-0.000121t}$ gives $t \approx 19{,}030$ years. (b) 13,300 years old. (c) 38,060 years old.

9. 2.73%

11. 2011

13. Let P denote the earth's population. $P(t) = Ce^{0.02t}$; $C = 3.5$; solve $P(t) = 20$ for t. The world's population will reach 20 billion in the year 2057.

15. 2024

17. 5.53 billion

19. Let $y(t)$ be the population at time t; $y(t) = \frac{1}{k}[r - Ce^{-kt}]$ with $k = 0.01$, $r = 15k = 0.15$; $C = 0.11$. Solve $y(t) = 8$ for t: $t \approx 45$ or the year 2020.

21. 2104

23. 8

25. Let P denote the population and $t = 0$ correspond to 1950. $P = 40e^{kt}$; $k = 0.00953$; $P(40) = 58{,}562$. The population in 1990 is estimated to be 58,562.

27. 4.066 billion

29. 158 years

31. (a) $L'(t) = k[100 - L(t)]$; $C = 100k$; $k = 0.0211$. Solve $L(t) = 80$ for t: $t \approx 76$ minutes. (b) 109 minutes.

33. 24.5 inches per tree
 −0.249 inch per tree
 −0.307 inch per tree

Review Exercises

1.

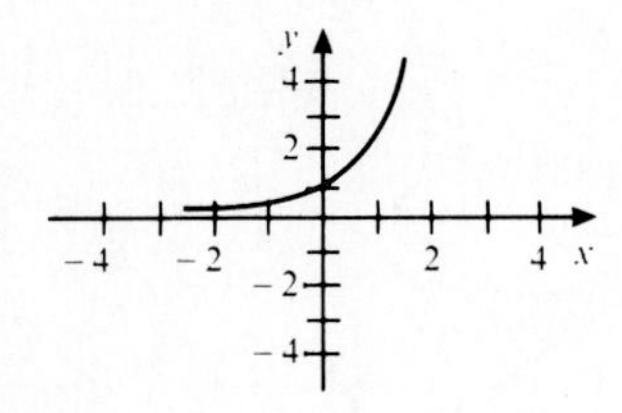

3.

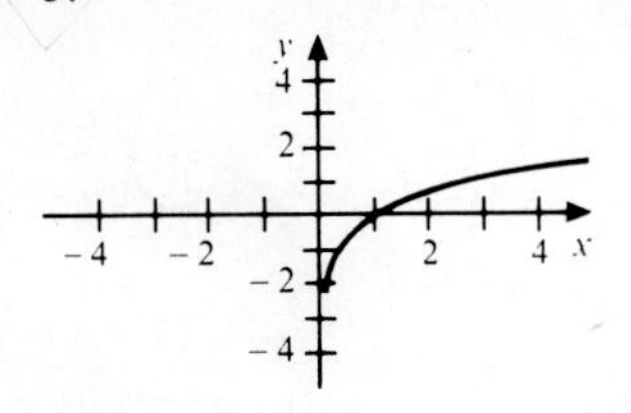

5. $\dfrac{\ln 30}{2\ln 8}$

7. $\frac{1}{2}(\ln 4 - 1)$

9. $\frac{1}{3}$

11. e^{12}

13. $\dfrac{5\ln 7}{(1 - \ln 7)} \approx -10.2859$

15. 0.43219

17. −4.5932

19. $2e^{2x}$

21. $4e^{-5x}(1 - 5x)$

23. $\dfrac{2x}{x^2 - 2}$

25. $\dfrac{3e^{3x}\ln 3x - \dfrac{e^{3x}}{x}}{(\ln 3x)^2}$

27. $\left(1 + \dfrac{1}{x}\right)e^{x + \ln x}$

29. $2x^{2x}(1 + \ln x)$

31. \$11,051.70

33. 7.92%

35. 13,725 years old

37. (a) $S(4) = 11{,}595$ (b) $S(8) = 13{,}445$ (c) 12.7 weeks

39. 6.75 years

Section 6.1

1. $F'(x) = f(x)$. To find another antiderivative of $f(x)$ add any constant to $F(x)$: $G(x) = x^2 + 1$

3. $G(x) = \frac{5}{2}x^2 - 3x - 1$

5. $G(x) = x^3 - 2x^{-1} + 13x + 1$

7. $G(x) = \dfrac{3x + 1}{2x - 1} + 1$

9. $G(x) = \dfrac{1}{2}\ln\left(\dfrac{x - 1}{x + 1}\right) + 1$

11. The two graphs are identical but are shifted vertically away from each other by 1.

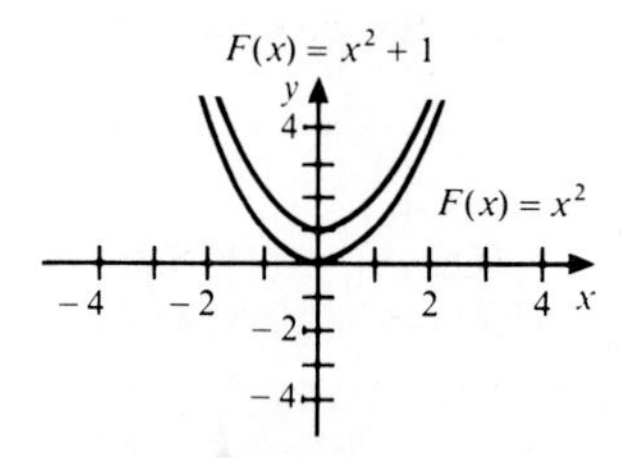

13. $F(x) = x^3 + 2x + 10$

15. $F(x) = \int (x^3 + x)\,dx = \int x^3\,dx + \int x\,dx = \frac{1}{4}x^4 + \frac{1}{2}x^2$

17. $F(x) = 4x - \frac{2}{3}x^3$

19. $F(x) = \frac{2}{3}x^{3/2} + x^3 + 6$

21. $F(x) = \int (4x^{-1/2} + x^2)\,dx = 4\int x^{-1/2}\,dx + \int x^2\,dx$
 $= 8\sqrt{x} + \dfrac{x^3}{3}$

23. $F(x) = \frac{3}{2}x^2 + 4\ln x$

25. $F(x) = 2e^{2x} + \frac{1}{3}e^{3x}$

27. $C(x) = \int (0.04x + 600)\,dx = 0.02x^2 + 600x + K$; $K = 468$
 $C(x) = 0.02x^2 + 600x + 468$; $C(100) = 60{,}668$

29. $C(x) = 0.01x^2 + 5x + 50$

31. $R(x) = -0.4x^3 + x^2 + 10x$

33. Use the power rule for integrals to get $\dfrac{x^4}{4} + C$.

35. $-\frac{1}{2}x^{-2} + C$
37. $5x + C$
39. $\int (2t^3 - 3t^2 + 5)\,dt = 2\int t^3\,dt - 3\int t^2\,dt + 5\int dt$
$= \frac{1}{2}t^4 - t^3 + 5t + C$
41. $-5u^{-1} + 2u^{1/2} + \frac{3}{8}u^{8/3} + C$
43. $\int \frac{1}{\sqrt[3]{v^2}}\,dv = \int v^{-2/3}\,dv = 3v^{1/3} + C$
45. $7 \ln x + C$
47. $3 \ln x + C$
49. $2e^{5t} + \frac{10}{3}t^3 + C$
51. $F(x) = \frac{3}{2}x^{2/3} + C$; $F(8) = 6$ gives $C = 0$; $F(x) = \frac{3}{2}x^{2/3}$
53. $F(x) = -2x^{-1/2} + \frac{4}{5}x^{5/2} + x^4 - 273.6$
55. $F(x) = \frac{3}{4}x^{4/3} + 2x - 4$
57. $F(x) = x^2 - 4x + C$; $F(0) = 0$ gives $C = 0$; $F(x) = x^2 - 4x$
59. $F(t) = \frac{1}{2}t^2 - \frac{2}{3}t^3 + t^4 + 4$
61. $F(x) = \ln x + e^x + e$
63. $f(x) = \frac{1}{3}x^3 + 2x + C$ represents all such functions.
65. $f(x) = 2x - \frac{1}{2}x^2 + C$
67. $f(x) = -\frac{1}{4}x^4 + C$
69. $\int x^2(2x^{1/2} + x)\,dx = \int (2x^{5/2} + x^3)\,dx = 2\int x^{5/2}\,dx + \int x^3\,dx = \frac{4}{7}x^{7/2} + \frac{1}{4}x^4 + C$
71. $\frac{2}{5}x^{5/2} + 2x^{1/2} + C$
73. $\frac{3}{2}x^2 + \ln x + C$

Section 6.2

1. $C(x) = 0.001x^2 + 4x + 4000$
3. Let $C(x)$ represent the cost function; then $C(x) = \int (-0.08x + 10)\,dx = -0.04x^2 + 10x + 6000$; $C(20) = -0.04(20)^2 + 10(20) + 6000 = 6184.00$; $C(21) = 6192.36$; the change in cost is \$8.36; the marginal cost is \$8.40 when $x = 20$.
5. $R(x) = 80x - 0.001x^2$
7. Let $R(x)$ represent the revenue function; then $R(x) = \int (-0.18x + 12)\,dx = -0.09x^2 + 12x$; $R(40) = -0.09(40)^2 + 12(40) = 336.00$; $R(41) = 340.71$; the change in revenue is \$4.71; the marginal revenue is \$4.80 when $x = 40$.
9. $C(31) - C(30) = \$28.17$
$M_C(30) = \$28.20$
11. $C(x) = 2\sqrt{x} + 0.001x + 1000$
13. (a) The cost function is given by $C(x) = \int (-x^2 + 80x + 20)\,dx = -\frac{1}{3}x^3 + 40x^2 + 20x + 23$.
(b) $C(21) = \$14{,}996$; $\frac{C(9)}{9} = \$355.56$ per unit; $\frac{C(21)}{21} = \$714.10$ per unit.
(c) At 9 units
15. (a) $R(x) = -\frac{1}{15}x^3 + 8x^2 + 13x + 946.67$
(b) $R(60) = 16{,}126.67$
$\frac{R(40)}{40} = \$250.00$
$\frac{R(60)}{60} = \$268.78$
(c) At 60
17. (a) 64 ft/sec
(b) 0 ft/sec
(c) 3 seconds
(d) 6 seconds
(e) 144 feet
(f) -96 ft/sec
19. (a) $v(t) = \int -9.7536\,dt = -9.7536t + 30$;
$s(t) = \int v(t)\,dt = -4.8768t^2 + 30t + 50$
(b) 7.5 seconds
21. -48 ft/sec
23. $s(t) = \frac{1}{2}at^2 + v_0 t$
$v(t) = at + v_0$
$a = 20$ ft/sec^2; $v_0 = 5$
25. (a) -10 ft/sec^2
(b) 5 seconds
(c) 125 ft
27. 320 ft/sec
29. (a) $s(t) = 16t^2$
(b) $s(3) = 144$ ft
(c) $s(5) = 400$ ft
(d) $s(10) = 1600$ ft

Section 6.3

1. For one subinterval, $3000 \times 7 = 21{,}000$ loaves.
For two subintervals, $(2800 \times 3.5) + (3000 \times 3.5) = 20{,}300$ loaves.
For seven subintervals, 16,600 loaves.
3. 192 ft
5. $A(x) = \frac{3}{2}x^2$; $A(5) = \frac{75}{2}$

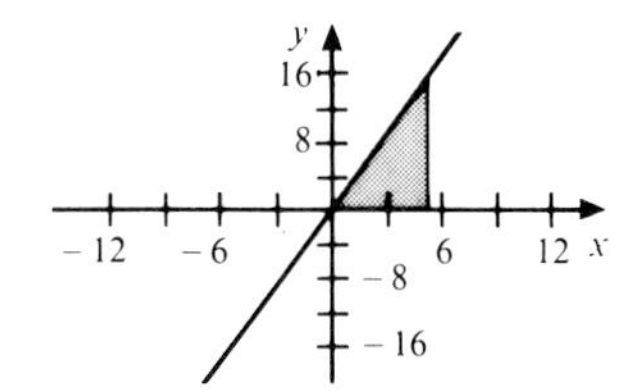

7. $A(x) = \int (4x + 1)\,dx = 2x^2 + x - 10$; $A(6) = 68$

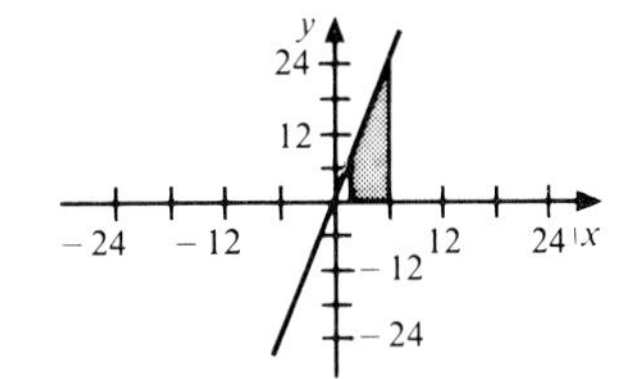

9. $A(x) = 5x - \frac{1}{2}x^2 + \frac{39}{2}$; $A(-1) = 14$

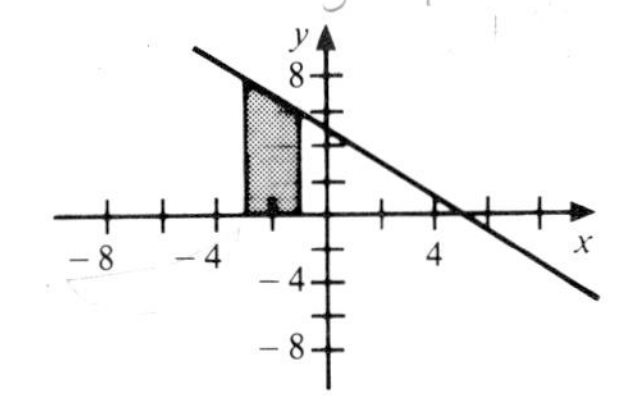

11. $A(x) = \ln x$; $A(e) = 1$

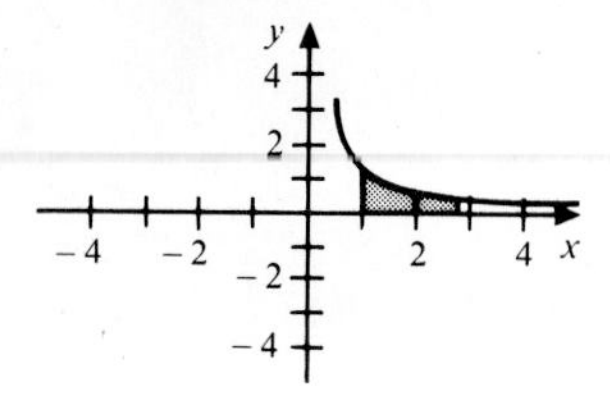

13. $A(x) = e^x - 1$; $A(1) = e - 1 \approx 1.72$

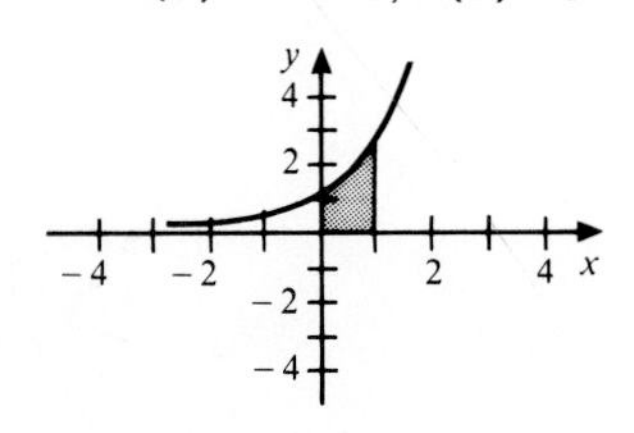

15. $A(x) = e^x + \frac{1}{2}x^2 - 1$; $A(1) = e - \frac{1}{2}$

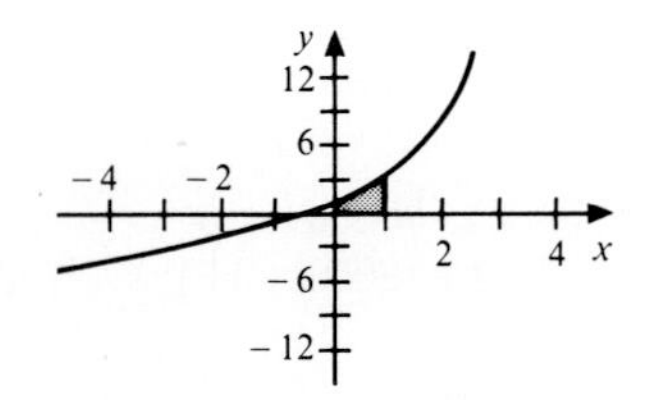

17. $G(x) = \int (2x + 1)\, dx = x^2 + x$
$G(4) - G(0) = 20$

19. $\frac{86}{3}$

21. $e^9 + 2$

23. $\frac{1}{2}(e^2 - 1)$

25. $G(x) = \int (-4 + e^x)\, dx = -4x + e^x$
$G(5) - G(3) \approx 120.33$

27. $\ln 2 - \frac{3}{8}$

29. 242.4 miles

Section 6.4

1. $\int_1^3 2x\, dx = x^2|_1^3 = [3^2] - [1^2] = 8$
3. $\frac{26}{3}$
5. 0
7. $\int_{-1}^{1} (x^2 - 2)\, dx = (\frac{1}{3}x^3 - 2x)|_{-1}^{1} = -\frac{10}{3}$
9. 1.36
11. −39.72
13. $\int_1^e (4 - \frac{3}{x})\, dx = (4x - 3\ln x)|_1^e = 3.87$
15. $\frac{1}{3}(e^6 - 1)$
17. 10.39
19. $\int_{-1}^{1} (e^x + e^{-x})\, dx = (e^x - e^{-x})|_{-1}^{1} = 4.70$
21. 0
23. 2
25. $\int_{-1}^{1} (6 + e^{-x})\, dx = (6x - e^{-x})|_{-1}^{1} = 14.35$
27. 21.75
29. 21

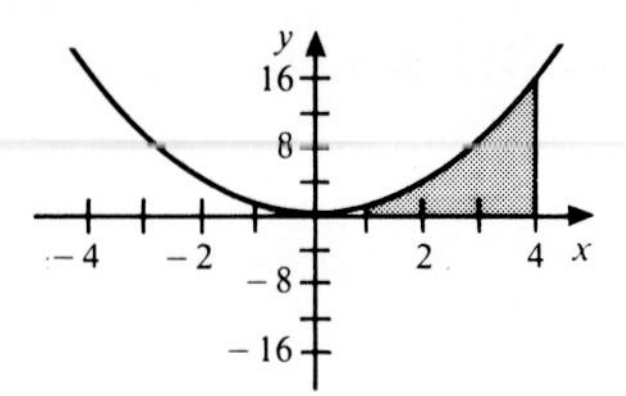

31. $\int_0^4 (5x + 2)\, dx = (\frac{5}{2}x^2 + 2x)|_0^4 = 48$

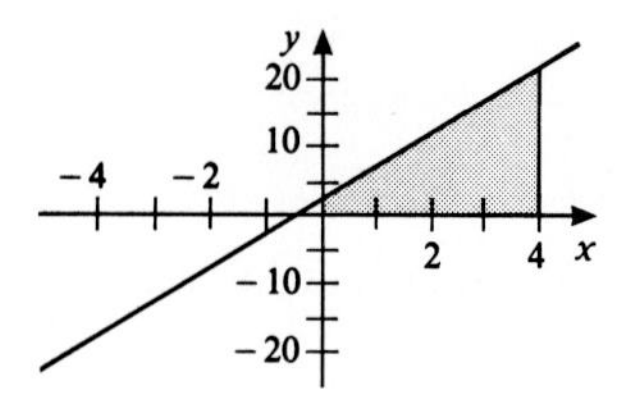

33. $\ln 3$

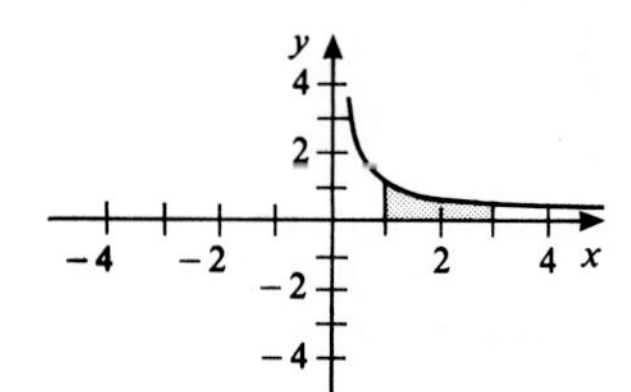

35. $6\frac{2}{3}$

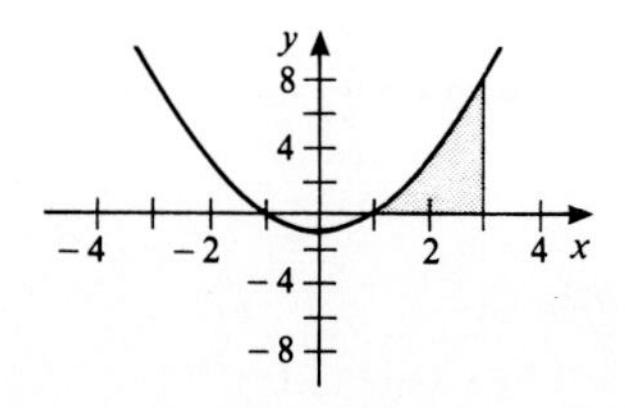

37. $\int_{-2}^{3} (4 - x)\, dx = (4x - \frac{1}{2}x^2)|_{-2}^{3} = \frac{35}{2}$

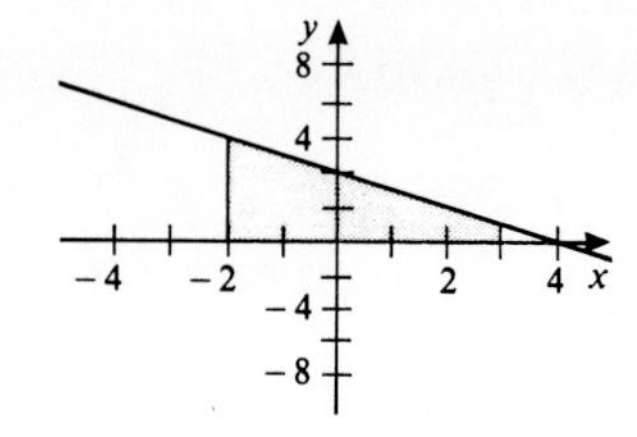

39. ln 4

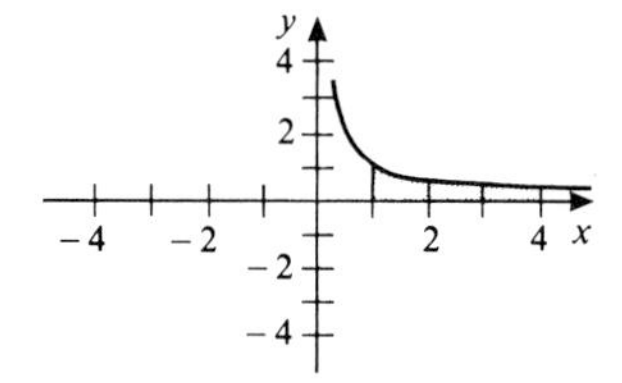

41. ln 4

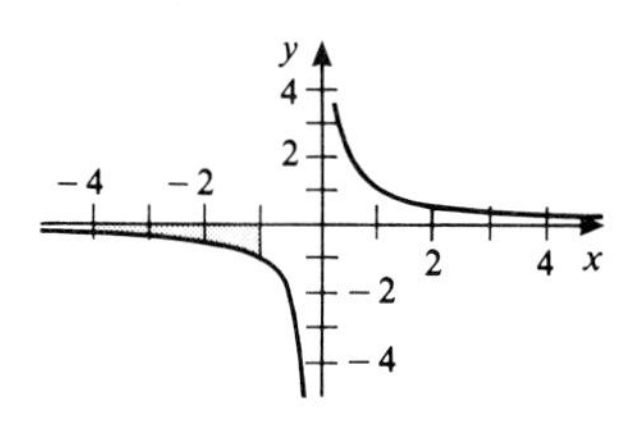

43. $\int_{-3}^{3} (4 - x^2)\, dx = (4x - \frac{1}{3}x^3)\Big|_{-3}^{3} = 6$

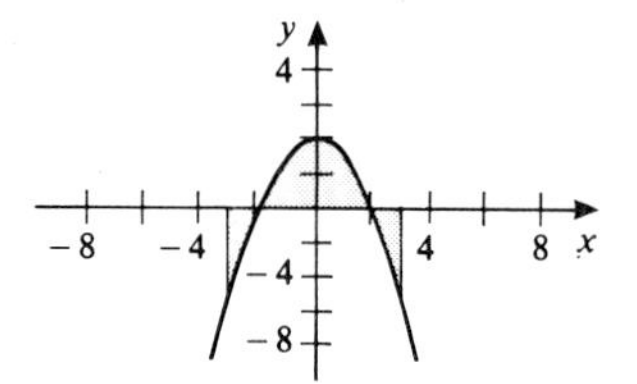

45. 6

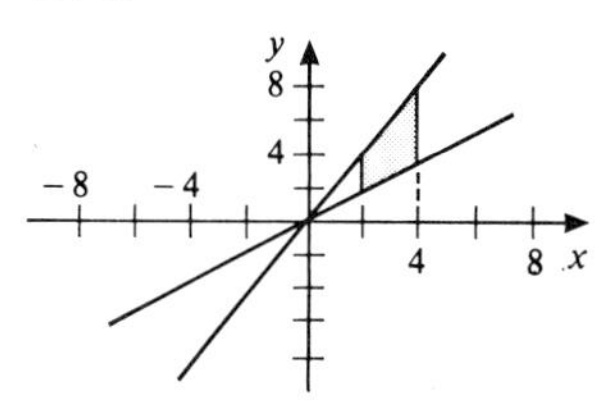

47. 7.568

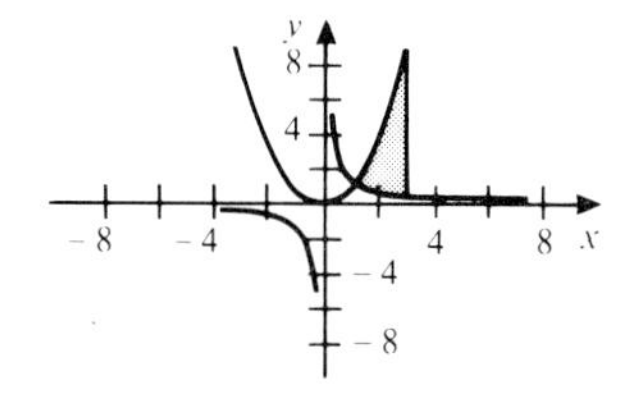

49. $A = \int_0^2 (e^{2x} - e^x)\, dx = \frac{1}{2}e^{2x} - e^x\Big|_0^2 \approx 20.41$

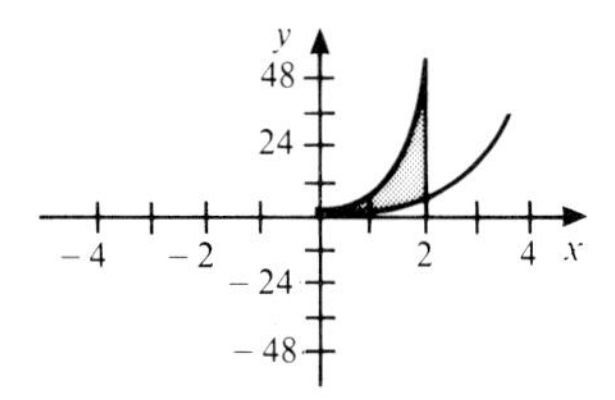

51. 5

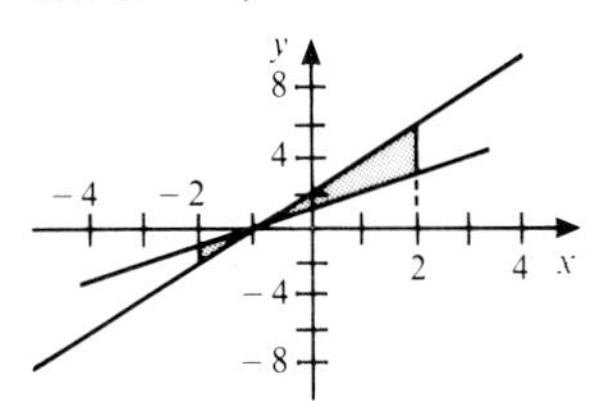

53. $\frac{1}{4}$

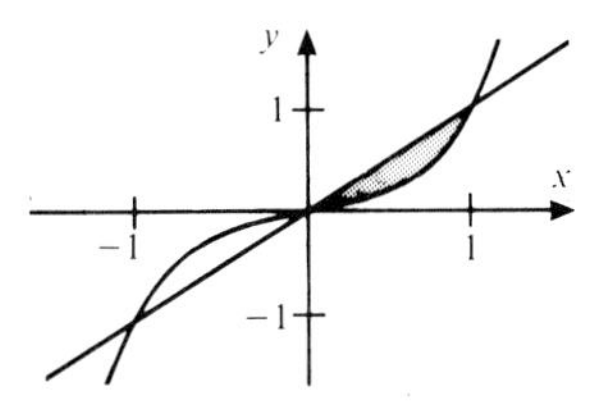

55. $A = \int_1^9 (2 - \frac{1}{x})\, dx = 2x - \ln x\Big|_1^9 \approx 13.80$

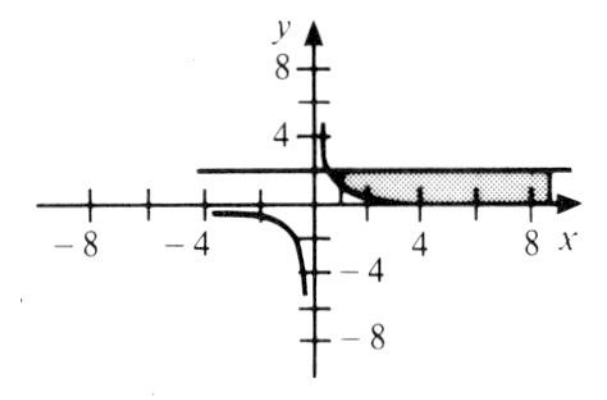

57. It represents the change in revenue from producing $x = a$ items per day to producing $x = b$ items per day.
59. It represents the difference in velocity of the object from time $t = b$ to time $t = c$.
61. Let $N(t)$ denote the number of calculators sold. Then $N(t) = \int S(t)\, dt$; $N(7) - N(0) = \int_0^7 S(t)\, dt = 742$. The company sells 74,200 calculators in the first 7 days.
63. 0.5846
65. $-\$45$
67. Let $I(t)$ denote the number of people interviewed. Then $I(t) = \int (30 + 2t)\, dt$; $I(10) - I(0) = 400$; 400 people will be interviewed in 10 weeks. They will not make it.

Section 6.5

1. Area $\approx (\frac{1}{1.5})0.5 + (\frac{1}{2})0.5 + (\frac{1}{3})1 + (\frac{1}{4})1 = \frac{7}{6}$
   Area $\approx (\frac{1}{1})0.5 + (\frac{1}{1.5})0.5 + (\frac{1}{2})1 + (\frac{1}{3})1 = \frac{5}{3}$

3. $\frac{32\pi}{3}$ 5. $\pi \ln 4 = 4.355$

7. $V = \int_1^4 \pi x\, dx = \frac{\pi x^2}{2}\Big|_0^4 = 8\pi$

9. $\frac{243\pi}{5}$ 11. $\frac{2\pi}{15}$

13. $\int_0^4 2\pi r(16 - r)\, dr = [16\pi r^2 - \frac{2}{3}\pi r^3]|_0^4 = 670{,}210$ people
    $\int_2^5 2\pi r(16 - r)\, dr = 810{,}530$ people

15. $-\frac{8}{3}$ 17. $\frac{11}{4}$ 19. 60.4°

21. The selling price is $b = 30$; the demand is $a = 35$. $\int_0^{35} (100 - 2x) - 30\, dx = 1225$; the consumer surplus is \$1225.

23. \$400 25. \$6666.67 27. \$5333.33

29. $S(x) = D(x)$ at $x = 3000$; $S(3000) = 34$; hence $a = 3000$, $b = 34$. $PS = \int_0^{3000} 34 - (4 + 0.01x)\, dx = 45{,}000$; the producer surplus is \$45,000.

31. \$6.08

Chapter 6 Review Exercises

3. $\frac{x^3}{3} - 2x + 2$ 5. $\ln|x| + 2$

7. $\frac{-1}{9x^6} + C$ 9. $\frac{2}{3}x^{3/2} - 2x^{1/2} + C$

11. -12 13. $-\frac{1}{4}$

15. $\frac{1}{2}(1 - e^{-6})$ 17. $\frac{1}{2}(e^8 - e^2 + \ln 4)$

19. 1886.58 gallons of water 21. 38 thousand

23. About 233 25. 64,298 cubic feet

27. 9 seconds 29. -240 ft/sec

31. 20 ft/sec^2

33. $s(t) = \frac{t^4}{12} - \frac{4t^{5/2}}{15} + 84t + 10$

35. $s(t) = t^2 - \frac{1}{12}t^4 - \frac{1}{3}t$ 37. 72 ft/sec

39. $\frac{17}{6}$ 41. $\frac{9}{2}$

43. $\frac{709}{12}$ 45. $\frac{3}{\ln 4} = 2.164$

Section 7.1

1. Let $u = x^2 + 3$; $\int (x^2 + 3)^{14}(2x)\, dx = \int u^{14}\, du = \frac{u^{15}}{15} + C = \frac{(x^2 + 3)^{15}}{15} + C$

3. $-\frac{(-4x^2 + 6)^{-4}}{4} + C$ 5. $-\frac{(2t^2 + 3t - 9)^{-3}}{3} + C$

7. $\frac{(2x + 4)^7}{14} + C$; use $u = 2x + 4$

9. $e^{x^2} + C$

11. $\frac{(5x^3 + 6x^2 + 2x - 4)^{24}}{24} + C$

13. $\frac{(-2t^3 + 5t - 6)^{50}}{50} + C$; use $u = -2t^3 + 5t - 6$

15. $\frac{1}{6} \ln (3x^2 + 5) + C$ 17. $\frac{-1}{9(3x + 1)^3} + C$

19. $\frac{1}{3} \ln (3e^w + 5) + C$; use $u = 3e^w + 5$

21. $e^{2x^2 + x} + C$

23. $\frac{(x^2 + 7)^{23}}{46} + C$

25. $\frac{(4t^2 + 6t + 3)^{34}}{68} + C$; use $u = 4t^2 + 6t + 3$

27. $-\frac{(3x^2 + 12x + 7)^{-14}}{84} + C$

29. $\frac{-1}{476(7x^4 + 14x + 3)^{34}} + C$

31. $\frac{1}{12}(x^2 + 1)^6 + C$; use $u = x^2 + 1$

33. 504 35. $1 - e^{-6}$ 37. 625; use $u = s^2 + 2s + 1$

39. $\frac{3\sqrt{2}}{2}$ 41. $\ln\frac{7}{3}$ 43. $\frac{1}{2} \ln \frac{5}{2}$; use $u = x^2 + 1$

45. 40 47. $-\frac{16}{3}$ 49. $1 - e^{-2}$

51. 20 53. $\frac{64}{21}$

55. 66.7 cm; use $u = t + 1$

57. \$5493.33

59. Use $u = kx$; the result is $x + C$ when $k = 0$.

Section 7.2

1. $xe^x - e^x + C$; start with

x	e^x
1	e^x

3. $\frac{1}{5}x(x - 3)^5 - \frac{1}{30}(x - 3)^6 + C$

5. $(5x - 4)^3e^x - 15(5x - 4)^2e^x + 150(5x - 4)e^x + 750e^x + C$

7. $-\frac{1}{18}(4 - 3x)(2 - 3x)^6 + \frac{1}{126}(2 - 3x)^7 + C$; start with

$4 - 3x$	$(2 - 3x)^5$
-3	$\frac{-1}{3 \times 6}(2 - 3x)^6$

9. $-x^2e^{-x} - 2xe^{-x} - 2e^{-x} + C$

11. $\frac{(2t + 1)^2(4t - 2)^{5/2}}{10} - \frac{(2t + 1)(4t - 2)^{7/2}}{35} + \frac{(4t - 2)^{9/2}}{315} + C$

13. $\frac{(2x^2 + 3x - 4)(6x + 5)^{7/2}}{21} - \frac{(4x + 3)(6x + 5)^{9/2}}{567} + \frac{4(6x + 5)^{11/2}}{18{,}711} + C$; start with

$2x^2 + 3x - 4$	$(6x + 5)^{5/2}$
$4x + 3$	$\frac{(6x + 5)^{7/2}}{6 \times (\frac{7}{2})}$

15. $\frac{2s^2}{5}(s - 1)^{5/2} - \frac{8s}{35}(s - 1)^{7/2} + \frac{16}{315}(s - 1)^{9/2} + C$

17. $\frac{x}{2}e^{2x} - \frac{1}{4}e^{2x} + C$

19. $-\frac{t}{2}e^{-2t} - \frac{1}{4}e^{-2t} + C$; start with

t	e^{-2t}
1	$-\frac{1}{2}e^{-2t}$

21. $\frac{1}{2}e^{-4x}(2x-1) + C$

23. $\frac{1}{4}(6x-11)e^{2x+3} + C$

25. $\frac{t^2e^{3t}}{3} - \frac{2te^{3t}}{9} + \frac{2e^{3t}}{27} + C$; start with

t^2	e^{3t}
$2t$	$\frac{1}{3}e^{3t}$

27. $\frac{e^{2x}}{4}(4x^2 + 6x - 9) + C$

29. $e^{3u+2}(u^2 + u - 1) + C$

31. $2x^3e^{x/2} - 12x^2e^{x/2} + 48xe^{x/2} - 96e^{x/2} + C$; start with

x^3	$e^{x/2}$
$3x^2$	$2e^{x/2}$

33. $\frac{e^{2t}}{4}(4t^2 + 2t + 1) + C$

35. $\frac{x^5 \ln x}{5} - \frac{x^5}{25} + C$

37. $\frac{(2x+3)^5 \ln(2x+3)}{10} - \frac{(2x+3)^5}{50} + C$; start with

$\ln(2x+3)$	$(2x+3)^4$
$\frac{2}{(2x+3)}$	$\frac{(2x+3)^5}{10}$

39. $-2p^{-2}\ln 5p - p^{-2} + C$

41. 1084

43. $\frac{54}{5}$; start with

t	$(2t+3)^{1/2}$
1	$\frac{1}{3}(2t+3)^{3/2}$

45. $\frac{749}{60}$

47. $2e^3$

49. $\frac{122}{27}e^{12} - \frac{26}{27}e^6$; start with

x^2	e^{3x}
$2x$	$\frac{1}{3}e^{3x}$

51. 14.68

53. 47.91

55. $\frac{209}{140}$; start with

x^3	$(x-1)^3$
$3x^2$	$\frac{1}{4}(x-1)^4$

57. 210.64
59. $\frac{1}{280}$
61. 106.24

63. 3363 heartbeats
65. 478,000 items

67. $s(t) = \int_0^2 6te^{-2t}\,dt \approx 1.362$ meters

69. 1

Section 7.3

1. $\frac{1}{2}$; $\int_1^x \frac{1}{t^3}\,dt = \frac{1}{-2x^2} + \frac{1}{2}$; $\lim_{x\to\infty} \frac{1}{-2x^2} = 0$

3. Diverges

5. 99.31

7. 1; $\int_0^t e^{-x}\,dx = -e^{-t} + 1 \to 1$ as $t \to \infty$

9. $\frac{1}{2e}$

11. Diverges

13. Diverges; $\int_2^t \frac{x}{x^2-1}\,dx = \frac{1}{2}\ln(t^2-1) - \frac{1}{2}\ln 3$; $\lim_{t\to\infty} \frac{1}{2}\ln(t^2-1)$ does not exist.

15. $\frac{1}{16}$

17. $\frac{3}{e^2}$

19. $\frac{1}{1156}$; $\int_3^x \frac{t}{(t^2+8)^3}\,dt = \frac{1}{-4(x^2+8)^2} + \frac{1}{1156} \to \frac{1}{1156}$ as $x \to \infty$

21. $-\frac{3}{8}$

23. $-\frac{1}{324}$

25. Diverges

27. -1

29. $-\frac{1}{219{,}501}$

31. $\frac{1}{98}$; $\int_{-\infty}^{\infty} \frac{|t|}{(t^2+7)^3}\,dt = \int_{-\infty}^{0} \frac{-t}{(t^2+7)^3}\,dt + \int_0^{\infty} \frac{t}{(t^2+7)^3}\,dt$

33. 0

35. (a) 500 gallons
(b) No, the integral diverges.

37. 66.7 acres

39. $p > 1$

Section 7.4

1. 15
3. 8.7311
5. 22.7126
7. (a) 131.3907
(b) 130.1429
9. 2.8554
11. 2.6082
13. About 34.8 cases
15. 112.5 burglaries
17. $\frac{255}{4}$
19. $\frac{64}{3}$
21. 13.8916
23. (a) 8.0871
(b) 8.0857
(c) 8.0856
25. 2.2103
27. 9.3063
29. 53.8638

Section 7.5

1. Use entry 1 with $a = 4$, $b = -9$: $\frac{1}{4}\ln|4x-9| + C$

3. $\frac{-1}{30(-3x^2+2)^5} + C$

5. $-\sqrt{-2x+1} + C$

7. Use $u = 2t$, then entry 6:
$\frac{t}{2}\sqrt{4t^2+12} + 3\ln|2t + \sqrt{4t^2+12}| + C$

9. $x + 6 - 6\ln|x+6| + C$

11. $\frac{x}{8}(2x^2-9)\sqrt{x^2-9} - \frac{81}{8}\ln|x + \sqrt{x^2-9}| + C$

13. Use entry 5 with $a = 8$, $b = 9$: $\frac{1}{4}\sqrt{8t+9} + C$

15. $\ln|x + \sqrt{x^2-1}| + C$

17. $\frac{-2}{9(3x-2)} + \frac{\ln|3x-2|}{9} + C$

19. Use entry 11 with $a = 2$:
$\frac{x}{8}(2x^2 - 4)\sqrt{x^2 - 4} - 2\ln|x + \sqrt{x^2 - 4}| + C$

21. $\frac{w^{18}}{324}[18 \ln w - 1] + C$
23. 2.3905
25. 0.729
27. 0.9303

Chapter 7 Review Exercises

1. $\frac{(x+3)^6}{6} + C$
3. $\frac{2(x^2-4)^{3/2}}{3} + C$
5. $2\sqrt{t^2 - 3t + 1} + C$
7. $\ln(e^x + 1) + C$
9. $\frac{1}{2}e^{x^2+1} + C$
11. $\frac{2x(3x+5)^{3/2}}{9} - \frac{4(3x+5)^{5/2}}{135} + C$
13. $2t + 4 - \ln|t+2| + C$
15. $\ln|2x - 5| + C$
17. $\frac{1}{3}(x^2+1)^{3/2} + C$
19. $\frac{(\ln s)^2}{2} + C$
21. $\frac{7}{3}$
23. 0.0878
25. ln 2
27. Converges to $\frac{1}{2}$
29. Diverges
31. $\frac{67}{60}$
33. 1.1
35. $\frac{2e^3 + 1}{9}$
37. 0.0722
39. $\ln|x + \sqrt{x^2 - 16}| + C$
41. $2360.00

Section 8.1

1. $f(2,3) = (2)^2 + (3)^2 - 7(2)(3) = -29$
$f(0,4) = (0)^2 + (4)^2 - 7(0)(4) = 16$
$f(3,0) = (3)^2 + (0)^2 - 7(3)(0) = 9$
3. 1188; −36
5. $\frac{22}{9}$; 2.8; $\frac{1}{3}$
7. Cost of producing x hardbound copies is $7.50x$. Cost of producing y paperback copies is $3.45y$. Cost of producing x hardbound copies and y paperback copies is $C(x, y) = 7.5x + 3.45y$.
9. $213,000
11. $254
13. $C(2,3) = 2(2)^2 + 17(2)(3)^2 + 140(3) + 12(2) + 200 =$ $958
15. $4200; $9400
17. 19

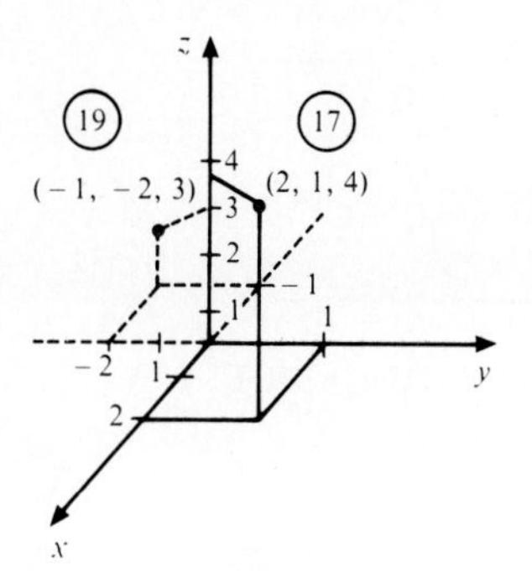

21.

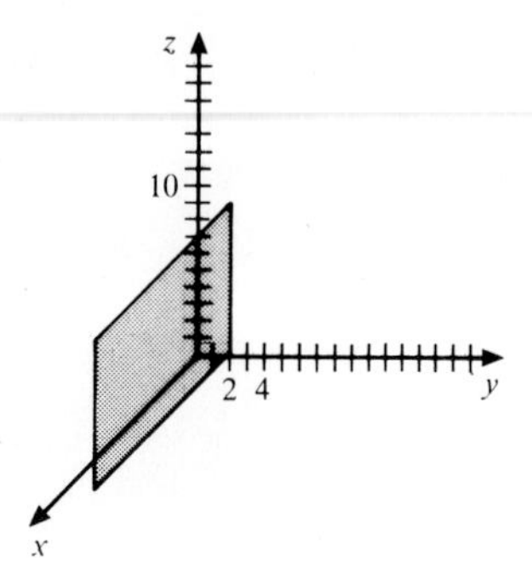

23.

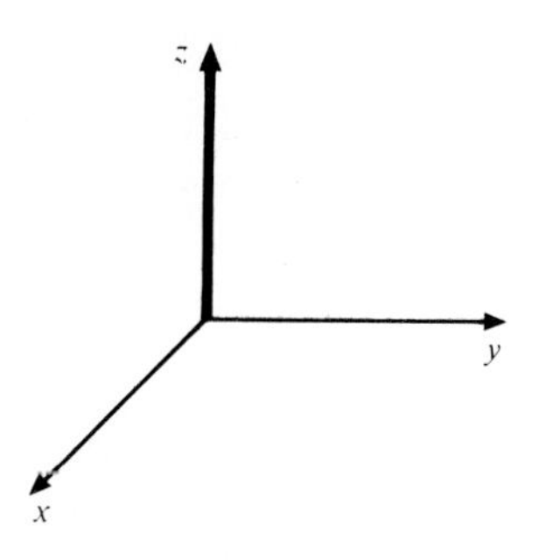

25.

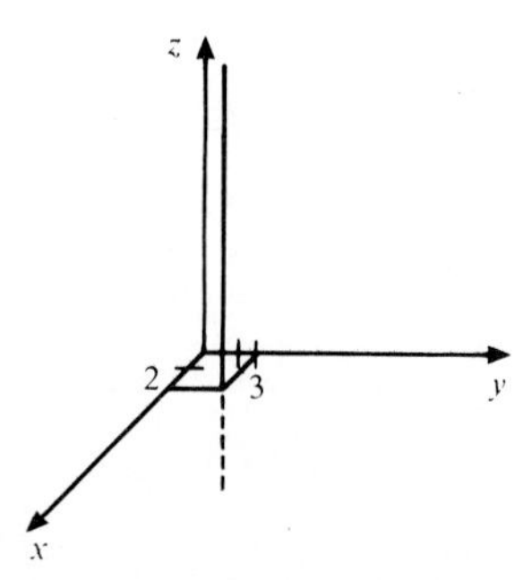

27.

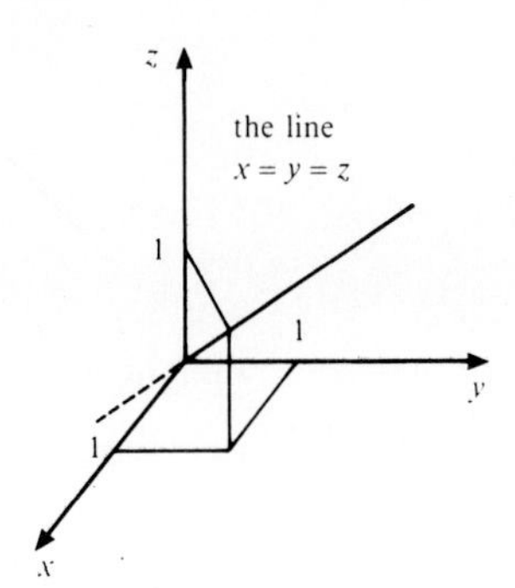

29.

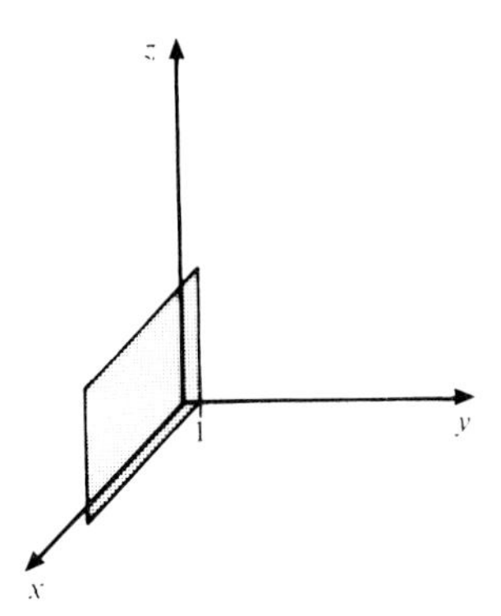

31.

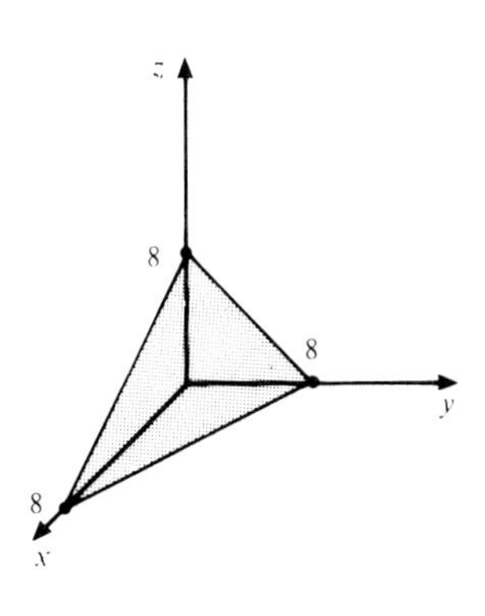

33.

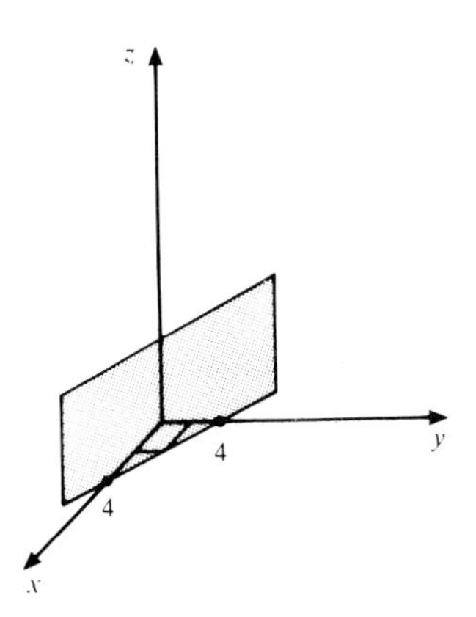

35.

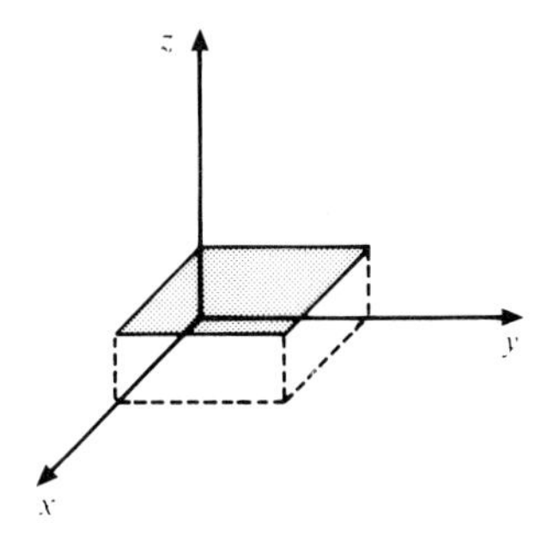

37.

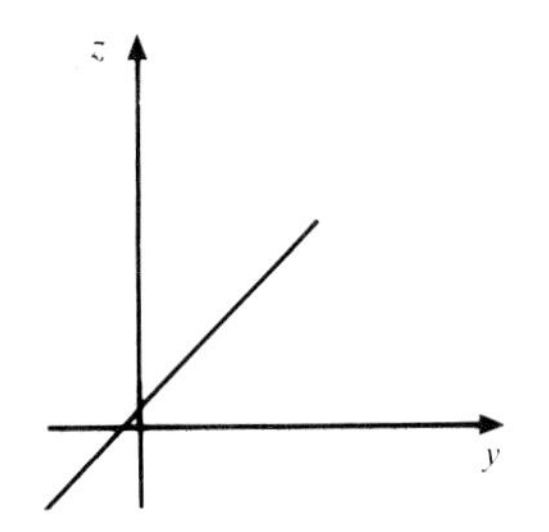

39.

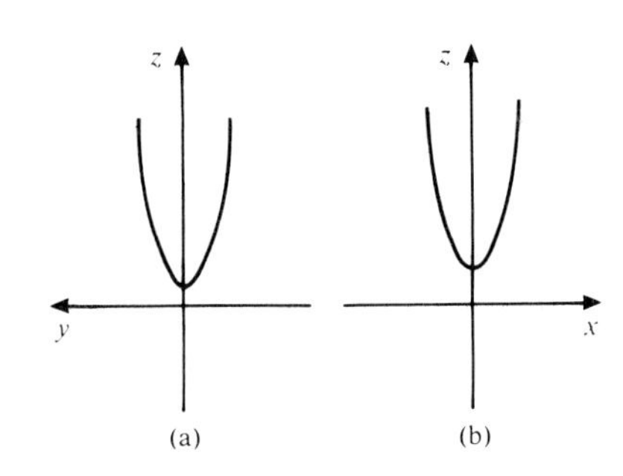

41.

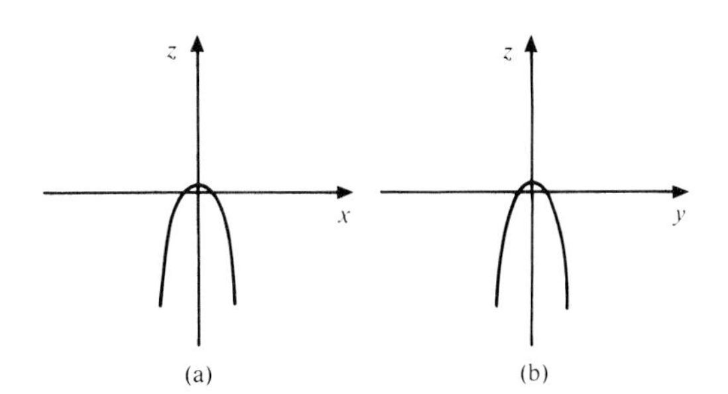

43.

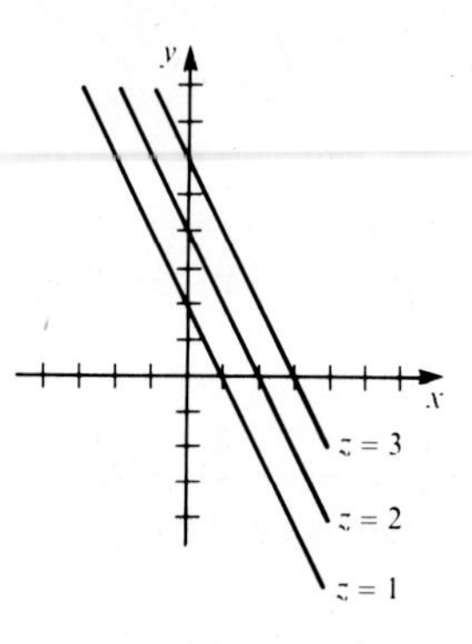

45.

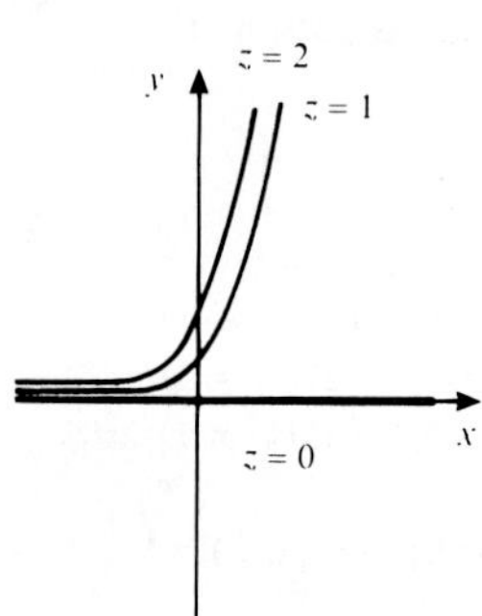

Section 8.2

1. Treating y as a constant, $\frac{\partial z}{\partial x} = 2x - y$. Evaluating at $(2, 2)$, we have $2(2) - 2 = 2$. Treating x as a constant, $\frac{\partial z}{\partial y} = 2y - x$.

3. $8x(x^2 + y^2)^3$
375,000

5. $15(3x + 2y - 4)^4$
65,610

7. $f_u(u, v) = \frac{\partial z}{\partial u} = 2u + v - 14; f_v(u, v) = -2v + u;$
$f_v(2, 3) = -2(3) + (2) = -4$

9. $\frac{2x^2 + 4y}{\sqrt{x^2 + 4y}}; 2x(x^2 + 4y)^{-1/2}; \sqrt{2}$

11. $8xz + 12y^2(x^2z^2 + 3xyz)(2xz^2 + 3yz)$
$4x^2 + 12y^2(x^2z^2 + 3xyz)(2x^2z + 3xy)$

13. Slope of $f(s, t)$ in the s-direction is $f_s(s, t) = \frac{s^2}{\sqrt{s^2 - t^2}} + \sqrt{s^2 - t^2}; f_s(5,3) = \frac{21}{4}$.

Slope in the t direction is $f_t(s, t) = \frac{-st}{\sqrt{s^2 - t^2}}$;

$f_t(5, 3) = -\frac{15}{4}$.

15. $-4; -7$

17. $\frac{2s}{t} - \frac{t^2}{(4 - s)^2}; \frac{-s^2}{t^2} - \frac{2t}{4 - s}; -5$

19. $\frac{\partial z}{\partial x} = 2x^2e^{x^2+y^2} + e^{x^2+y^2}; \frac{\partial z}{\partial y} = 2xye^{x^2+y^2}$

21. $\frac{2x}{x^2 + y^2}; \frac{4}{13}$

23. $\frac{1}{x}; 3y^2; 12$

25. $\frac{\partial p}{\partial x} = 0.016x + 0.03y; \left.\frac{\partial p}{\partial x}\right|_{(10,18)} = 0.70;$
$\frac{\partial p}{\partial y} = 0.03x + 0.004y; \left.\frac{\partial p}{\partial y}\right|_{(10,18)} = 0.372$

27. $\frac{19}{3}; -\frac{11}{27}$

29. $\frac{1}{2}; -\frac{1}{2}$

31. $f_x = 3x^2 - 3; f_y = 2y + 1$
Set these quantities equal to zero and solve for x and y: $x = 1, -1$ and $y = -\frac{1}{2}$; the points are $(1, -\frac{1}{2})$, $(-1, -\frac{1}{2})$.

33. $(-1, -\sqrt{2}), (-1, \sqrt{2}), (1, -\sqrt{2}), (1, \sqrt{2})$

35. $P_L = 3 + 2K; P_K = 4K + 2L; 558; 23; 64$
When $L = 11$ and $K = 12$, $P \approx 581$ (actual value is 585).
When $L = 10$ and $K = 13$, $P \approx 622$ (actual value is 628).

37. $f_x = -3 + 2x - y; f_x(10,15) = 2$

39. $5000y; 5000x$

41. When aired twice a day, sales increase by 10,000 tubes per day for every increase of 1 million viewers.

43. $R = a\frac{S}{V} = \frac{2a(l + 2r)}{\left(rl + \frac{4}{3}r^2\right)}; R_r = \frac{-2a\left(l^2 + \frac{8}{3}rl + \frac{8}{3}r^2\right)}{\left(rl + \frac{4}{3}r^2\right)^2}$; note that $R_r < 0$ so an increase in r produces a decrease in R.
$R_l = \frac{-4ar^2}{3\left(rl + \frac{4}{3}r^2\right)^2}$; since $R_l < 0$ an increase in l also produces a decrease in R.

45. (a) An increase from 800 hours to $800 + h$ hours will increase production by about $(49.3)(\frac{h}{100})$ tons.
(b) An increase in acreage from 4 to $4 + k$ will increase production by about $93.6k$ tons.

47. $2 + 6xy^2; 56; 6x^2y; 18$

49. $f_x = 2xy^4 - 3y; f_{xx} = 2y^4; f_{xx}(1,4) = 2(4)^4 = 512$
$f_{xy} = 8xy^3 - 3; f_{xy}(1,4) = 8(1)(4)^3 - 3 = 509$

51. $6y^5 - 60x^3y^2; -1626; 30xy^4 - 30x^4y; 2520$

53. $\frac{2}{x}; 2; \frac{1}{y}; \frac{1}{2}$

55. $\frac{\partial z}{\partial y} = 2x^4y - 5x^3y^4; \frac{\partial^2 z}{\partial x\, \partial y} = 8x^3y - 15x^2y^4$

57. $4xy + \frac{9}{4}y^2x^2(x^3 + y^3)^{-3/2}$

59. 0

61. $\frac{\partial z}{\partial y} = -2yx^2$; $\frac{\partial^2 z}{\partial x\,\partial y} = -4xy$

63. $12y^2 - 12x^2$

65. $\frac{4}{3}y(y - 3x + xy^2)^{-1/3} - \frac{2}{9}(1 + 2xy)(-3 + y^2)(y - 3x + xy^2)^{-4/3}$

67. $\frac{\partial z}{\partial y} = \frac{4xe^{xy}}{2x + e^{xy}}$; $\frac{\partial^2 z}{\partial x\,\partial y} = \frac{4e^{xy}(2x^2y + e^{xy})}{(2x + e^{xy})^2}$

69. $\frac{1}{y} + \frac{y^2}{x^3}$

71. $\frac{1}{y} + \frac{y^2}{x^3}$

73. $f_y = \frac{2y^3 - x^3}{2xy^2}$; $f_{yy} = \frac{(2xy^2)(6y^2) - (2y^3 - x^3)(4xy)}{4x^2y^4}$;
$f_{yy}(1,2) = \frac{9}{8}$

Section 8.3

1. $f_x = 2x = 0$ or $x = 0$
$f_y = 2y = 0$ or $y = 0$
The only point (0, 0).

3. (0, 0)

5. (0, 0), $(\frac{1}{2}, 0)$, $(-\frac{1}{2}, 0)$

7. $f_x = y(-2x + 4 - y) = 0$ so $y = 0$ or $-2x + 4 - y = 0$;
$f_y = x(-2y + 4 - x) = 0$ so $x = 0$ or $-2y + 4 - x = 0$.
Case 1: if $y = 0$, then $x(-2(0) + 4 - x) = 0$ or $x = 4$ or 0; (4, 0), (0, 0).
Case 2: If $x = 0$, then $y(-2(0) + 4 - y) = 0$ or $y = 4$ or 0; (0, 4), (0, 0).
Case 3: If $y \neq 0$, $x \neq 0$ then $-2x + 4 - y = 0$ and $-2y + 4 - x = 0$ yields $(\frac{4}{3}, \frac{4}{3})$.
The points are (0, 0), (0, 4), (4, 0), $(\frac{4}{3}, \frac{4}{3})$.

9. (0, 0)

11. Spend about \$4000 on promotion and set the price at \$5.00.

13. $R_w = 3w^2 - 6p$; $R_p = 12p - 12 - 6w$; these are zero at $w = 2$ and $p = 2$.

15. 9 barrels liquid fertilizer
6 barrels liquid insecticide

17. $x = 2\sqrt[3]{4}$; $y = 2\sqrt[3]{4}$; $z = 3\sqrt[3]{4}$

19. $s_y = 150 - y = 0$ or $y = 150$. Produce 150 cameras.

21. $6 \times 6 \times 3$

23. $x = 71$; $y = 50$

25. $p_x = 20 - 10xy = 0$ or $xy = 2$; $p_y = 20 - 5x^2 = 0$ or $x = 2$; if $x = 2$, then $xy = 2$ implies that $y = 1$ so the night shift produces 2000 and the day shift 1000 units.

27. 6 of the smaller size; 4 of the larger size.

29. 864 sq ft

31. $P_a = -8a - b + 20 = 0$; $P_b = -4b - a + 18 = 0$; solving simultaneously gives $a = 2$, $b = 4$. About \$2000 should be invested in variable a and about \$4000 in variable b.

Section 8.4

1. Algebraic Substitution:
$x = 2 - y$; $f(2 - y, y) = (2 - y)(y + 6) = -y^2 - 4y + 12$; $f_y = -2y - 4 = 0$ or $y = -2$; hence $x = 4$; the maximum is $f(4, -2) = 16$.
Lagrange Multiplier Method:
$g(x, y) = 2 - x - y$;
$F(x, y, \lambda) = x(y + 6) + \lambda(2 - x - y)$
$F_x = y + 6 - \lambda = 0$; $F_y = x - \lambda = 0$; $F_\lambda = 2 - x - y = 0$;
$\lambda = y + 6$; $\lambda = x$. Therefore, $x = y + 6$. The maximum is $f(4, -2) = 16$.

3. $\frac{9}{8}$

5. -49

7. $g(x, y) = 4 - 3y - 2x$; $F(x, y, \lambda) = x^2 + y^2 + \lambda(4 - 3y - 2x)$; $F_x = 2x - 2\lambda = 0$; $F_y = 2y - 3\lambda = 0$; $F_\lambda = 4 - 3y - 2x = 0$. The minimum is $\frac{208}{169}$ at $(\frac{8}{13}, \frac{12}{13})$.

9. $x = \frac{37}{6}$; $y = \frac{29}{6}$; $z = \frac{17}{6}$

11. Minimum at (3, 3, 4); Maximum at (−3, −3, −4)

13. $g(x, y, z) = 70 - 4x^2 - 2y^2 - z^2$; $F(x, y, z, \lambda) = 6x + 3y + 2z - 5 + \lambda(70 - 4x^2 - 2y^2 - z^2)$. The maximum occurs at (3, 3, 4) and the minimum occurs at (−3, −3, −4).

15. Maximum at $(2, 3\sqrt{2})$ and $(2, -3\sqrt{2})$
No minimum

17. 50 hours of lectures; 30 hours of recitations

19. $g(x, y) = 2x^2 + 8xy + 17y^2 - 1539$; $F(x, y, \lambda) = 2x + 5y + \lambda(2x^2 + 8xy + 17y^2 - 1539)$; $x = 21$, $y = 3$

21. Work 1.6 hours at one job and 6.4 hours at the other.

23. 22.9 units

25. $g(x, y) = 2400 - 200x - 400y = 0$; $S(x, y, \lambda) = \sqrt{x} + \sqrt{y} + \lambda(2400 - 200x - 400y)$; $x = 8$; $y = 2$

27. 1,410,453 units

29. (2, 4)

31. $\frac{\partial P}{\partial x}(25, 150) = 125(6)^{3/4}$; $\frac{\partial P}{\partial y}(25, 150) = 375(6)^{-1/4}$

33. About 110 pounds of pesticide x and about 390 pounds of pesticide y.

Section 8.5

1. $f_x = 2x + y + 3$; $f_x(1, -2) = 3$; $f_y = x$; $f_y(1, -2) = 1$;
$df = (3)(0.1) + (1)(0.5) = 0.8$

3. 0.7787

5. 130.0474

7. $f_x = \frac{(x + y)y - xy}{(x + y)^2}$; $f_y = \frac{(x + y)x - xy}{(x + y)^2}$;
$df = \frac{25(1) + 4(0.5)}{49} = \frac{27}{49}$

9. 1

11. Let $f(x, y) = x^{1/3}y^{1/4}$; $a = 216$; $b = 256$; $h = -6$ and $k = -6$.
$df = (0.0370)(-6) + (0.0234)(-6) = -0.3624$
$f(210, 250) \approx f(216, 256) + df = 24 - 0.3624 = 23.6376$; $210^{1/3}250^{1/4} \approx 23.6376$

13. 17.3125

15. 20.585

17. 0.793

19. 1.478

21. 10.4π

23. 0.00104

25. 3500

27. Let $x = 25$, $y = 13$, $a = 25$, $b = 12$ so $h = 0$ and $k = 1$. $dP = (450)(0) + (432.5)(1) = 432.5$; $P(25, 13) \approx P(25, 12) + dP = 3432.5$.
29. The maximum change in C is 0.2 and the maximum percent change in C is 0.5%.
31. Production would increase by 30 units.

Section 8.6

1. $6m + 4b = 12$; $14m + 6b = 23$; $y = x + 1.5$
3. $y = \frac{1}{2}x + 2$
5. $y = 0.1x + 1.1$
7. $-150m + 5b = 33$; $5500m - 150b = -950$; $y = -0.04x + 7.8$
9. $y = 3.45x - 7$
11. About 5.843 million dollars
13. $564m + 4b = 1149$; $79{,}910m + 564b = 162{,}197$; $y = 0.487x + 218.576$
15. (a) $y = 3.5x + 1$
    (b) 29
17. $y = 0.413x + 19$

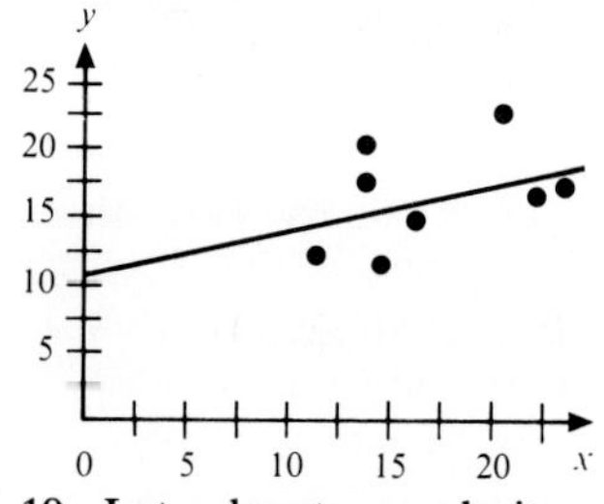

19. Let x denote popularity rank and y participation rank. Then $18m + 6b = 21$; $63.5m + 18b = 71$; $y = 0.8421x + 0.9737$.
21. $y = 0.000255x + 7.295$
23. $y = 0.000261x - 1.754$
25. See Example 8.34; for $y = kt + b$ the normal equations are: $16k + 5b = -4.6306$; $84k + 16b = -39.767$; $k = -0.7606$; $b = 1.5079$. $C = 4.5174$ and $k = -0.7606$.
27. $C = 1.807$
    $k = -0.3689$
29. $C = 5.770$
    $k = 0.6104$
31. Derive new data (t_i, y_i) from the given data (x_i, y_i) by $t_i = (x_i)^2$; the normal equations are: $51a + 5b = 77$; $963a + 51b = 1655$; $a = 1.9639$; $b = -4.6314$.
33. $\alpha = 0.0296$
    $m = 0.5178$

Section 8.7

1. 21; work from the inside, first evaluating $\int_2^5 xy\,dy = 10.5x$. Now use this function in the second integration, $\int_0^2 10.5x\,dx = 21$.
3. $e^3 - 2e^2 + e$
5. $-\frac{16}{35}$
7. 63.5; work from the inside, first evaluating
$$\int_{x^2}^{3x^2} x(y+1)^5\,dy = \frac{x(3x^2+1)^6}{6} - \frac{x(x^2+1)^6}{6};$$
in the second integration, divide into two separate integrals and use a u-substitution on each part.
9. $\frac{1}{4}(\frac{3}{e} - 1)$
11. 327.82
13. 2,221,528; work from the inside, first evaluating $\int_1^4 (y^2 \ln x + ye^{xy})\,dx = 4y^2 \ln 4 - 4y^2 + e^{4y} + y^2 - e^y$.
15. $\frac{768}{5}$
17. 0
19. $\frac{68}{3}$; work from the inside, first evaluating $\int_1^3 (x^2 + y^2)\,dy = 2x^2 + \frac{26}{3}$.
21. $\frac{217}{15}$
23. $e^2 - e - e^{-1} + e^{-2}$
25. -0.575
27. $\frac{2}{27}$
29. $\frac{1}{21}$

Review Exercises

1. $f(4, 2) = -4$
   $f(2, -1) = 7$
3.

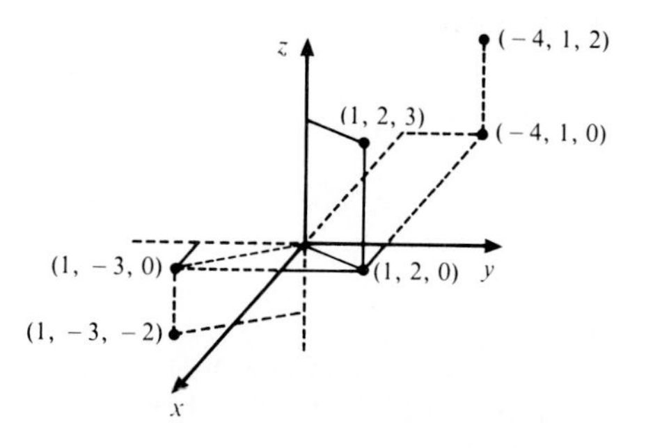

5.

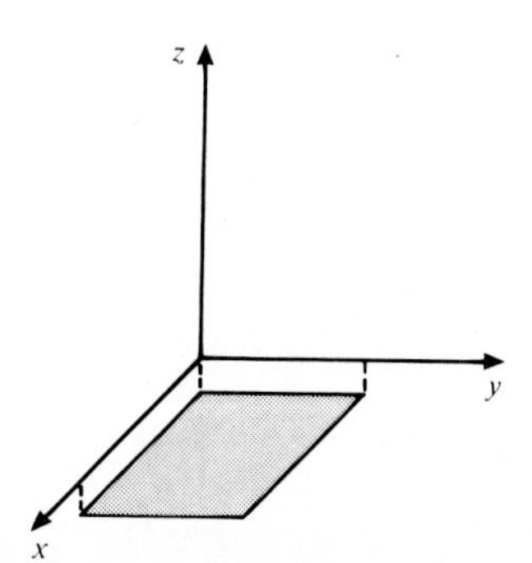

7.

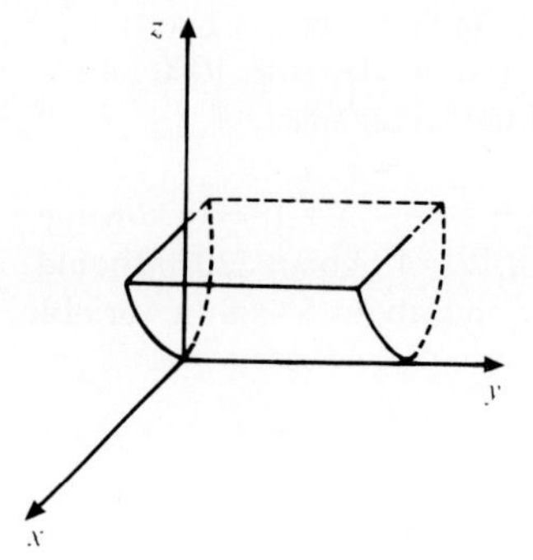

9. $\frac{3}{2}(x-y)^{1/2};\ -\frac{3}{2}(x-y)^{1/2};\ 3$

11. $\dfrac{2x}{x^2+y} - \dfrac{1}{x+y+1};\ \dfrac{1}{x^2+y} - \dfrac{1}{x+y+1}$

13. $\dfrac{\partial z}{\partial x} = yx^{y-1};\ \dfrac{\partial z}{\partial y} = x^y \ln x;\ \left.\dfrac{\partial z}{\partial x}\right|_{(2,-2)} = -\dfrac{1}{4}$

15. $6xy + 20x^3y^2;\ 3x^2 + 10x^4y;\ 23$

17. $4e^{2x-3y};\ -6e^{2x-3y};\ -6e^{-8}$

19. $f_{xx}(x, y) = \dfrac{2y}{x^3};\ f_{xy}(x, y) = -\dfrac{1}{x^2};\ f_{xy}(-1, 2) = -1$

21. Relative maximum at $(0, 0)$

23. Relative minimum at $(-1, -2)$

25. $(1, -3)$ is a relative minimum; $(-1, -3)$ is a saddle point.

27. 2 units of item A; 5 units of item B

29. \$3000 on newspaper advertising
\$1500 on TV advertising

31. 50	33. $\frac{293}{12}$	35. $y = 5.6538x - 5.4154$
37. 5	39. $-\frac{10}{3}$	41. $\frac{1}{70}$

Section A.1

1. 125	3. 64	5. $\frac{1}{2}$
7. 1	9. $\frac{1}{36}$	11. 64

13. $4^3 \times 4^2 = 4^{3+2} = 4^5 = 1024$

15. $\frac{1}{36}$ 17. 3

19. $(b^2a^2)^3 = (b^2)^3(a^2)^3 = b^6a^6$

21. 8000 23. 4096

25. $(a^2b^{1/2})^3(a^{-1/2}b^3)^{-2} = a^6b^{3/2}ab^{-6} = a^{6+1}b^{3/2-6} = a^7b^{-9/2}$

27. $\dfrac{2x}{(x-3)(1-x)}$ 29. $\dfrac{2(x^2+x+9)}{x(x+3)}$

31. $\dfrac{x}{x+2} + (x-1) = \dfrac{x}{x+2} + \dfrac{(x-1)(x+2)}{x+2}$
$= \dfrac{x + (x^2+x-2)}{x+2} = \dfrac{x^2+2x-2}{x+2}$

33. $\dfrac{3}{x}$

Section A.2

1. $2x - 2$	3. $x^2 + 2x + 1$	5. $x^2 - 1$

7. $(x+4)(x-2) = x^2 + 4x - 2x - 8 = x^2 + 2x - 8$

9. $x^2 + 3x - 4$ 11. $x^2 - 9$

13. $(x - \frac{1}{2})^2 = x^2 - \frac{x}{2} - \frac{x}{2} + \frac{1}{4} = x^2 - x + \frac{1}{4}$

15. $x = 1, 2$ 17. $x = 4, 3$

19. $2x^2 - 9x + 4 = (2x-1)(x-4) = 0$
$2x - 1 = 0$ hence $x = \frac{1}{2}$
$x - 4 = 0$ hence $x = 4$

21. No real roots 23. $x = \frac{1}{3}, -\frac{5}{2}$

25. $(x+8)(x+2)$ 27. $(x+8)(x-4)$

29. $3(x-3)(x-2)$ 31. $-2(x+3)(x+1)$

33. $(x-10)(x+5)$ 35. $3(x+\frac{2}{3})(x+1)$

37. $a = 4,\ b = 10,\ c = -6$

$$x = \frac{-10 \pm \sqrt{(10)^2 - 4(4)(-6)}}{2(4)} = \frac{-10 \pm \sqrt{196}}{8} = \frac{-10 \pm 14}{8}$$

$x = \frac{1}{2}, -3$

$4x^2 + 10x - 6 = 4(x - \frac{1}{2})(x + 3)$

Section A.3

1. True	3. False	5. False
7. False	9. <	11. <

13. Multiplication by a negative number reverses the direction of the inequality sign. Therefore, $ac > bc$ if $a < b$ and $c = -3$.

15. <	17. <	19. 4
21. $\frac{1}{2}$	23. 0	
25. $\lvert -4 \rvert + (-4) = 4 - 4 = 0$		27. 1

29

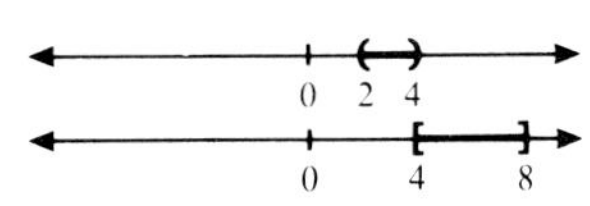

31.

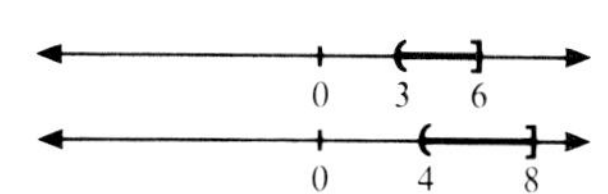

33.

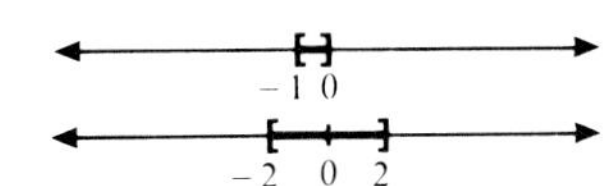

35.

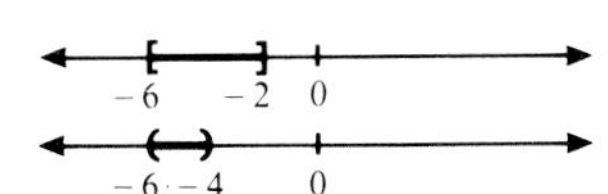

37.

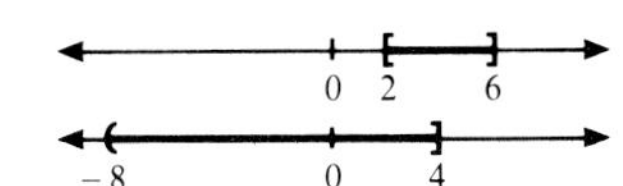

39. $(-2, 4)$

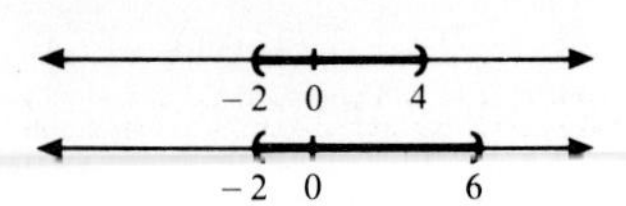

41. $[\frac{1}{2}, \frac{5}{2}]$

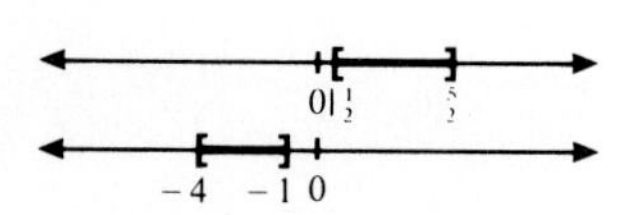

43. The absolute value of anything is never less than zero, and hence is greater than -2. So there are no values of x that satisfy the inequality.

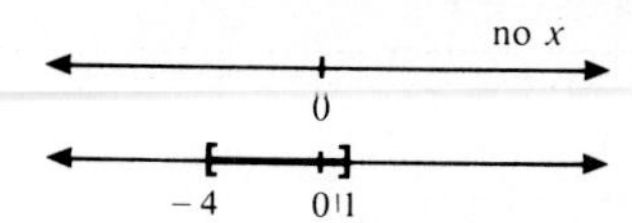

45. $(0, 6)$

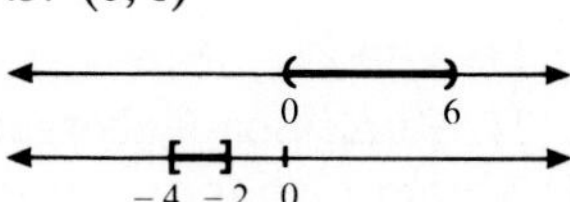

47. $|x-4| \leq 2$

49. $|x-3| < 5$

51. $|x-4.5| \leq 2.5$

CREDITS

Chapter 1: page 1, R. B. Hoit/Photo Researchers; page 6, David Powers/Stock, Boston; page 24, Runk/Schoenberger/Grant Heilman
Chapter 2: page 45, Dean Abramson/Stock, Boston; page 46, Tom McHugh/Photo Researchers; page 71, George H. Harrison/Grant Heilman; page 77, Richard Wood/Taurus
Chapter 3: page 97, Jerry Howard, Positive Images; page 114, Arthur Grace/Stock, Boston; page 132, Grant Heilman; page 138, Mark Antman/The Image Works
Chapter 4: page 147, Barbara Rios/Photo Researchers; page 170, Courtesy Harley-Davidson Co.; page 176, The Photo Works/Photo Researchers
Chapter 5: page 189, Runk/Schoenberger/Grant Heilman; page 193, Hazel Hankin/Stock, Boston; page 214, Culver Pictures; page 219, Stan Levy/Photo Researchers
Chapter 6: page 227, Earth Observation Satellite Co.; page 241, Paul Conklin
Chapter 7: page 271, Barrera/TexaStock; page 278, Michael D. Sullivan, TexaStock; page 288, Topham/The Image Works
Chapter 8: page 305, Ellis Herwig/Stock, Boston; page 328, Grant Heilman; page 353, Culver Pictures

INDEX

Absolute maximum-minimum, 129
on a closed interval, 130, 133
of a continuous function, 133
examples of, 129
existence conditions, 130, 133
finding, 131
on a finite interval, 133
theorem, 130
Absolute value,
as distance, 373
and inequalities, 373
notation for, 36, 373
of a real number, 36, 373
Absolute value function, 36, 92
graph of, 36, 92
Absorption constant, 220, 224
Absorption rate of medicine, 220, 224
Acceleration, 82, 84, 115, 236, 239, 240, 247, 257, 268, 277, 296
Account, interest-bearing, 192–193, 194, 200–201, 214–216, 223–224, 225–226
Accumulation of money, 214
formula, 214, 215
Accuracy,
of Riemann sums, 299
of Simpson's rule, 299
of the trapezoidal rule, 299
Advertising, 84, 85, 135, 166, 185, 225, 240, 306, 313, 323, 330, 331, 356, 364
Agriculture, 2, 75, 80, 110, 139, 142, 144, 179, 257, 268, 306, 307, 313, 320, 324, 330, 331, 332, 340, 364
Air monitoring, 221, 225
Algebra review, 367
Algebraic substitution, 332
solving optimization problems, 332
Altitude, 308
Animal metabolism rate, 353
Annual compounding of interest, 192, 195, 223
Anthropology, 93, 223
Antiderivative, 228
domain of, 230
Antidifferentiation, 230
Approximations,
and integrals, 240, 241, 242, 292, 296
with total differential, 395
Approximation theory, 346
Archaeology, 218, 223
Area,
maximizing, 139, 142, 146
minimizing, 139, 142, 146
net, 245, 250, 251
between two curves, 254
Area under a curve, 244–257
antiderivative, 245
approximation of, 245, 257, 258, 296
using integrals to calculate, 240, 241
of a nonnegative function, 245
and Riemann sum, 244
using Simpson's rule, 296
using the trapezoidal rule, 292
Assets, rate of change of, 104, 109
Asymptote(s), 148
horizontal, 149
to the natural logarithm function, 203
notation, 149
and rational functions, 151
vertical, 150
Athlete,
endurance and speed training of, 43, 143
workout of, 284, 286
Atmospheric pressure, 312
isobar, 312
Average,
rate of change,
of a cost function, 40
of a function, 38
of a linear function, 39
of production, 39, 71
rate of growth, 47
speed, 38, 81
temperature, 266
value of a function, 263
Axis (axes),
x, 11, 307
y, 11, 307
z, 307

Bacteria population, 41, 85, 95, 219, 223, 225, 268, 306, 323, 343, 351
Bacterial infection, 95
Base of an exponential function, 192
Batching problem, 174, 179, 188
Biology, 2, 10, 26, 30, 33, 37, 41, 46, 51, 84, 85, 95, 132, 219, 223, 224, 225, 253, 257, 268, 277, 291, 306, 323, 324, 328, 331, 340, 351, 353
Botany, 328, 331
Campaign fundraising, 43
Capacity, maximizing, 141
Capital, 319
marginal productivity of, 319, 323, 344, 346
Carbon-12, 218
Carbon-14, 218
Cartesian coordinate system, 11
Cash management, 177, 180
Census records, 93
Chain rule, 105–106
and implicit differentiation, 163
proof of, 110
Change,
in cost, 252
in orientation, 19
net, 252, 296
in price and demand function, 27, 181–182
in price and revenue function, 183
rate of. *See* Rate of change.
relative, 181
in supply and demand function, 27
in velocity, 296
Chemical reaction rate, 80
Chemistry, 80, 217, 218, 223, 224
City planning, 75
Class size related to fee, 80
Closed interval, 7, 374
CO_2 level, 221, 225
Cobb-Douglas production function, 319, 354, 355, 357
Codling moth, 33, 37, 132
Coefficient(s) of a polynomial, 31
leading, 31
College fund, 21, 23, 200
Community cooperation index, 357
Composition of functions, 105
Compound interest, 192–194, 215–216, 223–225
Concave downward, 115, 116, 125
Concave upward, 115, 122
Constant, 31
absorption, 220, 224
decay, 217
derivative of, 67
growth, 217
of integration, 233, 235, 246, 279
limit of, 55
of proportionality, 219
Constant multiple,

Constant multiple (*contd.*)
derivative of, 66
integral, 232
limit of, 55
Constrained optimization problem, 332
Constraint, 331
Constraint function, 332
Construction, 6, 143, 318, 330, 331, 336, 340
Consumer surplus, 264, 265, 266
Continuous,
and differentiable, 90
function, 87
graph of, 86
and Riemann sum, 247
polynomial, 87
rational function, 87
Continuous compounding of interest, 215, 223, 224, 225
Converging, 191, 287
Conversion,
of measurements, 24, 30
temperature, 24, 30
Coordinate,
x, 11
y, 11
Coordinate system,
Cartesian, 11
three-dimensional, 307, 308
Corner, 90
Costs(s),
of buying and operating a car, 26, 30
of increasing production, 252
of interviews, 30
ordering, 176, 179, 180, 181
plumbing, 30
rental car, 43
restaurant, 180, 268
shipping, 364
telephone, 30, 31
Cost function, 25, 40, 167, 168, 173, 174, 177, 180, 181, 187, 188, 234, 238, 252, 306, 313, 330, 331, 340
average rate of change of, 40
fixed, 25, 167, 234, 238
linear, 25
marginal. *See* Marginal cost function.
minimizing, 174, 180, 181, 188, 331, 340
production. *See* Production cost.
Crime rate, 330
Critical point, 77, 325
Critical value, 78, 120, 122
Crop yield, 144
Curve(s)
area between two, 264
area under, 244–247
approximation of, 245, 257, 258
using the antiderivative, 245
using Simpson's rule, 296
using the trapezoidal rule, 292
level, 310, 311, 337
line tangent to, 49
Curve fitting, 346
Cusp, 90
Customer receptivity, 331

Daily compounding of interest, 194, 215
Decay,
modeled by exponential function, 216
radioactive, 217, 223, 224
half-life, 217, 224
Decay constant, 217
Deceleration, 85, 236, 237, 239
Decreasing function, 75, 122, 125
Deer population program, 37
Defective parts, 313
Definite integral, 243
of a constant multiple, 250
definition of, 243
and net change, 252
properties of, 250
as a sum of integrals, 250
Degree of a polynomial, 31
Demand, 43
elastic, 184
inelastic, 184
price elasticity of, 182, 183, 184, 185, 188
unitary elastic, 184
Demand function, 10, 27, 28, 29, 43, 112, 165, 174, 186, 188, 264, 265, 266, 323
and change in price, 27, 181–182
and change in supply, 27
marginal, 323
Demographics,
census records, 93
community cooperation index, 357
crime rate, 330
education level and income level, 350, 358
household income, 30
income level and education level, 350, 358
mobility index, 357
park lands in city, 108
participation rank, 357
popularity rank, 357
population. *See* Population.
population density, 262, 266
poverty level income, 31
prison expenditures, 330
unemployment rate, 120
Density, population, 262, 266
Dependent variable, 3, 308
Derivative(s), 61
and chain rule, 105
of a constant, 67
of a constant multiple, 66
definition of, 63
of a difference, 66
evaluating, 62
of the exponential function, 195, 197
first derivative test, 122
of a function, 62
general power rule, 108
higher-order, 114
mixed partial, 321
of the natural logarithm function, 205, 206
notation, 62, 65
nth, 114
partial, 314, 315. *See also* Partial Derivatives.
power rule for, 65
product rule, 98
proof of, 103
quotient rule, 101
proof of, 104
rules, 66
second, 83, 114
second derivative test, 124
for two variables, 326, 327
second partial, 321
of a sum, 66
and tangent line, 62
third, 114
value of, notation for, 66
Difference,
derivative of, 66
limit of, 55
Difference quotient, 54, 91
Differentiable, 61, 89
and continuous, 90
function, 61
Differential, 73, 74, 273
definition of, 73
geometric definition of, 73
total, 341–342, 344
Differential equation, 216
Differentiation, 61
and the chain rule, 163
implicit, 162
logarithmic, 211
and optimization, 61
of partial derivatives, 317
and rate of change, 61
Discontinuous function, graphs of, 86, 88
Distance,
absolute value, 373
between two points, 22
stopping, for various speeds, 238
traveled, 38, 85, 237, 247, 286, 287
on the x- or y-axis, 11
Distance formula, 22
Diverging, 287
Domain,
of antiderivative, 230
of the exponential function, 191
of a function, 2, 7

of a function of two variables, 308, 312
rectangular region, 312
Double integral, 358, 360
integrand, 358
integration variables, 358
limits of, 359
as a Riemann sum, 360
and volume, 359
Drug, 306, 324, 330

e
as a limit, 213
the number, 195
Ecology, 223, 306, 320, 323
Economics, 30, 31, 43, 120, 138, 167–184, 192–194, 214–216, 223–226, 230, 234, 235, 238, 239, 240, 247, 252, 257, 264, 265, 266, 268, 277, 288, 291, 295, 299, 300, 304, 306, 313, 319, 323, 331, 340, 346, 350, 354, 356, 357, 358
Education level and income level, 350, 358
Efficiency,
engine, 144
worker, 357
Elastic demand, 184
unitary, 184
Elasticity, 182, 183, 184, 185, 188
price elasticity of demand, 182, 183, 184, 185, 188
price elasticity of supply, 185
Elastic supply, 185
Election campaign,
fundraising, 43
strategies, 113, 117, 143
snowballing, 113, 117
Electronics manufacturing, 102
Element,
of a Riemann sum, 260
of a solid of revolution, 260
End arrows, 151
Endpoint(s), 130
of a partition, 241
Engine efficiency, 144
Epidemic, 137, 143
Equal functions, 9
Equation(s),
differential, 216
of a line,
point-slope form, 17
slope-intercept form, 17
normal, 348
Reineke's, 222, 224
of a tangent line, 69
of a vertical line, 18
Equilibrium point, 29, 31, 265, 266
Equilibrium price, 29, 265, 266, 267
Error measurement,
and the total differential, 342, 344
and the trapezoidal rule, 294
Existence of absolute maximum-minimum, 130, 133
Exponential function(s), 191, 195
base, 192
and decay, 216
decay constant, 217
definition of, 192
derivative of, 195, 197
domain of, 191
graph of, 192, 197
and growth, 216
growth constant, 217
integral of, 232
and natural logarithm function, 202
properties of, 196
table of, 384–387
table of comparative values for, 196
Exponents,
irrational numbers as, 191
laws of, 366
properties of, 190, 367
rational numbers as, 191

Factoring, 369
Factoring a quadratic function, 34
Fahrenheit-to-Celsius conversion function, graph of, 25
Federal programs, 120, 135, 143, 330
Fee, relation to class size, 80
Finding absolute maximum-minimum, 131
Finding the vertex of a parabola, 32
Finding x-intercepts using the quadratic formula, 36
First derivative test, 122
Fixed cost, 25, 167, 234, 238
Forestry, 2, 27, 71, 74, 85, 222, 223, 224, 291, 300
Fractions, rules for computing with, 367
Fruit fly population, 46, 51, 84
Function(s), 2
absolute value, 36, 92
average rate of change of, 38
average value of, 263
chain rule, 106
Cobb-Douglas production, 319, 354, 355, 357
composition of, 105
concave downward, 115, 116, 125
concave upward, 115, 122
constraint, 332
continuous, 86, 87
cost. *See* Cost function; Marginal cost function.
critical point, 77, 325
critical value, 78, 120, 122
decreasing, 75, 122, 125
definition of, 2
demand. *See* Demand function.
derivative of, 62
differentiable, 61
discontinuous, 86, 88
domain of, 2, 7
equal, 9
exponential, 191, 195
Fahrenheit-to-Celsius conversion, graph of, 25
given explicitly, 162
given by a formula, 3
given implicitly, 162
given by a rule, 2
given by a table, 3
graph of, 11
maximum point on, 50
point on, 11
graphing checklist, 153
greatest integer, 88
increasing, 75, 122, 125
inflection point, 116
guidelines for finding, 119
inverse, 202
with jump, 86
Lagrange, 334
limit of, evaluating, 52
limit of a power of, 55
linear, 15, 31
average rate of change of, 39
linear cost, 25
logarithm, 200
maximum value of, 34
minimum value of, 34
natural logarithm, 202
negative, integral for, 245
nonlinear, in three dimensions, graph of, 308
notation, 4, 5
of two variables, 306
objective, 332
polynomial, 31, 68
postal rate, 3
postal scale, 58
price, 138, 149, 264
product of, derivative of, 198
production. *See* Production function; Marginal cost of production.
profit. *See* Profit function.
quandratic, factoring, 34
rate of change. *See* Rate of change.
rational, 53, 57, 87, 151
relative change of, 181
revenue. *See* Revenue function.
sign chart, 78, 117
slope of, 16
step, 89
supply. *See* Supply function.
temperature, 312
test value, 78, 117

Function(s) (*contd.*)
of two variables, 306
domain of, 308, 312
graph of, 307
notation for, 306
value of, 306
value, 5
volume, 49
zero of, 12
Functional notation, 4, 5
Fundamental Theorem of Calculus, 247

General power rule, 108
Geology, 311
Geometric definition,
of the derivative, 63
of the differential, 73
of tangent line as limit of secant lines, 49, 50, 69, 70
Geometric formulas, inside front cover
Geometric interpretation of partial derivatives, 315
Gerbil population, 42
Government, programs, 120, 135, 143, 330
regulation, 171, 172, 173, 174, 184–185
Graph(s),
of the absolute value function, 36, 92
of asymptotes, 148
of a concave downward function, 115, 116
of a concave upward function, 116
of continuous functions, 86
of a decreasing function, 75
of a demand function, 28
of demand and supply functions, 29
of discontinuous functions, 86, 88
of an exponential function, 192, 197
of the Fahrenheit-to-Celsius conversion function, 25
of a function, 11
maximum point on, 50
point on, 11
of two variables, 307
of the greatest integer function, 88
of an increasing function, 75
of a linear function, 15
in three dimensions, 308
maximum point on, 50
of the natural logarithm function, 202
of a nonlinear function in three dimensions, 308
of a parabola, 32
plotting points on, 11, 32
polynomial, 32
of relative maxima/minima, 325
slices of, 308
slope of, 16
of a sphere, 309
tangent line to, 70
Graphic representation of intervals, 8, 374
Graphing,
Cartesian coordinate system, 11
three-dimensional coordinate system, 307, 308
three-dimensional surface by slicing, 308, 310
Graphing checklist, 153
Grasshopper population related to temperature, 10
Greatest integer function, 88
Growth,
modeled by exponential function, 216
population, 219
rate of, 47, 328, 331
average, 47
Growth constant, 217

Half-life, 217, 224
Half-open interval, 374
Health, 95, 137, 138, 143, 220, 223, 224, 225, 284, 286, 306, 324, 330, 343
Higher-order derivative, 114
Highway safety, 208
Horizontal,
asymptote, 149
line, 17, 50
tangent line, 70
Household income, 30
Hyperbolic paraboloid, 326

Implicit differentiation, 162
and the chain rule, 163
Implicit form of a function, 162
Import fees, 173
Improper integral, 287
convergence, 287
divergence, 287
geometric interpretation of, 288
Income,
of a household, 30
poverty level, 31
salesperson's, 43
Income level and education level, 350, 358
Increasing function, 75, 122, 125
graph of, 75
Indefinite integral, 230
Independent variable, 3, 308
Inelastic demand, 184
Inelastic supply, 185
Inequalities,
and absolute value, 373
properties of, 373
Inequality symbols, definition of, 372
Infection, 95, 138, 143, 225
Infinity, limits at, 149
Inflection point(s), 116
guidelines for finding, 119
Initial condition, 228
Insect infestation, 46, 51, 84, 291, 340
Inspection of parts, 313
Instantaneous rate of change of a function, 51, 82
Instantaneous speed, 82
Integral(s), 230
antiderivative, 228
antidifferentiation, 230
and approximations, 240, 241, 242, 292, 296
area calculation, 240, 241, 245, 250, 254
of a constant multiple, 232
converging, 287
definite, 243. *See* also Definite integral.
diverging, 287
double, 358, 360. *See also* Double integral
of an exponential function, 232
improper, 287
geometric interpretation of, 288
indefinite, 230
iterated, 360, 361
of the natural logarithm function, 232
for a negative function, 245
notation for, 243, 248
overestimates of area, 247
partition, 241, 242, 244
of a polynomial, 232
of a power, 232
power rule for, 230
rectangular region, 241, 242
as a Riemann sum, 241, 242
of a sum, 232
table of, 300, 301, inside back cover
underestimates of area, 240, 241, 242
value of, 287
Integrand, 230, 249
of a double integral, 358
Integration, 228
antiderivative, 228, 230, 231
antidifferentiation, 230
constant of, 233, 235, 246, 279
of double integrals, 358, 359, 360
Fundamental Theorem of Calculus, 247
integrand, 230, 249
interval of, 249
limits of, 249
changing, 275
lower limit, 249
order of, 361
by parts, 278
formula, 278
table format, 281
by substitution, 272, 273
upper limit, 249
variables for double integrals, 358
Intercept(s),
x, 12
of a function, number of, 34
y, 12

Interest, 192, 214–216, 223, 225, 226
chart of various compounding periods, 193
compounded annually, 192, 215, 223
compounded continuously, 215, 223, 224, 225
compounded daily, 194, 215
compounded monthly, 215, 216
compounded quarterly, 193, 215, 225
Interest-bearing account, 192–193, 194, 200–201, 214–216, 223–224, 225–226
Interpolation, linear, 379
Intersection of two lines, 21
Interval(s), 7, 374
closed, 7, 374
graphic representation of, 8, 374
half-open, 374
of integration, 249
notation, 7, 374
open, 7, 374
Inventory model, 176
Inverse functions, 202
Investment, 74, 86, 138, 144, 187, 215, 223, 224, 225, 304
Iodine-131, 217
Irrational numbers as exponents, 191
Isobar, 312
Isotherm, 312
Iterated integral, 360
limits of, 361
notation for, 361
order of integration, 361
*i*th subinterval, 241, 242, 257–260, 263

Job creation program, 34, 37
Job satisfaction, 340
Jump, function with, 86

Labor, 257
marginal productivity of, 319, 323, 344, 346
Lagrange, Joseph Louis, 334
Lagrange function, 334
Lagrange multiplier, 334
solving optimization problems, 334
Lagrange multiplier theorem, 334
Laws of exponents, 366
Leading coefficient of a polynomial, 31
Learning, 2, 6, 76, 112, 223, 224, 356
Least squares method, 347, 352, 354
line of regression, 348
Left-hand limit, 58
Level curve, 310, 311, 337
Limit(s), 50, 52
of a constant, 55
of a constant multiple, 55
definition of, 52
determining, 55
of a difference, 55
and *e*, 213
evaluating, of a function, 52
existence of, 52
at infinity, 149
of integration, 249
changing, 275
for a double integral, 359
for an iterated integral, 361
left-hand, 58
lower, 249
notation, 52
one-sided, 58
of a polynomial, 56
of a power, 55
of a power of a function, 55
of a product, 55
of a quotient, 55
of a rational function, 53, 57
right-hand, 58
of a sum, 55
theorems, 55
upper, 249
Limiting process, 48
Line(s),
equation of
point-slope form, 17
slope-intercept form, 17
horizontal, 17, 50
intersection of two, 21
parallel, 20
perpendicular, 21
of regression, 348
least squares method, 347
secant, 49, 63, 69
and tangent lines, 49, 50, 69, 70
slope of, 16
tangent,
to the curve, 49
and derivative, 62
in three dimensions, graph of, 308
vertical, 17, 18, 93
equation of, 18
Linear cost function, 25
Linear function, 15, 31
average rate of change of, 39
graph of, 15
Linear interpolation, 379
Logarithm function, 200
natural, 202
Logarithmic differentiation, 211
Logarithms, natural, table of, 380–383
Lower limit of integration, 249

Manufacturing, 2, 48, 80, 81, 102, 135, 142, 144, 174, 180, 234, 288, 291, 319, 323, 331
costs, 2, 3, 10, 25, 80, 81
profit, 84, 85, 135, 142, 144
Marginal cost function, 168, 173, 187, 230, 234, 235, 238, 239, 252, 257, 268, 277
Marginal,
demand, 323
price, 257
productivity, 240, 257
of capital, 319, 323, 344, 346
of labor, 319, 323, 344, 346
of money, 336
profit, 268
revenue function, 169, 172, 173, 187, 234, 238, 239, 257, 277, 295, 299, 304
Marginal cost,
of production, 230, 234, 235, 238, 239, 268
of volume, 337
Mathematical models, 20
of cash management, 177, 180
of decay, 216
of growth, 216
of inventory, 176
of population, 351
Maximizing,
area, 139, 142, 146
capacity, 141
production, 72, 74, 85, 332, 335
profit, 74, 138, 142, 144, 169, 170, 171, 173, 174, 187, 328, 331, 364
rental income, 144
revenue, 140, 174, 187
volume, 49, 51, 80, 129, 136
yield, 320
Maximum,
absolute, 129
examples of, 129
relative, 121, 125, 325
examples of, 121
graphs of, 325
tangent plane at, 325
Maximum point on a graph, 50
Maximum value, 34, 50
of a parabola, 33
Max-min problems, 135
Maze experiments, 2, 224
Measurement conversion, 24, 25, 30
Medicine, 86, 220, 223, 224, 306, 324, 330
absorption constant, 220, 224
absorption rate, 220, 224
Metabolism, 353–354
Method of least squares, 347, 352, 354
line of regression, 348
Military spending, 143
Minimizing,
area, 139, 142, 146
costs, 174, 180, 181, 188, 331, 340
Minimum,
absolute, 129
examples, 129

Minimum (*contd.*)
relative, 121, 125, 325
examples of, 121
graphs of, 325
tangent plane to, 325
Minimum value, 34, 51
Missile's altitude recorder, 61
Mixed partial derivatives, 321
Mobility index, 357
Model(s), mathematical, 20, 177, 179, 180, 216, 351
Money, marginal productivity of, 336
Mouse-to-elephant function, 353–354
Multiplier, Lagrange, 334
solving optimization problems, 334
theorem, 334

Natural logarithm function, 202
and antiderivative, 231
asymptote to, 203
derivative of, 205, 206
and exponential function, 202
graph of, 202
integral of, 232
as inverse of exponential function, 202
properties of, 203, 205, 378
solving equations using, 204
table of, 380–383
use of tables of, 378–379
Negative,
function, integral for, 245
slope, 17
Net area, 245, 250, 251
Net change, 252, 296
in cost, 252
in velocity, 296
Newspaper advertising, 306, 313, 330, 364
Nonlinear function in three dimensions, graph of, 308
Normal equations, 348
Notation,
absolute value, 36, 373
asymptotes, 149
derivative, 62, 64, 114
value of, 66
function of two variables, 306
functional, 4, 5
integral, 243, 248
interval, 7, 374
iterated integral, 361
limits, 52
one-sided limits, 58
partial derivatives, 314, 315
partition, 241
Pythagorean theorem, 22
second partial derivatives, 321
summation, 258, 348
*n*th derivative, 114
Nuclear fallout, 217
Number of *x*-intercepts of a function, 34

Objective function, 332
Oil leak, 112, 145
One-sided limit, 58
Open interval, 7, 374
Optimal, amount of cash, 177, 178, 180
company size, 109, 112
order size, 176, 179, 180, 188
price, 171, 173
production level, 170, 171, 173
Optimization problem, 48, 80, 135, 324
constrained, 332
constraint, 331
and differentiation, 61
organizing, 136
solving by algebraic substitution, 332
solving by Lagrange multipliers, 334
Order of integration, 361
for an iterated integral, 361
Order size, optimal, 176, 179, 180, 188
Ordered pairs, 11
Ordering costs, 176, 179, 180, 181
Orientation, change in, 19, 23
Origin, 11

Pairs, ordered, 11
Parabola, 32
determining the direction of, 32
finding the vertex, 32
graph of, 32
maximum value of, 33
vertex of, 32
Paraboloid, hyperbolic, 326
Parallel lines, 20
Parklands in city, 108
Partial derivative(s), 314, 315
differentiation of, 317
geometric interpretation of, 315
mixed, 321
notation for, 314, 315
as a rate of change, 315
second, 321
value of, 315
Participation rank, 357
Partition, 241, 242, 244, 258
endpoints of, 241
notation for, 241
Riemann, 241, 242, 244, 258
Parts, integration by, 278
formula, 278
table format, 281
Perception, spatial, 19, 23
Perpendicular lines, 21
Pesticide/insecticide use, 46, 144, 330, 340
Pharmaceuticals, 84, 85, 330
Photosynthesis, 2
rate of trees, 26, 30
Physics, 38, 81, 82, 84, 85, 104, 115, 236, 237, 238, 239, 240, 247, 248, 253, 257, 268, 277, 286, 287, 291, 296, 345

Plane,
tangent, 324
at relative maxima-minima, 325
xy-, 308
xz-, 309
yz-, 308
Plotting points on a graph, 11, 32
Plumbing costs, 30
Point(s)
critical, 77, 325
distance between, 22
equilibrium, 29, 31, 265, 266
on the graph of a function, 11
maximum, 50
inflection, 116
guidelines for finding, 119
saddle, 325
of tangency, 49
Point-slope form of the equation of a line, 17
Politics, 135, 143
Pollution, 31, 217, 221, 224, 225, 306, 320, 323, 331, 340, 343, 345
Pollution index, 31, 320, 323, 343, 345
Polynomial, 31
coefficients of, 31
continuous, 87
degree of, 31
examples of, 31
factoring, 370
function, 31, 68
graph of, 32
integral of, 232
leading coefficient of, 31
limit of, 56
linear function, 31
quadratic, 31, 34
root of, 369
quadratic formula, 34, 369
finding *x*-intercepts using, 36
real solutions, 34, 369
solving by the quadratic formula, 369
x-intercepts of, 34, 36
Popularity rank, 357
Population, 10, 33, 37, 41, 42, 46, 51, 84, 85, 95, 108, 112, 145, 216, 219, 223, 224, 225, 262, 266, 268, 306, 323, 346, 351, 357, 364
density, 262, 266
growth, 219
rate of change of, 216, 223, 224
world, 223, 224
Positive slope, 17
Postal,
package limits, 340
rate function, 3, 86
scale function, 58
Poverty-level income, 31
Power,
integral of, 232

limit of, 55
Power rule,
for derivatives, 65
general, 108
for integrals, 230
Pressure, atmospheric, 312
isobar, 312
Price,
change in,
and demand function, 27, 181–182
and revenue function, 29
and supply function, 29
equilibrium, 29, 265, 266, 267
optimal, 171, 173
tax effect on, 171, 173
Price elasticity,
of demand, 182, 183, 184, 185, 186, 188
of supply, 186
Price function, 138, 144, 264
marginal, 257
Principal square root, 7
Prison expenditures, 330
Producer surplus, 267
Product column,
for Simpson's rule, 297
for the trapezoidal rule, 294
Product rule, 98, 369
limit of, 55
proof of, 103
for three functions, 100
Production cost, 25, 30, 40, 41, 42, 73, 74, 112, 143, 145, 235, 238, 239, 252, 277. *See also* Marginal cost of production.
Production function, 39, 71, 74, 85, 86, 170, 171, 252, 268, 288, 291, 295, 298–299, 300, 304, 306, 313, 319, 323, 330, 332, 333, 335, 340, 344, 346, 354, 355, 364
average rate of change in, 39, 71
Cobb-Douglas, 319, 354, 355, 357
maximizing, 72, 74, 85, 332, 335
wood, 71, 74, 85, 300
Production-inventory problem, 174, 179
Production level, optimal, 170, 171, 173
Production-storage problem, 174, 179, 188
Productivity, 240, 257, 319, 355, 357
marginal, 240, 257
of capital, 319, 323, 344, 346
of labor, 319, 323, 344, 346
of money, 336
Profit function, 42, 44, 74, 84, 85, 86, 135, 138, 142, 144, 148, 169, 170, 171, 173, 185, 187, 268, 328, 330, 331, 364
marginal, 268
maximizing. *See* Maximizing, profit.
Projectile, 236, 239, 240, 248, 253, 268
Proportionality, constant of, 219
Psychology, 2, 19, 23, 76, 80, 135, 143, 223, 224, 277, 331, 339, 340
Publishing, 143, 144, 313, 330
Pythagorean theorem, 22

Quadratic formula, 34, 369
finding x-intercepts using, 36
Quadratic function, factoring, 34
Quadratic polynomial, 31
root of, 369
Quarterly compounding of interest, 193, 215, 225
Quotient,
difference, 54, 90
limit of, 55
of two functions, derivative of, 101
Quotient rule, 101
proof of, 104

Radio advertising, 330
Radioactive decay, 217, 218, 223, 224
carbon-14, 218
carbon-12, 218
half-life, 217, 224
iodine-131, 217
radium, 217, 223
strontium-90, 224
Radium, 217, 223
Rate(s),
of absorption,
by bacterium, 323, 324
of medicine, 220, 224
of change. *See* Rate of change.
of growth, 47, 328, 331
average, 47
photosynthesis, 26, 30
production, 86
reaction, 306, 324
related, 108
unemployment, 120
Rate of change, 104, 109, 112, 114, 117–118, 145, 216, 218, 221, 222, 223, 224, 225, 226, 268, 286, 323
of assets, 104, 109
average, 38, 39, 40, 71
and differentiation, 61
instantaneous, of a function, 51, 82
partial derivative as, 315
of population, 216, 223, 224
of the slope of a function, 114
of volume, 112, 145, 268
Rational exponents, 191
Rational function(s),
and asymptotes, 151
continuous, 87
limit of, 53, 57
rule, 57
Reaction rate, 306, 324
Reading proficiency test, 112
Real number, absolute value of, 36, 373
Real solutions of a polynomial, 34, 369
Receptivity,
of customer, 331
of student, 77, 80
Rectangular region, 241, 242, 312
Regression. *See* Least squares method.
Regulation, government, 171, 172, 173, 174, 184–185
Reineke's equation, 222, 224
Related rates, 108
Relation of class size to fee, 80
Relative change, 181
Relative maximum-minimum, 121, 125, 325
examples of, 121
graphs of, 325
tangent plane at, 325
test for, 122, 124
Rental car costs, 43
Respiration rate, 41
Restaurant, costs, 180, 268
tax, 184, 185
Revenue function, 10, 31, 80, 167, 169, 172, 174, 187, 234, 238, 239, 264, 313, 330
and change in price, 183
marginal. *See* Marginal revenue function.
maximizing, 140, 174, 187
Revolution, solid of, 260, 266
element of, 260
volume of, 261
using washer method to find volume of, 266
Riemann partition, 241, 242, 244, 258
Riemann sum(s), 241, 258
accuracy of, 299
and area, 242
and continuous functions, 247
and double integral, 360
element of, 260
integral as, 241, 242
Right-hand limit, 58
Rise, 16
Rocket flight, 239
Rodent control, 41, 80
Root, 369
of a function, approximating by total differential, 342, 344
principal square, 7
of a quadratic polynomial, 369
square, 7
Rotating objects, 20
Rule, chain, 105–106
Run, 16

Saddle point, 325
Sales, 225, 238, 239, 240, 247, 248, 257, 266, 286, 300, 306, 313, 323, 356
Salesperson's income, 43

Secant line, 49, 63, 69
slope of, 49, 69
and tangent lines, 49, 50, 69, 70
Second derivative, 83, 114
Second derivative test, 124
for two variables, 326, 327
Second partial derivative, 321
Shade intolerant trees, 27
Shade tolerant trees, 27
Shipping costs, 364
Sign chart (for graphing a function), 78, 117
Simpson's rule, 296, 297
accuracy of, 299
to approximate area under a curve, 296
product column, 297
weights column, 297
Slice of a surface, 308, 315
Slicing, graphing a three-dimensional surface by, 308, 310
Slope,
of a function, 16
rate of change of, 114
of a line, 16
of the line tangent to the graph of a function, 51, 69
negative, 17
positive, 17
rise, 16
run, 16
of the secant line, 49, 69
of a tangent line, 51, 69
of a three-dimensional surface, 316, 317
undefined, 17
zero, 17, 50
Slope-intercept form of the equation of a line, 17
Snowballing campaign, 113, 117
Sociology, 30, 81, 135, 257, 300, 330, 349, 350, 357, 358
Solid of revolution, 260, 266
element of, 260
volume of, 261
using washer method to find volume of, 266
Solutions of a function, 34
Solving equations using the natural logarithm function, 204
Solving polynomials, equations
by factoring, 370
by the quadratic fomula, 369
Spatial perception, 19, 23
Speed,
average, 38, 81
chart of stopping distances, 238
instantaneous, 82
Speeding car, 345
Sphere, graph of, 309
Sports training, 43, 143, 284, 286
Square root, 7
principal, 7
Standard urea clearance function, 343
Statistics, 356
Step function, 89
Stopping distances for various speeds, 238
Straight line. *See* Linear function.
Strontium-90, 224
Student receptivity, 77, 80
Student test scores, 6, 356
Subinterval, ith, 241, 242, 257–260, 263
Substitution,
algebraic, 332
solving optimization problems by, 332
integration by, 272, 273
Sum,
of definite integrals, 250
derivative of, 66
integral of, 232
limit of, 55
Riemann, 241, 258
Summation notation, 258, 348
Supply,
change in, and demand function, 27
elastic, 185
inelastic, 185
price elasticity of, 185
Supply function, 29, 43, 265, 266
and change in price, 29
Surface, 308
slice of, 308, 315
tangent to, 324
Surplus
consumer, 264, 265, 266
producer, 267

Table,
of exponential functions, 384–387
of integrals, 300, 301, inside back cover
of natural logarithms, 380–383
Tangency, point of, 49
Tangent line,
to a curve, 49
and the derivative, 62
equation of, 69
horizontal, 70
and secant lines, 49, 50, 69, 70
slope of, 51, 69
vertical, 93
Tangent plane, 324
at relative maxima-minima, 325
Tangent to a surface, 324
Tax,
restaurant, 184, 185
effect on price, 171, 173
Telephone costs, 30
Television advertising, 135, 306, 313, 323, 364
Temperature, 10
average, 266
conversion, 24, 30
function, 312
Test value, 78, 117
Third derivative, 114
Three-dimensional coordinate system, 307, 308
Three-dimensional surface,
graphing of, by slicing, 308, 310
slope of, 316, 317
Topological map, 311
Total differential, 341
and approximating roots of functions, 342, 344
and error measurement, 342, 344
Trapezoidal rule, 292, 293
accuracy of, 299
to approximate area under a curve, 292
error measurement, 294
product column, 294
weights column, 294
x column, 293
Triple, 307

Undefined slope, 17
Unemployment rate, 120
Unitary elastic demand, 184
Upper limit of integration, 249

Vacation planning, 340
Value,
absolute, 36, 373
function, 36, 92
critical, 78, 120, 122
of the derivative, notation for, 66
of a function, 5, 12
average, 263
maximum, 34, 50
minimum, 34, 51
of two variables, 306
of the integral, 287
of a partial derivative, 315
test, 78, 117
Variable, 2
dependent, 3, 308
independent, 3, 308
Velocity, 82, 84, 104, 236, 239, 247, 253, 257, 268, 286, 291, 296
net change in, 296
Venture-capital investment, 74
Vertex of a parabola, 32
Vertical asymptote, 150
and rational functions, 151
Vertical line, 17, 18
equation of, 18
Vertical tangent line, 93
Volume,
and double integral, 359

of a solid of revolution, 261
Volume function, 49
marginal cost of, 337
maximizing, 49, 51, 80, 129, 136
rate of change of, 112, 145, 268

Washer method for finding the volume of a solid of revolution, 266
Water testing program, 253
Weather, 266, 312, 313
atmospheric pressure, 312
isotherm, 312
Weight as a function of height, 364
Weights column,
for Simpson's rule, 297
for the trapezoidal rule, 294
Welfare spending, 143, 330
Wildlife management programs, 37
Wood production, 71, 74, 85, 300
Worker efficiency, 357
World population, 223, 224

x-axis, 11, 307
x column for the trapezoidal rule, 293
x-coordinate, 11
z-intercept(s), 12
finding, using the quadratic formula, 36
of higher-degree polynomials, 36
number of, of a function, 34
xy-plane, 308
xz-plane, 308

y-axis, 11, 307
y-coordinate, 11
Yield, maximizing, 320
y-intercept, 12
yz-plane, 308

z-axis, 307
Zero
of a function, 12
slope, 17, 50